REVIEWS in MINER

MW00345816

VOLATILES IN MAGMAS

Editors:

Michael R. Carroll
University of Bristol

John R. Holloway
Arizona State University

Series Editor: **Paul H. Ribbe**
Department of Geological Sciences
Virginia Polytechnic Institute & State University
Blacksburg, Virginia 24061-0420

Mineralogical Society of America
Washington, D.C.

REVIEWS IN MINERALOGY

(Formerly: SHORT COURSE NOTES)

ISSN 0275-0279

Volume 30: *VOLATILES IN MAGMAS*

ISBN 0-939950-36-7

FOREWORD

Volatiles in Magmas is Volume 30 in the Reviews in Mineralogy series begun in 1974 by the Mineralogical Society of America (MSA). See the opposite page for a listing of available titles. Most of these books, including this one, have been produced in conjunction with short courses on their respective topics. The *Voltiles in Magmas* course was organized by Mike Carroll (Bristol University) and John Holloway (Arizona State University) and transpired at the Napa Valley Sheraton Hotel in California, December 2-4, 1994, just prior to the Fall Meetings of the American Geophysical Union in San Francisco.

Very special thanks are due Mike Carroll, who was responsible for most of the technical editing, and John Holloway, his co-editor. Most of the editing, revising, and proofing was, for the first time in our corporate experience, carried out through the Internet by e-mail with a high degree of success, with only minor frustrations. The crash of the Series Editor's hard disk at the last moment threatened disaster, but that is history of interest only to him and his skilled and concientious helpers, Brett Macey and Margie Sentelle.

Paul H. Ribbe
Series Editor
Virginia Tech, Blacksburg

EDITORS' INTRODUCTION

Volatile components, by which we mean those magma constituents which typically prefer to occur in the gaseous or super-critical fluid state, may influence virtually every aspect of igneous petrology. The study of volatile-bearing systems, both in nature and in the laboratory, has far exceeded the relative abundances of these components in igneous rocks, yet in many ways the words of Bowen (1928) are still broadly applicable:

> *"...to many petrologists a volatile component is exactly like a Maxwell demon; it does just what one may wish it to do." (Bowen, 1928, p. 282)*

What we hope to show in this volume are some areas of progress in understanding the behavior of magmatic volatiles and their influence on a wide variety of geological phenomena; in doing this it also becomes apparent that there remain many questions outstanding. The range of topics we have tried to cover is broad, going from atomistic-scale aspects of volatile solubility mechanisms and attendant effects on melt physical properties, to the chemistry of volcanic gases and the concentrations of volatiles in magmas, to the global geochemical cycles of volatiles. The reader should quickly see that much progress has been made since Bowen voiced his concerns about Maxwell demons, but like much scientific progress, answers to old questions have prompted even greater numbers of new questions.

This book has benefited from the efforts of many people: the authors themselves, most of whom managed to keep to deadlines; the many reviewers who provided valuable, detailed comments, often on short notice; the many graduate students at Bristol who served as guinea-pigs to help gauge the suitability of chapters for non-specialists; and the Series Editor, Paul Ribbe, and his staff, who actually made everything happen in the end. We thank all for their efforts. We also acknowledge secretarial support from Bristol University and Arizona State University, and MRC thanks Prof. A. Mottana and the Terza Università di Roma for a Visiting Professorship during the final, hectic month of preparation of this volume.

Michael R. Carroll
University of Bristol, UK

John R Holloway
Arizona State University, Tempe

DEDICATION TO C. WAYNE BURNHAM

Wayne Burnham became interested in volatiles in melts through his work on hydrothermal ore deposits (his Ph.D. dissertation), and his work with Dick Jahns on pegmatites. From the time he arrived at Penn State in 1955 he began experiments aimed at understanding H_2O in melts. The early experiments were on the effects of H_2O on phase equilibria in granitic systems, and studies of the more fundamental properties of H_2O solubility and melt viscosity quickly followed. In attempting to devise a method to explain H_2O solubility in melts Wayne realized that he needed to have a complete thermodynamic model for a hydrous melt. He knew that this meant measuring the P-V-T properties of a hydrous melt system, but he needed a highly precise volumometer which would work over the P-T range of silicic crustal melts. He decided that such a volumometer needed a working fluid with precisely known P-V-T properties and chose H_2O for the fluid. The first step then was to measure those properties over the range 1 bar to 8 kbars and 25° to 900°C. It took six years to complete that project before he and Nick Davis could measure the volumetric properties of hydrous albitic melts over the same P-T range. Wayne then generalized the thermodynamic model for H_2O solubility in hydrous albitic melts to bulk compositions spanning the range of common igneous rocks. Following that he developed a thermodynamic model for crystal-liquid equilibria which he continues to expand to include additional components. Using the thermodynamic model for H_2O in rhyolite, Wayne related the volatile content of magmas to the energetics of eruptive processes and breccia formation. Over the course of much of his career Wayne has published chapters in three successive volumes of The Geochemistry of Hydrothermal Ore Deposits (the latest one is currently in press), edited by his friend and colleague Hugh Barnes. Those chapters each deal with the interaction of the important volatiles in magmas. Wayne's research forms the basis of much of work described in this Volume, and we happily dedicate it to him and hope that he continues to provide us with his insights and enthusiasm.

John Holloway & Mike Carroll

Chronological listing of C. Wayne Burnham's papers

Burnham CW (1954) Contact metamorphism at Crestmore, California. Calif. Div. Mines Geol Bull. 170:61-70

Burnham CW (1954) The Crestmore Hills. California Div Mines Geol Bull.170, Guide No. 5:54-57

Burnham CW (1959) Contact metamorphism of magnesian limestones at Crestmore, California. Geol Soc Am Bull 70:879-920

Burnham CW (1959) Metallogenic provinces of the southwestern United States and northern Mexico. New Mexico Bur Mines Bull 65:1-76

Burnham CW (1962) Facies and types of hydrothermal alteration. Econ Geol 57:768-784

Burnham CW, Jahns RH (1962) A method for determining the solubility of water in silicate melts. Am J Sci 260:721-745.

Putman GW, Burnham CW (1963) Minor elements in igneous rocks, northwestern and central Arizona. Geochim Cosmochim Acta 27:53-106

Hamilton DL, Burnham CW, Osborn EF (1964) The solubility of water and effects of oxygen fugacity and water content on crystallization in mafic magmas. J Petrol 5:21-39

Anderson GM, Burnham CW (1965) The solubility of quartz in supercritical water. Am J Sci 263:494-511

Anderson GM, Burnham CW (1967) Reactions of quartz and corundum with aqueous chloride and hydroxide solutions at high temperatures and pressures. Am J Sci 265:12-27

Burnham CW (1967) Hydrothermal fluids at the magmatic stage. In: Geochemistry of Hydrothermal Ore Deposits, 38-76, Holt, Reinhart and Winston, New York

Holloway JR, Burnham CW, Millhollen GL (1968) Generation of H_2O-CO_2 mixtures for use in hydro-thermal experimentation. J Geophs Res 73:6598-6600

Burnham CW, Holloway JR, Davis NF (1969) The specific volume of water in the range 1000-8900 bars, 20 to 900° C. Am J Sci 267:70-95

Burnham CW, Holloway JR, Davis NF (1969) Thermodynamic properties of water to 1000°C and 10000 bars. Geol Soc Am Spec Paper 132: 96 p

Hsu LC, Burnham CW (1969) Phase relationships in the system $Fe_3Al_2Si_3O_{12}$–$Mg_3Al_2Si_3O_{12}$ at 2.0 kilobars. Geol Soc Am Bull 80:2293-2408

Jahns RH, Burnham CW (1969) Experimental studies of pegmatite genesis: I. A model for the derivation and crystallization of granitic pegmatites. Econ Geol 64:843-864

Burnham CW, Davis NF (1971) The role of H_2O in silicate melts: I. P-V-T relations in the system $NaAlSi_3O_8$-H_2O to 10 kilobars and 1000° C. Am J Sci 270:54-79

Millhollen GL, Wyllie PJ, Burnham CW (1971) Melting relations of $NaAlSi_3O_8$ to 30 kb in the presence of H_2O:CO_2 = 50:50 vapor. Am J Sci 271:473-480

Burnham CW (1972) The energy of explosive volcanic eruptions. Earth and Mineral Sci Bull 14:69-70

Holloway JR, Burnham CW (1972) Melting relations of basalt with equilibrium water pressure less than total pressure. J Petrol 13:1-29

Kilinc IA, Burnham CW (1972) Partitioning of chloride between a silicate melt and coexisting aqueous phase from 2 to 8 kilobars. Economic Geol 67:231-235

Eggler DH, Burnham CW (1973) Crystallization and fractionation trends in the system andesite-H_2O-CO_2-O_2 at pressures to 10 kb. Geol Soc Am Bull 84:2517-2532

Burnham CW (1974) $NaAlSi_3O_8$-H_2O solutions: a thermodynamic model for hydrous magmas. Bull Soc Fr Minéral Cristallogr 97:223-230

Burnham CW, Davis NF (1974) The role of H_2O in silicate melts: II. Thermodynamic and phase relations in the system $NaAlSi_3O_8$-H_2O to 10 kilobars, 700°C -1100°C. Am.J Sci 274: 902-940

Burnham CW (1975) Thermodynamics of melting in experimental silicate-volatile systems. Fortschr Mineral 52:101-118

Burnham CW (1975) Water and magmas: A mixing model. Geochim Cosmochim Acta 39:1077-1084

Burnham CW, Darken LS, Lasaga AC (1978) Water and magmas: Application of the Gibbs-Duhem Equation: A response. Geochim Cosmochim Acta 42:277-280

Flynn RT, Burnham CW (1978) An experimental determination of rare earth partition coefficients between a chloride containing vapor phase and silicate melts. Geochim. Cosmochim. Acta. 42: 685-701.

Burnham CW (1979) The importance of volatile constituents. In: The Evolution of the igneous rocks, 1077-1084, Princeton University Press, Princeton, NJ

Burnham CW (1979) Magmas and hydrothermal fluids. In: Geochemistry of hydrothermal ore deposits, 71-136, John Wiley & Sons, New York

Burnham CW, Lasaga AC (1979) Water and magmas: Another Reply. Geochim Cosmochim Acta 43: 643-647

Burnham CW, Ohmoto H (1980) Late-stage processes of felsic magmatism. Mining Geology (Japan) Special Issue, p 1-11

Burnham CW (1981) The nature of multicomponent aluminosilicate melts. In: Chemistry and Geochemistry of Solutions at High Temperatures and Pressures, p 197-229. Pergammon Press, New York.

Burnham CW (1981) Convergence and mineralization – Is there a relation? Geol Soc Am Memoir 154: 761-768.

Burnham CW (1981) Physiochemical constraints on porphyry mineralization. Arizona Geology Digest XIV:71-77

Burnham CW, Ohmoto H (1981) Late magmatic and hydrothermal processes in ore formation. In: Mineral Resources: Genetic Understanding for Practical Applications (Studies in Geophysics), 62-72, National Academy Press, Washington, DC

Boettcher AL, Burnham CW, Windom KE, Bohlen SR (1982) Liquids, glasses, and the melting of silicates to high pressure. J Geol 90:127-138

Anderson GM, Burnham CW (1983) Feldspar solubility and the transport of aluminum under metamorphic conditions. Am J Sci 283A: 283-297

Burnham CW (1983) Deep submarine pyroclastic eruptions, Econ Geol Monogr 5:142-148

Eggler DH, Burnham CW (1984) Solution of H_2O in diopside melts: a thermodynamic model. Contrib Mineral Petrol 85:58-66

Bodnar RJ, Burnham CW, Sterner SM (1985) Synthetic fluid inclusions in natural quartz. III. Determination of phase equilibrium properties in the system H_2O-NaCl to 1000°C and 1500 bars. Geochim Cosmochim Acta 49:1861-1873

Burnham CW (1985) Energy release in subvolcanic environments: implications for breccia formation. Econ Geol 80:1515-1522

Burnham CW, Nekvasil H (1986) Equilibrium properties of granitic magmas. Am Mineral 71:239-263

Nekvasil H, Burnham CW (1987) The calculated individual effects of pressure and water content on phase equilibria in the granite system. In Mysen BO (ed) Magmatic Processes: Physicochemical Principles. Geochemistry Soc Spec Publ 17:95-109

Burnham CW (1992) Calculated melt and restite compositions of some Australian granites. Trans Royal Soc Edinburgh: Earth Sciences 83:387-397.

Burnham CW (1994) Magmas and hydrothermal fluids. In: Geochemistry of Hydrothermal Ore Deposits, 3rd Edition, John Wiley & Sons (in press), New York.

VOLATILES IN MAGMAS

TABLE OF CONTENTS, VOLUME 30

Chapter 1   R. B. Symonds, W. I. Rose,
                                       G. J. S. Bluth & T. M. Gerlach

VOLCANIC-GAS STUDIES: METHODS, RESULTS, AND APPLICATIONS

Chapter 2   **P. D. Ihinger, R. L. Hervig & P. F. McMillan**

ANALYTICAL METHODS FOR VOLATILES IN GLASSES

Chapter 3   C. W. Burnham

DEVELOPMENT OF THE BURNHAM MODEL FOR PREDICTION OF H_2O SOLUBILITY IN MAGMAS

Chapter 4   P. F. McMillan

WATER SOLUBILITY AND SPECIATION MODELS

Chapter 5   J. G. Blank & R. A. Brooker

EXPERIMENTAL STUDIES OF CARBON DIOXIDE IN SILICATE MELTS: SOLUBILITY, SPECIATION, AND STABLE CARBON ISOTOPE BEHAVIOR

Chapter 6   **J. R. Holloway & J. G. Blank**

APPLICATION OF EXPERIMENTAL RESULTS TO C-O-H SPECIES IN NATURAL MELTS

Chapter 7   **M. R. Carroll & J. D. Webster**

SOLUBILITIES OF SULFUR, NOBLE GASES, NITROGEN, CHLORINE, AND FLUORINE IN MAGMAS

Chapter 8
         M. C. Johnson, A. T. Anderson, Jr.
             & M. J. Rutherford

PRE-ERUPTIVE VOLATILE CONTENTS OF MAGMAS

Chapter 9   **R. A. Lange**

THE EFFECT OF H_2O, CO_2 AND F ON THE DENSITY AND VISCOSITY OF SILICATE MELTS

xiii

DIFFUSION IN VOLATILE-BEARING MAGMAS

Chapter 12   A. Jambon

EARTH DEGASSING AND LARGE-SCALE GEOCHEMICAL CYCLING OF VOLATILE ELEMENTS

Chapter 1

VOLCANIC-GAS STUDIES:
METHODS, RESULTS, AND APPLICATIONS

Robert B. Symonds[1], William I. Rose[2], Gregg J. S. Bluth[2] and Terrrence M. Gerlach[1]

[1]*United States Geological Survey, Cascades Volcano Observatory*
5400 MacArthur Blvd.
Vancouver, Washington 98661 USA

[2]*Department of Geological Engineering, Geology, and Geophysics*
Michigan Technological University
Houghton, Michigan 49931 USA

INTRODUCTION

This chapter reviews several facets of the study of volcanic gases. The focus of the review is generally, but not exclusively on "high-temperature volcanic gases"—i.e., gases emitted at temperatures over 500°C from various sources at active volcanoes (erupting magma, lava flows, lava lakes, lava domes, eruption vents, fumaroles, fractures, etc.)— since these gases contain a substantial proportion of volatiles released directly from shallow magmas. Shallow magmas beneath active volcanoes release volatiles during both passive degassing and volcanic eruption. Passively degassing volcanoes often permit direct sampling of volcanic gases from ground-level sources and from volcanic plumes. Investigations of volcanic gases during eruptive degassing generally require remote sensing methods, including satellite-based methods. Because of the close tie to magma degassing, investigations of high-temperature volcanic gases provide important data for constraining the compositions, amounts, and origins of volatiles in magma. Volcanic gas data provide insights into magma degassing processes and critical information for evaluating volcanic hazards. In recent years, volcanic gas emissions have also received attention because of their effects on the atmosphere and climate, and as benchmarks for comparison with anthropogenic gas emissions.

SAMPLING AND IN-SITU MEASUREMENTS AT HIGH-TEMPERATURE SITES

Inherent risks and limitations

Health and safety hazards. Getting to high-temperature sites on volcanoes inevitably involves higher than normal risks regardless of the mode of transport (hiking, climbing, helicopter flight). At the site, there are serious risks from direct exposure to eruptions, explosions, lava spraying, and hot gases (Williams, 1993). Furthermore, it is virtually impossible to work at these sites without some exposure to corrosive and toxic gases, aerosols, and toxic trace metals.

Sampling bias. Unfortunately, sampling is only feasible at the Earth's surface, so all volcanic-gas samples represent low-pressure (~1 bar) discharges. A potentially more significant problem is that the samples may represent fractionated gases from very degassed magma.

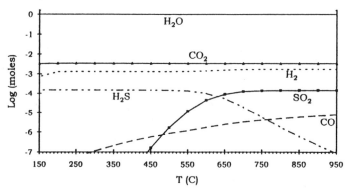

Figure 1. Calculated distribution of species for closed-system cooling of the August 11, 1960 gas sample from Showa-Shinzan's A-1 fumarole at 1 atm pressure. Gas analysis is from Mizutani and Sugiura (1982). For clarity, we exclude HCl (-3.4 log moles) and HF (-3.7 log moles) from the plot; all other species included in the calculations are shown.

Contamination. The ideal volcanic-gas sample would contain only magmatic volatiles. However, high-temperature volcanic gases are commonly mixtures of magmatic volatiles, air, meteoric steam, and gases from hydrothermal systems. It is often difficult to avoid some air contamination, because air can mix with magmatic gases at depth, in the vent, or during sampling. The common presence of meteoric water and hydrothermal fluids in volcanoes often leads to contamination of gas samples by meteoric steam and hydrothermal gases.

Reactions during cooling. For purposes of constraining magmatic-volatile compositions, volcanic-gas samples should be collected at magmatic temperatures. Unfortunately, most samples come from vents that are tens to hundreds of degrees Celsius cooler than the parental magma. During cooling, the concentrations of gas species can change (Fig. 1), although the changes are usually arrested by quenching of gas reactions at temperatures hotter than the vent temperature (Le Guern et al., 1982; Gerlach and Casadevall, 1986a).

Methods

Solution-filled collection bottles. The most common present-day method for sampling volcanic gases is to collect them in solution-filled bottles and analyze the mixtures in the laboratory (Giggenbach and Matsuo, 1991). This method was developed by Giggenbach (1975) and Giggenbach and Goguel (1989). A typical sampling setup is shown in Figure 2. A titanium or silica tube is inserted into the vent and attached to a dewared tube that minimizes condensation. The sampling train connects to an evacuated, pre-weighed sampling bottle partly filled with 4N NaOH solution. During sampling, the gas bubbles through the NaOH solution. H_2O condenses and the acid gases (CO_2, SO_2, H_2S, HCl, HF) are absorbed by the NaOH solution by the following reactions:

$$CO_2 + 2\ OH^-(aq) = CO_3^{2-}(aq) + H_2O \tag{1}$$
$$4\ SO_2 + 7\ OH^-(aq) = 3\ SO_4^{2-}(aq) + HS^-(aq) + 3\ H_2O \tag{2}$$
$$H_2S + OH^-(aq) = HS^-(aq) + H_2O \tag{3}$$
$$HCl + 2\ OH^-(aq) = 2Cl^-(aq) + H_2O \tag{4}$$
$$HF + 2\ OH^-(aq) = 2F^-(aq) + H_2O \tag{5}$$

The noncondensable gases (H_2, CO, CH_4, COS, N_2, Ar, O_2) collect in the headspace. In the laboratory, the bottles are reweighed to determine the weight gain during collection.

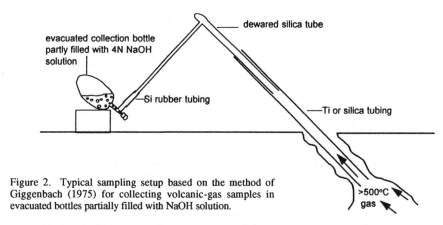

Figure 2. Typical sampling setup based on the method of Giggenbach (1975) for collecting volcanic-gas samples in evacuated bottles partially filled with NaOH solution.

The headspace gases are analyzed (normally by gas chromatography) to determine the molar amounts of each gaseous species collected. The solutions are analyzed by various techniques (e.g., wet-chemical methods, ion chromatography, selective ion electrode) for CO_3^{2-}, SO_4^{2-}, HS^-, total sulfur, Cl^-, and F^- to reconstruct the molar amounts of absorbed acid gases. H_2O is determined by difference (weight gain minus weight determined for non-H_2O gas species).

HS^- may oxidize to SO_4^{2-} after collection if the sample is contaminated with air. To preserve the original quantity of HS^-, some gas geochemists spike their NaOH solutions with Cd^{2+} or collect a second bottle filled with NH_4OH-$AgNO_3$ solution (Greenland, 1987b; Giggenbach and Matsuo, 1991). During collection, the HS^- reacts with the Cd^{2+} or Ag^+ to form CdS or AgS precipitate, which enables determination of the amount of H_2S by weighing the amount of sulfide precipitate or by determining the amount of Cd^{2+} or Ag^+ consumed by the precipitation reaction.

Another significant post-collection reaction is conversion of CO to formate (HCO_2^-):

$$CO + OH^-(aq) = HCO_2^-(aq) \qquad (6)$$

(Giggenbach and Matsuo, 1991). This reaction is controlled kinetically; CO loss depends on time and the normality of the NaOH solution, and is less of a problem when using higher normality (~4N) NaOH solutions and when collecting gases rich in acidic species (Giggenbach and Matsuo, 1991). Preliminary results suggest that CO loss follows the radioactive isotope decay equation; hence, the initial CO concentration in the gas may be determined by obtaining several CO analyses over time and extrapolating back to the time of collection (Giggenbach and Matsuo, 1991).

Samples obtained in solution-filled bottles can be affected by several alterations during collection. Condensation or re-evaporation of liquid in the sampling train due to temperature fluctuations below and above the H_2O boiling point may severely disturb H_2O, HCl, and SO_2 concentrations (Gerlach, 1980b; Giggenbach and Matsuo, 1991). Condensation can be minimized by using a dewared sampling train as shown in Figure 2. Spilling of solution from the sampling bottle during collection decreases the determined amount of H_2O but has little effect on the determined quantities of other gas species (Gerlach, 1993b). Chemical reactions between the volcanic gases and titanium sampling tubes produce excess H_2. Silica and mullite tubes do not react with high-temperature volcanic gases, but may be too fragile to use in vigorous vents (Greenland, 1987b; Giggenbach and Matsuo, 1991). Titanium tubing withstands rough field use and is probably unreactive in $\leq 400°C$ vents (Giggenbach and Matsuo, 1991). However, when used in $>400°C$ vents, titanium may react with the gases to produce H_2, especially if the gases contain $>0.1\%$ HCl (Greenland, 1987b; Giggenbach and Matsuo, 1991). The predicted reaction[1] between the titanium tube and the gas is as follows:

$$Ti(s) + 2 H_2O \Rightarrow TiO_2(s) + 2 H_2 \qquad (7)$$

Finally, minor amounts of sulfur may be lost by deposition of native sulfur prior to gas entry into the collection bottle (Matsuo, 1961; Gerlach, 1993b) by the reactions:

$$2 H_2S + SO_2 \Rightarrow 3 S(s) + 2 H_2O \qquad (8)$$

$$S_2 \Rightarrow 2 S(s) \qquad (9)$$

Native sulfur typically saturates at temperatures between 100° and 200°C (Bernard, 1985; Quisefit et al., 1989). Because temperatures of the gases often drop to ~100°C prior to entry into the collection bottle, loss of native sulfur is potentially a problem.

Pre-1975 collection techniques. Prior to 1975, most volcanic-gas samples were collected in flow-through or evacuated bottles without a solution to absorb the condensable gases, although notable exceptions are the 1954-1985 Showa-Shinzan samples (Nemoto et al., 1957; Mizutani 1962a,b; Mizutani and Sugiura, 1982). H_2O was not determined in many of the pre-1975 dry-bottle samples, although some analysts obtained estimates of the H_2O contents by determining the amount of condensate in the bottle (Giggenbach and Le Guern, 1976), by adding $CaC_2(s)$ and determining H_2O from the acetylene (C_2H_2) generated (Tazieff et al., 1972), by adding silica gel to the bottles to absorb H_2O (Huntingdon, 1973), or by condensing the H_2O prior to entry of gas into the bottle (Sigvaldason and Elisson, 1968). Dry-bottle samples are also affected by loss of sulfur from Reactions (8) and (9) and by the disproportionation of SO_2 in condensed H_2O:

$$4 SO_2 + 4 H_2O \Rightarrow H_2S + 6 H^+(aq) + 3 SO_4{}^{2-}(aq) \qquad (10)$$

Some pre-1975 collection techniques exposed the sampled gases to reactive metals (steel, Cu, Al) that caused copious H_2 generation from reaction with H_2O and HCl (Gerlach, 1979; Gerlach, 1980c; Giggenbach and Matsuo, 1991).

In-situ gas chromatography. Some investigators have obtained analyses of volcanic gases by injecting them directly into a field gas chromatograph (hereafter, FGC; Le Guern et al., 1982). This method can determine H_2O, H_2, CO_2, CO, SO_2, H_2S, and atmospheric gases, but HCl and HF must be determined by the bottle-sampling approach (Le Guern et al., 1982). This method potentially yields real-time data on gas composition before leaving the field. Successful application of FGC is limited to sites where it is possible to work for an extended period of time (hours to days). The bottle-sampling approach is more desirable than FGC for hazardous sites where an extended stay is impractical or unsafe and where the results are not needed immediately. Other disadvantages of FGC are relatively large errors in H_2O and the inability to measure HCl and HF.

Oxygen fugacity probes. Some workers have used solid electrolyte probes to measure the oxygen fugacity (fO_2) of the gas in the field (Gantes et al., 1983). The most common probe consists of an electrolyte core, generally made of Zr and Y oxides, coated by a porous noble metal such as Pt (Gantes et al., 1983). The exchange of oxygen ions between the electrolyte core and the noble metal coating fuels the oxidation process of reduced volcanic-gas species and produces a current that is a function of fO_2.

---

[1]Reaction predicted by reacting a small amount of Ti(s) with a range of volcanic-gas compositions at 400° to 1200°C using thermochemical equilibrium program, GASWORKS (Symonds and Reed, 1993). The results show that Ti(s) reacts to form rutile (TiO_2) in preference to 28 other Ti-bearing gases and solids.

EVALUATION OF VOLCANIC-GAS ANALYSES

General guidelines

We suggest several guidelines to aid in selecting and evaluating gas samples that are likely to contain a high proportion of magmatic volatiles. A high collection temperature (>500°C) is one of the best indicators that a volcanic-gas sample is mainly from a magmatic source. Samples obtained in solution-filled bottles or from in-situ FGC are in general preferable to samples collected in flow-though or evacuated bottles without caustic solution. After collection, gas samples can be evaluated in several ways. When available, stable isotope data ($\delta^{18}O$ and δD data on H_2O, $\delta^{13}C$ data on CO_2 and CH_4, and $\delta^{34}S$ data on total sulfur) are extremely valuable to help determine whether these constituents come from magma, meteoric fluids, sedimentary rocks or fluids within them, seawater, or from some other source (Allard, 1983; Taylor, 1986).For thermochemical evaluation of a gas sample, it is desirable to have analytical data for as many species as possible, although it is common for analyses to lack data for critical species (e.g., SO_2, H_2, H_2S, CO) either because no determinations were made or because of detection limitations. Finally, gas samples are preferred that contain <1% air contamination or that permit reliable removal of air-contamination effects.

The above guidelines should only be viewed as aids in selecting gas samples likely to be enriched in magmatic volatiles. Judgment is required, however, because the guidelines may exclude "good" samples. For instance, the 1918-19 J-series Kilauea samples display only minor modifications from their exhaled equilibrium state even though the samples contain up to 90% air and were collected in evacuated bottles without caustic solution (Gerlach, 1980a). Finally, some samples satisfy the above guidelines, but are nonetheless "bad" for other reasons. We now consider more systematic evaluation procedures.

Evaluation procedures

Air contamination. Air contamination is a common and sometimes serious problem with volcanic-gas samples. Of the three most abundant gases in air, Ar is unreactive and N_2 reacts very little with high-temperature volcanic gases (Giggenbach, 1980, 1987), but O_2 may react significantly with reduced-gas species. Air contamination commonly causes the N_2/Ar ratio of the sample to match approximately the value for air (83.6). N_2/Ar ratios that are much higher than 83.6 or that are about equal to 38.5 (air-saturated groundwater) suggest that N_2 and Ar may come from a deeper magmatic or sedimentary source, or from air-saturated-groundwater, respectively (Magro and Pennisi, 1991). Most gas analyses that contain more than 1% N_2 have N_2/Ar ratios close to air (Fig. 3). The amount of air in such samples can be determined with the equation:

$$air = N_2 + Ar + 0.268 \, N_2 \qquad (11)$$

where all terms are in moles and the term, $0.268 \, N_2$, estimates the original amount of oxygen before any reaction with the sample. Sometimes Ar is not determined separately, but is analyzed as O_2 + Ar. In this case, air contamination gives $N_2/(O_2 + Ar)$ about equal to the value for air (3.6), although the ratio will be greater if some O_2 reacted with the sample. The isotopes, ^{20}Ne, ^{22}Ne, ^{36}Ar, and ^{38}Ar, are atmospheric tracers that may also help discern air contamination in volcanic gases (Magro and Pennisi, 1991).

The severity of air contamination depends on how much atmospheric O_2 reacts with the sample. If all the N_2 in a sample comes from air, then the amount of reacted O_2, $(O_2)_r$, can be estimated with the equation:

$$(O_2)_r = 0.268 \, N_2 - (O_2)_m \qquad (12)$$

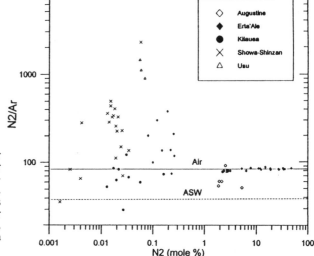

Figure 3. Plot showing N_2/Ar ratios versus total N_2 for selected analyses of high-temperature vol-canic gases. The solid and dashed lines show the N_2/Ar ratios of air and air saturated water (ASW), respectively. Sources of data are in Tables 3-5 (below).

where all terms are in moles and $(O_2)_m$ is the analyzed amount of O_2. Ideally, $(O_2)_r$ should account for all oxidation of reduced-gas species (Gerlach and Casadevall, 1986a). If they are determined to come from the atmosphere, N_2 and Ar are removed from gas analyses. O_2 is also removed from gas analyses, although if it has reacted with the reduced-gas species, the procedure is somewhat more complex (see below).

Equilibrium-disequilibrium analysis. Past investigations indicate that high-temperature volcanic gases initially approach a state of thermodynamic equilibrium (Ellis, 1957; Heald et al., 1963; Nordlie, 1971; Giggenbach and Le Guern, 1976; Gerlach, 1980a-d, 1993a,b; Le Guern et al., 1982; Gerlach and Casadevall, 1986a; Giggenbach, 1987). The strongest support for the equilibrium model of high-temperature volcanic gases is that some samples have equilibrium compositions (e.g., Le Guern et al., 1982; Gerlach and Casadevall, 1986a; Gerlach, 1993a,b) and the equilibrium compositions invariably imply equilibrium temperatures that are greater that the corresponding collection temperatures. These observations suggest that the equilibrium compositions represent the last equilibrium of the high-temperature volcanic gases before the gases stopped reacting (quenched) during cooling. Samples with "quenched equilibrium compositions", however, are relatively uncommon. Most analyses of samples collected before 1975 show evidence of severe disequilibrium (Gerlach, 1980a-d), and modern collection and analytical methods often give samples with mild to moderate disequilibrium (Gerlach, 1993b). To distinguish equilibrium and disequilibrium samples, we turn to thermodynamic modeling.

A flow chart for the equilibrium-disequilibrium determination is shown in Figure 4. We illustrate the procedure for SOLVGAS[2], an equilibrium computer program for gases (Symonds and Reed, 1993), although the same results can be achieved with other computer programs (e.g., Heald et al., 1963; Gerlach, 1980a, 1993b) or graphical methods (Greenland, 1987b). The procedure involves calculating correspondence temperatures (CTs) for the analyzed species to evaluate whether or not a sample is an equilibrium mixture. A CT is the temperature at which the calculated equilibrium molar abundance of

---

[2]The latest versions of programs SOLVGAS and GASWORKS, the GASTHERM thermochemical data base, and the accompanying manuals are now available for distribution. These are FORTRAN 77 programs and run on 80386- and 80486-level IBM-compatible personal computers, and also on IBM and VAX mainframe computers. They can be obtained for a small distribution cost from the senior author.

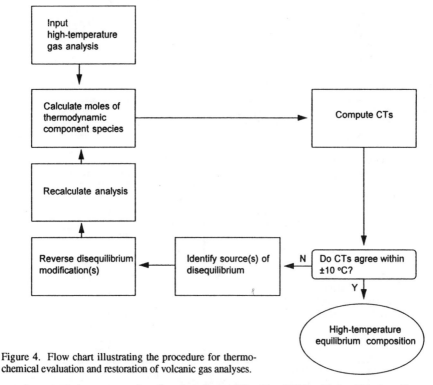

Figure 4. Flow chart illustrating the procedure for thermo-chemical evaluation and restoration of volcanic gas analyses.

species equals its measured molar abundance (Nordlie, 1971). If the CTs for all species agree within ±10°C, an acceptable limit for typical analytical error, the analysis is interpreted to be a quenched equilibrium composition (Gerlach, 1993b). If the CTs do not agree within ±10°C, the analysis is a disequilibrium composition.

Equilibrium-disequilibrium determinations normally include H_2O, CO_2, SO_2, H_2, H_2S, HCl, HF, and CO but exclude free O_2 and the amounts of N_2 and Ar indicated to be from air contamination. CH_4 and NH_3 are often excluded because they are almost always extraneous (see below). The analysis is recast in terms of thermodynamic component species, as shown in Table 1 for the August 11, 1960 sample from the A-1 vent at Showa-Shinzan (case without N_2, CH_4, and NH_3). SOLVGAS computes the distribution of gas species as a function of temperature using the mass balances of component species. In these calculations, only reactions between the selected analyzed species, H_2O, CO_2, H_2, HCl, HF, SO_2, CO, and H_2S in this case, are considered. Comparisons of the calculated moles of each species with the observed moles determine the CTs. We illustrate this graphically by plotting the logarithms of the calculated/observed moles versus temperature for each species, since this quantity equals zero at the CT but nonzero above and below the CT. The graphical (Fig. 5) and tabular (Table 1) results for the 1960 Showa-Shinzan sample show that the CTs for H_2O, H_2, SO_2, H_2S, CO_2, and CO converge at 735 ± 4°C. We do not report CTs for HCl and HF; their calculated and measured amounts are equal at all temperatures, since volcanic gases lack measurable amounts of other Cl- and F-bearing species that react with HCl and HF. The analysis is interpreted to be a quenched equilibrium composition because the divergence of CTs is less than ±10°C. If CTs diverge by more than ±10°C, the analysis is a disequilibrium composition, and the potential causes of disequilibrium are evaluated.

Table 1. Gas analysis, moles of component species, correspondence temperatures (CTs), and restored composition for the August 11, 1960 gas sample collected from the A-1 vent at Showa-Shinzan. The moles of component species are shown along with CTs are shown for cases with and without CH_4, NH_3, and N_2.

Species	Gas analysis[†] (moles)	With CH_4, NH_3, and N_2		Without CH_4, NH_3, and N_2		Restored composition (moles)
		Recast to component cpecies[‡] (moles)	CTs (°C)	Recast to component cpecies[‡] (moles)	CTs (°C)	
H_2O	99.39	99.41692	726	99.4176	739	99.39
CO_2	0.34	0.34074	903	0.34040	731	0.34
H_2	0.17	0.129835	720	0.12840	739	0.17
HCl	0.042	0.042	---	0.042	---	0.042
N_2	0.024	0.024025	194	---	---	---
HF	0.019	0.019	---	0.019	---	0.019
SO_2	0.014	---	740	---	739	0.014
H_2S	0.0008	0.0148	740	0.0148	739	0.0008
CO	0.00040	---	729	---	731	0.00040
CH_4	0.00034	---	210	---	---	6.4×10^{-12}
Ar	0.00016	---	---	---	---	---
O_2	0.0000	---	---	---	---	---
NH_3	0.00005	---	194	---	---	---
S_2	---	---	---	---	---	1.5×10^{-5}
$\log fO_2$						-14.38
T(°C)	722 (obs)					735 (calc)

[†] Mizutani and Sugiura (1962)
[‡] SO_2, CO, CH_4, and NH_3 are not component species, so they are re-expressed interms of the other species (see Symonds and Reed, 1993); Ar and O_2 are excluded.

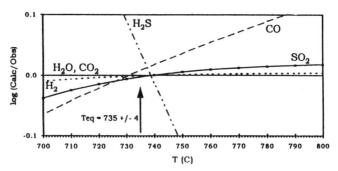

Figure 5. Log (calculated/observed) moles vs. temperature plot for a Showa-Shinzan gas sample collected by Mizutani and Sugiura (1982) from the A-1 vent, August 11, 1960. The CT for a species is the temperature at which the log (calculated/observed) moles equals zero. The plotted species are the only ones included in the calculations; we removed CH_4, NH_3, N_2, and O_2 from the original analysis before calculating the CTs. Note that the scales on the vertical axes conceal the tiny variations in the H_2O, CO_2, and N_2 lines.

Common causes of disequilibrium

There are many causes of disequilibrium in high-temperature volcanic-gas samples. Naturally occurring volatile compounds from non-magmatic sources (heated rock, hydrothermal systems, meteoric water, air, etc.) may contaminate a high-temperature gas initially at equilibrium and produce a disequilibrium mixture that never equilibrates fully before collection. A variety of collection-related factors (poor sampling design, equipment failures, operator errors, etc.) may cause modifications of high-temperature equilibrium gas compositions and result in an analysis with a disequilibrium composition. Inadequate stabilization of a sample after collection (leakage, post-collection reactions, etc.) may give a similar result. Finally, analytical errors may result in a disequilibrium analytical composition.

Oxidation of H_2, H_2S, or CO. The main cause of disequilibrium is oxidation of H_2, H_2S, or CO by the reactions:

$$H_2 + 1/2\, O_2 = H_2O \tag{13}$$
$$H_2S + 3/2\, O_2 = SO_2 + H_2O \tag{14}$$
$$CO + 1/2\, O_2 = CO_2 \tag{15}$$

Atmospheric oxidation of one of these species is potentially significant if $(O_2)_r$ (Eqn. 12) is greater than the concentration of that species. Nonequilibrium oxidation of H_2 to H_2O produces low CTs for H_2O, H_2, SO_2, and H_2S but high CTs for CO_2 and CO, whereas disequilibrium oxidation of H_2S to SO_2 or CO to CO_2 produce high CTs for H_2O, H_2, SO_2, and H_2S but low CTs for CO_2 and CO (Table 2).

Gains or losses of H_2O. Common sources of disequilibrium related to gains/losses of H_2O include condensation of H_2O in the intake line (Gerlach, 1980b), spilling of solution from the sampling bottle (Gerlach, 1993b), analytical errors (Gerlach, 1980b), and contamination of magmatic gases with unequilibrated meteoric steam (Symonds and Mizutani, in prep.). Condensate will form in the intake line if the gas drops below the boiling point of H_2O. Formation of condensate during collection can give samples with disequilibrium compositions from H_2O loss. However, sometimes increased gas flow during collection will cause condensate to re-evaporate and the sample may show disequilibrium from gains in H_2O. Spilling of solution from sampling bottles is a common problem. The main effect of spilling is disequilibrium related to H_2O loss, since water is determined by difference (Gerlach, 1993b). Disequilibrium can also result from analytical errors in determinations of H_2O concentrations. For instance, the 1971 Erta'Ale gas analyses display disequilibrium and contain highly variable amounts of H_2O because water was determined by an inefficient analytical method (conversion of H_2O to acetylene by CaC_2; Giggenbach and Le Guern, 1976). The H_2O analytical errors in consecutive samples from the same vent over a short period of time produce large variations in H/C while C/S ratios are relatively constant (Fig. 6). When H_2O concentrations are determined by difference, they may contain significant error if the mass of sample collected is close to the analytical error of the balance. Contamination of volcanic gases with unequilibrated meteoric steam prior to collection is yet another potential source of disequilibrium from H_2O gains. In these cases, δD and $\delta^{18}O$ data for H_2O may be a great help diagnosing mixing of meteoric and magmatic gases (Mizutani and Sugiura, 1982; Taylor, 1986). Gain of unequilibrated H_2O will produce low CTs for H_2O, H_2, SO_2, and H_2S but high CTs for CO_2 and CO (Table 2). In contrast, loss of H_2O without re-equilibration will produce high CTs for H_2O, H_2, SO_2, and H_2S but low CTs for CO_2 and CO (Table 2).

Extraneous CH_4 and NH_3. In high-temperature volcanic gases, the CTs for CH_4 and NH_3 are typically much lower than the CTs for other significant species. Samples with NH_3 may also have low CTs for N_2 because it is the only significant nitrogen species that reacts with NH_3. The Showa-Shinzan sample reported in Table 1 is an example of a gas analysis that has much lower CTs for N_2, CH_4, and NH_3 than for H_2O, CO_2, H_2, H_2S, SO_2, and CO. Low CTs for N_2, CH_4, and NH_3 arise from the relatively high concentrations of CH_4 and NH_3 in these samples, since their predicted equilibrium concentrations increase with decreasing temperature. This extraneous CH_4 and NH_3

probably comes from deep hydrothermal (Giggenbach, 1987) or sedimentary (Kodosky et al., 1991; Symonds and Mizutani, in prep.) gases that mix with the high-temperature magmatic gases without equilibrating with them. Sometimes, $\delta^{13}C$ data may support a sedimentary origin for CH_4 (Kodosky et al., 1991). Unequilibrated CH_4, NH_3, and N_2 are generally removed from the analysis, and the CTs are recalculated (Table 2).

Table 2. Summary of common disequilibrium modifications of volcanic-gas samples and restoration procedures to correct them.

Modification	Evidence	Restoration procedure
Extraneous CH_4 in sample	1-2	Remove CH_4 from the analysis. See Gerlach and Casadevall (1986a) and Kodosky et al. (1991).
Gains of unequilibrated H_2O	3-5	Remove a trial amount of H_2O from analysis, renormalize adjusted analysis, and recalculate CTs. Repeat procedure until all CTs converge. See Gerlach (1980a,b).
Nonequilibrium H_2O loss	4, 6-7	Add a trial amount of H_2O to the original analysis, renormalize adjusted analysis, and recalculate CTs. Repeat procedure until all CTs converge. See Gerlach (1980b), Symonds et al. (1990), and Gerlach (1993b).
Nonequilibrium oxidation of CO to CO_2	6, 8, 9	Convert a trial amount of CO to CO_2 in the original analysis by reversing the reaction: $CO + 1/2O_2 \Rightarrow CO_2$, renormalize the adjusted analysis, and recalculate CTs. Repeat procedure until all CTs converge. See Gerlach and Casadevall (1986a).
Nonequilibrium oxidation of H_2 to H_2O	3, 8, 10	Convert a trial amount of H_2O to H_2 in the original analysis by reversing the reaction: $H_2 + 1/2O_2 \Rightarrow H_2O$, renormalize adjusted analysis, and recalculate CTs. Repeat procedure until all CTs converge. See Le Guern et al. (1982).
Nonequilibrium oxidation of H_2S to SO_2	6, 8, 11	Convert trial amounts of H_2S to SO_2 by reversing the reaction: $H_2S + 3/2O_2 \Rightarrow SO_2 + H_2O$, renormalize adjusted analysis, and recalculate CTs. Repeat procedure until all CTs converge. See Gerlach (1980a).
Gains of unequilibrated H_2	6, 12	If relatively minor, identify and reverse reaction that produces excess H_2, renormalize adjusted analysis, and recalculate CTs. If major, estimate equilibrium gas composition from the mass balances of C, H, and S in the analysis, and from independent estimates of fO_2. See Gerlach (1979. 1980c).

(1) Very low CTs for CH_4 relative to all other species. (2) $\delta^{13}C$ data may indicate a sedimentary source for CH_4. (3) Low CTs for H_2, H_2O, SO_2, and H_2S relative to CTs for CO_2 and CO. (4) Simultaneous samples from the same vent will have widely variable H/C and H/S ratios but relatively constant C/S ratios. (5) δD and $\delta^{18}O$ data for contemporaneous fumarolic condensates may indicate mixing between magmatic and meteoric fluids if the excess H_2O is from a meteoric source. (6) High CTs for H_2, H_2O, SO_2, and H_2S relative to CTs for CO_2 and CO. (7) NaOH solution may have spilled from sampling bottle during collection. (8) Samples may show evidence of air contamination (e.g., >1% N_2, $N_2/Ar \approx$ atmospheric value of 83.6). (9) Simultaneous samples from the same vent may show variable $CO/(CO + CO_2)$ ratios. (10) Simultaneous samples from the same vent may show variable $H_2/(H_2 + H_2O)$ ratios. (11) Simultaneous samples from the same vent may show variable $H_2S/(H_2S + SO_2)$ ratios. (12) Samples probably reacted with Cu, Al, steel, or Ti collection materials.

Gains of H_2. Samples collected prior to 1975 may show disequilibrium from excess H_2 generated by reactions of H_2O and HCl with Cu, Al, or steel sampling materials, as discussed above (Gerlach, 1979, 1980c). Today, samples containing excess H_2 are rare because most are collected with relatively unreactive silica or titanium sampling tubes. However, recent work by Giggenbach and Matsuo (1991) suggests that titanium can react with >400°C volcanic gases to generate H_2 by Reaction (7), especially if the gases contain significant HCl. The symptoms of H_2 gains are anomalously low fO_2 values (Gerlach, 1979) and high CTs for H_2O, H_2, SO_2, and H_2S but low CTs for CO_2 and CO (Table 2).

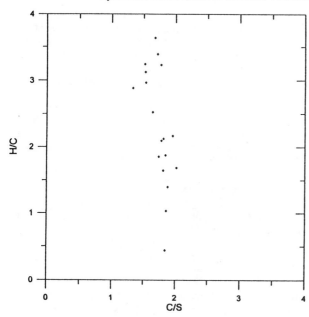

Figure 6. A plot of H/C versus C/S ratios for the 1971 Erta'Ale gas analyses from Giggenbach and Le Guern (1976). All samples were collected from a 1075°C vent over a two-hour period. H_2O errors cause H/C to vary over a wide range but do not affect C/S which is relatively constant (see text).

Retrieval of equilibrium compositions

We assume that high-temperature volcanic-gas samples with quenched equilibrium compositions represent better the volatiles degassed from shallow magma than do samples with disequilibrium compositions. It is therefore desirable to retrieve the initial equilibrium compositions of disequilibrium samples. Retrieval of equilibrium compositions is possible for some disequilibrium samples, but with varying degrees of reliability. If the samples have disequilibrium compositions caused predominantly by a single type of alteration or analytical error, as opposed to multiple or superimposed sources of disequilibrium, their equilibrium compositions often can be restored if the analyses are sufficiently complete (Gerlach and Casadevall, 1986a). Restoration makes no a priori assumptions of an initial equilibrium state but in fact tests the equilibrium hypothesis for high-temperature volcanic gases and yields the last equilibrium composition and temperature. "Restored equilibrium compositions" are considered about as reliable as quenched equilibrium compositions. "Apparent equilibrium compositions" are based on incomplete analyses and the assumption that the gases initially approached a state of chemical equilibrium (Gerlach and Casadevall, 1986a). Analytical data must be complete enough, however, to constrain both the last equilibrium temperature and the concentrations of the undetermined species. Apparent equilibrium compositions are less reliable than restored equilibrium compositions. "Estimated equilibrium compositions" are based on incomplete analyses and the assumption that the gases initially approached a state of chemical equilibrium (Gerlach and Casadevall, 1986a). However, for estimated compositions it is necessary to assume that the gases were in equilibrium at the collection temperature since the analytical data are insufficient for simultaneous determination of both the last equilibrium temperature and concentrations of the undetermined species. Estimated equilibrium compositions are the least reliable of the retrieved equilibrium compositions for high-temperature volcanic gases..

Removal of disequilibrium effects. There is a large body of literature (Gerlach 1979, 1980a-d, 1981, 1993a,b; Le Guern et al., 1982; Gerlach and Casadevall, 1986a; Symonds et al., 1990; Kodosky et al., 1991) that describes computer-based methods (e.g., using program SOLVGAS) for removing disequilibrium modifications from volcanic-gas analyses. These restoration procedures involve identifying the sources of disequilibrium, reversing their effects on the analytical data, recalculating the moles of component species, and recomputing the CTs to determine if the adjusted composition is closer to an equilibrium composition (Fig. 4). This process is repeated iteratively until all the CTs converge, or it is determined that a restored equilibrium composition cannot be obtained.

Table 2 summarizes the restoration procedures for the most common disequilibrium modifications of volcanic-gas analyses. For restorable samples that suffer from disequilibrium oxidation of H_2, H_2S, or CO, their restored equilibrium compositions are obtained by reversing Reactions (13), (14), or (15), respectively. For instance, if a sample suffers from disequilibrium oxidation of H_2, its restored equilibrium composition is obtained by converting the gained amount of H_2O to H_2 by reversing Reaction (13). This is done quickly on a computer by converting a trial amount of H_2O to H_2 in the original analysis, discarding the appropriate amount of oxygen, and recomputing the CTs. The procedure is repeated until the CTs converge to within ±10°C. Disequilibrium loss or gain of H_2O is treated in a similar fashion by adding or subtracting trial amounts of H_2O to/from the original analysis, depending on whether H_2O has been lost or gained, and computing CTs for the adjusted analysis. This process is repeated until the equilibrium amount of H_2O is found and the CTs converge.

Some gas analyses have multiple sources of disequilibrium. For instance, the 1974 Erta'Ale samples show evidence of gains or losses of H_2O, loss of sulfur due to reaction with the steel lead-in tubes, and atmospheric oxidation of reduced sulfur species (Gerlach, 1980b). Application of the numerical operations in Table 2 will eliminate only one source of disequilibrium so other methods must be used to remove multiple disequilibrium problems. One common approach is to estimate the last equilibrium composition using the analytical data for unmodified species and assume that the gas was last in equilibrium at the vent temperature (Gerlach, 1980a-c). For instance, Gerlach (1980b) obtains estimated equilibrium compositions for the 1974 Erta'Ale samples, all collected sequentially from the same vent, by the following steps: (1) compute the fO_2 of each sample at the vent temperature from equilibria (15) and analytical data for CO and CO_2, which show consistent CO/CO_2 values; (2) calculate the equilibrium amounts of H_2O for each sample from analytical data for H_2 and the computed fO_2 values; (3) add S to samples with relatively low S/C ratios; and (4) calculate the equilibrium distribution of S species from the computed fO_2 values and the mass balances for C, H, and S in each analysis.

Estimating concentrations of minor and trace species. Many volcanic gas analyses lack data minor and trace species such as COS, S_2, and magmatic O_2 and CH_4). Some of these species may be important in some high-temperature volcanic gases (COS, S_2, CH_4; Gerlach and Nordlie, 1975), and some define petrologically useful intensive parameters (fO_2, fS_2). The equilibrium amounts of COS, S_2, CH_4, and O_2 can be estimated from the following equilibria:

$$CO_2 + H_2S = COS + H_2O \qquad (16)$$
$$2\,H_2S = 2\,H_2 + S_2 \qquad (17)$$
$$CO_2 + 4\,H_2 = CH_4 + 2\,H_2O \qquad (18)$$
$$2\,H_2O = 2\,H_2 + O_2 \qquad (19)$$

combined with known equilibrium concentrations of H_2O, H_2, CO_2, and H_2S, and the following equations:

$$n_{COS} = \frac{(\phi_{CO_2} n_{CO_2})(\phi_{H_2S} n_{H_2S}) K_{15}}{\phi_{COS}(\phi_{H_2O} n_{H_2O})} \qquad (20)$$

$$n_{S_2} = \frac{N(\phi_{H_2S} n_{H_2S})^2 K_{16}}{\phi_{S_2}(\phi_{H_2} n_{H_2})^2 P} \tag{21}$$

$$n_{CH_4} = \frac{P^2(\phi_{CO_2} n_{CO_2})(\phi_{H_2} n_{H_2})^4 K_{17}}{\phi_{CH_4}(\phi_{H_2O} n_{H_2O})^2 N^2} \tag{22}$$

$$n_{O_2} = \frac{N(\phi_{H_2O} n_{H_2O})^2 K_{18}}{\phi_{O_2}(\phi_{H_2} n_{H_2})^2 P} \tag{23}$$

where n_i is the molar abundance of species i, ϕ_i is the fugacity coefficient of species i, P is the pressure in atmospheres, N is the total moles of gas, and K_{16}, K_{17}, K_{18}, and K_{19} are the equilibrium constants for Reactions (16) through (19), respectively, at pressure P and at the equilibrium temperature of the sample. Since pressure is atmospheric in collection environments (Stevenson, 1993) and the gases can be assumed to be ideal, the P and ϕ_i terms equal unity. For example, to calculate the equilibrium amount of S_2, n_{S_2} is first estimated from (21) with the initial value of N taken as the total moles of gas species (without S_2). The estimate for n_{S_2} is added to the molar amounts of species to calculate an improved value for N, which in turn gives a better estimate of n_{S_2}. This process is repeated until n_{S_2} and N converge. The interactive procedure can be neglected if n_i for the undetermined species is many orders of magnitude smaller than N.

EQUILIBRIUM COMPOSITIONS OF
HIGH-TEMPERATURE VOLCANIC GASES

In Tables 3-5, we compile the quenched, restored, apparent, and estimated equilibrium compositions of 136 high-temperature volcanic-gas samples from convergent-plate, divergent-plate, and hot-spot volcanoes (Tables 3-5). The tables include most published and some unpublished equilibrium compositions of >500°C volcanic gases. Table 6 lists additional sources of recent high-quality volcanic-gas data for samples collected at >500°C vents and that undoubtedly contain quenched and retrievable equilibrium compositions. Preliminary interpretation of these additional analyses suggests that they generally support our conclusions on volcanic gases as discussed below. Discussions of studies on <500°C volcanic gases are beyond the scope of this paper, since these low-temperature gases are less likely to contain significant amounts of magmatic volatiles.

Molecular compositions

For each sample in Tables 3-5, we report the 1-bar, mole% concentrations of H_2O, CO_2, SO_2, H_2, H_2S, and CO; many analyses also contain data for HCl and HF, and some include estimated amounts of S_2, COS, and SO. The tables also include the determined values for the equilibrium temperature and fO_2. The tables exclude data for minor amounts of N_2 and Ar that are available for many analyses because these species are often atmospheric in our compiled samples (e.g., Fig. 3).

H_2O, CO_2, and SO_2 are the dominant species in all samples. This is shown clearly in Figure 7 where all samples plot very close to the H_2O-XO_2 mixing line:

$$H_2O = 100 - XO_2 \tag{24}$$

where $XO_2 = SO_2 + CO_2$. Samples departing from the H_2O-XO_2 line have significant concentrations of one or more minor species, including H_2, H_2S, HCl, CO, and S_2. H_2 concentrations range from 0.01 to 3% H_2 (Tables 3-5), and are higher for higher temperatures (Fig. 1) and lower fO_2 values. The concentrations of H_2S and S_2 increase with rising amounts of total S, decreasing fO_2, and with decreasing temperature (Gerlach

Table 3. Equilibrium compositions, temperatures, and log fO_2 values of high-temperature and low-pressure (1 bar) volcanic gases from convergent-plate volcanoes. Concentrations of species reported in mole %; log fO_2 given as log bars.

Magma	Date	Number	T (°C)	log fO_2	H_2O	H_2	CO_2	CO	SO_2	H_2S	S_2	HCl	HF	COS	SO
Mount St. Augustine (Kodosky et al., 1991)															
andesite	7/79	79A3G	648	-17.54	97.23	0.381	1.90	0.0035	0.006	0.057	—	0.365	0.056	—	—
"	"	79A4Ga	746	-15.20	97.46	0.568	1.23	0.0053	0.171	0.326	—	0.157	<.003	—	—
"	"	79A4Gb	746	-15.37	97.41	0.690	1.12	0.0058	0.111	0.377	—	0.226	<.003	—	—
"	"	79A6G	721	-15.77	98.31	0.522	0.74	0.0026	0.045	0.128	—	0.239	<.003	—	—
Mount St. Augustine (Symonds et al., 1990)															
andesite	8/28/87	dome-2	870	-12.45	84.77	0.54	2.27	0.016	6.18	0.68	0.12	5.34	0.086	—	—
"	"	dome-3	"	-12.59	83.91	0.63	2.40	0.020	5.72	1.00	0.20	6.04	0.086	—	—
Mount St. Augustine (Symonds, Gerlach, and Iven, unpublished)															
andesite	7/6/89	Spine-1A	743	-15.20	96.83	0.54	1.49	0.0060	0.22	0.38	—	0.51	0.025	—	—
"	"	Spine-1B	764	-14.71	96.80	0.56	1.59	0.0072	0.29	0.32	—	0.41	0.028	—	—
"	"	Spine-1C	744	-15.18	96.93	0.54	1.39	0.0056	0.21	0.34	—	0.56	0.034	—	—
"	"	Spine-1D	775	-14.46	96.73	0.56	1.50	0.0072	0.33	0.29	—	0.55	0.030	—	—
"	"	Spine-1E	752	-14.99	96.94	0.54	1.45	0.0061	0.25	0.34	—	0.45	0.030	—	—
Mount Etna (Huntingdon, 1973; Gerlach, 1979)															
hawaiite	7/12/70	horn. 1 #11	1075	-9.47	27.71	0.30	22.76	0.48	47.70	0.22	0.76	—	—	—	0.06
"	7/12/70	" "	"	"	47.10	0.51	23.42	0.49	27.96	0.22	0.26	—	—	—	0.04
		#16A													
"	7/13/70	" #8	"	"	46.91	0.50	22.87	0.48	28.70	0.22	0.28	—	—	—	0.04
"	7/14/70	" #13	"	"	44.59	0.48	22.23	0.46	31.61	0.24	0.33	—	—	—	0.04
"	7/12/70	horn. 2 #10	"	"	47.26	0.51	26.06	0.54	25.18	0.20	0.21	—	—	—	0.03
"	7/12/70	" #17	"	"	46.80	0.50	24.55	0.51	27.14	0.21	0.25	—	—	—	0.03
"	7/12/70	" "	"	"	49.88	0.54	21.69	0.45	26.93	0.22	0.24	—	—	—	0.03
		#14A													
"	7/12/70	" "	"	"	53.69	0.57	20.00	0.42	24.85	0.22	0.21	—	—	—	0.03
		#12A													
"	7/13/70	" #5	"	"	49.33	0.53	24.32	0.51	24.86	0.20	0.21	—	—	—	0.03
"	7/13/70	" #7	"	"	43.09	0.46	24.00	0.50	31.34	0.23	0.33	—	—	—	0.04
"	7/13/70	" #6	"	"	49.91	0.54	23.56	0.49	25.05	0.21	0.21	—	—	—	0.03
"	7/13/70	" #9	"	"	48.27	0.52	19.71	0.41	30.48	0.25	0.31	—	—	—	0.04
"	7/14/70	" #15	"	"	47.82	0.51	17.08	0.36	33.54	0.27	0.38	—	—	—	0.04
"	7/13/70	horn. 3 #1	"	"	49.50	0.53	33.58	0.70	15.45	0.13	0.08	—	—	—	0.02
"	7/13/70	" #4	"	"	49.91	0.54	33.93	0.71	14.69	0.12	0.07	—	—	—	0.02

Table 3 (continued)

Magma	Date	Number	T(°C)	log fO₂	H₂O	H₂	CO₂	CO	SO₂	H₂S	S₂	HCl	HF	COS	SO
Gunung Merapi (Le Guern et al., 1982)															
andesite	1978	Mer 78-1	826	-13.70	95.30	0.87	3.31	0.03	0.25	0.24	0.003	—	—	—	—
"	"	Mer 78-2	806	-14.01	94.53	0.74	4.08	0.03	0.31	0.30	0.004	—	—	—	—
"	"	Mer 78-3	820	-13.79	93.64	0.81	4.54	0.04	0.50	0.46	0.010	—	—	—	—
"	"	Mer 78-4	900	-12.69	91.04	1.44	5.19	0.10	1.07	1.08	0.070	—	—	—	—
"	"	Mer 78-5	767	-14.86	95.83	0.71	3.26	0.02	0.06	0.12	0.0003	—	—	—	—
"	1979	Mer 79-1	895	-12.77	88.53	1.37	7.56	0.15	1.14	1.16	0.08	0.59	0.04	—	—
"	"	Mer 79-2	915	-12.49	88.87	1.54	7.07	0.16	1.15	1.12	0.08	0.59	0.04	—	—
Momotombo (Bernard, 1985)															
thol. basalt	12/80	MoMo-1	820	-13.55	97.11	0.70	1.44	0.0096	0.50	0.23	0.0003	2.89	0.259	—	—
"	"	MoMo-2	658	-16.52	97.90	0.17	1.47	0.0015	0.30	0.16	0.0009	2.68	0.240	—	—
"	"	MoMo-3	777	-14.35	97.30	0.50	1.44	0.0063	0.45	0.29	0.0040	2.86	0.256	—	—
Mount St Helens (Gerlach and Casadevall, 1986a)															
dacite	9/25/80	800925-710	802	-14.25	91.58	0.8542	6.942	0.06	0.2089	0.3553	0.0039	—	—	0.0008	—
"	9/16/81	CNRS	663	-16.76	98.52	0.269	0.913	0.0013	0.073	0.137	0.0003	0.089	—	0.00002	—
"	9/17/81	CNR	710	-15.77	98.6	0.39	0.886	0.0023	0.067	0.099	0.0002	0.076	0.03	1.8E-05	—
Poas (Delorme, 1983, Barquero, unpublished; Rowe, 1991)															
thol. basalt	6/17/81	P31	1002	-10.11	95.59	0.5997	0.8170	0.0083	1.877	0.0198	—	0.974	0.011	1.2E-06	—
"	6/18/81	P35	1010	-9.96	96.12	0.5889	0.7042	0.0071	1.622	0.0140	—	0.814	0.098	7.3E-08	—
"	6/18/81	P39	971	-10.53	96.04	0.5433	0.7289	0.0062	1.681	0.0224	—	0.874	0.102	1.1E-07	—
"	6/19/81	P44	989	-10.20	96.29	0.5240	0.7768	0.0066	1.511	0.0131	—	0.784	0.091	7.1E-08	—
"	6/28/81	P52	1007	-9.99	95.28	0.5639	0.9607	0.0093	2.004	0.0163	—	1.039	0.121	1.2E-07	—
"	6/28/81	P53	991	-10.25	95.86	0.5776	0.8097	0.0077	1.721	0.0194	—	0.894	0.104	1.1E-07	—
"	7/9/81	P60	965	-10.82	96.14	0.6659	0.7130	0.0073	1.535	0.4199	—	0.804	0.094	2.0E-07	—
"	8/13/81	P61	1044	-9.46	96.54	0.6030	0.8930	0.0099	1.232	0.0065	—	0.638	0.074	4.6E-08	—
"	8/13/81	P62	1045	-9.35	96.16	0.5416	0.9983	0.0100	1.450	0.0055	—	0.750	0.087	4.4E-08	—
"	10/83	P10-83	960	-10.42	97.08	0.3859	0.5281	0.0031	1.988	0.0113	—	—	—	3.9E-08	—

Table 3 (continued)

Magma	Date	Number	T(°C)	log fO_2	H_2O	H_2	CO_2	CO	SO_2	H_2S	S_2	HCl	HF	COS	SO
Showa-Shinzan (Nemoto et al., 1957; Mizutani, 1962a; Mizutani and Sugiura, 1982; Symonds and Mizutani, in prep.)															
dacite	9/8/54	A-1 vent	1015	-9.93	98.04	0.63	1.2	0.0129	0.043	0.0004	2.6E-07	0.053	0.024	--	--
"	8/7/57	"	791	-13.17	99.32	0.2	0.39	0.00068	0.021	0.0005	1.2E-07	0.047	0.022	--	--
"	8/13/58	"	720	-14.72	99.39	0.16	0.38	0.0004	0.01	0.0008	1.0E-07	0.044	0.019	--	--
"	7/7/59	"	750	-14.06	99.28	0.18	0.47	0.00065	0.012	0.0007	9.2E-08	0.039	0.02	--	--
"	8/11/60	"	735	-14.38	99.41	0.17	0.34	0.0004	0.014	0.0008	1.5E-07	0.042	0.019	--	--
"	8/5/61	"	811	-12.89	99.37	0.24	0.24	0.00056	0.02	0.0006	1.4E-07	0.088	0.03	--	--
"	8/7/62	"	718	-14.82	99.58	0.17	0.12	0.00013	0.016	0.0014	3.4E-07	0.089	0.028	--	--
"	8/7/63	"	701	-15.22	99.59	0.16	0.12	0.00012	0.0082	0.0011	1.3E-07	0.092	0.029	--	--
"	9/23/64	"	728	-14.56	99.67	0.17	0.056	0.000065	0.0093	0.0007	8.4E-08	0.062	0.027	--	--
"	10/5/73	"	698	-14.98	99.81	0.11	0.019	0.000013	0.0084	0.0004	3.2E-08	0.036	0.016	--	--
"	9/1/74	"	645	-16.27	99.83	0.083	0.032	0.000012	0.008	0.0008	6.4E-08	0.035	0.014	--	--
"	8/28/77	"	799	-13.75	99.46	0.48	0.013	0.000056	0.008	0.0023	4.7E-07	0.027	0.011	--	--
"	11/21/77	"	619	-17.11	99.86	0.085	0.014	0.000005	0.0049	0.0012	7.2E-08	0.021	0.012	--	--
"	10/8/78	"	588	-17.85	99.88	0.06	0.022	0.000005	0.0055	0.0014	8.1E-08	0.019	0.01	--	--
"	7/9/59	A-6A vent	656	-16.45	99.41	0.15	0.36	0.00027	0.0089	0.0037	5.6E-07	0.051	0.021	--	--
"	8/13/60	"	694	-15.24	99.49	0.13	0.29	0.00023	0.01	0.0009	1.1E-07	0.06	0.02	--	--
"	8/5/57	C-2 vent	759	-14.1	99.11	0.24	0.5	0.00096	0.033	0.0033	1.7E-06	0.078	0.031	--	--
"	8/10/58	"	663	-16.12	99.49	0.13	0.3	0.0002	0.011	0.0024	3.8E-07	0.047	0.018	--	--
"	7/1/59	"	650	-16.58	99.33	0.14	0.43	0.00029	0.01	0.0043	7.5E-07	0.057	0.023	--	--
"	8/9/60	"	630	-17.41	99.24	0.18	0.44	0.00035	0.0065	0.0109	1.7E-06	0.09	0.028	--	--
"	8/5/61	"	640	-16.82	99.64	0.13	0.15	0.000089	0.0077	0.0035	4.4E-07	0.047	0.018	--	--
"	8/5/62	"	601	-18.09	99.7	0.13	0.1	0.000052	0.0046	0.01	1.1E-06	0.032	0.019	--	--
Usu (Matsuo et al., 1982; Gerlach, unpublished)															
dacite	9/1/79	11	659	-16.92	95.8	0.273	3.024	0.00440	0.258	0.609	0.0052	0.0241	0.0116	0.00032	--
"	9/1/79	12	676	-16.58	96.1	0.329	2.641	0.00501	0.219	0.537	0.0042	0.160	0.0332	0.00026	--
"	9/28/79	16	656	-17.20	96.1	0.342	2.613	0.00468	0.142	0.714	0.0042	0.105	0.0208	0.00031	--
"	1/28/80	26	678	-16.45	97.3	0.303	1.696	0.0029	0.200	0.350	0.0022	0.160	--	0.00011	--
"	1/28/80	27	667	-16.70	97.0	0.285	1.864	0.0029	0.226	0.461	0.0033	0.126	0.019	0.00015	--

--- not determined or below detection

Table 4. Equilibrium compositions, temperatures, and log fO_2 values of high-temperature and low-pressure (1 bar) volcanic gases from divergent-plate volcanoes. Concentrations of species reported in mole %; log fO_2 given as log bars.

Magma	Date	Sample	T (°C)	log fO_2	H_2O	H_2	CO_2	CO	SO_2	H_2S	S_2	HCl	COS
Ardoukoba (Allard et al., 1979; Gerlach, 1981)													
thol. basalt	11/13/78	A89	1070	-10.17	78.71	1.73	3.78	0.16	12.94	1.57	1.07	—	—
" "	" "	A92	" "	" "	77.75	1.71	4.02	0.17	13.52	1.62	1.17	—	—
" "	" "	G82	" "	" "	76.05	1.68	3.39	0.14	15.38	1.80	1.51	—	—
Erta' Ale (Tazieff et al., 1972; Giggenbach and Le Guern, 1976; Gerlach, 1980b)													
thol. basalt	12/3/71	1032	1075	-10.12	69.41	1.57	17.16	0.75	9.46	1.02	0.59	—	0.02
" "	" "	1130	" "	" "	69.41	1.57	17.12	0.75	9.49	1.03	0.59	—	0.02
" "	" "	1131	" "	" "	69.45	1.57	17.92	0.79	8.77	0.95	0.50	—	0.02
" "	" "	1132	" "	" "	69.59	1.57	16.13	0.71	10.17	1.10	0.68	—	0.02
" "	" "	1133	" "	" "	69.60	1.57	17.58	0.77	8.93	0.97	0.52	—	0.02
" "	" "	1150	" "	" "	70.16	1.59	17.11	0.75	8.86	0.97	0.52	—	0.02
" "	" "	1151	" "	" "	68.36	1.55	18.88	0.83	8.87	0.94	0.52	—	0.02
" "	" "	1152	" "	" "	68.83	1.56	18.22	0.80	9.04	0.97	0.54	—	0.02
" "	" "	1153	" "	" "	70.88	1.60	17.47	0.77	7.95	0.88	0.41	—	0.01
" "	" "	1154	" "	" "	69.43	1.57	18.06	0.79	8.67	0.94	0.49	—	0.02
" "	" "	1155	" "	" "	68.72	1.55	18.81	0.83	8.63	0.92	0.49	—	0.02
" "	" "	1157	" "	" "	71.65	1.62	16.61	0.73	8.03	0.90	0.42	—	0.01
" "	" "	1214	" "	" "	69.55	1.57	18.10	0.80	8.53	0.92	0.48	—	0.01
" "	" "	1216	" "	" "	69.27	1.56	18.38	0.81	8.53	0.92	0.48	—	0.02
" "	" "	1224	" "	" "	69.99	1.58	17.93	0.79	8.31	0.91	0.45	—	0.02
" "	" "	1225	" "	" "	70.21	1.59	18.30	0.81	7.80	0.85	0.40	—	0.02
" "	" "	1226	" "	" "	66.99	1.51	18.57	0.82	10.29	1.07	0.70	—	0.02
" "	" "	1227	" "	" "	70.40	1.59	18.00	0.79	7.89	0.87	0.41	—	0.01
" "	1/23/74	910	1130	-9.16	77.24	1.39	11.26	0.44	8.34	0.68	0.21	0.42	—
" "	" "	911	" "	-9.29	76.05	1.57	12.52	0.56	7.59	0.93	0.31	" "	—
" "	" "	916	" "	-9.31	75.06	1.59	13.08	0.60	7.84	1.01	0.36	" "	—
" "	" "	920	" "	-9.21	78.28	1.48	10.92	0.45	7.51	0.72	0.21	" "	—
" "	" "	931	" "	-9.32	77.52	1.68	11.36	0.53	7.12	1.01	0.32	" "	—
" "	" "	932	" "	-9.32	76.52	1.66	11.94	0.56	7.46	1.05	0.35	" "	—
" "	" "	933	" "	-9.29	77.87	1.61	11.38	0.51	7.02	0.88	0.26	" "	—
" "	" "	934	" "	-9.25	77.47	1.53	11.51	0.49	7.47	0.82	0.25	" "	—
" "	" "	935	" "	-9.27	77.45	1.57	11.53	0.51	7.37	0.86	0.26	" "	—
" "	" "	936	" "	-9.36	78.19	1.76	11.22	0.55	6.46	1.05	0.31	" "	—
" "	" "	937	" "	-9.34	76.76	1.70	11.93	0.57	7.16	1.08	0.35	" "	—
Nyiragongo (Chaigneau et al., 1960; Gerlach, 1980d)													
nephelinite	1959	2	970	-12.0	43.50	1.29	48.55	2.20	2.02	1.72	0.62	—	0.09
" "	" "	12	1020	-11.3	45.90	1.59	45.44	2.72	2.30	1.41	0.55	—	0.08
" "	" "	13	960	-12.4	55.62	2.18	36.35	2.13	0.81	2.45	0.38	—	0.08
Surtsey (Sigvaldason and Elisson, 1968; Gerlach, 1980c)													
alk. basalt	10/15/64	12	1125	-9.80	81.13	2.80	9.29	0.69	4.12	0.89	0.25	—	—
" "	" "	13	" "	-9.78	81.98	2.77	9.79	0.71	3.15	0.64	0.13	—	—
" "	2/21/65	17	" "	-9.82	87.88	3.12	5.01	0.38	2.46	0.63	0.10	—	—
" "	" "	22	" "	-9.55	87.83	2.27	6.43	0.36	2.43	0.24	0.03	—	—
" "	" "	24	" "	-9.75	87.40	2.86	5.54	0.39	2.72	0.54	0.09	—	—
" "	3/31/67	29	" "	-9.11	91.11	1.42	3.31	0.11	2.81	0.06	0.01	—	—
" "	" "	30	" "	-9.30	92.47	1.79	1.21	0.05	3.27	0.14	0.02	—	—
" "	" "	31	" "	-9.27	92.46	1.74	1.31	0.05	3.74	0.15	0.02	—	—

--- not determined or below detection; HF and SO not determined in all samples

Table 5. Equilibrium compositions, temperatures, and log fO_2 values of high-temperature and low-pressure (1 bar) volcanic gases from hot-spot volcanoes. Concentrations of species reported in mole %; log fO_2 given as log bars.

Magma	Date	Sample	T (°C)	log fO_2	H_2O	H_2	CO_2	CO	SO_2	H_2S	S_2	HCl	HF	COS
Kilauea Summit Lava Lake (Shepherd, 1921; Jaggar, 1940; Gerlach, 1980a)														
thol. basalt	3/25/18	J8	1170	-8.38	37.09	0.49	48.90	1.51	11.84	0.04	0.02	0.08	—	—
"	3/13/19	J11	1100	-9.33	40.14	0.55	36.69	1.03	21.06	0.20	0.25	0.00	—	—
"	3/15/19	J13	1175	-8.40	69.29	1.01	17.82	0.62	10.93	0.08	0.03	0.21	—	—
"	3/16/19	J14	1100	-9.30	35.09	0.54	47.41	1.52	15.06	0.15	0.17	0.00	—	—
"	3/17/19	J16	1140	-8.84	60.42	0.87	23.21	0.74	14.31	0.14	0.07	0.21	—	—
"	"	J17	1085	-9.65	65.95	1.02	20.27	0.62	11.44	0.32	0.16	0.17	—	—
"	"	J18	1185	-8.45	58.14	1.03	21.76	0.93	17.46	0.17	0.13	0.32	—	—
Kilauea East Rift Zone (Gerlach, 1993a)														
thol. basalt	1/14/83	Pele 9	1010	-10.49	79.8	0.9025	3.15	0.0592	14.9	0.622	0.309	0.1	0.19	0.0013
"	"	Pele 4	997	-10.76	81.6	0.9929	3.80	0.0702	12.0	0.761	0.358	0.171	0.20	0.0016
"	"	Pele 6	1016	-10.54	80.0	1.059	3.55	0.0804	13.7	0.875	0.445	0.153	0.12	0.0022
"	1/15/83	Pele 12	935	-11.91	78.7	1.065	3.17	0.0584	11.5	3.21	1.89	0.167	0.20	0.0054
"	"	Pele 2	948	-11.55	78.5	0.9272	3.89	0.0594	13.4	1.70	1.12	0.148	0.19	0.0032
"	"	Pele 3	952	-11.41	80.7	0.8429	3.78	0.0617	12.5	1.32	0.540	0.162	0.17	0.0032
"	"	Pele 5	1003	-10.61	82.5	0.9370	3.56	0.0649	11.8	0.587	0.232	0.151	0.15	0.0013
"	"	Pele 7	1032	-10.19	80.4	0.9289	3.52	0.0784	14.0	0.511	0.197	0.174	0.19	0.0014
"	1/16/83	Pele 8	979	-11.00	76.2	0.8655	3.35	0.0587	17.2	1.22	0.808	0.169	0.18	0.0026
"	"	Pele 10	980	-10.92	78.8	0.8246	3.00	0.0496	15.6	0.902	0.469	0.173	0.17	0.0017

— not determined or below detection; SO not determined in all samples.

Table 6. Sources of high-quality volcanic-gas analyses from >500°C vents not listed in Tables 3-5.

Volcano	Tectonic setting	Date(s)	Reference
Kilauea	Hot Spot	1960	Heald et al. (1963)
"	"	1980-83	Greenland (1987a)
Klyuchevskoy	Convergent Plate	1988	Taran et al. (1991)
Mauna Loa	Hot Spot	1984	Greenland (1987b)
Momotombo	Convergent Plate	1982-85	Menyailov et al. (1986)
Satsuma-Iwojima	Convergent Plate	1962-67	Matsuo et al. (1974)
"	"	1990	Shinohara et al. (1993)
Tolbachik	Convergent Plate	1975-76	Menyailov & Nikitina (1980)
Usu	Convergent Plate	1978-80	Matsuo et al. (1982)
"	"	1985	Giggenbach & Matsuo (1991)
White Island	Convergent Plate	1971-88	Giggenbach (1987); Giggenbach & Sheppard (1989); Giggenbach & Matsuo (1991)

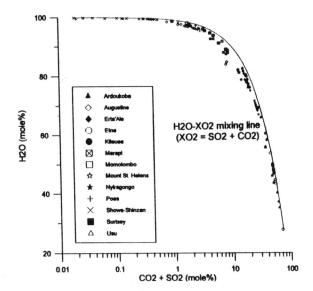

Figure 7. Concentrations of H_2O versus CO_2 + SO_2 for selected equilibrium volcanic-gas compositions. Most data lie close to the H_2O-XO_2 (XO_2 = CO_2 + SO_2) mixing line. Data from Tables 3,4,5. Filled symbols represent gases from divergent-plate and hot-spot volcanoes; others mark gases from convergent-plate volcanoes.

and Nordlie, 1975; Symonds et al., 1992), and are generally highest in samples from basalt (Tables 3-5). CO increases with temperature (Fig. 1) and total C (Gerlach and Nordlie, 1975), and is higher in samples from basalt (Tables 3-5). Samples from convergent plate volcanoes contain up to 6% HCl (e.g., Mount St. Augustine; Symonds et al., 1990; Kodosky et al., 1991).

Elemental compositions

Elemental compositions are calculated from the molar compositions in Tables 3-5 using the following equation (from Gerlach and Casadevall, 1986b):

$$e_j = \sum_{i=1}^{N} a_{ij} n_i \tag{23}$$

where e_j is the total moles of element j per 100 moles of sample, a_{ij} is the number of moles of element j in a mole of species i, n_i is the mole% concentration of species i (Tables 3-5), and N is the total number of species. For example, 100 moles of Kilauea sample J8 (Table 5) contains 75.32 moles H, 160.08 moles O, 50.41 moles C, 11.92 moles S, and 0.08 moles Cl. The elemental compositions provide a data base for graphical comparison of all samples in Tables 3-5, as shown in Figures 8 to 12.

Figure 8 shows a H-C-S ternary plot based on the equilibrium compositions of samples in Tables 3-5. In general, samples from convergent-plate volcanoes have proportionally more H than samples from divergent-plate and hot-spot systems (Fig. 8) and consequently have higher H/S and H/C ratios than samples from divergent-plate and hot-spot volcanoes (Fig. 9). These characteristics reflect much higher H_2O concentrations in convergent-plate samples.

C and S are most enriched in samples from divergent-plate and hot-spot volcanoes and least enriched in samples from convergent-plate volcanoes (Fig. 8). Mount Etna, an enigmatic convergent-plate volcano, also discharges C- and S-rich gases, probably because its alkaline basalt magma comes from a gas-rich source. The most C-rich samples come from the 1959 Nyiragongo and 1918-19 Kilauea summit lava lakes, whereas the most

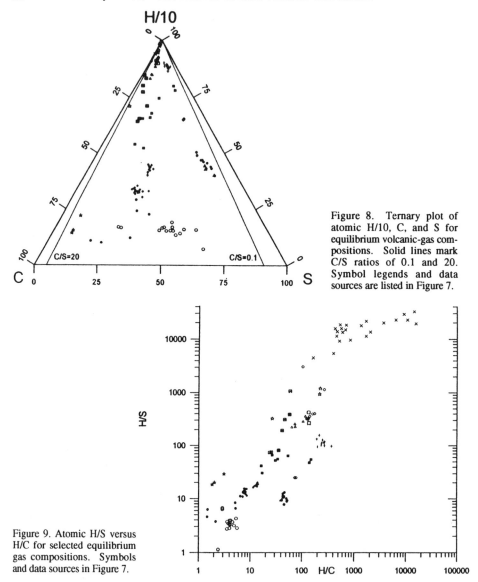

Figure 8. Ternary plot of atomic H/10, C, and S for equilibrium volcanic-gas compositions. Solid lines mark C/S ratios of 0.1 and 20. Symbol legends and data sources are listed in Figure 7.

Figure 9. Atomic H/S versus H/C for selected equilibrium gas compositions. Symbols and data sources in Figure 7.

S-rich samples come from flank lava flows at Mount Etna and from the east rift of Kilauea (Fig. 8). The C/S ratios of samples from convergent-plate volcanoes and divergent-plate and hot-spot volcanoes are roughly similar (Fig. 10). C/S ratios of volcanic gases depend largely on the extent of degassing of the magma. For instance, at Kilauea the C/S ratios decrease from 4 for samples from relatively undegassed summit lava lakes to about 0.15 for samples from partially degassed lavas of a flank eruption along the east rift zone.

Compared to samples from divergent-plate and hot-spot volcanoes, samples from convergent-plate volcanoes contain higher proportion of Cl in the C-S-Cl system (Fig. 11). Cl/S is clearly much higher in samples from convergent-plate volcanoes where it ranges from 0.03 to 10 than in samples from divergent-plate and hot-spot volcanoes where it

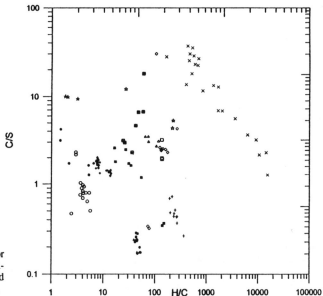

Figure 10. C/S versus H/C for selected equilibrium gas compositions. Symbol legends and data sources listed in Figure 7.

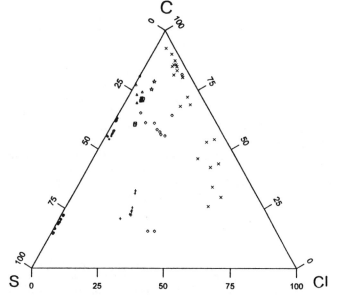

Figure 11. Ternary plot of atomic C, S, and Cl for equilibrium volcanic-gas compositions. Symbol legends, data sources listed in Figure 7.

ranges from 0.006 to 0.05. However, the Cl concentrations of divergent-plate and hot-spot volcanic gases are within the range of those for convergent plate volcanoes (Fig. 12).

The most likely explanation for the H_2O and Cl enrichments in convergent-plate gases (Figs. 8 and 11) is that their magmas are enriched in H_2O and Cl relative to hot-spot and divergent-plate magmas. This explanation accounts for the H_2O and Cl enrichments in convergent-plate gases for all magma types, it is consistent with the H_2O and Cl enrichments of subduction zone magmas, and it fits the modern conceptual tectonic

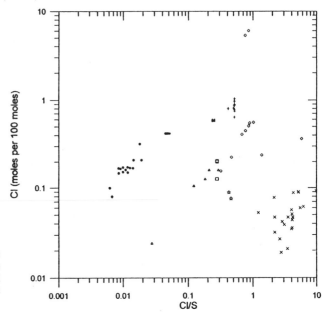

Figure 12. Atomic Cl versus Cl/S for selected equilibrium gas compositions. Symbol legends, data sources listed in Figure 7.

framework. Allard (1983) originally suggested that high-temperature gases from hot-spot and divergent-plate volcanoes contain mostly mantle-derived volatiles, whereas convergent-plate volcanoes discharge gases containing crustal and mantle components. Giggenbach (1992) used δD and $\delta^{18}O$ data of fumarolic condensates to argue that gas discharges from convergent-plate andesites contain subducted seawater. These observations are consistent with previous work (e.g., Gill, 1981; Ito et al., 1983; Stolper and Newman, 1994) which shows that magmas from convergent-plate volcanoes have higher pre-eruption H_2O and Cl contents (see Chapter 8). Subduction of hydrothermally altered oceanic slabs and sediments (e.g., Ito et al., 1983; Alt, 1994) beneath convergent-plate volcanoes is implicit in plate tectonic theory, and the subducted material provides an important source for volatiles associated with arc magmatism. Prograde metamorphic reactions in the oceanic crust and dewatering of the marine sediments drive H_2O, Cl, and other volatiles from the subducted slab into the overlying mantle wedge (e.g., Peacock, 1990), and these recycled volatiles enrich subduction zone magmas and their gas discharges (see Chapter 12 regarding volatile recycling).

Oxygen fugacities

The equilibrium fO_2 values of the samples are reported in Tables 3-5 and plotted as a function of inverse temperature in Figure 13. For comparison, Figure 13 shows that the calculated fO_2 values for the Mount St. Helens compositions (Table 3) are in excellent agreement with simultaneous in situ fO_2 measurements (Gerlach and Casadevall, 1986a). Similar agreement exists between equilibrium fO_2 values and simultaneous in situ fO_2 measurements at Merapi and Momotombo (Bernard, 1985). Simultaneous in situ fO_2 measurements are not available for samples from the other volcanoes.

Figure 13 shows that the fO_2 values lie between one log unit below quartz-fayalite-magnetite (QFM) to two log units above Ni-NiO (NNO), within the range expected for basic to silicic volcanic rocks (Carmichael, 1991). In general, the fO_2 values are broadly

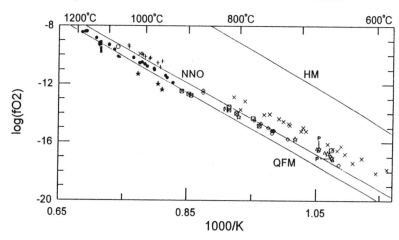

Figure 13. Plot of log fO_2 versus inverse temperature for selected equilibrium gas compositions (Tables 3-5). The quartz-fayalite-magnetite (QFM), Ni-NiO (NNO), and hematite-magnetite (HM) oxygen buffers are shown for reference. Symbol legends and data sources are listed in Figure 7. The plot also shows the September 16, 1981 oxygen fugacity probe data (P) for Mount St. Helens (Gerlach and Casadevall, 1986a).

similar to those expected for their respective lavas. For example, the fO_2's for Kilauea are within the range for Kilauea lavas (>1 log unit below to < 1 log unit above QFM; Carmichael, 1991); the fO_2's for Ardoukoba, Erta'Ale, and Surtsey, all divergent-plate basalts, are up to 1 log unit higher than the range for oceanic basalts (3 to <1 log units below QFM; Carmichael, 1991); the fO_2's for Merapi, Momotombo, and Poás are similar to those for other basaltic andesites (<1 log unit below to 1 log unit above NNO; Carmichael, 1991); the fO_2's for Augustine, Showa-Shinzan, and Usu are similar to other orthopyroxene-bearing silicic lavas (<1 log unit below to 1 log unit above NNO; Carmichael, 1991); and the fO_2's for Mount St. Helens are about 1 log unit below those for its hornblende-bearing dacite (Gerlach and Casadevall, 1986b). The general similarities between the fO_2's of the gases and their respective lava types suggests that lavas buffer fO_2 in high-temperature volcanic gases, although this has only been demonstrated at Kilauea (Gerlach, 1993a).

Pressure effects

To indicate the effects of pressure on gas compositions, we show the calculated equilibrium distribution of species from 1 to 100 bars for two gas samples: (1) sample 800925-710 from Mount St. Helens (Table 3) to represent SO_2-poor gases and (2) sample J8 from Kilauea (Table 5) to represent SO_2-rich gases. The calculations are done with program GASWORKS (Symonds and Reed, 1993) at the approximate magmatic temperatures of 1000° and 1170°C for Mount St. Helens and Kilauea, respectively (Gerlach and Casadevall, 1986b; Gerlach, 1980a). We treat the gases as ideal, which is reasonable for >600°C gas mixtures at < 500 bars (Holloway, 1977; Spycher and Reed, 1988), and consider only reactions between the most abundant species in the H-C-O-S system (H_2O, CO_2, H_2, SO_2, H_2S, CO, S_2, and COS); minor amounts of HCl and HF in these samples are neglected. The general effect of higher pressure is to increase H_2S, COS, and S_2, and to decrease SO_2, H_2, and CO (Fig. 14). The crossover between H_2S and SO_2 is at 18 bars for the Mount St. Helens case. A similar increase in H_2S is observed for the Kilauea case, but H_2 is not abundant enough in the J8 sample to cause H_2S to supersede SO_2 via Reaction (26). These pressure changes occur mostly in the first 50 bars and are the result oi volume effects that favor fewer moles of products than reactants in the following reactions (Gerlach and Casadevall, 1986b):

$$SO_2 + 3 H_2 \Rightarrow H_2S + 2 H_2O, \tag{26}$$

$$2 H_2 + CO + SO_2 \Rightarrow 2 H_2O + COS, \tag{27}$$

$$2 H_2 + SO_2 \Rightarrow 2 H_2O + 1/2 S_2. \tag{28}$$

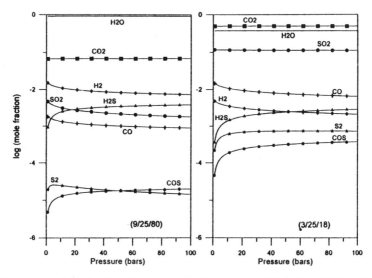

Figure 14. Calculated distribution of the most abundant species of H, O, C, and S as a function of pressure for (A) Mount St. Helens sample 800925-710 at 1000°C and (B) Kilauea sample J8 at 1170°C. Starting compositions at 1 bar pressure are tabulated in Tables 3-5.

SHALLOW MAGMA DEGASSING PROCESSES

Degassing at hot-spot and divergent-plate volcanoes

Kilauea volcano. Studies of high-temperature volcanic gases have contributed to understanding magma degassing processes at Kilauea volcano (Gerlach, 1980a, 1986, 1993a; Gerlach and Graeber, 1985; Greenland et al., 1985; Gerlach and Taylor, 1990). Sustained summit lava lake eruptions persisted at Kilauea during most of the previous century and on into the early 1920's and for brief periods in 1952 and 1967-1968. In essence, the roof of the summit magma chamber rose to the surface and a continuous resupply of magma sustained a lava lake at these times. Sustained summit lava lake eruptions emit relatively CO_2-rich, Type I volcanic gases (J series samples, Table 5, Fig. 15) directly to the atmosphere in a one-stage degassing process (Fig. 16a). The Type I gas samples (Table 5) plotted in Figure 15 include CO_2-rich samples like J-8, which was collected from a crack in a spatter cone adjacent to upwelling resupply magma, and several samples poorer in CO_2 that were emitted by partially degassed resupply magma at the margins of a summit lava lake.

Sustained summit lava lake eruptions have been comparatively uncommon in this century, when most eruptions have occurred along rift zones. The magma of rift zone eruptions is temporarily stored in the summit chamber prior to being discharged into the rift

Figure 15. Ternary plot of C, H, and S for Type I volcanic gases (circles) from the 1918-19 summit lava lake (J-series samples in Table 5), Type II volcanic gases (squares) collected in 1983 from the east rift zone of Kilauea (Pele samples in Table 5), and the CO_2-rich summit Chamber Gas composition (triangle). The degassing vector originating at J-8 (Table 5) illustrates that the CO_2-depleted, Type II gases lie on a continuation of the trend formed by Type I gases from partially degassed summit lava lake resupply magma. [By permission of the editor of Geochimica et Cosmochimica Acta; Gerlach (1993a), Fig. 3, p. 799.

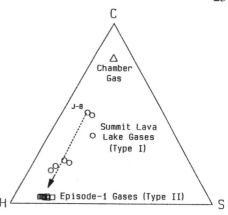

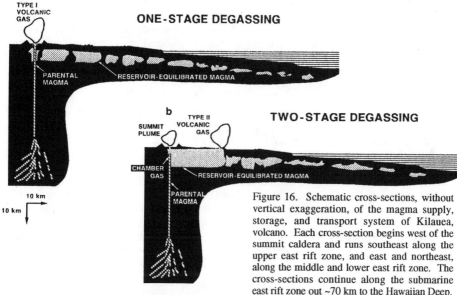

Figure 16. Schematic cross-sections, without vertical exaggeration, of the magma supply, storage, and transport system of Kilauea, volcano. Each cross-section begins west of the summit caldera and runs southeast along the upper east rift zone, and east and northeast, along the middle and lower east rift zone. The cross-sections continue along the submarine east rift zone out ~70 km to the Hawaiian Deep. For purposes of clarity the thickness of the feeder conduit for parental magma is greatly exaggerated and the summit chamber is somewhat enlarged. A plexus of dikes, sheets, and conduits is believed to constitute the summit chamber and east rift zone storage reservoirs shown schematically here as regions with high melt-to-rock ratios. (a) A one-stage degassing process during a continuously supplied summit lava lake eruption emitting Type I volcanic gases (J-series samples, Table 5, Fig. 15) derived from lava lake resupply magma. (b) A two-stage degassing process involving CO_2-rich chamber gas (Fig. 15) venting from new magma equilibrating in the summit reservoir and a rift zone eruption emitting type II volcanic gases (Pele samples, Table 5, Fig. 15) derived from summit reservoir-equilibrated magma. The chamber gas supplies volatiles to the summit hydrothermal system and to the summit fumaroles and plume. Eruptions of reservoir-equilibrated magma occur in the summit region and along the rift zone at subaerial and submarine sites. From Gerlach (1986).

zone complex (Fig. 16b). Equilibration and evolution of new magma injected into and temporarily stored in the summit chamber result in degassing of excess gas, which leads to the emission of a Chamber Gas containing >90% CO_2 (Fig. 15). Summit chamber degassing acts on all magma discharged into the reservoir system of the rift zone complex, and it is the first stage of a two-stage degassing process for magma that is subsequently erupted from the rift zone (Fig. 16b); the second stage occurs during the rift zone eruption itself (Fig. 16b). Summit chamber degassing removes so much CO_2 that, compared to Type I gases, rift zone eruptions produce distinctly CO_2-depleted, Type II volcanic gases, such as the Pele samples (Table 5, Fig. 15). The compositions of Chamber, Type I, and Type II gases plotted in Figure 15 show that Type I and Type II gases differ from each other mainly by their CO_2 content, and that the latter gases lie on a continuation of the trend formed by Type I gases from the partially degassed resupply magma of a summit lava lake (Fig. 15).

Divergent-plate volcanoes. Similar degassing processes may be common at divergent-plate volcanoes (Gerlach, 1989a, 1989b). For instance, in the axial volcanic fields of the Afar depression, there are similar differences in the compositions of gas samples from the sustained summit lava lake of Erta'Ale volcano and samples from a fissure eruption at Ardoukoba on the Asal Rift in 1978 (Table 4, Fig. 8). In this case also, the summit lava lake and rift zone volcanic gases differ from each other mainly by their CO_2 content, which is higher in the summit lava lake gases (Type I) compared to the rift zone eruption gases (Type II) (Table 4, Fig. 8), suggesting again that the magma of the rift eruption at Ardoukoba underwent a two-stage degassing process with the first stage consisting of CO_2-rich chamber degassing during prior storage in a crustal magma reservoir. Finally, there is evidence to suggest that mid-oceanic ridge basalts undergo an analogous two-stage degassing process in which the first stage of degassing is the release of CO_2-rich gas from the ridge magma reservoirs that overlie the principal mantle sources of partial melt (Gerlach, 1989b).

Generalizations. Chamber degassing fractionates the volatiles in the magma supplied from the mantle to hot-spot and divergent-plate crustal magma chambers. In general, it probably operates largely on the basis of volatile solubility in basaltic melts, removing most of the CO_2 and heavier rare gases (Ar, Kr, Xe, Rn), much of the He and Ne, some sulfur, and minor amounts of water and halogens (see also Chapters 6, 7 and 8). Chamber degassing also may be an effective mechanism for fractionating stable isotopes of volatile elements — e.g., carbon isotopes (Gerlach and Taylor, 1990). Therefore, Type I volcanic gases better represent the volatiles of new magma from the mantle, whereas Type II gases reflect fractionation by chamber degassing during crustal storage. Chamber degassing of CO_2-rich gas is responsible for the tendency of volcanic gases from hot-spot and divergent-plate basalts to separate into a high C/S group and a low C/S group (Fig. 8). The high C/S group consists of Type I gases and show somewhat variable C/S (~1-4), possibly reflecting different mantle sources. The low C/S group includes the Type II gases, which have comparatively restricted C/S values near 0.3, perhaps reflecting equilibration to similar low pressure conditions in crustal magma chambers.

Degassing at convergent-plate volcanoes

There are several examples of time-series studies of the degassing of magma-hydrothermal systems at convergent-plate volcanoes. In addition to those considered below for Showa-Shinzan, Mount St. Helens, and White Island volcanoes, there are studies of time-series gas sampling available for Momotombo volcano, Nicaragua (Menyailov et al., 1986) and Vulcano, Italy (Tedesco et al., 1991; Magro and Pennisi, 1991).

Showa-Shinzan dome, Japan. The 31 years of gas data from Showa-Shinzan is currently the longest set of time-series data on high-temperature fumarolic emissions (Mizutani, 1962b, 1978; Mizutani and Sugiura, 1982; Symonds and Mizutani, in prep.). Showa-Shinzan is a dacitic lava dome that formed during the 1943-45 eruption of Usu volcano, Japan. High-temperature fumaroles on the dome were sampled from 1954 to

1985, during which the dome degassed, cooled, and was invaded by meteoric vapor. Samples collected from the A-1 fumarole, the hottest vent, in 1954-78 (Table 3) and in 1985 (Symonds and Mizutani, in prep.) best represent the temporal evolution of the discharged volatiles during the 31-year period.

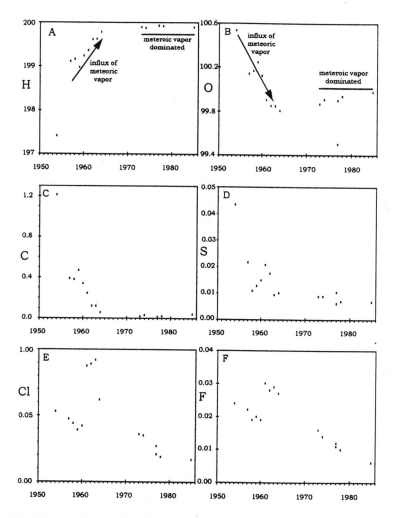

Figure 17. Plot showing moles of elemental hydrogen (A), oxygen (B), carbon (C), sulfur (D), chlorine (E), and fluorine (F) per 100 moles of gas for samples collected from the A-1 fumarole on Showa-Shinzan between 1954 and 1985 (Table 3). Arrows in (A-B) show schematically the inferred influx of meteoric H_2O into the A-1 gases from 1954 to 1964, whereas the solid lines in (A-B) mark the period from 1973 to 1985 when the A-1 gases were dominated by meteoric vapor.

Figure 17 shows the temporal evolution of elemental H, O, C, S, Cl, and F in the A-1 fumarole gases from 1954 to 1985. During this period, the equilibrium temperatures of the A-1 gases dropped from 1015 to 518°C. From 1954 to 1964 the A-1 gases became increasingly enriched in H but depleted in O, C, and S (Fig. 17A-D). After 1972, most samples contained >99.8% H_2O and contained nearly constant levels of H, O, C, and S (Fig. 17A-D). As suggested by Mizutani and Sugiura (1982), these trends reflect the escalating amounts of meteoric H_2O in these samples, which is especially pronounced from 1954 to 1964. After 1964, the A-1 fumarole gases were dominated by meteoric vapor. In contrast, concentrations of atomic Cl and F generally decreased from 1954 to 1960, increased sharply between 1961 and 1964, and decreased after 1964 (Fig. 17E-F). The general decrease in Cl and F from 1954 to 1985 represents the overall meteoric-water-dilution trend. However, the sharp increase in Cl and F between 1961 and 1964 may represent a late-stage halogen release from the magma. In essence, Cl and F partition from the magma much later than C and S, and these late-stage halogens overprint the overall meteoric-water-dilution trend during the 1961-64 period. Figure 18B shows that the Cl/S ratio increased five-fold from 1954 to 1964. Over the same interval, the C/S ratio decreased by a factor of five (Fig. 18A). In contrast, the 1954-64 Cl/F ratios were relatively constant (Fig. 18C).

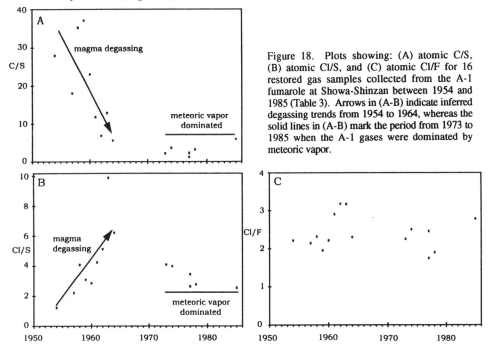

Figure 18.   Plots showing: (A) atomic C/S, (B) atomic Cl/S, and (C) atomic Cl/F for 16 restored gas samples collected from the A-1 fumarole at Showa-Shinzan between 1954 and 1985 (Table 3). Arrows in (A-B) indicate inferred degassing trends from 1954 to 1964, whereas the solid lines in (A-B) mark the period from 1973 to 1985 when the A-1 gases were dominated by meteoric vapor.

The 1015° to 518°C temperature drop of the equilibrium A-1 gases from 1954 to 1985 caused ratios of H_2O/H_2, CO_2/CO, and H_2S/SO_2 to increase (Fig. 19), chiefly because the equilibrium amounts of H_2S increased while those of H_2, SO_2, and CO decreased (Fig. 1). These temperature-induced changes reflect right shifts in Reaction (26) and this reaction:

$$SO_2 + 3\,CO + H_2O = H_2S + CO_2 \qquad (29)$$

Mount St. Helens, U.S.A.   The 1980-86 eruptions of Mount St. Helens provided an opportunity to study the evolution of gases discharging from a magma-hydrothermal system (Gerlach and Casadevall, 1986a,b). The volcano erupted explosively on 18 May 1980, followed by five smaller explosive eruptions in the summer of 1980 and the extrusion of a lava dome after the October 1980 explosive eruption. The dome grew by episodic exogenous and endogenous growth until 1986. Gases were collected from fumaroles on the crater floor and on the dome from September 1980 to September 1981 (Table 3; additional data in Gerlach and Casadevall, 1986a). Over this period, the H content of the

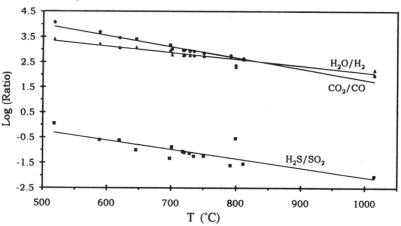

Figure 19. Plot showing ratios of H_2O/H_2 (triangles), CO_2/CO (circles), and H_2S/SO_2 (squares) as a function of temperature for the A-1 fumarole Showa-Shinzan gas samples (Table 3). Solid lines represent linear regression fits of the data. All correlation coefficients are significant at the 99% confidence level.

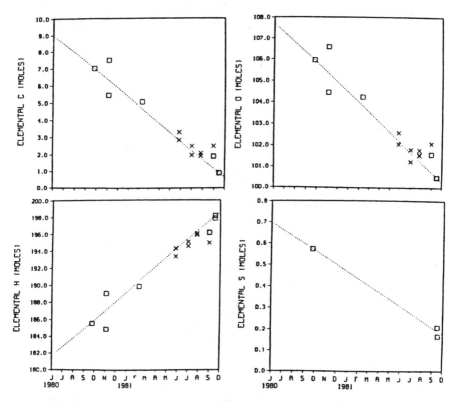

Figure 20. Elemental C, O, H, and S in moles per 100 moles of gas for 16 samples collected from Mount St. Helens, September 1980 to September 1981. The squares represent highest-quality restored and apparent compositions (Table 3) and Xs mark estimated compositions determined by Gerlach and Casadevall (1986a). The dashed line is the compositional trend for the 3 best samples reported in Table 3. [By permission of the editor of Journal of Volcanology and Geothermal Research; Gerlach and Casadevall (1986b), p. 150.]

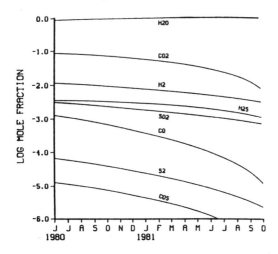

Figure 21. Equilibrium model of 1-atm fumarole gas compositions for Mount St. Helens between June 1980 and October 1981. [Used by permission of the editor of Journal of Volcanology and Geothermal Research, from Gerlach and Casadevall (1986b), Fig. 2, p. 152.]

gases increased linearly while the concentrations of C, O, and S decreased, also in a linear fashion (Fig. 20). These trends reflect the steady dilution of the magmatic gases by hydrothermal vapor over this time period. Estimates of the proportion of hydrothermal vapor in the Mount St. Helens gases range from 20 to 40% in June 1980 to 50 to 100% by October 1981. Like Showa-Shinzan, there are also temperature-related changes in the ratios of gas species. Linear changes in vent temperature and atomic compositions of fumarolic gases between September 1980 and September 1981 were used to constrain a 1-atm equilibrium model of the gases between June 1980 and October 1981 (Fig. 21).

White Island, New Zealand. White Island, an active andesitic-dacitiv volcano at the northeastern end of the Taupo volcanic zone, has emitted fumarolic gases for at least the past 150 years (Cole and Nairn, 1975). Regular fumarole gas sampling (about 2 samples per year from 6 to 10 fumaroles) commenced in 1971 and continues to the present day (Giggenbach and Sheppard, 1989). Over this time period, eruptions occurred in July 1971, between December 1976 and January 1982, and between December 1983 and February 1984. Between 1971 and 1984, samples from the Donald Mound fumarole, consistently the hottest vent, display cyclical variations in concentrations of HCl, H_2, SO_2, and H_2S that increase and decrease with fumarole temperature, although the decreases in H_2S during periods of lower temperatures are much larger than the corresponding decreases in SO_2 (Fig. 22). The concentration of CO_2 also varies with temperature, but somewhat less than the other gases (Fig. 22). Giggenbach and Sheppard (1989) suggest that fluctuating inputs of magmatic gas into the subvolcanic system trigger movements in the inferred subsurface brines (Fig. 23) that may account for the variations in gas compositions from the Donald Mound fumaroles. During periods of high magmatic-gas output between late 1982 until early 1983, magmatic gases vented freely from the magma to the Donald Mound fumaroles (Fig. 23). However, during periods of low magmatic gas flux, brines, inferred from the intermittent occurrence of highly mineralized acidic springs, invade the vapor-only zone (Fig. 23) and absorb HCl, SO_2, and other condensable gases. H_2S also decreases during the low-temperature periods, probably a consequence of subsurface precipitation and loss of native S. Subsequent reheating evaporates the brines and deposits sulfur, releasing some of the previously absorbed volatiles and sometimes producing higher-than-magmatic concentrations of HCl and total S. CO_2 probably fluctuates less than the HCl and SO_2 because it is less soluble in the brine (see below).

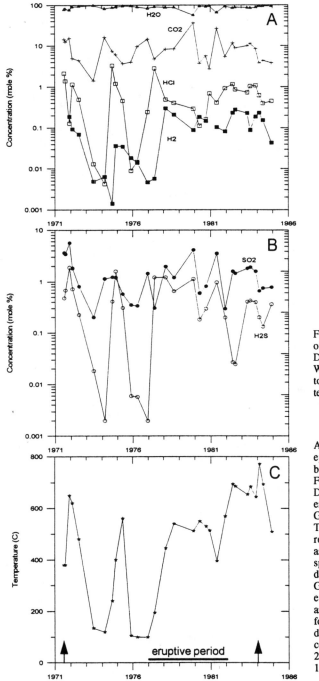

Figure 22. Temporal evolution of discharges from the hottest Donald Mound fumarole on White Island volcano from 1971 to 1984. Diagram shows the temporal evolution of

(A) H_2O, CO_2, HCl, and H_2;

(B) SO_2 and H_2S; and

(C) temperature.

Arrows mark short eruptive episodes in July 1971 and between December 1983 and February 1984; bar marks December 1976 to January 1982 eruptive period. All data from Giggenbach and Sheppard (1989). Total S and its oxidation state, n, recalculated as SO_2 and H_2S, assuming they are the dominant species, using the equations described by Giggenbach and Goguel (1989; p. 49); when n exceeded 4.0, we assigned an arbitrary value of 0.002 mole % for the concentration of H_2S. H_2 data unavailable for samples collected on July 29 and August 26, 1971, and on December 2, 1980.

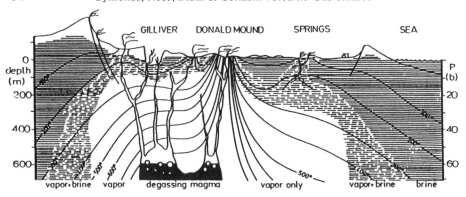

Figure 23. Schematic cross-section across the White Island crater showing inferred locations, relative to Donald Mound, of degassing magma, vapor-only zone, vapor-plus-brine zone, and brine-only zone for the high-temperature periods in 1972 and 1975. Diagram from Giggenbach and Sheppard (1989).

Gas/lava interactions during degassing

It is commonly assumed that, after discharging from magma, volcanic gases maintain homogeneous equilibrium with falling temperature until they quench. A test of this assumption with ten high-temperature gas samples (Pele samples, Table 5) reveals a more complex process for gases emitted from an erupting fissure in 1983 on the east rift zone of Kilauea (Gerlach, 1993a). The Pele analyses have log fO_2 values that fall on a tight linear trend ($r^2 = 0.982$) as a function of reciprocal (last equilibrium) temperature (Fig. 24a). Regression of the data yields the following equation:

$$\log fO_2 = -26,410\,(1/T) + 10.06 \qquad (29)$$

where fO_2 is in bars and T is in Kelvin; the equation is valid from 935 to 1032°C. Values of log fH_2, log fCO, log fS_2, log fH_2S, log fCl_2, and log fF_2 also follow linear trends as a function of reciprocal (last equilibrium) temperature (Fig. 24b-g). It can be shown that these linear trends do not represent the intrinsic closed-system trends expected if the gases maintained homogeneous equilibrium during cooling. Numerical simulation of closed-system (homogeneous equilibrium) cooling of each sample from high temperature to the last equilibrium temperature produces trends that depart from the observed linear trends for all species, except for F_2 (Fig. 24a-g). Open-system cooling involving oxygen transfer between the major gas species (H_2O, CO_2, SO_2) and lava, however, is consistent with the observed linear log fO_2-versus-reciprocal-temperature trend. Numerical simulations show, moreover, that the gas/lava oxygen transfer required to constrain log fO_2 by equation (29) during open-system cooling also produces the observed linear trends of all minor and trace species (Fig. 24a-g). Mass balance calculations suggest that redox reactions between the gas and ferrous/ferric iron in the lava are plausible mechanisms for the oxygen transfer. Oxygen transfer during cooling is variable—probably reflecting fluctuating rates of gas flow. Higher flow rates reduce the time available for gas/lava oxygen exchange and result in gases with higher last equilibrium temperatures; lower flow rates favor oxygen transfer down to lower temperatures. Gas/lava exchanges of other components (C,H,S) apparently are relatively insignificant during cooling; thus, multicomponent gas/lava heterogeneous equilibrium is not realized. Samples of higher-temperature—up to 1185°C—CO_2-rich volcanic gases from sustained summit lava lake eruptions (J-series samples, Table 5) also show buffering by gas/lava oxygen transfer. The two data sets give a tightly constrained log fO_2—relative to the nickel-nickel oxide (NNO) buffer—of NNO - 0.5(±0.05) for subaerially erupted Kilauea basalt from liquidus to solidus temperatures (Fig. 13). Thus,

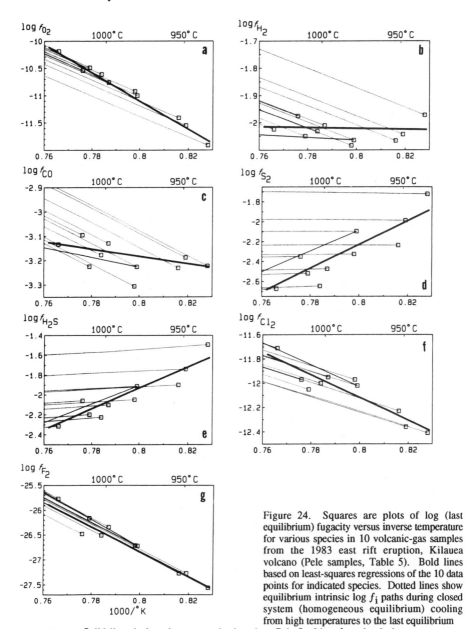

Figure 24. Squares are plots of log (last equilibrium) fugacity versus inverse temperature for various species in 10 volcanic-gas samples from the 1983 east rift eruption, Kilauea volcano (Pele samples, Table 5). Bold lines based on least-squares regressions of the 10 data points for indicated species. Dotted lines show equilibrium intrinsic log f_i paths during closed system (homogeneous equilibrium) cooling from high temperatures to the last equilibrium temperatures. Solid lines in b-g show examples based on Pele 8 of log f_i paths during open system (heterogeneous equilibrium) cooling with gas/lava oxygen transfer to constrain fO_2 by equation (29). All other samples also give open system curves parallel to the bold regression curves; to keep b-g legible, only the Pele 8 open system curve is shown. [Used by permission of the editor of Geochimica et Cosmochimica Acta, from Gerlach (1993a), Fig. 4, p. 800.]

the chemical equilibrium preserved in the Kilauea gas samples is a heterogeneous equilibrium involving gas/lava exchange of only oxygen, and the last equilibrium temperature is the temperature at which this exchange ceased. Preservation of this last equilibrium in the samples suggests that homogeneous reaction rates within the gas phase were slow compared to the time it took the gases to move from the last site of gas/lava equilibration to the site of collection. These findings contrast with the commonly held assumption that volcanic gases are released from lava in a state of chemical equilibrium and then continue equilibrating homogeneously with falling temperature until they quench. Although volcanic gases may often reflect closed-system equilibrium during cooling (Giggenbach, 1987), open system gas/lava equilibration may not be uncommon.

Trace-element vapor transport and deposition during degassing

Another use of high-temperature equilibrium compositions of volcanic gases is to constrain thermochemical models of the origin, transport, and deposition of trace elements in volcanic gases. In addition to H_2O, CO_2, SO_2, H_2, H_2S, HCl, HF, CO, S_2, and COS, high-temperature volcanic gases contain myriad trace elements (e.g., Na, Cu) that are transported as volatile species (e.g., NaCl, CuCl) or eroded-rock aerosols (Symonds et al., 1987, 1992; Quisefit et al., 1989). The volatile trace elements are of considerable interest for their atmospheric impacts (Buat-Ménard and Arnold, 1978), the origin of incrustations (colorful deposits that include both sublimates, precipitated directly from the gas, and solids that form by other processes) around volcanic vents (Stoiber and Rose, 1974), and their implications for the origin of some types of ore deposits (Krauskopf, 1957, 1964; Symonds et al., 1987; Bernard et al., 1990). Multicomponent chemical equilibrium calculations can help unravel the origin and transport of these trace elements, and help determine the origin of fumarolic incrustations. Central to this approach are tests for equilibrium between the volcanic gases (major and minor gases, trace elements) and the magma, sublimates, incrustations, and aerosols. Evidence for equilibrium among the major gases lends credence to thermochemical equilibrium calculations for trace elements.

Model results suggest that most trace elements in volcanic gases are degassed from magma as simple chlorides (e.g., $ZnCl_2$, KCl, AgCl) and other species (e.g., AsS, Hg, H_2MoO_4, H_2BO_3, H_2Se), and that the speciation of each volatile element depends on the bulk gas composition (Symonds et al., 1987, 1992; Quisefit et al., 1989; Symonds and Reed, 1993). The models also predict that Al, Ca, Mg, and other non-volatile rock-forming elements exist primarily as eroded-rock aerosols rather than volatile species in volcanic gases (Quisifit et al., 1989; Symonds et al., 1992; Symonds and Reed, 1993). A third result is that volcanic sublimates generally precipitate in the order of their equilibrium saturation temperatures, which explains the overall zoning of sublimates in volcanic fumaroles (Symonds et al., 1987; Quisefit et al., 1989; Symonds and Reed, 1993). wever, textures of volcanic sublimates indicate that this is a quasi-equilibrium process, ereby the gas starts precipitating sublimates at their equilibrium saturation temperature t fails to maintain equilibrium with the precipitating sublimates at cooler temperatures ymonds, 1993). A final result is that volcanic incrustations form by complex processes nvolving direct precipitation of sublimates from cooling volcanic gases, and reactions between volcanic gases, wall rock, and air (Getahun et al., 1994).

REMOTE SENSING OF GAS EMISSIONS

In addition to direct sampling, volcanic gases can be studied remotely by airborne and ground-based instruments and by satellite. Remote sensing offers great advantages in volcanic-gas work: 1. *Safety,* a point that cannot be overemphasized since the tragedy at Galeras, which occurred during attempts by scientists to collect direct gas samples at a central vent (Williams, 1993). Erebus Volcano in the Antarctic is another example of a volcano which has an open vent and plenty of active surficial degassing, but direct sampling is apparently impossible because of human-safety concerns. 2. *Expediency.* In other cases sampling is not done because suitable sites cannot be found free of substantial contamination of the gas by groundwater or the atmosphere. In these cases remote sensing measurements can often produce useful data. 3. *Gives flux data.* Remote sensing can determine fluxes of species whereas direct gas sampling alone cannot (see Chapter 12). 4. *Instantaneous analysis.* Remote sensing can determine concentrations or fluxes instantaneously in situ, rather than requiring transport (and reaction) of the samples between the sampling point and the laboratory. 5. *Potential economic advantages.* The opportunity for repetitive measurements of gas emissions at many or most volcanoes from satellite platforms exists as more extensive satellite based systems are developed. Since a satellite platform can potentially service many volcanoes, it could become a cost effective way to monitor volcanoes regularly and economically. 6. *Broad potential capability for many gas species.* It can in theory at least be applied to many gas species, and also to various types of particles or aerosols. This advantage is far from realized, however.

In spite of these advantages, remote sensing techniques have not been applied very widely at volcanoes. This is because of limited knowledge of the appropriate techniques, limited budgets for fieldwork at volcanoes, a lack of equipment suited for field work under the rugged field conditions, and the operational difficulties associated with the volcano/atmosphere interface.

Table 7. The ten most common volcanic gas species in volcanic plumes.

Column 1	Column 2			Column 3			Column 4		
Species	Estimated volcanic contribution to highly diluted plume			Ambient tropospheric concentration			Mixed plume concentration		
H_2O	10	- 20	ppm	40	- 40,000	ppm	40	- 40,000	ppm
CO_2	0.5	- 10	ppm	~	300	ppm	301	-	310 ppm
SO_2	1	- 2	ppm	0.1	- 70	ppb	1	-	2 ppm
HCl	0.1	- 2	ppm	~	1	ppb	0.1	-	2 ppm
H_2S	100	- 500	ppb	0.08	- 24	ppb	100	-	500 ppb
S_2	10	- 80	ppb		?			?	
H_2	5	- 40	ppb	540	- 810	ppb	545	-	850 ppb
HF	5	- 40	ppb		?			?	
CO	1	- 20	ppb	0.05	- 0.2	ppm	51	-	220 ppb
SiF_4	1	- 5	ppb		?			?	

Data are based upon homogeneous gas calculations (Symonds and Reed, 1993) and direct gas sampling at high temperature fumaroles. Dilution factors are constrained by direct data (Lazrus et al., 1979; Rose et al, 1980; Friend et al., 1982; Casadevall et al., 1984; Rose et al., 1985, 1988) which demonstrate factors of 10^4 to 10^5. Near vent concentrations may be higher due to lower dilution ratios. Plumes are assumed to have an SO_2 concentration of 1 to 2 ppm (Casadevall et al., 1981; Harris et al., 1981; Rose et al., 1985) for comparison with other species. Species ambient atmospheric levels are from Graedel (1978) and Schidlowski (1986). Column 4 = Column 2 + Column 3. From Andres and Rose (1994).

One way of describing the volcano/atmosphere interface is to compare the concentrations of various chemical species in volcanic gases with the background concentrations in the atmosphere (Table 7). Remote sensing is done from a distance and requires looking through the atmosphere at a gas emission plume. It is obviously more advantageous to make measurements when the concentration of the constituent in question is significantly above the background atmospheric concentrations. The two most abundant gas species in volcanic gases, H_2O and CO_2, are also found in the ambient atmosphere at significant concentrations. This partly explains why remote sensing of volcanoes is at a primitive level. The remote sensing of CO_2, for example, requires detecting and mapping a 1 to 5 ppm anomaly in a 300 ppm background, and doing it through a background that is usually many times the width of the plume. The lack of contrast between the plume and the atmosphere is influenced by the dilution of the volcanic plume by the atmosphere, generally by factors of 1000 to 100,000 or more (Rose et al., 1980), by the time that the remote sensor is applied. This happens within the first seconds of emission to the atmosphere, and means that volcanic plumes ingest large amounts of air and transport it upward. Only volcanic-gas species such as SO_2, which has a very low atmospheric background concentration, can be easily outlined in a plume by remote sensing. Examples of plume-concentration anomalies of SO_2 are shown in Figure 25.

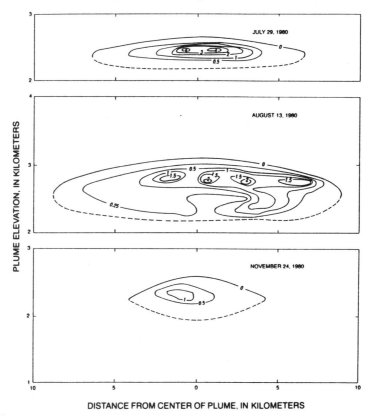

Figure 25. Interpretive cross-sections of SO_2 concentrations (ppm by volume) for three plumes from Mount St. Helens in 1980. Based on COSPEC traverses at 5 to 7 different altitudes in a fixed-wing aircraft. These demonstrate the relatively dilute nature of volcanic plumes. For comparison, undiluted gas leaving the magma in June 1980 contained ~0.3 mole % SO_2 (Gerlach and Casadevall, 1986b), so the plumes shown are diluted by factors of ~1000 to 12,000 at ambient pressures. [Used by permission of the editor of Journal of Volcanology and Geothermal Research; McGee (1992), Fig. 8, p. 280.]

The concentration of any volcanic-gas constituent can easily be overwhelmed by the atmosphere unless its abundance in volcanic gases greatly exceeds its ambient atmospheric concentration. Also if there is a gradient of concentration of some species in the atmosphere such that it decreases in concentration with altitude, a common situation for H_2O and CO_2, then ingestion of air at the base of the plume will result in a contrast between the plume and the surrounding atmosphere that will overwhelm and mask the volcanic component. On a typical summer day in temperate climates there is a higher H_2O content in air closer to the Earth's surface. Because a rising volcanic plume ingests large amounts of air at its base, the plume will be full of water vapor from the atmosphere. Any volcanic H_2O is swamped by atmospheric water vapor. Ingestion of lower altitude air can thus greatly complicate the utility of remote sensing, whether it is ground based or from an airborne or satellite platform. (see also Woods, 1993).

CORRELATION SPECTROMETER DETERMINATION OF SO_2 EMISSIONS

Instrumentation and methods

Of the three most abundant species in volcanic gases, H_2O, CO_2 and SO_2, the remote determination of SO_2 has enjoyed the greatest progress, mainly because of atmospheric problems discussed above with respect to H_2O and CO_2. Beginning in about 1971, an ultraviolet correlation spectrometer (COSPEC, made by Barringer, of Canada), which uses skylight as an ultraviolet source (Millán and Hoff, 1978) has been used at many volcanoes to measure the flux of SO_2 (Stoiber and Jepsen, 1973). This instrument has become a standard tool to make measurements at open-vent volcanoes (Stoiber et al., 1983), and a growing list of data sets has accumulated (see Andres et al., 1993; 1994). In all, the instrument has been used at approximately 50 volcanoes, far more than any similar measurement, and has been used as a tool to estimate the total SO_2 degassing to Earth's atmosphere from volcanoes (Berresheim and Jaschke, 1983; Stoiber et al., 1987). The COSPEC can be used in a number of ways. It can operate in an airplane, from an automobile or a boat, and also from a fixed point on a tripod. In general it works best when it is within a kilometer or two of the gas plume, because attenuation from a long path (>5 km) or from dust or aerosol is significant (Rose et al., 1985). At many volcanoes, aircraft measurements are the highest quality, but the cost of aircraft platforms is frequently too high to allow continuous measurements. The environmental conditions of each volcano (location of the active vent with respect to roads, atmospheric clarity, wind patterns, interfering SO_2 sources) dictate the way the COSPEC is used at each volcano, and the frequency of measurements reflects these as well as the resources of the local volcano observatory. The instrument cannot be used in low-light conditions (night, early morning or late evening in the tropics, anytime during high latitude winters). There are typically significant concentrations of aerosols in volcanic plumes (Fig. 26) and the attenuation effects of these aerosols on the COSPEC results are not well known. Also the signal can be saturated by thick plumes with high SO_2 burdens (Casadevall, 1985). The variety of possible styles of data collection (Fig. 27) reflect the environmental conditions as well as the scientific objectives of the measurements and the type of platform used. Sometimes, especially in aircraft surveys, only a few scans of the plume are made, whereas in other cases hundreds of data scans at intervals of minutes or even seconds are made (see Fig. 27H). There is interest in the interpretation of both long-term and short-term data sequences which measure fluxes of SO_2 at volcanoes (Table 8).

An important factor that influences the quality of data collected by COSPECs is the necessity of measuring the wind speed that is moving the plume or the rise speed of the plume. This is true because the COSPEC looks at a single angular path and is generally used to measure a transect across the plume, either horizontal or vertical, and this burden must be multiplied by a plume movement speed to get a flux (Stoiber et al., 1983). It has been shown by COSPEC users that uncertainty in the plume movement speed is usually the largest source of error in individual flux measurements, and that the high uncertainties (usually 20 to 50%) associated with COSPEC-determined fluxes are largely due to

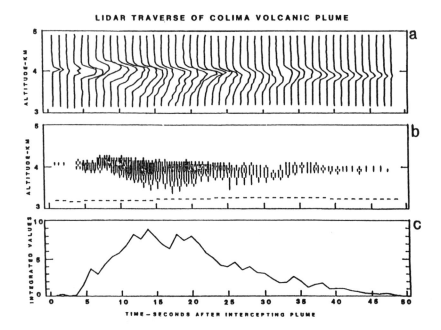

Figure 26. Lidar data collected on a traverse of a plume from Colima Volcano in February 1982 (Casadevall et al., 1984). This aircraft survey also did a COSPEC-based plume flux determination of 320 t/d, and the SO_2 absorption signal matched the lidar record in the traverses, which showed that aerosols in the plume tracked with the SO_2 in this case.

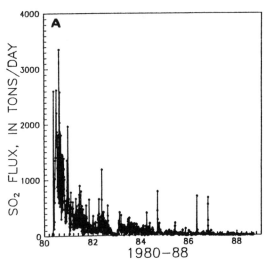

Figure 27. Examples of time dependent COSPEC data sets collected at volcanoes— see text for discussion of potential errors.

(A) Mount St. Helens, 1980-88. This is the most extensive data set of its kind, and shows a long term exponential decay since May 18, 1980. [By permission of the editor, Journal of Volcanology and Geothermal Research; McGee (1992), Fig.2, p. 271.]

imperfect wind speed measurements. Because aircraft have devices that can aid in active windspeed determinations, and because aircraft platforms allow easy adjustments of measurement distance and geometry of the transect, airborne COSPEC surveys are generally considered to be of higher quality. Their expense is higher, however, and this often means that various types of ground-based surveys are used instead. The principal

Table 8. Some examples of use of COSPEC data from the volcanological literature.

Volcano, date	Application	Reference
Fuego, 1978	COSPEC with particle and gas samplers	Lazrus et al. (1979)
Etna, 1978	SO$_2$ emissions and forecasting activity	Malinconico (1979)
St Helens, 1980	Low SO$_2$ before eruption	Stoiber et al. (1980)
Fuego, 1974-7	Excess SO$_2$ released	Rose et al. (1982)
Galunggung, 1982	SO$_2$ emissions and activity	Baddrudin (1983)
St Helens, 1980-2	Long term SO$_2$ and magmatic processes	Casadevall et al. (1983)
Colima, 1984	COSPEC with lidar data for particles	Casadevall et al. (1984)
Kilauea 1984-5	Summit versus rift SO$_2$	Greenland et al. (1985)
White Island 1984-5	COSPEC paired with direct gas data	Rose et al. (1986)
Masaya, 1980-2	Environmental hazards of plumes	Stoiber et al. (1986)
World, 1972-85	Estimating the annual volcanic flux	Stoiber et al. (1987)
Kilauea, 1985	Shallow magma processes, Pu'u O'o	Chartier et al. (1988)
Kilauea, 1979	SO$_2$ flux and earth tides	Connor et al. (1988)
Augustine, 1986	Scaling up degassing rates	Rose et al. (1988)
Kilauea, 1986	SO$_2$ related to magma supply rate	Andres et al. (1989)
Erebus, 1984-8	COSPEC paired with gas and particle samplers	Kyle et al. (1990)
Ruiz, 1985-8	Very large excess SO$_2$	Williams et al. (1990)
Etna, 1975-85	Long-term SO$_2$ average output	Allard et al. (1991)
Lascar, 1989	Excess SO$_2$ compared to surficial magma	Andres et al. (1991)
St. Helens, 1980-8	Exponential SO$_2$ decay after eruption	McGee (1992)
Stromboli, 1980-93	SO$_2$ flux and intrusive magma degassing budget	Allard et al. (1994)
Erebus, 1984-93	Shallow magma processes	Kyle et al. (1994)
Etna, 1987-94	Continuous COSPEC and high eruptive fluxes	Caltabiano et al. (1994)

advantage of ground-based platforms is that large data sets can be collected, which allows examination of the pattern of gas release in detail and enables more robust statistical averaging. In recent years the data collection and reduction processes have been automated (Kyle and McIntosh, 1989) and this has facilitated collection of large data sets which can be used for studies of surficial magmatic processes such as convection (Kyle et al., 1994).

SO$_2$ emissions and volcanic activity

Volcanoes release SO$_2$ at high rates during large explosive eruptions but these fluxes generally cannot be measured with COSPEC because large plume sizes, attenuation of ultraviolet radiation by ash and aerosols, and the geometric problems associated with traverses all cause difficulties. Instead the COSPEC is useful either during conditions of low-level explosive and effusive activity, continual open-vent degassing, or gas releases from shallow intrusions and domes. At any one time there are probably 50 to 100 fuming volcanoes in the world. The SO$_2$ emissions for these range from <10 to more than 5000 metric tonnes/day. The highest SO$_2$ emission rates measured at volcanoes by COSPECs were 32,000 t/d at Kilauea in 1984 (Greenland et al., 1985), 25,000 t/d at Etna in 1989 and 1991 (Caltabiano et al., 1994), and 24,000 t/d at Augustine in 1986 (Rose et al., 1988). Stoiber et al. (1983) broadly classified volcanic plumes based on their SO$_2$ flux as small (< about 200 t/d) moderate (200 to 1000 t/d) and large (>1000 t/d). Moderate and large SO$_2$ fluxes imply magma degassing because (1) SO$_2$ is an abundant magmatic gas, so magma degassing is its logical source, and (2) the main alternative source of these gaseous sulfur discharges, boiling of S-bearing hydrothermal fluids, seems unlikely, since boiling of hydrothermal fluids produces trivial amounts of SO$_2$ in comparison to other S-bearing gases and minerals, even when the fluids are heated to magmatic temperatures (Gerlach et al., 1995). A relationship between the rate of SO$_2$ release and earth tides has been shown at Kilauea (Fig. 27F; Conner et al., 1988).

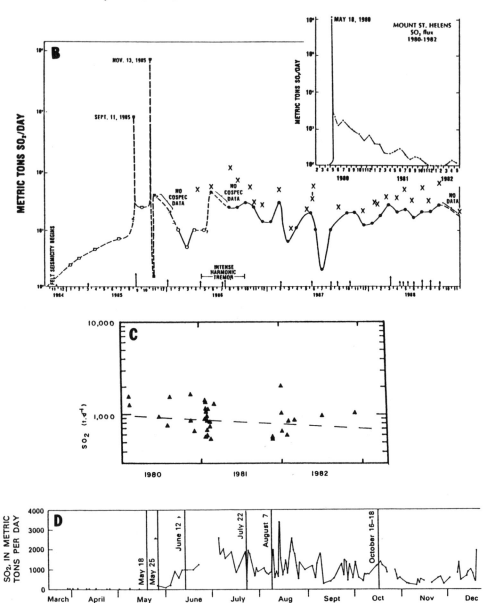

Figure 27 (continued).

(B) Nevado del Ruiz, 1984-88. Semilog plot of data collected at Ruiz, which shows a fairly steady emission rate of thousands of tons/day for years following the eruptions of 1985. Inset: data from Mount St. Helens for comparison. [Used by permission of the editor of Journal of Volcanology and Geothermal Research; Williams et al. (1990), Fig. 2, p. 56.]

(C) Masaya, 1980-82 (from Stoiber et al., 1986). Masaya is a basaltic shield characterized by lava lake and long term open vent degassing of about 1000 t/d.

(D) Mount St. Helens, 1980 (from Casadevall et al., 1981). The first several months of COSPEC data at Mount St. Helens, showing the relationship between explosive volcanic activity and the SO_2 flux.

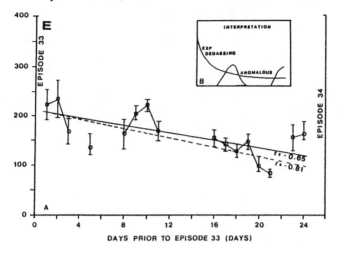

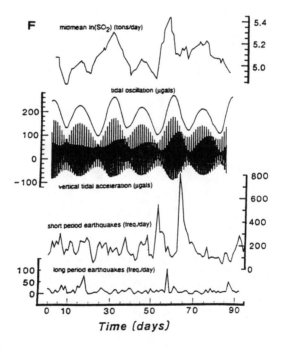

Figure 27 (continued):

E. Pu'u O'o, June 13-July 6, 1985. A daily record of degassing patterns of the ERZ vent during one repose period of 24 days. An exponential decrease with two anomalous periods was interpreted. [E and G by permission of the editor of Bulletin of Volcanology; Chartier et al. (1988), Figs. 9 and 11, respectively.]

F. Kilauea summit June 10-Sept 12, 1979 (from Connor et al., 1988). The daily fluxes are shown to correlate with vertical accelerations of earth tides.

SO$_2$ emission rates at volcanoes often correlate with eruptive activity. The COSPEC data sets from several volcanoes have shown a declining rate of SO$_2$ emission following a large eruption and coincidental with a decrease in volcanic activity. Such a pattern is best shown by the data set from Mount St. Helens, 1980-86 (McGee, 1992; Fig. 27A). On another scale, measurements made at the Puu O'o vent at Kilauea between fountaining phases also show a generally declining trend of emissions (Chartier et al., 1988; Fig 27E). The tendency for SO_2 emission rates to decline at inactive open vents can be supported by other examples, but there is the expectation that frequent measurements might detect increases in SO_2 emissions which correlate with and precede renewed activity. This was first noted by Malinconico (1979) at Etna Volcano; volcanologists have noted this kind of correlation on several occasions since (e.g., Fig. 27D; cf. Fig. 33, below). Precursory increases in SO_2 flux are not observed consistently, however.

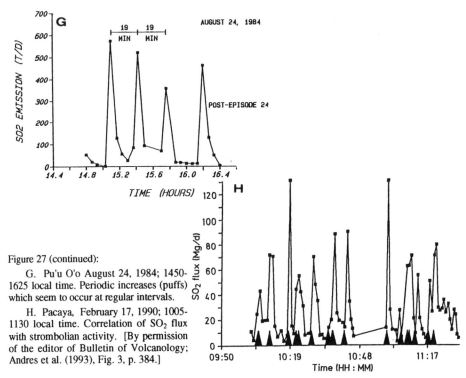

Figure 27 (continued):

G. Pu'u O'o August 24, 1984; 1450-1625 local time. Periodic increases (puffs) which seem to occur at regular intervals.

H. Pacaya, February 17, 1990; 1005-1130 local time. Correlation of SO_2 flux with strombolian activity. [By permission of the editor of Bulletin of Volcanology; Andres et al. (1993), Fig. 3, p. 384.]

The use of a COSPEC prior to a climactic eruption is of interest to volcanologists because the presence of SO_2 in a plume or eruption cloud is thought to be an indicator of near surface magma and an open vent, and thus gas emission measurements may provide a means of detecting changes associated with fracturing of the near surface rocks above a pressurized magma body. Before the Mount St. Helens eruption of 1980, the COSPEC measured generally very low SO_2 emission rates (Fig. 27D). By contrast, emission rates at Ruiz (Williams et al., 1986; 1990; Fig. 27B) were as high as several thousand tonnes/day before the main eruptive events. Similarly high SO_2 emissions were observed before the cataclysmic June 15, 1991 eruption at Mount Pinatubo (Daag et al., 1994). Such high (>1000 t/d) fluxes of SO_2 have come to mean that the volcano may be likely to erupt, although such observations are tempered by a lack of data on the background emissions of many volcanoes. Even with the uncertainties and the lack of many robust data sets, COSPEC gas emission measurements have been used on numerous occasions to try to help forecast eruptions, and following seismic monitoring, provide valuable information for volcanologists trying to understand evolving magmatic systems.

TOTAL OZONE MAPPING:
SPECTROMETER DETERMINATIONS OF SO_2 EMISSIONS

Introduction

As noted earlier, sampling of volcanic gases is often impractical and hazardous; another major limitation stems from the sheer size of large, powerful eruptions. There are no ground-based techniques that have the capability to fully analyze large eruption plumes, which often exceed thousands of square kilometers in area. Observing phenomena of this size can only be accomplished by instruments aboard Earth-orbiting satellites. A powerful tool in the remote sensing of volcanic events was discovered quite by accident, after the launch of the Total Ozone Mapping Spectrometer (TOMS) in 1978 on NASA's Nimbus-7 satellite. Originally designed for globally mapping ozone by measuring the Earth's ultraviolet albedo, TOMS measured anomalous ozone values over Mexico after the eruption of El Chichón in 1982. It turned out that erupted sulfur dioxide was absorbing in some of the same UV wavelengths as used for ozone, and Krueger (1983) developed an algorithm to quantify SO_2.

One advantage of the TOMS instrument is its ability to provide full spatial coverage of large eruptions of sulfur dioxide. Perhaps equally valuable are the nearly continuous global data that the original and subsequent TOMS instruments have produced, recording explosive volcanic SO_2 emissions since late 1978, with the potential to continue this data collection through the turn of the century. Combined with other techniques TOMS data have been used to place constraints on annual emplacement of SO_2 into the atmosphere (Bluth et al., 1993), and to evaluate volcanic eruption sulfur budgets (Bluth et al., 1994a). In addition to estimating SO_2 emissions from individual eruptions, TOMS data have also aided studies of atmospheric circulation using volcanic injections of SO_2 as natural tracers (Schoeberl et al., 1993).

Due to the dynamic nature of eruption plumes, quantification of sulfur dioxide from TOMS measurements is rarely straightforward. A good deal of this section is devoted to discussion of some of the fundamentals of the TOMS instrument, along with the basics of determining SO_2 tonnages and limitations of the technique. A short summary of TOMS data follows, with specific applications to studies of magma degassing. More detailed descriptions of the TOMS instrument and theory behind TOMS-based SO_2 measurement can be found in Krueger (1983; 1985).

Method

Basic parameters. The Nimbus-7 satellite was launched into a sun-synchronous orbit, making a complete observation of the Earth every day in 13.7 orbits. Each orbit consists of a series of 2800 km-wide scans of 35 pixels each, with pixel resolution ranging from 50 km at the orbit center (nadir) to about 200 km at the orbit edges. Each scan takes approximately eight seconds to complete, thus the data comprising most volcanic SO_2 emissions can be collected within a few minutes, essentially giving a snapshot view of eruption clouds. The Nimbus TOMS measures the reflected sunlight of the Earth's surface in six ultraviolet wavelengths. From this information, both total column ozone and sulfur dioxide are calculated, based on the decrease in reflected light due to absorption and scattering by these gases (Krueger, 1983). Two bands which are unaffected by ozone or sulfur dioxide are used to calculate surface albedo (which can include both the Earth and some types of water clouds) or reflectivity.

TOMS instruments. The Nimbus TOMS provided data from November 1978 to its demise in May 1993. A second TOMS instrument was launched in August 1991 in a joint

U.S.–U.S.S.R. venture on board the Meteor-3 satellite. Although it uses essentially the same UV wavelengths, the Meteor TOMS is in a higher altitude, precessing orbit compared to Nimbus, resulting in a slightly larger footprint and less coverage at high latitudes. A third TOMS instrument is aboard NASA's Earth Probe satellite, launched in late 1994. The Earth Probe orbit is identical to that of the Nimbus satellite, and has modified wavelengths resulting in improved discrimination of SO_2. Future launches of TOMS instruments, by various agencies, are planned through the turn of the century.

Basic technique. Column amounts of SO_2 (ideally, the amount of gas between the TOMS instrument and the Earth's surface) are calculated in units of milli-atm cm (also known as Dobson units), the one-dimensional thickness of the gas layer at STP. Typical SO_2 clouds erupted from volcanoes and detected by TOMS range from 20 to several hundred milli-atm cm. The column mass of SO_2 is obtained by multiplying the average column amount of an integration region by its area and a conversion factor.

Background levels of SO_2 are determined by examining the areas adjacent to the volcanic cloud. SO_2 tonnages are calculated for the cloud(s) simply by adding the column masses from all the pixels within the cloud, and accounting for any non-zero background. The timing of the TOMS data collection is noted; this information may be used to calculate SO_2 dispersion during the time between eruption and subsequent TOMS orbits. If the SO_2 cloud is cohesive enough so that tonnages may be calculated on successive days, data are used to extrapolate to the time of eruption and better estimate an original emitted tonnage.

Limitations. The TOMS minimum detectable SO_2 mass depends on eruption, background, and meteorological conditions. Under normal operating conditions for the average scan position area (approximately 5000 km^2), the minimum SO_2 amount considered above noise in one pixel is about 2.5 kilotons (kt). However, single high-value pixels are generally assumed to be random noise, and 2-3 adjacent high-value pixels are usually required to represent a statistically valid SO_2 cloud. Based on the majority of pixel areas and actual SO_2 clouds found within the TOMS database, a practical lower range of detection is about 5-10 kt. However, towards the outer edges of the TOMS scan (where the footprint areas increase to 40,000 km^2) the minimum cloud detection might be no better than 50 kt SO_2.

The maximum amount of SO_2 detectable by TOMS has not been determined. Because the TOMS instrument measures UV reflectance rather than transmittance, even an optically opaque cloud will produce a signal. A cross-section through such a cloud would presumably reveal that absorption (i.e., SO_2 concentration) values reach a plateau; however the thickest cloud, or highest level of SO_2 concentration, known to date (Mount Pinatubo 1991) did not produce such a pattern in the TOMS data.

As the TOMS instrument uses reflected light, the data are valid only in sunlit portions of the orbit. Due to available sunlight, large algorithm errors caused by increasing light path length and screening by ozone place practical limits on the detection of eruptions to about 70° latitude in the hemispheric summer and 50° in winter. This restraint can be significant because of the numerous eruptions from volcanoes in Alaska, Kamchatka, and Iceland. However, there is one advantage at high latitudes because subsequent orbits overlap. At 60° during the hemispheric summer, two or three consecutive observations (at roughly 100 minute intervals) of an SO_2 cloud are often available, instead of the usual single daily observation.

There is little known regarding the minimum and maximum altitudes at which TOMS can detect and measure SO_2 clouds. The TOMS SO_2 algorithm was designed for stratospheric measurements, and the SO_2 cloud altitude clearly has an effect on retrievals.

For example, UV absorption of the sulfur dioxide molecule decreases as temperature rises (McGee and Burris, 1987). Since atmospheric temperature increases with lowering altitude below the tropopause, SO_2 clouds in the lower troposphere should be harder to detect. Other factors that hamper low-altitude detection include the increase in UV light scattering with atmospheric thickness and the increasing likelihood of interference from UV-reflecting water clouds between TOMS and a low altitude SO_2 cloud. Despite these hindrances, TOMS has observed lower tropospheric clouds (e.g., Galunggung 1982, Mauna Loa 1984).

Sources of error. It is important to recognize that the TOMS instruments were not optimally designed to directly measure sulfur dioxide; rather, SO_2 quantities are inferred from UV reflectance data based on an algorithm and the unique conditions surrounding each volcanic eruption. The method of using TOMS to quantify volcanic SO_2 emissions involves three types of uncertainties: (1) the measurement error of a scan pixel; (2) the cloud tonnage error, and (3) the error in extrapolating back to an original emission. Implicit in TOMS tonnage estimates are a number of assumptions and corrections, unique to each volcanic eruption, which will be briefly discussed here.

(1) The accuracy of a single TOMS measurement of SO_2 relies on instrument calibration and operating conditions, errors in assigning absorption coefficients, and the physical and chemical conditions within and surrounding the sulfur dioxide cloud (e.g., Krueger et al., 1990). A (correctable) scan bias occurs in the TOMS data retrieval due to changing sensitivity of the instrument to SO_2 as TOMS scans from nadir to the orbit edges. Changes in the Earth's UV reflectance, from high altitude water clouds as well as high albedos over ice-covered areas, have been correlated to linear, correctable effects on TOMS SO_2. Coemitted tephra and the transformation of erupted sulfur dioxide into sulfuric acid aerosols can also affect SO_2 retrievals, but their precise effects are not well known; these effects are under investigation using radiative transfer calculations.

Because of the limited amount of ground-based SO_2 data on large eruptions and the discrepancies among instrument sensitivities to SO_2, there have been very few opportunities to compare TOMS measurements to other data. One such opportunity arose after the September 1992 eruption of Mount Spurr, Alaska when the emitted SO_2 cloud drifted over an operating Brewer instrument (a ground-based UV spectrophotometer designed to measure ozone) in Toronto. TOMS data taken during two overpasses agreed within 20% to the ground-based measurements (A. Krueger, personal communication).

(2) Evaluations of cloud tonnages are largely affected by background levels, determination of cloud boundaries, and movement of the cloud between TOMS overpasses. Background levels vary with respect to latitude, meteorological conditions, and operating conditions of the TOMS instrument; the variable nature of TOMS background requires careful evaluation in order to discriminate and quantify the SO_2 cloud. Determination of cloud boundaries is vital because the TOMS method works best for compact clouds, or "spikes" of SO_2, rather than discriminating between background (zero level) and cloud boundaries. For diffuse clouds, either due to the nature of the eruption or from dispersion of an originally compact emission, the tonnage uncertainty increases. Due to the size or location of the SO_2 cloud, or orbit overlap at high latitudes, it is not uncommon for an SO_2 cloud to overlap adjacent scans. In these cases cloud movement between TOMS overpasses must be assessed in order to avoid counting the same SO_2 region twice.

(3) When estimating the original amount of SO_2 emitted by an eruption, it is necessary to account for changes in SO_2 amount between the time of eruption and the satellite

observation(s). This calculation assumes that measured tonnages are minimum values because of both chemical conversion of SO_2 to H_2SO_4 and subsequent rainout, and physical dissipation of SO_2 at the cloud margins to below the TOMS detection limits. The loss rate of most volcanic SO_2 clouds (to TOMS observation), while driven by chemical reaction rates, is also highly dependent on cloud altitude. If the erupted SO_2 remains below the tropopause there is more rapid conversion to sulfuric acid and rainout than in the drier, relatively less reactive conditions in the stratosphere. With large stratospheric clouds the amount lost at cloud boundaries is small compared to the mass of the SO_2 cloud (e.g., total cloud tonnage decreased by roughly 10% a day during the first week of observation following the 1991 Mount Pinatubo eruption; Bluth et al., 1992). In contrast, tonnages of small, tropospheric sulfur dioxide clouds with a high perimeter to mass ratio are more affected by physical and chemical processes, and commonly decrease by up to 50% per day (e.g., Bluth et al., 1994a).

The uncertainty involved in estimating the original emission of SO_2 generally decreases with the number of available measurements; thus the more days an SO_2 cloud is detectable by TOMS, the better the extrapolation. Krueger et al. (1990) considered both instrument and typical cloud/background parameters to estimate a total error of ±30% for the 1985 Nevado del Ruiz eruption, a figure which is typically used regarding TOMS tonnages. However, this error is not necessarily applicable to all eruptions. Work in progress to improve accuracy estimates of the TOMS data includes comparison of TOMS data to those of other instruments such as the Brewer spectrometer, and radiative transfer modeling of sulfur dioxide, tephra, and aerosols.

Data

During its 14-year lifetime, the Nimbus-7 TOMS instrument detected over 100 different volcanic eruption clouds. TOMS quantifying capabilities have ranged over three orders of magnitude, from 5 kt (essentially its lower limit of detection) to 20 Megatons (Mt) emitted from Mount Pinatubo in 1991 (Bluth et al., 1992). Figure 28 shows the larger eruptions from the TOMS database; some eruption clouds have been combined to represent a single event. The Volcanic Explosivity Index, or VEI, allows categorization of volcanic eruptions, based mainly on the height of the eruption column and amount of emitted tephra (Newhall and Self, 1982). TOMS detected every highly explosive eruption which occurred during its operation (i.e., the 11 eruptions of VEI ≥ 4), approximately 33% of VEI 3 eruptions, and about 10% of VEI 2 eruptions, using the database of the Smithsonian's Global Volcanism Network (Smithsonian Institution/SEAN, 1989) from 1979 to 1985 as the reported ground truth. Calibration and algorithm development are ongoing for the TOMS on Meteor-3 and no eruption tonnages have been quantified using this instrument.

Applications

General correlations of SO_2 outgassing with eruption characteristics. Differences in volcanic degassing between non-arc (divergent plate and hot spot) and arc (convergent plate) tectonic regimes are fairly evident in the TOMS data. Non-arc volcanoes typically erupt more often than arc volcanoes, but are less numerous worldwide; arc eruptions are by far the most violent (Smithsonian Institution/SEAN, 1989). As shown in Figure 29, there is a general increase in outgassed SO_2 with explosivity for arc eruptions but non-arc activity is less dependent on VEI. Non-arc volcanoes typically degas more sulfur-rich magmas and emit relatively greater amounts of sulfur dioxide for a given VEI. When considering the two regimes on a global basis we found a 2:1 ratio of arc to non-arc outgassing in the TOMS SO_2 database—a ratio dominated by two arc eruptions, El Chichón and Mount Pinatubo (Bluth et al., 1993).

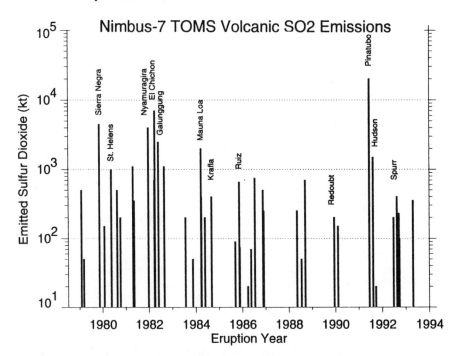

Figure 28. Inventory of TOMS SO_2 data from explosive eruptions, 1979-1992 (after Bluth et al., 1993). Most of the larger or well-known eruptions are labeled. Multiple eruptions from a single volcanic episode are combined, except for the three eruptions of Mount Spurr in 1992. The higher frequency of small eruptions observed prior to 1986 resulted from detailed searches through the TOMS database for all known eruptions using data from Smithsonian Institution/SEAN (1989).

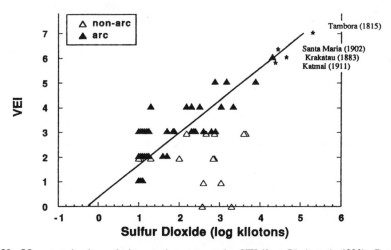

Figure 29. SO_2 outgassing by explosive eruptions compared to VEI (from Bluth et al., 1993). Data are separated by tectonic regimes to show relative contributions of volcanic activity as a function of explosivity. The TOMS SO_2-VEI regression line ($r^2 = 0.999$) is determined from average SO_2 for all VEI 4-6 eruptions between 1979 and 1992; VEI 2 and 3 averages are calculated using respective TOMS detection rates. Estimates of SO_2 emissions (*) from some historical, highly explosive eruptions are added for comparison.

Annual SO_2 emissions. In order to understand the impact of volcanism on the Earth's atmosphere, the eruption frequency and outgassing rates for the entire range of volcanic activity must be known. Non-explosive volcanoes (e.g., Etna, Kilauea) degas at relatively consistent rates, estimated at 9 Mt SO_2 annually (Stoiber et al., 1987). In contrast, forceful degassing characteristic of explosive eruptions (e.g., Mount Pinatubo) is notoriously sporadic and more difficult to quantify. Thus the total yearly volcanic flux of sulfur dioxide is not well constrained; previous ground-based estimates range from 1.5 to 50 Mt (see references within Bluth et al., 1993). The current man-made contribution is approximately 190 Mt/yr (Möller, 1984).

Using the TOMS data, Bluth et al. (1993) made the simplest estimate of annual $\overset{\bullet}{S}O_2$ output by explosive volcanism from the total explosive SO_2, 52 Mt, over 14 years of observation — an average of 4 Mt/yr. But Figure 28 demonstrates the rather sporadic distribution of the "average" output. The dataset is sensitive to the SO_2 emissions of a few relatively large eruptions. For example, one could argue that the 14-year TOMS database is simply dominated by two eruptions, El Chichón and Mount Pinatubo; removing these would halve the calculated annual SO_2 flux for this period.

Seeking another way to address the temporal distribution of global volcanism, Bluth et al. (1993) linked the TOMS SO_2 data to a longer-term dataset of eruption explosivity (VEI) and frequency. Although virtually useless in connecting individual eruptions to SO_2 emissions (in Fig. 29, note the 2-3 order of magnitude ranges of SO_2 observed over VEI 3 or 4 eruptions), a general relationship between average sulfur dioxide emission and VEI has been observed previously by Stoiber et al. (1987). Likewise in Figure 29, a TOMS SO_2/VEI relationship can be determined from average SO_2 amounts for each VEI level (VEI 2 and 3 averages reflect the TOMS' detection rates of these eruptions). The relationship of eruption frequency to VEI over the past 200 years (Simkin, 1993) can then be used to determine the annual eruptive activity for each VEI category. Multiplying the two (average SO_2 and number of eruptions from VEI 0 - 7) produces a total SO_2 flux from explosive volcanism of approximately 4 Mt, in agreement with the simple average of the total TOMS SO_2 database. This suggests that the dominance of the total explosive SO_2 output by a few large eruptions is not unique to the past 14 years. Most volcanic emissions are apparently generated from non-explosive degassing, and combining the non-explosive (9 Mt) and explosive (4 Mt) outputs yields an annual flux of 13 Mt SO_2 from volcanoes. Therefore, it appears that the average annual volcanic emission of SO_2 to the Earth's atmosphere comprises less than 10% of the present-day anthropogenic flux. Even a Mount Pinatubo-sized eruption is small compared to the current annual anthropogenic SO_2 contribution to the atmosphere.

Although much smaller in SO_2 mass output, volcanic activity produces a disproportionately large climatic impact compared to anthropogenic sources because explosive volcanism has the ability to inject sulfur directly into the stratosphere. In the TOMS database 65% of the total 52 Mt SO_2 was emitted by eruptions of VEI 4 and above (Bluth et al., 1993), which implies that stratospheric injection of sulfur dioxide by volcanoes averages roughly 2.5 Mt/yr.

Tandem use with COSPEC. In short-lived, powerful eruptions, such as those of Mount Pinatubo in June 1991, the explosive emissions of SO_2 are usually much greater than the non-explosive contribution. These emissions are best quantified by TOMS. In contrast, eruptions such as exhibited by Mauna Loa and Etna outgas essentially all their sulfur non-explosively, and are better evaluated by COSPEC. To accurately assess the total sulfur budget for moderate strength volcanism such as that exhibited by Galunggung in 1982, a combination of remote sensing techniques is most appropriate.

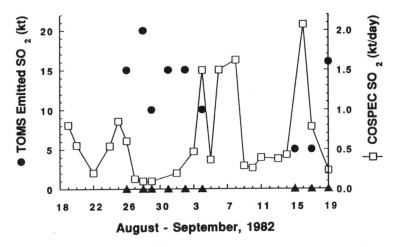

Figure 30. Comparison of TOMS and COSPEC SO₂ data for a coincident period of record of Galunggung eruptions in Indonesia, August 19 - September 19, 1982 (from Bluth et al., 1994a). Large Vulcanian eruptions reported by the Cikasasah Volcano Observatory are marked by filled triangles along the horizontal axis. Note that the TOMS and COSPEC data are plotted on different scales.

Galunggung awoke from 63 years of quiescence in April 1982. During its most violent period from April to October, the Cikasasah Observatory reported 88 eruptions (Katili and Sudradjat, 1984). Eruptions became less violent towards the end of September, but the volcano continued to erupt sporadically through January 1983. TOMS detected and measured 24 separate sulfur dioxide clouds from April 5 to September 19 (Bluth et al., 1994a). The rapid dispersion of the SO₂ clouds, altitude estimates of co-emitted ash clouds, and comparisons of cloud motion to tropospheric wind, suggest that most of the SO₂ remained below the tropopause. A total of just over 2000 kt SO₂ was produced primarily by the 24 explosive eruptions observed by TOMS, and to a smaller extent by the 64 explosive eruptions which were below the TOMS detection limit.

For the first time both TOMS and COSPEC were in operation during a coincident, 32-day period of explosive volcanism (Fig. 30). Bluth et al. (1994a) estimated that explosive outgassing of SO₂ (derived from TOMS) produced approximately six times the non-explosive emissions (estimated from COSPEC data; Badruddin, 1986). If this ratio remained relatively constant over the course of activity from April to September, approximately 350 kt would have outgassed non-explosively. COSPEC data indicate an additional 50 kt SO₂ were emitted by the less-violent activity from the end of September 1982 to the end of January 1983. Combining all the estimates of SO₂ emissions, Galunggung outgassed approximately 2500 kt of sulfur dioxide from both explosive and non-explosive activity between April 1982 to January 1983.

Excess *SO₂* from convergent-plate volcanoes

Volatile exsolution models for Kilauea basalt containing 0.3 wt % H_2O, 0.07 wt % S, and 0.02-0.05 wt % CO_2 suggest that vigorous exsolution of dissolved sulfur (as SO_2) occurs only after the magma has ascended to shallow depths (estimated to be about 150 m) where pressures are <30 bar (Gerlach, 1986). The observed amounts of SO_2 emitted from Kilauea basalt are roughly in balance with the amounts that would be expected from the dissolved sulfur content of the magma and the volumes of magma extruded and degassed

Table 9. Comparisons of erupted magma masses and masses of sulfur released. Modified from Andres et al. (1991).

Volcano, date	Erupted Magma Mass tonnes (10^6 g)	S mass tonnes ($= 10^6$ g)	Equiv % S wt %
Stromboli, 1980-93	1.9×10^5/yr[†]	1.1×10^5/yr[†]	58[†]
Etna, 1975	4.0×10^4	7.5×10^3	15
Láscar, 1989	3.7×10^6	2.7×10^5	6.9
Lonquimay, 1989	1.1×10^5	3.5×10^3	3.1
Ruiz, 1985	3.5×10^7	3.3×10^5	0.93
Fuego, 1974	2.2×10^8	1.6×10^6	0.72
El Chichón, 1982	1.2×10^9	7.0×10^6[‡]	0.58
Pacaya, 1972	2.6×10^4	1.3×10^2	0.5
Agung, 1963	5.0×10^9	0.6×10^7	0.12
St. Helens, 1980	7.4×10^8	0.3×10^6	0.04
Mauna Loa, 1984	5.4×10^8	1.2×10^5	0.02

[†] From Allard et al. (1994); [‡]slightly modified to reflect data of Bluth et al. (1993).

near the surface (Andres et al., 1989). These results are also likely to be true for eruptions of other tholeiitic basalts with similar low volatile contents, but they are at odds with SO_2 emissions depicted by TOMS and COSPEC data for eruptions of magma at convergent-plate volcanoes. For these systems, SO_2 emissions appear to be in excess of expected yields based on concentrations of dissolved sulfur, as represented by melt inclusions, and the quantity of erupted and near surface magma (Rose et al., 1982; Andres et al., 1991; Westrich and Gerlach, 1992; Allard et al., 1994; Gerlach et al., 1994, 1995; Gerlach and McGee, 1994). In many cases, the proportion of excess sulfur indicated by the SO_2 release is very large. Table 9 shows examples of the required sulfur contents of surficial coerupted magma that, if completely degassed, could account for the observed masses of released sulfur. The values are much higher than the sulfur content of the magma indicated by melt inclusions, except for Mauna Loa, which involves tholeiitic basalt likely to have a volatile content similar to that of Kilauea basalt. Most of the examples are for convergent plate volcanoes. The general lack of agreement in these cases shows that remote sensing of SO_2 is a poor way to estimate lava eruption rates and volumes for these systems, and that melt-inclusion methods are likely to underestimate sulfur yields during convergent plate volcanism (Devine et al., 1984; Palais and Sigurdsson, 1989) by large factors.

Comprehensive TOMS, COSPEC, and melt-inclusion data permit a comparison of SO_2 emissions based on remote sensing and melt inclusions for explosive volcanism at three convergent plate volcanoes. TOMS and COSPEC data from the climactic explosion of Mount St. Helens on May 18, 1980 to the last stages of quiescent dome degassing in September 1988 indicate a total SO_2 emission for the eruption of roughly 2 Mt (megaton) compared to a melt inclusion-based estimate for the eruption of 0.08 Mt (Gerlach and McGee, 1994). The total TOMS and COSPEC-based SO_2 emission estimate for the 1989-1990 eruption of Redoubt Volcano, Alaska, is approximately 1 Mt compared to a maximum emission estimate of 0.04 Mt from melt-inclusion data (Gerlach et al., 1994). The TOMS estimate for SO_2 emission during the 1991 eruption of Mount Pinatubo, Philippines, is about 20 Mt (Bluth et al., 1992), whereas the melt inclusions indicate no significant degassing of dissolved sulfur during magma ascent and eruption (Gerlach et al., 1995; see also Chapter 8). The sources of the excess sulfur are widely discussed and debated, but the remote sensing data for SO_2 and CO_2 and melt inclusion data for H_2O and CO_2 suggest that the Mount St. Helens dacite, the Redoubt andesites, and the Mount Pinatubo dacite were vapor-saturated at depth prior to eruption, and that sulfur in accumulated vapor provided the immediate source of excess sulfur for the SO_2 emissions (Westrich and

Gerlach, 1992; Gerlach et al., 1994; Gerlach and McGee, 1994; Wallace and Gerlach, 1994; Gerlach et al., 1995).

Detection of H_2S emissions. During the summer of 1992, Mount Spurr, Alaska ended a 39-year period of quiescence and emitted large SO_2 plumes during three separate eruptions June 27, August 18, and September 17 (Bluth et al., 1994b). After the August and September eruptions, an unusual pattern was observed: TOMS-measured SO_2 actually increased from the first to second day, before returning to a typical pattern of gradual cloud dispersion (Fig. 31). Both eruption plumes were observed after complete separation from the volcano, thus there was no possibility that the increase could have occurred from additional outgassing after the first TOMS overpass. An important implication for this retrieval pattern may be that significant H_2S emission occurred during the Spurr eruption.

Bluth et al. (1994b) first examined several other explanations. Measurement uncertainty was ruled out as the cause, based on the consistent SO_2 dispersion characteristics of the clouds after the initial days' low quantities. Neither interference from co-erupted ash, nor aerosol formation were consistent with the TOMS data. UV light scattering by the ash clouds, which Schneider et al. (1994) showed were coincident with the gas clouds, may have partially masked detection of sulfur dioxide, but TOMS reflectivity data showed no evidence of UV attenuation from ash particles. In addition, the slow fallout rate of micron-sized particles in the plume (Rose, 1993; Wen and Rose, 1994) indicated that any effect on TOMS measurements should have lasted for weeks, rather than only a single day. Radiative transfer modeling of ash and aerosol effects on TOMS retrievals suggest that the presence of either should cause measured SO_2 to increase, rather than be underestimated. Also, the aerosol effect should have increased with time, as more SO_2 converted to aerosol.

The eruption of H_2S, and oxidation to SO_2 subsequent to the first day, could explain the tonnage increases in the TOMS data. The amount of hydrogen sulfide required to produce the sulfur dioxide pattern observed in the TOMS data is roughly 75 kt H_2S for the August eruptions (a 3:1 SO_2:H_2S ratio), and 25 kt H_2S for the September eruption (a 6:1 ratio). Doukas and Gerlach (1994) report on evidence and a possible mechanism for hydrogen sulfide emissions from boiling of hydrothermal fluid within the volcano (see below). As discussed above under pressure effects, it is also possible that H_2S was a stable sulfur gas species in the magma at depth (Reaction 26, Fig. 14) and that rapid eruption and transport through the troposphere allowed a significant amount of it to reach the stratosphere and subsequently oxidize to SO_2. For example, Gerlach and Casadevall (1986b) suggested that sudden release of H_2S stabilized in magma at depth (Fig. 14) may have caused the high H_2S/SO_2 observed in Mount St. Helens plumes during explosive eruptions, including the May 18, 1980 climactic eruption plume (Casadevall et al., 1981; Hobbs et al., 1981), and that SO_2-related environmental effects of explosive volcanism may often require post-eruption oxidation of discharged H_2S.

The pattern of increasing SO_2 in this stratospheric cloud was also observed in the TOMS data immediately after the climactic May 18 eruption of Mount St. Helens in 1980. While the TOMS data cannot be used as definitive proof of H_2S gas emissions from either the Spurr or St. Helens eruptions, these results may open a new approach for investigating the degassing of magma-hydrothermal systems.

EMISSION RATES OF OTHER VOLCANIC GASES

Methods for measurement

Volcanologists have long recognized the value of remote sensing data on more species, but results of this sort are very sparse (Mori et al., 1993). An infrared gas spectrometer

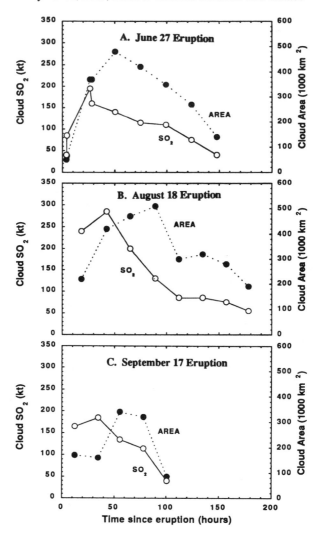

Figure 31. Comparison of cloud sulfur dioxide mass (open circles) and area (filled circles) as a function of time after eruptions of Mount Spurr, Alaska in 1992 (from Bluth et al., 1994b). (a) June 27 eruption. (b) August 18 eruption. (c) September 17 eruption.

(Miran) has been used to provide data on CO_2 fluxes of volcanic plumes at Mount St. Helens (Harris et al., 1981). This device is not a remote sensor, but requires direct sampling of the plume. This limitation means that the CO_2 concentrations of the plume must be mapped by aircraft traverses through the plume. As a result, the Miran has only been used in a few cases (Harris et al., 1981; Casadevall et al., 1987; Casadevall et al., 1994; Doukas and Gerlach, 1995). At Mount St. Helens Miran data on CO_2 were coupled with COSPEC determined SO_2 fluxes to allow comparison and ratioing of two species. Because most of the CO_2 may be lost from magmas before extensive degassing of SO_2— as discussed above for Kilauea volcano—there is a possibility that such data will be a more useful forecasting tool than SO_2 alone (Harris and Rose, 1994; Casadevall et al., 1994).

Methods of estimation

Combining flux measurements with compositional data on plumes and fumaroles gives estimated fluxes of additional species. For instance, by combining COSPEC SO_2 fluxes with direct gas samples that constrain the ratios of other species to SO_2, fluxes for all species in the samples can be estimated (e.g. Rose et al., 1986; McGee, 1992). When analytical data are not available for minor and trace species, thermochemical modeling can provide estimates of the appropriate ratios (Symonds et al., 1992). Combining the COSPEC with gas-filter sampling (Finnegan et al., 1989) also allows estimation of elemental fluxes (Phelan et al., 1982; Rose et al., 1988).

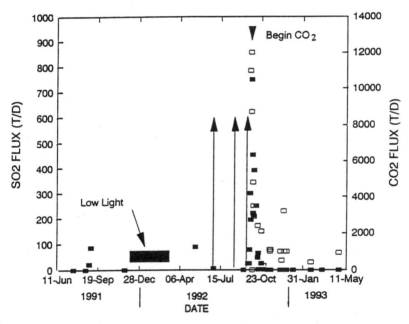

Figure 32. SO_2 and CO_2 emission-rate measurements for Crater Peak, Spurr Volcano, between 1991 and 1993. Solid squares and open squares give SO_2 (left vertical axis) and CO_2 (right vertical axis) fluxes, respectively. Arrows mark explosive eruptions. Solid bar indicates low light period when COSPEC measurements were not possible. From Doukas and Gerlach (1994).

Measuring both SO_2 and CO_2 emission rates–an example of the advantages

During the 1991-1993 activity of Crater Peak on Spurr Volcano, Doukas and Gerlach (1994) measured both SO_2 and CO_2 emission rates using COSPEC and Miran, respectively. Over this timeframe, there were three subplinian eruptions on June 27, August 18, and September 16-17, 1992. SO_2 measurements began on July 22, 1991 at the onset of precursory seismicity, whereas CO_2 fluxes were not initiated until September 25, 1992; flux measurements continued until April 24, 1993 (Fig. 32). Despite the elevated seismic activity and explosive eruptions, both indicating shallow intrusion of magma, SO_2 fluxes were generally <100 t/d during the 1991-1993 period, except during the eruptions themselves, as measured by TOMS (Bluth et al., 1994b), and between September 25 and October 10, shortly after the September eruption (Fig. 32). The <100 t/d fluxes are much lower than expected for a passively degassing volcano discharging moderately large plumes. Other active volcanoes discharging plumes of moderate size generally emit 200 to 1000 t/d of SO_2 (e.g., Mount St. Helens, Fig. 27A, McGee, 1992; Redoubt, Casadevall et al., 1994), as did Crater Peak between September 25 and October 10, 1992 (Fig. 32). In contrast, the CO_2 emission rates from late 1992 to early 1993 were generally between 1000 and 12,000 t/d, similar to other volcanoes emitting moderately large plumes (e.g., Redoubt, Casadevall, 1994; Mount St. Helens, Harris et al., 1981). The generally lower than expected SO_2 emission rates, normal CO_2 fluxes, and CO_2/SO_2 ratios between 10-100, much higher than expected for gases from undegassed convergent-plate magmas (Fig. 10), led Doukas and Gerlach (1994) to suggest that the gases released from shallow magma encountered liquid water (possibly as hydrothermal fluid) inside the volcano. They speculate that scrubbing of sulfur dioxide by hydrolysis reactions with water masked the SO_2 degassing, but boiling of hydrothermal fluid released CO_2 to the atmosphere. They also suggest that "drying out" of the channelways between September 25 and October 10, 1992 allowed a brief period of elevated SO_2 emissions. These results underscore the advantages to be gained from measuring emission rates of magmatic volatiles like CO_2, which are relatively insoluble in shallow magma and boiling water, for comparison with SO_2 emission rates.

GAS STUDIES TO ASSESS VOLCANIC HAZARDS

In addition to seismic and geodetic monitoring, gas studies play an important role in assessing volcanic activity. There are three basic gas-based approaches to assessing volcanic activity: (1) direct sampling of fumarolic or soil gases, (2) measuring gas emission rates, and (3) continuous monitoring of volcanic fumaroles and ambient air. The ideal gas program would integrate all three methods because each approach has advantages and disadvantages (Table 10). At some volcanoes, the gases vent into a crater lake and geochemical sampling of lake waters can detect changes in the degassing activity.

Direct gas sampling

Surveillance by direct gas sampling involves collecting gas samples from volcanic fumaroles at periodic intervals with the methods described above. It may also involve soil-gas sampling to detect diffuse degassing around the volcanic edifice. The time interval between sampling may vary depending on volcanic activity, logistical considerations, and available resources. Direct sampling of fumaroles is currently the only way to fully characterize the compositions of gases discharging from volcanoes. It is also the only way to collect isotopic data vital to constraining the origin(s) of individual gas species from potentially restless volcanoes (e.g., Sorey et al., 1993). Direct gas sampling can also capitalize on rare and urgent sampling opportunities because it requires a minimal amount of time and field equipment. Past studies have shown that direct gas sampling can determine the source(s) of the gases, including the presence or absence of subsurface magma (Allard et al., 1991b); if magma is present, these data can sometimes detect changes in its degassing state (Gerlach and Casadevall, 1986b; Tedesco et al., 1991). Chemical analyses of gases can also be used to estimate subsurface temperatures of volcanic-hydrothermal systems and to determine whether a system is heating up or cooling down (Tedesco and Sabroux, 1987). Moreover, direct gas sampling provides geochemical information on a volcano's hydrothermal system, which should be a part of an overall hazards assessment since a hydrothermal system may influence the eruptive style of a volcano.

Direct gas sampling is currently the most practical and cost effective way to geochemically characterize volcanoes with low gas fluxes. Surveillance of volcanoes with low gas fluxes requires methods that can detect magmatic gases in weak fumarolic

Table 10. Summary of advantages and disadvantages of gas geochemical methods to assess volcanic activity.

Advantages	*Disadvantages*
Direct Gas Sampling	
1. Characterizes complete gas composition. 2. Only way to collect isotopic data on gases. 3. Requires minimal time and equipment. 4. Best way to study low-flux volcanoes. 5. Can detect subsurface temperature changes. 6. Helps characterize volcano-hydrothermal systems. 7. Enhances volcanic emission studies. 8. Cheapest method.	1. Temporal resolution only as good as sampling frequency. 2. Labor intensive. 3. Each sample requires human access to potentially dangerous volcanic vents. 4. Weather dependent.
Measuring Gas Emission Rates	
1. Only way to measure fluxes of gas species. 2. Best spatial coverage of volcanic degassing. 3. Can be used during eruptions or during elevated volcanic activity when other methods are not feasible. 4. Only gas measurement method to estimate volumes of magma degassed. 5. Safest methods because it does not require human access to volcanic vents.	1. Currently, only SO_2 and CO_2 are measured. 2. Temporal resolution only as good as measurement frequency. 3. Labor intensive. 4. Weather dependent. 5. More expensive than gas sampling.
Continuous Monitoring	
1. Provides very frequent time-series data. 2. Can detect short-lived (hours to days) gas events. 3. Not weather dependent. 4. Safer than direct gas sampling. 5. Easy direct comparison with telemetered geophysical data on same time scale.	1. Ideal sensors for volcanoes not yet available 2. Monitoring equipment exposed to geologic hazards. 3. Servicing equipment requires human access to potentially dangerous volcanic vents. 4. More expensive than gas sampling.

emissions, including species like CO_2 and He that may pass through hydrothermal or meteoric water that is often present at such volcanoes. Recent work (Baubron et al., 1990; Allard et al., 1991a) shows that significant degassing of CO_2 also occurs on the flanks of volcanoes as diffuse soil-gas emissions, which suggests that periodic soil-gas sampling may detect changes in activity at volcanoes with low gas fluxes. Emission-rate methods (below) are often not sensitive enough to detect low gas fluxes in the early stages of a magmatic intrusion. Direct gas sampling has the sensitivity required to potentially detect such disturbances, including any changes in CO_2 and He.

One main disadvantage of assessing volcanic activity with gas samples is that each time series of samples requires another trip to the collection site. This approach is satisfactory when one can collect and analyze samples faster than the rate at which gas compositions are changing. However, when available resources, weather conditions, or other problems restrict the sampling frequency, the samples may miss rapidly changing chemical events. Moreover, the duration of some gas events are so short (e.g., several hours long; McGee and Sutton, 1994) that it is impractical to detect them by gas sampling.

Measuring gas emission rates

Remote sensing and direct sampling of plumes are the only ways of obtaining gas flux information, which provide the best gauge of overall volcanic degassing. Because measurements are made on volcanic plumes rather than in close proximity to volcanic vents, these are also the safest geochemical assessment methods and can even be used during some volcanic eruptions. Current technology enables measurement of SO_2 and CO_2 emission rates by airborne or ground-based methods, and SO_2 emissions by satellite (see above). Measurement intervals depend on volcanic activity, weather, and available resources, and like direct gas sampling, one major problem with emission-rate studies is that the temporal resolution of the data is only as good as the sampling frequency. Another significant disadvantage is that SO_2 and CO_2 are the only species that can currently be measured by remote sensing or direct sampling of plumes. However, new methods capable of measuring other species are under development (e.g., K. McGee, personal communication, 1994) and these may make remote sensing a more attractive and cost-effective way to measure volcanic emissions.

Continuous monitoring of fumaroles and ambient air

Continuous monitoring is accomplished by collecting very frequent gas-concentration data (e.g., every 10 minutes) with sensors placed inside a fumarole or in the air or soil around a fumarole. The data are telemetered back to a remote site for analysis and interpretation. This method has the advantages of being much safer than direct gas sampling and has excellent temporal control that can detect hourly changes in the compositions of gas emissions (Sutton et al., 1992; McGee and Sutton, 1994). For example, both reduced-gas monitoring and SO_2 flux measurements detected a large gas release prior to the October 1986 lava extrusion at Mount St. Helens (Fig. 33; McGee and Sutton, 1994). The chief disadvantage of continuous monitoring is that suitable sensor technology for volcanic environments is still to be developed. For instance, species-selective sensors that can withstand the acid-gas attack in volcanic fumaroles are not yet available (Sutton et al., 1992). However, new advances in environmental sensors have made it possible to monitor specific species in the air around volcanic vents (Sutton, 1990). Future improvements in sensor technology will make continuous monitoring more attractive.

Sampling of volcanic crater lakes

About 88 of the world's 619 active volcanoes contain near-summit crater lakes (Rowe and Varekamp, 1993). In such cases, periodic sampling of crater lakes has many of the advantages (numbers 2-8, Table 10) and disadvantages (2-4, Table 10) of direct gas sampling and some of the advantages (1-2, 4, Table 10) of measuring gas emission rates (Giggenbach, 1974; Giggenbach and Glover, 1975; Menyailov, 1975; Takano, 1987; Takano and Watanuki, 1990; Rowe et al., 1992a,b; Brantley et al., 1993; Christenson and Wood, 1993; Ágústsdóttir and Brantley, 1994; Delmelle and Bernard, 1994).

An integrated approach to surveillance

Table 11 outlines surveillance methods for an idealized eruption. The application of the various techniques is closely related to the level of activity. During early and late stages, direct sampling, if possible, is the method of choice. The best time for direct gas sampling is typically late, after the big eruptions, but while the vent is still open. COSPEC and Miran measurements of SO_2 and CO_2 become feasible usually at some point before the climactic eruption and continue throughout the eruptive cycle, except during large explosive

Table 11. A model of gas studies throughout an idealized eruption sequence.

Sequence of events	Timing	Vent characteristics	Sampling/Species	Expected Results
repose period	years prior	background degassing	Direct: H_2O, CO_2, SO_2, H_2, H_2S, HCl, HF, CO, He, isotopes†	-establish baseline gas compositions; -determination of sources of gases; -background hydrothermal activity
onset of activity	>1 year prior	minor degassing: interactions with groundwater, rock, atmosphere	Direct: H_2O, CO_2, SO_2, H_2, H_2S, HCl, HF, CO, He, isotopes†	-increases in magmatic-gas components; -CO_2 escape through diffuse cracks and soil; -elevated hydrothermal activity
ramping up of activity	1-12 months prior	increase in vent degassing; vent opening	Direct: H_2O, CO_2, SO_2, H_2, H_2S, HCl, HF, CO, He, isotopes†; COSPEC: SO_2; Miran: CO_2; Continuous Monitoring	-increases in magmatic-gas components; -compositions of magmatic gases; -fO_2 determination; -low-level SO_2 and CO_2 degassing rates; -periods of elevated degassing
paroxysmal eruption(s)	within months prior	open vent; minor, precursor eruptions	COSPEC: SO_2; Miran: CO_2; Continuous Monitoring; TOMS: SO_2	-low-level SO_2 and CO_2 degassing rates; -gas detection of intrusions; -periods of elevated degassing; -total SO_2 from explosive events
cataclysmic eruption	during	major explosive eruption	TOMS: SO_2	-total SO_2 degassed
recurring eruptions	days to months after	open vent; minor eruptions	TOMS: SO_2; COSPEC: SO_2; Miran: CO_2; Continuous Monitoring	-total SO_2 from explosive events; -low-level SO_2 and CO_2 degassing rates; -gas detection of intrusions; -periods of elevated degassing
ramping down of activity	months to years after	open vent degassing; rare eruptive activity	Direct: H_2O, CO_2, SO_2, H_2, H_2S, HCl, HF, CO, He, isotopes†; COSPEC: SO_2; Miran: CO_2	-evolution of magmatic-hydrothermal system; -compositions of magmatic gases; -fO_2 determination; -gas-rock interactions; -low-level SO_2 and CO_2 degassing rates

†Recommended isotopes for gas monitoring include δD and $δ^{18}O$ for H_2O, $δ^{13}C$ for CO_2, $δ^{34}S$ for total S, and $^3He/^4He$.

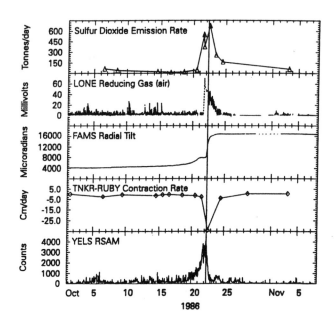

Figure 33. Plot of SO_2 emission rates, reducing-gas sensor data, tilt, electronic distance measurements (EDM), and seismic data at Mount St. Helens during October and November, 1986. Air reducing-gas data are from station LONE on the east side of the dome. The single datum on October 22 is shown enlarged and the dotted line represents the missing early portion of the reducing-gas peak. Tilt data are from FAMS (north side of dome); EDM data are from the line between TNKR (north crater floor) and RUBY (north side of dome); Seismicity is represented by hourly-averaged RSAM data from YELS (north crater floor). Vertical line represents the beginning of lava extrusion on 21 October 1986. Note that the increases in SO_2 fluxes and reducing-gas signals coincide with changes in tilt, EDM, and seismic data. From McGee and Sutton (1994).

eruptions. TOMS is useful (for SO_2) during the large explosive eruptions, but can't be used to measure lower-level emissions. The example shown in Figure 34 documents SO_2 emissions throughout 1980 at Mount St. Helens. Most of the data come from COSPEC observations, but the high spikes are TOMS data, taken during the major eruption of May 18. The ratio of explosive to non explosive degassing at Mount St. Helens in 1980 is roughly 5:1; ~1 Mt during the May 18 event and 220 Kt during non-eruptive days of 1980.

CONCLUSIONS

Volcanic gases contain H_2O, CO_2, SO_2, H_2, H_2S, HCl, HF, CO, S_2, COS, rare gases, and a number of trace-metal species. The main volcanic gases are H_2O, CO_2, and SO_2, although H_2S may be the dominant sulfur gas in equilibrium with some convergent-plate magmas at depth, and HCl may exceed total S in some convergent-plate volcanic gases. Gases discharging from convergent-plate volcanoes contain much more H_2O than gases from divergent-plate or hot-spot volcanoes, which have higher proportions of CO_2 and SO_2. Convergent-plate-volcano gases are also relatively enriched in HCl. These patterns reflect the tectonic origins of the respective magmas. Magmas from divergent-plate and hot-spot volcanoes derive volatiles primarily from the mantle, so they contain moderate amounts of C and S but are relatively depleted in H_2O and Cl compared to magmas from convergent-plate volcanoes, which contain volatiles from the subducted slab as well as the mantle, making them richer in H_2O and Cl.

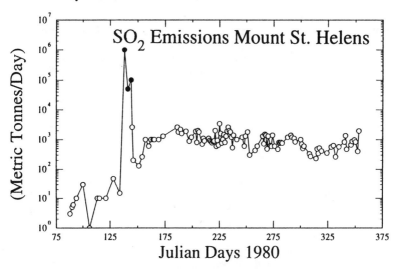

Figure 34. Combination of COSPEC (open circles) and TOMS (filled circles) data for Mount St. Helens in 1980. COSPEC data are from Casadevall et al. (1982); TOMS data are approximate. Highest spike of SO_2 emissions is on Julian day 138 (18 May), about 1 Mt, measured by TOMS.

The relative abundances of H_2, H_2S, and CO in volcanic gases depend on temperature, pressure, and bulk composition. For a given bulk atomic composition, the H_2/H_2O, SO_2/H_2S, and CO/CO_2 ratios increase with rising temperature and falling pressure. The abundances of H_2, H_2S, and CO also increase with the amounts of atomic H, H + S, and C, respectively, and with decreasing fO_2. The fO_2's of at least some high-temperature volcanic gases are apparently buffered by their respective lavas over a range of liquidus to solidus conditions. Thus, volcanic-gas data can provide independent constraints on fO_2 of magmas in addition to those provided by analyses of minerals and glasses in rocks.

Volcanic-gas studies contribute to the overall understanding of the storage and release of volatiles from magmas. One outstanding dilemma is that many andesitic-dacitic convergent-plate volcanoes emit more SO_2 during eruptions than can be explained by the amount of dissolved S released from erupted or near-surface magma. This also applies to CO_2 in the few cases where the appropriate CO_2 data exist. One emerging explanation is that many of these magmas may contain a separate gas phase at depth prior to eruption (see also Chapter 8). These andesitic-dacitic magmas also contain variable amounts of dissolved volatiles (mainly H_2O) that may exsolve from melt during ascent and eruption or at shallow depths by second boiling of magma (e.g., during crystallization of a dome). Most basaltic magmas also appear to contain dissolved volatiles and a separate gas phase. For instance, Kilauea magmas apparently saturate with CO_2-rich fluid at ~40 km depth, whereas most of the remaining volatiles exsolve at ≤150 m. This general feature of two-stage degassing may be common at hot-spot and divergent-plate volcanoes. Once a separate CO_2-rich fluid exists, it may scavenge other volatiles with especially low solubility (e.g., rare gases) from the melt.

Volcanic gases are one of several indicators used to understand and forecast volcanic activity. In particular, SO_2 fluxes correlate moderately well with volcanic activity, and have helped forecast some eruptions. However, SO_2-flux data have been much more successful at predicting the episodic eruptions or declining volcanic activity that follow an initial cataclysmic eruption than in predicting the first explosive eruption itself. This suggests that prior to the vent-clearing eruption, a hydrothermal or impermeable-rock barrier may prevent some or all of the SO_2 from reaching the surface. Recent work indicates that magmatic CO_2 and He are less affected by hydrothermal barriers and may therefore provide an improved basis for forecasting initial eruptions.

The principal atmospheric and global impacts of volcanic eruptions occur through stratospheric injection of SO_2, which converts to sulfuric acid aerosols that block incoming solar radiation and contribute to ozone destruction. On average, volcanoes emit 13 Mt of SO_2 per year, of which 4 Mt comes from explosive eruptions and 9 Mt is emitted by passive degassing. This represents only 5-10% of the annual anthropogenic SO_2 flux, but its climatic and atmospheric impacts are greater because much of the volcanic SO_2 is injected explosively into the stratosphere where it is not readily rained out, unlike the SO_2 discharged into the troposphere. Over the past 14 years, the eruptions of El Chichón and Mount Pinatubo dominate the explosive SO_2 flux, suggesting that most of the stratospheric injections come from infrequent large eruptions.

Future studies should expand the data on high-temperature equilibrium com-positions of volcanic gases. These studies should include thermochemical evaluation of analytical data and retrieval of last equilibrium compositions and temperatures where possible by methods reviewed above. High sampling priorities include gases from hot-spot and divergent-plate volcanoes, especially those with alkaline and evolved magmas. More time-series samples from multiple vents on the same volcano are also needed. In addition, more work is needed to develop sensors that detect specific gas species and resist corrosion by acidic volcanic gases so that the true potential of time-series monitoring of fumaroles can be realized. A high priority for plume studies is to measure CO_2 emission rates, as well as the emission rates of other gas species, in addition to SO_2. Past studies show that, in addition to SO_2 fluxes, ratios of gas species change with volcanic activity. Multiple-species flux data would provide data on both emission rates and the ratios of gas species. Satellite-based systems that can detect multiple species well into the troposphere are also desirable and may offer the most cost-effective approach for measuring volcanic-gas emissions when combined with in-situ or continuous sensor data in the future. Such systems may provide the only practical means for observing volcanic emissions on a global scale. Finally, we need additional TOMS measurements of SO_2 emissions from large explosive eruptions to improve understanding of the atmospheric and climatic impacts of these volcanic events.

ACKNOWLEDGMENTS

Timely and detailed reviews by Patric Allard, Alain Bernard, Jake Lowenstern, and Ken McGee greatly improved an earlier draft of this paper. Mark Reed suggested the method of graphically displaying correspondence temperatures. Funding for this project was provided in part by the U.S. Geological Survey Volcano Hazards and Geothermal Studies Program and the U.S. Geological Survey Global Change and Climate History Program.

REFERENCES

Ágústsdóttir AM, Brantley SL (1994) Volatile fluxes integrated over four decades at Grímsvötn volcano, Iceland. J Geophys Res 99:9505-9522.
Allard P (1983) The origin of hydrogen, carbon, sulphur, nitrogen and rare gases in volcanic exhalations: evidence from isotope geochemistry. In: Tazieff H, Sabroux JC (eds) Forecasting Volcanic Events. Elsevier, Amsterdam, 337-386.
Allard P, Tazieff H, Dajlevic D (1979) Observations of sea floor spreading in Afar during the November 1978 fissure eruption. Nature 279:30-33.

Allard P, Carbonnelle J, Dajlevic D, Le Bronec J, Morel P, Robe MC, Maurenas JM, Faivre-Pierre R, Martin D, Sabroux JC, Zettwoog P (1991a) Eruptive and diffuse emissions of CO_2 from Mount Etna. Nature 351:387-391.

Allard P, Malorani A, Tedesco D, Cortecci G, Turi B (1991b) Isotopic study of the origin of sulfur and carbon in Solfatara fumaorles, Campi Flegrei caldera. J Volcanol Geotherm Res 48:139-159.

Allard P, Carbonnelle J, Métrich N, Loyer H, Zettwooog P (1994) Sulphur output and magma degassing budget of Stromboli volcano. Nature 368:326-330.

Alt JC (1994) A sulfur isotopic profile through the Troodos ophiolite, Cyprus: primary composition and the effects of seawater hydrothermal alteration. Geochim Cosmochim Acta 58:1825-1840.

Andres RJ, Rose WI (1994) Remote sensing spectroscopy of volcanic plumes and clouds. In: McGuire WJ, Kilburn CRJ, Murray JB (eds) Monitoring Active Volcanoes: Strategies, Procedures, and Techniques. University College of London Press, London, England (in press).

Andres RJ, Kyle PR, Stokes JB, Rose WI (1989) Sulfur dioxide emissions from episode 48A, East Rift Zone eruption of Kilauea volcano, Hawaii. Bull Volcanol 52:113-117.

Andres RJ, Rose WI, Kyle PR, deSilva S, Francis P, Gardeweg M, Moreno Roa H (1991) Excessive sulfur dioxide emissions from Chilean volcanoes. J Volcanol Geotherm Res 46:323-329.

Andres RJ, Rose WI, Stoiber RE, Williams SN, Matiás O, Morales R (1993) A history of sulfur dioxide emission rate measurements from Guatemalan volcanoes. Bull Volcanol 55:379-388.

Andres RJ, Kyle PR, Chuan RL (1994) Sulfur dioxide, particle and elemental emissions from Mount Etna, Italy during July 1987. Geologishe Rundschau (in press).

Baubron JC, Allard P, Toutain JP (1990) Diffuse volcanic emissions of carbon dioxide from Vulcano Island, Italy. Nature 344:51-53.

Badruddin M (1986) Pancaran gas SO_2 pada Letusan G. Galunggung, 1982. In: Katili JA, Sudradjat A, Kumumadinata K (eds) Letusan Galunggung 1982-1983. Direktorat Vulkanologi, Direktorat Jenderal Geologi Dan Sumberdaya Mineral, Departemen Pertambangan Dan Energi, Bandung, 285-301 (in Indonesian).

Bernard A (1985) Les mécanismes de condensation des gaz volcaniques. PhD dissertation, University of Brussels, Belgium (in French).

Bernard A, Symonds RB, Rose WI Jr (1990) Volatile transport and deposition of Mo, W and Re in high temperature magmatic fluids. Appl Geochem 5:317-326.

Berreshim H, Jaeschke W (1983) The contribution of volcanoes to the global atmospheric sulfur budget. J Geophys Res 88:3732-3740.

Bluth GJS, Doiron SD, Schnetzler CC, Krueger AJ, Walter LS (1992) Global tracking of the SO_2 clouds from the June, 1991 Mount Pinatubo eruptions. Geophys Res Lett 19:151-154.

Bluth GJS, Schnetzler CC, Krueger AJ, Walter LS (1993) The contribution of explosive volcanism to global atmospheric sulphur dioxide concentrations. Nature 366:327-329.

Bluth GJS, Casadevall TJ, Schnetzler CC, Doiron SD, Walter LS, Krueger AJ, Badruddin M (1994a) Evaluation of sulfur dioxide emissions from explosive volcanism: the 1982-1983 eruptions of Galunggung, Java, Indonesia. J Volcanol Geotherm Res (in press).

Bluth GJS, Scott CJ, Sprod IE, Schnetzler CC, Krueger AJ, Walter LS (1994b) Explosive SO_2 emissions from the 1992 eruptions of Mount Spurr, Alaska. In: Keith T (ed) Spurr 1992. U S Geol Surv Bull (in press).

Brantley SL, Ágústsdóttir AM, Rowe GL (1993) Crater lakes reveal volcanic heat and volatile fluxes. GSA Today 3:173-178.

Buat-Ménard P, Arnold M (1978) The heavy metal chemistry of atmospheric particulate matter emitted by Mount Etna volcano. Geophys Res Lett 5:245-248.

Caltabiano T, Gennaro D, Romano R (1994) SO_2 flux measurements at Mount Etna (Sicily). J Geophys Res (in press).

Carmichael ISE (1991) The redox states of basic and silicic magmas: a reflection of their source region? Contrib Mineral Petrol 106:129-141.

Casadevall TJ, Johnston DA, Harris DM, Rose WI, Malinconico LL, Stoiber RE, Bornhorst TJ, Williams SN (1981) SO_2 emission rates at Mount. St. Helens from March 29 through December, 1980. In: Lipman PW, Mullineaux DR (eds) The 1980 eruptions of Mount St. Helens, Washington. U S Geol Surv Prof Paper 1250:193-200.

Casadevall T, Rose W, Gerlach T, Greenland LP, Ewert J, Wunderman R, Symonds R (1983) Gas emissions and the eruptions of Mount St. Helens through 1982. Science 221:1383-1385.

Casadevall TJ, Rose WI Jr, Fuller WH, Hunt WH, Hart MA, Moyers JL, Woods DC, Chuan RL, Friend JP (1984) Sulfur dioxide and particles in quiescent volcanic plumes from Poás, Arenal, and Colima volcanos, Costa Rica and Mexico. J Geophys Res 89:9633-9641.

Casadevall TJ, Stokes JB, Greenland LP, Malinconico LL, Casadevall JR, Furukawa BT (1987) SO_2 and CO_2 emission rates at Kilauea volcano, 1979-1984. In: Decker RW, Wright TL, Stauffer PH (eds) Volcanism in Hawaii. U S Geol Surv Prof Paper 1350:771-780.

Casadevall TJ, Doukas MP, Neal CA, McGimsey RG, Gardner CA (1994) Emission rates of sulfur dioxide and carbon dioxide from Redoubt volcano, Alaska during the 1989-1990 eruptions. J Volcanol Geotherm Res (in press).

Chaigneau M, Tazieff H, Febre R (1960) Composition des gaz volcaniques du lac de lave permanent due Nyiragongo (Congo belge). C R Acad Sci Paris, Ser D, 250:2482-2485.

Chartier TA, Rose WI, Stokes JB (1988) Detailed record of SO_2 emissions from Pu'u `O`o between episodes 33 and 34 of the 1983-86 ERZ eruption, Kilauea, Hawaii. Bull Volcanol 50:215-228.

Christenson BW, Wood CP (1993) Evolution of vent-hosted hydrothermal system beneath Ruapehu Crater Lake, New Zealand. Bull Volcanol 55:547-565.

Cole JW, Nairn IA (1975) Catologue of the active volcanoes of the world including solfatara fields—Part 22. New Zealand. Int'l Assoc Volcanol Chem Earth's Interior.

Connor CB, Stoiber RE, Malinconico, LL Jr (1988) Variation in sulfur dioxide emissions related to earth tides, Halemaumau crater, Kilauea volcano, Hawaii. J Geophys Res 93:14867-14871.

Daag A, Tubianosa B, Newhall C, Tungol N, Javier D, Dolan M, de los Reyes PJ, Arboleda R, Martinez M, Regalado MTM (1994) Monitoring sulfur dioxide emissions at Mount Pinatubo. In: Punongbayan RS, Newhall CG (eds) The 1991-1992 eruptions of Mount Pinatubo, Philippines. U S Geol Surv Prof Paper (in press).

Delmelle P, Bernard A (1994) Geochemistry, mineralogy, and chemical modelling of the acid crater lake of Kawah Ijen volcano, Indonesia. Geochim Cosmochim Acta 58:2445-2460.

Delorme H (1983) Composition chimique et isotopique de la phase gazeuse de volcans calcoalcalins: Amèrique Centrale et Soufrière de la Guadeloupe. Application à la surveillance volcanique. PhD disseration, University of Paris VII.

Devine JD, Sigurdsson H, Davis AN, Self S (1984) Estimates of sulfur and chlorine yields to the atmosphere from volcanic eruptions and potential climatic effects. J Geophys Res 89:6309-6325.

Doukas MP, Gerlach TM (1994) Results of volcanic gas monitoring during the 1992 eruption at Mt. Spurr. In: Keith T (ed) Spurr 1992. U S Geol Surv Bull (in press).

Ellis AJ (1957) Chemical equilibrium in magmatic gases. Am J Sci 255:416-431.

Finnegan DL, Kotra JP, Hermann DM, Zoller WH (1989) Use of ^{7}LiOH-impregnated filters for the collection of acidic gases and analysis by instrumental neutron activation analysis. Bull Volcanol 59:83-87.

Friend JP, Bandy AR, Moyers JL, Zoller WH, Stoiber RE, Torres AL, Rose WI, McCormick MP, Woods DC (1982) Research on atmospheric volcanic emissions: an overview. Geophys Res Lett 9:1101-1104.

Gantes M, Sabroux JC, Vitter G (1983) Chemical sensors for monitoring volcanic activity. In: Tazieff H, Sabroux JC (eds) Forecasting Volcanic Events. Elsevier, Amsterdam, 409-424.

Gerlach TM (1979) Evaluation and restoration of the 1970 volcanic gas analyses from Mount Etna, Sicily. J Volcanol Geotherm Res 6:165-178.

Gerlach TM (1980a) Evaluation of volcanic gas analyses from Kilauea Volcano. J Volcanol Geotherm Res 7:295-317.

Gerlach TM (1980b) Investigations of volcanic gas analyses and magma outgassing from Erta'Ale lava lake, Afar, Ethiopia. J Volcanol Geotherm Res 7:415-441.

Gerlach TM (1980c) Evaluation of volcanic gas analyses from Surtsey Volcano, Iceland, 1964-1967. J Volcanol Geotherm Res 8:191-198.

Gerlach TM (1980d) Chemical characteristics of the volcanic gases from Nyiragongo lava lake and the generation of CH_4-rich fluid inclusions in alkaline rocks. J Volcanol Geotherm Res 8:177-189.

Gerlach TM (1981) Restoration of new volcanic gas analyses from basalts of the Afar region: further evidence of CO_2-degassing trends. J Volcanol Geotherm Res 10:83-91.

Gerlach TM (1986) Exsolution of H_2O, CO_2, and S during eruptive episodes at Kilauea volcano, Hawaii. J Geophys Res 91:12177-12185.

Gerlach TM (1989a) Degassing of carbon dioxide from basaltic magma at spreading centers: I. Afar transitional basalts. J Volcanol Geotherm Res 39:211-219.

Gerlach TM (1989b) Degassing of carbon dioxide from basaltic magma at spreading centers: II. Mid-oceanic ridge basalts. J Volcanol Geotherm Res 39:221-232.

Gerlach TM (1993a) Oxygen buffering of Kilauea volcanic gases and the oxygen fugacity of Kilauea basalt. Geochim Cosmochim Acta 57:795-814.

Gerlach TM (1993b) Thermodynamic evaluation and restoration of volcanic gas analyses: An example based on modern collection and analytical methods. Geochem J 27:305-322.

Gerlach TM, Casadevall TJ (1986a) Evaluation of gas data from high-temperature fumaroles at Mount St. Helens, 1980-1982. J Volcanol Geotherm Res 28:107-140.

Gerlach TM, Casadevall TJ (1986b) Fumarole emissions at Mount St. Helens Volcano, June 1980 to October, 1981: degassing of a magma-hydrothermal system. J Volcanol Geotherm Res 28:141-160.

Gerlach TM, Graeber EJ (1985) Volatile budget of Kilauea volcano. Nature 313:273-277.

Gerlach TM, McGee KA (1994) Total sulfur dioxide emissions and pre-eruption vapor saturated magma at Mount St. Helens, 1980-1988. Geophys Res Lett (in review).

Gerlach TM, Nordlie BE (1975) The C-O-H-S gaseous systems, Part II: temperature, atomic composition, and molecular equilibria in volcanic gases. Am J Sci 275:377-394.

Gerlach TM, Taylor BE (1990) Carbon isotope constraints on degassing of carbon dioxide from Kilauea volcano. Geochim Cosmochim Acta 54:2051-2058.

Gerlach TM, Westrich HR, Casadevall TJ, Finnegan DL (1994) Vapor saturation and accumulation in magmas of the 1989-1990 eruption of Redoubt volcano, Alaska. J Volcanol Geotherm Res 62 (in press).

Gerlach TM, Westrich HR, Symonds RB (1995) Pre-eruption vapor saturation in magma of the climactic Mount Pinatubo eruption: Source of the giant statospheric sulfur dioxide cloud. In: Punongbayan RS, Newhall CG (eds) The 1991-1992 eruptions of Mount Pinatubo, Phillipines. U S Geol Surv Prof Paper (in press).

Getahun A, Reed MH, Symonds R (1994) Mount St. Augustine volcano fumarole wall rock alteration: mineralogy, zoning, and numerical models of its formation process. J Volcanol Geotherm Res (in review).

Giggenbach WF (1974) The chemistry of Crater Lake, Mt. Ruapehu (New Zealand) during and after the 1971 active period. New Zealand J Sci 17:33-45.

Giggenbach WF (1975) A simple method for the collection and analysis of volcanic gas samples. Bull Volcanol 39:15-27.

Giggenbach WF (1980) Geothermal gas equilibria. Geochim Cosmochim Acta 44:2021-2032.

Giggenbach WF (1987) Redox processes governing the chemistry of fumarolic gas discharges from White Island, New Zealand. Appl Geochem 2:143-161.

Giggenbach WF (1992) Isotopic shifts in waters from geothermal and volcanic systems along convergent plate boundaries and their origins. Earth Planet Sci Lett 113:495-510.

Giggenbach WF, Glover RB (1975) The use of chemical indicators in the surveillance of volcanic activity affecting the crater lake on Mt. Ruapehu, New Zealand. Bull Volcanol 39:70-81.

Giggenbach WF, Goguel RL (1989) Collection and analysis of geothermal and volcanic water and gas discharges. DSIR Chemistry Rept CD 2401.

Giggenbach WF, Le Guern F (1976) The chemistry of magmatic gases from Erta'Ale, Ethiopia. Geochim Cosmochim Acta 40:25-30.

Giggenbach WF, Matsuo S (1991) Evaluation of results from second and third IAVCEI field workshops on volcanic gases, Mt Usu, Japan, and While Island, New Zealand. Appl Geochem 6:125-141.

Giggenbach WF, Sheppard DS (1989) Variations in the temperature and chemistry of White Island fumarole discharges 1972-85. New Zealand Geol Surv Bull 103:119-126.

Gill JB (1981) Orogenic andesites and plate tectonics. Springer-Verlag, Berlin.

Graedel TE (1978) Chemical compounds in the atmosphere. Academic Press, New York.

Greenland LP (1987a) Hawiian eruptive gases. In: Decker RW, Wright TL, Stauffer PH (eds) Volcanism in Hawii. U S Geol Surv Prof Paper 1350:781-790.

Greenland LP (1987b) Composition of gases from the 1984 eruption of Mauna Loa volcano. In: Decker RW, Wright TL, Stauffer PH (eds) Volcanism in Hawii. U S Geol Surv Prof Paper 1350:781-790.

Greenland LP, Rose WI, Stokes JB (1985) An estimate of gas emissions and magmatic gas content from Kilauea volcano. Geochim Cosmochim Acta 49:125-129.

Harris DM, Rose WI (1994) Dynamics of carbon dioxide emissions, magma crystallization and magma ascent: hypothesis, theory and application to Mount St. Helens. Bull Volcanol (in review).

Harris DM, Sato M, Casadevall TJ, Rose WI Jr, Bornhorst TJ (1981) Emission rates of CO_2 from plume measurements. In: Lipman PW, Mullineaux DR (eds) The 1980 eruptions of Mount St. Helens, Washington. U S Geol Surv Prof Paper 1250:201-207.

Heald EF, Naughton JJ, Barnes IL Jr (1963) Use of equilibrium calculations in the interpretation of volcanic gas samples. J Geophys Res 68:545-557.

Hobbs PV, Radke LF, Eltgroth MW, Hegg DA (1981) Airborne studies of the emissions from the volcanic eruptions of Mount St. Helens. Science 211:816-818.

Holloway JR (1977) Fugacity and activity of molecular species in supercritical fluids. In: Fraser DG (ed) Thermodynamics in Geology. Reidel Publishing Co, Dordrecht, Holland, 161-181.

Huntingdon AT (1973) The collection and analysis of volcanic gases from Mount Etna. Philos Trans R Soc London, Ser A, 274:119-128.

Ito E, Harris DM, Anderson AT Jr (1983) Alteration of oceanic crust and geologic cycling of chlorine and water. Geochim Cosmochim Acta 47:1613-1624.

Jaggar TA (1940) Magmatic gases. Am J Sci 238:313-353.

Katili JA, Sudradjat A (1984) Galunggung: The 1982-1983 eruption. Report of the Volcanological Survey of Indonesia, Republic of Indonesia.

Kodosky LG, Motyka RJ, Symonds RB (1991) Fumarolic emissions from Mount St. Augustine, Alaska,

1979-1984: degassing trends, volatile sources, and their possible role in eruptive style. Bull Volcanol 53:381-394.

Krauskopf KB (1957) The heavy metal content of magmatic vapor at 600°C. Econ Geol 52:786-807.

Krauskopf KB (1964) The possible role of volatile metal compounds in ore genesis. Econ Geol 59:22-45.

Krueger AJ (1983) Sighting of El Chichón sulfur dioxide clouds with the Nimbus 7 Total Ozone Mapping Spectrometer. Science 220:1377-1379.

Krueger AJ (1985) Detection of volcanic eruptions from space by their sulfur dioxide clouds. American Institute of Aeronautics and Astronautics, AAIA 23rd Aerospace Sciences Meeting, AIAA-85-0100, January 14-17, Reno, Nevada, 1-5.

Krueger AJ, Walter LS, Schnetzler CC, Doiron SD (1990) TOMS measurement of the sulfur dioxide emitted during the 1985 Nevado del Ruiz eruptions. J Volcan Geotherm Res 41:7-15.

Kyle PR, McIntosh WC (1989) Automation of a correlation spectrometer for measuring volcanic SO_2 emissions. New Mexico Bur Mines Mineral Res Bull 131:158.

Kyle PR, Meeker K, Finnegan D (1990) Emission rates of sulfur dioxide, trace gases and metals from Mount Erebus, Antarctica. Geophys Res Lett 17:2125-2128.

Kyle PR, Sybeldon LM, McIntosh WC, Meeker K, Symonds R (1994) Sulfur dioxide emission rates from Mount Erebus, Antarctica. In: Kyle P (ed) Volcanological Studies of Mount Erebus, Antarctica. Antarctic Res Series, Washington, DC (in press).

Lazrus AL, Cadle RD, Gandrud BW, Greenberg JP, Huebert BJ, Rose WI (1979) Trace chemistry of the stratosphere and of volcanic eruption plumes. J Geophys Res 84:7869-7875.

Le Guern F, Gerlach TM, Nohl A (1982) Field gas chromatograph analyses of gases from a glowing dome at Merapi volcano, Java, Indonesia, 1977, 1978, 1979. J Volcanol Geotherm Res 14:223-245.

Magro G, Pennisi M (1991) Noble gases and nitrogen: mixing and temporal evolution in the fumarolic fluids of Vulcano, Italy. J Volcanol Geotherm Res 47:237-247.

Malinconico LL (1979) Fluctuations in SO_2 emissions during recent eruptions of Etna. Nature 278:43-45.

Matsuo S (1961) On the chemical nature of fumaorlic gases of volcano, Showashinzan, Hokkaido, Japan. J Earth Sci Nagoya Univ 9:80-100.

Matsuo S, Suzuoki T, Kusakabe M, Wada H, Suzuki M (1974) Isotopic and chemical compositions of volcanic gases from Satsuma-Iwojima, Japan. Geochem J 8:165-173.

Matsuo S, Ossaka J, Hirabayashi J, Ozawa T, Kimishima K (1982) Chemical nature of volcanic gases of Usu volcano in Japan. Bull Volcanol 45:261-264.

McGee K (1992) The structure, dynamics and chemical composition of non-eruptive plumes from Mount St. Helens. J Volcanol Geotherm Res 51:269-282.

McGee K, Sutton AJ (1994) Eruptive activity at Mount St. Helens, Washington, USA, 1984-1988: a gas geochemistry perspective. Bull Volcanol (in press).

McGee TJ, Burris J Jr (1987) SO_2 absorption cross sections in the near U.V. J Quant Radiative Transfer 37:165-182.

Menyailov IA (1975) Prediction of eruptions using changes in compositions using changes in compositions of volcanic gases. Bull Volcanol 39:112-125.

Menyailov IA, Nikitina LP (1980) Chemistry and metal contents of magmatic gases: the new Tolbachik volcanoes case (Kamchatka). Bull Volcanol 43:197-205.

Menyailov IA, Nikitina LP, Shapar VN, Pilipenko VP (1986) Temperature increase and chemical change of fumarolic gases at Momotombo volcano, Nicaraqua, in 1982-1985: are these indicators of a possible eruption? J Geophys Res 91:12199-12214.

Millán MM, Hoff RM (1978) Remote sensing of air pollutant by correlation spectroscopy—instrumental response characteristics. Atmos Environ 12:853-864.

Mizutani Y (1962a) Chemical analysis of volcanic gases. J Earth Sci Nagoya Univ 10:125-134.

Mizutani Y (1962b) Origin of lower temperature fumarolic gases at Showashinzan. J Earth Sci Nagoya Univ 10:135-148.

Mizutani Y (1978) Isotopic compositions of volcanic steam from Showashinzan volcano, Hokkaido, Japan. Geochem J 12:57-63.

Mizutani Y, Sugiura T (1982) Variations in chemical and isotopic compositions of fumarolic gases from Showashinzan volcano, Hokkaido, Japan. Geochem J 16:63-71.

Möller D (1984) Estimation of the global man-made sulphur emission. Atmos Environ 18:19-27.

Mori T, Notsu K, Tohjima Y, Wakita H (1993) Remote detection of HCl and SO_2 in volcanic gas from Unzen volcano, Japan. Geophys Res Lett 20:1355-1358.

Nemoto T, Hayakawa M, Takahashi K, Oana S (1957) Report on the geological, geophysical, and geochemical studies of Usu volcano (Showashinzan). Geol Surv Japan Rep 170:1-149.

Newhall CG, Self S (1982) The volcanic explosivity index (VEI): An estimate of explosive magnitude for historical volcanism. J Geophys Res 87:1231-1238.

Nordlie BE (1971) The composition of the magmatic gas of Kilauea and its behavior in the near surface environment. Am J Sci 271:417-463.

Palais JM, Sigurdsson H (1989) Petrologic evidence of volatile emissions from major historic and prehistoric volcanic eruptions. In: Berger A, Dickinson RE, Kidson JW (eds) Understanding Climate Change. Am Geophys Union Monograph 52:31-53.

Peacock SM (1990) Fluid processes in subduction zones. Science 248:329-337.

Phelan JM, Finnegan DL, Ballantine DS, Zoller WH, Hart MA, Moyers JL (1982) Airborne aerosol measurements in the quiescent plume of Mount St. Helens: September, 1980. Geophys Res Lett 9:1093-1093.

Quisefit JP, Toutain JP, Bergametti G, Javoy M, Cheynet B, Person A (1989) Evolution versus cooling of gaseous volcanic emissions from Momotombo Volcano, Nicaragua: Thermochemical model and observations. Geochim Cosmochim Acta 53:2591-2608.

Rose WI (1993) Comment on "another look at the calculation of fallout tephra volumes" by Judy Fierstein and Manuel Nathenson. Bull Volcanol 55:372-374.

Rose WI, Chuan RL, Cadle RD, Woods DC (1980) Small particles in volcanic eruption clouds. Am J Sci 280:671-696.

-Rose WI Jr, Stoiber RE, Malinconico LL (1982) Eruptive gas compositions and fluxes of explosive volcanoes: Budget of S and Cl emitted from Fuego volcano, Guatemala. In: Thorpe RS (ed) Orogenic Andesites and Related Rocks. Wiley and Sons, New York, NY, 669-676.

Rose WI, Symonds RB, Chartier T, Stokes JB, Brantley S (1985) Simultaneous experiments with two correlation spectrometers at Kilauea and Mount St. Helens. EOS Trans Am Geophys Union 66:1142.

Rose WI, Chuan RL, Giggenbach WF, Kyle PR, Symonds RB (1986) Rates of sulfur dioxide and particle emissions from White Island volcano, New Zealand, and an estimate of the total flux of major gaseous species. Bull Volcanol 48:181-188.

Rose WI, Heiken G, Wohletz K, Eppler D, Barr S, Miller T, Chuan RL, Symonds, RB (1988) Direct rate measurements of eruption plumes at Augustine volcano: a problem of scaling and uncontrolled variables. J Geophys Res 93:4485-4899.

Rowe GL (1991) The acid crater lake system of Poás volcano, Costa Rica: geochemistry, hydrology, and physical characteristics. PhD Dissertation, Pennsylvania State University, University Park, PA

Rowe GL, Varekamp JC (1993) A call for contributions to the IWGCL crater and caldera lake database. International Working Group on Crater Lakes (IWGCL) Newsletter 6:54-61.

Rowe GL Jr, Brantley SL, Fernandez M, Fernandez JF, Borgia A, Barquero J (1992a) Fluid-volcano interaction in an active stratovolcano: the crater lake system of Poás volcano, Costa Rica. J Volcanol Geotherm Res 49:23-51.

Rowe GL Jr, Ohsawa S, Takano B, Brantley SL, Fernandez JF, Barquero J (1992b) Using crater lake chemistry to predict volcanic activity at Poás volcano, Costa Rica. Bull Volcanol 54:494-503.

Schidlowski M (1986) The atmosphere. In: Hutzinger O (ed) The Natural Environment and the Biogeochemical Cycles. Springer Verlag, New York.

Schneider DJ, Rose WI, Kelley L (1994) Tracking of the 1992 Crater Peak/Spurr volcano eruption clouds using AVHRR. In: Keith T (ed) Spurr 1992. U S Geol Surv Bull (in press).

Schoeberl MR, Doiron SD, Lait LR, Newman PA, Krueger AJ (1993) A simulation of the Cerro Hudson SO_2 cloud. J Geophys Res 98:2949-2955.

Shepherd ES (1921) Kilauea gases, 1919. Hawaiian Volcano Obs Bull 9:83-88.

Shinohara H, Giggenbach WF, Kazahaya K, Hedenquist JW (1993) Geochemistry of volcanic gases and hot springs of Satsuma-Iwojima, Japan: following Matsuo. Geochem J 27:271-285.

Sigvaldason GE, Elisson G (1968) Collection and analysis of volcanic gases at Surtsey, Iceland. Geochim Cosmochim Acta 32:797-805.

Simkin T (1993) Terrestrial volcanism in space and time. Ann Rev Earth Planet Sci 21:427-452.

Smithsonian Institution/SEAN, 1989, Global Volcanism 1975-1985. Prentice-Hall, Englewood Cliffs, New Jersey, and American Geophysical Union, Washington, DC.

Sorey ML, Kennedy BM, Evans WC, Farrar CD, Suemnicht GA (1993) Helium isotope and gas discharge variations associated with crustal unrest in Long Valley Caldera, California, 1989-1992. J Geophys Res 98:15871-15889.

Spycher NF, Reed MH (1988) Fugacity coefficients of H_2, CO_2, CH_4, H_2O and of H_2O-CO_2-CH_4 mixtures: A virial equation treatment for moderate pressures and temperatures applicable to calculations of hydrothermal boiling. Geochim Cosmochim Acta 52:739-749.

Stevenson DS (1993) Physical models of fumarolic flow. J Volcanol Geotherm Res 57:139-156.

Stoiber RE, Jepsen A (1973) Sulfur dioxide contributions to the atmosphere by volcanoes. Science 182:577-578.

Stoiber RE, Rose WI Jr (1974) Fumarolic incrustations at active Central American volcanoes. Geochim Cosmochim Acta 38:495-516.

Stoiber RE, Malinconico LL Jr, Williams SN (1983) Use of the Correlation Spectrometer at Volcanoes. In: Tazieff H, Sabroux JC (eds) Forecasting Volcanic Events. Elsevier Science Publishers, Amsterdam, 425-444.

Stoiber RE, Williams SN, Huebert BJ (1986) Sulfur and halogen gases at Masaya Caldera complex, Nicaragua: total flux and variations with time. J Geophys Res 91:12215-12231.

Stoiber RE, Williams SN, Huebert BJ (1987) Annual contribution of sulfur dioxide to the atmosphere by volcanoes. J Volcanol Geotherm Res 33:1-8.

Stolper E, Newman S (1994) The role of water in the petrogenesis of Mariana trough magmas. Earth Planet Sci Lett 121:293-325.

Sutton AJ (1990) Chemical sensors for volcanic gases with a compilation of commercial availability. U S Geol Surv Open-File Rept 90-44.

Sutton AJ, McGee KA, Casadevall TJ, Stokes JB (1992) Fundamental volcanic-gas-study techniques: an integrated approach to monitoring. In: Ewert JW, Swanson DA (eds) Monitoring volcanoes: techniques and strategies used by the staff of the Cascades Volcano Observatory, 1980-90. U S Geol Surv Bull 1966:181-188.

Symonds R (1993) Scanning electron microscope observations of sublimates from Merapi volcano, Indonesia. Geochem J 26:337-350.

Symonds RB, Reed MH (1993) Calculation of multicomponent chemical equilibria in gas-solid-liquid systems: Calculation methods, thermochemical data and applications to studies of high-temperature volcanic gases with examples from Mount St. Helens. Am J Sci 293:758-864.

Symonds RB, Rose WI, Reed MH, Lichte FE, Finnegan DL (1987) Volatilization, transport and sublimation of metallic and non-metallic elements in high temperature gases at Merapi Volcano, Indonesia. Geochim Cosmochim Acta 51:2083-2101.

Symonds RB, Rose WI, Gerlach TM, Briggs PH, Harmon RS (1990) Evaluation of gases, condensates, and SO_2 emissions from Augustine Volcano, Alaska: the degassing of a Cl-rich volcanic system. Bull Volcanol 52:355-374.

Symonds RB, Reed MH, Rose WI (1992) Origin, speciation, and fluxes of trace-element gases at Augustine volcano, Alaska: insights into magma degassing and fumarolic processes. Geochim Cosmochim Acta 56:633-657.

Takano B (1987) Correlation of volcanic activity with sulfur oxyanion speciation in a crater lake. Science 235:1633-1635.

Takano B, Watanuki K (1990) Monitoring of volcanic eruptions at Yugama crater lake by aqueous sulfur oxyanions. J Volcanol Geotherm Res 40:71-87.

Taran YA, Rozhkov AM, Serafimova EK, Esikov AD (1991) Chemical and isotopic compositions of magmatic gases from the 1988 eruption of Klyuchevskoy volcano, Kamchatka. J Volcanol Geotherm Res 46:255-263.

Taylor BE (1986) Magmatic volatiles: isotopic variations of C, H, and S. In: Valley JW, Taylor HP, O'Neil JR (eds) Stable isotopes in high temperature geological processes. Rev Mineral 16:185-225.

Tazieff H, Le Guern F, Carbonnelle J, Zettwoog P (1972) Etude chimique des fluctuations des gaz éruptifs de volcan Erta'Ale (Afar, Ethiopia). C R Acad Sci Paris, Ser D, 274:1003-1006.

Tedesco D, Sabroux JC (1987) The determination of deep temperatures by means of the CO-CO_2-H_2-H_2O geothermometer: an example using fumaroles in the Campi Flegrei, Italy. Bull Volcanol 49:381-387.

Tedesco D, Toutain JP, Allard P, Losno R (1991) Chemical variations in fumarolic gases at Vulcano Island (Southern Italy): seasonal and volcanic effects. J Volcanol Geotherm Res 45:325-334.

Wallace PJ, Gerlach TM (1994) Magmatic vapor source for SO_2 released during volcanic eruptions: evidence from Mount Pinatubo. Science 265:497-499.

Wen S, Rose WI (1994) Retrieval of sizes and total masses of particles in volcanic clouds using AVHRR bands 4 and 5. J Geophys Res 99:5421-5431.

Westrich HR, Gerlach TM (1992) Magmatic gas source for the stratospheric SO_2 cloud from the June 15, 1991, eruption of Mount Pinatubo. Geology 20:867-870.

Williams SN (1993) Galeras volcano, Colombia: perspectives from a researcher and survivor. Geotimes 38-6:12-14.

Williams SN, Stoiber RE, Garcia NP, Londoño AC, Gemmell JB, Lowe DR, Connor CB (1986) Eruption of the Nevado del Ruiz volcano, Colombia, on 13 November 1985: gas flux and fluid geochemistry. Science 233:964-967.

Williams SN, Sturchio NC, Calvache MLV, Mendez RF, Londoño AC, García NP (1990) Sulfur dioxide from Nevado del Ruiz volcano, Colombia: total flux and isotopic constraints on its origin. J Volcanol Geotherm Res 42:53-68.

Woods A (1993) Moist convection and the injection of volcanic ash into the atmosphere. J Geophys Res 98:17627-17636.

Chapter 2

ANALYTICAL METHODS FOR VOLATILES IN GLASSES

Phillip D. Ihinger

Department of Geology and Geophysics
Yale University
New Haven, Connecticut 06511 USA

Richard L. Hervig

Center for Solid State Science
Arizona State University
Tempe, Arizona 85287 USA

Paul F. McMillan

Department of Chemistry and Biochemistry
Arizona State University
Tempe, Arizona 85287 USA

INTRODUCTION

Our understanding of the behavior of volatiles in silicate melts has been closely linked to the development of the experimental and analytical methods that have been applied to investigate them. The techniques vary widely in character, ranging from bulk extraction methods, to site- and species-specific spectroscopic studies. Some familiarity with these techniques and their range of application is essential to the investigator who wishes to pursue or evaluate studies of volatile behavior in magmatic systems. The goal of this chapter is to review the analytical techniques that are most routinely applied to volatile-bearing systems of geological interest. These techniques include bulk vacuum extraction, Secondary Ion Mass Spectroscopy (SIMS), and vibrational spectroscopy (principally Fourier transform infrared (FTIR) absorption for dissolved volatile analysis, and Raman spectroscopy for fluid inclusions). We present here a concise description of each technique, along with a discussion of their advantages and disadvantages, especially when applied to studies of volatiles dissolved in silicate melts.

Overview and historical development of analytical techniques

A variety of analytical methods has been applied to the study of the volatile contents of geological materials. The principal methods can be broadly divided into four classes: (1) bulk extraction techniques, (2) energetic particle bombardment techniques, (3) vibrational spectroscopic techniques, and (4) phase equilibrium studies. Methods (1), (2), and (4) have proven highly successful for determining the total volatile contents of silicate melts and glasses. However, these methods reveal little about the abundances of individual volatile-bearing *species*, and as such contribute little to understanding the structural role of volatiles in molten silicates. Analytical methods that fall within classes (1) and (2), however, are readily applicable to determining volatile concentrations in natural samples and have been especially successful in illuminating the degassing behavior of volatiles in igneous environments (see Chapters 7, 8).

The spectroscopic techniques (class (3)) are ideally suited for determining the concentrations of particular volatile species, but in the absence of good theoretical models for absolute peak intensities, these methods must rely on prior calibration of standard samples via bulk techniques. Analytical methods within classes (1) and (4) are sufficiently accurate to generate standards for methods within classes (2) and (3). Methods that comprise class (4) include a variety of experiments involving melt-vapor systems, such as determining freezing point depression by observing the onset of melting, or determining solubility through the appearance of dimples in capsules (e.g., Burnham and Jahns, 1962). These indirect methods for volatile determination involve series of time-intensive experiments, and so are not ideally suited to routine determination of volatile contents of natural or synthetic samples; therefore, they will not be discussed in this chapter. We present here a brief overview as well as a summary of the historical development of methods (1)-(3) as applied to the study of volatiles in silicate melts.

Bulk extraction techniques. Bulk extraction techniques involve subjecting the sample of interest to high temperatures in order to physically remove all volatile elements, and subsequently analyzing the exsolved gas and/or the refractory residue. The Penfield method (Penfield, 1894) provided quick and easy measurements of the volatile content of geological samples. The method involves. measuring the loss on ignition of hydrous samples heated within glass tubes. Tilden (1896) performed the first systematic study of the release of volatiles from natural rocks. Following the earlier definitive conclusions about the structural state of volatiles in minerals (Davy, 1824; Brewster, 1826; Sorby and Butler, 1869), Tilden suggested that volatiles evolved on heating magmatic rocks were "wholly inclosed in cavities." Chamberlin (1908) analyzed 112 natural rocks (including fresh volcanic rocks) and concluded, contrary to Tilden, that volatiles released upon heating rock powders *in vacuo* represented magmatic components not associated with gas cavities. Instead, he proposed that the gas evolved at high temperatures was either chemically bonded to the glasses and solids or physically "occluded" within the rock in the molecular state.

Goranson (1931, 1936) was the first to quantitatively apply bulk extraction techniques to the study of volatiles in glasses quenched from melts of geological composition. He experimentally determined the solubility of water in both natural (rhyolite) and synthetic (albite, orthoclase, orthoclase-silica) melts using the weight-loss on ignition technique, after equilibrating vapor saturated melts at pressures up to 4 kbars and temperatures up to 1200°C. After removing excess vapor by piercing and heating charges to 110°C, he combusted the entire hydrous assemblage at high temperature and atmospheric pressure. The dissolved water content was assumed to be equal to the difference in weight before and after the combustion process, with a minor correction made for some diffusive loss of indigenous water during the 110°C heating step. Shepherd (1938) applied Goranson's quantitative methods to accurately determine the volatile content of natural obsidians collected from around the world.

The "weight-loss" technique has been subsequently modified in many ways. Kennedy (1950), on measuring the solubility of water in pure silica, measured only the amount of water released during the 110°C heating step. By knowing the amount of water loaded into the charge before the experiment, the concentration of silica in the vapor phase before quenching, the amount of water the colloidal silica scavenged from the vapor upon precipitation, and the total anhydrous weight of the silica starting material, he calculated the total water content of the glass by evaporative weight difference. Hamilton et al. (1964) applied the same technique to melts of basaltic and andesitic composition. These workers ignored the solute correction, but applied a new correction to account for the loss of H_2 through the platinum capsule walls generated by the oxidation of ferrous to ferric iron in the melt during the experiment. A simplified version of this weight-loss technique (ignoring both corrections) was adopted by Oxtoby and Hamilton (1978a,b,c,d; also Hamilton and Oxtoby, 1986) for solubility determinations on natural and synthetic aluminosilicate glasses. Yoder, Stewart, and Smith (1956) applied another variation of the "weight-loss" technique, in which water contents of feldspathic melts were determined by completely crystallizing hydrated glasses and measuring the difference in the weight between their anhydrous crystalline products and their hydrous glassy experimental run products.

Khitarov et al. (1959) applied a further modification to the weight-loss technique; rather than assuming that all weight lost on ignition was due to water released from the glass, these workers measured the quantity of water actually released on ignition by collecting the released water in absorbent anhydrone, for which the water content was measured by weight difference. Their results on rhyolitic compositions at 900°C differed from Goranson's (1931) by up to 2.0 weight percent absolute, for samples prepared under identical conditions.

The standard method by which released water is collected and chemically analyzed was first developed by Fischer (1935), and is commonly termed Karl Fischer titration. In this method, evolved water reacts quantitatively with coulometrically generated iodine. The relative standard deviation for this technique is less than 5% for low water content samples (~0.1 wt % total water; Westrich, 1987), and less than 2% for higher water content glasses (~1.5 wt % total water; Turek et al., 1976). The relatively rapid (samples are completely analyzed within thirty minutes), simple, and inexpensive nature of this technique has led to its routine use for determining the water content of silicate glasses (e.g., Holz et al., 1994; Behrens, 1994).

Several analytical techniques have been used in conjunction with bulk extraction techniques to characterize vapors that evolve on fusion of silicate glasses in vacuum. For example, volume manometry (e.g., Craig, 1953) and pressure manometry (Harris, 1981; Brett et al., 1987) have been routinely applied to determine the absolute abundance of volatiles within glassy samples. Gas chromatographs (e.g., Moore et al., 1970; Brey and Green, 1976; Pichavant, 1981), quadrupole mass spectrometers (e.g., Killingley and Muenow, 1975; Delaney et al., 1978; Byers et al., 1986), and Knudsen Cell mass spectrometers have been attached to vacuum extraction lines to distinguish the relative concentrations of gaseous species exsolved from silicate samples (see below, Analytical Methods Involving Bulk Extraction, for more details of these techniques).

Light element analyzers that employ these techniques are now commercially available for determining the volatile contents of geological materials. For example, the Carlo Erba Strumentazione (trademark of Montedison, Milan, Italy) is a C, H, S, and N analyzer that uses chromatographic columns to separate out volatile species. In this device, the sample is heated to ~1000°C in a tin crucible and is placed in contact with a Cr_2O_3 catalyst. The exothermic reaction of the catalyst with the tin results in a combustion reaction at about 1700°C, ensuring complete evolution and oxidation of volatiles within the sample, and these oxidized species are carried to the chromatographic columns using helium as a carrier gas. Accuracies for N_2 gas range from ±10% for samples of 0.01% total N to ±0.2% for samples with 10 % total N content. These analyses take about 5 minutes per sample and require sample sizes ranging from 0.5 to 100 mg, depending on the concentrations of the volatiles of interest.

The LECO Corporation (St. Joseph, MI) also manufactures light element analyzers capable of determining the volatile contents of geological materials. Separate carbon and sulfur (model # CS-444), nitrogen and oxygen (model # TC-436), and hydrogen (model # RH-404) analyzers are available at prices ranging from \$50,000 to \$65,000. These instruments have quoted accuracies of 0.1 ppm C and S (or 0.5% of total C and S), 0.5 ppm O, 1.0 ppm N (or 1% of total O_2 and N_2), and 0.05 ppm H (or 1% of total H). Both resistance and induction furnaces are employed in these analytical devices. Their detection methods include IR absorption for C, S, and O_2 and thermal conductivity for N_2 and H_2. The precision of an older LECO analyzer (model # SC-132) was reported by Krom and Berner (1983) to be 0.7 % absolute on replicate extractions of calcite crystals, and the analyses were accurate to within 4% of the total carbon in the sample. They noted that sample sizes as low as 34 mg were sufficient to yield reasonable results in sediment samples containing 2.55 wt % total carbon.

State-of-the-art vacuum extraction techniques have been developed largely because of the intense interest in the isotopic composition of volatiles in geological materials. Determining the isotopic composition of dissolved volatiles requires the preparation of pure gas samples for which the vacuum extraction line is ideally suited.

The isotopic composition of water in erupted magmas was first investigated by Friedman and Smith (1958). They examined water released from both the water-rich perlitic over-growths and the water-poor rhyolitic cores of obsidians, to demonstrate the isotopic distinction between inherent magmatic water and water subsequently added during hydration at low temperatures. High vacuum extraction lines have subsequently been used in conjunction with mass spectrometers to determine the volatile content and isotopic composition of unaltered plutonic rocks (e.g., Godfrey, 1962; O'Neil et al., 1977; O'Neil and Chappel, 1977; Masi et al., 1981; Craig and Lupton, 1976; Nabelek et al., 1983; Brigham and O'Neil, 1985), fresh continental volcanic rocks (e.g., Taylor et al., 1983; Newman et al., 1988; Anderson and Fink, 1989), and fresh oceanic rocks (e.g., Moore, 1970; Sakai et al., 1971; Pineau and Javoy, 1983; Des Marais and Moore, 1984; Kyser and O'Neil, 1984; Mattey, et al., 1984; Poreda, 1985; Poreda et al., 1986; Blank, et al., 1986; Dobson and O'Neil, 1987; Garcia et al., 1989). Vacuum extraction techniques have proven useful for many applications, and will be described in greater detail below.

Energetic particle techniques. A range of techniques involving energetic particles has been successfully applied to the study of volatile contents in silicate glasses. These techniques include electron microprobe, ion microprobe, and nuclear reaction techniques. They are advantageous in that they require only a small amount of sample, and are extremely precise. Their chief disadvantage is that they require accurate calibration standards, and that they are generally incapable of distinguishing individual volatile-bearing species concentrations.

The ion microprobe (or secondary ion mass spectrometer, SIMS) is most commonly applied to analytical problems in the semiconductor industry. However, the high sensitivity afforded by mass spectrometric analysis and the good lateral resolution (usually ranging from a few to ~30 micron diameter incident ion beam) make it a useful tool for geochemists. The interested reader is referred to Williams (1985; 1992; references therein) for a review of the physical principles of SIMS. With respect to volatile-containing quenched melts, SIMS has been successfully applied to the analysis of H (Delaney and Karsten, 1981; Steele, 1986; Hervig et al., 1989), C (Pan et al., 1991; Thibault and Holloway, 1994), F (Hervig et al., 1989; Dunbar and Hervig, 1992a,b), and Cl (Hervig et al., 1989).

The development of layered synthetic microstructure (LSM) crystals, used as dispersing elements in X-ray spectrometers, has allowed the quantitative application of the electron microprobe to the study of light volatile elements within silicate glasses (e.g., Wood et al., 1985; Nicolosi et al., 1986; Huang et al., 1989). These crystals can be manufactured with specific d values optimized for the light elements of interest. For example, WSi crystals ($2d \sim 60$ angstroms) are ideally suited for the study of F, O, N, and C (Potts and Tindle, 1989; Nash, 1992; McGuire et al., 1992), and MoB_4C ($2d \sim 150$ angstroms) crystals are used for the study of B and Be (McGee, et al. 1991). These crystals have replaced lead stearate (organic multilayered) crystals that have traditionally been applied to the study of light elements in silicate glasses. Some of the analytical details of electron microprobe applications to light elements can be found in the following references: C – Mysen et al. (1976), Mathez and Delaney (1981); S – Carroll and Rutherford (1985); Cl – Harris and Anderson (1984); Ar – White et al. (1989), Carroll and Stolper (1991).

High energy beam nuclear reaction techniques have also been applied with some success to the study of volatiles in silicate melts (e.g., see Courel et al., 1991). These include the measurement of hydrogen (Lee et al., 1974; Mosbah et al., 1991), fluorine (Mosbah, 1991), nitrogen (Mosbah et al., 1993), chlorine (Metrich et al., 1993) and bulk carbon (Mathez et al., 1984; Fine et al., 1985).

Vibrational spectroscopic techniques. Researchers in the glass sciences have long applied spectroscopic methods to the quantitative determination of volatiles in silicate glasses (e.g., Harrison, 1947; Schölze, 1960, 1966: Dodd and Fraser, 1966; Ernsberger, 1977; Pearson et al., 1979; Bartholomew et al., 1980; Wu, 1980; Takata et al., 1981; Shelby, 1987). Keller and Pickett (1954) were the first to apply vibrational spectroscopic techniques to materials of direct geological interest. They observed significant absorptions at 3500 cm^{-1} and 1630 cm^{-1} in the infrared spectrum of natural volcanic glasses and assigned these to the presence of hydroxyl and molecular water species. Evidence for the presence of molecular water species in a silicate melt was provided by Orlova (1964), who analyzed infrared spectra of albitic melts quenched to hydrous glasses. Ostrovskiy et al. (1964) used this technique to investigate the influence of cation size on the solubility of water in alkali melts. They distinguished the two hydrous species (bound hydroxyl and molecular H_2O), and observed Henrian behavior for the molecular species based on their experimental results at 2 and 4 kbars.

The near-IR absorption technique has been refined and developed by Stolper and coworkers in a series of studies of water speciation and solubility within melts and glasses of geological interest, and has now become a routine method for rapid and reliable analysis of hydroxyl species, molecular H_2O, and total dissolved water (Stolper, 1982a; Newman et al., 1988; Silver and Stolper, 1989; Stolper, 1989; Silver et al., 1990; Ihinger, 1991; Pan et al., 1991; Pawley et al., 1992; Blank et al., 1993; Dixon et al., 1994). The technique has also proven to be especially effective for the determination of the water contents within melt inclusions as small as 25 μm in diameter (e.g., Anderson et al., 1989; Skirius et al., 1990; Bacon et al., 1992; Lowenstern, 1993; Stolper and Newman, 1994), and is easily applied to diffusion studies (Stanton et al., 1985; Stanton, 1989; Zhang et al., 1991; Zhang and Stolper, 1991). Such spectroscopic studies have contributed greatly to our understanding of the solubility mechanisms of water in silicate melts (see Chapter 4).

Infrared spectroscopic studies have been applied with equal success to the investigation of dissolved CO_2 in natural and synthetic silicate glasses. Early studies in the glass sciences used infrared spectroscopy to investigate the incorporation of CO_2 in silicate glasses (Schölze, 1966). Powder transmission methods, in which the ground sample is mixed with KBr (or some other mounting medium), then pressed into a disc for infrared transmission spectroscopy, has been used qualitatively to explore the solubility mechanism of CO_2 in mafic and synthetic aluminosilicate melts (Mysen, 1976; Mysen et al., 1976; Brey, 1976; Taylor, 1990). These workers found evidence for both molecular CO_2 and carbonate CO_3^{2-} species. Fine and Stolper (1985; 1986) have calibrated characteristic absorption bands assigned to molecular CO_2 (at 2350 cm^{-1}) and carbonate species (two bands between 1400 and 1600 cm^{-1}) for aluminosilicate glass compositions, and have quantitatively investigated the solubility and speciation of these carbon-bearing species in quenched silicate melts via infrared absorption on doubly polished plates of known thickness. This technique has subsequently been applied to the determination of carbon contents in natural samples from a variety of localities (e.g., Dixon et al., 1988; Newman et al., 1988; Stolper and Newman, 1994), and to a range of experimental run products (Fogel and Rutherford, 1990; Pan et al., 1991; Pawley et al., 1992; Blank et al., 1993). Later in this chapter, we describe in detail the theory and application of vibrational spectroscopy to the study of volatiles in silicate glasses.

Raman scattering spectroscopy has been used extensively in the structural study of volatile-containing melts and glasses (e.g., Sharma, 1979; Sharma et al., 1979; Mysen et al., 1980; Mysen and Virgo, 1980a,b; 1986a,b; Rai et al., 1983; McMillan et al., 1983; 1993: see chapter 5), but has not yet been reliably calibrated for the quantitative analyses of these species. The Raman technique could be applied to the analysis of dissolved volatile species, based on its demonstrated success in the analysis of molecular species in fluid inclusions (Dhamelincourt et al., 1979; Roedder, 1984; Pasteris et al., 1988; Dubessy et al., 1989). It would be quite easy to extend these calibrations to species dissolved in volatile-containing glasses, to provide an alternate analytical technique for situations where infrared absorption becomes difficult. One obvious example is in the analysis of compositions with very high dissolved CO_2 contents, which require extremely thin doubly polished samples to be prepared for IR analysis, because of the intensity of the infrared absorption (R. Brooker, unpublished data). Raman spectroscopy would permit easy analysis of such samples, from the strong characteristic peak for symmetric stretching of the CO_3^{2-} group (Sharma, 1979; Sharma et al., 1979; Mysen et al., 1980; Mysen and Virgo, 1980a,b; 1986a,b: see below). In addition, Raman spectroscopy gives site-specific information on dissolved volatiles which is not easily available from IR studies: for example, the Raman band observed near 900 cm^{-1} in hydrous aluminosilicate glasses may well be a vibration of OH species associated with Al sites (see Sykes and Kubicki, 1993; McMillan et al., 1993: also Chapter 4 for discussion of this assignment). Calibration of the intensity of this band would permit quantification of the concentration of these species as a function of total water content.

Nuclear magnetic resonance (NMR) techniques have also been applied to the quantitative study of H- and C-bearing volatiles in silicate glasses (e.g., Müller-Warmuth et al., 1965; Schölze, 1966; Belton, 1979; Bartholomew and Schreurs, 1980; Bray and Holupka, 1984; Eckert et al., 1987; 1988; Farnan et al., 1987; Kohn et al., 1989b, 1991). The technique and its application to geological samples has been reviewed by Kirkpatrick (1988) and Stebbins (1988). In the NMR experiment, the sample is placed in an external magnetic field, which causes nuclear spins to become preferentially aligned. The isotopes ^1H, ^2H and ^{13}C (also ^{19}F, ^{10}B, ^{11}B, ^{17}O), either at natural abundance, or artificially enriched, are convenient NMR nuclei. If the sample is irradiated with a radio frequency pulse, the sample absorbs energy as the spins are tipped. The precise frequency of the radio frequency energy absorbed is dependent upon the nucleus of interest and the magnitude of the applied magnetic field. For example, at a field of 8.45 T, ^{13}C resonates near 90.5 MHz, and ^1H at 360 MHz, so that different nuclei are well separated in the experiment. The precise resonance frequency for each nucleus studied is highly dependent on the local environment for that atom or ion, so NMR spectroscopy is a sensitive structural probe for dissolved volatile species and framework ions (e.g., Farnan et al., 1987; Eckert et al., 1988; Kirkpatrick, 1988; Stebbins, 1988; Kohn et al., 1989a,b; 1991; 1992: see Chapter 5). NMR spectroscopy for samples in the condensed phase is complicated because the various orientation-specific interactions are not averaged, with the result that NMR lines are usually broad. This is usually overcome by the technique of "magic angle sample spinning" (MAS), in which the sample is rapidly spun about an axis at a "magic angle" (usually 54.71°) to the applied magnetic field (Kirkpatrick, 1988).

The amount of radio frequency energy absorbed is directly proportional to the number of nuclei excited in the experiment, so that integration of the observed signal directly gives the concentration of the species contributing to that resonance, with no further calibration. For this reason, NMR spectroscopy should provide a powerful, site-specific quantitative tool for the analysis of dissolved volatiles in silicate glasses. However, various factors related to the nature of the NMR experiment, including both instrumental parameters and the specific properties of the nucleus under investigation, do not yet permit such a simple, direct, precise quantification of the NMR signal for most samples of geological interest (although these questions are being actively addressed: see e.g., Eckert et al., 1988; Kohn et al., 1991). NMR techniques are particularly advantageous in that they require minimal sample preparation and are in principle non-destructive, although the sample must generally be ground for MAS studies. The application of this family of techniques to quantitative, site-specific studies of volatiles in silicate melts holds much promise for the future.

ANALYTICAL METHODS INVOLVING BULK EXTRACTION

General description of the technique

The most direct method for determining the concentration of volatile components in silicate glasses involves fusion of the sample at high temperature, under vacuum, in a glass or metal-walled extraction line. The gas evolved on fusion is cryogenically purified so that the volatile species of interest can be isolated and measured. Several comprehensive and authoritative works describing the physical basis of technique and measurement in vacuum have been published (e.g., Martin and Hill, 1948; Dushman, 1962; Turnbull et al., 1962; Diels and Jaeckel, 1966; Yarwood, 1967; Redhead et al., 1968; Berman, 1985). The vacuum line consists of a devolatilizing chamber, separate oxidizing and reducing furnaces, a cryogenic trap, and any analytical device (including volume manometer, pressure manometer, gas chromatograph, isotope ratio mass spectrometer, quadrupole mass spectrometer, and/or any of a variety of chemical reactors) used to measure the amount of different chemical species in the gaseous sample (Fig. 1). The precision of this technique is variable, and depends mainly on the total amount of extracted gas and the care taken in reducing the background associated with each measurement.

The devolatilizing chamber consists of a metal crucible sealed within a quartz glass reaction vessel, placed within a high-temperature fusion furnace. Because H_2 gas can diffuse out of the line through heated walls, radio-frequency induction furnaces are often used to generate the temperatures necessary for fusion. The crucible (usually made of platinum or molybdenum) is placed in the axis of a water-cooled copper coil (the inductor), and is heated as an alternating electric current within the copper coil induces an oscillating electric field within the metal crucible (which acts as the susceptor). The high temperatures attained within such induction furnaces can be measured quite accurately above ~700°C

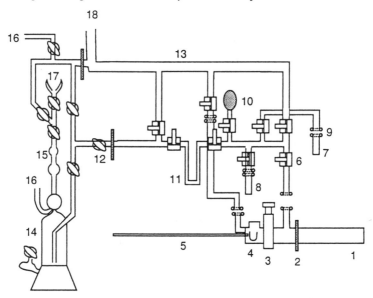

Figure 1. Schematic diagram of the vacuum extraction line; 1) evacuated quartz reaction vessel, 2) glass to metal seal, 3) gate valve, 4) evacuated loading chamber, 5) rotative magnetic insertion rod, 6) air-actuated vacuum valve, 7) oxidation furnace, 8) reduction furnace, 9) VCR coupling, 10) vacuum gauge, 11) cryogenic trap, 12) glass stopcock, 13) vacuum arc remelted stainless steel manifold, 14) Toepler pump, 15) calibrated volume, 16) to mercury manometer, 17) exit port to analytical apparatus, 18) to turbo molecular vacuum pump.

with the use of a suitably calibrated optical pyrometer. However the lack of sufficient thermal radiation at temperatures lower than this, and the uncertainties due to the interaction of metallic thermocouple wires with the oscillating electromagnetic field, make accurate low temperature calibration of induction furnaces difficult. For this reason, conventional metal-wound resistance furnaces have been used for high precision, low temperature heating steps of volatile-bearing material.

Other sample heating techniques that avoid heating of the chamber walls involve CO_2 laser and infra-red imaging techniques. The laser technique has proven particularly successful in the extraction of rare gases from geologic materials, and can be used for sample areas down to several tens of μm in dimension. Large thermal gradients developed during irradiation of large samples require caution in applying this teachnique (e.g., see Sharp, 1990). Infra-red image furnaces utilize gold-lined, parabolic-shaped mirrors to reflect radiation emitted from heated tungsten elements and generate an axial-symmetric hot-zone. The quartz glass walls of the extraction line are relatively transparent to the IR radiation, so that H_2 diffusion through the wall is minimized. These furnaces have extremely fast response times, and can be reproducibly controlled throughout complicated heating schedules (P. Ihinger and M. Davis, unpublished results). Another advantage is that they are cheaper and more compact than induction furnaces, so that their use in the Earth sciences will continue to grow.

Before samples are loaded into the extraction line, empty crucibles are preheated for several hours at higher temperatures than those to be encountered during the extraction procedure. This ensures the removal of any volatiles within the crucible, and also serves to remove any adsorbed volatiles from the surface of the quartz reaction vessel. If high precision is not required, time can be saved by loading the weighed sample directly into the pre-heated reaction vessel in air. The blank contribution of average-sized vessels exposed to air is typically less than 3 μmoles water and ~0.25 μmoles CO_2 and is reproducible to within 3% of these values (Ihinger, 1991). When small amounts of water (<10 μmoles) are to be exsolved, the hydrous sample is carefully weighed and loaded into the reaction vessel in a N_2 or Ar atmosphere dry box. Reaction vessels vary in size and surface area, and the blank for each vessel must be determined prior to use. Epstein and Taylor (1970) report blanks as small as .02 μmoles for H_2O and CO_2, and these values are readily reproduced if adequate precautions are taken while loading the sample in the dry box. The blank contribution of the chamber walls can also be eliminated by inserting a gate valve between the devolatilizing chamber and an evacuated loading chamber (Fig. 1). Glassy samples with low total water contents can be preheated at temperatures between 110° and 150°C to dislodge adsorbed water molecules on the surface of the sample grains; however, water-rich samples have been known to lose non-negligible amounts of indigenous molecular water even at these low temperatures (e.g., Goranson, 1936; Zhang et al., 1991).

The heating schedule used in bringing samples up from room temperature to high temperature depends upon the volatile to be extracted and the composition of the host glass. In general, samples are brought up to temperature at rates slow enough to prevent the rapid release of evolved gas that results in the ejection of glassy or partly molten material from the crucible (a volcanic eruption, on a microscopic scale). Samples are then held at temperatures slightly below fusion temperature so that diffusive length scales remain small for as long as reasonably possible. As described below, high silica samples with high water contents necessitate unusually long extraction times at lower temperatures wo ensure complete volatile yields. The samples are then brought to maximum extraction temperature (up to 1400°C) until gas is no longer evolved.

Volatiles may exsolve from silicate materials in either oxidized or reduced form, and separate oxidizing and reducing agents are necessarily an integral part of the extraction line. For example, the presence of significant ferrous iron in the devolatilizing sample will result in a non-trivial fraction of H_2 exsolving along with H_2O. Hydrogen gas is converted to a condensable form via reaction with an oxidizing agent (typically copper oxide held at 500-600°C; the lower temperatures reduce the time required for adsorption of the reduced gas molecules onto the CuO surface but increase the time required for oxidation: see Dushman, 1962). Thus, noncondensable gas is first passed through an oxidizing furnace so that reduced H, C, and S bearing species can be collected and analyzed. Condensable gas is frozen into a liquid nitrogen trap (-196°C) so that the remaining non-condensable fraction can be isolated and characterized by techniques such as manometry, gas chromatography, and quadrupole mass spectrometry.

The condensed gas sample is then subjected to a series of variable temperature steps in order to separate individual gas species and quantitatively measure their abundances. The condensed gas is separated by sequentially pumping away evolved fractions of gas at higher trap temperatures, using either continuously variable temperature cryogenic traps (Des Marais, 1978), or by subjecting the condensed gas to fixed temperatures with a variety of standard freezing point slurries. For example, CO_2 is condensed by bathing the trap in a mixture of dry-ice and alcohol at ~-85°C. Evolved gas is collected by passing through a Toepler pump into a calibrated volume for manometric analysis. A small analytical uncertainty in the manometric measurement is associated with reading the height of the column of mercury (< 1% of the total volume of the sample). Gas samples can then be sent directly to auxiliary analytical devices, such as a gas chromatograph, quadrupole mass spectrometer, or isotope ratio mass spectrometer, for further characterization.

Because H_2O molecules readily adsorb onto glass and metal walls at room temperature, evolved water samples are converted to H_2 gas before further analysis. Pure water is converted to H_2 via two commonly applied techniques. The procedure of Bigeleisen et al. (1952) involves passing water vapor over hot (750°C) uranium. Application of this technique in isotopic studies is limited by a "memory effect", most probably associated with the formation of U-hydrides within the furnace. The magnitude of the "memory effect" must be calibrated by analyzing a series of samples with known isotopic composition and sample size (e.g., Ihinger, 1991). This effect can further be minimized by prior flushing of the line with a sample of known isotopic composition that is similar to the "expected" isotopic composition of an unknown sample. A further disadvantage in using uranium as a reducing agent is that depleted uranium (still radioactive) is becoming harder to find commercially, and deal with bureaucratically.

An alternative reduction technique was introduced by Friedman (1953), and later developed and refined by Schiegl and Vogel (1970), Coleman et al. (1982), Kendall and Coplen (1985), Dubois (1985), Tanweer et al. (1988), Coplen et al. (1991), and Vennemann and O'Neil (1993). This technique involves reduction of H_2O with a Zn alloy at a temperature slightly below its fusion temperature, and has proven useful for water reduction in routine analyses. The disadvantages of this technique are centered around the "art" of manufacturing a zinc alloy that is suitable for H_2 production, and the problems associated with production of very small amounts of H_2. Other reduction catalysts, such as Mg metal, have been applied with some success toward the production of hydrogen vapor; however, there is clearly a need for further research and development of alternate efficient reducing catalysts.

Application to volatile contents in glasses

Hydrogen. Glassy samples are prepared for analysis by gently crushing them to small grain sizes in order to minimize the diffusion length of bound volatiles during high temperature extraction. Chips with high silica contents that are larger than 1mm in diameter require unusually long extraction times for complete yields. On the other hand, Newman et al. (1986) used hydrogen isotope studies to show that grain sizes smaller than 150 μm in diameter are associated with a measurable contribution of adsorbed water onto the surfaces of obsidian chips. Thus, for water analyses, glassy samples should be sieved, and only grain size fractions between 150 μm and 500 μm should be used in volatile extractions.

In Figure 2, the release rate of water from a rhyolitic glass with 1.2 wt % water as a function of the temperature of extraction is shown (Sally Newman, unpublished data). The details of the release rate of water from each glass vary from sample to sample, depending upon the water content and the size of the chips. However, the bimodal release trend shown in Figure 2 has been observed for all water-bearing glasses with more than a few tenths of a wt % total water. Figure 2 shows that a significant amount of water is released at ~500°C. Zhang et al. (1991) report that, at these temperatures, the diffusion rate of the molecular species is at least fifty times greater than that for the hydroxyl species. Thus, the peak at lower temperature represents the release of the molecular water fraction in the glass. A second peak of evolved water occurs at ~900°C, corresponding to the release of the hydroxyl component of dissolved water in the glass.

It is necessary to completely extract the molecular water fraction of dissolved water at temperatures less than ~450°C to prevent the nucleation of tiny bubbles within the glass (Ihinger, 1991). The bubbles serve as storage sites for diffusing hydroxyl species at higher temperatures. The high viscosity of silica-rich samples allows the integrity of the bubbles to be maintained, thus preventing the escape of exsolved water to the vacuum line.

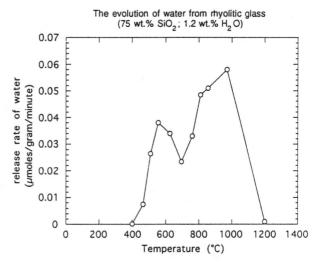

Figure 2. The release rate of water from a hydrous rhyolitic glass (1.2 wt % total dissolved water. The peak at ~500 °C represents the evolution of water molecules from the glass. Peak at ~900°C represents the release of hydroxyl species (see text). Unpublished data courtesy of Dr. S. Newman, California Institute of Technology

Table 1.

Bakeout procedure for water-rich silicate glasses that is used to prevent formation of bubbles at high temperatures during vacuum extraction. This procedure should be followed only for the accurate determination of total water content (less than 1% absolute) or in determining the isotopic composition of dissolved volatiles.

Approximate temperature		Duration
110°C	for	10 minutes
150°C	for	20 minutes
200°C	for	10 hours
350°C	for	12 hours
500°C	for	1 hour
650°C	for	30 minutes
800°C	for	5 hours
850°C	for	3 hours
900°C	for	1-5 hours
1000°C	for	30 minutes
1100°C	for	30 minutes
1200°C	for	2.5 hours
Total		36 to 40 hours

Care must be taken to slowly extract the dissolved water from these Si-rich glasses; extraction times greater than 40 hours were needed to completely extract the water from water-rich rhyolitic samples. Table 1 shows a typical heating schedule needed to successfully prevent the formation of bubbles during the extraction procedure.

Other volatiles. Carbon and sulfur can exist in multiple oxidation states in silicate materials, and care must be taken in collecting all exsolving species. In general, the same procedure as that described for H-bearing glasses is followed. Ueda and Sakai (1984) used a modified "Kiba reagent" (dehydrated Sn^{2+}-bearing phosphoric acid) to decompose basaltic glasses at temperatures greater than 150°C. In this process, dissolved sulfate, sulfide, and carbonate are released from the glass as SO_2, H_2S, and CO_2, respectively. These species are readily separable using the cryogenic techniques described above, and their relative abundances are accurately measured by manometric methods. The speciation of sulfate and sulfide into different gas phase species means that the oxidation state of sulfur within the glass can be determined. The above discussion regarding slow heating rates also applies to extracting other volatiles from water-rich, high silica samples.

Knudsen Cell Mass Spectrometry (KCMS) allows the quantitative analysis of volatile elements with masses of 1 to 200 atomic mass units (amu: 1 amu = 1.67×10^{-27} kg) (Knudsen, 1909). Equilibrium partial pressures of volatile elements are determined by measuring the flux of various vapor species through a minute aperture into an evacuated volume. This technique is particularly advantageous because it permits simultaneous determination of the absolute amounts (with standard deviation ~ 7%) of the many volatiles released from a single sample. Details of this technique have been described by Grimley (1967), and results on the volatile content of mafic glasses can be found in Killingley and Muenow (1975) and Delaney et al. (1978). Fraser and Rammensee (1987) have used KCMS to determine the volatility of Na and K in molten silicate systems, and Rosenlieb et al. (1992) have used the technique to measure the solubility of noble gases in albite melt.

Gas chromatography has also been used in conjunction with vacuum fusion techniques to determine the concentrations of more complex gaseous molecules such as CO and CH_4 (e.g., Moore et al., 1970; Eggler et al., 1979). In addition, Harris (1981) has developed an elegant method for measuring the abundances of the "noncondensable" gas species, H_2, CO, N_2, Ar, CH_4, and others, by using the characteristic vapor pressures of these gas species within a known volume. The estimated errors for this technique are on the order of 10% of their relative concentrations.

ANALYTICAL METHODS INVOLVING
SECONDARY ION MASS SPECTROMETRY (SIMS)

Introduction

Secondary ion mass spectrometry, (SIMS, or ion microprobe analysis) represents a microanalytical technique that has been successfully applied to characterizing the volatile element content of glasses and crystals (Hinthorne and Anderson, 1975; Jones and Smith, 1984; Steele, 1986, Delaney and Karsten, 1981; Hervig et al., 1989). The instrument is capable of measuring H, C, F, S, and Cl at levels from parts per million to several weight percent.

In SIMS, a beam of primary ions at an impact energy of ~3-20 keV is focused to a spot on a sample - preferably a polished, flat sample. Each incoming ion knocks off, or "sputters", 1 to ~12 atoms from the surface of the sample and some (1 to 10%) of these sputtered atoms are ionized in the process (see Williams, 1985; 1992 for reviews of the physical process). In general, secondary ions are accelerated into a mass spectrometer where they are selected for energy and mass, and detected by an electron multiplier (faraday cups are often used to measure secondary ion signals in excess of ~10^6/s). Variables in this technique include: (1) the polarity and species of the primary ion beam (most commonly O_2^+, O^-, Ar^+, Cs^+), (2) the impact energy, current and diameter of the primary beam, (3) the polarity of the secondary ion beam, and (4) the energies of the secondary ions detected.

SIMS is useful in geochemistry in general because of its high sensitivity to much of the periodic table, and because of its good lateral resolution (typical primary beam diameters of 1 to 30 µm, depending on several parameters) and excellent depth resolution (100 to 500 Å). In addition, isotope ratios of certain elements can be determined. Application of this technique to the measurement of volatile elements in quenched silicate melts is the topic of this section.

General description of SIMS techniques

The discussion of how analyses are obtained refers specifically to the Cameca IMS 3f instrument at Arizona State University (ASU). There should be no significant differences between this and other Cameca 3f-6f SIMS instruments, but it is likely that analyses with the SHRIMP, Isolab (Fisons), Atomika, Phi, Riber, AEI, and new Cameca 1270 instruments may be somewhat different.

Primary beam characteristics. Most geochemists study insulating phases, which can lead to problems with sample charging when the phase of interest is bombarded by a charged particle beam. Remedies to charging have been discussed in detail by Werner and Morgan (1976). The simplest technique is to apply a conducting coat to the sample (e.g., C, Au, Au-Pd), and sputter with a negative oxygen beam. While the developing crater may build up a negative charge, secondary electrons produced during sputtering "hop" to the coated area at the edge of the crater, relieving this excess negative charge. An auxiliary electron flood gun can be used in conjunction with positive primary ion beams if O^- cannot be used, but extreme care is required if this approach is taken for isotope ratio measurements (Hervig et al., 1992).

The size of the primary ion beam is adjusted to fit the problem of interest, with the minimum diameter for O^- from most duoplasmatron sources being on the order of a few microns. If more current is required (to increase the count rate for the element of interest), the beam diameter increases. Most new secondary ion mass spectrometers are equipped with a primary beam mass filter that allows only ^{16}O to strike the sample. Eliminating the primary mass filter will allow all like-charge ions to strike the sample, with the possibility that H^- (or OH^-) will be an abundant component in the beam, driving up the H secondary ion signal from nominally dry samples. However, in the absence of a primary mass filter, careful attention to the purity of the O_2 gas entering the primary ion source can minimize this problem (G. Layne, Institute of Meteoritics, University of New Mexico, pers. comm., 1991). Typical conditions used at ASU on a Cameca IMS 3f are 1nA $^{16}O^-$ with 17 keV impact energy into an approximately 10 to 20 µm diameter spot.

Secondary ion characteristics. The designs of secondary ion optics on different model ion probes vary greatly. This discussion will concentrate on the Cameca design as represented by the IMS 3f–6f models. A positive or negative potential (~4500 V) is placed on the sample to accelerate either positive or negative secondary ions into the mass spectrometer. Positive secondary ions are most commonly analyzed by geochemists because of the large number of interesting elements that are accessible, including alkali metals, alkaline earths, and transition metals. Negative secondary ions are more useful in the study of carbon, halogens and group V-VI elements. Hydrogen shows high ion yields using either polarity. In the Cameca instruments, the energy of the secondary ion beam can be selected. In general, the electrostatic analyzer and energy bandpass (see Fig. 3) is set to

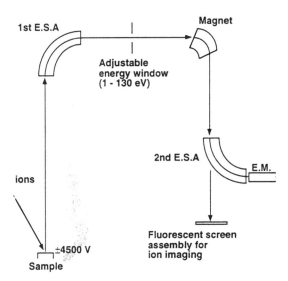

Figure 3. Schematic of Cameca IMS 3f ion microscope. A beam of ions at ~30° to the sample normal is directed at an area of interest. Positive or negative ions (depending on the polarity of the sample voltage) are accelerated along the sample normal into the mass spectrometer. The first electrostatic analyzer (ESA) selects the secondary ion beam for energy, with an adjustable slit set to allow ions with a range of initial kinetic energies (from 1 to 130 eV) into the magnet. After mass selection in the magnet, the ion beam follows a second ESA into the counting system (most commonly, an electron multiplier operating in pulse counting mode). The second ESA can be grounded, and the mass-analyzed ion beam then will pass through a hole in the ESA and strike a microchannel plate/fluorescent screen assembly to allow direct imaging of the distribution of the selected isotope leaving the sample surface. This use of the SIMS as a "chemical microscope" facilitates the location of areas for analysis.

allow secondary ions with energies of 4500 eV into the magnetic sector of the instrument. If the sample potential is dropped from 4500 to 4400 V without changing the conditions in the electrostatic sector, then only those secondary ions which are ejected with 100 eV excess kinetic energy will be detected. The elimination of low-energy secondary ions from the mass spectrum is commonly referred to as "energy filtering". Ions ejected with several tens of eV are less abundant than lower energy ions (see energy spectrum for $^{28}Si^+$ in Fig. 4), but cluster ions, particularly those consisting of 3 or more atoms, can be eliminated from the mass spectrum, depending on the secondary ion energy selected. Furthermore, it is suggested that matrix effects are smaller for these high energy ions than low energy ions (e.g., Ottolini et al., 1993).

In the Cameca SIMS design, raw secondary ion count rates can vary, even on a homogeneous phase, as the sample is moved. This is probably related to the fact that a very strong extraction field is above the sample, and that as the sample edge is approached, the field lines are distorted somewhat. Regardless of the cause, secondary ion *ratios* are little affected, if at all, so most analyses are given in terms of normalized intensities. For example, a calibration for H_2O might plot the H^+/Si^+ ion ratio normalized to the silica abundance: $[H^+/Si^+] *SiO_2$ vs. H_2O contents for several phases. This calculation assumes that the secondary ion count rate for silicon varies linearly with abundance. This assumption is probably acceptable for small variations in silica concentration (e.g., from basalt to andesite compositions), but must be viewed warily when studying phases with silica abundances much different than standards (Shimizu, 1986; Hinton, 1990). Depending on

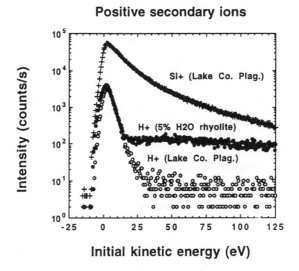

Positive secondary ions

Figure 4. Energy spectra for H^+ and $^{28}Si^+$ sputtered from Lake County, Oregon, plagioclase (An_{60}) and H^+ from ME35D, an experimentally hydrated rhyolitic glass (5.1 wt % H_2O by Karl Fisher titration, courtesy H.R. Westrich, Sandia National Laboratory). Primary beam was $^{16}O^-$ emitted from a duoplasmatron ion source at -12.5 kV accelerating potential. Note high secondary ion intensities at low initial kinetic energies, and the gradual decrease of secondary Si^+ ions with increasing initial kinetic energies. This is in sharp contrast to the abrupt change in H^+ intensities with increasing kinetic energy, indicating two modes of production of H^+ ions: electron stimulated desorption (ESD) caused by the abundant secondary electrons generated during primary ion impact gives the low energy peak while sputtering processes produce the higher energy ions.

the matrix being studied, other elements can be used for normalizing purposes (e.g., $^{18}O^-$ when studying negative secondary ions).

Applications to volatile contents in silicate glasses: Hydrogen

H analysis using positive secondary ions. One important parameter in SIMS measurements for H is the distribution of H^+ ions as a function of initial kinetic energy sputtered from a hydrous phase. Such energy spectra are shown in Figure 4 (see also Fig. 5 from Deloule et al., 1991a) for H-rich and (nominally) H-free phases. We observe a very intense peak for H ions with a few eV excess kinetic energy for both samples (these data were obtained shortly after inserting the sample, so the vacuum near the sample was ~3 x 10^{-7} torr). Ions with >~20 eV initial kinetic energy are approximately constant with increasing energy, but the hydrous rhyolite has ~10x the H^+ count rate. The high intensity at low energies shows that analyses of H using positive ions can be quite sensitive. However, the high intensity recorded for H in the H-free phase indicates that background signals can be a serious problem in quantitative analysis.

Our early attempts at calibrating the H signal on hydrous glass standards using low energy secondary ions gave apparent water concentrations of over 1 wt % in blank glasses (Hervig and Williams, 1988). If, however, only secondary H^+ ions with >50 eV excess kinetic energy were admitted into the mass spectrometer, the background signal was decreased by a factor of >10 (Hervig and Williams, 1988). It is known that abundant secondary electrons are generated during sputtering. During positive ion analysis, these electrons are attracted back to the sample and their impact is favorable for desorbing surface H atoms as positive ions. This process is known as electron stimulated desorption (ESD: Williams, 1981; Williams et al., 1983). Ions generated by ESD have a characteristic narrow energy spectrum, and lack the characteristic high energy tail shown by sputtered ions (compare the H^+ energy spectra with that of $^{28}Si^+$ in Fig. 4). It seems most probable that the low energy peak observed for H^+ ions on nominally H-free samples relates to the desorption of surface atoms by electron impact, while the ions with >~20eV excess kinetic energy are generated by collisions of the primary ion beam with the sample, as for the Si^+ ions. Thus, low energy H^+ ions may be strongly controlled by a combination of the frequency with which gas phase hydrogen lands on (and sticks to) the sample and its secondary electron yield.

Because of the strong dependence of secondary electron yield on chemistry, it might be expected that H^+ calibrations based on low energy H^+ ions would be subject to large matrix effects. This was observed by Steele (1986) using an AEI ion probe, where low energy secondary ions were preferentially accepted into the mass spectrometer to maximize count rates. Hinthorne and Anderson (1975) did not observe large matrix effects on an ARL instrument, but the study by Delaney and Karsten (1981) using an ARL ion probe did find a significant difference between ion yields of H between rhyolite and basalt. It is possible to tune the ARL to bias against low-energy secondary ions, and perhaps different tuning is responsible for the different magnitudes of matrix effects observed.

While many researchers (ASU; Paillat et al., 1992; Sisson and Layne, 1993; Webster et al., 1993; Ottolini et al., 1994) use H^+ ions with >20 eV excess kinetic energy for quantitative analyses, Deloule et al. (1991a) emphasized improving vacuum near the sample to reduce the contribution of adsorbed H to the low energy mass spectrum. They found that if the sample remains in the SIMS vacuum chamber for 3 to 4 days, surface H has diminished to a point low enough to allow precise measurements of D/H ratios in hydrous phases while detecting low energy ions. When sample changes need to be more frequent, the use of higher energy ions will allow analyses to be obtained for H_2O fairly rapidly.

The calibration of the ASU SIMS for hydrogen in rhyolite and basaltic glasses is shown on Figure 5a,b. For this plot, only positive secondary ions with 75±20 eV excess kinetic energy were allowed into the mass spectrometer. The data for basalt span 5 years of analysis sessions; during this period, the background intensity for hydrogen varied somewhat, but the slope of the calibration curve remained constant at ~0.2. Results for a suite of experimentally hydrated phonolitic glasses are also shown on Figure 5a (characterized by FTIR, provided by M. Carroll, University of Bristol). For rhyolite, the calibration curve has been somewhat more variable over time. Most analysis sessions from 1990 to mid-1993 produced calibrations with slopes near 0.3. However, one session in early 1994 gave a rhyolite calibration with a slope near 0.2. Subsequent sessions on rhyolites and a suite of pantellerite glasses have returned slopes near 0.3.

The rhyolite calibrations hint that the 5% H_2O glasses are the most variable. One source of variation in the hydrogen calibration curves may be the primary current density. It is conceivable that H in some glass compositions is mobile under the conditions of impact by a high energy ion beam (and the resulting charge build-up in the crater). If so, the mobility might be expected to increase with increasing density of deposited charge on the sample (see e.g., Clark et al., 1978; Webster, 1990). A rigorous test of current density on H ion yields is difficult, but a semi-quantitative measurement at ASU (Hervig, unpublished data) revealed that the value of $(H^+/Si^+)*SiO_2$ on a rhyolite glass with 5 wt % H_2O increased from 1.4 to 2.1 as the current density was decreased by ~40x. This covers a wide range of values on Figure 5b. Hinton (1990) warned that varying primary current density could dramatically change the secondary ion yields for particular elements. It would appear that for H analysis of rhyolites, it is important to calibrate the instrument at the same conditions used to analyze unknowns. Calibration for H in basalts seems much less sensitive to primary beam conditions.

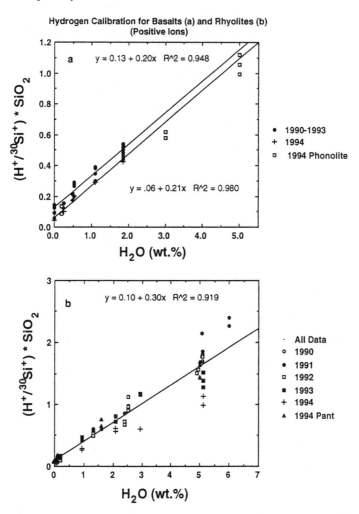

Figure 5. (a) Calibration of the secondary ion mass spectrometer (SIMS) for H in basalt. Bulk analyses of samples from Delaney and Karsten (1981; provided by J. Karsten) by hydrogen extraction. Vertical axis is $H^+/^{30}Si^+ * SiO_2$ (wt fraction). Secondary ions with 75 ± 20 eV excess kinetic energy were detected. ● indicate measurements taken during analysis sessions from 1990 to 1993 (5 different sessions). + indicate most recent analysis session (1994) with improved vacuum in the sample chamber (note the constant slope of the calibration since 1990). Open boxes represent hydrated phonolite glasses (characterized by FTIR and provided by M. Carroll). (b) Calibration for H in rhyolite. Standards were experimentally hydrated and characterized by either Karl Fisher titration or FTIR (KFT from H.R. Westrich at Sandia National Lab and FTIR from E. Stolper at Caltech and J. Barclay, University of Bristol). Vertical axis as in (a). Secondary ions with 75 ± 20 eV excess kinetic energy were detected. Note the greater variability of the high water standards (>3 wt %). Also note general colinearity of pantellerite and metaluminous rhyolites.

Negative secondary ions

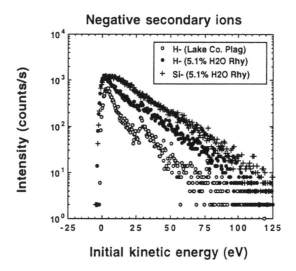

Figure 6. Energy spectra for H^- and $^{28}Si^-$ sputtered from ME35D, an experimentally hydrated (5.1% H_2O by Karl Fisher titration) rhyolite glass and H^- from Lake Co. plagioclase (An_{60}). Primary beam was $^{16}O^-$ emitted from a duoplasmatron ion source at -12.5 kV accelerating potential. Note high secondary ion intensities at low initial kinetic energies, and the gradual decrease of secondary silicon and hydrogen ions with increasing initial kinetic energies. The negative ion energy spectra of silicon and hydrogen are very similar, in contrast to positive ion energy spectra (Fig. 4).

H analysis using negative secondary ions. The energy spectra for H^- sputtered from the same phases as in Figure 4 are shown in Figure 6. The negative ion spectra are distinct from those for positive ions in that they resemble very closely the spectrum for Si^-. When negative ions are analyzed, secondary electrons are still generated, but they are immediately repelled by the negative potential on the sample, and are not as likely to cause desorption of surface hydrogen. Note the secondary peak for H^- ions sputtered from the Lake Co. plagioclase sample at ~30 eV excess kinetic energy (also to a lesser degree on the rhyolite glass). During analysis, the samples charged up by ~30V. The charging in the crater was compensated by adding -30V to the sample potential, but at the edge of the crater, near the remaining conducting coat, there is no charging, and we observe the sputtering of H^- ions with an apparent energy difference of 30 eV compared to the center of the crater. Si^- ions do not show the same secondary peak, as no Si is adsorbed to the conducting coat.

A hydrogen calibration can be obtained using negative secondary ions, but as suggested by the energy spectra in Figure 6, background signals can be a serious problem. An example is shown on Figure 7, for some of the hydrous basaltic and rhyolite glasses depicted in Figure 7. Negative secondary ions with 50±20 eV were used to construct this plot. As for positive ions, the slope of the rhyolite calibration is greater than for basalt (by nearly identical factors of 1.5; Figs. 5,7). The effect of varying secondary ion energy and sample composition on hydrogen calibrations using negative secondary ions has not been examined.

Hydrogen detection levels. Anecdotal evidence suggests that it is the mounting medium that controls the background for H analyses. For example, the lowest water backgrounds we have observed (≤0.1 wt % H_2O) were on samples embedded and polished

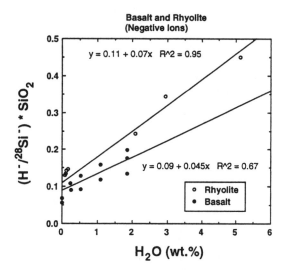

Figure 7. Calibration for H in basalt and rhyolite glasses using negative secondary ions with 50±20 eV excess kinetic energy. H count rate normalized to that for $^{28}Si^-$. Standards characterized as in Fig. 3.

in orthodontic resin (for infrared analysis). For SIMS analysis, these wafers were stuck to a glass slide with a drop of "super glue" (cyanoacrylate). More commonly, our epoxy-rich standard grain mounts give ~0.3 wt % H_2O backgrounds, while polished thin sections of rocks give backgrounds closer to 0.2 wt % H_2O. Leaving the sample in the vacuum chamber for several days will significantly lower the H background, if such time is available (e.g., 1994 basalt and phonolite calibrations shown on Fig. 5a; Deloule et al., 1991a). Another technique to reduce the H background from the vacuum is to greatly increase the intensity of the primary ion beam, which, in effect, removes H-bearing gas contaminants from the sample surface (by sputtering) as fast as or faster than they can be deposited. The limit of this technique occurs when further increases in primary ion current do not increase the current *density* in the crater. Yurimoto et al. (1989) used high primary beam currents to obtain detection levels near 5 ppm (atomic) for H in quartz at the expense of using a 100 μm diameter primary ion beam. Very low background levels can also be achieved when experiments are conducted with D_2O instead of H_2O. The background signal for D is nearly absent, allowing very low levels of D in synthetic, nominally anhydrous phases to be determined (Rovetta et al., 1989, Pawley et al., 1993). When negative secondary ions are studied, we observe higher backgrounds for H_2O than for positive ions, but no rigorous comparison has been made by us.

Matrix effects. Besides the glasses described above, we have studied the minerals kaersutite, tremolite, biotite, staurolite, and cordierite over the past five years using the conditions for positive ion analysis described above. All of these phases fall very close to the calibration defined by the basaltic glasses in Figure 5a with the exception of the tremolite (closer to the rhyolite calibration) and the biotite (very much below the basaltic calibration). Until more examples of these two phases are tested we would suggest that for compositions from 28 up to 59 wt % SiO_2 (the silica content of the phonolite in Fig. 5a) a calibration based on basalt glass will provide reasonably accurate analyses. The rhyolite (and pantellerite) glasses at 70 to 76 wt % SiO_2 show higher sensitivities (higher H^+/Si^+ per wt % H_2O) than basalts, but we do not know where in composition-space the calibration changes, or if the change is sudden or gradual. For analysis of high-silica glasses, it would be best to obtain standards as close as possible to unknowns.

Precision. The data in Figure 5 indicate that H^+ calibrations are reproducible in that repeated analysis sessions generally give the same slope. Specifically, we can calculate the error in the slope for a particular calibration. For example, the 1994 basalt calibration gives a slope of 0.206 ± 0.013 (one standard error), a precision of about 6%. For the 1993 rhyolite calibration, we calculate a slope of 0.249 ± 0.022, approximately a 10% uncertainty. The percentage uncertainty in the rhyolite slope is cut in half if the glass containing 5 wt % H_2O is removed from the fit. Thus, precision of H_2O analyses on rhyolite glass by SIMS is significantly improved when water contents are less than ~3 wt %. This may be related to dominance of (mobile?) molecular H_2O in water-rich glasses.

D/H measurements by SIMS. D/H ratios measured by Deloule et al. (1991a) used positive, low-energy secondary ions to maximize sensitivity. This technique is useful (Deloule et al., 1991b; Watson et al., 1994), but is difficult to apply to compositions other than represented by standards because of the large matrix effects (instrumental fractionation varied by several tens of per mil over a range of amphibole solid solutions). It is known that ions generated by electron stimulated desorption are subject to very large isotope fractionation effects (Madey et al., 1970; Houston and Madey, 1982), suggesting that there is a significant change in the secondary electron yields for the samples studied by Deloule et al. (1991a). Analyses using negative secondary ions (minimal contribution by secondary electron-generated ions) may reduce this effect, but at the expense of isotope ratio precision.

Applications to volatile contents in silicate glasses: Carbon

Carbon is difficult to analyze using positive secondary ions because ion yields are low, and $^{24}Mg^{2+}$ at the same nominal mass/charge value as $^{12}C^+$ must be resolved. Much greater sensitivity can be achieved with negative secondary ions (and multiply charged elemental negative ions are not stable). Pan et al. (1991) synthesized CO_2-bearing basaltic and basanitic glasses and characterized them by total carbon analysis. These samples provided a SIMS calibration with identical C ion yields for basaltic and basanitic glasses. Later analyses on bulk-analyzed leucitites (Thibault and Holloway, 1994) and one minette (synthesized by Y. Thibault, now at the C. Scarfe Laboratory at the University of Alberta) showed colinearity with basaltic and basanitic samples. The leucitite + minette + basanite + basalt calibration is shown on Figure 8a. A suite of acmitic ($NaFeSi_2O_6$) glasses was also studied by SIMS (synthesized by Y. Thibault; Fig. 8a). The sensitivity is ~1/2 that for leucitites. Several other samples are shown on Figure 8b, with the leucitite calibration given for scale. CO_2-bearing albitic and albite$_{95}$–anorthite$_5$ glasses were synthesized using ^{13}C-bearing starting materials and were characterized by NMR (provided by R. Brooker, University of Bristol). The two rhyolitic glasses (provided by R.A. Fogel) were characterized by Fourier transform infrared spectroscopy on normal isotopic abundance samples (Fogel and Rutherford, 1990). The data for the rhyolitic and Ab$_{95}$An$_5$ glasses are approximately colinear, and this trend has slightly lower slope than the acmitic glasses in Figure 8a. The albitic glasses suggest a slope which is about 20% that of the leucitite composition. There is a general trend of decreasing sensitivity to C with increasing amounts of C dissolved as molecular CO_2 species, but the acmite glasses provide an exception. The matrix effects for C analyses by SIMS are poorly understood at present.

Carbon detection levels. The background for carbon (detected as negative secondary ions) is determined by a combination of the sample mounting medium, and back streaming of pump oil. Increasing the primary ion current (and as a result, the primary ion beam spot size) decreases the contribution of C from the vacuum, until a point is reached where larger currents do not translate into higher current densities. Rastering a ~5 nA primary beam over an area ~30 μm on a side for ~5 minutes before returning to a stationary

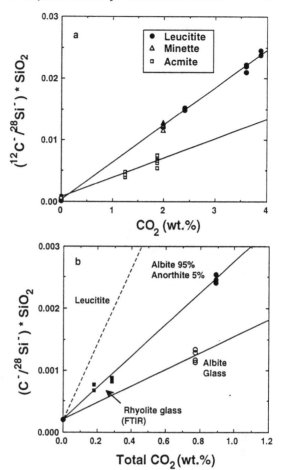

Figure 8. (a) Calibration for carbon in experimentally carbonated leucitite, minette, and acmite glasses (characterized by total carbon analysis). The $^{12}C^-/^{28}Si^-$ ratio is multiplied by the wt. fraction SiO_2 in the glass. The slope of this line is indistinguishable from the calibration for basalt and basanite (Pan et al., 1991). Acmite glasses show lower sensitivity for C analysis by a factor of ~2.

(b) Calibration for carbon in rhyolite glass (calibrated by FTIR; Fogel and Rutherford, 1990), albite glass, and $Ab_{95}An_5$ glass, both calibrated by nuclear magnetic resonance (NMR) spectroscopy on samples synthesized with pure ^{13}C. Vertical axis represents secondary ion intensity ratios for either $^{13}C^-$ or $^{12}C^-$, depending on the sample. The leucitite calibration from (a) is shown for comparison. Samples from Drs. Y. Thibault (Chris Scarfe Lab, University of Alberta), R. Brooker (University of Bristol), and R.A. Fogel (American Museum of Natural History).

beam for analysis gave the minimum background levels at ASU, corresponding to 0.1 to 0.3 wt % CO_2 (Pan et al., 1991; Hervig, 1992). However, background levels as low as 0.02 wt % CO_2 have been reported on synthetic albite glass on another Cameca SIMS instrument (Ian Hutcheon, Caltech, now at Lawrence Livermore National Laboratories, pers. comm., 1991).

Precision. The precision of SIMS measurements on silicic glasses has not been tested at ASU, but repeated measurements of our bulk-analyzed basanites and leucitites has given the same slope (within 10%) on our calibration plots since 1990.

Applications to volatile contents in silicate glasses: Halogens

Fluorine. Fluorine can be analyzed using either positive or negative secondary ions. Positive F ions are surprisingly intense at low secondary ion energies, the result of the efficiency with which secondary electrons can desorb F ions from the sample surface (see Williams, 1992). When high energy ions are studied, the sensitivity for F drops severely, in concert with its high ionization potential (1 count/s $^{19}F^+$ ~ 200 ppm F for 1 nA O^- and secondary ions with 75 ± 20 eV). Ottolini et al. (1994) suggested that matrix effects for

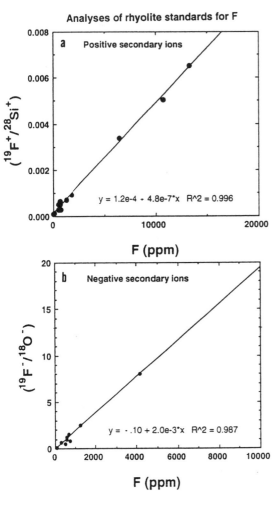

Figure 9. Analyses of rhyolite standards for fluorine. (a) calibration with positive secondary ions with 75±20 eV excess kinetic energy normalized to the $^{30}Si^+$ count rate. Data include analyses from 1991 and 1994. The analytical sessions are not distinguishable. (b) calibration with negative secondary ions with 50±20 eV excess kinetic energy normalized to the $^{18}O^-$ count rate. Note that the vertical axes are not normalized for silica or oxygen abundance as the rhyolitic glasses represent a limited range of compositions. Standards characterized by ion specific electrode (most by H.R. Westrich, Sandia National Labs).

high energy (100±25 eV) F^+ ions were small based on their analyses of amphiboles, micas, kornerupines, and topaz. One problem with measuring F^+ is that $^{18}OH^+$ is not completely eliminated by energy filtering and can be a significant interference for low-F, high H glasses. A calibration curve for F^+ in rhyolite glasses is shown on Figure 9a. For this calibration, we have combined analyses taken in 1991 and 1994.

Negative secondary ions of F are much more intense than positive ions; 1 count/s F^- ~<1ppm for 1 nA O^- and secondary ions with 50±20 eV. A calibration on many of the same rhyolite glasses shown in Figure 9a is displayed in Figure 9b. We have analyzed one F-rich amphibole and three different apatites. The results for the crystals were colinear and the slope of the resulting calibration curve was greater by a factor of 1.3 than that for the rhyolite glasses. The extent of matrix effects for this element remains to be investigated.

Chlorine. Ottolini et al. (1994) found that the signal for high energy (100±25 eV) Cl^+ was about half that for F^+, limiting the use of positive ion analysis to levels of tens of ppm Cl. No calibration curve was presented in this work. Much higher sensitivities are

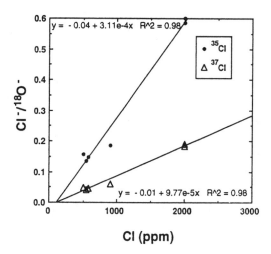

Figure 10. Calibration for chlorine in rhyolitic glasses. Negative secondary ions of $^{35}Cl^-$ and $^{37}Cl^-$ with 50 ± 20 eV excess kinetic energy normalized to the $^{18}O^-$ count rate plotted against Cl determined by electron probe. The ratio of the slopes is within 2% of the standard terrestrial Cl isotope ratio. Note the negative intercept. Regardless of which isotope is used for the SIMS calibration, there is a constant difference of ~120 ppm Cl between SIMS (low) and electron probe (higher abundances) indicating a minor background signal in the electron probe.

observed for negative secondary ions; our calibration on the same rhyolite glasses as in Figure 9b is shown in Figure 10. We measured both isotopes of Cl, and the slopes for the two calibrations are different by a factor of 0.315 ± 0.003, very similar to the terrestrial $^{37}Cl/^{35}Cl$ ratio of 0.319. For the other calibrations we have described, the intercepts have either been close to zero (e.g., F) or positive (e.g., H, C). The latter calibrations indicate an instrumental background for the species studied. For Cl, the intercept is negative, indicating that the nominal analyses of the standards are too high. Both isotopes imply that the standards are uniformly high by ~120 ppm Cl. Because the standards are characterized by the electron probe, this should not be surprising, as the level of detection for Cl is on this order.

Sensitivity and precision. Sensitivity for halogens can be very good when using negative secondary ions; measurements of less than 1 ppm of either F or Cl should be feasible. However, at these low levels, contamination problems may be important. We have found that F and Cl signals can be significant on nominally anhydrous, halogen-free minerals in thin section, presumably because the epoxy in which they are embedded is halogen-bearing and is smeared across the sample surface during polishing. Rastering the primary ion beam over a small area (e.g., 50 x 50 microns) for a few minutes seems to virtually eliminate this contamination. Reproducibility has been good for fluorine; note the combination of data sets from 1991 and 1994 in Figure 9a. We also note that we have been getting the same $^{19}F^-/^{18}O^-$ ratio (within 10%) on Durango apatite since 1990.

Applications to volatile contents in silicate glasses: Sulfur

Whereas SIMS has been commonly used for S isotope ratio measurements on sulfides (e.g., Eldridge et al., 1989), the measurement of S in volcanic glasses by SIMS has been limited. This is because the electron probe is adequate for many naturally occurring compositions (e.g., Devine et al., 1984). For high-silica compositions, however, the solubility of sulfur diminishes to the point that electron probe analyses are at or below easily achieved detection levels. SIMS analyses for S in rhyolites are difficult because there are serious interferences on S from oxygen dimers ($^{16}O_2^-$ and $^{16}O^{18}O^-$ on $^{32}S^-$ and $^{34}S^-$, respectively) and because of the slightly greater problem of sample charging when negative secondary ions are analyzed. An attempt to measure sulfur in a suite of high-silica glasses is shown on Figure 11a. Only ions with 50 ± 20 eV excess kinetic energy were

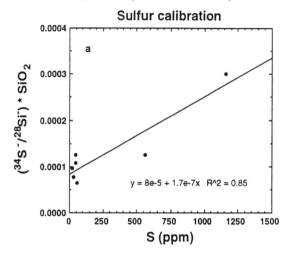

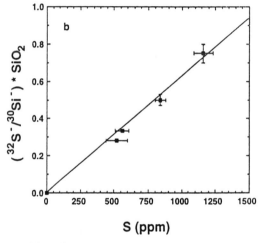

Figure 11. (a) Calibration for S in high silica glasses using an $^{16}O^-$ primary beam and energy filtering (detecting secondary ions with 50±20 eV excess kinetic energy) in an attempt to reduce O_2^- interferences on the sulfur isotopes. Analyses were made at ASU.

(b) Calibration for S in high-silica glasses using a Cs^+ primary ion beam and high mass resolution to eliminate O_2^- interferences on sulfur. Analyses were made on the Cameca IMS 4f at CANMET Ottawa, Canada. X-axis error bars reflect the uncertainty in the accepted values for the standards as provided by the National Institute of Standards and Technology (formerly NBS) and the Society of Glass Technology certificates (NBS 621, SGT 5) and Palais and Sigurdsson (1989; NBS 610 and NBS 620). Y axis error bars reflect the 1 sigma of replicate analyses on each standard. The precision of the instrumental ratio measurements far exceeds the apparent heterogeneity of the standard materials.

allowed into the mass spectrometer in an attempt to minimize the contribution of the O_2 dimer on ^{34}S. The plot shows a significant background signal (judging by the positive intercept), and the two high sulfur glasses are not well-correlated with the other samples or the origin. The correlation was not improved by increasing the initial kinetic energy of the secondary ions detected. Detecting ^{32}S instead of ^{34}S gave S^-/Si^- ion ratios for the low-S glasses that were even higher than the high S glasses! Clearly, the use of $^{16}O^-$ as a primary ion beam combined with energy filtering to remove O_2 interferences is inappropriate for this analytical problem. Most SIMS analysts in the semiconductor industry routinely use a primary ion beam of Cs^+ as it has been shown to dramatically increase the yields of negative secondary ions. However, when applied to insulators, the Cs^+ beam causes the crater to charge up positively, resulting in a very low and unstable secondary ion current. Cameca has included a novel electron gun design with some of its instruments sold in the past 7 years that effectively compensates the charge build-up on insulators during Cs^+ bombardment. This allows secondary ion intensities to be measured to a precision approaching 1%. These analysis conditions were tested on a suite of high-silica glasses giving the calibration results shown on Figure 11b (analyses provided by Graham Layne, at the Institute of Meteoritics at the University of New Mexico). Analyses were made on the Cameca IMS 4f at CANMET Ottawa (Canada). The interfering oxygen dimer ions were eliminated from the mass spectrum by closing the entrance slits to the mass spectrometer until the peak at $^{32}S^-$ could be clearly resolved from $^{16}O_2^-$. This required a mass resolving power of 18 parts in 32000, easily achievable in the Cameca IMS instruments. Note the near-zero intercept, and the good colinearity of the standards, two of which are also shown in Figure 11a.

Applications to volatile contents in silicate glasses: Other elements

The SIMS technique is sensitive to much of the periodic table (see Hinton, 1990 for relative positive ion yields). While we have discussed the potential for microanalyses of common volatile elements, under certain P-T-X conditions normally lithophile elements can be preferentially partitioned into the vapor phase and so may be considered as potentially volatile elements (e.g., Li, B, Rb, Y, Nb, Cs, and Ce; Pichavant, 1981; London et al., 1989; Webster et al., 1989). These elements can all be detected at the ppm level or below by SIMS. Of these elements, boron is probably the most easily transferred into vapor phases in natural systems. Very precise SIMS measurements of boron crystal/liquid partition coefficients and B isotope ratios in silicate melts are presented in Chaussidon and Libourel (1993) and Chaussidon and Jambon (1994), respectively.

Summary of SIMS

Secondary ion mass spectrometry can provide information on several volatile elements as well as quantitative analyses of trace elements ranging from Li to U, all on the same ~≤20 μm diameter spot. Knowledge of both trace and volatile elements on a microscale can be very useful when tracing the evolution of fluids from magmas (e.g., in trapped melt inclusions vs. matrix glass). In addition, isotope ratio measurements of selected elements are possible (Zinner, 1989; Deloule et al., 1991a; Chaussidon and Jambon, 1994). As opposed to FTIR, no speciation information is available from SIMS, and the sample is lost during analysis (although the amount consumed ranges from <1 to 10 ng). In addition, precision of volatile element analyses tends to be large (±10% or greater), especially for H_2O in water-rich glasses. Thus, electron probe analyses for water at high water contents (e.g., ≥5 wt % H_2O) may have comparable errors (Nash, 1992). Application of SIMS microanalytical techniques to volatile elements in magmas would be improved by a better understanding of matrix effects and improved vacuum near the sample.

ANALYTICAL METHODS INVOLVING VIBRATIONAL SPECTROSCOPY

Introduction

Vibrational spectroscopic techniques, including infrared (IR) transmission and Raman scattering spectroscopy, have been used extensively both in structural studies and in quantitative analyses of geologically important glasses and melts containing volatile species, including H_2O, CO_2, F and H_2 (e.g., Keller and Pickett, 1954; Orlova, 1964; Ostrovskiy et al., 1964; Schölze, 1966; Mysen, 1976; Mysen et al., 1976; Brey, 1976; Sharma, 1979; Sharma et al., 1979; Mysen et al., 1980; Mysen and Virgo, 1980a,b; 1985; 1986a,b; Stolper, 1982a,b; Rai et al., 1983; McMillan et al., 1983; 1993; Fine and Stolper, 1985; 1986; Dixon et al., 1988; Newman et al., 1988; Luth, 1988a,b; Silver and Stolper, 1989; Stolper, 1989; Silver et al., 1990; Taylor, 1990; Fogel and Rutherford, 1990; Ihinger, 1991; Pan et al., 1991; Pawley et al., 1992; Blank et al., 1993; Dixon et al., 1994; Stolper and Newman, 1994).

The technique of choice for analysis of volatile (H- and C-bearing) species in silicate glasses is infrared absorption spectroscopy. Both the mid-IR region (using light with wavelengths (λ) between approximately 2.5 to 7 μm, corresponding to a wavenumber ($\bar{v}$ = 1/λ) range of 4000 to 1400 cm^{-1}) and the near-IR (1.25 to 2.7 μm: 8000 to 3700 cm^{-1}) region are used, depending on the characteristic vibrational frequencies of the volatile species of interest (Stolper, 1982a,b; Fine and Stolper, 1985; 1986).

Raman spectroscopy has not yet been applied to the quantitative determination of volatiles dissolved in silicate melts or glasses, but has been used extensively in the study of fluid inclusions (Dhamelincourt et al., 1979; Roedder, 1984; Pasteris et al., 1988; Dubessy et al., 1989; McMillan et al., 1994), and in qualitative studies of the speciation of H- , C- and F-bearing volatiles in geologically relevant glass compositions (Verweij et al., 1977; Sharma, 1979; Sharma et al., 1979; Mysen et al., 1980; Mysen and Virgo, 1980a,b; 1985; 1986a,b; Rai et al., 1983; McMillan et al., 1983; 1993; Luth, 1988a,b). It would take little effort to develop the quantitative application of Raman spectroscopy to cases where FTIR analysis is difficult, or does not provide sufficient information on individual species, and there are obvious opportunities for worthwhile overlap between both techniques. In particular, the combination of micro-FTIR and micro-Raman analyses can give detailed quantitative information on volatiles dissolved in the silicate melt and on the composition of the coexisting fluid phase, from analysis of fluid inclusions and adjacent glass (Konijnendijk and Buster, 1977; Barres et al., 1987; Wopenka et al., 1990; Pawley et al., 1992).

Both infrared and Raman techniques are non-destructive, although some sample preparation may be necessary. Both can be used as microbeam techniques, either by aperturing the sample area to be studied (IR), or by focusing the incident beam through an optical microscope (micro-Raman), so that samples or areas down to a few μm in dimension can be analyzed. In micro-IR spectroscopy, the incident light is focused on the sample using reflecting optics, and apertures are used to further select the region of interest.

Before describing the application of these techniques to specific geochemical systems, it is useful to review the nature of the characteristic vibrational modes used for volatile analysis.

Vibrational properties of important volatile species

OH and H₂O. Water determination via infrared spectroscopy relies on quantitative infrared absorption studies of characteristic vibrations of hydrous species: O-H stretching, HOH bending, or their combinations and overtones. The diatomic OH molecule has a classical harmonic frequency at $v_0 = 1.065 \cdot 10^{14}$ s^{-1}, corresponding to a wavenumber aperture of 3550 cm^{-1}. (In the spectroscopic literature, vibrational "frequencies" are most often expressed in wavenumber units, cm^{-1}, with $\bar{v} \equiv c / v$, where c is the speed of light in cm/s. The wavelength of the infrared light absorbed is given by $\lambda = c/v = c\,\bar{v}$, and corresponds to 2.817 μm for the free OH absorption). The value of the IR absorption frequency is determined by the molecular force constant k and the reduced mass of the molecule

$$m_{OH}^{*} = \frac{m_{o} m_{H}}{m_{o} + m_{H}}, \tag{1}$$

via the classical equation

$$v_{o} = \frac{1}{2\pi c} \sqrt{\frac{k}{m^{*}}}. \tag{2}$$

This vibration is infrared active because there is a change in the dipole moment (μ) of the OH molecule during the vibration, evaluated at the equilibrium bond distance (r_0):

$$\left(\frac{d\mu}{d\Delta r}\right)_{r_o} \neq 0. \tag{3}$$

Δr is the change in O-H bond length during the vibrational mode (Herzberg, 1950; McMillan, 1985).

In a quantum mechanical model of the molecular vibration, the quantized vibrational energies are given by

$$E_n = (n + \tfrac{1}{2})h\sqrt{\frac{k}{m^*}}. \tag{4}$$

h is Planck's constant (6.626×10^{-34} J·s), and n is a vibrational quantum number: n = 0, 1, 2, 3, etc. The energy levels for such a *harmonic oscillator* are equidistant, separated by $\Delta E = h v_0$, and vibrational excitations can only take place between adjacent energy levels, $\Delta n = \pm 1$. The transition between energy levels with n = 0 and n = 1 is known as the *fundamental* vibrational transition of the molecule, and is accompanied by absorption of radiation with energy $\Delta E = h v_0$.

Because the hydrogen atom is so light, it undergoes large displacements during its vibrational motion. In addition, the force constant for O-H stretching is quite large, so that the spacing between energy levels is high. For these reasons, it becomes particularly important to consider the effects of *anharmonicity*, or departure of the molecular potential energy function from a parabola (Herzberg, 1950; McMillan, 1985), when discussing the vibrational properties of O-H groups. Because of the anharmonicity of the molecular potential, the quantized vibrational energies become

$$E_n = (n + \tfrac{1}{2})h v_e - (n + \tfrac{1}{2})^2 h v_e x_e + (n + \tfrac{1}{2})^3 h v_e y_e - ..., \tag{5}$$

in which the v_e, x_e and y_e are anharmonicity constants, with $v_e \gg v_e x_e \gg v_e y_e$ (Herzberg, 1950; Huber and Herzberg, 1979). Anharmonicity permits the appearance of *overtone* transitions, or transitions in which $\Delta n = \pm 2, \pm 3$, etc., in the infrared spectrum. This is important for volatile analysis, because the near-IR method for water determination relies on detection of the first overtone of the O-H stretching vibration, near 7000 cm^{-1} (Stolper, 1982a,b) (Fig. 12). For anharmonic vibrations, the spacing between adjacent vibrational levels decreases with increasing quantum number n, so that the overtone transitions do not occur at exact multiples of the fundamental frequency, but at decreasingly smaller intervals (Herzberg, 1950; Strens, 1966; Burns and Strens, 1966; Stone and Walrafen, 1982; McMillan, 1985). For example, the maximum in the fundamental O-H infrared absorption band in rhyolitic obsidians occurs at 3560-3580 cm^{-1}, but the first overtone appears at 7018-7080 cm^{-1} (Stolper, 1982a,b; Newman et al., 1986) (Fig. 12).

The O-H stretching frequency in condensed samples is affected by hydrogen bonding (Ryskin, 1974; Rossman, 1988a). The O-H bond is quite polar, with the hydrogen atom carrying a small net positive charge ($\delta+$), and a small negative charge on the oxygen ($\delta-$). For this reason, the hydrogen atom of an OH group is attracted to any adjacent negative centers, such as other oxygens to which it is not bonded. The resulting hydrogen bond lowers the frequency of the O-H stretching vibration, in some cases by several hundred or even as much as 2000 cm^{-1} (Ryskin, 1974; Rossman, 1988a). The effects of this can be seen in the O-H stretching vibrations of dissolved H-bearing volatile species. For SiO_2 samples with OH content near 1200 ppm, the O-H stretching from silanol (SiOH) groups shows a single sharp peak at 3670 cm^{-1} which is quite symmetric (McMillan and Remmele, 1986: see Chapter 4). This indicates that, in this case, the OH groups are sufficiently removed from adjacent oxygen atoms that little or no hydrogen bonding takes place.

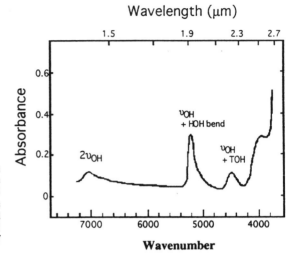

Figure 12. The near-IR transmission spectrum (displayed in absorbance units) of a hydrous albite glass (6.85 wt % H_2O), showing the overtone bands associated with stretching and bending of molecular water and T-OH groups. Redrawn from Figure 4 of Stolper (1982a).

However, samples with higher water contents (greater than several thousand ppm) show an O-H stretching vibration which is significantly broadened, is shifted to lower wavenumber, and is asymmetric on its lower frequency side (McMillan and Remmele, 1986; see Chapter 4). The O-H stretching vibration in aluminosilicate glasses also shows evidence for hydrogen bonding (Stolper, 1982a,b; McMillan et al., 1983; 1993; Newman et al., 1986). Changes in the degree of hydrogen bonding can have a significant effect on the magnitude of the infrared absorption coefficient, which determines the strength of the O-H absorption band (see Paterson, 1982).

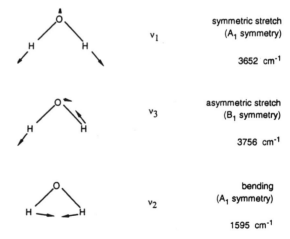

Figure 13. Sketch of the three vibrational modes for molecular H_2O, with the gas phase frequencies. The symmetries of the vibrations with point group C_{2v} are also shown.

Molecular H_2O has three vibrational modes (v_1, v_2 and v_3) which correspond to symmetric O-H stretching, HOH bending, and asymmetric stretching, respectively (Fig. 13). All are infrared and Raman active. The infrared absorption spectrum of gaseous H_2O contains intense bands at 3756 and 3652 cm^{-1} (unresolved) due to the O-H stretching fundamentals, and a medium intensity absorption at 1595 cm^{-1}, due to the bending vibration (Herzberg, 1945). A band is also observed at 3151 cm^{-1} due to the first overtone

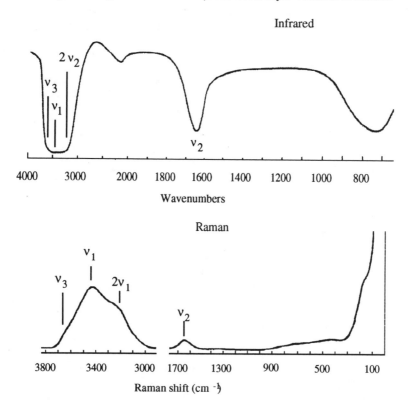

Figure 14. Infrared and Raman spectra of liquid water, redrawn from Figures 1 and 2 of Walrafen (1964).

of the bending mode ($2v_2$) (Herzberg, 1945). Because the moments of inertia of the molecule are small, the rotational constants are large, and rotational structure is observed on these vibrational bands under even moderate resolution, in the gas phase. In liquid water, or water molecules trapped in a matrix, the rotational structure is lost, and the fundamental vibrations appear as broadened bands, shifted from their gas phase positions (Fig. 14). In liquid water at room temperature and pressure, the v_2 bending vibration is observed at 1630 cm^{-1}, and the v_1 and v_3 vibrations appear at 3450 and 3615 cm^{-1} (Walrafen, 1964).

Because of the small mass of hydrogen, most of the vibrational displacement in hydrous molecules concerns the H atom, and the O-H stretching frequencies for species containing the OH group remain approximately constant in the absence of significant hydrogen-bonding interactions. For this reason, the fundamental O-H stretching vibrations of molecular H_2O and groups such as SiOH or AlOH in condensed aluminosilicates all occur at approximately 3300 to 3700 cm^{-1}, close to the value for the isolated OH molecule (see Chapter 4).

In quantitative IR studies of hydrous glasses, most interest has been focused on the first overtone region (Stolper, 1982a,b). This includes a band near 7100 cm^{-1}, due to the first overtone of O-H stretching, and is common to both molecular H_2O dissolved in the glass and bound OH groups, a band near 5230 cm^{-1} which is due to the O-H stretching + HOH bending combination of molecular H_2O, and a band near 4500 cm^{-1} due to bound hydroxyl (TOH) (Stolper, 1982a,b) (Fig. 12). The position of this band requires that the

OH groups in the glass must be bound to a species with a vibration near 1000 cm^{-1}; either an Si-O or Al-O bond, or a TOT linkage (Newman et al., 1986; Pichavant et al., 1992; McMillan et al., 1993). Consistent with this assignment, the infrared and Raman spectra of hydrous SiO_2 glasses contain a peak at 970 cm^{-1}, due to stretching or bending of Si-OH groups in the glass (Stolen and Walrafen, 1976; Hartwig and Ran 1977; Mysen and Virgo 1986a; McMillan and Remmele 1986) (see Chapter 4). A similar feature appears near 900 cm^{-1} in the spectra of hydrous aluminosilicate glasses, which could be due to either a terminal AlOH species, or to a protonated AlOSi linkage (Kohn et al., 1989a; 1992; McMillan et al., 1993; Sykes and Kubicki, 1993) (Chapter 4).

If this 900 cm^{-1} band is in fact due to a protonated AlOSi linkage, and the water dissolution model proposed by Kohn et al. (1989a) is correct (see Chapter 4), there should also be $Na^+...OH^-$ species present in the hydrous glass, presumably as hydration complexes (Pichavant et al., 1992). The vibrations contributing to the 4500 cm^{-1} combination band then only represent one half of the OH species present, with the result that the determined value for the molar absorptivity (see below) is in error by a factor of two. It is interesting that there is an overtone band near 4000 cm^{-1} in the infrared spectra of hydrous aluminosilicate glasses (Stolper, 1982a,b; Newman et al., 1986) (Fig. 12) which has not been satisfactorily assigned. This could be due to the combination of the O-H stretching vibration with that of a metal-oxygen stretching (e.g., Na-O stretching) within such a complex (Pichavant et al., 1992). This observation would not, however, affect either the conclusions or the analytic results obtained by Stolper and colleagues, because the molar absorptivity values obtained in those studies are all internally consistent, and return correct values for the total water content.

CO, CO_2 and CO_3^{2-}. Carbon-oxygen species form the second major class of volatiles analyzed routinely by quantitative FTIR spectroscopy. The diatomic molecule CO has its characteristic absorption at 2145 cm^{-1}, and is visible in infrared and Raman spectra. From micro-Raman studies, this species is known to be present in CO-CO_2 fluid inclusions (Bergman and Dubessy, 1984). However, it is unlikely that molecular CO is an important dissolving species in silicate melts, based on the experimental findings of Pawley et al. (1992) on basaltic liquids in equilibrium with CO-CO_2 fluids.

Carbon dioxide forms a linear symmetric molecule in the gas phase. The expected vibrational modes include the symmetric (v_1) and asymmetric (v_3) C-O stretching modes, and a doubly degenerate OCO bending mode (v_2). The symmetric stretching vibration v_1 is only Raman active, whereas v_3 and v_2 are only infrared active. As expected, the infrared spectrum of CO_2 shows two strong absorption bands due to the v_3 and v_2 vibrations, at 2349 and 667 cm^{-1}, respectively (Herzberg, 1945). However, the observed Raman spectrum shows *two* strong peaks in the C=O stretching region, near 1388 and 1286 cm^{-1} (Fig. 15). These appear due to *Fermi resonance* between the v_1 fundamental and the first overtone of the bending vibration, $2v_2$ (Herzberg, 1945). The normally weak $2v_2$ overtone gains a large component of the character of the v_1 vibration through the Fermi coupling, and becomes strongly Raman active (Herzberg, 1945). The frequencies of both components are affected by this vibrational coupling, in that both move away from the positions of their unperturbed components. The result is two new vibrational modes (denoted v_+ and v_-), each of which has a partial character of both the v_1 and $2v_2$ components.

The relative intensities and positions of the v_+ and v_- components of the Fermi doublet of CO_2 is critically dependent on the degree of v_1-$2v_2$ coupling, which is extremely sensitive to the molecular geometry and to any intermolecular interactions. The relative

Figure 15. Raman spectrum of CO_2 vapor in the C=O stretching region from a fluid inclusion in quartz. The two strongest peaks are the high-(v_+) and low-frequency (v_-) components of the v_1-$2v_2$ Fermi doublet of normal isotopic CO_2 used in quantitative analysis. Weaker peaks due to the Fermi doublet of $^{13}CO_2$ are also apparent (Howard-Lock and Stoicheff, 1971). An additional weak peak at 1409 cm^{-1} is due to a "hot band" transition from the first excited state of the bending vibration (the 010 → 110 transition; Herzberg, 1945).

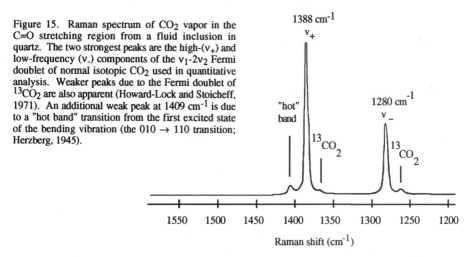

Raman shift (cm^{-1})

positions and intensities of the Raman doublet have been studied extensively in the gas, liquid, fluid and solid states of CO_2, as well as in aqueous solution (Davis, 1972; Garrabos, 1980; Hanson, 1980; Hanson, 1981; Wang, 1973; Wright, 1973; Wright, 1974). This is particularly important in studies of CO_2 in fluid inclusions, which may be subject to considerable internal pressure, increasing their density and affecting the intensity calibration of the Raman signal (Wopenka and Pasteris, 1986, 1987; Pasteris et al., 1988; Dubessy et al., 1989).

Along with molecular CO_2, the carbonate group (CO_3^{2-}) is the primary C-bearing species in carbonated silicate and aluminosilicate glasses (Fine and Stolper, 1985; 1986; Stolper et al., 1987). The vibrational modes of the free ion, which has trigonal planar symmetry, are well known in aqueous solution: a Raman-active symmetric stretch (v_1) at 1064 cm^{-1}, an IR-active (v_2) bending vibration at 879 cm^{-1}, a doubly degenerate asymmetric stretch (v_3) at 1415 cm^{-1}, and a doubly degenerate bending mode (v_4) at 680 cm^{-1} (Fig. 16). These last two are both IR and Raman active (Herzberg, 1945; White, 1974). In quantitative infrared studies of C-bearing aluminosilicate glasses, most attention

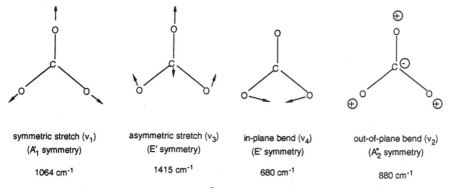

Figure 16. The vibrational modes of the free CO_3^{2-} anion, with trigonal (D_{3h}) symmetry. The frequencies given are those measured in aqueous solution. The totally symmetric stretching vibration is only Raman active, the out-of-plane bend (v_2) only appears in the infrared spectrum, and the other vibrations are infrared and Raman active. Distortion of the molecule results in splitting of the asymmetric stretch (and the in-plane bend) into two components, and the symmetric stretch may become weakly infrared active.

has been focused on the asymmetric stretching vibration of molecular CO_2 near 2350 cm^{-1}, and the v_3 vibrations of carbonate species in the 1350 to 1600 cm^{-1} region (Fine and Stolper, 1985; 1986) (Fig. 17) The asymmetric stretching of CO_3^{2-} species is usually perturbed by interaction with metal cations, or with the Si or Al framework cations in the glass, which removes the degeneracy of this vibration, and valuable information regarding speciation can be obtained from the degree of splitting of the v_3 modes (Fine and Stolper, 1985; 1986; R. Brooker, unpublished data) (Fig. 17). The symmetric stretching vibration gives rise to a characteristic strong peak in the Raman spectra of carbonate-bearing glasses (Sharma, 1979; Sharma et al., 1979; Mysen and Virgo, 1980; Rai et al., 1983) (Fig. 18), but this has not yet been calibrated for quantitative studies.

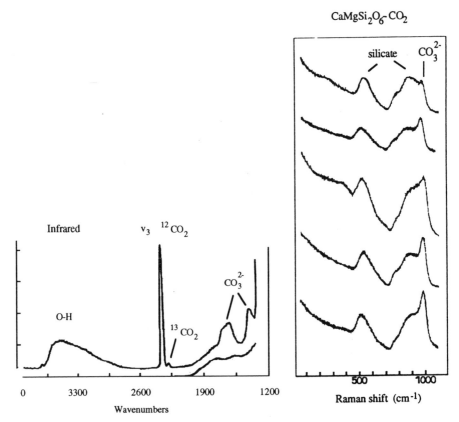

Figure 17 (left). Infrared transmission spectrum of a doubly polished section of albite glass containing dissolved carbon species, redrawn from Figure 1 of Stolper et al. (1987). The CO_3^{2-}/CO_2 ratio is determined from the relative absorbances of the asymmetric stretch bands of CO_2 and CO_3^{2-}. Note the splitting of the CO_3^{2-} asymmetric stretch, indicating distortion of the anion due to coordination to cationic species in the glass. There is also structure on the higher frequency component, indicating that more than one type of CO_3^{2-} group is present. A weak feature due to the asymmetric stretching of $^{13}CO_2$ is observed. In addition, a broad band due to O-H stretching is present near 3500 cm^{-1}, allowing the simultaneous determination of water content of the volatile-bearing glass.

Figure 18 (right) Raman spectra of diopside glasses containing 1 to 5 wt % dissolved CO_2, showing the appearance of the strong v_1 symmetric stretching vibration of CO_3^{2-} groups dissolved in the glass. This could be used for quantitative determination of carbonate species. Redrawn from Figure 6 of Mysen and Virgo (1980).

Quantitative vibrational spectroscopy

Infrared absorption: theory. Because this is the primary technique used in determination of volatile species concentration via FTIR spectroscopy, it is important to understand the theoretical basis for the method.

When a beam of infrared light with incident intensity I_0 (this is the *irradiance*) is passed through a sample of thickness d (which is conventionally given in cm), it is attenuated at the vibrational frequency v_i by interaction with the absorbing species in the sample. The transmitted intensity I_t is given by:

$$I_t = I_o e^{-\alpha d} \tag{6}$$

The constant α is known as the *absorption coefficient* associated with the infrared active vibrational mode with frequency v_i, and its magnitude reflects the degree of infrared activity of that mode. Within a harmonic model, the theoretical expression for the net absorption coefficient for a collection of N molecules per unit volume is given by (Steele, 1971):

$$\alpha = \frac{1}{d} \ln \frac{I_o}{I_t} = \frac{N\pi}{3m * c} \left(\frac{d\mu}{dr} \right)^2_{r_o} \tag{7}$$

Here m* is the reduced mass of the molecule, n is the quantum number of the lower vibrational energy level, v is the vibrational frequency, and c is the speed of light. The magnitude of the absorption coefficient also depends upon the square of the change in molecular dipole moment during the vibrational displacement, as defined above. In this expression, α is independent of temperature and frequency, for a given vibrational mode (Steele, 1971). This is no longer strictly true for the vibrations of real anharmonic molecules, for which α becomes temperature dependent, and the vibrational transitions for even a single molecule are spread over a band of frequencies.

For quantitative analysis applications, it is usual to work in decadic logarithms, and the corresponding intensity expression becomes

$$I_t = I_o 10^{-Kd} \tag{8}$$

The decadic absorption coefficient is K, with $\alpha = 2.303 \cdot K$. For quantitative analysis of a sample containing a uniform concentration of the absorbing species of c moles per liter, the corresponding expression is the Bouguer-Beer-Lambert law:

$$I_t = I_o 10^{-c\varepsilon d} \tag{9}$$

The parameter ε is known as the *extinction coefficient*, with $\varepsilon = K/c$, in units of liter/mol·cm. The transmittance of the infrared beam through the sample is given by

$$T^{sample} = \frac{I_t^{sample}}{I_o} \tag{10}$$

usually expressed as a percentage. In an FTIR experiment (see below), it is usual to run a blank spectrum to account for the spectral response of the spectrometer system, and absorption by any species in the sample compartment environment (e.g., water vapor in the air, etc.). The transmittance of the blank is

$$T^{blank} = \frac{I_t^{blank}}{I_o} \tag{11}$$

The *absorbance* due to the species of interest is then given by

$$A = \log_{10} \frac{T_{blank}}{T_{sample}} \qquad (12)$$

It can be seen that, if the sample does not absorb at a given wavelength, the transmittance through the blank and the sample are the same, and the absorbance is zero. If the sample absorbs strongly, such that the transmittance through the sample is only 1% of the blank, the absorbance is 2. It is not recommended to carry out quantitative work with such low transmission values. Slight variations in the thickness of the sample, resulting in minor changes in the % transmission, or non linearity in the detector response at low transmission levels, can lead to large errors in the measured absorbance. Typical quantitative IR analyses for volatile species are obtained with absorbances of 0.1 to 1.8, corresponding to transmittances of 80 to 15%.

In general, an infrared absorption peak is not infinitely sharp, but has some finite width and shape. This occurs because of anharmonicity of the individual vibrations, and because there is usually some range in atomic environments contributing to the signal. The measured absorption coefficient will then be a function of frequency (v) across the finite width of the absorption band. To account for this, it is usual to replace the extinction coefficient with an *integrated molar absorptivity*:

$$\varepsilon^* = \int_{band} \varepsilon(v)dv \qquad (13)$$

obtained by integration over the surface of the experimentally measured band. If the vibrational "frequency" scale is in wavenumbers, the units of ε^* are liters mol^{-1} cm^{-2}, often simplified to cm/mmol, commonly known as *"darks"* in the spectroscopic literature.

There is as yet no good method for the absolute prediction of infrared absorption coefficients for the vibrational modes of species of interest in analytical geochemical studies, and so ε or ε^* must be determined empirically for each vibrational mode of each species of interest. In practice, this is carried out by preparing (or otherwise obtaining) series of standard samples with known concentrations of the absorbing species, and measuring the peak height or integrated absorbance in the region of interest as a function of sample thickness. This results in a series of empirical extinction coefficients or molar absorptivities, for determination of species concentration in suites of unknown samples sufficiently similar to the standards that the calibration is applicable. Details of the empirical calibration of ε and ε^* for volatiles of interest are discussed below.

In general, there is an orientation dependence to the infrared absorption coefficient, pending on the direction and state of polarization of the incident infrared beam, relative to e orientation of the molecular vibration dipole in the sample. This forms the basis for sing polarized infrared spectroscopy as a sensitive tool to determine the crystallographic orientation of specific structural groups, such as OH, in minerals (Paterson, 1982; Aines and Rossman, 1984; Rossman, 1988a). In order to compare the magnitude of the extinction between different minerals and glasses, it is important to take this orientation dependence into account. Paterson (1982) has suggested the definition of an idealized extinction coefficient $\varepsilon_{\parallel}$ (or corresponding molar absorptivity, $\varepsilon^*_{\parallel}$) in which all of the OH dipoles in the sample are forced to lie parallel to the electric vector of a plane-polarized incident infrared beam. This "parallel" extinction coefficient is obtained from the measured ε value by dividing by an orientation factor, γ, obtained from the geometric distribution of dipoles in the sample (Paterson, 1982):

$$\varepsilon_\parallel = \frac{\varepsilon}{\gamma} \qquad (14)$$

For a completely isotropic distribution of dipoles, as found in a volatile-bearing glass, the orientation factor is $\gamma = 0.333$.

From comparison of values for the O-H stretching fundamental obtained for different hydrous minerals and glasses, the magnitude of the extinction coefficient or molar absorptivity is inversely correlated with the O-H stretching frequency, from 10,000 to over 100,000 cm^{-2} per mol H atoms per liter (Paterson, 1982), and a value near 350,000 cm^{-2} per mol H per liter is reached for OH bonded to octahedral silicon in stishovite (Pawley et al., 1993). The precise reasons for this variation are not yet well understood, but could be related to effects of hydrogen bonding (Paterson, 1982), or to the strength of the adjacent Si-O bond in stishovite.

Practical application of the technique. The basic instrumentation required for volatile analysis is a suitable spectrometer. Reliable quantitative analyses via infrared or near-infrared absorption can be conveniently carried out using conventional scanning instruments (Stolper, 1982a; Newman et al., 1986), in which the incident beam is passed through the sample, then dispersed by a prism or grating via a slit on to the detector. The infrared transmission or absorbance signal is often read out directly on chart paper. In order to improve the signal to noise ratio in an experiment of this type, it is necessary to slow down the scan speed, or to open the slits, which reduces resolution. Most modern infrared instruments, however, utilize the Fourier transform technique, which provides a number of advantages for quantitative spectroscopy.

Instead of scanning the wavelengths of light transmitted by the sample, the entire wavelength range emitted by the infrared source, passed through the sample, and collected by the detector is modulated to give a signal intensity fluctuating in time, $I(t)$. This is then converted via Fourier transform (FT) techniques using a bench top computer attached to the spectrometer, to obtain the spectrum in the frequency $(I(\nu))$ domain. A description of the FTIR technique applied to geological samples is given by McMillan and Hofmeister (1988), and detailed accounts of the theory and practice of FTIR spectroscopy appear in the books by Hadni (1967) and Griffiths and de Haseth (1986).

Fourier transform spectrometers are based on the Michelson interferometer (Fig. 19). Light from the infrared source is passed into a beamsplitter, which separates the light into two beams of equal intensity. One beam is directed on to a fixed mirror, and the second beam on to a moving mirror. Both mirrors reflect the light back into the beamsplitter, where the two beams are recombined and sent to the sample. Because one of the mirrors is moving periodically along the axis of the light beam, constructive or destructive interference occurs when the two beams are recombined at the beamsplitter, depending upon the instantaneous position of the moving mirror (Fig. 19). The result is a regularly fluctuating incident beam of light, with a time dependent intensity $I_0(t)$, being sent to the sample. If there is no sample present in the light path, the light beam reaching the detector contains all of the wavelengths emitted by the source. The fluctuating light intensity is recorded by the detector and stored in a digital computer as an interferogram. The desired result is usually a spectrum of intensity as a function of frequency, which is obtained by carrying out the time Fourier transform of the function $I(t)$ into the inverse time, or frequency domain (Hadni, 1967; Griffiths and de Haseth, 1987). When no sample is present, the result is a blank spectrum of the emitted light of the detector, multiplied by the sensitivity curve of the detector, and the transmission function of the instrument. Small

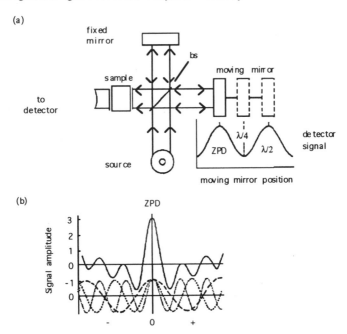

Figure 19. Schematic showing the principle of operation of the Michelson interferometer, as used in FTIR spectroscopy. (a) Light from the infrared source is separated into two beams by the beamsplitter (bs). One beam is reflected by a fixed mirror, and the other by a moving mirror. The two reflected beams are recombined at the beamsplitter, and are sent through the sample chamber to the detector. If the source emits a single fixed frequency, the resulting I(t) is a sinusoidal function due to constructive and destructive interference of the recombined reflected beams. The maximum intensity occurs when the distance to the moving mirror is the same as that to the fixed mirror (zero path difference, or ZPD). The minimum occurs when the paths differ by $\lambda/4$, where λ is the wavelength of the incident light. (b) When the source contains several (or many) different wavelength components, the resulting I(t) is a Fourier sum of sine waves, which all add constructively at the ZPD point (shown here are three such sine waves, corresponding to a source emitting three different wavelengths of light). The intensity maximum at the ZPD is known as the "centerburst" of the interferogram, and serves to center the interferometer. In experimental studies, the instrument is aligned to maximize the intensity at the centerburst, for optimal performance. If a sample is placed in the beam, its infrared absorption will remove specific wavelengths, or particular sine wave components, from the interferogram. Fourier transformation of the signal from the time domain (I(t)) to the frequency domain (I(1/t) or I(v)) results in the infrared absorption spectrum.

spikes due to absorption of radiation by atmospheric water or CO_2 may also be present. This blank spectrum is stored in the computer. The experiment is repeated with a sample in the beam path. This time, specific wavelengths are absorbed by the infrared active vibrations of the sample, and the resulting interferogram is different. The sample interferogram is Fourier transformed, and ratioed to the stored blank background. The result is an FTIR spectrum, in the spectral range of interest.

With the FTIR technique, the time for a complete scan of the spectral range is determined by the time for the moving mirror to complete one displacement cycle, which is on the order of a few tenths or hundredths of a second. Within an experimental time of several minutes, many hundred scans are averaged, permitting a dramatic improvement in the signal to noise factor. This means that high quality spectra from small sample areas with low volatile concentrations can be obtained in a fraction of the time which would be necessary using a scanning instrument. In addition, the spectra are automatically stored on computer, which greatly facilitates subsequent spectral manipulations. Further, the constant scanning rate associated with the moving mirror and the subsequent time Fourier transform requires an absolute frequency reference, provided by a small laser mounted inside the FTIR bench, which means that measured peak positions are exact.

For IR spectroscopy, samples can be mounted inside the main bench and areas of interest selected by appropriate apertures, or a micro-IR attachment can be used. In an IR microscope, the incident beam is focused on the sample using a parabolic mirror, and the transmitted light is collected using a reflecting optics (Cassegrain) objective. The principal advantages of the micro-IR are that (a) it is easy to visualize the sample using white light optics before the experiment, (b) the focused IR beam allows more efficient sampling and collection of the transmitted beam, and (c) the detector placed after the IR microscope is usually optimized for small sample areas. The signal to noise ratio in the IR experiment is partly determined by the degree of coverage of the active area of the detector by the image of the sample. Detectors used in the main sample bench commonly have large active areas (several tens to hundreds of mm^2), to collect the maximum signal passing through a bulk sample. Detectors designed for IR microscopes have much smaller active areas, to account for the much smaller sample area irradiated.

The source, beamsplitter and detector used in an FTIR analysis experiment depend on the spectral region of interest (Hadni, 1967; McMillan and Hofmeister, 1988). Determination of CO-bearing species is carried out using the C-O stretching vibrations in the 1350 to 2500 cm^{-1} infrared region of the spectrum, and a "standard" mid-IR instrument configuration with a heated ceramic ("Globar") source, KBr beamsplitter and HgCdTe (MCT: the range of the detector is determined by the precise composition of this material) detector is ideal. In experiments in this spectral range, information is also simultaneously obtained on trace (0.01 to 0.5 wt %) amounts of OH species dissolved in the glass, via the absorption at 3500 cm^{-1} which has been calibrated for this purpose (Fine and Stolper, 1985; 1986; Newman et al., 1986; Pan et al., 1991; Pawley et al., 1992; Dixon et al., 1994) (Fig. 17). Determination of hydrous species at higher species concentrations is usually made from the characteristic overtone bands in the near-IR region of the spectrum, between 4000 and 7500 cm^{-1} (Stolper, 1982a,b; Newman et al., 1986; Silver and Stolper, 1989; Silver et al., 1990; Blank et al., 1993) (Fig. 12). Although the Globar source still has appreciable intensity in this spectral region, and can be used, a tungsten-halogen lamp source is much brighter and often gives better results. In principle, a KBr beamsplitter could also still be used in this region, but in practice, the coating (usually Ge) applied to protect the alkali halide gives rise to interference fringes which obscure parts of the spectrum. For this reason, a quartz or CaF_2 beamsplitter is usually used for near-IR analysis of hydrous species. There are several detectors which cover this region, including a broad band MCT detector (which could be the same as that used for the mid-IR determinations), or a PbSe detector (Pawley et al., 1992, 1993). For careful work with low water contents, a liquid-N_2 cooled InSb detector is extremely sensitive in this spectral range (Dixon et al., 1994).

It is useful to note that MCT detectors mounted in micro-FTIR attachments are usually narrow band detectors chosen for especially high sensitivity (with D* values, which are used to compare the relative sensitivity of different detectors, on the order of 10^{10} to 10^{11}, or even higher). These permit high quality results to be obtained on extremely small samples or areas (down to approximately 5 μm: McMillan et al., 1991), with, for example, carbonate concentrations on the order of 100 ppm (Pawley et al., 1992). The advent of infrared absorption studies using synchrotron radiation should also allow the determination of extremely low concentrations of dissolved volatile species, perhaps down to the ppm range (Meade et al., 1994).

In selecting small sample areas, the minimum aperture dimension is generally fixed by the wavenumber range of interest. For example, 1500 cm^{-1} (for carbonate analysis) corresponds to a light wavelength of 6.7 μm, so that sample dimensions down to approximately twice this value can be studied without fringes from diffraction of the infrared beam by the aperture interfering with the spectrum. In practice, even if interference fringes do begin to appear, these form a regular pattern in frequency space, and so correspond to a single point in the FTIR interferogram. This can be removed by computer subtraction before Fourier transformation of the data. Studies in the near-IR region permit even smaller apertures to be selected, on the order of 1 to 2 μm for the first overtone of O-H stretching near 7000 cm^{-1}. The principal limitation in these experiments becomes the intensity of the source radiation and the sensitivity of the detector used.

For quantitative infrared absorbance measurements, doubly polished sections of the volatile-containing glass are prepared, usually by manual polishing. Sample chips are mounted in epoxy (an orthodontic resin works well), and glued to microscope slides using an adhesive soluble in acetone (e.g., Crystalbond 509, supplied by Aremco Products, Inc.) (Newman et al., 1986). Doubly polished sections are prepared by grinding and polishing with alumina and diamond paste or diamond-impregnated paper (water and cutting oil used as lubricants: alcohol tends to soften the resin). It is best to finish with 0.3 μm abrasive: a good surface polish makes it easier to locate features on or within the samples for choice of regions for analysis (a few surface scratches can help in "mapping" the surface). Samples can be stored in air after polishing. Because the volatile content is measured in the bulk sample, traces of adsorbed atmospheric water or CO_2 do not affect the results. However, special care should be taken with samples with a high alkali content (or some germanates), which react substantially with atmospheric water and CO_2. These should be stored in a desiccator or dry box.

The final sample thickness depends on the volatile content and the extinction coefficient of the peaks to be measured. For example, the extinction coefficient for the fundamental O-H stretching vibration lies between 50 and 100 L/cm·mol (Newman et al., 1986). For a sample containing 5 wt % water, the sample thickness would be required to be less than 40 μm to have transmission greater than 1% (absorbance less than 2). For this reason, it is not usually convenient to use the fundamental O-H stretching vibration for routine water determination at high concentrations. The molar absorptivities of the overtone bands of hydrous species lie in the range between 0.1 and 2 L/cm·mol (Table 2). For water determination using these bands, samples ranging in thickness from several tens to several thousand μm (0.1 cm = 1000 μm) are prepared. Concentrations of carbon-oxygen bearing species are usually on the order of 0.1 wt % or less. The extinction coefficients for C-O stretching vibrations lie in the range between 200 and 1000 L/cm·mol (Table 2), and samples studied for CO_2 and CO_3^{2-} determination have thicknesses on the order of 30 to 400 μm (Fine and Stolper, 1985; 1986; Newman et al., 1986; Pan et al., 1991; Pawley et al., 1992; Dixon et al., 1994). The principal problem with analysis of samples containing extremely high carbonate concentrations is that, because of the large extinction coefficients for the characteristic vibrations, the samples must be so thin that thickness measurements can introduce substantial error into the volatile determination. For this reason, it would be useful to explore the application of infrared reflectance spectroscopy (Hadni, 1967; McMillan, 1985; McMillan and Hofmeister, 1988), or quantitative Raman scattering (see below), to the analysis of carbonate rich samples.

Table 2. Compilation of molar absorptivity values for characteristic infrared absorption bands of OH and H_2O species dissolved in silicate glasses. See text for discussion.

OH and H_2O Species

(top value: extinction coefficient ε (L/mol-cm))
(bottom value: integrated molar absorptivity ε^* (L/mol-cm^2))

	1.41 μm (7100 cm⁻¹) OH	1.41 μm (7100 cm⁻¹) H₂O	1.91 μm (5200 cm⁻¹) H₂O	2.22 μm (4500 cm⁻¹) OH	2.53 μm (4000 cm⁻¹) OH	2.53 μm (4000 cm⁻¹) H₂O	2.82 μm (3550 cm⁻¹) OH	2.82 μm (3550 cm⁻¹) H₂O	6.13 μm (1630 cm⁻¹) H₂O	Compositional range
Stolper (1982a)	0.21 / 70	0.21 / 70	1.76 / 300	0.98 / 200	0.95 / 220	0.95 / 220	67 / 32,000	67 / 32,000	- / -	basalt-rhyolite, albite
Newman et al. (1986)	0.320±0.008 / 96	0.184±0.012 / 83	1.61±0.05 / 248±24	1.73±0.02 / 341±25	1.14 / 290	1.07 / 350	100±2 / 44,000±1000	56±4 / 26,300±2200	55±2 / 2640±200	rhyolite
Dixon et al. (1988)	- / -	- / -	- / -	- / -	- / -	- / -	63±5 / -	63±5 / -	- / -	basalt
Dobson et al. (1989)	0.26±0.01 / -	- / -	- / -	- / -	- / -	- / -	88±2 / -	- / -	- / -	rhyolite
Silver and Stolper (1989)	0.20±0.002 / 59±1	0.20±0.002 / 59±1	1.67±0.06 / 268±9	1.13±0.04 / 219±8	- / -	- / -	70±2 / 35,000±500	70±2 / 35,000±500	49±2 / 3081±143	albite (natural and synthetic)
Silver et al. (1990)	Jadeite / Ca-aluminosilicate (E2) / Orthoclase / Anorthite		1.13 / 1.07±0.04 / 1.87±0.07 / 1.50±0.15	1.12 / 0.85±0.03 / 1.43±0.05 / 1.20±0.59	- / -	- / -	- / -	- / -	- / -	orthoclase, jadeite, Ca-aluminosilicates
Ihinger and Stolper (1994)	- / -	- / -	1.86±0.05 / -	1.50±0.05 / -	- / -	- / -	80±4 / -	- / -	- / -	rhyolite
Dixon et al. (1994)	- / -	- / -	0.62±0.07 / -	0.67±0.03 / -	- / -	- / -	- / -	- / -	25±3 / -	basalt
Cocheo (1994)	- / -	- / -	0.56±0.04 / -	0.58±0.02 / -	- / -	- / -	- / -	- / -	- / -	basanite

Table 2, continued

Compilation of molar absorptivity values for characteristic infrared absorption bands of CO_2 and CO_3^{2-} species dissolved in silicate glasses. See text for discussion.

CO_2 and CO_3^{2-} Species

(top value: extinction coefficient ε (L/mol·cm))
(bottom value: integrated molar absorptivity ε^* (L/mol·cm²))

	2350 cm⁻¹ CO_2	1500-1700 cm⁻¹ CO_3^{2-}	1350-1430 cm⁻¹ CO_3^{2-}	
Fine and Stolper (1985)	945±45 25,200±1200	200±15 24,100±1900	235±20 16,800±1500	SiO₂-NaAlSi₂O₆ glasses
Fine and Stolper (1986)		375±20 69,500±3000	375±20 69,500±3000	basalt; Ca-Mg silicates Note: only for CO_2 < 0.8 wt%
Stolper et al. (1987)		199±17 27,300±2300	235±20 16,300±1400	albite
Thibault and Holloway (1994)		355 -	- -	Ca-rich leucitite

After polishing, the sample thickness is carefully measured via micrometric techniques, or by visual determination under a calibrated microscope. The former technique is better, due to the often ragged sample edge (Stolper, 1982a; Fine and Stolper, 1985; 1986; Cocheo, 1994). The measurement error via the micrometric technique is usually stated to be within 2 to 3 μm. It is obvious that this can introduce substantial error for the thinnest samples. In our micro-FTIR experiments, we have also calibrated the microscope stage height to give the sample thickness at the point of measurement, by focusing on the top and bottom of the sample (e.g., McMillan et al., 1991; Pawley et al., 1993).

Before using the technique for quantitative analysis, a set of extinction coefficients for the characteristic species bands of interest, appropriate for the type of sample to be studied, must be established (Stolper, 1982a; Fine and Stolper, 1985; 1986; Newman et al., 1986). These are developed by obtaining a suite of standard samples for which the volatile concentrations have been determined by another, reliable, quantitative method (see the first section of this chapter).

Stolper (1982a) obtained a suite of nineteen natural and synthetic aluminosilicate glasses, with compositions ranging from basaltic to rhyolitic and albitic, and water contents determined by a variety of methods ranging from 0.063 to 6.85 wt %. The densities (ρ) of these samples were determined experimentally, or calculated from previous density measurements on related hydrous glass compositions. Samples were then doubly polished to appropriate thickness, and infrared absorption measurements carried out. The mid-IR peaks at 3550 cm⁻¹ (2.82 μm) was assigned to the fundamental O-H stretching, and the near-IR peaks at 7100 cm⁻¹ (1.41 μm), 5230 cm⁻¹ (1.91 μm) and 4500 cm⁻¹ (2.22 μm) assigned to the first overtone of the O-H stretch (OH + H₂O), the stretch + bend combination of molecular H₂O, and a combination vibration of bound hydroxyl groups (OH), respectively. The assignment of a further band at 3950 cm⁻¹ (2.53 μm) is still not yet completely resolved (see above: also Newman et al., 1986; Pichavant et al., 1992).

Stolper (1982a) noted that the intensity of this band appeared to scale with total water content, rather than those of H_2O or hydroxyl species.

Stolper (1982a) then first plotted the known water content (c) against the absorbance (A), measured as the maximum peak height after background subtraction, divided by the product of the sample thickness (d) and density (ρ), and obtained a straight line. If the molar absorptivity ε for each sample was the same, the slope of this line is equal to ε. The expression used is

$$c = \frac{18.02 \times A}{d\rho\varepsilon} \tag{15}$$

This relates the water content, expressed as the weight per cent H_2O contained within the glass, to the absorbance, the sample thickness in cm, the density in g/L, and the molar absorptivity in L/mol·cm. The value of ε obtained by Stolper (1982a) for the fundamental O-H stretching vibration was 67 L/mol·cm, with a scatter of approximately ±10%. Stolper (1982a) compared this value with previous determinations in the literature, for a range of different (synthetic) glass compositions. Although the value of ε does show considerable dependence on bulk sample composition, much of the variation is for samples with compositions of no direct geologic interest, and the value determined by Stolper (1982a) showed no obvious systematic variation from rhyolitic to basaltic compositions.

The infrared spectra of hydrous binary alkali and alkaline earth silicate glasses show strong features at much lower wavenumber, associated with the presence of strongly hydrogen bonded OH groups (Schölze, 1966; Stolper, 1982a; McMillan and Remmele, 1986). These strongly affect the values of ε determined for these glasses. It is possible that future calibrations of the spectroscopic technique for some highly peralkaline geological compositions might have to take this into account.

For comparison with the molar absorptivities of minerals, glasses and aqueous solution, the ε value determined by Stolper (1982a) must be divided by the appropriate orientation factor ($\gamma = 0.333$) to obtain $\varepsilon_\parallel$ Paterson (1982), then divided by two to account for the fact that Stolper's (1982a) concentration is expressed in terms of H_2O molecules, rather than H atoms. This returns a value of 100 L/mol.cm per H atom for $\varepsilon_\parallel$, consistent with the correlation of Paterson (1982) (see also Rossman, 1988a).

Stolper (1982a) then carried out the same procedure as above to determine the molar absorptivities for the characteristic bands at 1.41, 1.91, 2.22 μm, and for the 2.53 μm band which is not currently used in water determination. The 1.91 μm band is used to determine the concentration of molecular water species dissolved in the glass, and that at 2.22 μm gives the hydroxyl concentration (Fig. 12). The 1.41 μm band is used to return the total water concentration. The values obtained are listed in Table 2.

Stolper (1982a) had noted the potential for error in the "known" water solubility values for his standard samples. Newman et al. (1986) recalibrated the infrared technique, using a suite of natural rhyolitic obsidians which were carefully analyzed "in-house", using vacuum extraction and high precision manometry (see first section of this chapter). The D/H ratio was also investigated via mass spectrometry to discriminate between adsorbed and dissolved water. Substantial errors were found in the stated water content of "known" samples. The revised bulk water contents were used to establish a more reliable set of ε values for rhyolite compositions (Table 2). First, the molar absorptivities for the 1.92 μm (5200 cm^{-1}) and 2.22 μm (4500 cm^{-1}) bands were determined, assuming that these represented the totality of the molecular water and hydroxyl species in the glasses. To demonstrate the self-consistency of the technique, the sum of the OH and H_2O species obtained from the absorbances of these characteristic bands returned the total water content, as measured by bulk extraction (Newman et al., 1986).

The mid-IR spectrum was measured in the region of 1600 cm^{-1}, to obtain an absorption coefficient for the fundamental H_2O bending vibration at 1630 cm^{-1} (6.13 µm). The molar absorptivity for this band was found by searching for a best fit between the measured absorbance of the fundamental and its overtone at 5200 cm^{-1} (Table 2). Newman et al. (1986) noted that the molar absorptivity of the OH stretching fundamental at 3570 cm^{-1} decreased with increasing total water content, and suggested that this might be due to the differing proportion of H_2O molecules and hydroxyl groups contributing to the OH stretch. By modeling the observed absorbance of this band as a sum of contributions from the two species, Newman et al. (1986) obtained different ε values for both OH-containing species (Table 2). Newman et al. (1986) likewise obtained separate molar absorptivities for the OH stretching overtone, and for the band near 4000 cm^{-1}. Dobson et al. (1989) noted that the extractions of Newman et al. (1986) were incomplete in that small amounts of water remained in glassy run products after extraction. Dobson et al. (1989) succeeded in degassing their samples by holding the glassy chips at temperatures just below their softening point for durations up to eight hours. Ihinger and Stolper (unpublished results) have shown that additional water molecules can be trapped in bubbles in water-rich rhyolitic samples, thus further preventing the complete extraction of volatiles for manometric analysis. They have demonstrated that slowly increasing the temperature between 200 and 400°C will eliminate the formation of "frothy" run products and provide complete yields for manometric analysis. Their measurements further refine the ε value for the bands at 3570 cm^{-1}, 4500 cm^{-1}, and 5200 cm^{-1} in high silica rhyolitic glass (Table 2).

Silver and Stolper (1989) obtained additional ε values for natural and synthetic albite samples, but did not distinguish between the hydroxyl and molecular water contributions to the 3570, 4000 and 7100 cm^{-1} bands. In fact, at this point, the technique is now so well established that it is only really necessary to calibrate the 5200 cm^{-1} and 4500 cm^{-1} bands to obtain the H_2O and OH concentrations, and then sum these to obtain the total water content. This was done by Silver et al. (1990), who measured molar absorptivities for these bands for several sodium, potassium and calcium aluminosilicate compositions (Table 2), and found excellent agreement with the total water contents obtained by bulk measurement techniques. At this point, it is useful to note that, although Stolper (1982a), Newman et al. (1986) and Silver and Stolper (1989) have obtained and reported values for the integrated molar absorptivities ε* of the water bands (Table 2), it is only really necessary for reliable water analysis of the samples studied to date to use the measured peak height at the band maximum, and the extinction coefficient ε. The values determined by Silver and Stolper (1989) and Silver et al. (1990) for model compositions serve to give an indication of the effect of changing composition on the molar absorptivities, at high silica content.

Dixon et al. (1988) calibrated the fundamental OH stretching vibration for a series of tholeiitic basalts, and found an ε value of 63 L/mol·cm, close to that obtained by Stolper (1982a). Identical values of 61 L/mol·cm and 60.6±1.8 L/mol·cm were determined for basaltic compositions by Pandya et al. (1992) and Danyushevsky et al. (1993), respectively. This value is lower than that found for the hydroxyl OH stretching vibration for rhyolites (100 L/mol·cm), but is similar to that for the molecular water OH fundamental (56 L/mol·cm) (Newman et al., 1986) (Table 2). At the low water contents studied, the dissolved water is primarily present as hydroxyl groups (Stolper, 1982b; Dixon et al., 1988; 1991; 1994), so that this effect reflects a compositional dependence of ε between rhyolite and basalt, rather than a simple average of the OH and H_2O values obtained by Newman et al. (1986).

Dixon et al. (1994) have investigated the compositional dependence of the molar absorptivities of the 5200 cm^{-1} and 1630 cm^{-1} bands (molecular water) and the 4500 cm^{-1} band (hydroxyl groups) in a series of mid-ocean ridge basalts, and compared these with values for various silicate compositions (Table 2). Cocheo (1994) has recently carried out a similar calibration for a seafloor basanite from Hawaii (PA Cocheo and JR Holloway, in prep), and finds similar values to those for basalt (Table 2). These investigations provide an appropriate set of molar absorptivities for analysis of water content and speciation in basaltic composition glasses.

The molar absorptivities summarized in Table 2 (see also Dixon et al., 1994) also provide an indication of the compositional variation in ε for the different characteristic species vibrations, which should permit an assessment of the confidence to have in the technique when it is applied to compositions for which no calibration is yet available. However, in such cases, the surest way is to simply calibrate the appropriate molar absorptivities for the melt composition of interest, if this does not correspond sufficiently closely to an available calibration.

At this point it is worth mentioning one problem with the application of the technique to work on basalts, or other systems with high iron content. Ferrous iron species in tetrahedral and octahedral coordination give rise to optical absorption bands centered at approximately 9500 and 5500 cm^{-1} (Bell et al., 1976; Goldman and Berg 1980; Rossman, 1988b). These provide a broad baseline underneath the vibrational overtones, which must be subtracted for quantitative analysis (Dixon et al., 1994). This can be achieved either by fitting the baseline spectrum with a sum of Gaussian bands (Dixon et al., 1994), or simply by tracing the form of the baseline with a french curve (Cocheo, 1994; Cocheo and Holloway, in prep). The former method has the advantage that it can be quantified and easily reproduced; however, the second allows for some "artistic impression" in obtaining the best fit. The error introduced by both is likely of the same order, however. Unfortunately, these features can not simply be removed by subtracting the spectrum of a water free sample recorded under the same conditions, because changes occur in the "baseline" spectrum due to removal of the water.

In an extension of the quantitative IR technique to carbon-bearing species, Fine and Stolper (1985) obtained molar absorptivities for the 1350-1430 cm^{-1} and 1600-1660 cm^{-1} bands of CO_3^{2-} species and the 2350 cm^{-1} band of molecular CO_2, dissolved in glasses along the SiO_2-$NaAlSi_2O_6$ join. The carbon content of most of the glasses was assumed to correspond to the nominal concentration loaded into the capsule: the validity of this assumption was verified for four of the samples by analysis via the $^{12}C(d,p_0)^{13}C$ nuclear reaction technique (see earlier). The molar absorptivities obtained for the three characteristic vibrational modes are listed in Table 3. These values have been applied to measurements of the dissolved CO_2 content and speciation in natural obsidians (Newman et al., 1988). In analyses of CO_2 and CO_3^{2-} species, Equation (15) is used, but the molar mass of water (18.02) is replaced by that for carbon dioxide (44.01 g/mol). This then returns the concentration of the carbon-bearing species in wt %, expressed as dissolved CO_2. Because CO_2 concentrations are often very low, this number is usually multiplied by 10^6 to give the CO_2 concentration in ppm.

Fine and Stolper (1986) next calibrated the infrared technique for a suite of natural basaltic glasses, and several synthetic Ca-Mg silicate and aluminosilicate glasses. These authors observed a shift in the higher frequency component of the bending doublet of CO_3^{2-} compared with the sodium aluminosilicate glasses, down to approximately 1515 cm^{-1}. None of the glasses, save one with 1.29 wt % total CO_2 content, contained any observable molecular CO_2, unlike the sodium aluminosilicate series. The molar absorptivity for the carbonate bands was determined to be 375 L/mol·cm, and Fine and Stolper (1986) suggested that this value should only be used for total CO_2 contents ≤ 0.8 wt %, for similar bulk compositions. This extinction coefficient has been used in subsequent studies of CO_2 solubility in basaltic liquids (Stolper and Holloway, 1988; Dixon et al., 1991; Pan et al., 1991; Pawley et al., 1992).

Stolper et al. (1987) undertook a more detailed study of the solubility and speciation of CO_2 dissolved in albitic melt. The principal 2350 cm^{-1} band is asymmetric and can be fit by two Gaussian components. These do not change when the sample is frozen to liquid nitrogen temperature, so these correspond to two different environments for molecular CO_2 groups dissolved in the glass. The symmetric stretching band of $^{13}CO_2$ molecules is clearly visible in their spectrum (Fig. 17), with a molar absorptivity determined to be 11.7±1 L/mol·cm. The carbonate band in the 1600-1700 cm^{-1} region clearly contains two unresolved components (Fig. 17). These correspond to carbonate groups in different chemical environments within the glass, with different degrees of structural perturbation through interaction with the Na$^+$ ions or the aluminosilicate framework. Stolper et al. (1987) obtained a molar absorptivity for this composite band which differed only slightly from that of Fine and Stolper (1985). R. Brooker (unpublished results) has recently carried out an extensive series of studies at Arizona State University, which demonstrate that there are indeed two (and for some compositions, more) carbonate components in these glasses, and that the relative intensities of their contributions to the infrared spectrum change with silica content and metal oxide:alumina ratio. This needs to be further explored in future work, as the different types of carbonate sites will have different molar absorptivity values.

Fogel and Rutherford (1990) applied the extinction coefficient determined by Fine and Stolper (1985) for molecular CO_2 in sodium aluminosilicates to measure the CO_2 content in natural rhyolite. In a similar study, Blank et al. (1993) developed a new value for the extinction coefficient for this band in rhyolite, but this was not published. Thibault and Holloway (1994) have recently examined the solubility of carbon dioxide in a calcium-rich leucitite composition. They determined an ε value of 355 L/mol·cm for the carbonate band at 1515 cm^{-1}, slightly lower than that obtained by Fine and Stolper (1986) for Ca-Mg glasses.

Quantitative Raman spectroscopy. Although Raman scattering spectroscopy has not yet been applied to the quantitative study of volatile species dissolved in silicate melts, it has been used extensively in structural studies of volatile-containing glasses (Fig. 16). In addition, quantitative Raman studies are now an essential tool in the chemical and physical characterization of fluid inclusions, which may represent fluids coexisting with a volatile-bearing silicate melt phase at depth. Micro-Raman spectroscopy provides a rapid, non-destructive method for qualitative and quantitative analysis of species present in the gas and liquid phases of individual fluid inclusions. Usually a rapid scan over characteristic frequency ranges is sufficient to detect the presence or absence of particular molecular species (Dhamelincourt et al., 1979; Bergman and Dubessy, 1984; Pasteris et al., 1988). Along with microthermometry, this is often sufficient to characterize the composition of the inclusion (Dubessy et al., 1989). It is often found that the wavenumber of the characteristic Raman line of a particular molecular species is shifted from the gas phase value, which gives additional useful information on the density or internal pressure within the fluid inclusion (Dubessy et al., 1989; Chou et al., 1990). It is thus useful to review the nature of Raman spectroscopy, for potential future applications to a broader range of problems associated with volatile determination in quenched melts of geological interest.

In a Raman scattering experiment, a beam of laser light is passed through the sample, and the scattered light is analyzed by a spectrometer and sensitive detection system. The energy of the incident light ($E_0 = h\nu_0$) is slightly raised (anti-Stokes) or lowered (Stokes scattering) through an inelastic interaction with the vibrational modes (ν_i) of the sample, and weak satellite peaks (at $\nu_0 \pm \nu_i$) appear at frequencies on either side of the incident laser line (Long, 1977; McMillan, 1985). A Raman signal is observed if the molecular polarizability (α) is changed during the vibration, so that the incident light induces an oscillating induced dipole μ_{ind} in the molecule.

The intensity of Raman scattering induced by an incident beam of irradiance I_0 at the sample is expressed in terms of the Raman scattering cross section σ, defined by

$$\sigma(v) = \frac{\Phi(v)}{NI_o(v)} \tag{16}$$

Φ is the total scattered power (in J s^{-1}), integrated over all scattering directions and polarization states of the radiation, and there are N molecules per unit volume in the region sampled. For a given polarization state of the incident and scattered radiation, the total scattered intensity entering the spectrometer at a given frequency for all molecules in the volume sampled is

$$I_s = \frac{d\Phi}{d\Omega} = NI_o\left(\frac{d\sigma}{d\Omega}\right) \tag{17}$$

where the solid angle Ω is determined by the collection and spectrometer optics. The quantity $d\sigma/d\Omega$ is known as the *differential scattering cross section*. Ignoring polarization effects, the quantum mechanical expression for the Raman scattered intensity for a collection of diatomic molecules with vibrational frequency ν_i at temperature T is

$$I_s = NI_o\frac{d\sigma}{d\Omega} = NI_o\frac{(v_o - v_i)^4}{2\varepsilon_o^2 c^4 m^* v_i(1 - e^{-\frac{hv_i}{kT}})}\left(\frac{d\alpha}{d\Delta r}\right)^2 \tag{18}$$

for Stokes scattering. It is seen that, unlike infrared absorption coefficients, the intensity of Raman scattering depends implicitly upon the temperature and the wavelength (frequency) of excitation, even within the harmonic approximation. For vibrations of polyatomic molecules, or if the polarization properties of the incident and scattered radiation are considered, the term in $(d\alpha/d\Delta r)^2$ in this expression is replaced by appropriate elements of the polarizability tensor (Long, 1977; McMillan, 1985).

Absolute values of the differential Raman scattering cross section for the vibrational modes of a few gas phase molecules have been measured (Schrötter and Klöckner, 1979; Bernstein, 1982; Kiefer, 1982). However, for most cases, differential Raman cross sections are commonly measured and reported relative to a standard for which the absolute cross section is well known, such as N_2 gas (Schrötter and Klöckner, 1979). The resulting *relative normalized differential Raman cross sections* for a range of molecules have been compiled by Schrötter and Klöckner (1979) and Bernstein (1982), and these values are commonly used in fluid inclusion analysis (Dhamelincourt et al., 1979; Bergman and Dubessy, 1984; Pasteris et al., 1988; Dubessy et al., 1989). The relative concentrations of species a and b within an inclusion are obtained by ratioing their Raman intensities:

$$\frac{N_a}{N_b} = \frac{I_s^a}{I_s^b} \frac{\left(d\sigma/d\Omega\right)_b}{\left(d\sigma/d\Omega\right)_a}, \tag{19}$$

in which N_a/N_b represents the relative numbers of species a and b, usually expressed as a percentage or mole fraction in the scattering volume, V. This approach eliminates the need for independent determination of V, which would be required for an absolute determination of species concentration. In a Raman scattering experiment, the scattering volume is a function of both the incident laser focusing optics and the collection optics, which define an approximate cylinder several μm^3 in volume, within which the exciting radiation is concentrated (Pasteris, 1988).

In recent fluid inclusion studies, more reliable values for the relative normalized differential Raman cross sections of different gas phase species have been adopted (Schrötter and Klöckner, 1979; Dubessy, 1989). In addition, it has been recognized that these should be corrected by a frequency- and temperature-dependent factor, arising from the frequency and temperature dependence of the Raman cross section (Dubessy et al., 1989). This results in a correction to the relative differential cross sections for species a and b, $(d\sigma/d\Omega)_a/(d\sigma/d\Omega)_b$, which depends on their characteristic Raman frequencies relative to a particular excitation frequency, at a given temperature. Further, the vibrational modes of species in fluid inclusions, particularly CO_2, are affected by the density and by interaction with solvent molecules, resulting in changes in the Raman cross sections relative to dilute gas phase molecules (Pasteris et al., 1988; Dubessy et al., 1989; Chou et al., 1990).

A successful fluid inclusion analysis depends on the assumption that the instrumental response is linear over the entire wavelength range studied, which is demonstrably not true for commercial Raman instruments. Especially the grating reflectivity and detector efficiency are strongly dependent on wavelength, and also on the polarization state of the light passing through the spectrometer. One method to correct for this is to determine an intensity calibration factor as a function of wavelength, by passing light from a standard calibration lamp through the instrument. The use of instrument-specific "Raman quantification factors" has also been suggested to replace the Raman scattering cross sections for fluid inclusion studies (Wopenka and Pasteris, 1987; Pasteris et al., 1988).

From this discussion, it would be relatively easy to establish a working set of Raman scattering cross sections for various volatile species of interest dissolved in silicate glasses. Small known quantities of a standard such as N_2 or CH_4 could be dissolved in the glass (either as truly dissolved molecular species, or as homogeneously distributed submicroscopic bubbles) to provide an initial internal calibration, then one of the Raman bands of the silicate framework could serve as the standard in later work. Application of this technique could permit quantification of species concentrations not easily measured by infrared spectroscopy, such as carbonate contents in very high carbon content samples, or the species associated with the 900 cm^{-1} band in hydrous aluminosilicate compositions (McMillan et al., 1993).

REFERENCES

Aines RD, Rossman GR (1984) Water in minerals: A peak in the infrared? J Geophys Res 89:4059-4072
Anderson AT, Newman S, Williams SN, Druitt TH, Skirius C, Stolper E (1989) H$_2$O, CO$_2$, Cl and gas in Plinian and ash flow Bishop rhyolite. Geology 17:221-225

Anderson SW, Fink JH (1989) Hydrogen-isotope evidence for extrusion mechanisms of the Mount St Helens lava dome. Nature 341:521-523

Bacon CR, Newman S, Stolper E (1992) Water, CO_2, Cl, and F in melt inclusions in phenocrysts from three Holocene explosive eruptions, Crater Lake, Oregon. Am Mineral 77:1021-1030

Barres O, Burneau A, Dubessy J, Pagel M (1987) Application of micro-FT-IR spectroscopy to individual hydrocarbon fluid inclusions. Appl Spectroscopy 41:1000-1008

Bartholomew DR, Butler BL, Hoover HL, Wu CK (1980) Infrared specta of a water-containing glass. J Am Ceram Soc 63:481-485

Bartholomew RF, Schreurs JWH (1980) Wide-line NMR study of protons in hydrosilicate glasses of different water content. J Non-crystalline Solids 38-39:679-684

Behrens H (1994) Measurement of solubilities of water in melts of albitic ($NaAlSi_3O_8$) and orthoclasic ($KAlSi_3O_8$) compositions. Eur J Mineral, submitted

Bell DR, Rossman GR (1992) Water in Earth's mantle: the role of nominally anhydrous minerals. Science 255:1391-1396

Bell PM, Mao HK, Weeks RA (1976) Optical spectra and electron paramagnetic resonance of lunar and synthetic glasses: a study of the effects of controlled atmosphere, composition and temperature. Proc 7th Lunar Sci Conf:2543-2559

Belton PS (1979) An ^{1}H pulsed N.M.R. study of some amorphous silicate hydrates. J Chem Technol Biotechnol 29:19-25

Bergman SC, Dubessy J (1984) CO_2-CO fluid inclusions in a composite peridotite xenolith for upper mantle oxygen fugacity. Contrib Mineral Petrol 85:1-13

Berman A (1985) Total Pressure Measurements in Vacuum Technology. Academic Press, Inc, Orlando, Florida

Bernstein HJ (1982) Raman intensities of gases In: Vibrational Intensities in Infrared and Raman Spectroscopy, WB Person, G Zerbi (eds), Elsevier, Amsterdam. p 258-265

Bigeleisen J, Perlman ML, Prosser HC (1952) Conversion of hydrogenic materials to hydrogen for isotopic analysis. Anal Chem 24:1356-1357

Blank JG, Delaney JR, Des Marais DJ (1986) Carbon in basaltic glass from the Juan de Fuca Ridge. EOS Trans Am Geophys Union 67:1253

Blank JG, Stolper EM, Carroll MR (1993) Solubilities of carbon dioxide and water in rhyolitic melt at 850°C and 750 bars. Earth Planet Sci Lett 119:27-36

Boulos EN, Kreidl NJ (1972) Water in glass: a review. J Canadian Ceram Soc 41:83-90

Bray PJ, Holupka R (1984) Proton resonance in natural glasses. J Non-Cryst Solids 67:119-126

Brett R, Evans HT, Gibson EK, Hedenquist JW, Wandless M, Sommer MA (1987) Mineralogical studies of sulfide samples and volatile concentrations of basalt glasses from the southern Juan de Fuca Ridge. J Geophys Res 92:11,373-11,379

Brewster D (1826) On the existence of two new fluids in the cavities of minerals, which are immiscible, and possess remarkable physical properties. Trans R Soc Edinburgh 10:1-41

Brey G (1976) CO_2 solubility and solubility mechanisms in silicate melts at high pressures. Contrib Mineral Petrol 57:215-221

Brey G, Green DH (1976) Solubility of CO_2 in olivine melilitite at high pressures and the role of CO_2 in the earth's upper mantle. Contrib Mineral Petrol 55:217-230

Brigham RH, O' Neil JR (1985) Genesis and evolution of water in two-mica pluton: a hydrogen isotope study. Chem Geol 49:159-177

Burns RG, Strens RGJ (1966) Infrared study of the hydroxyl bands in clinoamphiboles. Science 153:890-892

Burnham CW, Jahns RH (1962) A method for determining the solubility of water in silicate melts. Am J Sci 260:721-745

Byers CD, Garcia MO, Muenow DW (1986) Volatiles in basaltic glasses from the East Pacific Rise at 21°N: implications for MORB sources and submarine lava flow morphology. Earth Planet Sci Lett 79:9-20

Carroll MR, Rutherford MJ (1985) Sulfide and sulfate saturation in hydrous silicate melts. J Geophys Res 90:C601-C612

Carroll MR, Stolper EM (1991) Argon solubility and diffusion in silica glass: implications for the solution behavior of molecular gases. Geochim Cosmochim Acta 55:211-225

Chamberlin RT (1908) The gases in rocks. Publication No. 106 Carnegie Inst of Wash Washington, DC, 80pp

Chase B (1987) Fourier transform Raman spectroscopy. Anal Chem 59:881-889

Chaussidon M, Jambon A (1994) Boron content and isotopic composition of oceanic basalts: Geochemical and cosmoshemical implications. Earth Planet Sci Lett 121:277-283

Chaussidon M, Libourel G (1993) Boron partitioning in the upper mantle: An experimental and ion probe study. Geochim Cosmochim Acta 57:5053-5062

Chou I-M, Pasteris JD, Seitz JC (1990) High-density volatiles in the system C-O-H-N for the calibration of a laser Raman microprobe. Geochim Cosmochim Acta 54:535-543

Clark GJ, White CW, Allred DD, Appleton BR, Tsong IST (1978) Hydrogen concentration profiles in quartz determined by a nuclear reaction technique. Phys Chem Minerals 3:199-211

Coleman ML, Shepherd TJ, Durham JJ, Rouse JE , Moore GR (1982) Reduction of water with zinc for hydrogen isotope analysis. Anal Chem 54:993-995

Coplen TB, Wildman JD, Chen J (1991) Improvements in the gaseous hydrogen-water equilibration technique for hydrogen isotope ratio analysis. Anal Chem 63:910-912

Courel P, Trocellier P, Mosbah M (1991) Nuclear reaction microanalysis and analysis of light elements in minerals and glasses. Nucl Instr Methods Phys Res 54B:429-432

Craig H (1953) The geochemistry of the stable carbon isotopes. Geochim Cosmochim Acta 3:53-92

Craig H, Lupton JE (1976) Primordial neon, helium, and hydrogen in oceanic basalts. Earth Planet Sci Lett 31:369-385

Danyushevsky LV, Falloon TJ, Sobolev AV, Crawford AJ, Carroll M, Price RC (1993) The H_2O content of basalt glasses from Southwest Pacific back-arc basins. Earth Planet Sci Lett 117:347-362

Davis AR, Oliver BG (1972) A vibrational spectroscopic study of the species present in the CO_2-H_2O system. J Solution Chem 1:329-339

Davy H (1822) On the state of water and aeriform matter in cavities found in certain crystals. Phil Trans Part II:367-376

Delaney JR, Karsten JL (1981) Ion microprobe studies of water in silicate melts: Concentration-dependent water diffusion in obsidian. Earth Planet Sci Lett 52:191-202

Delaney JR, Muenow DW, Graham DG (1978) Abundance and distribution of water, carbon and sulfur in the glassy rims of submarine pillow basalts, Geochim Cosmochim Acta 42:581-594

Delhaye M, Dhamelincourt P (1975) Raman microprobe and microscope with laser excitation. J Raman Spectroscopy 3:33-43

Deloule E, France-Lanord C, Albarede F (1991a) D/H analysis of minerals by ion probe. In: Taylor HP Jr, O'Neil JR, Kaplan IR (eds) Stable Isotope Geochemistry: A Tribute to Samuel Epstein, Geochemical Society, Special Publication No. 3, 53-62

Deloule E, Albarede F, Sheppard SMF (1991b) Hydrogen isotope heterogeneities in the mantle from ion probe analysis of amphiboles from ultramafic rocks. Earth Planet Sci Lett 105:543-553

Des Marais DJ (1978) Variable temperature cryogenic trap for the separation of gas mixtures. Anal Chem 50:1405-1406

Des Marais DJ , Moore JG (1984) Carbon and its isotopes in mid-oceanic basaltic glasses. Earth Planet Sci Lett 69:43-57

Devine JD, Sigurdsson H, Davis AN (1984) Estimates of sulfur and chlorine yield to the atmosphere from volcanic eruptions and potential climatic effects. J Geophys Res 89:6309-6325

Dhamelincourt P, Bény JM, Dubessy J, Poty B (1979) Analyse d'inclusions fluides à la microsonde MOLE à effet Raman. Bull Minéral 102:600-610

Diels K, Jaeckel R (1966) Leybold Vacuum Handbook. Pergamon Press, Oxford

Dingwell DB, Harris DM, Scarfe CM (1984) The solubility of H_2O in melts in the system SiO_2-Al_2O_3-Na_2O-K_2O at 1 to 2 Kbars. J Geol 92:387-395

Dixon JE, Stolper EM, Delaney JR (1988) Infrared spectroscopic measurements of CO_2 and H_2O glasses in the Juan de Fuca Ridge basaltic glasses. Earth Planet Sci Lett 90:87-104

Dixon JE, Clague DA, Stolper EM (1991) Degassing history of water, sulfur, and carbon in submarine lavas from Kilauea volcano, Hawaii. J Geol 99:371-394

Dixon JE, Stolper EM, Holloway JR (1994) An experimental study of water aand carbon dioxide solubilities in mid-ocean ridge basaltic liquids. J Petrol, in press

Dobson PF, Epstein S, Stolper EM (1989) Hydrogen isotope fractionation between coexisting vapor and silicate glasses and melts at low pressure. Geochim Cosmochim Acta 53:2723-2730

Dobson PF , O'Neil JR (1987) Stable isotope compositions and water contents of boninite series volcanic rocks from Chichi-jima, Bonin islands, Japan. Earth Planet Sci Lett 82:75-86

Dodd DM , Fraser DB (1966) Optical determinations of OH in fused silica. J Appl Phys 37:3911

Dubessy J, Poty B, Ramboz C (1989) Advances in C-O-H-N-S fluid geochemistry based on micro-Raman spectrometric analysis of fluid inclusions. Eur J Mineral 1:517-534

Dubois A-D (1985) An accurate zinc charcoal reduction system for D/H measurements of water and cellulose. Bull Soc Belg Geol 94:113-115

Dunbar NW, Hervig RL (1992a) Petrogenesis and volatile stratigraphy of the Bishop Tuff: evidence from melt inclusion analysis. J Geophys Res 97:15129-15150

Dunbar NW, Hervig RL (1992b) Volatile and trace element composition of melt inclusions from the Lower Bandelier Tuff: Implications for magma chamber processes and eruptive style. J Geophys Res 97:15151-15170

Dushman S (1962) Scientific Foundations of Vacuum Technique. John Wiley and Sons, Inc, New York

Eckert H, Yesinowski JP, Stolper EM, Stanton TR, Holloway J (1987) The state of water in rhyolitic glasses: a deuterium NMR study. J Non-crystalline Solids 93:93-114

Eckert H, Yesinowski JP, Silver LA, Stolper EM (1988) Water in silicate glasses: quantitation and structural studies by ^{1}H solid echo and MAS-NMR methods. J Phys Chem 92:2055-2064

Eggler DH, Mysen B, Hoering TC, Holloway JR (1979) The solubility of carbon monoxide in silicate melts at high pressures and its effect on silicate phase relations. Earth Planet Sci Lett 43:321-330

Eldridge CS, Compston W, Williams IS , Walshe JL (1989) Sulfur isotope analyses on the SHRIMP ion microprobe. In: Shanks WS, Criss RE (eds) New frontiers in stable isotopic research: Laser probes, ion probes, and small sample analysis. US Geol Surv Bull 1890:163-174

Epstein S, Taylor HP (1970) The concentration and isotopic composition of hydrogen, carbon and silicon in Apollo 11 lunar rocks and minerals. Geochim Cosmochim Acta, Supp 1, 2:1085-1096

Ernsberger FM (1977) Molecular water in glass. J Am Ceram Soc 60:91-92

Farnan I, Kohn SC, Dupree R (1987) A study of the structural role of water in hydrous silica glass using cross-polarization magic angle spinning NMR. Geochim Cosmochim Acta 51:2869-2873

Fine GJ, Stolper EM (1985) The speciation of carbon dioxide in sodium aluminosilicate glasses. Contrib Mineral Petrol 91:105-121

Fine GJ, Stolper E (1986) Carbon dioxide in basaltic glasses: concentrations and speciation. Earth and Planet Sci Lett 76:263-278

Fine G, Stolper E, Mendenhall MH, Livi RP, Tombrello TA (1985) Measurement of the carbon content of silicate glasses using the $^{12}C(d,P_0)^{13}C$ nuclear reaction In: Microbeam Analysis-1985 San Francisco Press 241-245

Fischer K (1935) A new method for the analytical determination of the water content of liquids and solids. Angew Chem 48:394-396

Fogel RA, Rutherford MJ (1990) The solubility of carbon dioxide in rhyolitic melts: A quantitative FTIR study. Am Mineral 75:1311-1326

Fraser DG, Rammensee W (1987) Determination of the mixing properties of granitic and other aluminosilicate melts by Knudsen Cell Mass Spectrometry. In: Mysen BO (ed) Magmatic Processes: Physicochemical Principles The Geochemical Society Special Publication No. 1 401-410

Friedman I (1953) Deuterium content of natural water and other substances. Geochim Cosmochim Acta 4:89-97

Friedman I , Smith RL (1958) The deuterium content of water in some volcanic glasses. Geochim Cosmochim Acta 15:218-228

Garcia MO, Muenow DW, Aggrey KE, O'Neil JR (1989) Major element, volatile, and stable isotope geochemistry of Hawaiian submarine tholeiitic glasses. J Geophys Res 94:10525-10538

Garrabos Y, Tufeu R, Le Neindre B, Zalczer G, Beysens D (1980) Rayleigh and Raman scattering near the critical point of carbon dioxide. J Chem Phys 72:4637-4651

Godfrey JD (1962) The deuteruim content of hydrous minerals from the East-Central Sierra Nevada and Yosemite National Park. Geochim Cosmochim Acta 26:1215-1245

Goldman DS, Berg JI (1980) Spectral study of ferrous iron in Ca-Al-borosilicate glass at room and melt temperatures. J Non-Crystalline Solids 38&39:183-188

Goranson R (1931) The solubility of water in granite magmas. Am J Sci 22:481-502

Goranson RW (1936) Silicate-water systems: the solubility of water in albite-melt. Trans Am Geophys Union 17:257-259

Griffiths PR, de Haseth JA (1986) Fourier Transform Infrared Spectroscopy. John Wiley and Sons, New York.

Grimley RT (1967) Mass spectrometry In: Margrave, JL (ed) The Characterization of High Temperature Vapors Wiley 195-243

Hadni A (1967) Essentials of Modern Physics Applied to the Study of the Infrared. Pergamon Press, Oxford

Hamilton DL, Burnham CW, Osborn EF (1964) The solubility of water and effects of oxygen fugacity and water content on crystallization in mafic magmas. J Petrol 5:21-39

Hamilton DL, Oxtoby S (1986) Solubility of water in albite-melt determined by the weight-loss method. J Geol 94:626-630

Hanson RC, Bachman K (1980) Raman studies of pressure tuning of the Fermi resonance in solid CO_2. Chem Phys Lett 73:338-342

Hanson RC, Jones LH (1981) Infrared and Raman studies of pressure effects on the vibrational modes of solid CO_2. J Chem Phys 75:1102-1112

Harris DM (1981) The microdetermination of H_2O, CO_2, and SO_2 in glass using a 1280°C microscope vacuum heating stage, cryopumping, and vapor pressure measurements from 77 to 273 K. Geochim Cosmochim Acta 45:2023-2036

Harris DM, Anderson AT (1984) Volatiles H_2O, CO_2, and Cl in a subduction related basalt. Contrib Mineral Petrol 87:120-128

Harrison AJ (1947) Water content and infrared transmission of simple glasses. Am Ceram Soc J 30:362-366

Hervig RL (1992) Ion probe microanalyses for volatile elements in melt inclusions. EOS Trans Am Geophys Union 73:367

Hervig RL, Dunbar N, Westrich HR, Kyle P (1989) Pre-eruptive water content of rhyolitic magmas as determined by ion microprobe analyses of melt inclusions in phenocrysts. J Volcan Geotherm Res 36:293-302

Hervig RL, Williams P (1988) SIMS microanalyses of minerals and glasses for H and D. In: Benninghoven A, Huber AM, Werner HW (eds) Secondary Ion Mass Spectrometry, SIMS VI. J. Wiley & Sons, New York, 961-964

Hervig RL, Williams P, Thomas RM, Schauer SN, Steele IM (1992) Microanalyses of oxygen isotopes in insulators by secondary ion mass spectrometry. Int J Mass Spec Ion Proc 120:45-63

Herzberg G (1945) Molecular Spectra and Molecular Structure II Infrared and Raman Spectra of Polyatomic Molecules. Van Nostrand Reinhold, New York

Herzberg G (1950) Molecular Spectra and Molecular Structure I Spectra of Diatomic Molecules. Van Nostrand-Reinhold, New York

Hinthorne JR, Anderson CA (1975) Microanalysis of fluorine and hydrogen in silicates with the ion microprobe mass analyzer. Am Mineral 60:143-147

Hinton RW (1990) Ion microprobe trace-element analysis of silicates: Measurement of multi-element glasses. Chem Geol 83:11-25

Holtz F, Behrens H, Dingwell DB, Johannes W (1994) Water solubility in haplogranitic melts: Compositional, pressure, and temperature dependence. Am Mineral, in press

Houston JE, Madey TE (1982) Core-level processes in the electron-stimulated desorption of CO from the W(110) surface. Phys Rev B26:554-566

Howard-Lock HE, Stoicheff BP (1971) Raman intensity measurements of the Fermi diad v_1, $2v_2$, in $^{12}CO_2$ and $^{13}CO_2$. J Mol Spectroscopy 37:321-326

Huang TC, Fung A, White RL (1989) Recent measurements of long-wavelength X-rays using synthetic multilayers. X-ray Spectrometry 18:53-56

Huber KP, Herzberg G (1979) Molecular Spectra and Molecular Structure IV Constants of Diatomic Molecules. Van Nostrand Reinhold, New York

Ihinger PD (1991) An experimental study of the interaction of water with granitic melt. PhD thesis, California Institute of Technology

Jones AP, Smith JV (1984) Ion probe analysis of H, Li, B, F, and Ba in micas with additional data for metamorphic amphibole, scapolite, and pyroxene. N Jb Mineral Mh 1984:228-240

Kadik AA, Lebedev TB (1968) Temperature dependence of the solubility of water in albite melt at high pressures. Geochem Int 5:1172-1181

Karsten JL, Holloway JR, Delaney JR (1982) Ion microprobe studies of water in silicate melts: temperature-dependent water diffusion in obsidian. Earth Planet Sci Lett 59:420-428

Keller WD, Pickett EE (1954) Hydroxyl and water in perlite from Superior, Arizona. Am J Sci 252:87-98

Kendall C, Coplen TB (1985) Multisample conversion of water to hydrogen by zinc for stable isotope determination. Anal Chem 57:1437-1440

Kennedy GC (1950) A portion of the system silica-water. Econ Geol 45:629-653

Khitarov NI, Lebedev EB, Rengarten EV (1959) The solubility of water in basaltic and granitic melts. Geokhimiya 5:479-492

Kiefer W (1982) Experimental techniques in Raman spectroscopy In: Person WB, Zerbi G (eds) Vibrational Intensities in Infrared and Raman Spectroscopy Amsterdam, Elsevier 239-257

Killingley JS, Muenow DW (1975) Volatiles from Hawaiian submarine basalts determined by dynamic high temperature mass spectrometry. Geochim Cosmochim Acta 39:1467-1473

Kirkpatrick, RJ (1988) MAS NMR spectroscopy of minerals and glasses. In: Spectroscopic Methods in Mineralogy and Geology, FC Hawthorne (ed), Rev Mineral 18:341-403

Knudsen M (1909) Die Molekularströmung der Gase durch Offnungen unde die Effusion. Ann Phys 28:999-1016

Kohn SC, Dupree R, Golam Mortuza M (1992) The interaction between water and aluminosilicate magmas. Chem Geol 96:399-409

Kohn SC, Dupree R, Smith ME (1989a) A multinuclear magnetic resonance study of the structure of hydrous albite glasses. Geochim Cosmochim Acta 53:2925-2935

Kohn SC, Dupree R, Smith ME (1989b) ^1H NMR studies of hydrous silicate glasses. Nature 337:539-541

Kohn SC, Brooker RA, Dupree R (1991) ^{13}C MAS NMR: a method for studying CO_2 speciation in glasses. Geochim Cosmochim Acta 55:3879-3884

Konijnendijk WL, Buster JHJM (1977) Raman-scattering measurements of silicate glasses containing sulphate. J Non-Crystalline Solids 23:401-418

Krom MD, Berner RA (1983) A rapid method for the determination of organic and carbonate carbon in geological samples. J Sed Petrol 53:660-663

Kyser TK, O'Neil JR (1984) Hydrogen isotope systematics of submarine basalts. Geochim Cosmochim Acta 48:2123-2133

Lee RR, Leich DA, Tombrello TA, Ericson JE, Friedman I (1974) Obsidian hydration profile measurements using a nuclear reaction technique. Nature 250:44-47

Leshin LA, Stolper EM, Eckert H (1990) Water in alkali silicate glasses: A MAS NMR and FTIR study. VM Goldschmidt Conf Progr Abstr 2:61

London D, Hervig RL, Morgan VI GB (1988) Melt-vapor solubilities and elemental partitioning in peraluminous granite-pegmatite systems: experimental results with Macusani glass at 200 MPa. Contrib Mineral Petrol 99:360-373

Long DA (1977) Raman Spectroscopy. McGraw-Hill, New York

Lowenstern JB (1993) Evidence for a copper-bearing fluid in magma erupted at the Valley of Ten Thousand Smokes, Alaska. Contrib Mineral Petrol 114:409-421

Luth RW (1988a) Effects of F on phase equilibria and liquid structure in the system $NaAlSiO_4$-$CaMgSi_2O_6$-SiO_2. Am Mineral 73:306-312

Luth RW (1988b) Raman spectroscopic study of the solubility mechanisms of F in glasses in the system CaO-CaF_2-SiO_2. Am Mineral 73:297-305

Madey TE, Yates JTJ, King DA, Uhlaner CJ (1970) Isotope effect in electron stimulated desorption: oxygen chemisorbed on tungsten. J Chem Phys 52:5213-5220

Martin LH, Hill RD (1948) A Manual of Vacuum Practice. Cambridge University Press, London

Masi U, O'Neil JR, Kistler RW (1981) Stable isotope systematics in Mesozoic granites of central and northern California and southwestern Oregon. Contrib Mineral Petrol 76:116-126

Mathez EA, Blacic JD, Beery J, Maggiore C, Hollander M (1984) Carbon abundances in mantle minerals determined by nuclear reaction analysis. Geophys Res Lett 11:947-950

Mathez EA, Delaney JR (1981) The nature and distribution of carbon in submarine basalts and peridotite nodules. Earth Planet Sci Lett 56:217-232

Mattey DP, Carr RH, Wright IP, Pillinger CT (1984) Carbon isotopes in submarine basalts. Earth Planet Sci Lett 70:196-206

McGee JJ, Slack JF, Herrington CR (1991) Boron analysis by electron microprobe using MoB_4C layered synthetic crystals. Am Mineral 76:681-684

McGuire AV, Francis CA, Dyar MD (1992) Mineral standards for electron microprobe analysis of oxygen. Am Mineral 77:1087-1091

McMillan PF (1985) Vibrational spectroscopy in the mineral sciences. In: Microscopic to Macroscopic: Atomic Environments to Mineral Thermodynamics. SW Kieffer and A Navrotsky (eds) Rev Mineral 14:9-63

McMillan P, Remmele RL (1986) Hydroxyl sites in SiO_2 glass: a note on infrared and Raman spectra. Am Mineral 71:772-778

McMillan PF, Hofmeister AM (1988) Infrared and Raman spectroscopy. In: Spectroscopic Methods in Mineralogy and Geology, FC Hawthorne (ed), Rev Mineral 8:99-159

McMillan PF, Jakobsson S, Holloway JR, Silver LA (1983) A note on the Raman spectra of water-bearing albite glasses. Geochim Cosmochim Acta 47:1937-1943

McMillan P, Akaogi M, Sato R, Poe B, Foley J (1991) Hydroxyl groups in β-Mg_2SiO_4. Am Mineral 76:354-360

McMillan PF, Stanton TR, Poe BT, Remmele RR (1993) A Raman spectroscopic study of H/D isotopically substituted hydrous aluminosilicate glasses. Phys Chem Minerals 19:454-459

McMillan PF, Dubessy J, Hemley RJ (1994) Applications to earth science and environment. In: Raman Microscopy, J. Corset (ed), Academic Press, to be published

McMillan PW, Chlebik A (1980) The effect of hydroxyl ion content on the mechanical and other properties of soda-lime-silica glass. J Non-cryst Solids 38:509-514

Meade C, Reffner JA, Ito E (1994) Synchrotron infrared absorbance of hydrogen in $MgSiO_3$ perovskite. Science 264:1558-1560

Metrich N, Clocchiatti R, Mosbah M (1993) The 1989-1990 activity of Etna magma mingling and ascent of H_2O-Cl-S-rich basaltic magma evidence from melt inclusions. J Volcan Geotherm Res 59:131

Moore CB, Gibson EK, Larimer JW, Lewis CF, Nichiporuk W (1970) Total carbon and nitrogen abundances in Apollo 11 lunar samples and selected chondrites and basalts. Geochim Cosmochim Acta Suppl 1:431

Moore JG (1970) Water content of basalt erupted on the ocean floor. Contrib Mineral Petrol 28:272-279

Mosbah M, Bastoul A, Cuney M (1993) Nuclear microprobe analysis of ^{14}N and its application to the study of ammonium-bearing minerals. Nucl Inst Methods Phys Res 77:450

Mosbah M, Clocchiatti R, Tirira J (1991) Study of hydrogen in melt inclusions trapped in quartz with a nuclear microprobe. Nucl Inst Methods Phys Res 54:298

Mosbah MN (1991) PIGME fluorine determination using a nuclear microprobe with application to glass inclusions. Nucl Inst Methods Phys Res 58:227

Moulson AJ, Roberts JP (1960) Water in silica glass. Trans Brit Ceram Soc 59:388-399

Mysen BO (1976) The role of volatiles in silicate melts: solubility of carbon dioxide and water in feldspar, pyroxene, and feldspathoid melts to 30 kb and 1625°C Am J Sci 276:969-996

Mysen BO, Virgo D (1980) Solubility mechanisms of carbon dioxide in silicate melts: a Raman spectroscopic study. Am Mineral 65:885-899

Mysen BO, Virgo D (1980) Solubility mechanisms of water in basalt melt at high pressures and temperatures: NaCa $AlSi_2O_7$-H_2O as a model. Am Mineral 65:1176-1184

Mysen BO, Virgo D (1985) Interaction between fluorine and silica in quenched melts on the joins SiO_2-AlF_3 and SiO_2-NaF determined by Raman spectroscopy. Phys Chem Minerals 12:77-85

Mysen BO, Virgo D (1986a) Volatiles in silicate melts at high pressure and temperature: 1 Interaction between OH groups and Si^{4+}, Al^{3+}, Ca^{2+}, Na^+, and H^+. Chem Geol 57:303-331

Mysen BO, Virgo D (1986b) Volatiles in silicate melts at high pressure and temperature. 2 Water in melts along the join $NaAlO_2$-SiO_2 and a comparison of solubility mechanisms of water and fluorine. Chem Geol 57:333-358

Mysen BO, Eggler DH, Seitz MG, Holloway JR (1976) Carbon dioxide in silicate melts and crystals. Part I solubility measurements. Am J Sci 276:455-479

Mysen BO, Virgo B, Harrison WJ, Scarfe CM (1980) Solubility mechanisms of H_2O in silicate melts at high pressures and temperatures: A Raman spectroscopic study. Am Mineral 65:900-914

Nabelek PI, O'Neil JR, Papike JJ (1983) Vapor phase exolution as a controlling factor in hydrogen isotope variation in granitic rocks: the Notch Peak granitic stock, Utah. Earth Planet Sci Lett 66:137-150

Nash WP (1992) Analysis of oxygen with the electron microprobe: Applications to hydrated glass and minerals. Am Mineral 77:453-456

Newman S, Stolper EM, Epstein S (1986) Measurement of water in rhyolitic glasses: calibration of an infrared spectroscopic technique. Am Mineral 71:1527-1541

Newman S, Epstein S, Stolper EM (1988) Water, carbon dioxide, and hydrogen isotopes in glasses from the ca 1340 AD eruption of the Mono Craters, California: Constraints on degassing phenomena and initial volatile content. J Volcan Geotherm Res 35:75-96

Nicolosi JA, Broven JP, Merlo D, Jenkins R (1986) Layered synthetic microstructures for long wavelength X-ray spectrometry. Optical Eng 25:964-969

O'Neil JR, Chappell BW (1977) Oxygen and hydrogen isotope relations in the Berridale batholith. J Geol Soc Lond 133:559-571

O'Neil JR, Shaw SE, Flood RH (1977) Oxygen and hydrogen isotope compositions as indicators of granite genesis in the New England Batholith, Australia. Contrib Mineral Petrol 62:313-328

Orlova GP (1964) Solubility of water in albite melts-under pressure. Int Geol Rev 6:254-258

Ostrovskiy IA, Orlova GP, Rudnitskaya YS (1964) Stoichiometry in the solution of water in alkali-aluminosilicate melts. Doklady Akad Nauk SSR 157:149-151

Ottolini L, Botazzi P, Vannucci R (1993) Quantification of lithium, beryllium, and boron in silicates by secondary ion mass spectrometry using conventional energy filtering. Analyt Chem 65:1960-1964

Ottolini L, Botazzi P, Zanetti A (1994) Quantitative analysis of hydrogen, fluorine, and chlorine in silicates using energy filtering. In: SIMS IX, Proc Ninth Int Conf Secondary Ion Mass Spectrometry, in press

Oxtoby S, Hamilton DL (1978a) The discrete association of water with Na_2O and SiO_2 in NaAl silicate melts. Contrib Mineral Petrol 66:185-188

Oxtoby S, Hamilton DL (1978b) Solubility of water in melts of Na_2O-Al_2O_3-SiO_2 and K_2O-Al_2O_3-SiO_2 systems. In: WS Mackenzie (ed) Progress in Experimental Petrology, vol 4, Natural Environment Research Council, Manchester, England. p 33-36

Oxtoby S, Hamilton DL (1978c) Water in plagioclase melts. In: WS Mackenzie (ed) Progress in Experimental Petrology, vol 4, Natural Environment Research Council, Manchester, England. p 36-37

Oxtoby S, Hamilton DL (1978d) Calculation of the solubility of water in granitic melts. In: WS Mackenzie (ed) Progress in Experimental Petrology, vol 4, Natural Environment Research Council, Manchester, England. p 37-40

Paillat O, Elphick SC, Brown WL (1992) The solubility of water in $NaAlSi_3O_8$ melts: a re-examination of Ab-H_2O phase relationships and critical behavior at high pressures. Contrib Mineral Petrol 112:490-500

Palais JM, Sigurdsson H (1989) Petrological evidence of volatile emissions from major historic and pre-historic volcanic eruptions. In: Geophysical Monographs No. 15: Contribution of Geophysics to Climate Change Studies. Am Geophys Union, Washington DC, p 31-53

Pan V, Holloway JR, Hervig RL (1991) The pressure and temperature dependence of carbon dioxide solubility in tholeiitic basalt melts. Geochim Cosmochim Acta 55:1587-1595

Pandya N, Muenow DW, Sharma SK (1992) The effect of bulk composition on the speciation of water in submarine volcanic glasses. Geochim Cosmochim Acta 56:1875-1883

Pasteris JD, Wopenka B, Seitz JC (1988) Practical aspects of quantitative laser Raman microprobe spectroscopy for the study of fluid inclusions. Geochim Cosmochim Acta 52:979-988

Paterson MS (1982) The determination of hydroxyl by infrared absorption in quartz, silicate glasses and similar materials. Bull Minéral 105:20-29

Pawley AR, Holloway JR, McMillan PF (1992) The effect of oxygen fugacity on the solubility of carbon-oxygen fluids in basaltic melt. Earth Planet Sci Lett 110:213-225

Pawley AR, McMillan PF, Holloway JR (1993) Hydrogen in stishovite, with implications for mantle water content. Science 261:1024-1026

Pearson AD, Pasteur GA, Northover WR (1979) Determination of the absorptivity of OH in a sodium borosilicate glass. J Mat Sci 14:869-872

Penfield SL (1894) On some methods for the determination of water. Am J Sci 48:30-37

Pichavant M (1981) An experimental study of the effect of boron on a water-saturated haplogranite at 1 kbar pressure: geological applications. Contrib Mineral Petrol 76:430-439

Pichavant M, Holtz F, McMillan P (1992) Phase relations and compositional dependence of H_2O solubility in quartz-feldspar melts. Chem Geol 96:303-319

Pineau F, Javoy M (1983) Carbon isotopes and concentrations in mid-oceanic ridge basalts. Earth Planet Sci Lett 62:239-257

Poreda R (1985) Helium-3 and deuterium in back-arc basalts: Lau-Basin and the Mariana Trough. Earth Planet Sci Lett 73:244-254

Poreda R, Schilling J-G, Craig H (1986) Helium and hydrogen isotopes in ocean-ridge basalts north and south of Iceland. Earth Planet Sci Lett 78:1-17

Potts PJ, Tindle AG (1989) Analytical characteristics of a multilayer dispersion element ($2d = 60$Å) in the determination of fluorine in minerals by electron microprobe. Mineral Mag 53:357-362

Rai CS, Sharma SK, Muenow DW, Matson DW, Byers CD (1983) Temperature dependence of CO_2 solubility in high pressure quenched glasses of diopside composition. Geochim Cosmochim Acta 47:953-958

Redhead PA, Hobson JP, Kornelsen EV (1968) The Physical Basis of Ultrahigh Vacuum. Chapman and Hall Ltd, London

Richet P, Roux J, Pineau F (1986) Hydrogen isotope fractionation in the system H_2O-liquid $NaAlSi_3O_8$: New data and comments on D/H fractionation in hydrothermal experiments. Earth Planet Sci Lett 78:115-120

Roedder E (1984) Fluid Inclusions. Rev Mineral 12, 644 p

Rosasco CJ, Roedder E (1979) Application of a new Raman microprobe spectrometer to non-destructive analysis of sulphate and other ions in individual phases in fluid inclusions in minerals. Geochim Cosmochim Acta 43:1907-1915

Rosasco GJ, Roedder E , Simmons JH (1975) Laser-excited Raman spectroscopy for non-destructive partial analysis of individual phases in fluid inclusions in minerals. Science 190:557-560

Rosasco GJ, Simmons JH (1974) Investigation of gas content of inclusions in glass by Raman scattering spectroscopy. Am Ceram Soc Bull 53:626-630

Roselieb K, Rammensee W, Büttner H, Rosenhauer M (1992) Solubility and diffusion of noble gases in vitreous albite. Chem Geol 96:241-266

Rossman GR (1988a) Vibrational spectroscopy of hydrous components. In: Spectroscopic Methods in Mineralogy and Geology, FC Hawthorne (ed), Rev Mineral 18:193-206

Rossman GR (1988b) Optical spectroscopy. In: Spectroscopic Methods in Mineralogy and Geology, FC Hawthorne (ed), Rev Mineral 18:207-243

Rovetta MR, Blacic JD, Hervig RL, Holloway JR (1989) An experimental study of hydroxyl in quartz using infrared spectroscopy and ion microprobe techniques. J Geophys Res 94:5840-5850

Russell LE (1957) Solubility of water in molten glass. J Soc Glass Technology 41:304-317T

Ryskin YI (1974) Vibrations of protons in minerals: hydroxyl, water and ammonium. In: The Infrared Spectra of Minerals, VC Farmer (ed) London, Mineral Soc 137-181

Sakai H, Smith JW, Kaplan IR, Petrowski C (1971) Microdeterminations of C, N, S, H, He, metallic Fe, $\delta^{13}C$, $\delta^{15}N$, and $\delta^{34}S$ in geologic samples. Geochem J 10:85-96

Schiegl WE, Vogel JC (1970) Deuterium content of organic matter. Earth Planet Sci Lett 7:307-313

Schölze H (1960) Zur frage der unterscheidung zwischen H_2O-moleculen und OH-Gruppen in Gläsern und Mineralen Naturwiss 47:226-227

Schölze H (1966) Gases and water in glass: Parts I, II, III. The Glass Industry 47:546-551; 622-628; 670-674

Schrötter HW, Klöckner HW (1979) Raman scattering cross sections in gases and liquids In: Raman Spectroscopy of Gases and Liquids, A Weber (ed), Springer-Verlag, Berlin. p 123-166

Sharma SK (1979) Structure and solubility of carbon dioxide in silicate glasses of diopside and sodium melilite composition at high pressures from Raman spectroscopic data. Carnegie Inst Wash Yearb 78:532-537

Sharma SK, Hoering TC, Yoder HS (1979) Quenched melts of akermanite compositions with and without CO_2 - characterization by Raman spectroscopy and gas chromatography. Carnegie Inst Wash Yearb 78:537-542

Sharp Z (1990) A laser-based microanalytical method for the *in situ* determination of oxygen isotope ratios of silicates and oxides. Geochim Cosmochim Acta 54:1353-1357

Shelby JE (1987) Quantitative determination of the deuteroxyl content of vitreous silica. Commun Am Ceram Soc 70:C-9 - C-10

Shepherd ES (1938) The gases in rocks and some related problems. Am J Sci 35-A:311-351

Shimizu N (1986) Silicon-induced enhancement in secondary ion emission from silicates. Int'l J Mass Spec Ion Proc 69:325-338

Silver LA, Stolper EM (1989) Water in albitic glasses. J Petrol 30:667-709

Silver LA, Ihinger PD, Stolper EM (1990) The influence of bulk composition on the speciation of water in silicate glasses. Contrib Mineral Petrol 104:142-162

Sisson TW, Layne GD (1993) H_2O in basalt and basaltic andesite glass inclusions from four subduction-related volcanoes. Earth Planet Sci Lett 117:619-635

Skirius CM, Peterson JW, Anderson AT (1990) Homogenizing rhyolitic inclusions from the Bishop Tuff. Am Mineral 75:1381-1398

Sorby HC, Butler PJ (1869) On the structure of rubies, sapphires, diamonds, and some other minerals. Proc R Soc 17:291-303

Stanton TR (1989) High pressure isotopic studies of the water diffusion mechanism in silicate melts and glasses. PhD Dissertation, Arizona State University.

Stanton TR, Holloway JR, Hervig RL, Stolper EM (1985) Isotope effect on water diffusivity in silicic melts: an ion microprobe and infrared analysis. Eos Trans Am Geophys Union 66:1131

Stebbins J (1988) NMR spectroscopy and dynamic processes in mineralogy and geochemistry. In: Spectroscopic Methods in Mineralogy and Geology, FC Hawthorne (ed), Rev Mineral 18:405-429

Steele D (1971) Theory of Vibrational Spectroscopy. W B Saunders, Philadelphia

Steele IM (1986) Ion probe determination of hydrogen in geologic samples. N Jb Mineral Mh 1986:193-202

Stewart DB (1958) The system $CaAl_2Si_2O_8-SiO_2-H_2O$. Geol Soc Am Bull 69:1648

Stolen RH, Walrafen GE (1976) Water and its relation to broken bond defects in fused silica. J Chem Phys 64:2623-2631

Stolper EM (1982a) Water in silicate glasses: an infrared spectroscopic study. Contrib Mineral Petrol 81:1-17

Stolper E (1982b) The speciation of water in silicate melts. Geochim Cosmochim Acta 46:2609-2620

Stolper EM (1989) Temperature dependence of the speciation of water in rhyolitic melts and glasses. Am Mineral 74:1247-1257

Stolper EM, Holloway JR (1988) Experimental determination of the solubility of carbon dioxide in molten basalt at low pressure. Earth Planet Sci Lett 87:397-408

Stolper E, Fine G, Johnson T, Newman S (1987) Solubility of carbon dioxide in albitic melt. Am Mineral 72:1071-1085

Stolper EM, Newman S (1994) The role of water in the petrogenesis of Mariana trough magmas. Earth Planet Sci Lett, in press

Stone J, Walrafen GE (1982) Overtone vibrations of OH groups in fused silica optical fibers. J Chem Phys 76:1712-1722

Strens RG (1966) Infrared study of cation ordering and clustering in some (Fe,Mg) amphibole solid solutions. Chem Commun 15:519-520

Sykes D, Kubicki JD (1993) A model for H_2O solubility mechanisms in albite melts from infrared spectroscopy and molecular orbital calculations. Geochim. Cosmochim. Acta 57:1039-1052

Takata M, Acocella J, Tomozawa M, Watson EB (1981) Effect of water content on the electrical conductivity of $Na_2O \cdot 3SiO_2$ glass. J Am Ceram Soc 64:719-724

Tanweer A, Hut G, Burgman JO (1988) Optimal conditions for the reduction of water to hydrogen by zinc for mass spectrometric analysis of the deuterium content. Chem Geol 73:199-203

Taylor BE, Eichelberger JC, Westrich HR (1983) Hydrogen isotopic evidence of rhyolitic magma degassing during shallow intrusion and eruption. Nature 306:541-545

Taylor BE, Westrich HR (1985) Hydrogen isotope exchange and water solubility in experiments using natural rhyolite obsidian. EOS Trans Am Geophys Union 66:387

Taylor W (1990) The dissolution mechanism of CO_2 in aluminosilicate melts - infrared spectroscopic constraints on the cationic environment of dissolved $[CO_3]^{2-}$. Eur J Mineral 2:547-563

Thibault Y, Holloway JR (1994) Solubility of CO_2 in a Ca-rich leucitite: effects of pressure, temperature, and oxygen fugacity. Contrib Mineral Petrol 116:216-224

Tilden WA (1896) An attempt to determine the condition in which helium and the associated gases exist in minerals. Proc R Soc 59:218-224

Turek A, Riddle C, Cozens BJ, Tetley NW (1976) Determination of chemical water in rock analysis by Karl Fischer titration. Chem Geol 17:261-267

Turnbull AH, Barton RS, Rivière JC (1962) An Introduction to Vacuum Technique. George Newnes Ltd, London

Ueda A , Sakai H (1984) Sulfur isotope study of Quaternary volcanic rocks from the Japanese Islands Arc. Geochim Cosmochim Acta 48:1837-1848

Uys JM, King TB (1963) The effect of basicity on the solubility of water in silicate melts. Trans Met Soc AIME 227:492-500

Vennemann TW, O'Neil JR (1993) A simple and inexpensive method of hydrogen isotope and water analyses of minerals and rocks based on zinc reagent. Chem Geol 103:227-234

Verweij H, van den Boom H, Breemer RE (1977) Raman scattering of carbonate ions dissolved in potassium silicate glasses. J Am Ceram Soc 60:529-534

Walrafen GE (1964) Raman spectral studies of water structure. J Chem Phys 40:3249-3256

Watson LL, Hutcheon ID, Epstein S, Stolper EM (1994) Water on Mars: Clues from deuterium/hydrogen and water contents of hydrous phases in SNC meteorites. Science 265:86-90.

Webster JD (1990) Partitioning of F between H_2O and CO_2 fluids and topaz rhyolite melt. Contrib Mineral Petrol 104:424-438

Webster JD, Holloway JR, Hervig RL (1989) Partitioning of lithophile trace elements between H_2O and $H_2O + CO_2$ fluids and topaz rhyolite melt. Econ Geol 84:116-134

Webster JD, Taylor RP, Bean C (1993) Pre-eruptive melt composition and constraints on degassing of a water-rich pantellerite magma, Fantale volcano, Ethiopia. Contrib Mineral Petrol 114:53-62

Werner HW, Morgan AE (1976) Charging of insulators by ion bombardment and its minimization for secondary ion mass spectrometry (SIMS) measurements. J Appl Phys 47:1232-1242

Westrich HR (1987) Determination of water in volcanic glasses by Karl-Fischer titration. Chem Geol 63:335-340

White WB (1974) The carbonate minerals. In: The Infrared Spectra of Minerals, VC Farmer (ed), Mineral Soc, London, p 227-284

White BS, Brearley M, Montana A (1989) Solubility of argon at high pressures. Am Mineral 74:513-529

Williams P (1981) Ion-stimulated desorption of positive halogen ions. Phys Rev B 23:6187-6190

Williams P (1985) Secondary ion mass spectrometry. Ann Rev Mater Sci 15:517-548

Williams P (1992) Quantitative analysis using sputtering techniques: secondary ion and sputtered neutral mass spectrometry. In: Briggs D, Seah MP (eds) Practical Surface Analysis (second edition). John Wiley & Sons, New York, p 177-228

Williams P, Reed DA, Morgan AE (1983) Ion microscopy using electron-desorbed ions. Appl Surface Sci 16:345-350

Wood JL, Grukpido NJ, Har KL, Flessa SA, Dadin AM, Deem JE, Ferris DH (1985) Measured X-ray performance of synthetic multilayers compared to calculated effects of layer imperfection. Soc Photo-Optical Instrumentation Engineers (SPIE) Proc 563:238-244

Wopenka B, Pasteris J (1986) Limitations to quantitative analysis of fluid inclusions in geological samples by laser Raman spectroscopy. Appl Spectroscopy 40:144-151

Wopenka B, Pasteris JD, Freeman JJ (1990) Analysis of individual fluid inclusions by Fourier transform infrared and Raman microspectroscopy. Geochim Cosmochim Acta 54:519-533

Wright RB, Wang CH (1973) Density effect on the Fermi resonance in gaseous CO_2 by Raman scattering. J Chem Phys 58:2893-2895

Wright RB, Wang CH (1974) Effect of density on the Raman scattering of molecular fluids. II. Study of intermolecular interaction in CO_2. J Chem Phys 61:2707-2710

Wu CK (1980) Nature of incorporated water in hydrated silicate glasses. J Am Ceram Soc 63:453-457

Yarwood J (1967) High Vacuum Technique. William Clowes and Sons, Ltd, London

Yoder HS, Stewart DB, Smith JR (1956) Ternary feldspars. Carnegie Inst Wash Year Book 57:189-191

Yurimoto H, Kurosawa M, Sueno S (1989) Hydrogen analysis in quartz crystals and quartz glasses by secondary ion mass spectrometry. Geochim Cosmochim Acta 53:751-755

Zhang Y, Stolper EM, Wasserburg GJ (1991) Diffusion of water in rhyolitic glasses. Geochim Cosmochim Acta 55:441-456

Zhang Y, Stolper EM (1991) Water diffusion in a basaltic melt. Nature 351:306-309

Zinner E (1989) Isotopic measurements with the ion microprobe In: Shanks WS, Criss RE (eds) New frontiers in stable isotopic research: Laser probes, ion probes, and small sample analysis. US Geol Surv Bull 1890:145-162

CHAPTER 3

DEVELOPMENT OF THE BURNHAM MODEL
FOR PREDICTION OF H_2O SOLUBILITY IN MAGMAS

C. Wayne Burnham

Geology Department
Arizona State University
Tempe, Arizona 85287 USA

HISTORICAL PERSPECTIVE

N. L. Bowen, in his classic treatise on *The Evolution of the Igneous Rocks* (1928), was of the opinion that small amounts of a volatile, such as H_2O, would have correspondingly small effects on the liquidus phase relations appropriate to the "dry" melts in silicate systems, as indicated by his admonition (his Chapter XVI): "To many petrologists a volatile is exactly like a Maxwell demon; it does exactly what one may wish it to do." Only three years later Goranson (1931) embarked upon a series of experimental studies which culminated seven years later (Goranson, 1938) and showed, for the first time, that a relatively small weight percent of H_2O in albite and sanidine melts does, indeed, behave almost like a "Maxwell demon" in its effect upon liquidus phase relations in these systems. To his credit, Bowen realized the significance of Goranson's findings and later did much to establish the importance of "small amounts" of H_2O in the origin of granites (Bowen and Tuttle, 1950; Tuttle and Bowen, 1958). The work embodied in this latter publication constituted the *coup de grace* to the transformist school of the origin of granites, and virtually all definitive work since 1958, whether field or laboratory oriented, has supported their main conclusions regarding the magmatic origin of granites.

Perhaps the main reason that Bowen did not more fully appreciate the critical role of H_2O in the origin of granites at an earlier time was the lack, at that time, of an understanding of how silicate melts and H_2O interact. His early position, prevalent at the time, appears to have been based on the assumption that H_2O, like other simple oxides such as Na_2O, affects liquidus equilibria merely by simple dilution. Goranson's (1938) experimental results clearly showed, however, that the addition of a relatively small weight percent H_2O has a disproportionately large effect on the liquidus temperature of albite. At 2.0 kbar H_2O pressure and 1100 K, for example, albite melt is saturated with only 6.1±0.3 wt % H_2O (Burnham and Jahns, 1962, Fig. 8), but the liquidus temperature is approximately 300°C lower than in the anhydrous melt at the same pressure (Burnham and Davis, 1974, Fig. 18). Were this large effect to be attributed to dilution, alone, the cryoscopic relations (Burnham, 1992, Fig. 1) require that the 6.1 wt % H_2O lower the activity of the *ab* component of the melt to 0.25, but, owing to differences in gram formula weight between H_2O (18.02 g) and $NaAlSi_3O_8$ (262.2 g), the most this amount of H_2O could lower a_{ab}^m by dilution is to 0.50. Thus, only approximately half the lowering of the liquidus temperature at 2.0 kbar can be attributed to dilution; the other half may be attributed to the mechanism by which H_2O dissolves in silicate melts, as discussed below.

In an effort to resolve this discrepancy, as well as many other puzzling aspects of the role of H_2O in granitic magmas, experimental research was undertaken to determine the thermodynamic properties of H_2O in $NaAlSi_3O_8$ melts. Melts of albite composition were chosen with the presumption that the behavior of H_2O in them is representative of that in melts of granitic composition, as attested by close similarities in H_2O solubilities and in its effects on melting relations. Before the results of these experiments on hydrous melts, in the form of P-V-T measurements at pressures to 10 kbars encountered in the earth's crust (Burnham and Davis, 1971), could be used to derive therefrom other thermodynamic parameters, however, it was necessary to know the standard-state thermodynamic properties of pure H_2O throughout the high temperature-pressure region of interest. An extensive experimental effort was mounted in the mid 1960s, therefore, to determine, for the first time, the thermodynamic properties of H_2O to 1273 K and 10 kbars (Burnham et al., 1969). These data, combined with those on the partial molal volumes (Burnham and Davis, 1971) and solubilities of H_2O in $NaAlSi_3O_8$ melts (Burnham and Jahns, 1962; Orlova, 1962) provide the data required to completely determine the partial molal thermodynamic properties of H_2O in these melts (Burnham and Davis, 1974). The volumetric measurements also provided the data base necessary for the first quantitative evaluation of the mechanical (PΔV) energy released in the exsolution of H_2O from a magma, as in explosive volcanic eruptions and igneous breccia formation (Burnham, 1972, 1983, and 1985).

THE THERMODYNAMIC PROPERTIES OF H₂O IN GRANITIC MELTS

Notation

$a_i{}^m$ = activity of component i in melt m relative to melt of pure i.

$f_w{}^m$ = fugacity of $H_2O(w)$ in melt m.

$f_w{}^o$ = fugacity of pure H_2O at P and T.

$G_w{}^o$ = molal Gibbs free energy of pure H_2O at P and T.

$k_w{}^{mi}$ = Henry's-law analogue constant for H_2O in melt m of i composition.

μ_i = chemical potential of component i in the system.

$X_i{}^m$ = mole fraction of i in melt m.

The fundamental criterion for chemical equilibrium is that the partial molal Gibbs free energy of each substance *(i)*, or its chemical potential *(μ_i)*, be the same in every phase at equilibrium. Hence any change in an independent variable that leads to a change in μ_i in one phase must produce precisely the same change in μ_i in all phases, if equilibrium is to be maintained. In terms of the three independent variables of interest here - P, T, and X_i - the total differential, $d\mu_i$, in any given phase may be expressed as

$$d\mu_i = (\partial \mu_i / \partial P)_{T,Xi}\, dP + (\partial \mu_i / \partial T)_{P,Xi}\, dT + (\partial \mu_i / \partial X_i)_{P,T}\, dX_i \qquad (1)$$

or, from the first and second laws of thermodynamics and the Lewisian definition of activity (a_i), $d\mu_i \equiv RTd\ln a_i$, the partial derivatives in Equation (1) may be replaced by their equivalents to yield

$$d\mu_i = V_i dP - S_i dT + RT(\partial \ln a_i / \partial X_i)_{P,T} dX_i \qquad (2)$$

where V_i and S_i are the partial molal volume and entropy of component i, respectively, and the last partial derivative gives the rate of change of μ_i with respect to composition.

The integral of Equation (2) for $i = w$ with respect to pressure, at constant temperature and $X_w{}^m$, yields an expression for the chemical potential of H_2O in the melt $(\mu_w{}^m)_{T,Xw}$ relative to some unspecified reference pressures where $\mu_w{}^m$ is known. These reference pressures were chosen at several points along isotherms on the H_2O saturation surface, where the melt is in equilibrium with almost pure H_2O and $G_w{}^o$, the molal Gibbs free energy of H_2O (Burnham et al., 1969), is known (cf. Burnham and Davis, 1974, Fig. 1). One of these isotherms (1173 K) is shown in Figure 1 below. As a consequence of this choice, the isothermal dependence of $\mu_w{}^m$ on $X_w{}^m$, the mole fraction of H_2O in the melt, could be readily evaluated. Having thus obtained isothermal values for $\mu_w{}^m$ as a function of P and $X_w{}^m$, it became a simple matter to compare isobaric and isoplethal values at different temperatures and thereby obtain values for $\mu_w{}^m$ as a function of all three independent variables, P, T, and $X_w{}^m$.

Although the chemical potential relationships are of fundamental importance, they are of limited utility for present purposes, because, as $X_w{}^m$ approaches zero, $\mu_w{}^m$ approaches minus infinity. To circumvent this inconvenience, Burnham and Davis (1974) first converted chemical potentials into fugacities $(f_w{}^m)$ through the identity

$$d\mu_w{}^m \equiv RT \, d\ln f_w{}^m \tag{3}$$

and found that not only did $f_w{}^m$ approach zero as $X_w{}^m$ approaches zero, as required, but that it varied essentially linearly with the square of the mol fraction $(X_w{}^m)^2$ up to $X_w{}^m = 0.5$. This same relationship holds, of course, if fugacities are converted into activities $(a_w{}^m)$ through the relationship

$$(\mu_w{}^m)_{P,T,Xw^m} - (G_w{}^o)_{P,T} = RT \, ln \, (f_w{}^m/f_w{}^o)_{P,T} = RT \, ln \, a_w{}^m \tag{4}$$

where $G_w{}^o$ and $f_w{}^o$ are the Gibbs free energy and fugacity of pure H_2O, respectively, at P and T (Burnham, 1975a, Figs. 1-5).

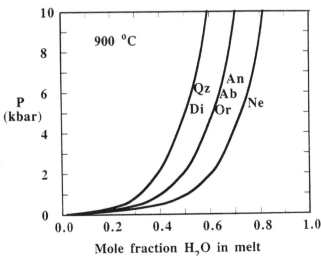

Figure 1. The calculated saturation mol fraction of H_2O (solubility) in feldspar (Ab,An,Or) and nepheline (Ne), as well as quartz and diopside (Qz,Di), melts at 900°C (1173 K).

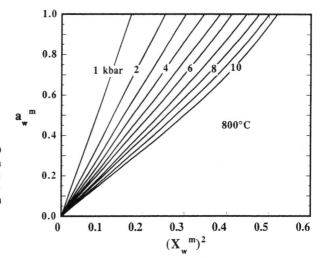

Figure 2. The activity of H_2O ($a_w{}^m$) in silicate melts as a function of the square of its mol fraction $(X_w)^2$ and pressure (kbar) at $800^\circ C$, as presented in Burnham (1974).

This linear relationship between $f_w{}^m$ or $a_w{}^m$ and $(X_w{}^m)^2$ for values of $X_w{}^m \leq 0.5$, as shown in Figure 2, was not unexpected, as Hamilton et al. (1964) found that the solubility of H_2O in andesitic and basaltic melts tended to vary linearly with the square root of total pressure $(P_t{}^{1/2})$. It was thus interpreted by Burnham and Davis (1974) and Burnham (1975a,b) to indicate that up to one mol of H_2O per mol of $NaAlSi_3O_8$ melt dissolves mainly by a dissociation-type reaction to produce two mols of products, presumably OH^-. Accordingly, the activity of H_2O was interpreted to obey, within the limits of experimental error on solubilities, a Henry's-Law analogue relationship for a dissociated solute

$$(a_w{}^m)_{P,T,Xw^m \leq 0.5} = k_w{}^{ma} (X_w{}^m)^2 \tag{5}$$

where $k_w{}^{ma}$ is a Henry's-Law analogue constant for melts of feldspar composition.

At values of $X_w{}^m > 0.5$, as shown in Figure 2, the linear relationship between $a_w{}^{hm}$ and $(X_w{}^m)^2$ no longer holds, but the constant slopes of the $\mu_w{}^m$ vs. $X_w{}^m$ isobars at $X_w{}^m > 0.5$, which are independent of P (Burnham and Davis, 1974, Figs. 2–6), lead to the relationship (Burnham, 1975a):

$$(a_w{}^m)_{P,T,Xw^m > 0.5} = 0.25 k_w{}^{ma} \exp[(6.52 - 2667/T)(X_w{}^m - 0.5)] \tag{6}$$

Equations (5) and (6), together with values for $\ln k_w{}^{ma}$ from

$$\ln k_w{}^{ma} = 5.00 + (\ln P)(4.481 \times 10^{-8} T^2 - 1.51 \times 10^{-4} T - 1.137)$$

$$+ (\ln P)^2(1.831 \times 10^{-8} T^2 - 4.882 \times 10^{-5} T + 4.656 \times 10^{-2})$$

$$+ 7.80 \times 10^{-3}(\ln P)^3 - 5.012 \times 10^{-4}(\ln P)^4 + T(4.754 \times 10^{-3}$$

$$- 1.621 \times 10^{-6} T) \tag{7}$$

where P and T are in units of bars and K, respectively, are all that is required to evaluate $a_w{}^m$ as a function of P,T, and $X_w{}^m$ in feldspar melts under the conditions of interest to granite petrology. Conversely, because the aqueous phase is more than 99 mol % H_2O at melt saturation below 10 kbars, $a_w{}^m$ may be set to unity and Equations (5) and (6) solved for X_w^m, the saturation mol fraction of H_2O in the melt (Fig. 1). Graphs of $ln\ k_w^{ma}$ as a function of ln P and T are presented in Figure 3 for the 973 and 1373 K isotherms, from which it is apparent that the temperature dependence of $ln\ k_w^{ma}$ becomes very small at pressures of 10 kbar and higher (10^{-5} K^{-1} or less).

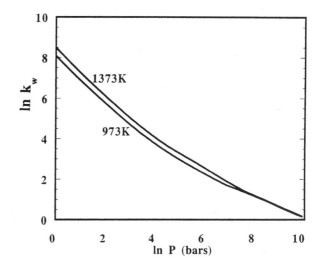

Figure 3. The natural logarithm of the Henrys-Law analogue constant (k_w^{ma}) for H_2O in albite melts versus the natural logarithm of pressure (in bars). Curves shown are the 973 K and 1373 K isotherms.

Equation (7) was obtained by least-squares fit to data taken from Figure 9.2 of Burnham (1981) which, in turn, were obtained from the data embodied in Figures 8–12 of Burnham and Davis (1974). Uncertainties in the values of $ln\ k_w^{ma}$ obtained from Equation (7), which are directly applicable only to melts of feldspar composition, range from ±0.2 at 0.5 kbar to ±0.08 at 10 kbar, based on the estimates of Burnham (1981, p. 228) for the uncertainties in the experimental H_2O solubility data used (±0.03 in X_w^m). In melts of other compositions Equation (7) must be modified as indicated below.

In Si_4O_8 (qz) melts, for example, Burnham (1981) showed that the expression for $ln\ k_w^{ma}$ could be made applicable simply by the addition of a constant 0.47, which gives rise to the lower solubilities shown in Figure 1. Thus Equation (7) can be made applicable to qz melts simply by replacing $ln\ k_w^{ma}$ with $ln\ k_w^{mqz}$ and adding 0.47 to the first term, making it 5.47. The same procedure may be adopted for melts of pyroxene composition - to a close approximation - despite the fact that the experimental data on diopside melts analyzed by Eggler and Burnham (1984) suggest a mechanism of solution which may be somewhat different from that in aluminosilicate melts. This caveat notwithstanding, application of the Gibbs-Duhem relation to the saturation values of X_w^m (cf. Fig. 1) calculated on the assumption that $ln\ k_w^{mdi} = ln\ k_w^{mqz} = ln\ k_w^{ma} + 0.47$, coupled with the cryoscopic equations for diopside ($Ca_{1.33}Mg_{1.33}Si_{2.67}O_8$), yield good agreement with the experimental data of Eggler and Rosenhauer (1978) on the H_2O-saturated solidus of diopside at pressures of five kilobars and higher. These results, coupled with those from the system Si_4O_8-H_2O, suggest that the factor 0.47 arises from the interaction of H_2O with the Al-containing complexes in melts of feldspar composition. Evidence in support of this correlation is forthcoming from melts which contain more than one gram atom of Al per eight gram atoms of oxygen (Dingwell et al., 1984).

In melts of nepheline composition ($Na_2Al_2Si_2O_8$), for example, the two Al per eight oxygens should lead to a $ln\ k_w^{mne}$ that is 0.47 lower than $ln\ k_w^{ma}$. In fact, the much higher solubilities of H_2O in nepheline melts determined by Boettcher and Wyllie (1969) and by Burnham (ca. 1960, unpublished experimental results), as indicated in the calculated curves of Figure 1, are fully accounted for, within experimental error, by this factor. Moreover, the dramatic lowering of the H_2O-saturated solidus temperature of nepheline from approximately 1800 K at atmospheric pressure to 963 K at 10 kbar, as also demonstrated by Boettcher and Wyllie (1969), can be accounted for only if the high-pressure melts contain the much larger amounts of H_2O indicated in Figure 1. Similarly, the H_2O-saturated solidus of leucite ($K_{1.333}Al_{1.333}Si_{2.667}O_8$) is lowered substantially relative to that of sanidine ($KAlSi_3O_8$) at a given pressure (Goranson, 1938; Spengler, 1965), in part because $ln\ k_w^{mlc}$ is 0.157 lower ($0.333 \cdot 0.47 = 0.157$) than $ln\ k_w^{ma}$ and X_w^m is therefore correspondingly greater at a given a_w^{hm}, such as at saturation. The very steep negative initial slope of the leucite liquidus curve in hydrous melts of sanidine composition, however, is due also in part to a rapidly decreasing extent of speciation (incongruency) to produce lc with increasing H_2O pressure.

From the immediately foregoing discussions, it is apparent that to use Equations (5) and (6) for the purpose of calculating the mol fraction of H_2O (X_w^m) in a melt for a given a_w^{hm}, it is necessary first to recast the chemical composition of the melt into mineral-like components—analogous to the CIPW norm—using the eight-oxygen mole convention (cf. Burnham, 1981a, Table 9.2; Burnham and Nekvasil, 1986, Table 3). In this context, the method proposed by Burnham (1975a,b; 1979a) for computing the "albite-equivalent" mass of melt results in variable formula weights of individual melt components and corresponding discontinuities at $X_w^m = 0.5$, hence it cannot be used in calculations involving application of the Gibbs-Duhem equation and, therefore, should be abandoned. Next, the appropriate k_w^{mi} is used to compute X_w^{mi} in each melt component, as in Figure 1, then these X_w^{mi}'s are weighted by the corresponding X_i^{am}'s and summed. The appropriate $ln\ k_w^{mi}$ to use in these computations is given, to a close approximation for all silicate components, by the relation

$$ln\ k_w^{mi} = ln\ k_w^{ma} + 0.47(1 - N_{Al}^{mi}) \tag{8}$$

where $ln\ k_w^{ma}$, which is applicable to all three feldspar-like components, is given by Equation (7), and N_{Al}^{mi} is the number of hydroxy-aluminate complexes per 8-oxygen mole of component i. In albite and anorthite melts, for example,

$N = 1$ [$NaAlO(OH)^+ \cdot Si_3O_6(OH)^-$ and $CaAl_2O_3(OH)^+ \cdot Si_2O_4(OH)^-$],

whereas in nepheline melts $N = 2\{2[NaAlO(OH)^+ \cdot SiO_2(OH)^-]\}$.

Although the solubilities of H_2O in Si_4O_8 melts calculated from Equations (5) to (7) are within the uncertainty of those determined experimentally for pressures up to approximately seven kilobars, the calculated solubility curve shown in Figure 1 does not exhibit the marked inflection toward much higher H_2O contents of the melt phase that are found experimentally as the "second critical endpoint" (complete miscibility) is approached at ~10 kbar (Kennedy et al., 1962). This discrepant behavior was interpreted by Burnham (1981) to indicate that most of the excess H_2O (15 to 17 wt %) is not actually dissolved in the melt. This interpretation, which implies that much of the H_2O forms an emulsion with an $Si_4O_8 \cdot H_2O$ solution, is supported by the fact that the S-L-V curve for quartz, as shown by Kennedy et al. (1962), does not exhibit a down-temperature inflection at pressures above 7 kilobars, as it must if this much additional H_2O was in true melt solution and, therefore, markedly lowered a_{qz}^{hm}. Other silicate systems in which H_2O solubilities appear to be positively temperature dependent at elevated pressures and temperatures above approximately 1000°C are similarly interpreted.

REFERENCES

Boettcher AL, Wyllie PJ (1969) Phase relationships in the system $NaAlSiO_4$-SiO_2-H_2O to 35 kilobars pressure. Am J Sci 267:875-909

Bowen NL (1928) The Evolution of the Igneous Rocks. Princeton University Press, Princeton, NJ, 332 p

Bowen NL, Tuttle OF (1950) The system $NaAlSi_3O_8$-H_2O. J Geol 58:489-511

Burnham CW (1972) The energy of explosive volcanic eruptions. Earth Mineral Sci (The Pennsylvania State University) 41:69-70

Burnham CW (1974) $NaAlSi_3O_8$-H_2O solutions: a thermodynamic model for hydrous magmas. Bull Soc fr Minéral Cristallogr 97:223-230

Burnham CW (1975a) Thermodynamics of melting in experimental silicate-volatile systems. Fortschr Mineral 52:101-118

Burnham CW (1975b) Water and magmas: A mixing model. Geochim Cosmochim Acta 39:1077-1084

Burnham CW (1979a) The importance of volatile constitutents. In: The Evolution of the Igneous Rocks, Princeton University Press, Princeton, NJ, p 1077-1084

Burnham CW (1979b) Magmas and hydrothermal fluids. In: Geochemistry of Hydrothermal Ore Deposits. John Wiley and Sons, New York, p 71-136

Burnham CW (1981) The nature of multicomponent aluminosilicate melts In: Chemistry and Geochemisry of Solutions at High Temperatures and Pressures. Phys Chem Earth 13&14:191-227

Burnham CW (1983) Deep submarine pyroclastic eruptions, Econ Geol Monogr 5:142-148

Burnham CW (1985) Energy release in subvolcanic environments: implications for breccia formation. Econ Geol 80:1515-1522

Burnham CW (1992) Calculated melt and restite compositions of some Australian granites. Trans Royal Soc Edinburgh: Earth Sciences 83:387-397

Burnham CW (1994) Magmas and hydrothermal fluids. Geochemistry of Hydrothermal Ore Deposits, John Wiley and Sons, New York, in press

Burnham CW, Davis NF (1971) The role of H_2O in silicate melts: I. P-V-T relations in the system $NaAlSi_3O_8$-H_2O to 10 kilobars and 1000°C. Am J Sci 270:54-79

Burnham CW, Davis NF (1974) The role of H_2O in silicate melts: II. Thermodynamic and phase relations in the system $NaAlSi_3O_8$-H_2O to 10 kilobars, 700°C -1100°C. Am J Sci 274:902-940

Burnham CW, Holloway JR, Davis NF (1969) Thermodynamic properties of water to 1000°C and 10000 bars. Geol Soc Am Spec Paper 132:1-96

Burnham CW, Jahns RH (1962) A method for determining the solubility of water in silicate melts. Am J Sci 260:721-745

Burnham CW, Nekvasil H (1986) Equilibrium properties of granitic magmas. Am Min 71:239-263

Dingwell DB, Harris DM, Scarfe CM (1984) The solubility of H_2O in melts in the system SiO_2-Al_2O_3-Na_2O-K_2O at 1 to 2 kbars. J Geol 92:387-395

Dingwell DB, Webb SL (1990) Relaxation in silicate melts. Eur J Min 2:427-449

Eggler DH (1974) Effect of CO_2 on the melting of peridotite. Carnegie Inst Wash Yrbk 73:215-224

Eggler DH, Burnham CW (1984) Solution of H_2O in diopside melts: a thermodynamic model. Contrib Mineral Petrol 85:58-66

Eggler DH, Rosenhauer M (1978) Carbon dioxide in silicate melts: II. Solubilities of CO_2 and H_2O in $CaMgSi_2O_6$ (diopside) liquids and vapors at pressures to 40 kb. Am J Sci 287:64-94

Goranson RW (1931) The solubility of water in granite magmas. Am J Sci 22:481-582

Goranson RW (1938) Silicate-water systems: Phase equilibria in the $NaAlSi_3O_8$-H_2O and $KAlSi_3O_8$-H_2O systems at high temperatures and pressures. Am J Sci 35A:71-91

Hamilton DL, Burnham CW, Osborn EF (1964) The solubility of water and effects of oxygen fugacity and water content on crystallization in mafic magmas. J Petrol 5:21-39

Kennedy GC, Wasserburg GJ, Heard HC, Newton RC (1962) The upper three-phase region in the system SiO_2-H_2O. Am J Sci 260:501-521

Orlova GP (1962) The solubility of water in albite melts. Int'l Geol Rev 6:254-258.

Spengler CJ (1965) The upper three-phase region in a portion of the system $KAlSi_2O_6$-SiO_2-H_2O at water pressures from two to seven kilobars. PhD dissertation, The Pennsylvania State University, University Park, PA

Stolper E (1982) The speciation of water in silicate melts. Geochim Cosmochim Acta 46:2609-2620

Tuttle OF, Bowen NL (1958) Origin of granite in the light of experimental studies in the system $NaAlSi_3O_8$-$KAlSi_3O_8$-SiO_2-H_2O. Geol Soc Am Memoir 74:153 p

Chapter 4

WATER SOLUBILITY AND SPECIATION MODELS

Paul F. McMillan

Department of Chemistry and Biochemistry
Arizona State University
Tempe, Arizona 85287 USA

INTRODUCTION

Water is the most important magmatic volatile, both in terms of its abundance, and in its influence on melt properties and crystallization pathways. Since the general realization of its importance, in the early part of this century (Bowen, 1928; Ingerson, 1960), there have been on-going efforts to understand and quantify the extent of water incorporation in aluminosilicate liquids at high pressures and temperatures, and its effect on magma rheology and crystal-liquid phase relations. There have also been many attempts to model the nature of the water-melt interaction, and the number and types of hydrous species present in the molten aluminosilicate. These models are essential for construction of realistic thermodynamic models of water-containing magmas, and for understanding the microscopic details of water dissolution and diffusion processes. Despite extensive studies, there remain major questions to be resolved regarding the mode and even the extent of water dissolution as a function of pressure, temperature and bulk melt composition. In addition, the advent of in-situ studies at high pressures and temperatures, coupled with a better understanding of relaxation behavior in molten silicates in general, are beginning to permit a more sophisticated approach to understanding the dynamic nature of hydration processes in the high temperature liquids.

The aim of the present chapter is to give a general overview of the present state of affairs in understanding water solubility, including a brief summary of current controversies in the field. It is hoped that this may help stimulate further work, especially with the design of critical experiments to help resolve these points of contention, as "water" moves further from its status of "Maxwell demon" in hydrous magmas (Bowen, 1928; Burnham, 1979), and the details of its fascinating chemistry in high temperature silicates begin to be revealed.

Early work: low pressure solubility

The extent and mechanism of incorporation of hydrous species in molten and glassy silicates has long been of interest in both geochemistry and the glass sciences. In several series of experiments carried out at or near room pressure, workers in the glass sciences investigated the isobaric solubility of water as a function of the partial pressure of water in the vapor phase (Tomlinson, 1956; Russell, 1957; Kurkjian and Russell 1958; Moulson and Roberts 1961; Schölze 1966; Bedford 1975; Coutures and Peraudeau 1981). These results showed that the amount of water incorporated in the glass or melt varied with the square root of the partial pressure of water vapor:

$$S \propto \sqrt{p_{H_2O}} \tag{1}$$

This observation was used to support a water dissolution model involving dissociation of H_2O into two hydroxylated species in the glass or melt:

$$H_2O(v) + O^{2-}(m) \Leftrightarrow 2OH^-(m), \tag{2}$$

consistent with spectroscopic studies of the hydrous glasses (Adams and Douglas, 1959; Schölze et al., 1962; Müller-Warmuth, 1965a,b; Schölze, 1966).

The effect of pressure: "square root relationships"

From the pioneering experimental work of Goranson (1931), it was known that the maximum water solubility in silicate magmas increases rapidly with increasing total pressure. In geochemical studies covering several decades, water solubilities have been measured in many natural and synthetic melt compositions over a wide range of pressures and temperatures (Jahns and Burnham, 1958; Tuttle and Bowen, 1958; Khitarov et al., 1959, 1968; Khitarov, 1960; Ingerson, 1960; Burnham and Jahns, 1962; Orlova, 1962; Kennedy et al., 1962; Friedman et al., 1963; Hamilton et al., 1964; Ostrovskiy et al., 1964; Kadik and Lebedev, 1968; Burnham and Davis, 1971, 1974; Kadik et al., 1972; Eggler, 1972; Hodges, 1973, 1974; Oxtoby and Hamilton, 1978a,b,c,d, 1979; Voigt et al., 1981; Day and Fenn, 1982; Dingwell et al., 1984; Hamilton and Oxtoby, 1986; McMillan et al., 1986; Pichavant, 1987; McMillan and Holloway, 1987; Silver et al., 1990; Pichavant et al., 1992; Paillat et al., 1992; Holtz et al., 1992; Blank et al., 1993; Dixon et al., 1994; Cocheo, 1994).

Guided by the work on hydrous glasses at low pressure, the solubility data from such experiments at high pressure have often been plotted against the square root of the water pressure, or water fugacity, in the fluid (Hodges, 1973, 1974; Hamilton et al., 1964; Mysen, 1977; Oxtoby and Hamilton, 1978a; Eggler and Rosenhauer, 1978). In plots of this type, linear relations between the solubility (S_{H_2O}, equal to the mole fraction or activity of hydrous species dissolved in the melt) and $\sqrt{p_{H_2O}}$ or $\sqrt{f_{H_2O}}$ have been taken to support the heterogeneous equilibrium (2), involving dissociation of the H2O species, as the water dissolution mechanism at high pressure (Burnham and Davis, 1974; Bedford, 1975; Rosenhauer and Eggler, 1975; Mysen, 1977; Eggler and Burnham, 1984) (Fig. 1).

However, a linear relation between S_{H_2O} and $\sqrt{f_{H_2O}}$ should not in fact be obtained in such high pressure solubility experiments, even if the dissolution did occur via Reaction (2) (Goranson, 1937; Wasserburg, 1957; Stolper, 1982a; McMillan and Holloway, 1987). Unlike the atmospheric pressure data, the high pressure solubility measurements are not isobaric, but instead represent polybaric points on the water saturation surface. Any equilibrium expression for the water dissolution reaction along the saturation boundary must then contain a pressure dependent exponential term, involving the change in partial molar volumes of the reacting species

$$(\Delta \overline{V}^{reaction} = \overline{V}_{hydrous\,melt} - \overline{V}_{dry\,melt} - \overline{V}_{H_2O}): \; e^{-\frac{\int \Delta \overline{V}^{reaction} dP}{RT}}$$

(Wasserburg, 1957; Stolper, 1982a; McMillan and Holloway, 1987).

Although most synthetic and natural aluminosilicate melts studied to date do not show any simple linear relation between S_{H_2O} and $\sqrt{f_{H_2O}}$ above approximately 500 bars pressure (Oxtoby and Hamilton, 1978a; McMillan and Holloway, 1987), as expected, linear "square root" relationships have in fact been found for particular (unrelated) aluminosilicate melt compositions, at least up to several (6-8) kbar pressure (Hamilton et al., 1964; McMillan et al., 1986; McMillan and Holloway, 1987) (Fig. 1). From the previous discussion, observation of such a linear relation in these polybaric saturation experiments does not give any direct information on the speciation associated with the water dissolution reaction, but indicates that some cancellation of terms in the

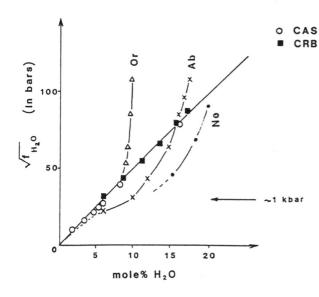

Figure 1. Plots of square root of water fugacity in fluid versus molar water solubility for selected melt compositions (water mole fractions calculated on a one-oxygen molar basis for the dry melt: see McMillan and Holloway, 1987). Note that Columbia River basalt (CRB: Hamilton et al., 1964) and a calcium aluminosilicate melt composition (CAS: McMillan et al., 1986) show linear root fugacity-water solubility relations, whereas compositions such as albite (Ab), orthoclase (Or) and nepheline (Ne) (Oxtoby and Hamilton, 1978a) do not.

thermodynamic expression for the dissolution reaction has occurred. In fact, if the mechansim for the dissolution reaction were known, along with an equation of state for the dry melt, the the variation of S_{H_2O} with $\sqrt{f_{H_2O}}$ would provide valuable information on the reaction volume, and the molar volume of the hydrous melt, at high pressure and temperature.

In their plots of water solubility versus $\sqrt{p_{H_2O}}$ or $\sqrt{f_{H_2O}}$, Hodges (1974) and Oxtoby and Hamilton (1978a) considered that their data represented linear regions separated by breaks in slope, indicative of changes in the water solubility mechanism at particular pressures or water contents. However, these solubility data could be equally well represented by continuous solubility curves. Such continuous changes in slope of water solubility relations would be expected due to variations in the partial molar volumes of the aqueous fluid or the hydrous melt species with pressure, and need not reflect any changes in the water dissolution mechanism.

Burnham's model for water dissolution

Burnham and Davis (1971, 1974) used the results of P–V–T measurements on two compositions in the albite-water system to construct isobaric sections relating $a_{H_2O}^{fluid}$ and the mole fraction of water dissolved in the melt, $X_{H_2O}^{melt}$. This work is summarized in Chapter 3 of this volume. The results suggested a linear relationship between water activity and the square of the mole fraction of dissolved water $\left(X_{H_2O}^{melt}\right)^2$ for $X_{H_2O}^{melt} < 0.5$, with positive deviations from this relation occurring at higher water contents. These observations were used to construct a model for water dissolution in albitic melts. For $X_{H_2O}^{melt} < 0.5$, it was proposed that the molecular H_2O component dissociated by exchanging one hydrogen with the Na^+ charge-balancing the aluminosilicate network:

$$NaAlSi_3O_8 + H_2O \Leftrightarrow AlSi_3O_7(OH) + NaOH \ . \tag{3}$$

Although this reaction as written does not necessarily imply any breakage of TOT (T = Si, Al) linkages in the melt (Kohn et al., 1989b, 1992), the cartoon used to depict the

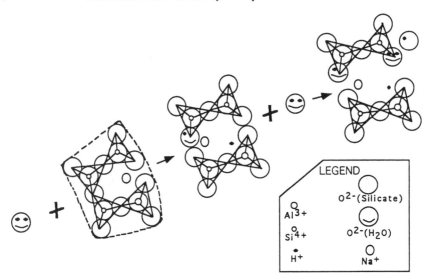

Figure 2. Cartoon depicting water dissolution mechanism in albite ($NaAlSi_3O_8$) melt, proposed by Burnham (1975). The figure shows an H_2O molecule (happy face) reacting with the $NaAlSi_3O_8$ unit, breaking a TOT linkage, and exchanging Na^+ with H^+. The first step corresponds to the reaction $NaAlSi_3O_8 + H_2O \Rightarrow AlSi_3O_7(OH) + NaOH$. The TOT (T = Al, Si) linkage is broken, and the charge-balancing Na^+ is exchanged by H^+. The second step occurs at higher water content and corresponds to the reaction $AlSi_3O_7(OH) + nH_2O \Rightarrow AlSi_3O_{7-n}(OH)_{2n+1}$ in which the remaining TOT linkages are hydrolyzed. Redrawn from Burnham (1975).

reaction by Burnham (1975a, 1979) does involve rupture of intertetrahedral linkages (Fig. 2). At higher H_2O contents, with $X^{melt}_{H_2O} > 0.5$, there is no Na^+ available for such exchange reactions, and the TOT linkages are hydrolyzed:

$$AlSi_3O_7(OH) + nH_2O \iff AlSi_3O_{7-n}(OH)_{2n+1} \qquad (4)$$

(Burnham, 1974, 1975a,b, 1979, 1981; Eggler and Burnham, 1984). In these models for water dissolution, no molecular water was presumed to be present as a species dissolved in the melt.

Molecular water as a dissolving species

Early infrared spectroscopic studies had in fact indicated the presence of molecular (i.e., undissociated) H_2O as a dissolving species in hydrous silicate glasses (Keller and Pickett, 1954; Orlova, 1962; Ostrovskiy et al., 1964; Schölze, 1966). In addition, the presence of substantial amounts of molecular water in high pressure aluminosilicate melts was suggested from solubility measurements and theoretical considerations (Wasserburg, 1957; Shaw, 1964; Hodges, 1973, 1974; Mysen, 1977; Eggler and Rosenhauer, 1978). This is now supported by many modern infrared and NMR studies on high water content glasses (Ernsberger, 1977; Bartholomew et al., 1980; Bartholomew and Schreurs, 1980; Wu, 1980; Takata et al., 1981; Stolper, 1982a,b, 1989; McMillan and Remmele, 1986; Newman et al., 1986; Farnan et al., 1987; Eckert et al., 1987, 1988; Silver and Stolper, 1989; Kohn et al., 1989a,b, 1992; Silver et al., 1990; Kummerlen et al., 1992). Mysen et al. (1980) had suggested from Raman spectroscopic results that no molecular H_2O was present in hydrous albite glasses, based on the absence of the HOH bending band near 1600 cm^{-1}. However, McMillan et al. (1983) showed that this band is in fact present in hydrous albite glass samples, although its Raman intensity is low. This assignment was accepted in subsequent Raman spectroscopic studies of hydrous glasses (Mysen and Virgo, 1985, 1986a,b).

Stolper (1982a,b) carried out systematic quantitative infra-red absorption studies of water incorporation in a range of natural and synthetic aluminosilicate glasses, and demonstrated the presence of molecular water species at high water contents (Fig. 3). At water contents below approximately 1 wt %, a dissolution mechanism involving only hydroxyl (-OH) groups is a good approximation, consistent with previous studies of glasses prepared at or near room pressure (Schölze, 1966; Coutures and Peraudeau, 1981). However, for higher water contents corresponding to solubilities at higher pressure, both hydroxyl and molecular H_2O groups were found to be present in the hydrous glasses, and the ratio of molecular water to hydroxyl groups increased rapidly with increasing water content (Fig. 3). By carrying out infrared experiments at liquid nitrogen temperature, where any bulk liqud water would be frozen to crystalline ice, Stolper (1982b) showed that the molecular H_2O was not simply present as sub-microscopic bubbles of trapped fluid, but should be considered as a true dissolved species in the melt. This near-infrared absorption technique has now been fully calibrated for routine quantitative analyses of hydrous species in aluminosilicate glass (Newman et al., 1986), has been extended to a wide range of natural and synthetic compositions (Silver and Stolper, 1989; Silver et al., 1990; Blank et al., 1993; Dixon et al., 1994), and has been applied to study speciation in water diffusion studies (Stanton et al., 1985; Stanton, 1989; Zhang and Stolper, 1991; Zhang et al., 1991).

Based on their findings, Stolper (1982a,b) and Silver and Stolper (1985) proposed a thermodynamic model for water dissolution in aluminosilicate melts, involving both hydroxyl (OH) and molecular H_2O in the dissolution mechanism. Reaction (2) above becomes a homogeneous equlibrium:

$$H_2O\,(m) \quad + \quad O^{2-}\,(m) \quad \Leftrightarrow \quad 2\,OH^-\,(m) \tag{5}$$

with an equilibrium constant K_1, and is preceded by a heterogeneous dissolution of molecular water in the melt:

$$H_2O\,(fl) \quad \Leftrightarrow \quad H_2O\,(m), \tag{6}$$

with equilibrium constant K_2. The regular solution model derived from these observations and its application to hydrous aluminosilicate melts is discussed in detail by Silver and Stolper (1985) and Silver et al. (1990), and in Chapter 6 of this volume.

This mechanism for water incorporation, involving both hydroxyl groups and undissociated molecular water, is more consistent with the spectroscopic data for hydrous glasses than is the speciation model proposed by Burnham (Burnham and Davis, 1971, 1974; Burnham, 1974, 1975a,b). However, Stolper (1982a) has compared the activity-composition relations for hydrous albitic melts calculated by both methods, and found a close correspondence between the two. If this observation can be extended to other cases of geologic interest, it would be extremely useful. The actiyity-composition relation derived by Burnham is algebraically very simple ($a_{H_2O}^{fluid} = \left(X_{H_2O}^{melt}\right)^2$ for $X_{H_2O}^{melt} < 0.5$), and has now been incorporated into an extensive thermodynamic data base and model for phase behavior in multicomponent systems (Burnham, 1981; Burnham and Nevkasil, 1986). If it does provide a useful working approximation to the true activity-composition relations of hydrous melts, there is no reason to abandon the simple Burnham form for the more accurate Stolper-Silver relations, unless the simpler form breaks down. The range of validity of the simple Burnham activity-composition relation should continue to be tested, for practical applications in igneous petrology.

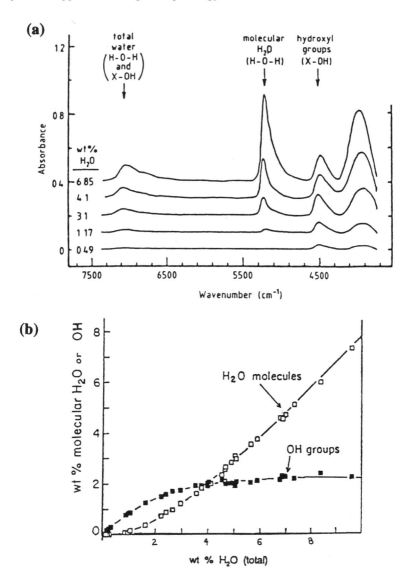

Figure 3. (a) Near infrared absorption spectra for albitic glass with variable amounts of dissolved water, showing the characteristic overtone and combination bands associated with molecular H_2O and bound hydroxyl groups (redrawn from Stolper, 1982b), and (b) the variation in relative proportion of H_2O molecules and OH groups dissolved in hydrous albite glass as a function of total water content (redrawn from Silver and Stolper, 1989).

Critical behavior of aluminosilicate-water systems at high pressure and temperature

Kennedy et al. (1962) observed critical behavior in the system SiO_2-H_2O at high pressure and temperature, with complete mutual solubility in the fluid phase above ~10 kbar at near 1000°C. These authors suggested that a similar phenomenon might also occur in the SiO_2-$KAlSi_3O_8$-$NaAlSi_3O_8$-H_2O system. Burnham and Jahns (1962) noted that the low temperature solubility curve of for albite melt showed a distinct flattening to higher solubility values above around 5 to 7 kbar. This did not agree with the water solubilities measured for albite at higher temperatures by S. Oxtoby (Oxtoby and Hamilton, 1978a; Hamilton and Oxtoby, 1986; McMillan and Holloway, 1987). However, the data of Oxtoby and Hamilton were obtained by weight loss measurements on quenched samples, and are likely to represent minimum values of the solubility above around 5 kbar (Paillat et al., 1992). Kadik and Lebedev (1968) have discussed the difficulty in obtaining solubilities for albite melt above a total water content of 8 to 10 wt %, from experiments on quenched glasses. Their results indicated that, above this water content, the melt does not retain its dissolved water, but a large amount is released as bubbles during the quench. This suggests that quench experiments cannot be used to obtain reliable water solubilities for samples with high melt water contents, and that phase equilibrium determinations, including the "dimple" technique employed by Burnham and Jahns (1962), must be used.

Combination of available solubility determinations below around 10 kilobars with measurements at pressures up to 30 kbar for albite and diopside melts (Eggler, 1973; Rosenhauer and Eggler, 1975; Eggler and Rosenhauer, 1978; Oxtoby and Hamilton, 1978a,b) also indicate some curvature which would support supercritical behavior (McMillan and Holloway, 1987).

In their studies of albite melting under hydrothermal conditions, Goldsmith and Jenkins (1985) noted that, above 14 kbar, the melting curve showed a distinct flattening, and the quenched product consisted of a fine-grained amorphous powder. Scanning electron microscopy showed that this powder consisted of rounded globules 2 to 5 μm in diameter, which could represent aluminosilicate component deposited from the fluid on quenching. This observation would suggest that the albite-water system had become supercritical above around 14 kbar at 600-700°C.

Paillat et al. (1992) have carried out a detailed study of the $NaAlSi_3O_8$-H_2O system. These workers used gel techniques to obtain homogeneous amorphous $NaAlSi_3O_8 \cdot nH_2O$ starting materials with high water contents, and analyzed the quenched glassy matrix after the high P-T run using the ion microprobe, in an attempt to minimize problems with volatile loss during quench from the high water content samples. They observed that the water solubility curved strongly to large $X_{H_2O}^{melt}$ values, on the order of 10 to 14 wt % for pressures of 5 to 6 kbar at temperatures between 900° and 1400°C, consistent with the early measurements of Burnham and Jahns (1962). From their data on the dependence of solubility on pressure and temperature, Paillat et al. (1992) constructed a P-T phase diagram in the system $NaAlSi_3O_8$-H_2O, with complete mutual solubility of $NaAlSi_3O_8$ and H_2O in the fluid above 1400°C at 9 kbar, and an increase in the solvus temperature with decreasing pressure (Fig. 4). It is likely that such critical behavior is in fact exhibited by many hydrous silicate and aluminosilicate systems, and more studies are needed to evaluate the implications for geologically important melts and fluids.

Stolper (1982a) noted that, because water solubility at high pressures should be dominated by solvation of molecular H_2O as the primary component, there will be little

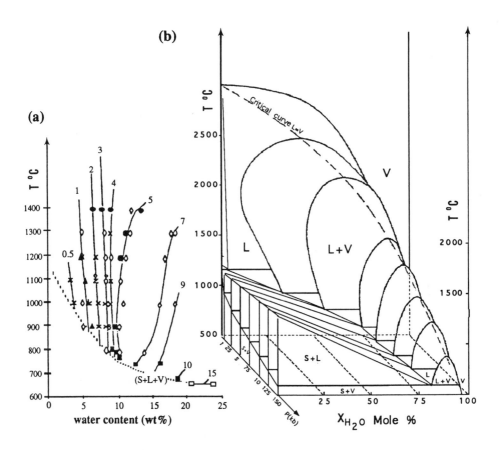

Figure 4   (a) Variation of water content with T for different pressures (marked in kilobars at ends of curves), compiled by Paillat et al. (1992) from various sources of experimental data. Note change from retrograde to prograde solubility with increasing pressure, at approximately 5 kbar. (b) Critical behaviour in the system $NaAlSi_3O_8$–H_2O, as suggested by Paillat et al. (1992), based on solubility determinations and thermodynamic considerations. Redrawn from Paillat et al. (1992).

compositional dependence of water solubility at high pressures. This appears to be borne out by the existing results of water solubility experiments at pressures in the 20 to 30 kbar range, and correlates with the difficulty of quenching hydrous melts with high water contents. Whereas water solubility shows a strong compositional dependence between room pressure and a few kilobars pressure, such widely different melt compositions as Fo, Di, Ab, An and En have solubilities within a few per cent of each other at high pressure (McMillan and Holloway, 1987). The rapidly increasing importance of the molecular water component in the hydrous melt phase may provide the precursor for the homogeneous supercritical hydrous aluminosilicate fluid encountered at high pressures and temperatures, as suggested by the results of Kennedy et al. (1962), Goldsmith and Jenkins (1985) and Paillat et al. (1992).

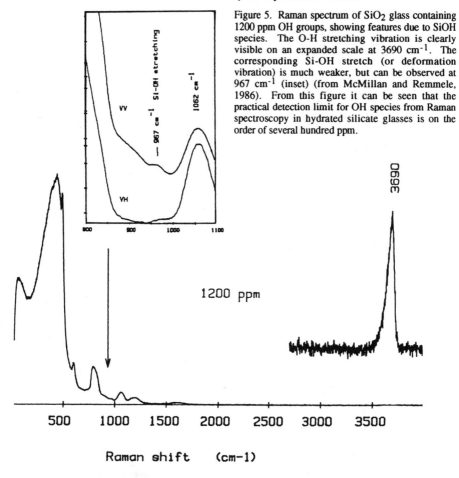

Figure 5. Raman spectrum of SiO_2 glass containing 1200 ppm OH groups, showing features due to SiOH species. The O-H stretching vibration is clearly visible on an expanded scale at 3690 cm⁻¹. The corresponding Si-OH stretch (or deformation vibration) is much weaker, but can be observed at 967 cm⁻¹ (inset) (from McMillan and Remmele, 1986). From this figure it can be seen that the practical detection limit for OH species from Raman spectroscopy in hydrated silicate glasses is on the order of several hundred ppm.

Raman shift   (cm-1)

MICROSCOPIC MODELS FOR WATER DISSOLUTION

Although significant progress has been made in this area, many important questions remain to be resolved regarding the mode of incorporation of H-bearing species, both in aluminosilicate glasses quenched from hydrous liquids, and especially in magmas at high pressures and temperatures.

SiO_2-H_2O

Even for hydrous SiO_2 glass, detailed assignment of the structural species present remains problematic. The infrared and Raman spectra of silica glasses containing small amounts of dissolved water (several hundred to thousands of ppm) show a single, asymmetric peak at 3670 to 3700 cm⁻¹, due to the O-H stretching vibration of weakly hydrogen-bonded OH groups (Fig. 5). In a study of SiO_2 glasses reacted with H_2 or D_2, a weak peak at 970 cm⁻¹ (for the H-bearing sample) was identified via its isotope shift (-30 cm⁻¹) as due to a stretching or deformation vibration of SiOH groups (Hartwig and Rahn, 1977; van der Steen and van den Boom, 1977). The same peak is observed in silica glasses containing dissolved water (Mysen and Virgo, 1985, 1986a; McMillan and

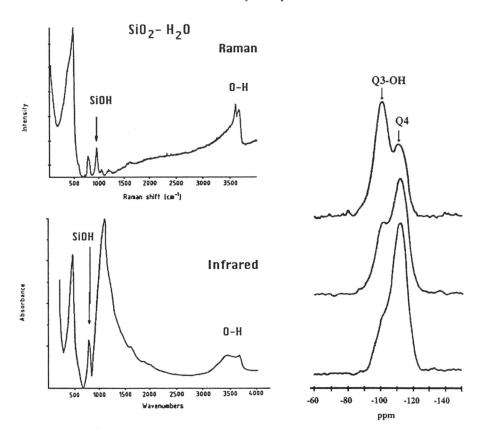

Figure 6 (left). Raman (top) and infrared (bottom) spectra of SiO_2 glass containing ~6% H_2O, showing features due to SiOH species and molecular H_2O. The bending vibration of molecular water at 1630 cm^{-1} is present in both spectra, but is weak, and is partly obscured by overtone features from the SiO_2 glass framework (from McMillan and Remmele, 1986). The molecular water contribution to the O-H stretching is clearly visible as the broad infrared band at 3450 cm^{-1}.

Figure 7 (right). ^{29}Si-^1H cross polarized (CP) MAS NMR spectrum of hydrous silica glass, showing peak due to Si-OH (Q^3-OH) species in addition to the fully polymerized Q^4 species. The three spectra correspond to different CP contact times, which optimize the exchange of spin between the two nuclei. Scale is in ppm relative to TMS (tetramethylsilane) standard. Redrawn from Farnan et al. (1987).

Remmele, 1986). Huffman and McMillan (1985) also observed a peak at 885 cm^{-1} in the Raman spectra of amorphous SiO_2 samples prepared by chemical vapor deposition, which could correspond to silicate tetrahedra with two OH groups (=Si(OH)$_2$ or Q^2(OH)$_2$ species). As the water content of SiO_2 glasses is increased, the O-H stretching vibration broadens and its maximum shifts to lower wavenumber, consistent with an increase in hydrogen bonding (Fig. 6) (Mysen and Virgo, 1985, 1986a; McMillan and Remmele, 1986). At the same time, a broad band near 3450 cm^{-1}, and a weak peak near 1630 cm^{-1}, appear in the spectra due to the presence of molecular water. An additional sharp feature at 3598 cm^{-1} is also observed in the Raman spectra of high water content silica samples (Mysen and Virgo, 1986a; McMillan and Remmele, 1986). This corresponds to the O-H stretching vibration of a species which has not yet been identified.

These infrared and Raman spectroscopic observations are generally consistent with the results of NMR studies on hydrous SiO_2. Farnan et al. (1987) have used the technique of cross-polarization (CP) magic angle spinning (MAS) NMR, in which magnetization is transferred from a proton (^{1}H) to a neighbouring [29]Si nucleus, to demonstrate the presence of SiOH groups ($Q^3(OH)$), with a characteristic resonance at -100 ppm relative to the TMS (trimethylsilane) standard in the CP-MAS spectrum (Fig. 7). For a sample containing 8.7 wt % water, a small peak was also present at -91 ppm, which was assigned to $=Si(OH)_2$ ($Q^2(OH)_2$) groups in the glass. Kohn et al. (1989b) used the MAS technique to study the proton environments directly via ^{1}H NMR in the same glasses. For the SiO_2 sample containing 2.5 wt % water, they observed a narrow peak at 3.1 ppm relative to TMS, assigned to SiOH species, and a broader feature centered at 4.2 ppm due to molecular H_2O. For the sample with higher water content (8.7 wt %), four resonances were observed. The detailed assignment of these was not clear, but it was suggested that molecular water might be present, hydrogen-bonded to adjacent bridging SiOSi oxygen linkages, and a neighboring SiOH group (Kohn et al. 1989b).

Binary silicate systems

Within binary alkali and alkaline earth silicates, water solubility appears to increase with decreasing silica content, both at low (≤ 1 atm) pressures and in high pressure studies, although much work remains to be done on these systems (Schölze, 1966; Coutures and Peraudeau, 1981; McMillan and Holloway, 1987). The solubility also appears to increase with alumina content along the binary SiO_2-Al_2O_3 join (Coutures et al., 1980; Coutures and Peraudeau, 1981; McMillan and Holloway, 1987). These generalizations of the compositional dependence of water solubility are obviously complicated at high pressures and temperatures by the onset of critical behavior in SiO_2-H_2O, and perhaps also in other silicate systems (McMillan and Holloway, 1987; Paillat et al., 1992).

There have been extensive infrared studies of alkali and alkaline earth silicate glasses prepared at low pressures. These studies are often complicated by the reaction of the samples with atmospheric water, which can give rise to molecular water features in the spectra due to adsorbed H_2O (Chen and Park, 1981; Pandya et al., 1994). In addition to the O-H stretching vibration at 3600 cm^{-1}, a band appears near 2800 cm^{-1} in the infrared spectra of a number of binary silicate glasses SiO_2-M_2O-$M'O$ (M = Li, Na, K; M' = Ca, Ba) (Fig. 8) (Schölze, 1966). This feature has been assigned to O-H stretching of SiOH groups involved in intramolecular hydrogen bonding, presumably to non-bridging oxygens associated with the depolymerized silicate tetrahedra (Fig. 8; McMillan and Remmele, 1986). The appearance of such strongly hydrogen-bonded species could help rationalize the observed increase in water solubility with decreasing silica content along binary silicate joins (Coutures and Peraudeau, 1981; McMillan and Holloway, 1987), through increased stabilization of hydroxylated species via hydrogen bonding interactions. However, this argument would not explain why the water solubility in alkaline earth silicate melts is apparently so much lower than in corresponding alkali silicates (Coutures and Peraudeau, 1981; McMillan and Holloway, 1987).

There have been several infrared and Raman spectroscopic studies of alkali and alkaline earth silicate glasses containing higher water contents, prepared by quenching hydrous liquids from high pressures and temperatures (Mysen et al., 1980; McMillan and Remmele 1986; Mysen and Virgo 1986a). The spectra all show evidence for the

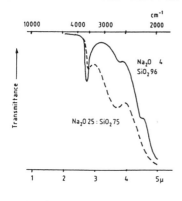

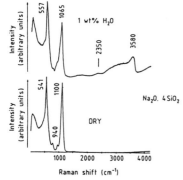

Figure 8. Infrared and Raman spectra of hydrous sodium silicate glasses. The band at 2800 cm^{-1} (and also perhaps that at 2350 cm^{-1}) is due to O-H stretching of hydrogen bonded O-H...⁻O--Si groups (see McMillan and Remmele, 1986). The weak features in the 1550-1650 cm^{-1} region of the Raman spectra are due to HOH bending of molecular H$_2$O species (1630 cm^{-1}), along with overtone vibrations of the silicate framework (~1570 cm^{-1}). The infrared spectra were redrawn from Schölze (1966) by McMillan and Holloway (1987).

presence of molecular water, through observation of the weak bending vibration at 1630 cm^{-1} (Mysen et al. 1980; McMillan and Remmele 1986; Mysen and Virgo, 1986a). The changes in the 900-1200 cm^{-1} region are complicated, and bands have not yet been definitively assigned to Si-O⁻ stretching, or Si-OH stretching or deformation vibrations. Likewise, the high frequency O-H stretching region is complex, with several broad bands corresponding to a range of hydrogen-bonded OH-containing species (Mysen et al., 1980; McMillan and Remmele, 1986; Mysen and Virgo, 1986a).

Kohn et al. (1989b) have used ^{1}H MAS NMR to investigate the hydrogen environments in alkali and alkaline earth disilicate glasses, and concluded that strongly hydrogen-bonded Si-OH species and molecular H$_2$O were present, consistent with the vibrational data, but no more detailed interpretation could be made. Kummerlen et al. (1992) used a range of NMR techniques (^{29}Si and ^{1}H MAS, ^{1}H-^{29}Si CP-MAS, and combined rotation and multiple pulse sequence (CRAMPS) for ^{1}H) to investigate dry and hydrous Na$_2$Si$_4$O$_9$ glasses. These workers observed that the proportion of fully polymerized (Q^4) silicate species decreased with addition of water. They also suggested that the proportion of Q^2 species increased, based on the observed increase in intensity of a peak near -82 to -85 ppm in the CP-MAS spectra. However, these workers did not assign any peaks in their spectra to Si-OH species, in contrast to Farnan et al. (1987) for SiO$_2$-H$_2$O samples, although such species must be present in the hydrated alkali silicate glass. The ^{1}H CRAMPS spectra clearly show two principal resonances, at 0 ppm and 7.5 ppm, relative to liquid H$_2$O. The former is assigned to molecular H$_2$O species in the glass, and the second to Si-OH species. Analysis of the molecular H$_2$O signal showed that these groups do not rotate freely within the glass, consistent with the conclusions of Kohn et al. (1989b).

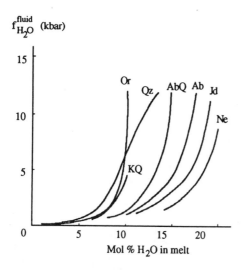

Figure 9. Water solubility (given in mole%, on a one-oxygen molar basis for the dry melt: see McMillan and Holloway, 1987) as a function of silica content for aluminosilicate liquids along the SiO_2 - $NaAlSiO_4$ (Qz, AbQ, Ab, Jd, Ne) and SiO_2 - $KAlSi_3O_8$ (Qz, KQ, Or) joins. Compiled by McMillan and Holloway (1987).

Hydrous aluminosilicate glasses

The mechanism of water dissolution in aluminosilicate glasses and melts is even less well understood, and has been the subject of some discussion in the recent literature (McMillan and Holloway, 1987; Kohn et al., 1989a, 1992, 1994; Sykes and Kubicki, 1993, 1994). For melts along the join SiO_2-$NaAlSiO_4$, water solubility appears to increase regularly with decreasing silica content, which might suggest an association between the hydrous component and Al-bearing species in the melt (Fig. 9) (Oxtoby and Hamilton, 1978a,b; McMillan and Holloway, 1987). However, this observation could be complicated by (a) the methods used to measure water solubility in those studies, and (b) the occurrence of supercritical behavior (Kennedy et al., 1962; Paillat et al., 1992). It is quite likely that the water solubilities reported by Oxtoby and Hamilton (1978a,b) represent minimum values, especially for runs at high temperature and pressure. In addition, it appears that water solubility does not increase with aluminate component along the join SiO_2-$KAlSi_3O_8$ (Oxtoby and Hamilton, 1978b; Voigt et al., 1981; McMillan and Holloway, 1987; Holtz et al., 1992; Pichavant et al., 1992) (Fig. 9).

In Raman and infrared spectroscopic studies of hydrous aluminosilicate glasses along the SiO_2-$NaAlSiO_4$ join, a new peak is observed to appear near 900 cm^{-1} on adding water to the system (Fig. 10). The relative intensity of this band increases with increasing water content, at least up to water contents of a few weight per cent (Mysen et al., 1980; McMillan et al., 1983; Mysen and Virgo, 1986b; Sykes and Kubicki, 1993). Freund (1982) and Remmele et al (1986) suggested that this peak might be due to an Al-OH stretching or bending vibration in the hydrous glass, by analogy with the assignment of the 970 cm^{-1} peak of hydrous SiO_2 glass. However, Mysen et al. (1980a) and Mysen and Virgo (1986b) observed no H/D isotope shift for this 900 cm^{-1} band, and also noted that the frequency of the band varied with glass composition. These workers concluded that it could not be due to OH-containing species, but was due an Si-O$^-$ non-bridging oxygen vibration of Q^2 sites in the hydrous glass (Mysen and Virgo, 1986b). McMillan et al. (1993) carried out a careful Raman spectroscopic study of H/D substituted sodium aluminosilicate glasses, and confirmed the lack of an isotope shift for the 900 cm^{-1} band, within experimental error (Fig. 11). However, McMillan et al. (1993) and Sykes and

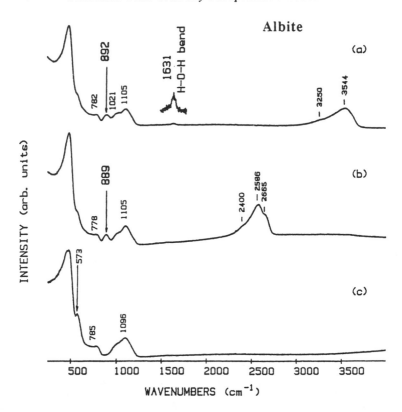

Figure 10. Raman spectra of dry and hydrous albite glasses ((a) $NaAlSi_3O_8$-H_2O; (b) $NaAlSi_3O_8$-D_2O; (c) dry), showing the O-H stretch in the 3200 to 3700 cm^{-1} region (2300 to 2700 cm^{-1} for O-D stretching) and the weak band near 1630 cm^{-1} due to the bending vibration of molecular water dissolved in the glass. The corresponding vibration of molecular D_2O occurs near 1200 cm^{-1}, and is obscured by the aluminosilicate framework bands in this region. The band near 890 to 900 cm^{-1} which appears on addition of water is indicated by an arrow (from McMillan et al., 1993).

Kubicki (1993) have argued that the band could in fact be due to a T-OH stretching or deformation vibration of hydrated aluminate species in the glass, and that the negligible H/D isotope shift is due to vibrational mode coupling. This suggestion is supported by the results of ab initio vibrational frequency calculations on hydrated aluminate fragments (McMillan et al. 1993; Sykes and Kubicki 1993).

Kohn et al. (1989a) carried out a multinuclear MAS NMR study of hydrated albitic glasses, investigating the behavior of the ^1H, ^{29}Si, ^{23}Na and ^{27}Al nuclei as a function of water content. They also carried out CP-MAS experiments, to probe the transfer of spin from ^1H nuclei to nearby ^{29}Si or ^{27}Al. In contrast to the SiO_2-H_2O glasses studied by Farnan et al. (1987), the ^1H-^{29}Si CP-MAS spectra showed no clear evidence for the presence of SiOH groups, although the spectra did change slightly compared with the single pulse ^{29}Si spectra. This suggested that protons were present in the vicinity of the silicon atoms within the structure, but not as terminal SiOH units (Kohn et al. 1989a). The ^{27}Al spectra also showed a surprising result. The spectra became narrower and more symmetric with increasing water content, consistent with a decrease in the nuclear

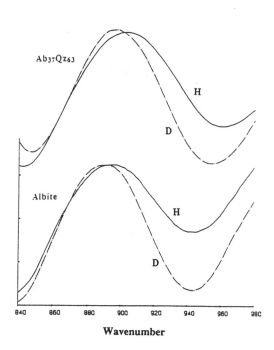

Figure 11. Detail in the 840-980 cm^{-1} region of the Raman spectra of H$_2$O- and D$_2$O-containing sodium aluminosilicate glasses (NaAlSi$_3$O$_8$, and a sample along the NaAlSi$_3$O$_8$-SiO$_2$ join). Although there is apparent slight shift in the peak maximum of the 890-900 cm^{-1} band, which could be interpreted as a small H/D isotope shift, this could simply be due to changes in the baseline arising from modifications in the higher frequency bands. The 890-900 cm^{-1} band could still be assigned to an AlOH vibration, however, due to vibrational coupling effects (from McMillan et al., 1993).

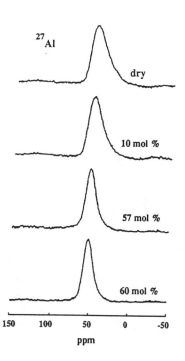

Figure 12 . ^{27}Al MAS NMR spectra for hydrous albite glasses, redrawn from Kohn et al. (1989a). Water contents are given in mole%. Scale is in ppm relative to Al^{3+} in aqueous nitrate solution. Magic angle spinning speeds were all approximately 7 kHz for these spectra.

quadrupolar coupling constant and also the chemical shift range, and suggesting that the Al environments become more regular with increasing water content (Fig. 12). This result has been borne out in further experiments on hydrous glasses along the SiO_2-$NaAlSiO_4$ join (Kohn et al., 1992) and on $NaAlSi_3O_8$ gels (Schmelz and Stebbins, 1993). Once more, there was no clear evidence for AlO-H bonding from [1]H-[27]Al CPMAS experiments, although slight shifts of the peak maximum were observed (Kohn et al., 1989a, 1992). The lack of an obvious signal which could be assigned to terminal Al-OH species, along with the observation of narrowed lines for the hydrous glasses, led Kohn et al. (1989a) to conclude that no TOT bond breaking occurred via the water dissolution mechanism. Instead, they have proposed a new model for water solubility in aluminosilicate melt compositions, based on exchange of the metal cation "charge-balancing" the bridging oxygen in SiOAl or AlOAl linkages by a proton (Fig. 13).

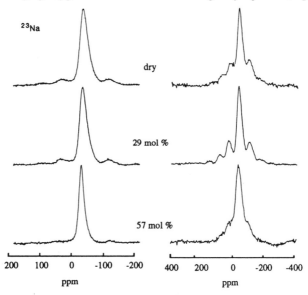

Figure 13. ^{23}Na MAS NMR spectra for hydrous albite glasses from the same series as in the previous Figure, obtained at two magnetic field strengths (8.45 T on left: 4.7 T on right). Redrawn from Kohn et al. (1989a). Scale is in ppm relative to Na^+ in aqueous solution. Combining the results from the two fields indicates that the [23]Na isotropic chemical shift (situated near the left hand edge of the resonance) becomes less negative (i.e., approaches 0 ppm) with increasing water content. The quadrupolar coupling constant, which describes the asymmetry of the site, varies little until approximately 30 mole % H_2O, then increases sharply to more than twice its original value (Kohn et al., 1989a).

In this model, the TOT linkage is not broken, but is simply weakened by attachment of hydrogen to the bridging oxygen. The observed changes in properties of hydrous aluminosilicate glasses with increasing water content, such as decrease in viscosity and increase in electrical conductivity (Burnham, 1979; Stolper, 1982a), could be explained within this model by the weakened T-O and O-H bonds (Kohn et al., 1989a, 1992; Pichavant et al., 1992). The model is also consistent with the ^{23}Na MAS NMR observations of Kohn et al. (1989a, 1992). With increasing water content, the [23]Na chemical shift becomes more positive, moving toward the value for Na^+ in aqueous solution (0 ppm) (Fig. 13). The mean quadrupolar coupling constant increases, implying that individual Na sites are more distorted, and there is a smaller range in chemical shift for the hydrous glasses (Kohn et al., 1989a, 1992). Kohn et al. (1989a, 1992) interpreted these results to indicate that water (both OH⁻ and molecular H₂O) is associated with Na⁺ ions in the hydrated glasses, consistent with their structural model (Fig. 14).

Within this new model, the 900 cm⁻¹ band observed in the Raman and infrared spectra is due to a stretching or deformation vibration of the hydrated T(OH)T linkages (Kohn et al., 1992; Pichavant et al., 1992; McMillan et al., 1993). It is interesting that, if the model is correct, it implies that one half of the OH groups in the glass are not associated with the T(OH)T linkages but are present within hydrated complexes of the type Na⁺(OH⁻)(H₂O)ₙ (Fig. 14), so do not give rise to a 900 cm⁻¹ vibration, and do not contribute to the 4500 cm⁻¹ combination band in the near-IR spectrum (Kohn et al., 1992; Pichavant et al., 1992). If this is the case, then the determined molar absorptivity for the 4500 cm⁻¹ band obtained by Stolper (1982a,b) is in error by a factor of two; however, this has no effect on the conclusions drawn by Stolper and co-workers regarding water speciation, because the intensity of the 4500 cm⁻¹ band still scales with OH content of the glass.

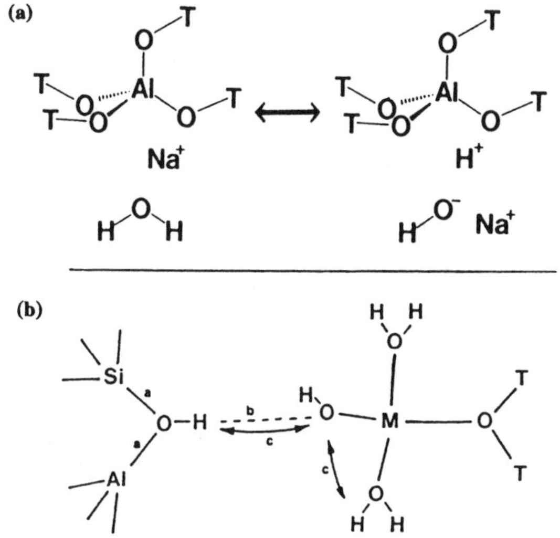

Figure 14 (a). Mechanism proposed by Kohn et al. (1989a) for exchange between molecular water and hydroxyl groups in hydrous albite glasses, consistent with their NMR observations, and vibrational spectroscopic data. Redrawn from Kohn et al. (1989). The AlOT bond is not broken during the hydration process, but is protonated as one of the H⁺ ions in the H₂O molecule is replaced by Na⁺, which moves from its original site charge-balancing the AlOT linkage. This mechanism permits the local Al site symmetry to be little affected by water dissolution, and provides an alternative to Burnham's scheme for alkali ion exchange, without requiring TOT linkages to be broken. (b). Modification of the Kohn et al. (1989a) dissolution model, suggested by Pichavant et al. (1992). The metal ion (M⁺) is coordinated by bridging oxygen (TOT), molecular water (H₂O) and OH⁻. One TOT linkage is protonated. In this picture, the Si-O and Al-O bonds are weakened ("a"), hydrogen bonding is present ("b"), and rapid proton exchange may occur ("c"). From Pichavant et al. (1992).

Pichavant et al. (1992) obtained water solubilities in the SiO₂-NaAlSi₃O₈ and SiO₂-KAlSi₃O₈ systems via phase equilibrium determinations, and confirmed the observations of Oxtoby and Hamilton (1978a) and Voigt et al. (1981), that water solubility in potassium-bearing systems is substantially lower than that in corresponding sodium aluminosilicates. They noted that this could not be ascribed to a difference in the OH/H₂O speciation, from the measurements of Silver et al. (1990). Within the context of the model proposed by Kohn et al. (1989, 1992), these workers suggested that the solubility in potassium aluminosilicate melts might be limited by the size of hydrated $K(OH)(H_2O)_n$ clusters within the aluminosilicate framework at high water content, when most of the water is dissolved as H_2O groups (Pichavant et al., 1992).

Sykes and Kubicki (1993) have criticized the interpretation of Kohn et al. (1989, 1992), based mainly on results of ab initio molecular orbital calculations for hydrated aluminosilicate clusters and infrared spectroscopic observations, and have re-interpreted the NMR results of Kohn et al. (1989, 1992). These authors suggest that water dissolution does in fact proceed by rupturing TOT linkages, most likely contained within highly strained three-membered T_3O_9 rings (Kubicki and Sykes, 1993), to give terminal AlOH units, and perhaps also some proportion of SiOH in the hydrous glass (Sykes and Kubicki, 1993). Sykes and Kubicki (1993) suggest that the ensemble of infrared, Raman and NMR spectroscopic data is more consistent with their model, than with that of Kohn et al. (1989a, 1992), which does not involve formation of terminal AlOH. This has led to a lively exchange in the recent literature. Kohn et al. (1994) have criticized the re-interpretation of their NMR data, have pointed out some apparent weaknesses in the arguments of Sykes and Kubicki (1993), and have re-stated their confidence in the dissolution mechanism involving hydration of TOT linkages. Sykes and Kubicki (1994) have replied with a re-statement of their own position, that ab initio calculations do not support the presence of protonated TOT linkages in the glass, and that the spectroscopic data remain consistent with linkage breaking to form terminal Al-OH groups.

It is probably most fair to conclude at this stage that there is no single definitive piece of evidence for or against either model, although those workers closest to the field presumably have a preferred view of the water dissolution mechanism. It is quite possible that neither model is completely correct in all its aspects, and that the true situation incorporates elements of both. We still do not have clear knowledge of the number of distinguishable types of protons present in the hydrous aluminosilicate glass, nor of the nature of the oxygen environment as a function of water content. Elucidation of the number and types of cation and anion sites in such complex multicomponent systems as $Na_\alpha Al_\beta Si_\gamma H_\delta O_\varepsilon$ is a difficult task, which is obviously far from complete, and may never be resolved in an absolute sense. In any case, it is highly unlikely that the proton environments in the low temperature glass actually represent the instantaneous positions in the hydrous liquid, at high pressures and temperatures.

IN-SITU STUDIES AT HIGH TEMPERATURE AND PRESSURE, AND EFFECTS OF RELAXATION AND PROTON EXCHANGE

Aines et al. (1983) carried out in-situ near-IR studies on hydrous albitic glasses to temperatures above the glass transition, and demonstrated that both molecular H_2O and OH groups were still present, with an OH/H_2O ratio which depended slightly on temperature. There have been several further experimental studies carried out which demonstrate a temperature- or quench rate-dependence to the H_2O/OH ratio in the hydrous glasses, in addition to the effects of bulk composition and total water content (Stolper et al., 1983; Silver, 1988; Stolper, 1989; Ihinger 1991; Zhang et al. 1994).

It is important to recognize that there are at least two widely different time scales which are likely to be important in understanding the kinetics and equilibration of water dissolution processes. One involves any proton exchange reactions between hydrated species within the glass or melt (Eckert et al., 1987; Kohn et al., 1989a, 1992; Pichavant et al., 1992). These reactions are likely to be rapid compared with laboratory quench rates, but slower than the vibrational timescale, for temperatures below approximately 1200°C (Keppler and Bagdassarov, 1993). For example, such proton exchange reactions in hydrous silicate liquids and glasses might be compared with H^+ transfer processes in aqueous solution and other protonated solvents, which occur with rates on the order of 10^{-7} to 10^{-12} s, at room temperature (Robinson et al. 1986; Lee et al. 1986).

The second, slower, timescale is that for equilibration of the dissociation reaction involving molecular water and hydroxyl groups ($H_2O(m) + O^{2-}(m) \Leftrightarrow 2\ OH^-(m)$) (Stolper, 1989; Dingwell and Webb, 1990; Zhang et al., 1994), with or without accompanying TOT bond breaking (Kohn et al., 1989a, 1992; Pichavant et al., 1992; Sykes and Kubicki, 1993). This reaction timescale is most likely to couple with that of structural re-equilibration of the aluminosilicate melt network (Dingwell and Webb, 1990).

Stolper and colleagues have carried out extensive series of experiments to explore the effects of water content, bulk melt composition, run temperature and quench rate on the OH/H_2O ratio determined by equilibrium (5) (Stolper, 1982a,b, 1989; Silver and Stolper, 1988; Silver et al., 1990; Zhang et al., 1994). In particular, Silver (1988), Silver and Stolper (1988), and Silver et al. (1990) provide data on the variation of the OH/H_2O ratio with water content at quench rates which differ by approximately two orders of magnitude (200°C/min and 200°C/s). Dingwell and Webb (1990) suggested that the equilibration kinetics of Reaction (5) would be determined by the timescale for structural relaxation (viscous flow) of the hydrous aluminosilicate melt. They used Shaw's (1972) method for estimating viscosities of dry and hydrous silicate liquids to construct relaxation curves (log τ vs 1/T) for rhyolite melts with 1, 3 and 5 wt % total H_2O (Fig. 15). The glass transition, or the temperature at which the speciation according to Reaction (1) is frozen in, is given by the point at which the relaxation time (τ) is equal to the quench rate. Dingwell and Webb (1990) used this approach to correct the H_2O/OH speciation data of Silver (1988) to the fictive temperature (T_g), corresponding to the glass transition for that water content and quench rate from the hydrous liquid. The glass transition temperature depends strongly on water content, dropping from approximately 1080 K for 1 wt % total water (200°C/s quench) to 790 K for 5 wt % water, and is lowered for a slower quench rate (approximately 950 K for 1 wt % water at 200°C/min, and 710 K for 5 wt %) (Dingwell and Webb 1990). Dingwell and Webb found that, for a given total water content, the proportion of water as OH groups in the melt increased rapidly with temperature, with an enthalpy change for Reaction (5) of 25±5 kJ/mol.

Stolper (1989) investigated the temperature dependence of water speciation in a series of equilibration experiments at room pressure, for obsidian glasses containing 0.6 to 1.7 wt % total water. Most of the runs were carried out below 600°C, and for all except two runs (1.89 wt %, 900°C; 1.75 wt %, 700°C), were below the glass transition temperature determined by the relaxation curves of Dingell and Webb (1990). In this case, the run temperature can be taken as the fictive temperature, and equilibration is achieved when the run duration is equal to or greater than the relaxation time (10^5 to 10^6 s for 1 wt % water and 400°C). In apparent contrast to Dingwell and Webb (1990), Stolper (1989) found a small temperature dependence of the OH/H_2O ratio, with ΔH for Reaction (1) of 6.1 kJ/mol. It could be argued that the difference between the two analyses arises because one (Stolper 1989) refers to the glassy state, in which the re-equilibration kinetics of Reaction (5) are unaccompanied by large-scale structural relaxation, and the other (Dingwell and Webb 1990) refers to the supercooled liquid above Tg, in which both structural relaxation and local OH/H_2O re-equilibration are present. In fact, in their recent analysis of relaxation times in this system, Zhang et al.

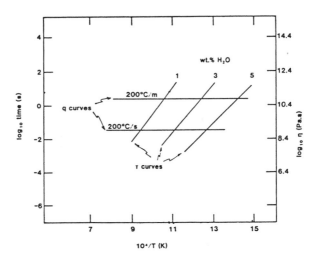

Figure 15. Relaxation (τ) curves for hydrous melts, redrawn from Dingwell and Webb (1990). These curves represent the logarithm of the time for structural relaxation, calculated from Shaw's (1972) model for viscosity (given at right in Pa.s) of a rhyolitic composition with 1, 3, and 5 wt % H_2O, versus reciprocal temperature. These τ curves represent the time necessary to achieve structural relaxation in the hydrous aluminosilicate liquid at a given temperature. The horizontal lines drawn at given values of $\log_{10}$ time are the "q" curves, determined by the experimental quench rate (200°C/min and 200°C/s, in the experiments of Silver, 1988). The intersection of the τ and q curves gives the temperature at which the supercooled liquid falls out of equilibrium; i.e., can no longer relax during the timescale determined by the quench: this intersection defines the glass transition temperature for that composition and quench rate.

(1994) have demonstrated that equilibration of Reaction (5) can not be represented by a single simple relaxation in the vicinity of the glass transformation. However, the values of the equilibrium constant ln K_1 as defined by Stolper (1989) differ by almost a factor of ten from those calculated by Dingwell and Webb (1990), and also from those reported by Zhang et al. (1994), for the same experimental data. This is the most likely source of the difference between values obtained for ΔH in the two studies.

Zhang et al. (1994) have recently carried out another extensive series of equilibration experiments to investigate the relaxation timescale of Reaction (5) as a function of water content. All runs were well below the glass transition temperatures defined by Dingwell and Webb (1990), except for samples with 2.36 wt % total water run at 550° and 600°C. Zhang et al. (1994) suggested that it was important to distinguish between a fictive temperature for structural relaxation, determined by the structural relaxation time for viscous flow ($T_f \equiv T_g$), and an apparent equilibration temperature, T_{ae}, determined by the timescale for equilibration of Reaction (5). They considered the reaction quotient, Q, determined by Reaction (5):

$$Q = \frac{[OH^-]^2}{[H_2O][O^{2-}]},$$
(6)

equilibrated natural obsidian or synthetic hydrous albite at high or low temperature, and observed the return of Q to equilibrium ($Q \equiv K_1$, the equilibrium constant for Reaction (5)).

Zhang et al. (1994) concluded that, although the approach of Dingwell and Webb (1990) was correct in spirit, it was quantitatively incorrect because the time scale for re-equilibration of the dissolution Reaction (5) (τ_r) was not equal to that for structural relaxation (τ_s) ; i.e., $T_{ae} \neq T_g$. However, the results are in fact remarkably close, if it is recognized that the relaxation timescale used by Dingwell and Webb (1990) was taken from the empirical viscosity parametrization of Shaw (1972): the fact that τ_r and τ_s agree to within an order of magnitude in fact strongly suggests that both timescales are the same, at least at high temperatures.

Zhang et al. (1994) then examined the possibility of a rapid re-equilibration of Reaction (1) at low temperature, which might affect the OH/H_2O speciation. They took a sample which had been equilibrated at 500°C, then cooled it in 50°C decrements at 5 minute intervals, and examined the OH/H_2O ratio (ln Q). As expected, Q decreased between 450° and 400°C, then remained unchanged on further cooling. Zhang et al. (1994) took this result to indicate that there was no low temperature re-equilibration component to Reaction (1).

However, when taken together with the structural relaxation considered by Dingwell and Webb (1990), a different picture emerges. The data of Zhang et al. (1994) demonstrate that the Reaction (5) falls out of equilibrium (i.e., Q is frozen in) at 400°C, for a cooling rate of 50°C/5 min, or 0.17°C/s. This defines a sampling time scale of 0.17^{-1} = 6 s. However, the curves drawn by Dingwell and Webb (1990) indicate that the structural relaxation, as determined by shear viscosity, for a rhyolite containing 1 wt % water falls out of equilibrium at 650°C (T_g) for the same quench rate (0.17°C/s = 10°C/min), a much higher temperature. This could well indicate that relaxation of the overall structure of the hydrous silicate melt in response to applied shear stress occurs on a longer time scale than that for local re-equilibration of the speciation Reaction (5), if in fact the viscosity prediction is accurate enough for this comparison. This will be important to test in future work. Such behavior would imply that the OH/H_2O speciation is affected by two distinct mechanisms: a local equilibration of the OH/H_2O ratio, and an overall relaxation of the entire aluminosilicate melt structure in response to applied external stress. In the high temperature limit, above the glass transition determined by the structural relaxation, both timescales are the same. However, at the structural glass transition, the structural relaxation is frozen, although the local OH/H_2O re-equilibration can still occur, until it becomes fixed at some lower temperature. In the language of glassy and liquid state relaxation theory, the local OH/H_2O exchange would be considered a "β" relaxation process, distinct from the primary structural, or "α", process (McCrum et al., 1991). This behavior could well be responsible for the complex relaxation behavior observed by Zhang et al. (1994) for samples equilibrated at high temperature, for short heating periods, due to the interaction between the two timescales. This important and interesting topic obviously deserves further study.

Finally, it is important to distinguish between the local OH/H_2O re-equilibration suggested above and rapid dynamic proton transfer processes which may well occur in the hydrous glass, even at temperatures close to room temperature. Solid state NMR studies have indicated that the molecular H_2O species in hydrous glasses undergo rotation about their two-fold axes on a time scale of $\sim 10^{-5}$ s at room temperature (Eckert et al., 1987: see also Kohn et al., 1989b). At temperatures above 100°C, the ^2H NMR studies of Eckert et al. (1987) suggested that hydrogen (deuterium) exchange between OH groups and molecular H_2O might begin to occur. The recent infrared spectroscopic results of Keppler and Bagdassarov (1993) suggest that, at temperatures on the order of 1300°C, proton hopping between oxygens in the melt might begin to occur, on a vibrational timescale (10^{-14} s). This would be completely decoupled from the primary structural relaxation timescale, as is the timescale for conductivity of alkali cations in silicate glasses (Dingwell and Webb, 1990).

CONCLUSION

From this survey, it is obvious that much progress has been made in measuring and understanding the extent, mechanisms, and implications of water dissolution in magmatic liquids, but also that much fundamental work remains to be done. Sufficient solubility determinations have still not been made to clearly define the dependence of solubility on bulk melt composition, and we have only begun to realize the potential importance of supercritical behavior in hydrous aluminosilicate systems at high pressures and temperatures. More experiments are needed to fully define the high P-T phase diagrams of hydrous aluminosilicates, to assess for which systems a fluid phase rich in aluminosilicate is present. The importance of atomic or ionic diffusion has long been recognized in determining geologic processes, but the kinetics of species equilibration and structural relaxation are only beginning to receive equal attention. There is now an opportunity for detailed study of the diffusional, re-equilibration and relaxation timescales in hydrous aluminosilicate liquids and glasses, which will yield fundamental information on the detailed mechanism of the processes, and practical insights into the nature of magmatic phenomena ranging from eruption dynamics (ascent rates, vesiculation) to cooling and crystallization pathways.

REFERENCES

Adams RV, Douglas RW (1969) Infra-red studies on various samples of fused silica with special reference to the bands due to water. J Soc Glass Tech 43:147T-158T

Aines RD, Silver LA, Rossman GR, Stolper EM, Holloway JR (1983) Direct observation of water speciation in rhyolite at temperatures up to 850°C. Geol Soc Am Abstr Programs 15:512

Anderson GM, Burnham CW (1983) Feldspar solubility and the transport of aluminum under metamorphic conditions. Am J Sci 283A:283-297

Bartholomew RF, Schreurs JWH (1980) Wide-line NMR study of protons in hydrosilicate glasses of different water content. J Non-Cryst Solids 38&39:679-684

Bartholomew RF, Butler BL, Hoover HL, Wu CK (1980) Infrared spectra of a water-containing glass. J Am Ceram Soc 63:481-485

Blank JG, Stolper EM, Carroll MR (1993) Solubilities of carbon dioxide and water in rhyolitic melt at 850°C and 750 bars. Earth Planet Sci Letts 119:27-36

Bedford RG (1975) The solubility of water in molten silicates. Glass Technol 16:20-24

Bowen NL (1928) The Evolution of the Igneous Rocks. Princeton Univ Press, New Jersey

Burnham CW (1974) $NaAlSi_3O_8$-H_2O solutions: a thermodynamic model for hydrous magmas. Bull Soc fr Min Cristallogr 97:223-230

Burnham CW (1975a) Water and magmas: a mixing model. Geochim Cosmochim Acta 39:1077-1084

Burnham CW (1975b) Thermodynamics of melting in experimental silicate-volatile systems. Fortschr Min 52:101-118

Burnham CW (1979) The importance of volatile constituents. In: Yoder HS Jr (ed) The Evolution of the Igneous Rocks: Fiftieth Anniversary Perspectives. Princeton Univ Press, p 439-482

Burnham (1981) The nature of multicomponent aluminosilicate melts. In: DT Rickard, FE Wickman (eds) Chemistry and Geochemistry of Solutions at High Temperatures and Pressures, Phys Chem Earth 13&14: 191-227

Burnham CW, Jahns RH (1962) A method for determining the solubility of water in silicate melts. Am J Sci 260:721-745

Burnham CW, Davis NF (1971) The role of H_2O in silicate melts I. P-V-T relations in the system $NaAlSi_3O_8$-H_2O to 10 kilobars and 1000°C. Am J Sci 270:54-79

Burnham CW, Davis NF (1974) The role of H_2O in silicate melts: II. Thermodynamic and phase relations in the system $NaAlSi_3O_8$-H_2O to 10 kilobars, 700° to 1000°C. Am J Sci 274:902-940

Burnham CW, Nevkasil H (1986) Equilibrium properties of granite pegmatite magmas. Am J Sci 71:239-263

Chen H, Park JW (1981) Atmospheric reaction at the surface of sodium disilicate glass. Phys Chem Glasses 22:39-42

Cocheo PA (1994) The solubility of water in basanitic melts at low pressures. MS thesis, Arizona State University, Tempe, AZ

Coutures JP, Devauchelle JM, Munoz R, Urbain G (1980) Dissolution of water vapor in SiO_2-Al_2O_3 melts. Rev Int Hautes Temp Réfractaires 17:351-361

Coutures JP, Peraudeau G (1981) Examen de la dissolution de l'eau dans les silicates liquides [$P(H_2O)\leq 1$ atm]. Rev Int Hautes Temp Réfractaires 18:321-346

Day HW, Fenn PM (1982) Estimating the P-T-X_{H_2O} conditions during crystallization of low calcium granites. J Geol 90:485-507

Dingwell DB, Webb SL (1990) Relaxation in silicate melts. Eur J Min 2:427-449

Dingwell DB, Harris DW, Scarfe CM (1984) The solubility of H_2O in melts in the system SiO_2-Al_2O_3-Na_2O-K_2O at 1 to 2 kilobars. J Geol 92:387-395

Dixon JE, Stolper EM, Holloway J (1994) An experimental study of water and carbon dioxide solubilities in mid-ocean ridge basalts. J Petrol, in press.

Eckert H, Yesinowski JP, Stolper EM, Stanton TR, Holloway J (1987) The state of water in rhyolitic glasses. A deuterium NMR study. J Non-Cryst Solids 93:94-114

Eckert H, Yesinowski JP, Silver LA, Stolper EM (1988) Water in silicate glasses: Quantitation and structural studies by ^{1}H solid echo and MAS-NMR methods. J Phys Chem 92:2055-2064

Eggler DH (1972) Water-saturated and undersaturated melting relations in a Paricutin andesite and an estimate of water content in the natural magma. Contrib Min Petrol 34:261-271

Eggler DH (1973) Role of CO_2 in melting processes in the mantle. Carnegie Inst Wash Yearb 72:457-467

Eggler DH, Rosenhauser M (1978) Carbon dioxide in silicate melts. II. Solubilities of CO_2 and H_2O in $CaMgSi_2O_6$ (diopside) liquids and vapors at pressures to 40 kilobars. Am J Sci 278:64-94

Eggler DH, Burnham CW (1984) Solution of H_2O in diopside melts: a thermodynamic model. Contrib Min Petrol 85:58-66

Ernsberger FM (1977) Molecular water in glass. J Am Ceram Soc 60:91

Farnan I, Kohn SC, Dupree R (1987) A study of the structural role of water in hydrous silica glass using cross-polarisation magic angle spinning NMR. Geochim Cosmochim Acta 51:2869-2873

Freund F (1982) Solubility mechanisms of H_2O in silicate melts at high pressures and temperatures: a Raman spectroscopic study: discussion. Am Min 67:153-154

Friedman I, Long W, Smith RL (1963) Viscosity and water content of rhyolite glass. J Geophys Res 68:6523-6535

Goldsmith JR, Jenkins DM (1985) The hydrothermal melting of low and high albite. Am Min 70:924-933

Goldsmith JR, Petersen JW (1990) Hydrothermal melting behavior of $KAlSi_3O_8$ as microcline and sanidine. Am Min 75:1362-1369

Goranson RW (1931) The solubility of water in granitic magmas. Am J Sci 22:481-502

Goranson RW (1937) The osmotic pressure of silicate melts. Am Min 22:485

Hamilton DL, Oxtoby S (1986) Solubility of water in albite melt determined by the weight-loss method. J Geol 94:626-630

Hamilton DL, Burnham CW, Osborn EF (1964) The solubility of water and effects of oxygen fugacity and water content on crystallization in mafic magmas. J Petrol 5:21-39

Hartwig CM, Rahn LA (1977) Bound hydroxyl in vitreous silica. J Chem Phys 67:4260-4261

Hodges FN (1973) Solubility of H_2O in forsterite melt at 20 kilobars. Carnegie Inst Wash Yearb 72:495-497

Hodges FN (1974) The solubility of H_2O in silicate melts. Carnegie Inst Wash Yearb 73:251-255

Holtz F, Behrens H, Dingwell DB, Taylor RP (1992) Water solubility in aluminosilicate melts of haplogranite composition at 2 kbar. Chem Geol 96:289-302

Huffman M, McMillan P (1985) A Raman and infrared study of chemically vapor deposited amorphous silica. J Non-Cryst Solids 76:369-379

Ihinger PD (1991) An experimental study of the interaction of water with granitic melt. PhD dissertation, California Inst Technology, Pasedena, CA

Ingerson E (1960) The water content of primitive grantic magma. Am Min 35:806-815

Jahns RH, Burnham CW (1958) Experimental studies of pegmatite genesis: the solubility of water in granitic melts. Geol Soc Am Bull 69:1544-1555

Kadik AA, Lebedev YB (1968) Temperature dependence of the solubility of water in an albite melt at high pressures. Geochem Int 5:1172-1181

Kadik AA, Lukanin OA, Lebedev YB, Korovughkina EY (1972) Solubility of H_2O and CO_2 in granite and basalt melts at high pressures. Geochem Int 9:1041-1051

Keller WD, Pickett EE (1954). Hydroxyl and water in perlite from Superior, Arizona. Am J Sci 252:87-98

Keppler H, Bagdassarov NS (1993) High-temperature FTIR spectra of H_2O in rhyolite melt to 1300°C. Am Min 78:1324-1327

Kennedy GC, Wasserburg GJ, Heard HC, Newton RC (1962) The upper three-phase region in the system SiO_2-H_2O. Am J Sci 260:501-521

Khitarov NI (1960) Relations between water and magmatic melts. Geochem 7:699-703

Khitarov NI, Lebedev EB, Rengarten EV, Arsen'va RV (1959) The solubility of water in basaltic and granitic melts. Geokhimya 5:479-492

Khitarov NI, Kadik AA, Lebedev YB (1968) Solubility of water in a basalt melt. Geochem Int 5:667-674

Kohn SC, Dupree R, Smith ME (1989a) A multinuclear magnetic resonance study of the structure of hydrous albite glass. Geochim Cosmochim Acta 53:2925-2935

Kohn SC, Dupree R, Smith ME (1989b) Proton environments and hydrogen-bonding in hydrous silicate glasses from proton NMR. Nature 337:539-541

Kohn SC, Dupree R, Golam Mortuza M (1992) The interaction between water and aluminosilicate magmas. Chem Geol 96:399-409

Kohn SC, Smith ME, Dupree R (1994) Comment on "A model for H_2O solubility mechanismsn in albite melts from infrared spectroscopy and molecular orbital calculations" by D. Sykes and J.D. Kubicki. Geochim Cosmochim Acta 58:1377-1380

Kubicki JD, Sykes D (1993) Molecular orbital calculations of vibrations in three-membered aluminosilicate rings. Phys Chem Minerals 19:381-391

Kummerlen J, Merwin LH, Sebald A, Keppler H (1992) Structural role of H_2O in sodium-silicate glasses - results from Si-29 and H-1 NMR spectroscopy. J Phys Chem 96:6405-6410

Kurkjian CR, Russell LE (1958) Solubility of water in molten alkali silicates. J Soc Glass Technol 42:130T-144T

Lee J, Robinson GW, Webb SP, Philips LA, Clark JH (1986) Hydration dynamics of protons from photon initiated acids. J Am Chem Soc 108:6538-6542

Luth RW, Mysen BO, Virgo D (1987) Raman spectroscopic study of the solubility behavior of H_2 in the system Na_2O-Al_2O_3-SiO_2-H_2. Am.Min.72:481-486

McCrum NG, Read BE, Williams G (1991) Anelastic and Dielectric Effects in Polymeric Solids. Dover Pub., New York

McMillan P, Jakobsson S, Holloway JR, Silver LA (1983) A note on the Raman spectra of water-bearing albite glasses. Geochim Cosmochim Acta 47:1937-1943

McMillan P, Remmele RL (1986) Hydroxyl sites in SiO_2 glass: a note on infrared and Raman spectra. Am Min 71:772-778

McMillan PF, Peraudeau G, Holloway JR, Coutures JP (1986) Water solubility in a calcium aluminosilicate melt. Contrib Min Petrol 94:178-182

McMillan P, Holloway JR (1987) Water solubility in aluminosilicate melts. Contrib Min Petrol 97:320-332

McMillan PF, Stanton TR, Poe BT, Remmele RL (1993) A Raman spectroscopic study of H/D isotopically substituted hydrous aluminosilicate glasses. Phys Chem Minerals 19:454-459

Moulson AJ, Roberts JP (1961) Water in silica glass. Trans Faraday Soc 57:1208-1216

Müller-Warmuth vW (1965a) Magnetische Resonanz in Gläsern. Teil I: Einführung und Grundlagen. Glastech Ber 38:121-133

Müller-Warmuth vW (1965b) Magnetische Resonanz in Gläsern. Teil II: Ergebnisse.Glastech Ber 38:405-414

Mysen BO (1977) The solubility of H_2O and CO_2 under predicted magma genesis conditions and some petrological and geophysical implications. Rev. Geophys. Space Phys. 15:351-361

Mysen BO (1983) The solubility mechanisms of volatiles in silicate melts and their relations to crystal-andesite liquid equilibria. J Volcanological Geothermal Res 18:361-385

Mysen BO (1990) Interaction between water and melt in the system $CaAl_2O_4$-SiO_2-H_2O. Chem Geol 88:223-243

Mysen B, Virgo D (1980) Solubility mechanism of water in basalt at high pressures and temperatures: $NaCaAlSi_2O_7$-H_2O as a model. Am Min 65:1176-1184

Mysen BO, Virgo D (1982) Solubility mechanisms of H_2O in silicate melts at high pressures and temperatures: a Raman spectroscopic study: reply. Am Min 67:155

Mysen BO, Virgo D (1985) Raman spectra and structure of fluorine- and water-bearing silicate glasses and melts. In: Advances in Materials Characterization II. RL Snyder, RA Condrate, PF Johnson (eds). Plenum Pub Co, New York. p 43-55

Mysen BO, Virgo D (1986a) Volatiles in silicate melts at high pressure and temperature. 1. Interaction between OH groups and Si^{4+}, Al^{3+}, Ca^{2+}, Na^+ and H^+. Chem Geol 57:303-331

Mysen BO, Virgo D (1986b) Volatiles in silicate melts at high pressure and temperature. 2. Water in melts along the join $NaAlO_2$-SiO_2 and a comparison of solubility mechanism of water and fluorine. Chem Geol 57:333-358

Mysen BO, Virgo D, Harrison WJ, Scarfe CM (1980) Solubility mechanism of H_2O in silicate melts at high pressures and temperatures: a Raman spectroscopic study. Am Min 65:900-914

Newman S, Stolper EM, Epstein S (1986) Measurement of water in rhyolitic glasses: calibration of an infrared spectroscopic technique. Am Min 71:1527-1541

Orlova GP (1962) The solubility of water in albite melts. Int Geol Rev 6:254-258

Ostrovskiy IA, Orlova GP, Rudnitskaya YS (1964). Stoichiometry in the solution of water in alkali-aluminosilicate melts. Doklady Akad Nauk SSR 157:149-151

Oxtoby S, Hamilton DL (1978a) The discrete association of water with Na_2O and SiO_2 in NaAl silicate melts. Contrib Min Petrol 66:185-188

Oxtoby S, Hamilton DL (1978b) Solubility of water in melts of Na_2O-Al_2O_3-SiO_2 and K_2O-Al_2O_3-SiO_2 systems. In: WS Mackenzie (ed) Progress in Experimental Petrology, vol 4, Natural Environment Research Council, Manchester, England. p 33-36

Oxtoby S, Hamilton DL (1978c) Water in plagioclase melts. In: WS Mackenzie (ed) Progress in Experimental Petrology, vol 4, Natural Environment Research Council, Manchester, England. p 36-37

Oxtoby S, Hamilton DL (1978d) Calculation of the solubility of water in granitic melts. In: WS Mackenzie (ed) Progress in Experimental Petrology, vol 4, Natural Environment Research Council, Manchester, England. p 37-40

Oxtoby S, Hamilton DL (1979) The solubility of water in alkali-alumino-silicate melts to 8 kbars. In: KD Timmerhaus and MS Barber (eds) High-Pressure Science and Technology, Proc 6th AIRAPT Conf, vol 2, Plenum Press, New York. p 153-158

Paillat O, Elphick SC, Brown WL (1992) The solubility of water in $NaAlSi_3O_8$ melts: a re-examination of Ab-H_2O phase relationships and critical behaviour at high pressures. Contrib Min Petrol 112:490-500

Pandya N, Muenow DW, Sharma SK, Sherriff BL (1994) The speciation of water in hydrous alkali silicate glasses. J Non-Cryst Solids (in press)

Pichavant M (1987) Effects of B and H_2O on liquidus phase relations in the haplogranite system at 1 kbar. Am Min 72:1056-1070

Pichavant M, Holtz F, McMillan P (1992) Phase relations and compositional dependence of H_2O solubility in quartz-feldspar melts. Chem Geol 96:303-319

Remmele R, Stanton T, McMillan P and Holloway J (1986) Raman spectra of hydrous glasses along the quartz-albite join. EOS Trans Am Geophys Union 67:1274

Robinson GW, Thistlethwaite PJ, Lee J (1986) Molecular aspects of ionic hydration reactions. J Phys Chem 90:4224-4233

Rosenhauer M, Eggler DH (1975) Solution of H_2O and CO_2 in diopside melt. Carnegie Inst Wash Yearb 74:474-479

Russell LE (1957) Solubility of water in molten glass. J Soc Glass Technol 41:304T-317T

Schmelz CE, Stebbins JF (1993) Gel synthesis of an albite ($NaAlSi_3O_8$) glass. Geochim Cosmochim Acta 57:3949-3960

Schölze H (1966) Gases and water in glass: Parts one, two and three. The Glass Industry 47:546-551; 622-628; 670-674

Schölze H, Mulfinger HO, Franz H (1962) Measurement of the physical and chemical solubility of gases in glass melts (He and H_2O). In: Advances in Glass Technology. Am Ceram Soc Plenum Press. p 230-248

Shaw HR (1964) Theoretical solubility of H_2O in silicate melts: quasi-crystalline models. J Geol 72:601-617

Shaw HR (1972) Viscosities of magmatic silicate liquids: an empirical method of prediction. Am J Sci 272:870-893

Silver LA (1988) Water in silicate glasses. PhD dissertation, California Inst Tech, Pasedena, CA

Silver L, Stolper E (1985) A thermodynamic model for hydrous silicate melts. J Geol 93:161-177

Silver L, Stolper E (1989) Water in albitic glasses. J Petrol 30:667-709

Silver LA, Ihinger PD, Stolper E (1990) The influence of bulk composition on the speciation of water in silicate glasses. Contrib Min Petrol 104:142-162

Stanton TR (1989) High pressure isotopic studies of the water diffusion mechanism in silicate melts and glasses. PhD dissertation, Arizona State University, Tempe, AZ

Stanton TR, Holloway JR, Hervig RL, Stolper EM (1985) Isotope effect on water diffusivity in silicic melts: an ion microprobe and infrared analysis. Eos Trans Am Geophys Union 66:1131

van der Steen, GHAM, van den Boom H (1977) Raman spectroscopic study of hydrogen-containing vitrous silica. J Non-Cryst Solids 23:279-286

Stolper E (1982a) The speciation of water in silicate melts. Geochim Cosmochim Acta 46:2609-2620

Stolper E (1982b) Water in silicate glasses: an infrared spectroscopic study. Contrib Min Petrol 81:1-17

Stolper E (1989) The temperature dependence of the speciation of water in rhyolitic melts and glasses. Am Min 74:1247-1257

Stolper E, Silver LA, Aines RD (1983) The effects of quenching rate on the speciation of water in silicate glasses. EOS Trans Am Geophys Union 65:339

Sykes D, Kubicki JD (1993) A model for H_2O solubility mechanisms in albite melts from infrared spectroscopy and molecular orbital calculations. Geochim. Cosmochim. Acta 57:1039-1052

Sykes D, Kubicki JD (1994) Reply to the Comment by S.C. Kohn, M.E. Smith, and R. Dupree on "A model for H_2O solubility mechanisms in albite melts from infrared spectroscopy and molecular orbital calculations". Geochim Cosmochim Acta 58:1381-1384

Takata M, Acocella J, Tomozawa M, Watson EB (1981) Effect of water content on the electrical conductivity of $Na_2O.3SiO_2$ glass. J Am Ceram Soc 64:19-724

Tomlinson JW (1956) A note on the solubility of water in molten sodium silicate. J Soc Glass Technol 4:25T-31T

Tuttle OF, Bowen NL (1958) Origin of granite in the light of experimental studies in the system $NaAlSi_3O_8$-$KAlSi_3O_8$-SiO_2-H_2O). Geol Soc Am Mem 74:1-154

Voigt DE, Bodnar RJ, Blencoe JG (1981) Water solubility in melts of alkali feldspar composition at 5 kilobars, 950^oC. Eos Trans Am Geophys Union 62:428

Wasserburg GJ (1957) The effects of H_2O in silicate systems. J Geol 65:15-23

Wu CK (1980) Nature of incorporated water in hydrated silicate glasses. J Am Ceram Soc 63:453-457

Zhang Y, Stolper EM (1991) Water diffusion in a basaltic melt. Nature 351:306-309

Zhang Y, Stolper EM, Wasserburg GJ (1991) Diffusion of water in rhyolitic glasses. Geochim Cosmochim Acta 55:441-456

Zhang Y, Stolper EM, Ihinger PD (1994) Kinetics of the reaction $H_2O+ O = 2OH$ in felsic glasses: Preliminary results. Am Min, in press

Chapter 5

EXPERIMENTAL STUDIES OF CARBON DIOXIDE IN SILICATE MELTS: SOLUBILITY, SPECIATION, AND STABLE CARBON ISOTOPE BEHAVIOR

Jennifer G. Blank

Institut de Physique du Globe de Paris
Laboratoire de Systèmes Dynamiques et Géologiques
4 Place Jussieu, 75 252 Paris Cedex 05 France

Richard A. Brooker

University of Bristol
Department of Geology, Wills Memorial Building
Queens Road, Bristol BS8 1RJ UK

INTRODUCTION

Experimental studies of volatiles in silicate melts provide a framework for the interpretation of observations made of natural magmatic systems. Their collective aim an understanding of the behavior of volatiles in melts and the role they play in magmas. As most of the magmas erupted at the earth's surface have degassed to a large extent, experimental studies offer a unique view to volatile-related processes occurring at depth.

Volatile solubility experiments typically involve exposing a silicate melt to a high pressure fluid phase at high (super-liquidus) temperatures. The melts are quenched to glasses and then analyzed for dissolved volatile contents (Chapter 2). Use of spectroscopic techniques can provide important information on the nature of the volatile species dissolved in the melts. More importantly, the combination of solubility and speciation information allows formulation of thermodynamic models and provides insight into possible dissolution mechanisms and related changes in the physical properties of the melt. Studies of the partitioning of stable isotopes of volatile components between vapor[1] and melts offer another source of information about volatile speciation and its effect on melt structure. The stable isotope systematics allow volatiles to be used as tracers of geologic environments and processes such as partial melting, degassing, and magma mixing.

Although CO_2 is the second most abundant volatile emitted at volcanic centers (see Chapters 1 and 12), its relatively low concentrations in erupted lavas (typically <200 ppm [0.02 wt %] in submarine basalts erupted on the seafloor, «30 ppm in rhyolitic glasses) and the ubiquity of C-bearing materials as potential contaminants have historically made analysis of CO_2 in silicate glasses difficult. Early experiments with CO_2 and silicate melts at moderate pressures (<1600 bars, e.g., Morey and Fleischer, 1940; England and Adams, 1951; Wyllie and Tuttle, 1959) produced glasse with levels of dissolved CO_2, below the detection limits of available analytical techniques, and therefore relied on indirect

---

[1] The term "vapor" is used throughout this chapter due to its common usage in the literature. The term "fluid" is more correct as volatiles are super-critical in most experiments (cf. Holloway, 1981).

measurements to determine solubilities. Some of the analytical difficulties became apparent following the work on lunar soils in the mid-1970s; lunar samples exposed to the terrestrial atmospheric environment yielded significantly greater quantities of carbon which was later demonstrated to be surficial contamination (e.g., Des Marais, 1978). With recent developments of microbeam analytical techniques, determinations of dissolved carbon species and abundances are now possible on a routine basis to the ppm level, and these techniques promise even higher resolution in the near future (e.g., Wright and Pillinger, 1989).

This chapter presents a general overview of the experimental and spectroscopic work pertaining to the behavior of carbon dioxide in synthetic (analog) and natural silicate melt compositions. After a brief description of methodology, we summarize relevant laboratory studies, addressing (1) solubility measurements; (2) speciation; (3) dissolution mechanisms; and (4) stable isotope fractionation. Practical applications of some of the material discussed here are presented in Chapter 6.

EXPERIMENTAL AND ANALYTICAL METHODS

Experiments

The type of apparatus used for determination of CO_2 solubilities depends on the pressure range of interest. At pressures from a few hundred bars to several kilobars, either cold-seal (T < ~1200°C for TZM, Williams, 1968; Kerrick, 1987) or internally heated (T up to ~1400°C, Yoder, 1950; Holloway, 1971) pressure vessels have been used. At higher pressures, piston cylinder devices (P to 35 kbar, T to 1750°C, e.g., Mysen et al., 1976) have been used extensively (for illustrations and developmental histories of this equipment, refer to Holloway and Wood, 1988). As cold-seal and internally-heated pressure vessels have large heat capacities and cool slowly, they are often modified to increase the quench rate of the sample, usually by moving the capsule rapidly from the heated area of the apparatus to a water-cooled environment (e.g., Ihinger, 1991; Holloway et al., 1992; Roux and Lefèvre, 1992). Other experimental apparatus such as the diamond (or sapphire) anvil cell may soon permit in-situ spectroscopic studies at high pressures and temperatures (e.g., Herrmannsdorfer and Keppler, 1992; Bassett et al., 1993; Franz et al., 1993).

Samples are loaded into noble metal capsules, and although a sample can be exposed to CO_2 gas used as a pressurizing medium, it is more common to seal a CO_2 source inside the capsule along with the sample material. Sources of CO_2 are usually a carbonate or oxalate material (such as $BaCO_3$, $CaCO_3$, K_2CO_3, $MgCO_3$, Na_2CO_3 or $Ag_2C_2O_4$, e.g., Mysen et al., 1975; Stolper et al., 1987; Pan et al., 1991) which releases CO_2 upon heating. Mixed CO_2-H_2O fluids can be generated with mixtures of water + silver oxalate ($Ag_2C_2O_4$) or carbonate (e.g., Holloway et al., 1968; Mysen, 1976), or anhydrous oxalic acid ($H_2C_2O_4$) or oxalic acid dihydrate ($H_2C_2O_4 \cdot 2H_2O$) ± water (see Holloway et al., 1968). CO_2-H_2O fluids diluted with N_2 can also be produced by use of guanidine nitrate ($CH_5N_3 \cdot HNO_3$) or ammonium oxalate ($[NH_4]_2C_2O_4 \cdot H_2O$) (Kesson and Holloway, 1974). Any C-O fluid in equilibrium with C will be a mixture of CO-CO_2 and graphite is sometimes included along with the CO_2 source for this purpose (see Pawley et al., 1992).

In the perfect solubility experiment, a sample is held at an accurately-determined high P and T under closed-system conditions. In practice, however, experimental capsules are not closed, and they are subject in particular to diffusive loss or gain of H_2 which can readily alter the composition of the vapor inside an experimental charge. Precautions are routinely taken to protect the sample against changes in the hydrogen mass balance within a capsule, often through use of a double-capsule assembly, with an outer "chemical buffer" (such as hematite or pyrex) to prevent hydrogen exchange to and from the inner capsule (e.g., Chou, 1987; Rosenbaum, in press). At the end of the experiment, the concentration or total quantity of the dissolved volatile components (and perhaps any excess vapor) of interest should be sufficient to enable their accurate and precise determination. A record of the vapor composition at the end of an experiment may be a representative indication of the vapor phase during a run (e.g., Blank et al., 1993a), but it is usually very difficult to measure and the vapor may be susceptible to changes during quenching when mixed-volatile species are involved (Jakobsson and Oskarsson, 1994).

One of the assumptions implicit in most studies is that the quenched glass ("run product") obtained at the end is a true record of the conditions during the experiment (i.e., there is no change in speciation upon quenching, there are no crystals formed or volatiles exsolved, and there is no interaction with the capsule or other materials). In actual experiments, these conditions are not always realized, and experimentalists are continuously driving towards the ideal experiment by inventing new and creative experimental techniques. Details of experimental methodology can be found in references mentioned above as well as in other studies, many of which are summarized as tables in the Appendix.

Analysis

Methods used for analysis of dissolved CO_2 in silicate melts are described in detail in Chapter 2. Absolute methods commonly used are manometry and commercial bulk carbon analyzers. Other techniques rely heavily on calibrations that may introduce errors in accuracy and are often compositionally dependent. However, the infrared spectroscopic technique is particularly sensitive and measurements can be performed on materials (such as natural samples) with very low dissolved CO_2 concentrations, and spectra can be obtained for very small samples (tens of microns in diameter).

One of the most significant challenges in comparing measurements of CO_2 concentrations in silicate glasses is the lack of quantitative agreement between different analytical techniques. For example, published CO_2 determinations among overlapping suites of natural tholeiitic glasses differ by up to two orders of magnitude (Des Marais, 1986). β-track autoradiography in particular has been criticized, and Tingle and Aines (1988) have shown that this analytical method applied to albitic glasses (with 0.5 to 2.0 wt % CO_2, synthesized at 20 kbar and 1450°C) yielded values 10 to 50% in excess of the amount of CO_2 initially added to the sample. In contrast, infrared measurements on the same (undersaturated) samples show a 1:1 correlation. Values obtained with the β-track method for CO_2-saturated diopside were 10% lower than the amount of CO_2 loaded into a sample charge (Tingle, 1987). These errors may be dependent on factors such as composition and density, or other complexities (e.g., nonlinear exposure response; B.J. Wood, pers. comm., 1994) associated with the β-track method which have become apparent since the initial work of Mysen and coworkers.

It is clear that a comparison of CO_2 contents obtained using different analytical techniques is desirable, and to date Pan et al. (1991) and Thibault and Holloway (1994) have reproduced CO_2 solubilities in tholeiitic and Ca-leucititic melts using FTIR spectroscopy, SIMS, and LECO bulk analysis.

SOLUBILITY MEASUREMENTS

Detailed reviews of CO_2 solubility measurements can be found in Mysen (1977) and Dingwell (1986), and Spera and Bergman (1980) and Holloway (1981; 1987) for C-O-H volatiles and thermodynamic modeling of their behavior. These and other reviews are listed in the Appendix (Table A1). Selected solubility data are presented in this section.

Early work

While the earliest measurements of CO_2 solubility in silicate melts were made on commercial glass compositions (Pearce, 1964; Faile and Roy, 1966), Wyllie and Tuttle (1959) are generally acknowledged as the first workers to determine experimentally volatile solubilities in geologically relevant (granitic) melts (at pressures up to ~2 kbar) with a mixed CO_2-H_2O fluid. Studies in the early 1970s by Kadik and his colleagues (Table A2) focused on CO_2 solubilities in natural rhyolitic, andesitic, and basaltic melts at pressures of 1 to 5 kbar. Although there is some scatter in their CO_2 solubility data (determined by weight loss), these workers observed that the solubility of CO_2 was approximately a factor of ten lower than that of water in these melt compositions (see Kadik et al., 1972).

Selected data

Published data on CO_2 solubility in silicate melts span a range of pressure and temperature for natural (e.g. andesite, basalt, dacite, Ca-leucitite, rhyolite) and synthetic (e.g. albite, anorthite, diopside, jadeite, sodamelilite) melt compositions. The majority of these studies were conducted at pressures above 5 kbar. Although such high pressure data are relevant to phase equilibria studies and melt generation in source regions, they require considerable extrapolation to the P-T conditions of other important aspects of volatile behavior in magmas, in particular volcanic degassing. Initially, limitations of analytical techniques dictated that high-pressure conditions be used in order to produce concentrations of dissolved CO_2 above detection limits. As analytical techniques have improved, lower pressures studies have become practical.

Representative results from CO_2 solubility studies conducted over a range of P and T are presented below, and spectroscopic quantification of the carbon species which constitute the total dissolved CO_2 (i.e., a combination of molecular CO_2 + carbonate) is discussed in the next section.

Pressure dependence. Without exception, the effect of increasing P_{CO_2} is to increase the amount of CO_2 dissolved in the melt (Fig. 1). Shilobreyeva and Kadik (1990) observed a linear relation over pressures from 0.5 to 5.0 kbar for basaltic, andesitic, and rhyodacitic melt compositions. However, Fogel and Rutherford (1990) and Blank et al. (1993a) examined CO_2 solubilities in rhyolitic melts in more detail and observed nonlinear behavior above ~1 kbar. This deviation from ideal behavior reflects the increased importance of the molar volume term with increasing pressure (see Chapter 6). As the CO_2 solubility deviates from a linear trend, the rate of increase in solubility is reduced (note that the logarithmic scale in Fig. 1 exaggerates this).

Temperature dependence. On the basis of data from recent studies (presented in Figs. 2 and 3), the solubility of CO_2 decreases with increasing temperature in most silicate melt compositions, regardless of speciation of dissolved CO_2 (molecular CO_2 or carbonate). Notable exceptions are basaltic melt, which exhibits negligible temperature dependence between 1300° and 1600°C (Pan et al., 1991), and the nepheline composition,

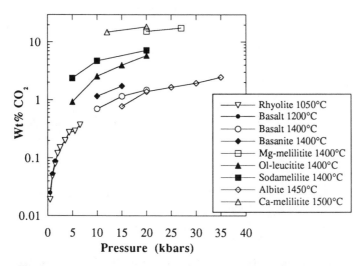

Figure 1. The increase in CO_2 solubility in silicate melts as a function of pressure. Data sources: basalt (15 kbar and 1400°C, Pan et al., 1991; 1200°C, Pawley et al., 1992), albite (Stolper et al., 1987), diopside (Rai et al., 1983), ol-leucitite (Thibault and Holloway, 1994), nepheline (Brooker et al., in prep.), sodamelilite (Mattey et al., 1990), basanite (Pan, pers. comm.), rhyolite (Fogel and Rutherford, 1990), phonolite, andesite, Ca- and Mg-rich melilitite (Brooker and Holloway, in prep.). Note log scale of y-axis.

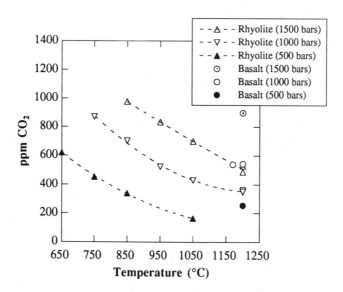

Figure 2. The negative temperature dependence of CO_2 solublity in rhyolitic melts at relatively low pressures (up to 1.5 kbar). Basalt data are included to show the higher CO_2 solubility in this composition at lower pressures (not apparent in Fig. 1). Melt compositions and pressures indicated in legend. Data from: Fogel and Rutherford (1990) and Blank et al. (1993a)—rhyolite; Dixon et al. (in prep.) and Pan et al. (1991)—basalt. Dissolved CO_2 contents were determined using FTIR spectroscopy. Rhyolite data of Fogel and Rutherford (1990) were corrected using a molar absorptivity coefficient of 1066 (1 mol^{-1} cm^{-2}) for molecular CO_2 (Blank, 1993).

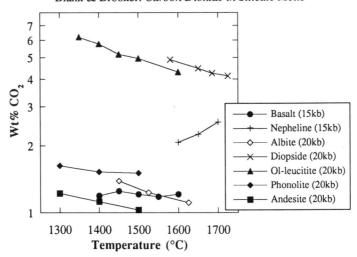

Figure 3. The variable influence of temperature at high pressure (15 to 20 kbar) on CO_2 solubility in a range of silicate melts. Melt compositions and pressures indicated in legend. Data from: Pan et al. (19910 —basalt; Stolper et al. (1987)—albite; Rai et al. (1983)—diopside, Thibault and Holloway (1994) (Ol-leucitite); Brooker et al. (in prep.)—nepheline; Mattey et al. (1990)—sodamelilite; Pan (pers. comm.) —basanite; Brooker and Holloway (in prep.)—phonolite and andesite. Note logorithmic scale on *y*-axis.

which has a positive temperature dependence (Brooker and Holloway, in prep.). The positive temperature dependence of CO_2 solubility in the nepheline melts (in which all dissolved CO_2 is detected as carbonate) is similar to that of the minor carbonate component observed in albitic melts (Stolper et al., 1987) and reflects similarities between these types of dissolved carbonate groups which are discussed later.

Mysen and coworkers (e.g., Mysen et al., 1975; Holloway et al., 1976; Mysen, 1976; Mysen et al., 1976; Mysen and Virgo, 1980a; Mysen and Virgo, 1980b) have made a large number of CO_2 solubility determinations on a number of silicate compositions synthesized under a wide range of P-T conditions. Unfortunately, carbon contents of the quenched melts were determined using [14]C β-track autoradiography, and as mentioned in the previous section, there may be substantial errors associated with this method. When compared with other data, the β-track results also appear to show opposing trends in ɪperature dependence (see Brey, 1976; Rai et al., 1983; Stolper et al., 1987). In Figure results of β-track measurements are compared with results for similar compositions termined (and in some cases reproduced) using a range of other techniques (FTIR ɹectroscopy, ion probe, and bulk extraction methods). Because of these discrepancies, ɴis substantial body of β-track data must be treated with caution. Models relying on these earlier data (e.g., Spera and Bergman, 1980) may need to be modified as new data become available.

Compositional dependence. There is a general increase in CO_2 solubility with decreasing mole percent SiO_2, but in detail there are significant variations among melts with similar silica contents (Fig. 1). For instance, in Figure 3, albite and andesite have solubility trends which cross the basalt data whereas the phonolite composition accommodates a higher CO_2 solubility at all temperatures. The relation between CO_2 solubility and melt composition is poorly constrained; some of the important factors are discussed in the Dissolution Mechanism section later in the chapter.

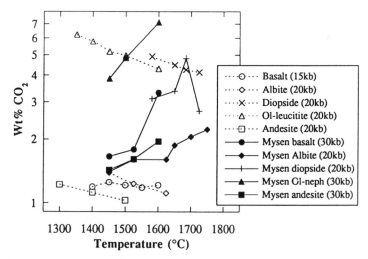

Figure 4. Comparison of data from Figure 3 with selected data of Mysen and co-workers. Obvious discrepancies exist between the β-track data of Mysen and co-workers and the data collected using other analytical techniques. Melt compositions indicated in legend. Open symbols are data from Figure 3. Solid symbols represent data from Mysen et al. (1975)—andesite, basalt and ol-nephelinite; Mysen and Virgo (1980b) —diopside; Mysen (1976), and Mysen and Virgo (1980a)—albite.

The effect of water. As water often dominates the volatile budget of natural magmas, it is important to determine its influence on CO_2 solubility. Blank et al. (1993a) and Dixon et al. (in press) documented ideal- (Henrian-) like mixing behavior at relatively low pressures (<1000 bars) for rhyolite with up to 3.3 wt % H_2O and for basalt with up to 2 to 3 wt % H_2O (see Chapter 6). Whether Henrian mixing prevails at high pressures is unknown. Larger proportions of dissolved water, as well as non-ideal behavior in the vapor phase, may contribute to non-Henrian solubility behavior of CO_2 in the melt. High-pressure solubility data is from Mysen and coworkers, and suggests a maximum in CO_2 solubility at a molar $CO_2/(CO_2 + H_2O)$ volatile ratio of ~0.6 for andesite and basalt, and ~0.7 for albite, jadeite and nepheline compositions (see Fig. 5a, also, Mysen et al., 1975; Mysen, 1976; Mysen et al., 1976). Mysen (1976) proposed that formation of OH groups depolymerizes the melt, thus allowing more carbonate to form by reaction with nonbridging oxygens (see section on Dissolution Mechanisms, this chapter), resulting in an increase in CO_2 solubility. At higher water activities, the reduced fugacity of CO_2 (diluted by the H_2O) leads to lower CO_2 solubilities. Uncertainties regarding the β-track analytical method have been discussed above, and Stolper et al. (1987) saw no evidence for an increase in CO_2 solubility with increasing amounts of (accidental) water (up to 0.8 wt %) in their study of CO_2 solubility in albitic melt. Recent models of water solubility in albitic melt suggest that the formation of OH may not involve depolymerization of the melt (Kohn et al., 1989), and Kohn and Brooker (1994) have also shown that the proportion of CO_2 as carbonate in quenched albitic melts is actually reduced to a minimum at the $CO_2/(CO_2+H_2O)$ ratio of ~0.8, whereas Mysen and coworkers report a CO_2 maximum (compare Figs. 5a,b). Kohn and Brooker (1994) note that these changes in speciation could be a function of the change in melt viscosity and quench rate (due to the presence of water). While it remains unclear whether this feature is accompanied by a change in CO_2 solubility, a CO_2 minimum rather than a maximum at a $CO_2/(CO_2+H_2O)$ ratio of 0.8 might be expected from the data of Kohn and Brooker (1994).

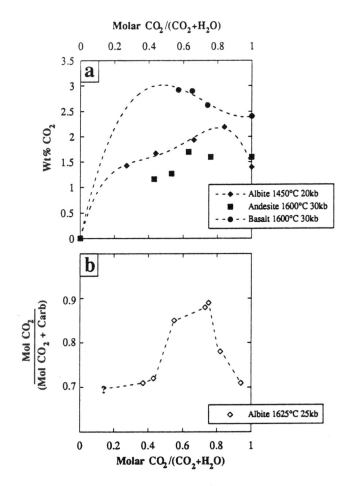

Figure 5. The effect of water on CO_2 solubility in silicate melts. The molar $CO_2/(CO_2+H_2O)$ reflects the ratio of volatiles loaded into an experimental charge. (a) Addition of small amounts of water leads to an ease in CO_2 solubility, particularly in albitic melts. Nephelinitic and jadeitic compositions (not wn) exhibit trends similar to those of albite. Data for albitic melts from Mysen (1976); data for basaltic andesitic melts, Mysen et al. (1975). (b) Change in $CO_2/(CO_2+CO_3^{2-})$ ratio in albitic melt as a ction of volatile composition, as determined using NMR spectroscopy. Data from Kohn and Brooker)94), with slight revision.

SPECIATION

Spectroscopic techniques have been used extensively to characterize the carbon species dissolved in silicate melts. Such information has been essential in developing thermo-dynamic descriptions of CO_2 solubility and also in evaluating volatile dissolution mechanisms. Infrared, Raman, and NMR spectroscopic studies of CO_2-bearing silicate glasses show that CO_2 dissolves in two general forms, as molecules of CO_2 and as carbonate groups (CO_3^{2-}), and that the speciation is a strong function of melt structure and available cations such as Na^+, Ca^{2+}, and Mg^{2+} (cf. Fine and Stolper, 1985).

Vibrational spectroscopy

Much of our understanding of the interaction between CO_2 vapor and silicate liquids derives from vibrational spectroscopy studies. Both molecular CO_2 and carbonate have been identified in a range of natural and synthetic melt compositions, and the proportions of these species show a high degree of compositional dependence (Brey and Green, 1976; Mysen et al., 1976; Fine and Stolper, 1985; Fine, 1986) It has been shown (e.g., Fine and Stolper, 1985; Fine, 1986) that proportions of dissolved species are independent of the total CO_2 concentration (in a selection of compositions with pure CO_2). The inverse is true for water speciation (see Chapter 4). Details of vibrational spectra have also provided information on the interaction of species with the melt structure, allowing speculation on mechanisms for dissolution (Sharma, 1979; Mysen and Virgo, 1980a; Mysen and Virgo, 1980b; Taylor, 1990).

Theory. There are many reviews of vibrational spectroscopy which refer to the range of structural information available from this technique (e.g., Herzberg, 1945; Cotton, 1971; Farmer, 1974, in particular Chapter 1; Harris and Bertolucci, 1978; McMillan and Hess, 1988; McMillan and Hofmeister, 1988). In this section we cover briefly a select number of concepts related specifically to carbon species to illustrate the significance of various aspects of the spectra shown.

Vibrational spectra result from the absorption of certain frequencies of electromagnetic radiation that correspond to the frequencies of vibrational translations of a molecule (see McMillan and Hess, 1988). Internal vibrations produced when the atoms of a molecule move relative to one another can be described in terms of components in three mutually perpendicular axes; their number and type for a molecule of a given symmetry are predicted by group theory. In general, a molecule will have 3N-6 vibrations, where N is the number of atoms, and -6 accounts for (i) the 3 cases in which all the atoms move by the same amount in the same direction along x,y,z; and (ii) the 3 cases in which the molecule rotates around one of the three axes (these are not vibrations as they do not involve relative displacement of atoms). Each vibration has a descriptive name and a v_n symbol (here following the convention of Herzberg, 1945); see Tables 1 and 2.

Vibrations may differ among molecules of like composition if their symmetry is not identical. This is due to the interaction of the vibration with different arrangements of electron charge in neighboring bonds. The relation between bond vibrations and electron distribution is significant as it determines whether a vibration will be detectable in the IR or Raman spectrum (or both). For the vibration of a bond to absorb in the infrared spectrum, it must change the dipole of the electron charge distribution; a change in polarizability is required for Raman activity (McMillan and Hofmeister, 1988). Minor overtones and combination bands can sometimes be seen in spectra at frequencies approximately equal to the sum and products of the fundamental frequencies (McMillan and Hess, 1988), and there are also bands related to isotopic substitutions (e.g., $^{13}C^{16}O_2$, $^{14}C^{16}O_2$, etc.).

The abundance and types of vibrations (IR or Raman active) have been used to interpret the structure of molecules or groups of molecules. Strongly bonded (highly covalent) molecules such as CO_2 and CO_3^{2-} retain the character of an isolated molecule when incorporated into other structures, simplifying their identification and quantitative analysis. Subtle changes from their isolated configurations provide information on the structural environment of these dissolved species that can be used to model the way in which they are incorporated into a silicate melt. Below follows a brief review of the structure and vibrations predicted for isolated CO_2 and CO_3^{2-} molecules.

The CO_2 molecule. A consequence of the linear symmetry ($C_{2\infty}$) of the CO_2 molecule is that one of the molecule's rotations (about the long axis of the molecule) will not translate any atoms, and thus the number of internal vibrations is 3N-5. There are therefore 4 internal vibrational modes; these are illustrated in Figure 6 and described in Table 1.

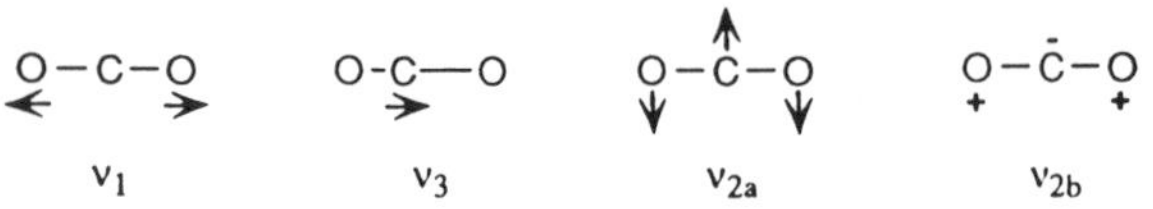

Figure 6. Internal vibrational motions of molecular CO_2. Motions in the z direction are indicated by + and −. See Table 1.

Table 1. Internal vibrational modes of the CO_2 molecule.
Fundamental frequencies given are for the molecule in its gaseous state.

Herzberg notation	v_1	v_3	v_{2a}	v_{2b}
Type of vibration	symmetric stretch	asymmetric stretch	symmetric bend	symmetric bend
Frequency (cm^{-1})	1337	2349	667	667
Activity	Raman	IR	IR	IR

Although the two v_2 symmetric-bending motions are different vibrations, they have the same frequency (and are said to be degenerate) and hence there are two rather than three peaks in the IR spectrum of the CO_2 molecule. The CO_2 molecule has a center of symmetry, and a rule of mutual exclusion states that no vibration can occur in both the Raman and IR spectra (e.g., Herzberg, 1945), and the Raman spectrum should have just the one v_1 peak at 1337 cm^{-1}. However, the Raman spectrum of gaseous CO_2 has two peaks (the Fermi doublet) rather than one. This phenomenon occurs because the first overtone of the IR-active v_2 peak coincides with the predicted position of v_1 ($2v_2 \approx 2 \times 667$ ≈ 1334 cm^{-1}). The existence of small perturbations causes the two energy levels to resonate and interact. The wavefunctions of the two states become mixed via aharmonic coupling and the IR active mode gains some Raman character to become Raman active (without resonance there would actually be a small $2v_2$ peak 50 times less intense than v_1, Howard-Lock and Stoicheff, 1971). As the energies are similar they repel each other giving peaks above and below the unperturbed positions.

The carbonate molecule. 3N-6 translations for the isolated carbonate ion produce the 6 internal vibrations represented in Figure 7 and described in Table 2. If the oxygen atoms associated with the carbonate ion are indistinguishable, then the triangular planar molecule has D_{3h} symmetry. However, if one oxygen is dissimilar to the others, for example in the way it is coordinated to a neighboring cation or through isotopic substitution, then the symmetry is lowered to C_{2v}. The effects of this change of symmetry are also distinguished in Table 2.

The two asymmetric stretching modes and the two in-plane bending modes are degenerate for D_{3h}, but if this symmetry is lowered to C_{2v} (or C_s) then the degenerate vibrations assume different character and symmetry to yield two different frequencies.

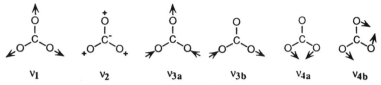

Figure 7. Internal vibrational motions of CO_3^{2-} ion. See Table 2.

Table 2. Internal vibrational modes of the carbonate ion. Fundamental frequencies listed are for an isolated ion (Kujumzelis, 1938) with D_{3h} symmetry *

Herzberg notation	v_1	v_2	v_{3a}	v_{3b}	v_{4a}	v_{4b}
Type of vibration	symmetric stretch	out-of-plane bend	asymmetric stretch		in-plane bend	
Frequency (cm^{-1})	1063	879	1415	1415	680	680
D_{3h} symmetry species	A_1'	A_2''	E'	E'	E'	E'
Activity	Raman	IR	IR+Raman		IR+Raman	
C_{2v} symmetry species	A_1	B_1	A_1	B_2	A_1	B_2
Activity	IR+Raman	IR+Raman	IR+Raman		IR+Raman	

* Note: For D_{3h} symmetry, the first overtone $2v_2$ (≈ 1760 cm^{-1}) is Raman-, not IR-active. [See Poulet and Mathieu (1970) and Verweij et al. (1977).]

Both the v_3 and v_4 vibrations will now produce two resolvable spectral peaks (a doublet), and it has been suggested (Nakamoto et al., 1957) that the difference in wavenumber between the two asymmetric stretching vibrations (Δv_3) is directly proportional to the distortion of the carbonate structure (i.e., the difference in environment among the carbonate oxygens). While the effect of interactions between ionic species can reduce the carbonate symmetry and produce slight degrees of distortion (e.g., calcite $D_3 \rightarrow$ aragonite C_s), the distortion caused by covalent coordination with the silicate melt network is thought to be much stronger and the Δv_3 split can be increased if two, rather than one, of the carbonate oxygens are coordinated in this way (for an example of 80 and 275 cm^{-1} splits caused by this mechanism, see Nakamoto et al., 1957). The significance of single peaks and v_3 doublet splits of ~250 and 80 cm^{-1} will become apparent when actual spectra are presented (Figs. 8 and 9). It should be noted that a loss of degeneracy gives two v_3 and two v_4 peaks, but v_1 and v_2 remain as single peaks, whereas the presence of two distinct D_{3h} carbonate groups could double the total number of peaks. In the following discussion, a distinction is made between different carbonate groups, but it is noted that each group may not have a unique configuration.

Sample spectra. Many spectra for a variety of CO_2-bearing natural and synthetic glass compositions have been published. Examples used in this section illustrate the range of spectral features that have been observed.

1. The $NaAlO_2$-SiO_2 join. Many of the earlier CO_2 solubility studies used melt compositions that plot along the $NaAlO_2$-SiO_2 join as simple analogs for granitic melts. Figure 8 illustrates the mid-IR spectra of glasses quenched from CO_2-bearing melt compositions on this join. Below 1200 cm^{-1}, the IR spectra of these fully-polymerized

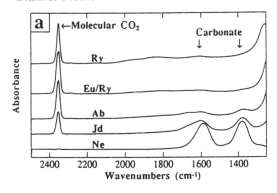

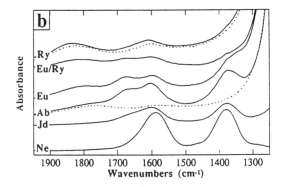

Figure 8. IR spectra for compositions on the $NaAlO_2$-SiO_2 join. (a) Mid-IR absorbance spectra for doubly-polished plates showing dissolved CO_2 as CO_2 molecules and carbonate. Compositions in order of increasing SiO_2 content are: Ne = $NaAlSiO_4$; Jd = $NaAlSi_2O_6$; Ab = $NaAlSi_3O_8$; Eu = $NaAlSi_4O_{10}$ (a composition close to the 1 atm silica-albite eutectic); Eu/Ry = $NaAlSi_5O_{12}$; Ry = $NaAlSi_6O_{14}$ (a synthetic 'rhyolite' with a Si:Al ratio similar to that of the rhyolite studied by Fogel and Rutherford (1990) which also showed no evidence of any carbonate species). Data from Brooker et al. (in prep.); the Jd, Ab and Eu spectra are consistent with Fine and Stolper (1985) and the Ne spectrum resembles a powder spectrum of Taylor (1990). Samples contain 1.0 to 2.2 wt % dissolved CO_2 and were prepared at 15 to 25 kbar, 1600°C. Spectra are scaled to approximately equal thickness. (b) Enlargement of the region of the carbonate doublet to illustrate the change in relative proportions of peaks at 1600 to 1680 cm^{-1}. Ab, Eu, Eu/Ry and Ry are exaggerated vertically by a factor of 3 relative to other spectra. Dashed lines are for spectra of volatile-free Ab and Ry samples prepared at 20 kbar and 1600°C. The broad background bands (most prominent in Ry) that appear as SiO_2 content increases are probably overtones of fundamental Si-O vibrations.

melts are dominated by peaks related to the aluminosilicate structure. These tend to obscure some of the carbon-related peaks, and the only vibrations resolvable are those assigned to the ν_3 asymmetric stretching of carbonate (1375 to 1660 cm^{-1}) and the ν_3 asymmetric stretching of molecular CO_2 (2352 cm^{-1}). Mysen and Virgo (1980a) have published Raman spectra of albite that show the Fermi doublet of molecular CO_2 (1272 and 1376 cm^{-1}) and the ν_1 symmetric stretching of carbonate at about 1075 cm^{-1}.

The molecular CO_2 peak in the IR spectra varies little in shape and position for a large range of compositions. As noted by Fine and Stolper (1985), the ν_3 2352 cm^{-1} peak position is distinct from the spectrum of gaseous CO_2, which is sometimes present in glasses in pressurized bubbles. The IR ν_3 in gaseous CO_2 occurs at 2348 cm^{-1}, but due to rotational effects a doublet appears at 2340 and 2360 cm^{-1} (e.g., Brey, 1976; Blank, 1993). Dissolved molecular CO_2 remains unaffected by cooling to liquid nitrogen temperatures (Stolper et al., 1987), suggesting that molecules are isolated from each other (i.e., there is no condensation). Minor bands related to molecular CO_2 are present at 3711 cm^{-1} (v_1 and v_3 combination band), 2287 cm^{-1} (v_3 $^{13}CO_2$) and 2229 cm^{-1} (v_3 $^{14}CO_2$) (cf. spectrum in Tingle and Aines, 1988) .

Fine and Stolper (1985) have assigned the IR peaks at ~1375-1680 cm^{-1} to the v_3 antisymmetric vibrational modes of distorted carbonate groups, which would be represented by a single peak (at ~1415 cm^{-1} for D_{3h} symmetry) if all C-O bonds were equivalent. Stolper et al. (1987) proposed the existence of two carbonate groups to account for the two components that can be distinguished at 1610 and 1680 cm^{-1} among the higher SiO_2 compositions. The 1680-1375 cm^{-1} split would therefore represent a more distorted group than the split at 1610-1375 cm^{-1}. Although the intensities of the peaks are not directly proportional to the concentration of species (each peak is related to concentration by its own molar absorptivity coefficient; Chapter 2), it is obvious that the relative proportion of carbonate decreases towards SiO_2-rich compositions and the relative proportion of the more distorted carbonate component increases over this range (see Fig. 8). IR spectra of CO_2-bearing glasses whose compositions fall along on the $NaAlSi_3O_8$-$CaAl_2Si_2O_8$ join exhibit a reduction in the molecular CO_2/carbonate ratio moving along the join from albite to anorthite, decreasing until no dissolved molecular CO_2 is detected at an Ab:An mixture of ≈ 1:1. There also appears to be a reduction in the magnitude of the split in the v_3 doublet, or perhaps the addition of a less-distorted carbonate component.

Stolper et al. (1987) demonstrated that in albitic glasses quenched from a range of P and T there is a slight decrease in the molecular CO_2/carbonate ratio with increasing pressure and a more significant decrease with increasing temperature. However, the speciation of dissolved volatiles may be susceptible to change during quenching (see discussion regarding water species in Chapter 4), and differing fictive temperatures may be responsible for these trends rather that responses to pressure and temperature at equilibrium.

2. Natural melt compositions. Relative proportions of carbonate and molecular CO_2 in natural silicate glasses are highly-dependent on glass composition. The most obvious differences between IR spectra of natural glasses (Fig. 9) and those of glasses from the $NaAlO_2$-SiO_2 join are the magnitudes of the split of the carbonate doublet (Δv_3), which is reduced from ~225 and 300 cm^{-1} to ~80 cm^{-1}, and the reduction from two distinct carbonate groups to one. Published spectra of basalts (e.g., Fine and Stolper, 1986; Pan et al., 1991), ol-melilitite (Brey, 1976) and a Ca-rich leucitite (Thibault and Holloway, 1994) are very similar to the nephelinite in Figure 9, as are the spectra of the depolymerized synthetic compositions sodamelilite (Mattey et al., 1990), diopside, and a calcium aluminosilicate (Fine and Stolper, 1986; Fine, 1986).

Raman spectra have been published for sodamelilite, diopside and akermanite compositions with high concentrations of carbonate (for examples of spectra, refer to Sharma, 1979; Sharma et al., 1979; Mysen and Virgo, 1980a). In these spectra, single v_1 and $2v_2$ carbonate peaks appear at ~1075 cm^{-1} and ~1730 cm^{-1}, respectively , but the v_3 appears as a doublet similar to that observed in IR spectra (Sharma, 1979). The Raman spectra of CO_2-bearing $K_2Si_{1.5}O_4$ - K_2SiO_3 glasses do not show a split in the v_3 peak (Verweij et al., 1977), in contrast to spectra of other (divalent cation-bearing) depolymerized compositions.

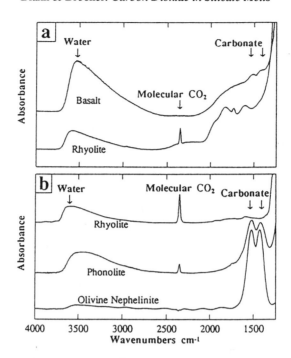

Figure 9. Representative IR spectra of natural CO_2-bearing silicate melt compositions. (a) Comparison of speciation of CO_2 dissolved in rhyolitic and basaltic melts in spectra normalized to 100-μm thickness: basaltic glass (WOK28-3; Sally Newman, unpublished) is from the Lau Basin and contains 100 ppm dissolved CO_2 (as carbonate) determined by the magnitude of the carbonate absorption bands at 1515 cm^{-1} and 1430 cm^{-1} (Fine and Stolper, 1986); rhyolitic glass (GB1-6P; Blank, 1993) contains approximately the same amount (115 ppm) of dissolved CO_2 (as CO_2 molecules, absorption band at 2350 cm^{-1}). Note the greater sensitivity of the IR spectra to molecular CO_2 rather than CO_3^{2-} absorptions. (b) IR spectra for olivine nephelinite (4.55 wt % CO_2), phonolite (1.63 wt % CO_2) and rhyolite (~0.45 wt % CO_2), all normalized to 20-μm thickness (all spectra from Brooker, unpublished).

NMR spectra

[13]C Nuclear Magnetic Resonance (NMR) spectroscopy can provide useful information concerning the bonding environments and speciation of carbon dissolved in melts (for detailed presentation of NMR spectroscopic method, refer to Kirkpatrick, 1988).

NMR spectroscopy is much more sensitive to subtle differences in the nature of the carbonate environments than are other spectroscopic methods. Another positive feature of this technique is that the relative proportions of species present are directly proportional to their spectral peak heights (i.e., species-dependent calibration is not required) and in principle, quantification of the spectra could lead to very reliable measurements of the total dissolved carbon concentration (see Kohn et al., 1991). Because the natural abundance of ^{13}C is low (constituting only ~1.1% of the total C in natural samples), ^{13}C MAS NMR is presently only practical for experimental samples made with ^{13}C-enriched materials. Also, NMR cannot be performed on Fe-bearing samples (although amounts <1% may be permissible), thus restricting NMR studies to simple analogs and more-evolved natural melt compositions.

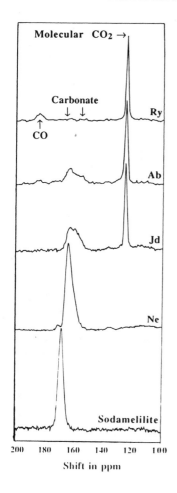

NMR spectroscopy has been applied recently to studies of CO_2 solubility in silicate melt (Kohn et al., 1991; Kohn and Brooker, 1994) and ^{13}C MAS NMR spectra for selected $NaAlO_2$-SiO_2 compositions are seen in Figure 10. Kohn et al. (1991) have assigned the 125 ppm peak in CO_2-bearing albitic glass to molecular CO_2 (the shift for gaseous CO_2 is at 124.2 ppm relative to the TMS standard). The other broad band in the spectrum has been assigned to carbonate groups and can be fit by at least two components for the $NaAlO_2$-SiO_2 compositions. The molecular CO_2 and carbonate bands change intensity with composition in a manner similar to those in IR spectra, and Kohn et al. (1991) suggest that the most distorted IR carbonate component (the one with the largest split, the most distant from undistorted carbonate) correlates with the NMR peak shifted the furthest distance from 170 ppm (^{13}C position for low-distortion crystalline carbonate). The less distorted component appears to decrease in relative intensity with increasing SiO_2. Changes in molecular CO_2/carbonate ratios are generally consis-tent with the IR spectroscopic determinations of Fine and Stolper (1985). As with the IR spectra, the sodamelilite composition (also in Fig. 10) appears to show one carbonate group, which is closer to the crystalline carbonate shift than the $NaAlO_2$-SiO_2-join carbonates, in agreement with its lesser distortion.

Figure 10. ^{13}C MAS NMR spectra for compositions on the $NaAlO_2$-SiO_2 join (from Brooker et al., in prep.; cf. Kohn et al., 1991) and sodamelilite $NaCaAlSi_2O_7$ (Kohn et al., 1991). Peak at 185 ppm is due to dissolved CO. Shifts are relative to TMS standard.

Under reducing (low f_{O_2}) conditions, dissolved CO may be present in some CO_2-silicate melt systems. Certain reduced (graphite-bearing) ^{13}C samples (see Ab and Ry in Fig. 10) have an NMR peak at 185 ppm. This is near the position for gaseous CO and this feature has been assigned to this species by Kohn et al. (1991). This peak is observed only in compositions that dissolve large quantities of CO_2 as molecular species and some CO solubility might be expected as the molecule is smaller than CO_2 and so should fit easily into available "holes" or sites in the silicate melt structure. CO solubility is below detection in basalt and Ca-leucitite compositions (see Pawley et al., 1992; Thibault and Holloway, 1994); these compositions also lack detectable CO_2. However, Eggler et al. (1979) have inferred that CO is soluble in diopside melt in which only carbonate has been detected (Mysen and Virgo, 1980a).

DISSOLUTION MECHANISMS

Several dissolution mechanisms for CO_2 in silicate melts have been proposed on the basis of vibrational (Mysen and Virgo, 1980a,b; Fine and Stolper, 1985; Taylor, 1990) and NMR spectra (Kohn et al., 1991) and also through the use of molecular orbital (MO) modeling techniques (deJong and Brown, 1980; Kubicki and Stolper in press). A large proportion of the detailed work has focused on synthetic compositions, and albitic melts in particular. It is important to be aware that the relevance of this composition to natural melts may be somewhat limited in light of the differences between the carbonate spectra of albitic and natural compositions (Figs. 8 and 9).

Proposed dissolution mechanisms are linked to existing concepts of melt structure. In one model for melt structure, cations are classified into two broad categories, tetrahedrally-coordinated network formers (referred to as 'T') such as Si and Al, and network modifiers such as Na, K and Ca. Network-forming tetrahedra are linked at their corners by oxygen atoms (referred to as 'bridging oxygens', BO); oxygens not shared between tetrahedra are termed non-bridging oxygens (NBO). Fully-polymerized melts lack non-bridging oxygens (NBO/T = 0) while fully-depolymerized melts contain only isolated SiO_4^{4-} tetrahedra (NBO/T = 4). Most melts are a mixture of structures and thus an average NBO/T can be calculated (see Mysen, 1986). These structures can also be described by Q^n notation, where n refers to the number of bridging oxygens around a network-forming atom (Engelhardt and Michel, 1987). Using this terminology, fully polymerized structures are Q^4, and isolated tetrahedra are Q^0.

Environments of molecular CO_2

Features of the NMR spectra of dissolved molecular CO_2 suggest that the molecules are constrained 'rigidly' and not free to rotate in a random fashion (see Kohn et al., 1991). As with the v_3 IR peak, the NMR peak at 125 ppm is also shifted slightly from the position for gaseous CO_2 (124.2 ppm). Theoretical calculations by Kubicki and Stolper (in press) predict existence of a 'bent' (177°) CO_2 molecule (first suggested by deJong and Brown, 1980) with a long-range interaction between the carbon and an oxygen in the aluminosilicate framework. This interaction is too weak for the molecule to have the characteristics of a carbonate group; it may exist as a structure intermediate between molecular CO_2 in the vapor and carbonate in the melt. An extreme weakness of any interaction between dissolved CO_2 molecules and the silicate matrix appears to be consistent with the ionic porosity model of Carroll and Stolper (1993); for example, solubility of CO_2 as molecules in rhyolite mimics the trend defined by 'inert' noble gas atoms. The ionic porosity model is based on the concept of a distribution of 'holes' in the melt that nonreactive volatile species can occupy, and it may be the availability of such 'holes' that has a major control on the amount of molecular CO_2 dissolved in melts (cf. Taylor, 1990). Depolymerization of melt structure or an increase in the number of cations 'filling' the framework could contribute to the decreasing amount of molecular CO_2 from rhyolitic to basaltic melt (in addition to the increased number of reactive non-bridging or free oxygens). Cation-filling of available holes could also contribute to variations in the proportion of molecular CO_2 in melts along the $NaAlO_2$-SiO_2 join. However, the reduction of molecular CO_2 as $CaAl_2Si_2O_8$ is substituted for $NaAlSi_3O_8$ on the albite-anorthite compositional join (Brey, 1976) indicates that the increase in Al/Si ratio is more important in these polymerized compositions.

Carbonate in $NaAlO_2$-SiO_2 compositions

In a study of albitic (and anorthitic) glasses, Mysen and Virgo (1980a) fitted Raman spectra with a number of peaks and interpreted the (non-unique) solution as indicating the formation of a Ca- or Na-carbonate complex, with the resultant non-charge-balanced Al being expelled from the framework, resulting in depolymerization. Taylor (1990) also interpreted spectra of CO_2-bearing nepheline and sodamelilite glasses as evidence

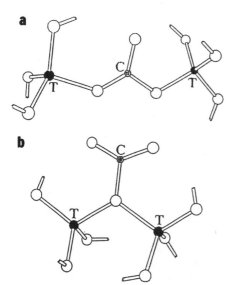

Figure 11. Configurations of a 'network carbonate': (a) originally suggested by Fine and Stolper (1985); (b) additional configuration proposed by Kohn et al. (1991). T = Si or Al tetrahedrally-coordinated by oxygen. Both illustrations adapted from Kubicki and Stolper (in press).

for the formation of this non-tetrahedral AlO_6. However, Taylor disagreed with the metal-carbonate complex model of Mysen and Virgo (1980b) as there is no indication of a single absorption peak, characteristic of the symmetrical, undistorted, ionic CO_3^{2-} group (which occurs near 1450 cm^{-1} for simple crystalline carbonates, see Verweij et al., 1977). Kohn et al. (1991) examined CO_2-saturated albite and nepheline composition glasses, using ^{27}Al NMR spectroscopy, and showed that all detectable Al remains in tetrahedral coordination with dissolution of CO_2. Kohn et al. (1991) favored a mechanism involving either of the 'network carbonate' configurations in Figure 11 that does not require Al-expulsion. The MO simulations of Kubicki and Stolper (in press) support the idea that configuration "a" (Fig. 11) is the stable form of carbonate in these melts and has the correct spectral characteristics; configuration "b" would be unstable as a carbonate group, but it could exist with a long and weak framework-oxygen to carbon bond, to give the bent molecular CO_2, mentioned in the previous section. The assignment of T's in Figure 11 to various combinations of Al and Si, and their roles in describing the spectra features, are discussed in detail in Fine and Stolper (1985), Taylor (1990), Kohn et al. (1991) and Kubicki and Stolper (in press). Kubicki and Stolper (in press) also calculate that the IR v_3 split can be heavily influenced by the proximity of the Na^+ cation and the importance of the alkali cation has been demonstrated by Brooker et al. (1992), who showed that both the doublet splits decreases as the alkali is substituted (K < Na < Li) in both albitic and jadeitic compositions. The substitution of Ca for Na also appears to reduce the split of the doublet (see Brey, 1976).

Carbonate in other compositions

Published IR and Raman spectra of nearly all synthetic and natural melt compositions (e.g., Sharma, 1979; Sharma et al., 1979; Mysen and Virgo, 1980a; Fine, 1986; Mattey et al., 1990; Taylor, 1990; Pan et al., 1991; Thibault and Holloway, 1994) show a v_3 carbonate doublet with a split of only 70 to 100 cm^{-1}. Apart from the compositions on the SiO_2–$NaAlO_2$ join, the only other exceptions are the $K_2Si_{1.5}O_4$–K_2SiO_3 compositions of Verweij et al. (1977) which do not show any obvious split in the v_3 peak. The 70 to 100 cm^{-1} split is also seen in crystalline scapolite (Papike, 1964), and Papike and

Stephenson (1966) suggest that the doublet results from the distorted nature of this *single* carbonate site. The bimodal release pattern of CO_2 on heating of CO_2-bearing diopsidic glass led Rai et al. (1983) to conclude that the doublet represented two distinct carbonate sites, but Sharma and others (1979) concluded that the v_3 doublet in diopside, sodamelilite and akermanite represents a single distorted carbonate, as there are only single v_1 and $2v_2$ peaks in the Raman spectra. As discussed in the NMR section, the single symmetric peak in the NMR spectra of a sodamelilite glass (Fig. 10) appears to confirm that the IR doublet represents one carbonate group, the degree of distortion being intermediate between that of carbonate in $NaAlO_2$–SiO_2 glasses and simple crystalline carbonates. The structure of carbonate in natural melt compositions could be similar to the configuration in Figure 11a, but with only one of the carbonate oxygens coordinated to the (alumino)silicate network. This rearrangement of the coordination could result in the reduction of the Δv_3 split from ~250 down to 80 cm^{-1} (cf. Nakamoto et al., 1957).

Mysen and Virgo (1980b) have interpreted the peak-fit Raman spectra of CO_2-sodamelilite ($NaCaAlSi_2O_7$; NBO/T = 0.67) and CO_2–diopside ($CaMgSi_2O_6$; NBO/T = 2) glasses to indicate that the melt becomes more polymerized with CO_2 dissolution. CO_2 is considered to react with a NBO (rather than a BO) to form a metal-carbonate complex. The oxygen reactivity and the polymerization of the melt conflict with the Mysen and Virgo (1980a) model for albite (see Taylor, 1990), and such differences could have implications for expected changes in melt viscosities, which generally increase as the degree of polymerization increases (see Chapter 9). Sharma (1979) also interpreted features of Raman spectra as indicating polymerization accompanying CO_2 dissolution in diopside and sodamelilite compositions, but did not favor the formation of a 'metal-carbonate complex' as in the Mysen and Virgo (1980a) model, due to the distorted nature of the carbonate. The lack of spectroscopic evidence for the 'metal-carbonate complex' is reviewed in detail by Taylor (1990) and the only compositions which could be interpreted as showing such species are the K-silicates of Verweij et al. (1977), which are very distinct from natural and divalent cation-bearing compositions.

Although viscosity measurements could provide some constraints for solubility mechanism models, data are limited. A slight decrease in viscosity has been reported for dissolution of CO_2 in both albite (Brearley and Montana, 1989) and sanidine (White and Montana, 1990) melt compositions, possibly supporting a depolymerizing dissolution mechanism for these fully polymerized melts. Alternatively, inclusion of 'network carbonate' may weaken the framework structure to give the same effect. More high quality data on the viscosities of CO_2-bearing melts are required before it is possible to use viscosity arguments to infer aspects of CO_2 dissolution mechanisms.

Holloway et al. (1976) demonstrated a positive correlation between CO_2 solubility and the Ca/(Ca+Mg) ratio at a given SiO_2 content (NBO/T) in meta- and orthosilicate melts. Along the Mg_2SiO_4-$MgCaSi_2O_6$ join, CO_2 solubility might be expected to decrease as the SiO_2 content increases (NBO/T decreases), but the increased Ca/(Ca+Mg) appears to be more important and an increase in solubility is observed. Holloway et al. (1976) also found changes in the temperature dependence of CO_2 solubility in some compositions, but it should be noted that [14]C β-track analysis was used, and the results must be subject to the same caution mentioned early in this chapter. Taylor (1990) suggests that the results of Holloway et al. (1976) can be explained in terms of disproportionation, where the relative proportions of Q species are to some extent a function of the modifying cation present; the increase in CO_2 solubility with the substitution of Ca^{2+} by Mg^{2+} also being observed when Na^+ is replaced by Li^+ (deJong et al., 1981). The relatively high charge to radius ratio cations ($Mg^{2+} > Ca^{2+}$ and $Li^+ > Na^+$) tend to produce a bimodal distribution of extreme Si environments, such as Q^4 and Q^0 (e.g., $2 Q^2 = Q^4 + Q^0$) and if the resulting species are less reactive to CO_2, then solubility is lowered.

Thermodynamic models

In very general terms, the formation of carbonate may be described by the following heterogeneous equilibrium (Fine and Stolper, 1985; Stolper and Holloway, 1988; Pan et al., 1991; Thibault and Holloway, 1994), first suggested by Pearce (1964);

$$CO_2 \text{ (fluid)} + O^{2-} \text{ (melt)} = CO_3^{2-} \text{ (melt)} \tag{1}$$

where O^{2-} represents oxygen ions in the melt which are able to react with CO_2. A solubility constant can be written from this equation and if the temperature and pressure dependence of the reaction is known, values of the volume change (ΔV), and enthalpy change (ΔH) associated with CO_2 dissolution can be calculated (see Chapter 6 for specific examples). However, the available data suggest that ΔV and ΔH may differ significantly among different melt compositions. For example ΔH (for additional detail, refer to Stolper and Holloway,, 1988) varies from near zero for a tholeiitic basalt (Pan et al., 1991), to -28.2 kJ/mol for an Ca-leucitite (Thibault and Holloway, 1994). This demonstrates the limitation of Equation (1) in describing dissolution, in that the constants derived cannot be transferred between compositions. Fine and Stolper (1985) suggest that there are a variety of reactive oxygens, each with its own solubility constant, and that proportions of these vary in response to melt compositions (see below). It has been suggested that NBO are more reactive to CO_2 than BO (Fine and Stolper, 1985) and that a modified reaction such as

$$CO_2 \text{ (fluid)} + 2 O^- \text{ (NBO)} = O^\circ \text{ (BO)} + CO_3^{2-} \tag{2}$$

is more appropriate. If the carbonate groups are assumed to mix ideally with NBOs in the melt (thus allowing the activity of carbonate to be expressed as the mole fraction), Equation (2) predicts a positive correlation between CO_2 solubility (as carbonate) and the activity of NBO (e.g., Dingwell, 1986).

In an attempt to introduce a thermodynamic approach to the problem, and at the same time avoid uncertainty in the type of reaction involved, Fine and Stolper (1985; 1986) suggested the following solubility reaction;

$$CO_2 \text{ (molecular)} + O_r^{2-} = CO_3^{2-} \tag{3}$$

where O_r^{2-} represents a distinguishable, reactive oxygen species (e.g., BO, NBO or 'free' oxygen), and an equilibrium constant can be written as:

$$K = \frac{(a_{CO_3^{2-}})}{(a_{CO_2mol})(a_{O_r})} \tag{4}$$

If the dissolution of CO_2 is considered as the mixing of CO_2 molecules with C-free melt, then each oxygen type has a different value of K. Highly-reactive oxygens would have a high value of K, while those inert to CO_2 would have a K value approaching zero. If two melts differ in concentration of oxygen species, then the melt with the higher ratio of reactive oxygens would have the higher ratio of carbonate/CO_2. This would explain the change in the carbonate/CO_2 ratio on the $NaAlO_2$–SiO_2 join (Fig. 8) if reactive oxygens could be considered to vary monotonically with the concentration of sodium in the melt, and if K for the reaction between molecular CO_2 and a BO between two Si tetrahedra is zero. This would also imply that the CO_2 dissolved in silica would be entirely in molecular form. Fine and Stolper (1985) consider the value of K in the presence of different cations, with a lower value for the reaction between CO_2 molecules and O_r associated with sodium than for O_r associated with calcium (i.e., $(CaCO_3)^\circ$ is stabilized, relative to CaO and CO_2, to a greater extent than $(Na_2CO_3)^\circ$, to NaO^- and CO_2).

While it appears that there may be a covalent association of carbonate with the network structure, it is also clear that the association of a cation is important. This suggests that the stabilities (and hence solubilities) of carbonate in melts is related to a combination of suitable cations and available reactive oxygens (possibly NBO), although for a given polymerization, the cation could be more important in creating the reactive oxygens (disproportionation) than in stabilizing the carbonate structure. An increase in the number of cations and/or the break-up of the network might also lead to a reduction of available sites (or holes) for molecular CO_2. It is clear that a more systematic examination of CO_2 solubility in relation to changes in melt structure and in particular cation substitution is needed before a model of the compositional dependence can be developed with an associated thermodynamic description of pressure and temperature effects.

STABLE ISOTOPE FRACTIONATION

Introduction

The atomic percent abundances of the two stable isotopes of carbon are 98.9% (^{12}C) and 1.1% (^{13}C) (see Walker et al., 1984), and their $^{13}C/^{12}C$ ratio can be determined to a very high level of precision, even when the total amount of carbon present is very small. Variations in the stable isotope ratios of carbon in melts may reflect source material, contamination, and fractionation. The magnitude of fractionation depends on factors such as temperature, speciation, chemical environment, and kinetic effects (cf. Faure, 1986; O'Neil, 1986). While equilibrium fractionations of stable isotopes among phases approach zero at very high temperature (Urey, 1947), they may still remain significantly large at magmatic temperatures, to produce measurable isotopic variation resulting from gain or loss of volatiles. Consequently, experimental stable isotope studies of C-bearing fluids provide an important means of understanding fluid-melt-rock interactions. Measurements of stable isotope ratios of C in magmatic gases, erupted glasses, and melt inclusions in phenocrysts are useful tools for interpreting the volatile history of natural magmas (e.g., Gerlach and Taylor, 1990; Sarda and Graham, 1990; Blank et al., 1993b).

A variety of carbon compounds are found in igneous rocks: (1) carbonate minerals and C-bearing gases in fluid inclusions; (2) elemental carbon in the form of graphite or diamonds; (3) mixtures of organic molecules and (4) carbides. The stable isotope composition of carbon in the oxidized form (carbonates and CO_2 gas) differs markedly from that in the reduced states (graphite, carbides, and organic molecules). Reduced carbon, strongly depleted in ^{13}C, has $\delta^{13}C$ values[1] typically between -20‰ and -30‰ (e.g., Feux and Baker, 1973; Hoefs, 1980). Oxidized carbon has distinctly different $\delta^{13}C$ values ranging from approximately +3‰ to -18‰ relative to the PeeDee Belemnite (PDB) isotopic standard.

Essential to the interpretation of natural variations of stable isotope ratios is knowledge of the magnitude and temperature dependence of the equilibrium isotopic fractionations between species, and this section summarizes and discusses current best-estimates of the values for carbon in silicate melt-CO_2 systems. A more detailed presentation of terminology relevant to stable isotope studies is presented in Chapter 6 of this volume.

---

[1] Where $\delta^{13}C = \{ (R - R_{std})/R_{std} \} \times 10^3$ and $R = (^{13}C/^{12}C)$, R_{std} is the corresponding ratio in a standard reference material.

Experimental fractionation studies

The experiments required to determine the partitioning behavior of ^{13}C between CO_2 vapor and silicate melt are notoriously difficult, and as a result there are few published data. Experimental studies of carbon isotope fractionation between coexisting CO_2 vapor and silicate melt (Δ_{v-m}) have focused on a few natural (basalt and rhyolite) and synthetic (tholeiite and sodamelilite) compositions. Published data are summarized in Table 3 and discussed in more detail below.

Table 3. Measured carbon isotope fractionations between coexisting CO_2 vapor and silicate melt

System	$\Delta_{\text{vapor-melt}}$ (‰)	T (°C)	Speci- ation	Reference
CO_2-basalt	4.3 (±0.3)	1120-1280	CO_3^{2-} *	Javoy et al., 1978
CO_2-basalt	4.2 (±0.4)	1200	CO_3^{2-}	Trull et al., 1991
CO_2-sodamelilite	2.4 (±0.2)	1200-1400	CO_3^{2-}	Mattey et al., 1990
CO_2-basalt (*syn., Fe-free*)	2.0 (±0.2)	1200-1400	CO_3^{2-}	Mattey, 1991
CO_2-rhyolite	0.0 (±0.2)	800-1200	CO_2	Blank & Stolper, 1993

* Inferred

CO$_2$-basalt. Experimental studies (e.g., Javoy et al., 1978; Mattey, 1991; Trull et al., 1991) have determined that CO_2 vapor is enriched in ^{13}C by ~2 to 4‰ relative to carbonate in coexisting basaltic melt Such fractionation information has been used to interpret the observed variations in ^{13}C/^{12}C ratios in submarine basaltic glasses (e.g., Des Marais and Moore, 1984; Bottinga and Javoy, 1989; Javoy and Pineau, 1991; Blank et al., 1993b; MacPherson and Mattey, 1994) and fumarole gases (e.g., Gerlach and Taylor, 1990), and also to estimate the initial carbon content of undegassed basaltic magma and calculate the flux of CO_2 from mid-ocean ridge volcanism to the atmosphere (e.g., Javoy et al., 1983; Des Marais, 1985; Sarda and Graham, 1990).

Javoy et al. (1978) determined experimentally the carbon isotopic fractionation between vapor and basaltic melt at 1120° to 1280°C and 7 to 8.4 kbar using two brands of oxalic acid with $\delta^{13}C$ values of -24.1 and -16.6‰. They reported Δ_{v-m} values of 4.0 to 4.6‰ that closely resemble calculated fractionations for CO_2-CO (Richet et al., 1977) and CO_2-graphite (Bottinga, 1969) at magmatic temperatures. These fractionation values appear to be too large given that the principal dissolved species in basaltic melt is carbonate (e.g., Fine and Stolper, 1986), and the calculated carbon isotopic fractionation for CO_2-carbonate should be no higher than 1 to 2‰ at magmatic temperatures (Ohmoto and Rye, 1979). Javoy et al. (1978) also report unusually high and variable CO_2 solubilities in their glassy run products, up to an order of magnitude higher than those predicted by Stolper and Holloway (1988) for the same pressure and temperature. Javoy et al. (1978) noted that the presence of an unidentified, reduced species of carbon (graphite?) may have affected their results, and thus the reported C isotope fractionation values are somewhat suspect. Trull et al. (1991) obtained similar fractionation results, and their measured CO_2 solubilities increased systematically with the P_{CO_2} of their experiments.

A smaller Δ_{v-m} value between 1.8‰ and 2.2‰ was determined experimentally by Mattey (1991) for basaltic melt held at 1200° to 1400°C and 5 to 20 kbar. These results are comparable to those from similar experiments performed on sodamelilite ($NaCaAlSi_2O_7$), an Fe-free basaltic melt analog in which dissolved carbon also occurs as carbonate ions (Mattey et al., 1990). However, Mattey and coworkers could not account for the mass loss of carbon during their experiments but showed that variable mass loss and shifts in $\Delta^{13}C_{(initial\ vapor-final\ vapor)}$ did not seem to affect their measured $\Delta_{vapor-melt}$ fractionations.

CO_2-rhyolite. Isotopic fractionation is predicted to be smaller for the CO_2-rhyolite system than for CO_2-basalt, as the dissolved species is molecular CO_2 rather than carbonate. Blank and Stolper (1993b) equilibrated natural, CO_2-free rhyolitic glasses with CO_2 gases having five different $^{13}C/^{12}C$ ratios. Experiments were conducted at 800° to 1200°C and at pressures ranging from 250 to 1444 bars. At the end of an experiment, vapor coexisting with rhyolitic melt was collected and analyzed for its $^{13}C/^{12}C$ ratio. The rhyolitic glasses were combusted on a vacuum line using a stepped-heating procedure, and CO_2 was extracted and $^{13}C/^{12}C$ ratios were determined for direct comparison with the vapor. Individual glass chips were also analyzed for dissolved CO_2 using FTIR spectroscopy. Blank and Stolper (Blank, 1993; Blank and Stolper, 1993) detected no measurable difference in the isotopic composition of carbon in the vapor and coexisting rhyolitic melts and concluded that the vapor-melt fractionation is close to zero over the range of temperatures of their study. If the fractionation is indeed approximately zero, then variations in the isotopic composition of C in natural rhyolites will not be due to CO_2 degassing. However, formation of other C-bearing phases (e.g., carbon-rich films) accompanied by significant fractionation of C isotopes could affect the composition of natural samples.

Intermediate compositions

The basalt and high-silica rhyolite samples represent two end-members in the natural spectrum of igneous rocks with respect to silica content. At the pressures of the isotopic experiments, all dissolved CO_2 has been detected as carbonate in basalt and as molecules of CO_2 in rhyolite (e.g., Fine and Stolper, 1986; Fogel and Rutherford, 1990; Pan et al., 1991). Presently, a systematic evaluation of the speciation of dissolved carbon dioxide in silicate melts of intermediate composition remains to be conducted, but andesitic and phonolitic melts have been observed to contain both molecular CO_2 and CO_3^{2-} (see Fig. 10, also Fogel and Rutherford, 1990; Brooker and Holloway in prep.). If the relative proportions of CO_2/CO_3^{2-} increase moving from mafic to silicic natural compositions, the corresponding $\Delta_{CO_2\text{-silicate melt}}$ would be expected to decrease with increasing silica contents.

On the basis of the experimental results presented above, the CO_2 vapor in equilibrium with basaltic melt is enriched in ^{13}C by ~2 to 4‰, whereas the CO_2 vapor in equilibrium with a rhyolitic melt has the same $\delta^{13}C$ value. Studies of $CO_2(g)$-$CO_2(l)$ and $CO_2(g)$-$CO_3^{2-}(aq)$ have shown a similar trend in carbon isotope fractionation. Grootes et al. (1969) report that the carbon isotopic fractionation between gaseous and liquid CO_2 [$CO_2(g)$-$CO_2(l)$] at low temperatures is nearly zero. Vogel et al. (1970) observed a small fractionation (~1‰) between gaseous CO_2 and CO_2 dissolved in water at 0° to 60°C whereas similar studies of CO_2 vapor and carbonate ion in solution have found 8 to 12‰ fractionation at temperatures ≤ 40°C (e.g., Lesniak and Sakai, 1989). Theory would predict that CO_2 vapor would be enriched in ^{13}C relative to the liquid state, because binding energy tends to favor the heavier isotope in symmetric molecules (Grootes et al., 1969).

Only CO_2 dissolved in molecular form has been detected in rhyolitic melts quenched from pressures of up to ~7 kbar (Fogel and Rutherford, 1990). However, Stolper et al. (1987) noted that the relative proportion of CO_3^{2-} as total dissolved CO_2 increases slightly with increasing pressure in albitic melt and therefore it might become significant in rhyolitic melts at high pressure. An obvious study for future work is the comparison of the manner in which $\Delta_{vapor-melt}$ and carbon speciation evolve in rhyolitic melt as a function of increasing pressure.

SUMMARY

While considerable progress has been made towards the formulation of a theoretical and empirical description of the observed interaction of carbon dioxide and silicate melts, many issues remain to be addressed. In particular, there are no predictive models to describe CO_2 solubility over the compositional range of natural melts, especially in terms of their variable water contents. This limits our understanding of volatile behavior during such processes as partial melting and fractionational crystallization. As these processes may be important in leading to vapor saturation, such modeling is an important aspect of volcanology. More systematic experiments are required to develop a compositionally-based model which can incorporate a thermodynamic description of pressure and temperature dependence. Such a model combined with observations of volcanic gas compositions and isotopic systematics at volcanic centers would provide an especially powerful tool to interpret processes occurring at inaccessible depths in the earth's interior.

Future experimental work must include development of analytical standards in order to arrive at a consensus among CO_2 measurements obtained using different analytical techniques. A comprehensive understanding of the atomistic aspects of C dissolution mechanisms is required if we hope to develop general theoretical models capable of predicting solubilities in natural melts over a range of pressures and temperatures.

ACKNOWLEDGMENTS

This chapter was improved through careful reviews by Mike Carroll and Simon Kohn and useful comments by Tina Clough, Patrick Dobson and Alison Pawley. JGB thanks Julie Shemeta and Jeff Meredith for assistance with network access and acknowledges the hospitality of Claude Jaupart and financial support from the French government (CROUS).

APPENDIX

Table A1. Reviews of CO_2 ($\pm$ H_2O) solubilities in silicate melts

Year	Author(s)	Comments
1964	Pearce	Lists studies pre-1964
1977	Mysen	General review, dissolution models
1977	Eggler & Holloway	Theory, phase equilibria
1979	Wyllie	General review, geologic context
1980	Spera & Bergman	Thermodynamic treatment of CO_2
1981	Holloway	General review, solubilities
1986	Dingwell	General review, CO_2 + other volatiles
1987	Holloway	General review of major volatiles (Rev Mineral v 17)

Table A2

EXPERIMENTAL STUDIES OF CO_2 EQUILIBRATED WITH SILICATE MELTS

Year	Authors	System [1]	Apparatus [2]	T (°C)	P (kbars) [3]	Analytical Method [4]
Synthetic Magma Compositions						
1932	Eitel & Weyl	CO_2-Na_2O-SiO_2	PC	1100	1.25	Weight loss
1940	Morey & Fleischer	CO_2-H_2O-K_2O-SiO_2	CS	340-424	P_{CO_2} to ≈25 bars	Weight loss
1951	England & Adams	CO_2-Kimberlite	CS	up to 1290	1 atm-0.16 kbar	Depression of Liquidus
1959	Wyllie & Tuttle	CO_2-Gran-Ab, -Ab-Or0, H_2O	CS	730-930	≈.07-2.14	mb
1964	Pearce	CO_2-Na-silicate melts	CS	850-1400	1 atm	vf/manometry
1966	Faile & Roy	CO_2-$K_2Si_4O_9$, -B_2O_3	CS	750-800	4	weight loss (mb), manometry
1971	Milhollen et al.	CO_2-Ab-H_2O	IHPV/PC	760-1000	1-30	weight loss
1974	Kesson & Holloway	CO_2-Ab-H_2O-N_2	IHPV	951-1058	4.14-4.85	weight loss
1976	Brey	CO_2-Ol Mel, -Ab-An	PC	1450-1650	30	gc, IR
1976	Brey	CO_2-Ol Mel	PC	1400-1520	5-40	electron microprobe shortfall, Hewlett-Packard CHN analyzer
1976	Holloway et al.	CO_2-CaO-MgO-SiO_2	PC	1500-1700	20/30	β-track
1976	Mysen et al.	CO_2-Ab-An	PC	1400-1700	5-40	β-track, gc, IR (KBr pellets)
1976	Mysen	CO_2-H_2O-Ab-Jd-Ne	PC	1450/1525/1625	10-30	β-track
1977	Verweij et al.	CO_2-K-silicate	1 atm furnace	1000-1200	1 atm	Raman
1978	Eggler & Rosenhauer	CO_2-Di-H_2O	PC	1180-1680	to 40	DTA
1979	Eggler et al.	CO_2-various silicate melts-CO	PC	1525-1700	20/30	β-track, gc
1979	Eggler & Kadik	CO_2-Ab-H_2O	PC	900-1100	3-20	β-track
1980	Mysen & Virgo	CO_2-Ab, -An	PC	1450-1700	10-25	Raman, β-track
1980	Mysen & Virgo	CO_2-Di, -Sm	PC	1275-1725	10-25	Raman, β-track
1981	Kadik et al.	CO_2-Ab-SiO_2	IHPV	1250/1350	2-3/16,17,26,35	β-track
1982	Bohlen et al.	CO_2-Ab-H_2O	PC	630-1040	5-20	depression of liquidus
1983	Rai et al.	CO_2-Di	PC	1580-1725	20	quadrapole ms/Raman
1984	Boettcher	CO_2-SiO_2-H_2O	PC	1000-1650	7-27.5	wt loss (mb), petrographic analysis
1985	Fine & Stolper	CO_2-$NaAlO_2$-SiO_2-join	PC	1400-1560	15-33	IR. Understaurated runs.
1987	Boettcher et al.	CO_2-Ab, -An, -San	PC	1160-1590	10-25.5	wt loss (mb), petrographic analysis
1987	Stolper et al.'	CO_2-Ab	PC	1450-1625	15-30	FTIR
1988	Tingle & Aines	CO_2-Ab	PC	1450	20	β-track, IR
1990	Mattey et al.	CO_2-Sm	PC	1200-1400	5-30	SHMS

Table A2, continued

Natural Magma Compositions

1972	Kadik et al.	CO_2-Gran, -Bas, $\pm H_2O$	IHPV	1200	1-3	weight loss
1975	Brey & Green	CO_2-Ol Mel	PC	1400	30	electron microprobe shortfall/IR
1975	Mysen et al.	CO_2-Ol Mel, -Neph, -Bas, $\pm H_2O$	PC	1400-1650	15-30	β-track
1986	Jakobsson & Holloway	CO_2-Ab, -Basan, -Bas, COH fluids	PC	1000/1100/1200	10-25	IR, Raman, thermogravimetry, quadrupole ms
1988	Stolper & Holloway	CO_2-Basalt	IHPV	1200	0.10-1.50	FTIR (low curve; high P run not well-equilibrated)
1990	Fogel & Rutherford	CO_2-Rhyolite	TZM/IHPV	950/1050/1150	0.50-6.60	FTIR (used jadeite epsilon; must corr 15%)
1990	Shilobreyeva & Kadik	CO_2-Bas, -And, -Dac, -Rhy	IHPV	1250	0.50-3.00	β-track
1991	Pan et al.	CO_2-Basalt	RQIHPV/PC	1200/1300-1600	1,10,15	FTIR/SIMS/LECO determinator
1991	Mattey	CO_2-Basalt	PC	1200-1400	5-20	SHMS
1991	Trull	CO_2-Basalt	IHPV	1200°	1-2	SHMS
1992	Pawley et al.	CO_2-Basalt-CO	RQ-IHPV	1200°C	0.503-1.503	vacuum line (mb) for vapor/ FTIR
1993	Blank et al.	CO_2-Rhyolite, $\pm H_2O$	RQTZM	850	0.75	FTIR
1994	Thibaut & Hollaway	CO_2-Ca-rich Leucitite	RQIHPV/PC	1200-1600	1-20	FTIR/SIMS/LECO bulk analysis
in press	Dixon et al.	CO_2-Basalt (comp. of S&H 1988)	RQIHPV	1200	<1	FTIR

Notes:

1 Abbreviations are the following: Gran = Granitic melt; Ab = Albite; Ol Mel = Olivine Melilitie; An = Anorthite; Jd = Jadeite; Ne = Nephelinite; K-Silicate = K-rich Silicate Composition; Di = Diopside; San = Sanidine; Sm = Sodamelilite; Ne Syen = Nepheline Syenite; Bas = Basalt; Basan = Basanite; And = Andesite; Dac = Dacite; Rhy = Rhyolite.

2 Types of apparatus include: PC = Piston Cylinder; CS = Cold Seal (Nickel) Pressure Vessel; IHPV = Internally-Heated Pressure Vessel; TZM = Titanium/Zirconium/Molybdenum Alloy Pressure Vessel; RQ = Rapid-Quench (e.g., RQCS = Rapid-Quench Cold Seal Pressue Vessel).

3 Pressure in kbars unless otherwise noted. Entries with greater number of significant figures were reported in bars in the literature.

4 Analytical techniques include: MB = mass balance; VF/Manometry = vacuum fusion/manometry; GC = gas chromatography; (FT)IR = (Fourier Transform) Infrared Spectroscopy; β-Track = [14]C autoradiography; Raman = Raman Spectroscopy; DTA = Differential Thermal Analysis; Quadrapole MS = Quadrapole Mass Spectrometry; SHMS = Stepped-Heating Mass Spectrometry; SHM = Stepped-Heating/Manometry; SIMS = Secondary Ion Mass Spectrometry; LECO = LECO Bulk Carbon Analyzer.

Table A3
Publications concerning CO_2 structure and dissolution mechanisms in silicate melts.

Year	Authors	System ($CO_2 +$ ___) [1]	Principal Analytical Technique
1976	Brey	Ol Mel, Ab-An	Gas Chromatography, IR Spectroscopy
1977	Verweij et al.	K-Silicate	Raman Spectroscopy
1979	Sharma	Di, Sm	Raman Spectroscopy
1979	Sharma et al.	Akermanite	Raman Spectroscopy
1980	deJong & Brown	$H_6Si_2O_7$	MO calculations
1983	Mysen	And	Raman Spectroscopy
1985	Fine & Stolper	$NaAlO_2$-SiO_2 join	IR Spectroscopy
1989	Brearley & Montana	Ab+Sm	Viscosity determintaions
1990	White & Montana	San	Viscosity determinations
1990	Taylor	Sm, Ne	IR Spectroscopy and review
1991	Kohn et al.	Ab, Ne, Sm	^{13}C NMR Spectroscopy
1980a	Mysen & Virgo	Ab, An	Raman Spectroscopy
1980b	Mysen & Virgo	Di, Sm	Raman Spectroscopy
in press	Kubicki & Stolper	$NaAlO_2$-SiO_2 join	MO calculations

[1] Abbreviations used: Ol Mel = Olivine Melilitite; K-Silicate = K-rich Silicate; Di = Diopside; Sm = Sodamelilite; And = Andesite; Ab = Albite; San = Sanidine; Ne = Nepheline; An = Anorthite.

REFERENCES

Bassett WA, Shen AH, Bucknum M, Chou IM (1993) Hydrothermal studies in a new diamond-anvil cell up to 10 GPa and from -190°C to 1200°C. Pure Appl Geophys 141:487-495

Blank JG (1993) An Experimental Investigation of the Behavior of Carbon Dioxide in Rhyolitic Melt. PhD dissertation, California Institute of Technology, Pasadena, CA

Blank JG, Delaney JR, Des Marais DJ (1993b) The concentration and isotopic composition of carbon in basaltic glasses from the Juan de Fuca Ridge. Geochim Cosmochim Acta 57:875-887

Blank JG, Stolper EM (1993) The partitioning of ^{13}C between coexisting CO_2 vapor and rhyolitic melt. EOS Trans Am Geophys Union 74:347-8

Blank JG, Stolper EM, Carroll MR (1993a) Solubilities of carbon dioxide and water in rhyolitic melt at 850°C and 750 bars. Earth Planet Sci Lett 119:27-36

Boettcher AL, Luth RW, White BS (1987) Carbon in silicate liquids: the system $NaAlSi_3O_8$-CO_2, $CaAl_2Si_2O_8$-CO_2, and $KAlSi_3O_8$.CO_2. Contrib Mineral Petrol 97:297-304

Boettcher AL (1984) The system SiO_2-H_2O-CO_2: melting, solubility mechanisms of carbon, and liquid structure to high pressures. Am Mineral 69:823-833

Bohlen SR, Boettcher AL, Wall VJ (1982) The system albite-H_2O-CO_2: a model for melting and activities of water at high pressures. Am Mineral 67:451-462

Bottinga Y (1969) Calculated fractionation factors for carbon and hydrogen isotope exchange in the system calcite-carbon dioxide-graphite, methane-hydrogen-water vapor. Geochim Cosmochim Acta 33:49-64

Bottinga Y, Javoy, M (1989) MORB degassing: Evolution of CO_2. Earth Planet Sci Lett 95:215-225

Brearley M, Montana A (1989) The effect of CO_2 on the viscosity of silicate liquids at high pressure. Geochim Cosmochim Acta 53:2609-2616

Brey G (1976) CO_2 solubility and solubility mechanisms in silicate melts at high pressures. Contrib Mineral Petrol 57:215-221

Brey G, Green DH (1975) The role of CO_2 in the genesis of olivine melilitite. Contrib Mineral Petrol 49:93-103

Brey GP, Green DH (1976) Solubility of CO_2 in olivine melilitite at high pressures and role of CO_2 in the earth's upper mantle. Contrib Mineral Petrol 55:217-230

Brooker RA, Holloway JR (in prep.) Polymerization, carbonate environment and CO_2 solubility in natural melt compositions

Brooker RA, Holloway JR, Kohn SC, McMillan PF (in prep.) CO_2 dissolution in $Na_2O-Al_2O_3-SiO_2$ melts, Part I: Speciation and solubility along the $NaAlO_2-SiO_2$ join

Brooker RA, McMillan PF, Holloway JR (1992) The structural environment of C-N-O species dissolved in aluminosilicate melts by FTIR spectroscopy. EOS Trans Am Geophys Union 73:619

Carroll MR, Stolper EM (1993) Noble gas solubilities in silicate melts and glasses: New experimental results for argon and the relationship between solubility and ionic porosity. Geochim Cosmochim Acta 57(23/24):5039-5051

Chou I-M (1987) Oxygen buffer and hydrogen sensor techniques at elevated pressures and temperatures. In: GC Ulmer, HL Barnes (eds) Hydrothermal Experimental Techniques. John Wiley & Sons, New York, p 61-99

Cotton FA (1971). Chemical Applications of Group Theory. New York, John Wiley & Sons

deJong BHWS, Brown GE (1980) Polymerization of silicate and aluminate tetrahedra in glasses, melts, and aqueous solutions - II. The network modifying effects of Mg^{2+}, K^+, Na^+, Li^+, H^+, OH^-, F^-, Cl^-, H_2O, CO_2, and H_3O^+ on silicate polymers. Geochim Cosmochim Acta 44:1627-1642

deJong BHWS, Keefer KD, Brown GE, Taylor CM (1981) Polymerization of silicate and aluminate tetrahedra in glasses, melts and aqueous solutions - III. Local silicon environments and internal nucleation in silicate glasses. Geochim Cosmochim Acta 45:1291-1308

Des Marais DJ (1978) Carbon, nitrogen, and sulfur in Apollo 15, 16, and 17 rocks. Proc Lunar Planet Sci Conf 9:2451-2467

Des Marais DJ (1985) Carbon exchange between the mantle and the crust, and its effect upon the atmosphere: today compared to Archean time. In: The Carbon Cycle and Atmospheric CO_2: Natural Variations Archean to Present. Geophys Monogr 32:602-611, Am Geophys Union, Washington, DC

Des Marais DJ (1986) Carbon abundance measurements in oceanic basalts: the need for a consensus. Earth Planet Sci Lett 79:21-26

Des Marais DJ, Moore JG (1984) Carbon and its isotopes in mid-oceanic ridge basaltic glasses. Earth Planet Sci Lett 69:43-57

Dingwell DB (1986) Volatile solubilities in silicate melts. In: CM Scarfe (ed) Short Course in Silicate Melts. Mineral Assoc Canada, p 93-129

Dixon JE, Stolper EM, Holloway JR (in press) An experimental study of water and carbon dioxide solubilities in mid-ocean ridge basaltic liquids. J Petrol

Eggler DE, Holloway JR (1977) Partial melting of peridotite in the presence of H_2O and CO_2: Principles and review. Oregon Dept Geol Mineral Indust Bull 96:15-36

Eggler D, Rosenhauer M (1978) Carbon dioxide in silicate melts: II. Solubilities of CO_2 and H_2O in $CaMgSi_2O_6$ (Diopside) liquids and vapors at pressures to 40 kb. Am J Sci 278:64-94

Eggler DE, Mysen BO, Hoering TC, Holloway JR (1979) The solubility of carbon monoxide in silicate melts at high pressures and its effect on silicate phase relations. Earth Planet Sci Lett 43:321-330

Eggler DH, Kadik AA (1979) The system $NaAlSi_3O_8-H_2O-CO_2$ at pressure: I. Compositional and thermodynamic relations of liquids and vapors coexisting with albite. Am Mineral 64:1036-1048

Eitel W Weyl W (1932) Residuals in the melting of commercial glasses. J Am Ceram Soc 15:159-166

Engelhardt G, Michel D (1987). High-Resolution Solid-State NMR of Silicates and Zeolites. New York, John Wiley & Sons, 485 p

England JL and Adams LH (1951) Effect of carbon dioxide on melting behavior of natural rocks. Carnegie Inst Wash Yrbk 50:49

Faile SP, Roy,= DM (1966) Solubilities of Ar, N_2, CO_2 and He in glasses at pressures to 10 kbar. J Ceram Soc 49:638-643

Farmer VC (1974). The Infrared Spectra of Minerals. Mineral Soc Monogr 4, 539 p

Faure G (1986). Principles of Isotope Geology, 2nd Ed. 2nd ed., New York, John Wiley & Sons, 589 p

Feux AN, Baker DR (1973) Stable carbon isotopes in selected granitic, mafic, and ultra mafic igneous rocks. Geochim Cosmochim Acta 37:2509-2521

Fine G (1986) Carbon dioxide in synthetic and natural silicate glasses. PhD dissertation, California Institute of Technology

Fine G, Stolper EM (1985) The speciation of carbon dioxide in sodium aluminosilicate glasses. Contrib Mineral Petrol 91:105-121

Fine GJ, Stolper EM (1986) Dissolved carbon dioxide in basaltic glasses: Concentrations and speciation. Earth Planet Sci Lett 76:263-278

Fogel RA, Rutherford MJ (1990) The solubility of carbon dioxide in rhyolitic melts: a quantitative FTIR study. Am Mineral 75:1311-1326

Franz JD, Dubessy J, Mysen B (1993) An optical-cell for Raman-spectroscopic studies of supercritical fluids and its application to the study of water to 500°C and 2000 bars. Chem Geol 106:9-26

Gerlach TM, Taylor BE (1990) Carbon isotope constraints on degassing of carbon dioxide from Kilauea Volcano. Geochim Cosmochim Acta 54:2051-2058

Grootes PM, Mook WG, Vogel JC (1969) Isotopic fractionation between gaseous and condensed carbon dioxide. Zeit Physik 221:257-273

Harris DC, Bertolucci MD (1978). Symmetry and Spectroscopy. Oxford, Oxford University Press, 550 p

Herrmannsdorfer G, Keppler,H (1992) A sapphire anvil cell for high pressure, high temperature Raman measurements of hydrous silicate melts. EOS Trans Am Geophys Union 73:599

Herzberg C (1945). Molecular Spectra and Molecular Structure, II. Infrared and Raman Spectra of Polyatomic Molecules. New York, Van Nostrand Reinhold Company, 632 p

Hoefs J (1980). Stable Isotope Geochemistry. 2nd ed., New York, Springer-Verlag, 208 p

Holloway JR (1971) Internally heated pressure vessels. In: GC Ulmer (ed) Research Techniques for High Temperature and Pressure. Springer, New York, p 217-257

Holloway JR (1981) Volatile interactions in magmas. In: RC Newton, A Navrotsky, BJ Wood (eds) Thermodynamics of Melts and Minerals. Advances in Physical Geochemistry I, Springer-Verlag, New York, p 273-293

Holloway JR (1987) Igneous fluids, thermodynamic modelling of geological materials: Minerals, fluids and melts. Rev Mineral 17:182-186

Holloway JR, Burnham CW, Millhollen GL (1968) Generation of H_2O-CO_2 mixtures for use in hydrothermal experimentation. J Geophys Res 73:6598-6600

Holloway JR, Mysen BO, Eggler DH (1976) The solubility of CO_2 in liquids on the join CaO-MgO-SiO_2-CO_2. Carnegie Inst Wash Yrbk 75:626-631

Holloway JR, Dixon JE, Pawley AR (1992) An internally heated, rapid-quench, high-pressure vessel. Am Mineral 77:643-646

Holloway JR, Wood BJ (1988). Simulating the Earth: Experimental Geochemistry. Boston, Unwin Hyman, 196 p

Howard-Lock HE, Stoicheff BP (1971) Raman intensity measurements of the Fermi diad v_1, $2v_2$ in $^{12}CO_2$ and $^{13}CO_2$. J Mol Spec 37:321-326

Ihinger PD (1991) The interaction of water with granitic melt. PhD dissertation, California Institute of Technology, Pasadena, CA

Jakobsson S, Holloway JR (1986) Crystal-liquid experiments in the presence of a C-O-H fluid buffered by graphite + iron + wustite: Experimental method and near-liquidus relations in basanite. J Volcanol Geotherm Res 29:265-291

Jakobsson S, Oskarsson N (1994) The system C-O in equilibrium with graphite at high pressure and temperature: An experimental study. Geochim Cosmochim Acta 58:9-17

Javoy M, Pineau F (1991) The volatiles record of a "popping" rock from the Mid-Atlantic Ridge at 14°N: Chemical and isotopic composition of gas trapped in the vesicles. Earth Planet Sci Lett 107:598-611

Javoy M, Pineau F, Allègre CJ (1983) Carbon geodynamic cycle--reply. Nature 303:731

Javoy M, Pineau F, Liyama,I (1978) Experimental determination of the isotopic fractionation between gaseous CO_2 and carbon dissolved in tholeiitic magma. Contrib Mineral Petrol 67:35-39

Kadik AA, Lukanin OA, Lebedev YB, Kolovushkina,EY (1972) Solubility of H_2O and CO_2 in granite and basalt melts at high pressures. Geochem Int 9:1041-1050

Kadik AA, Shilobreyeva SN, MV, Akhmanova Slutskiy AB, Korobkov VI (1981) Solubility of CO_2 in melts of acid composition for the case of the albite-silica system (65:35). Geokhimiya 1:63-70

Kerrick DM (1987) Cold-seal systems. In: GC Ulmer, HL Barnes Jr. (eds) Hydrothermal Experimental Techniques. Wiley, New York, p 293-323

Kesson SE, Holloway JR (1974) The generation of $N_2-CO_2-H_2O$ fluids for use in hydrothermal experimentation, II. Melting of albite in a multispecies fluid. Am Mineral 59:598-603

Kirkpatrick RJ (1988) MAS NMR spectroscopy of minerals and glasses. Rev Mineral 18:341-403

Kohn SC, Brooker RA (1994) The effect of water on the solubility and speciation of CO_2 in aluminosilicate glasses along the join $NaAlO_2-SiO_2$. Abstracts of the Goldschmidt Conference.

Kohn SC, Brooker RA, Dupree, R (1991) ^{13}C MAS NMR: A method for studying CO_2 speciation in glasses. Geochim Cosmochim Acta 55:3879-3884

Kohn SC, Dupree R, Smith ME (1989) A multinuclear magnetic resonance study of the structure of hydrous albite glasses. Geochim Cosmochim Acta 53:2925-2935

Kubicki JD, Stolper EM (in press) Structural roles of CO_2 and $[CO_3]^{2-}$ in fully-polymerized, sodium aluminosilicate melts and glasses. Geochim Cosmochim Acta

Kujumzelis TG (1938) Uber die schwingungen und die strukture der XO_3-ionen. Z Physik 109:586-597

Lesniak PM, Sakai H (1989) Carbon isotope fractionation between dissolved carbonate (CO_3^{2-}) and CO_2 (g) at 25° and 40°C. Earth Planet Sci Lett 95:297-301

MacPherson C, Mattey D (1994) Carbon-isotope variations of CO_2 in central Lau Basin basalts and ferrobasalts. Earth Planet Sci Lett 121:263-276

Mattey DP (1991) Carbon dioxide solubility and carbon isotope fractionation in basaltic melt. Geochim Cosmochim Acta 55:3467-3473

Mattey DP, Taylor WR, Green DH, Pillinger CT (1990) Carbon isotopic fractionation between CO_2 vapor, silicate and carbonate melts: an experimental study to 30 kbar. Contrib Mineral Petrol 104:492-505

McMillan PF, Hess AC (1988) Symmetry, group theory and quantum mechanics. Rev Mineral 18:11-61

McMillan PF, Hofmeister AF (1988) Infrared and Raman Spectroscopy. Rev Mineral 18:99-159

Millhollen GL, Wyllie PJ, Burnham CW (1971) Melting relations of $NaAlSi_3O_8$ to 30 kb in the presence of $H_2O:CO_2$ = 50:50. Am J Sci 271:473-480

Morey GW, Fleischer,M (1940) Equilibrium between vapor and liquid phases in the system CO_2-H_2O-K_2O-SiO_2. Geol Soc Am Bull 51:1035-1058

Mysen BO (1976) The role of volatiles in silicate melts: Solubility of carbon dioxide and water in feldspar, pyroxene, and feldspathoid melts to 30 kb and 1625°C. Am J Sci 276:969-996

Mysen BO (1977) The solubility of H_2O and CO_2 under predicted magma genesis conditions and some petrological and geophysical implications. Rev Geophys Space Phys 15(3):351-361

Mysen BO (1983) The solubility mechanisms of volatiles in silicate melts and their relations to crystal-andesite liquid equilibria. J Volcanol Geotherm Res 18:361-385

Mysen BO, Arculus RJ, Eggler DH (1975) Solubility of carbon dioxide in natural nephelinite, tholeiite and andesite melts to 30 kbar pressure. Contrib Mineral Petrol 53:227-239

Mysen BO, Eggler DH, Seitz MG, Holloway JR (1976) Carbon dioxide solubilities in silicate melts and crystals. Part I. Solubility measurements. Am J Sci 276:455-479

Mysen BO, Virgo D (1980a) The solubility behavior of CO_2 in melts on the join $NaAlSi_3O_8$-$CaAl_2Si_2O_8$-CO_2 at high pressure and temperatures: A Raman spectroscopic study. Am Mineral 65:1166-1175

Mysen BO, Virgo D (1980b) Solubility mechanisms of carbon dioxide in silicate melts: a Raman spectroscopic study. Am Mineral 65:885-899

Nakamoto K, Fujita J, Tanaka S, Kobayashi M (1957) Infrared spectra of metallic complexes, IV. Comparison of the infrared spectra of unidentate and bidentate metallic complexes. J Chem Phys 79:4904-4908

O'Neil JR (1986) Theoretical and experimental aspects of isotopic fractionation. Rev Mineral 16:1-40

Ohmoto H, Rye RO (1979) Carbon and sulfur in ore bodies. In: Barnes (ed) Hydrothermal Ore Deposits. John Wiley & Sons, New York, p 509-567

Pan V, Holloway JR, Hervig RL (1991) The pressure and temperature dependence of carbon dioxide solubility in tholeiitic basalt melts. Geochim Cosmochim Acta 55:1587-1595

Papike JJ (1964) The Structure of Scapolite. PhD dissertation, University of Minnesota

Papike JJ, Stephenson NC (1966) The crystal structure of mizzonite, a calcium- and carbonate-rich scapolite. Am Mineral 51:1014-1027

Pawley AR, Holloway JR, McMillan P (1992) The effect of oxygen fugacity on the solubility of carbon-oxygen fluids in basaltic melt. Earth Planet Sci Lett 110:213-225

Pearce ML (1964) Solubility of carbon dioxide and variation of oxygen ion activity in soda-silicate melts. J Ceram Soc 47:342-347

Poulet H, Mathieu JP (1970). Vibrational Spectra and Symmetry of Crystals. Paris, Gordon & Breach, 338 p

Rai CS, Sharma SK, Meunow DW, Matson DW, Byers CD (1983) Temperature dependence of CO_2 solubility in high pressure quenched glass of diopside composition. Geochim Cosmochim Acta 47:953-958

Richet P, Bottinga Y, Javoy M (1977) A review of hydrogen, carbon, nitrogen, oxygen, sulphur, and chlorine stable isotope fractionation among gaseous molecules. Ann Rev Earth Planet Sci 5:65-110

Rosenbaum JM (in press) Stable isotope fractionation between coexisting carbon dioxide and calcite at 900°C. Geochim Cosmochim Acta

Roux J, Lefèvre A (1992) A fast-quench device for internally heated pressure vessels. Eur J Mineral 4:279-281

Sarda P, Graham,D (1990) Mid-ocean ridge popping rocks: implications for degassing at ridge crests. Earth Planet Sci Lett 97:268-289

Sharma SK (1979) Structures and solubility of carbon dioxide in silicate glasses of diopside and sodium melilitite composition at high pressures from raman spectrscopic data. Carnegie Inst Wash Yrbk 78:532-537

Sharma SK, Hoering, TC Yoder, HS (1979) Quenched melts of akermanite composition with and without CO_2 - characterization by Raman spectroscopy and gas chromatography. Carnegie Inst Wash Yrbk 78:537-542

Shilobreyeva SN, Kadik AA (1990) Solubility of CO_2 in magmatic melts at high temperatures and pressures. Geochem Int 27:31-41

Spera FJ, Bergman SC (1980) Carbon dioxide in igneous petrogenesis: I. Aspects of the dissolution of CO_2 in silicate liquids. Contrib Mineral Petrol 74:55-66

Stolper EM, Fine GJ, Johnson T, Newman S (1987) The solubility of carbon dioxide in albitic melt. Am Mineral 72:1071-1085

Stolper EM, Holloway JR (1988) Experimental determination of the solubility of carbon dioxide in molten basalt at low pressure. Earth Planet Sci Lett 87:397-408

Taylor WR (1990) The dissolution mechanism of CO_2 in aluminosilicate melts—infrared spectroscopic constraints on the cationic environment of disolved $(CO_3)^{2-}$. Eur J Mineral 2:547-563

Thibault Y, Holloway JR (1994) Solubility of CO_2 in a Ca-rich leucitite: effects of pressure, temperature and oxygen fugacity. Contrib Mineral Petrol 116:216-224

Tingle TN (1987) An evaluation of the carbon-14 beta track technique: Implications for solubilities and partition coefficiens determined by beta track mapping. Geochim Cosmochim Acta 51:2479-2987

Tingle TN, Aines RD (1988) Beta track autoradiography and infrared spectroscopy bearing on the solubility of CO_2 in albite melt at 2GPa and 1450°C. Contrib Mineral Petrol 100:222-225

Trull T, Pineau F, Bottinga Y, Javoy M (1991) Experimental study of CO_2 bubble growth and $^{13}C/^{12}C$ isotopic fractionation in tholeiitic melt. 4th Silicate Melt Workshop: Program/Abstracts, 19-23 March, 1991, p 7

Urey HC (1947) The thermodynamic properties of isotopic substances. J Royal Chem Soc 1:562-581

Verweij H, van der Boom H, Breemer RE (1977) Raman scattering of carbonate ions dissolved in potassium silicate glass. J Am Ceram Soc 60:529-532

Vogel JC, Grootes PM, Mook WG (1970) Isotopic fractionation between gaseous and dissolved carbon dioxide. Z Physik 230:225-238

Walker FW, Miller DG, Feiner F (1984). Chart of the Nuclides. 13th edn, 59 p.

White BS, Montana A (1990) The effect of H_2O and CO_2 on the viscosity of sanidine liquid at high pressure. J Geophys Res 95:15683-15693

Williams DW (1968) Improved cold seal pressure vessels to operate to 1100°C at 3 kilobars. Am Mineral 53:1765-1769

Wright IP, Pillinger CT (1989) Carbon isotopic analysis of small samples by use of stepped-heating extraction and static mass spectrometry. U S Geol Surv Bull 1890:9-34

Wyllie PJ (1979) Magmas and volatile components. Am Mineral 64:469-500

Wyllie PJ, Tuttle OF (1959) Effect of carbon dioxide on the melting of granite and feldspars. Am J Sci 257:548-655

Yoder HS Jr (1950) Stability relations of grossularite. J Geology 58:221-253

Chapter 6

APPLICATION OF EXPERIMENTAL RESULTS TO C-O-H SPECIES IN NATURAL MELTS

John R. Holloway

Departments of Chemistry and Geology
Arizona State University
Tempe, AZ 85287-1604 USA

Jennifer G. Blank

Institut du Physique du Globe
Laboratoire de Dynamiques des Systèmes Géologiques
4 Pl Jussieu, Cedex 05, Paris 75252 France

INTRODUCTION

The goal of this chapter is to provide a practical overview of experimental results pertaining to CO_2 and H_2O solubilities and stable isotope partitioning behavior in natural silicate melt compositions. It is geared to those wishing to apply experimental data, and models derived from such data, to calculation of solubilities and melt-vapor stable isotope partitioning in natural systems in which CO_2, H_2O and CO are the dominant fluid species, and CO_2 and H_2O are the major volatile components dissolved in the melt.

The behavior of C-O-H volatiles in magmatic systems depends strongly on the relative stabilities of the different molecular species present (e.g., CO, CO_2, H_2O, CH_2), and thus we will first consider speciation in the C-O-H system as a function of oxygen fugacity (f_{O_2}), pressure, temperature, and bulk composition. Following, we provide simple descriptions of thermodynamic models useful for calculating water and carbon dioxide solubilities in silicate melts equilibrated with pure H_2O or CO_2 fluid phases, and we show how these models can be extended to calculation of the solubilities of H_2O and CO_2 in melts equilibrated with a mixed-volatile fluid phase. Finally, we summarize the experimentally-determined stable isotope partitioning behavior of H and C between fluid and species dissolved in silicate melts, and show briefly how such data may be used to provide insight into magma degassing processes.

Because discrepancies have been documented for different analytical techniques, particularly for CO_2 measurements (see Chapter 2), we restrict our treatment for most compositions to glasses in which H_2O and CO_2 contents were measured using infrared spectroscopy or secondary ion mass spectrometry.

OXIDATION STATE AND FLUID PHASE SPECIATION IN THE C-O-H SYSTEM AT IGNEOUS TEMPERATURES AND HIGH PRESSURES

Fluid-phase species in the C-O-H system are discrete molecules, and their relative abundance is commonly given in terms of mole fraction (of each species, i) in the fluid (X_i^{fl}); most important are H_2O, CO_2, CO, H_2, and CH_4. The sum of all molecular species comprises the bulk composition of the fluid, which can be expressed as atom fractions of C, O and H. The range of equilibrium fluid compositions in the C-O-H system is shown in

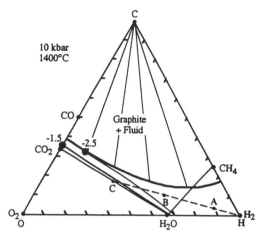

Figure 1. The C-O-H system at 10 kbar and 1400°C, calculated as described in text. The composition of graphite-saturated fluids are indicated by the heavy curve on the ternary diagram. The two straight line segments sub-parallel to the CO_2-H_2O join are fO_2 isobars (lines of constant fO_2), calculated by fixing fO_2 and decreasing carbon activity; their log fO_2 units relative to the Ni-NiO (NNO) buffer (O'Neill, 1987) are denoted by the numbers next to the large, filled circles. The dashed segment labeled A-B-C, a "hydrogen reaction line," illustrates how fluid composition changes by addition of hydrogen to (move toward A) or sub-traction of hydrogen from (move away from A) the system. Point C bisects the CO_2-H_2O join.

Figure 1. Geologically-important fluid compositions lie in the region bounded by the graphite saturation surface (heavy curve, concave towards C apex in Fig. 1) and the CO_2-H_2O join. Fluid compositions lying below the CO_2-H_2O join contain free oxygen (as molecular O_2) and hence have oxygen fugacities much greater than those commonly found in natural magmatic systems; therefore, magmatic fluids have compositions lying only above the CO_2-H_2O join. Likewise, fluids rich in methane and hydrogen occur only at fO_2 values lower than those commonly found in magmatic systems (Carmichael, 1991).

Bulk fluid compositions and species concentrations in fluids can be calculated using several thermodynamic approaches. The accuracy of the calculations depends on how well we can describe the free energy of each species as a function of pressure and temperature, and the mixing properties of each species in the fluid. The degree to which we can constrain these two variables, and their use in fluid speciation calculations, are discussed below, followed by a brief evaluation of the effects of C-O-H equilibria on experimental measurements in silicate rock systems. Calculations similar to those we describe can be found in the literature, beginning with French and Eugster (1965), French (1966) and Eugster and Skippen (1967), and more recently, Holloway and Reese (1974), Frost (1979) and Holloway et al. (1992).

The thermodynamic data

The major species in the C-O-H system under magmatic conditions at crustal and upper-mantle pressures are CO_2, CO, CH_4, H_2, H_2O and graphite (diamond is not considered here because it is not stable at any of the pressures at which solubility determinations have been made). The thermodynamic data for these species fall into three categories: (1) free energies of formation at one atmosphere as a function of temperature for all species (which ultimately require enthalpies of formation and heat capacities); (2) molar volume for the crystalline phase (graphite); and (3) fugacities of the molecular fluid species as a function of pressure, temperature and fluid composition.

Standard state properties for C-O-H species are given in compilations such as Chase et al. (1985). Heat capacities as a function of temperature for the gaseous species are based on statistical mechanical fits to spectroscopic data and have relative uncertainties equivalent to about ±3% in free energy units. Standard free energies at 298 Kelvin (K) include an uncertainty derived from the calorimetric determination of enthalpy, which in turn has an associated uncertainty of ±0.05% to ±0.5%, depending on the species. None of these uncertainties has a significant effect on the calculations described below. Gaseous C-bearing species are referenced to graphite, which has minimal errors associated with its measured standard state properties (because it is in elemental form at 1 atm and all temperatures). The molar volume of graphite is well known under the pressure and temperature conditions of interest (1 atm to 30 kbar; 600° to 1600°C).

Fugacities of pure fluids such as H_2O or CO_2 are based on pressure-volume-temperature (P-V-T) measurements as shown in the defining equation for fugacity (e.g., Anderson and Crerar, 1993):

$$\ln f_{P_2} - \ln f_{P_1} = \int_{P_1}^{P_2} \frac{V}{RT} dP = (\Delta G_{P_2} - \Delta G_{P_1}) / RT \tag{1}$$

where f_{P_1} and f_{P_2} are the fugacity of the species at pressures P_1 and P_2, V is molar volume, R is the ideal gas constant, T is the temperature (K), and ΔG_{P_1} and ΔG_{P_2} are the free energies of formation at P_1 and P_2.

P-V-T measurements are technically difficult at high P and T and the quality and P-T range of available P-V-T data varies greatly among the species of interest. The applicable pressure and temperature ranges of existing data for C-O-H gaseous species are listed in Table 1. The range for H_2O and CO_2 is considerably greater than those for the minor species. Data for CO and CH_4 are limited by the thermodynamic instabilities of these species, as discussed in more detail below. The limited range of data for H_2 reflects the virtual impossibility of containing H_2 quantitatively at high P-T conditions due to its high diffusion rate in all possible container materials.

Supplementary to P-V-T measurements are data for phase equilibrium reactions such as:

$$NiO_{(bunsenite)} + C_{(graphite)} = Ni_{(metal)} + CO_2_{(fluid)} \tag{2}$$

in which the free energies of the crystalline phases are well known at the P and T of interest (Ulmer and Luth, 1991). Such data provide useful checks on the validity of fugacities calculated for conditions beyond the limits of direct P-V-T measurements.

Table 1. Upper limits of available P-V-T data and estimated accuracy of fit by MRK equation-of-state for selected volatile species.

Species	Max T (°C)	Max P (bars)	MRK fit [1]
CO_2	900	8000	± 5
CO	300	10000	± 5
CH_4	200	6800	± 5
H_2	600	2500	± 3
H_2O	900	8000	± 3

[1] Numbers in this column indicate the relative percent deviation between values calculated using an empirical fit to the data and the actual data.

Trends defined by existing P-V-T data can be extrapolated and interpolated using an equation-of-state (EOS), provided that the algebraic form of the EOS is known to be valid over the P-T range of interest. Two types of equations that frequently meet this requirement are currently in use. One useful type of EOS uses the corresponding states approximation (see discussion in Prausnitz et al., 1986). Corresponding state equations are based on the observation that when P, V and T values for a molecular species are "reduced" to P_r, V_r and T_r by dividing the actual P, V, T values by their critical point values (P_c, V_c and T_c), the reduced quantities are approximately the same (e.g., Prausnitz et al., 1986). Special treatment (resulting in adjustment of P_c and T_c values) is required for H_2 to account for quantum mechanical effects related to its small mass, and for H_2O, which requires a special fit to take into consideration its large dipole moment (Prausnitz et al., 1986).

Semi-theoretical equations such as the Redlich-Kwong and its modified versions contain only a small number of adjustable parameters that are independent of P and T. Modified Redlich-Kwong (MRK) equations which use a constant valued b parameter (a measure of molecular volume), provide acceptable fits to P-V-T data at lower pressures but are limited inherently at pressures above ~10 to 30 kbar because the compressibility of molecules is not taken into account. Some forms of the modified Redlich-Kwong equation have a b parameter that does vary as a function of P and T (Kerrick and Jacobs, 1981), thereby providing better a more accurate fit to the data at high pressures; as a result, however, these equations contain many more adjustable parameters.

The version of the modified Redlich-Kwong (MRK) referred to in the remainder of this chapter is that used by Holloway (1981, 1987):

$$P = \frac{RT}{(V-b)} - \frac{a_0 + a_1T + a_2T^2}{V(V+b)\sqrt{T}} \tag{3}$$

in which V is the molar volume, a_i (i = 0, 1, 2) are coefficients describing inter-molecular attraction, and b is a parameter related to the size of the molecule of interest. For non-polar molecules, the value of a_1 and a_2 is zero.

The approximate fit of the MRK to the species of interest here is shown in Table 1. In the case for H_2 and H_2O, the P-V-T data together with either type of EOS provide a relative accuracy of ~±3% over the P-T range of interest (10^{-3} to 30 kbar, 700° to 1500°C). Recent measurement of the molar volume of H_2O at pressures up to 30 kbar in the igneous temperature range (Brodholt and Wood, 1994) confirm that the simple version of the MRK with parameters given in Holloway (1981; 1987) predicts the measured values to within this uncertainty. Uncertainties in CO_2 fugacity are probably less than ±5% (Mader and Berman, 1991). CO and CH_4 are simple molecules, with CH_4 being spherical and non-polar, whereas CO is nearly non-polar and should behave in a fashion similar fashion to that of N_2; the fugacities of both species should thus be predicted fairly well by either a Redlich-Kwong or corresponding state EOS. Uncertainties in CO and CH_4 fugacities at high P-T conditions are difficult to evaluate due to lack of experimental data, although experimental measurement of $CO-CO_2$ fluid compositions in equilibrium with graphite at 7.5 to 20 kbar, 1100° to 1500°C (Jakobsson and Oskarsson, 1994) suggests that use of the Redlich-Kwong parameters given in Holloway (1987) result in underestimation of CO fugacities by about 20%.

Fugacities of C-O-H species may be affected strongly by nonideal behavior in fluid mixtures, with deviations so large as to cause phase separation at low T (e.g., Holloway, 1984). The low temperature data show a progressive shift toward ideality with increasing

T, but data for activity-composition relations among the species are not available for the P-T range of interest here. The available data, combined with calculated mixing parameters lead to the conclusion that positive deviations of less than 10% relative will occur at T >700°C at pressures up to 4 kbar and >900°C at 10 kbar (Joyce and Holloway, 1993). In the calculations presented here it is assumed that the fluid species mix ideally.

Fluid speciation calculations

Calculation of fluid speciation in the two component C-O system at magmatic temperatures is relatively straightforward. A maximum of two phases exist, graphite and fluid. According to the phase rule, the variance is either zero or one at fixed P and T, depending on whether the system is graphite saturated. Graphite-saturated fluids are invariant at fixed P and T, and consequently both the CO/CO_2 ratio and f_{O_2} are known. Graphite under-saturated fluids are univariant at constant P and T, and knowledge of either the CO/CO_2 ratio or the f_{O_2} completely defines the system.

Calculation of the abundance of fluid species in the 3-component C-O-H system is somewhat more involved. This system can also have a maximum of two phases, graphite and fluid. At fixed P and T, graphite-saturated fluids are univariant, and graphite under-saturated fluids are divariant. Specifying the f_{O_2} in graphite-saturated fluids fixes the fluid composition and *vice versa*. Equilibrium calculations have been conducted using either an equilibrium constant approach (French, 1966; Frost, 1979; Holloway, 1987) or a global free-energy minimization (Dayhoff et al., 1967; Gordon and McBride, 1971; Holloway and Reese, 1974). The former approach involves fixing carbon activity, either at unity by imposing graphite saturation, or at values less than one for graphite-undersaturated fluids; in either scenario, the oxygen fugacity is fixed. Figure 1 illustrates typical results. The fluid compositions (in terms of atomic C, O, and H) saturated in graphite form the "graphite saturation surface", indicated by the heavy curve (Fig. 1).

Calculations using the free energy minimization technique are well-suited for determining fluid species concentration as a function of P and T for a fixed atomic composition of the system. Fixing the atomic composition of a system at constant P and T completely specifies the concentration of equilibrium species and f_{O_2}. Free energy minimization programs are also well adapted for calculating minor and trace species concentrations. Examples of the results of such calculation are given in Table 2. Note the low concentrations of species other than CO_2 , CO, CH_4, H_2, H_2O at magmatic conditions.

Implications for experiments

Oxidizing conditions: The H_2O-CO_2 join. Nearly-binary H_2O-CO_2 fluids are stable over a wide range of f_{O_2} (Fig. 2). The conditions range from unrealistically high values of f_{O_2} (more than several log units above the Ni-NiO (NNO) buffer) to values more typical of magmatic systems (~0-1 log unit below NNO). However, many high pressure phase equilibria and volatile solubility experiments are subject to loss or gain of H_2 due to its rapid hydrogen diffusion through capsule walls and its ubiquitous presence in pressure vessels. The consequences of such diffusive flux can be seen for a fictive composition "B", chosen to lie at the intersection of the C-H_2O join with a segment connecting H_2 to the midpoint of the CO_2-H_2O join (Fig. 1). Loss of H_2 from a fluid initially at point B will result in decreased H_2O activity and oxidation, as well as decreased Fe^{3+}/Fe^{2+} in iron-bearing melts. Gain of H_2 will result in reduction, decreased CO_2 activity, and increased CO and/or CH_4 activities in the fluid phase. This is illustrated in Figure 2, which shows

Table 2. Relative abundances of major and minor gas species calculated for "Composition B" at 10 kbars and 1400°C as denoted in Figure 2.

Species	Mole Fraction[1]
H_2O	0.56780
H_2	0.17994
CH_4	0.11296
CO	0.08197
CO_2	0.05498
C_2H_4	0.00141
CH_2O	0.00089
CH_3	0.00004
HCO	0.00001
C_2H_2	0.00001

[1] Calculated as described by Holloway and Reese (1974).

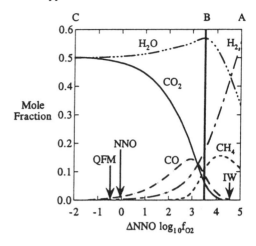

Figure 2. Binary plot of fluid compositions along the line segment A-C in Figure 1 illustrating the change in fluid composition as a function of f_{O_2} (relative to NNO), which increases moving from point A (near the H apex) to point C (the 1:1 $H_2O:CO_2$ composition). The vertical line shows the fluid composition at point B on the C-H_2O join in Figure 1. Note that fluids more oxidizing than NNO-0.5 (equivalent to QFM) are essentially pure $H_2O + CO_2$. NNO f_{O_2} values are from O'Neill (1987).

how speciation of fluid with initial composition at B will change with H_2 loss or addition, along the dashed (hydrogen reaction) line A-B-C in Figure 1. Although H_2 loss through a capsules wall is very common in a range of experiments, H_2 gain occurs frequently in CO_2-rich experiments and causes formation of H_2O. H_2 gain can also facilitate loss of iron into the capsule wall in iron-bearing melts as a result of a lowered f_{O_2} (although the loss of iron causes a small increase in Fe_2O_3 which will partially offset the H_2 gain; Holloway et al., 1992). Successful execution of experiments to measure solubilities of "pure" CO_2 or H_2O fluids, or of CO_2 and H_2O fluid mixtures must be done at oxygen fugacities greater than ~NNO-0.5 to avoid the presence of significant amounts of reduced species. As seen in Figure 2, the relative concentrations of H_2O in the fluid is less sensitive to change in f_{O_2} than is CO_2.

Reducing conditions: Limits on CO, H_2, and CH_4. Experimental measurement of CO and CH_4 solubility is impeded by their partial decomposition at magmatic conditions. CO reaches maximum concentrations in the C-O system at high T and low P when the fluid is saturated with graphite. As shown in Figure 3, the CO/CO_2 ratio in the fluid declines rapidly with increasing P.

High methane concentrations are found only in strongly-reducing fluids in which H_2O is often present in significant amounts, hindering separation of methane solubility from that of the highly soluble H_2O. However, the limited available data for the solubility of CH_4 in basaltic melt in equilibrium with methane-rich fluids suggest very low methane solubilities (Holloway and Jakobsson, 1986). In contrast, Taylor and Green (1988) conducted peridotite melting experiments in the presence of reduced C-O-H fluids and suggested that methane was significantly soluble on the basis of solidus temperatures.

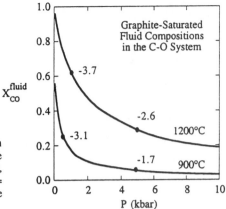

Figure 3. Graphite-saturated fluid compositions in the C-O system as a function of P and T. The mole fraction of CO in the fluid X_{CO}^{fluid} is shown (similarly, the mole fraction of CO_2 in the fluid = $X_{CO_2}^{fluid} = 1 - X_{CO}^{fluid}$). Numbers next to solid circles are log fO_2 values relative to NNO (O'Neill, 1987).

There are no thermodynamic limitations to the generation of fluids with very high hydrogen concentrations, but such fluids would be extremely reducing (Fig. 2). Very strongly reducing conditions reduce the silicon in SiO_2 to metal. For this reason the solubility of H_2 in silicate melts is unknown. However, it seems probable that H_2 solubility is at least as great as that of argon, which has solubility similar to that of CO_2 in silica-rich melts (White et al., 1989; Carroll and Stolper, 1993; see Chapter 7).

Graphite-H_2O experiments. Use of graphite capsule-liners in high-pressure melting experiments is common but care must be taken in such experiments to control the volatile species activities. Some attempts have been made to add H_2O to such experiments either as water or a hydrous mineral in the starting material. It is important to note that in this case the volatiles will include all components of the C-O-H system and that if hydrogen or oxygen fugacities are not controlled carefully there can be a significant decrease in the amount of water present in the experimental capsule both through diffusive loss of hydrogen and/or reaction of H_2O with graphite to produce CO_2; such effects are discussed in detail by Holloway et al. (1992).

PRACTICAL CALCULATION OF H_2O SOLUBILITY IN ROCK MELTS

Experimentally-determined equilibrium volatile solubility curves provide information useful for interpretation of measured volatile contents of erupted silicate glasses. Water solubilities in rock composition melts can be done via two distinct methods which we will refer to as the Burnham or the Stolper models. The approach outlined in the Stolper model requires fitting of thermodynamic parameters to available solubility data for individual rock compositions. The result is an accurate fit to the data, but the Stolper model does not allow direct calculation of H_2O solubilities of compositions other than those whose volatile solubility data sets have been considered. In contrast, the Burnham model is not fit to a specific data set for a natural melt compositions but can be used to calculate H_2O solubility in virtually any silicate melt composition. A fundamental difference between the two models is that the Burnham model is based principally on direct P-V-T measurements on hydrous albitic melts, thus allowing direct estimation of the partial molar volume of H_2O in melts. Burnham and Davis (1974) found a convincing change in the partial molar volume of H_2O in albitic melt with pressure and temperature, whereas the Stolper model, in contrast, assumes a constant molar volume which is independent of P and T. These differences are not important in calculating solubilities, but are of considerable importance in calculating melt densities as discussed in Chapter 9 of this volume. The Stolper model as usually applied is not as versatile as the Burnham model, but is likely to be more accurate for the compositions to which it has been applied, particularly as it is based on a wider range of experimental water solubility data and on direct physical measurements of the hydrous species in glasses quenched from high temperature melts. In principle, the Stolper model can be expanded to encompass compositional variations (Silver and Stolper, 1989) but this remains to be done.

The Burnham model

Calculation of H_2O solubility using the Burnham model involves recasting the melt composition of interest in terms of 8-oxygen normative components, as discussed in Chapter 3 and as illustrated in the Appendix to this chapter. Given a melt composition and a choice of P and T, the Burnham model requires calculation of a value of $\ln k_w^{ma}$, which is a Henry's-Law analog constant for melts of feldspar composition; this calculated $\ln k_w^{ma}$ value is then adjusted to account for the effects of non-feldspar components on water solubility. The general sequence of steps can be summarized as follows (see also Appendix and Chapter 3):

1. Use the following equation (Eqn. 7, Chapter 3) to calculate $\ln k_w^{ma}$ for a chosen P (kbar) and XT (hundreds of Kelvins).

$$\ln k_w^{ma} = 5.00 + (\ln P)(4.481 \times 10^{-8} \, XT^2 - 1.51 \times 10^{-4} \, XT - 1.137)$$
$$+ (\ln P)^2 (1.831 \times 10^{-8} \, XT^2 - 4.882 \times 10^{-5} \, XT + 4.656 \times 10^{-2})$$
$$+ 7.80 \times 10^{-3} (\ln P)^3 - 5.012 \times 10^{-4} (\ln P)^4 + XT(4.754 \times 10^{-3} - 1.621 \times 10^{-6} \, XT) \quad (4)$$

2. Use the following expression (a generalized form of Eqn. 8 from Chapter 3) to calculate $\ln k_w^{rock\ melt}$ for the melt composition of interest:

$$\ln k_w^{rock\ melt} = \ln k_w^{ma} + 0.47 \cdot (1 - \{(X_{ab} + X_{or} + X_{an}) - 2 \cdot X_{ne}\} \quad (5)$$

where the mole fractions (X_i) are of eight-oxygen normative components in the melt composition in question.

3. Find X_w^m from the equations below which are obtained from (Eqns. 5 and 6 in Chapter 3):

$$X_{w^m} = \left[\frac{a_w}{k_w^{rockmelt}}\right]^{1/2} \quad (6a)$$

$$X_{w^m} = 0.5 + \ln\left[\frac{a_w}{0.25 * k_w^{rockmelt}}\right] \Big/ \left[\frac{6.52 - 2667}{T}\right] \quad (6b)$$

where a_w = the activity of H_2O in the system and T is in Kelvins. If $X_w^m > 0.5$ as calculated by Equation (6a), then re-calculate a value using Equation (6b).

4. Convert the mole fraction of H_2O to wt % H_2O with the following algorithm:

$$wt\%_{H_2O}^{melt} = \left[\frac{18.02 \cdot X_{w^m}^m}{18.02 \cdot X_{w^m}^m + (1 - X_{w^m}^m) \cdot FW_{eight}}\right] \quad (7)$$

where FW_{eight} is the 8-oxygen formula weight (g/mol) of the volatile-free melt, X^m is the mole fraction of H_2O in the melt (also calculated on a 8-oxygen basis, and 18.02 is the formula weight of H_2O. Values of $k_w^{rock\ melt}$ and examples of typical values of FW_{eight} for some common rock compositions are given in Table 3 along with a sample calculation.

Table 3. Eight-oxygen formula weights and K-factors used to calculate H_2O solubiity in melts of a variety of compositions with the Burnham method.

Rock Composition	FW_{eight}	K-factor
basanite	297	0.14
alkalic basalt	296	0.19
tholeiitic basalt	292	0.24
andesite	276	0.17
dacite	265	0.15
rhyolite	261	0.19
quartz	240	0.47
albite	262	0
nepheline	284	-0.47

Sample calculation, using the Burnham model, of H_2O solubility in the rhyolite composition given in Table 4 at a pressure of 1000 bars and a temperature of 850°C with a "pure" H_2O fluid ($a_w^{fl} = 1.0$):

1. From Equation (4): $\ln k_w^{ma} = 1.7967$

2. Calculate the eight-oxygen norm, or choose a value of the K-factor from Table 3 (for rhyolite, this value is 0.203). The corresponding eight-oxygen normative mineral mole fractions are:

   $ab = 0.352$; $or = 0.238$; $an = 0.025$; $qz = 0.386$.

3. From Equation (5),
   $\ln k_w^{rock\ melt} =$
   $\ln k_w^{ma} + 0.47 \times (1-\{(X_{ab} + X_{or} + X_{an}) - 2 \times X_{ne}\}$

   or: $\ln k_w^{rock\ melt} = \ln k_w^{ma} + \text{K-factor} = 1.7967 + 0.1900$
   $= 1.9867$,

   and $k_w^{rock\ melt} = 7.2914$.

4. Calculate X_w^m using Equation (6a). $X_w^m > 0.5$. If $X_w^m > 0.5$ then use Equation (6b) to calculate X_w^m. In this example,
   $X_w^m = (1.0/7.2914)^{0.5} = 0.3703$.

5. Use Equation (7) to calculate the solubility in wt % H_2O (from Table 3, $FW_8 = 261$):

   $$\text{wt \% } H_2O = 100 \times \frac{18.02\ (0.3703)}{18.02\ (0.3703) + (1-0.3703)261}$$

   $= 3.90$

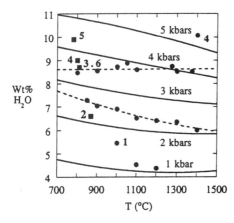

Figure 4. Comparison of the Burnham model predictions with experimentally-determined H_2O solubilities in albitic melts. Symbols represent the following data: filled squares (Burnham and Jahns, 1962); filled circles (Hamilton, 1987) except the two points at 1400°C from (Paillat et al., 1992). Data points falling far from the normal trend are labeled to indicate pressure in kbar. Solid lines were calculated with the Burnham model as described in the text. Dashed lines are least-squares fits to the data of Hamilton et al. (1964) at 2 and 4 kbar.

The accuracy of the Burnham model can be evaluated by using it to calculate H_2O solubility in melt compositions for which experimental measurements of H_2O solubility exist (Fig. 4). As described in Chapter 3 of this volume, the model is based on, and strongly influenced by, the P-V-T measurements of hydrous albitic melts (Burnham and Davis, 1971; 1974). Of the data shown, only those of Burnham and Jahns (1962) were available at the time the albite model was formulated. Precise and accurate measurements of H_2O solubility are difficult, especially at high H_2O concentrations, and different studies may yield differences of 10 to 20% or more for nominally the same experimental conditions (e.g., McMillan and Holloway, 1987; see also Chapter 4). Comparison of measured water solubilities in albitic melt from a variety of sources and the predictions of the Burnham model shows that in general the model is in reasonable agreement with experimental findings.

Application of Burnham's model to natural melts requires use of Equation (4) to account for the effect of bulk composition on solubility. Figure 5 shows water solubility data for several natural and haplo- rock compositions, and the calculated solubilities based on the Burnham model. In some cases the model underestimates the experimental data, especially at higher pressures, whereas for silicic and basaltic magmas it overestimates the data. The model most nearly approximates the silicic compositions (compare model with granite minimum composition EA1.2K in Figure 5A, high-silica rhyolite composition in Fig. 5B), while it is not nearly as accurate for andesitic compositions (Fig. 5C). The fit to experimental data for basaltic compositions is not as good as for rhyolitic melt at low pressure. The recent measurements on basaltic and rhyolitic compositions (Figs. 5B and 5D) are based on infrared measurements which have high precision at these relatively low H_2O concentrations. There is clearly need for improvement in the modeling of compositional effects, and for additional highly accurate experimental solubilities on rock compositions. A summary of calculated H_2O solubilities for a wide range of rock compositions is shown in Figure 6 where the solubility is presented as the *difference* relative to albitic melt.

An issue of importance to volcanic degassing calculations is the effect of temperature on H_2O solubility at low pressures. The Burnham model predicts a moderate decrease in solubility with increasing temperature, and this is in agreement with the results of experimental studies on albitic (Fig. 4) and haplogranitic (Fig. 7) melts. However, there have been no systematic studies of H_2O solubility as a function of temperature at pressures below 2 kbar, a pressure range important to volcanic degassing.

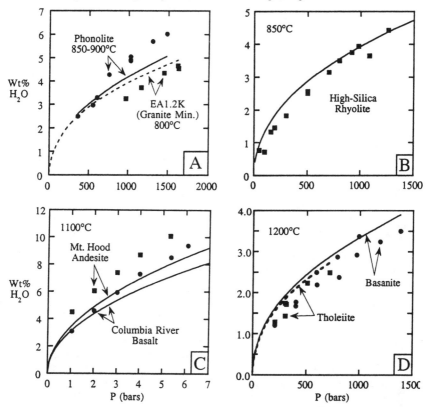

Figure 5. Comparison of measured H_2O solubilities as a function of pressure with those predicted using the Burnham model. (A) phonolite (Carroll and Blank, 1994) and a granite minimum melt (Dingwell et al., 1984). (B) Rhyolite data (Silver et al., 1989) at 850°C (only rapid-quench data are shown). (C) Columbia River tholeiitic basalt and Mount Hood andesite (Hamilton et al., 1964). (D) Mid-ocean ridge basalt (Dixon et al., 1994) and basanite (composition in Table 3; Cocheo and Holloway, 1993).

The Stolper model

A thermodynamic model for H_2O solubility in silicate melts was developed by Stolper and coworkers to account for the presence of both hydroxyl (OH) and molecular H_2O groups in aluminosilicate glasses. The model has been applied to results of quantitative infrared spectroscopy measurements (Chapters 2 and 4), and it provides activity-composition relations and solubility parameters specific to a given melt composition. In applying the model it was assumed initially that the concentrations of OH and H_2O species measured in quenched glasses are representative of those dissolved in the silicate melts at high temperature. It is now realized that H_2O speciation may be affected to varying degrees (depending on the total water content in the melt) during quenching; Zhang et al. (1991) derived an algorithm to correct for these changes in speciation, although the issue of speciation itself is one of some debate (see Chapter 4).

Although earlier qualitative observations of molecular H_2O and hydroxyl species in silicate glasses had been made (see summary in Chapter 3), Stolper (1982a,b) provided the

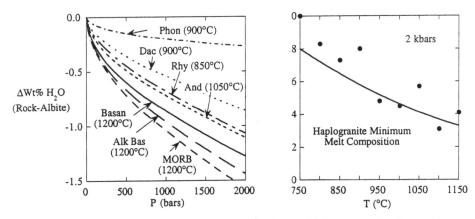

Figure 6 (left). Calculated H_2O solubilities using the Burnham model for a range of melt compositions compared to H_2O solubility in albitic melt.

Figure 7 (right). An example of the H_2O solubility:effect of temperature on H_2O solubility in silicate melts, ilustrating, for the haplogranite-water system at 2 kbar, the gradual decrease from 6 to ~5.4 wt % water as the temperature increases by 400°C from 750° to 1150°C. H_2O solubility data for a haplogranite minimum melt composition from Holtz et al. (1992). The water solubility curve calculated using Burnham's model is included for comparison.

first quantitative determinations of species concentrations as a function of the total dissolved H_2O concentration. He proposed a thermodynamic model based on ideal mixing of O, OH and molecular H_2O which provided an adequate fit to the data available at that time. Silver and Stolper (1985) expanded on the original ideal mixing model by incorporating temperature effects on solubility and speciation, and they showed that the model provided a reasonable fit to H_2O-saturated melting experiments for albite, diopside and silica compositions at low to moderate pressures; at P > ~8 kbar the model significantly underestimated solubilities.

The ideal mixing model was superseded by the introduction of three more complex models (Silver and Stolper, 1989), primarily because the ideal mixing model could not fit newly obtained speciation data for high total H_2O contents. Silver and Stolper showed that any of the three models could provide a good fit to all available speciation data for albitic glasses, but they preferred the regular solution model due to its relative simplicity, although they noted that the model with several distinguishable types of O and OH sites could potentially be developed into a general model to describe the effects of melt composition on water solubility.

The regular solution model is explained in detail in Silver and Stolper (1989) and Stolper (1989), and it has been used to model water solubility and speciation in rhyolitic (Silver et al., 1990; Blank et al., 1993b) and basaltic compositions (Dixon et al., in press). The model describes the homogeneous equilibrium between dissolved molecular water, hydroxyl groups, and melt oxygen species

$$H_2O_{(melt)} = O_{(melt)} + 2\,OH_{(melt)}$$

and can be reduced to the following expression, with the empirical parameters A', B', and C' derived from fitting of spectroscopically-measured H_2O-speciation data:

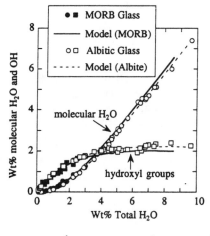

Figure 8. Comparison of measured and calculated H_2O speciation in albitic and basaltic glasses from quenched melts. Curves are best fits to the measured infrared spectroscopic data for OH and molecular H_2O using Equations (8) and (10) with data and model parameters from Table 4. Modified after Dixon et al. (in press).

$$-\ln\left(\frac{\left(X_{OH}^m\right)^2}{\left(X_{H_2O,mol}^m\right)\left(1 - X_{OH}^m - X_{H_2O,mol}^m\right)}\right) = A' + B' X_{OH}^m + C' X_{H_2O,mol}^m \tag{8}$$

The fit of this model to water speciation in albitic and basaltic glasses quenched from high temperature melts is shown in Figure 8.

Equation (8) allows evaluation of the homogeneous equilibrium between different hydrous species in water-bearing quenched melts, but in order to calculate water solubilities as a function of P and T it is necessary to consider the heterogeneous equilibrium between melt and coexisting aqueous fluid phase. This can be expressed by the reaction

$$H_2O \text{ (fluid)} = H_2O \text{ (melt)} \tag{9}$$

In the Stolper model, molecular H_2O is the species used to evaluate thermodynamic activity-composition relations. The dependence of the activity of molecular H_2O in fluid-saturated melt on total pressure and H_2O fugacity and temperature is given by the following expression for the equilibrium in Equation (9) (Silver and Stolper, 1985; 1989):

$$X_{H_2O}^m(P,T) = X_{H_2O}^{0,m}(P_0,T_0)\frac{f_{H_2O}(P,T)}{f^0{}_{H_2O}(P_0,T_0)}\exp\left\{\frac{-V_{H_2O}^{0,m}(P - P_0)}{RT_0} + \int_{T_0}^{T}\frac{\Delta H_{H_2O}^0(T,P_0)}{RT^2}dT\right\} \tag{10}$$

where $X_{H_2O}^m(P,T)$ is the activity of water in the melt at pressure P (bars) and temperature T (°K); $X_{H_2O}^{0,m}(P_0,T_0)$ is the activity of water in the melt in equilibrium with pure aqueous fluid at the reference pressure (P_0) and temperature (T_0); $f_{H_2O}(P,T)$ is the fugacity of H_2O in the coexisting fluid and $f^0{}_{H_2O}(P_0,T_0)$ is the fugacity of pure H_2O in the fluid at the P_0 and T_0 ; R is the gas constant (chosen to agree with the units used for P and V, e.g., 83.14 bar cm^3 mol^{-1} K^{-1}); $V_{H_2O}^{0,m}$ is the molar volume of water in the melt in its standard state, and is assumed to be a constant and equal to the partial molar volume of water in the melt; and $\Delta H_{H_2O}^0(T,P_0)$ is the heat of solution of a mole of H_2O in the melt and has only been determined in the case of albite melt (Silver and Stolper, 1985). In most cases, Equation (10) is applied to isothermal data, and the enthalpy term omitted. The partial molar volume of H_2O in the melt is usually assumed independent of P and H_2O concentration, the exception is again albite (Silver and Stolper, 1989).

Application of Equation (10) to experimental data assumes that molecular water in the melt obeys Henry's law; i.e.,

$$a_{H_2O}^m \propto a_{H_2O,mol}^m \propto X_{H_2O,mol}^m \qquad (11)$$

Validity of the Henrian approximation can be tested by plotting water fugacity against the mole fraction of molecular water in glasses quenched from melts equilibrated with fluid at constant total pressure. The variation in water fugacity at constant total pressure is achieved by varying the composition of the fluid (Blank et al., 1993b; Dixon et al., in press). An example is shown in Figure 9A. In this case, the volume-dependent term drops out of Equation (10) and activity is directly proportional to fugacity, so if the concentration of molecular water is proportional to water fugacity, it must also be proportional to activity.

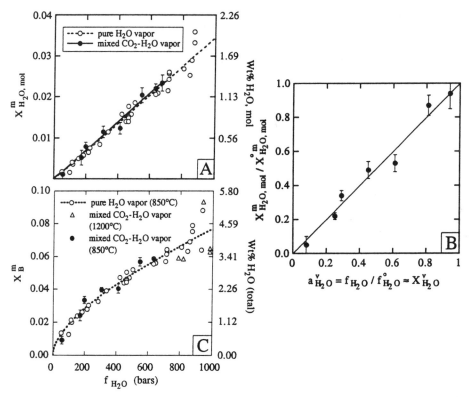

Figure 9. (A) The ratio of the mole fraction of molecular H_2O in samples of hydrous rhyolitic glass to the reference value plotted against the ratio of H_2O fugacity in the fluid to its value in the reference state. Experimental data fall on the 1:1 line, obeying Henry's law. (B) Plot of 100 times the mole fraction of molecular H_2O versus H_2O fugacity for rhyolitic glass samples equilibrated with pure H_2O and CO_2-H_2O fluids. The straight line fits were used in conjunction with Equation (10) to calculate $X_{H_2O}^{o,m}(P_o, T_o)$. Both figures after Blank et al. (1993). (C) Total dissolved water (OH$^-$ and molecular H_2O) content of vapor-saturated rhyolitic melt (expressed as X_B, calculated on a single-oxygen basis) versus water fugacity. Fugacities were calculated using MRK EOS (see Appendix). Symbols are: 850°C pure H_2O vapor results (open circles; Silver et al., 1990); 850°C mixed H_2O-CO_2 data (closed circles; Blank et al., 1993); 1200°C mixed CO_2-H_2O data (open triangles; Kadik et al., 1972). 850°C data were obtained with FTIR. Errors for 850°C mixed-volatile data are 2σ; errors for pure water data are comparable. Data from Kadik et al. (1972) were aquired using bulk analytical techniques. Dashed curve is the best-fit concentration of H_2O in rhyolitic melt coexisting with pure H_2O vapor (where f_{H_2O} varies due to changes in total pressure) and was calculated by the method outlined in this chapter using the thermodynamic parameters listed in Table 4.

A similar test can be carried out using solubility data from runs equilibrated with fluid over a range of pressures, provided that the range of total pressures is sufficiently small to allow neglecting the pressure-dependent term in Equation (10). An example is shown in Figure 9B. For example, at 1200°C with $V_{H_2O}^{o,m} = 12$ cm³/mol, the exponential term in Equation (10) increases from 1.1 at 1 kbar, to 1.2 at 2 kbar, to 1.6 at 5 kbar. The correction for molar volume is thus <10% relative in terms of H_2O solubility at P < 1 kbar, which is usually within experimental uncertainty. Comparison of isobaric and polybaric data sets in Figure 9B shows they are in fact identical within experimental uncertainty.

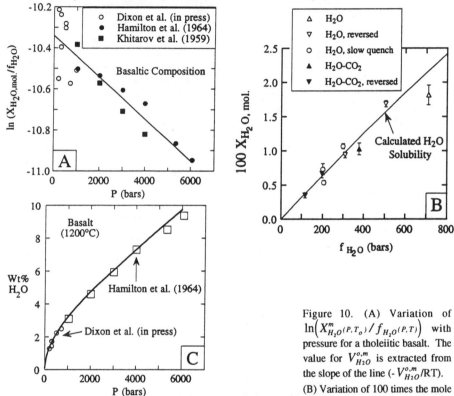

Figure 10. (A) Variation of $\ln\left(X_{H_2O}^m(P,T_o)/f_{H_2O}(P,T)\right)$ with pressure for a tholeiitic basalt. The value for $V_{H_2O}^{o,m}$ is extracted from the slope of the line ($-V_{H_2O}^{o,m}/RT$).
(B) Variation of 100 times the mole fraction of molecular H_2O in basaltic glasses with changing H_2O fugacity. (C) Total H_2O content of basaltic melts as a function of pressure. All data are for basaltic melts equilibrated with pure H_2O fluids. The curve is calculated from the parameters listed in Table 4. Extrapolation of this model to one atmosphere yields an H_2O solubility of 0.11 wt %, in good agreement with an experimentally determined value of 0.10-0.11 wt % for a tholeiitic melt (Baker and Grove, 1985). Figures after Dixon et al. (in press).

Determination of $V_{H_2O}^{o,m}$ can be accomplished by plotting $\ln\left(X_{H_2O}^m/f_{H_2O}\right)$ vs. P as shown in Figure 10A on which the slope of the isotherm is $-V_{H_2O}^{o,m}/RT$ (Dixon et al., in press). The results are well described by a single straight line, corresponding to a constant $V_{H_2O}^{o,m}$ of 12 ± 1 cm³/mol, although the scatter in the low pressure data is too great to determine if there is a decrease in slope and thus in the partial molar volume of water at low pressures as has been proposed for rhyolitic melts by Silver and Stolper (1989). Note that molar volume determinations require high pressure solubility data (the large scatter in Fig. 10A at low pressures is typical of all such plots). A consequence of this numeric constraint is that molar volume determinations by this technique are applicable to high concentrations of H_2O in melts, but not low concentrations (see Chapter 9).

Water-basalt. The Stolper model has been applied to basaltic compositions by Dixon et al. (in press). Figure 10B illustrates the relationship between molecular water content and water fugacity as determined by their H_2O- and H_2O-CO_2-saturated experiments.

Dixon et al. (in press) used Equation (10) to fit their data and those of Khitarov et al. (1959) and Hamilton et al. (1964), assuming that $a_{H_2O}^m \propto X_{H_2O,mol}^m$ and that the dependence of molecular water on total water content can be described by extrapolation of the regular solution based on low pressure data. The resulting calculated solubilities, based on fit parameters given in Table 4, are compared with measured solubilities in Figure 10C.

Water-rhyolite. The calculated water solubility in rhyolitic melt coexisting with pure H_2O fluid, where f_{H_2O} varies due to changes in total pressure, is shown as a dashed curve in Figure 9C. This curve was calculated following the method of Stolper and co-workers using bulk water data from Silver and Stolper (1989) and Karsten et al. (1982). Equilibrium molecular water contents were calculated from bulk water data using the regular solution model parameters given by Zhang et al. (1991; cf. caption to their Fig. 6). Thermodynamic parameters derived from the H_2O calculations are listed in Table 4.

Table 4. Experimentally-determined model parameters for H_2O (wt %) solubility in natural rock melt compositions [1].

	Rhyolite [2]	**Tholeiite** [3]
ln K°	-10.31 (±0.04)	10.32
ΔH° (kJ/mol)	-35.8 (±4.6)	-
ΔV° (cm³/mol)	-2.5 (±3)	12 (±1)
P° (bars)	0.0001	0.0001
T° (°C)	850	1200
SiO_2	77.7	50.8
TiO_2	0.07	-
Al_2O_3	13	13.7
FeO*	0.38	12.4
MgO	0.05	6.67
CaO	0.52	11.5
Na_2O	4.08	2.68
K_2O	4.19	0.15
P_2O_5	-	0.19
S	-	0.15

[1] Parameters were derived using data from solubility experiments and CO_2 fugacities calculated using MRK parameters (Holloway, 1987) up to 0.4 GPa and parameters from Saxena & Fei (, 1987) at higher pressures. (Use of other values of f_{CO_2} with these parameters will produce erroneous solubility results.) Also used in the calculations: FW_{one} = 32.5 for rhyolite, 36.6 for the mafic rocks. Numbers in parentheses are 2σ deviations from a linear regression using solubility data and calculated fugacities. See text for discus-sion and description of calculations. P° and T° are the refer-ence pressures and temperatures. [2] Blank et al. (1993b). [3] Dixon et al. (in press).

Calculating H_2O solubility. Detailed procedures are given by Silver and Stolper (1985) and Dixon et al. (in press) who include a useful table of calculated results. The following is a summary of the steps involved (also see the Appendix to this chapter):

1. The parameters required are:

   a. Regular solution parameters; those for albite are often used A'= 0.403, B'= 15.333, C'= 10.894

   b. $X_{H_2O}^{o,m}{}_{(P_o,T_o)}$; $V_{H_2O}^{o,m}$; and for polythermal calculations, $\Delta H_{H_2O}^o{}_{(T,P_o)}$ (Table 4).

   c. FW_{one}, the single oxygen melt formula weight (Table 4).

2. Select a pressure and temperature and calculate $f_{H_2O}(P,T)$ ($f_{H_2O}^o{}_{(P_o,T_o)}$ is 1 bar).

3. Calculate $X_{H_2O}^m{}_{(P,T)}$ using Equation (10).

4. Calculate X_{OH}^m by solving iteratively Equation (8). One way to do this is calculate the right and left sides of the equations and vary X_{OH}^m until the two sides are equal (see program listing in Appendix).

5. Calculate the total mole fraction of H_2O in the melt, $X_b = X_{H_2O}^m{}_{(P,T)} + 1/2\ X_{OH}^m$.

6. Calculate the weight percent H_2O in the melt using Equation (7) with FW_{one} instead of FW_{eight}.

CO_2 SOLUBILITY IN ROCK MELTS

The only carbon species observed in basaltic glasses is the carbonate anion (CO_3^{2-}) and in rhyolitic glasses the only species detected is molecular CO_2. The Spera-Bergman thermodynamic treatment (Spera and Bergman, 1980) is simplified compared to the case of water because in the case of either basaltic or rhyolitic compositions only one species need be considered.

CO_2 in basaltic melts

Fine and Stolper (1986) showed that carbonate solubility can be expressed by the simple thermodynamic relationship (see also Chapter 5):

$$K = \frac{X_{CO_3^{2-}}^{melt}}{X_{O^{2-}}^{melt} \bullet f_{CO_2}} \tag{12}$$

where
$$K = K^o \exp\left[\frac{-\Delta V^o(P - P^o)}{RT} - \frac{\Delta H^o}{R}\left(\frac{1}{T} - \frac{1}{T^o}\right)\right] \tag{13}$$

with K^o as the equilibrium constant at a reference temperature and pressure, and other symbols as defined previously. The quantities ΔV^o and ΔH^o are treated as independent of pressure and temperature and refer to the partial reaction:

$$O^{2-\,melt} = CO_3^{2-\,melt} \tag{14}$$

where f_{CO_2} is the fugacity of CO_2 at the P and T of interest, and $X_{O^{2-}}^{melt}$ is defined by:

$$X_{O^{2-}}^{melt} = 1 - X_{CO_3^{2-}}^{melt} \tag{15}$$

Experiments on CO_2 solubility in basaltic melts can be illustrated using the approach taken for a basanite composition (Table 5). In this case the source of CO_2 was silver

Table 5. Experimentally determined model parameters for CO_2 solubility (wt %) in a variety of melt compositions. [1]

	Rhyolite [2]	Tholeiite [3]	Basanite [4]	Leucitite [5]
ln K°	-14.44 (±0.02)	-14.83	-14.32	-13.36
ΔH° (kJ/mol)	-27.1 (±2.0)	5.20 (±4.30)	-13.1 (±13.9)	-28.15 (±4.24)
ΔV (cm³/mol)	28.1 (±1.6)	23.14 (±1.03)	21.72 (±1.27)	21.53 (±0.42)
P° (bars)	1	1000	1000	1000
T°(°C)	850	1200	1200	1200
SiO_2	78	49	45	44
TiO_2	0.07	2.3	2.9	2.7
Al_2O_3	13	13	15	13
FeO^*	0.4	11	12	9
MgO	0.05	10	8	9
CaO	0.5	11	8	14
Na_2O	4.1	2.2	4.3	3.2
K_2O	4.2	0.5	1.1	3.5
P_2O_5	-	0.1	0.8	0.8

[1] Parameters were calculated using a method analgous to that for Table 4. Numbers in parentheses are 2σ deviations from a linear regression using solubility data and calculated fugacities. P° and T° are the reference pressure and temperature. See text for discussion and description of calculations. Data from [2]Blank et al. (1993b), [3]Pan et al. (1991), [4](Pan and Holloway, submitted) and [5]Thibault and Holloway (1994).

oxalate which was sealed, along with powdered basanite, in platinum capsules pre-saturated with iron. Samples were run in a piston-cylinder apparatus, except for one which was run at 1200°C and one kbar in an argon pressurized, internally-heated vessel. The latter point is used as the reference value in Equation (13). The high pressure data are shown in Figure 11, along with the calculated fit using Equation (13); fit parameters are given in Table 5. The scatter in the data shown in Figure 11 indicates the experimental uncertainty. Experiments on CO_2 solubility in tholeiitic (MOR) basalt and olivine leucitite were done in a similar manner. In all cases the data were fit to Equation (13) using CO_2 fugacities calculated by the modified Redlich-Kwong (Holloway, 1987) at pressures up to 4 kbar and the corresponding state equation of Saxena and Fei (1987) at higher pressures.

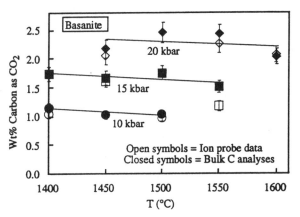

Figure 11. Experimentally-determined CO_2 solubilities in basanite 3048 from Pan et al. (submitted). Open symbols represent concentrations determined by SIMS, filled symbols indicate concentrations determined by bulk analysis using a LECO bulk carbon analyzer. Error bars are 67% confidence intervals (1σ). Circles indicate 10-kbar, squares 15-kbar, and diamonds 20-kbar experiments. The symbol size corresponds roughly to the size of the error bar in most cases.

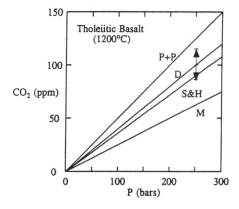

Figure 12. Comparison of measured CO_2 solubilities in tholeiitic (MORB) basalts. All at approximately 1200°C. Data from: P+P (Pan et al., 1991; Pawley et al., 1992), D (Dixon et al., in press), S&H (Stolper and Holloway, 1988), and M (Mattey, 1991), who used a synthetic haplobasalt with significantly lower iron than typical MORB's). The error bar represents the estimated ±15% relative uncertainty in the analyses (after Dixon et al., in press).

Small differences in CO_2 solubilities measured in tholeiitic (MOR) basalts by different workers are shown in Figure 12. These differences are mainly a result of the low carbon concentration at low CO_2 pressures and consequent analytical difficulties (see Chapter 2). For this reason, CO_2 solubilities based on high pressure experiments, where the carbon concentrations are more than an order of magnitude greater (Pan et al., 1991) are probably more precise, but the accuracy of low pressure solubilities derived from modeling of such data is determined by the accuracy of the thermodynamic model (i.e., assuming constant ΔH and ΔV) and the equation of state for CO_2.

Rhyolitic melts

Modeling of CO_2 solubility in rhyolitic melts, which apparently contain only molecular CO_2 (Fogel and Rutherford, 1990; Blank et al., 1993), is similar to that described for carbonate in mafic melts. It is probable that melts with compositions intermediate between tholeiitic basalt and rhyolite will contain both molecular CO_2 and carbonate ion. Such cases (andesite-dacite?) will require expressions analogous to those for water described previously (see Stolper et al, 1987 for a treatment of both carbon species in albitic melt).

A general reaction to describe the interaction between gaseous and dissolved CO_2 is

$$CO_2 \text{ (fluid)} = CO_2 \text{ (melt)} \tag{16}$$

with the equilibrium constant given by

$$K = \frac{a^m_{CO_2}}{a^v_{CO_2}} \approx \frac{X^m_{CO_2}}{f_{CO_2}/f^o_{CO_2}} \tag{17}$$

where $a^m_{CO_2}$ and $a^v_{CO_2}$ are the activities of carbon dioxide in the melt and fluid, f_{CO_2} is the fugacity of CO_2 in the fluid, and $f^o_{CO_2}$ is the fugacity of pure CO_2 at a reference P and T. The equation of CO_2 activity with its mole fraction ($X^m_{CO_2}$) implies that Henry's law is valid for CO_2 in rhyolitic melt, and this is supported by experimental data on CO_2 solubility up to $f_{CO_2} \approx 1300$ bars (see Fig. 13A). Deviations in measured solubilities from Henry's law at higher pressures may be accounted for by incorporating a term for the partial molar volume of dissolved CO_2 (e.g., Prausnitz et al, 1986),

$$\left(\frac{\partial \ln f_{CO_2}}{\partial P}\right)_T = \frac{V^{o,m}_{CO_2}}{RT} \tag{18}$$

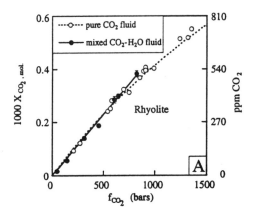

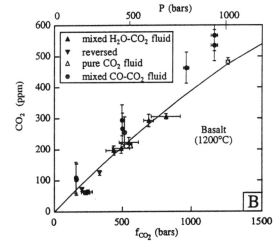

Figure 13. Experimentally-determined CO_2 solubilities for melts in equilibrium with pure CO_2, mixtures of CO_2 and H_2O, and mixtures of CO_2 and CO. (A) Rhyolite (Blank et al., 1993). (B) Tholeiitic basalt. Pure CO_2 (Stolper and Holloway, 1988), model and mixed CO_2-H_2O (Dixon et al., in press), and mixed CO_2 and CO (Pawley et al., 1992).

Likewise, the effect of temperature on CO_2 solubility may be accounted for by incorporation of an enthalpy of solution term for dissolution of CO_2 in melt

$$\left(\frac{\partial \ln f_{CO_2}}{\partial T}\right)_P = \frac{\Delta H^o_{CO_2}}{RT^2} \tag{19}$$

where $\Delta H^o_{CO_2}$ is the enthalpy of the reaction (Eqn. 16). Henry's law, incorporating changes due to pressure and temperature, can thus be written as:

$$\ln\left\{\frac{f_{CO_2}(P,T)}{X^m_{CO_2}(P,T)}\right\} = \ln\left\{\frac{f^o_{CO_2}(P_o,T_o)}{X^m_{CO_2}(P_o,T_o)}\right\} + \frac{V^{o,m}_{CO_2}}{RT}(P-P_o) - \frac{\Delta H^o_{CO_2}}{R}\left(\frac{1}{T^2} - \frac{1}{T^2_o}\right) \tag{20}$$

where P_0 and T_0 refer to a reference pressure and temperature. Note that $V^{o,m}_{CO_2}$ and $\Delta H^o_{CO_2}$ are assumed to be independent of temperature and pressure. Given a set of solubility determinations it is possible to fit the data by multiple linear regression to extract $V^{o,m}_{CO_2}$, $\Delta H^o_{CO_2}$, and $X^m_{CO_2}(P_o,T_o)$. These values can then be used to calculate CO_2 solubility as a function of P and T.

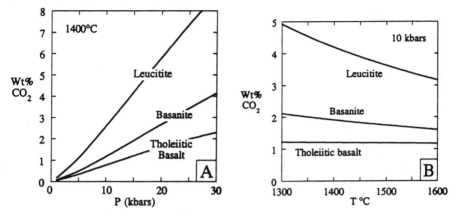

Figure 14 . The solubility of CO_2 in melts of basaltic composition, calculated using the parameters given in Table 5 and CO_2 fugacities calculated with the MRK (Holloway, 1981; 1987) and the corresponding state equation of Saxena and Fei (1987). (A) Variation of CO_2 with pressure. Note the strong effect of melt composition; generally-speaking, CO_2 solubility increases with the degree of melt depolymerization (cf. Chapter 5, this volume). (B) The effect of temperature on CO_2 solubility at a constant pressure of 15 kbar. Note the strong effect of melt composition on the temperature dependence.

Effects of composition on CO_2 solubility in rock melts

CO_2 solubility measurements have been parameterized with the above models for rhyolitic, tholeiitic (basaltic), basanitic, and olivine leucitite melts. The parameters are given in Table 5. The value of $K°$ is proportional to CO_2 solubility at the reference P and T. The value of $\Delta H°$ is a measure of the effect of T on solubility and $\Delta V°$ the effect of P on solubility. The value for the molar volume of CO_2 dissolved in rhyolitic melt is similar to the value of the b parameter in the Redlich-Kwong equation of state (29.7 cm^3/mol), and also very similar to the value obtained by Stolper et al. (1987) for the molar volume of molecular CO_2 dissolved in albitic melt (28.6±0.5 cm^3/mol). The volume changes for carbonate dissolution in basaltic melts are significantly smaller than the molar volume for CO_2 in rhyolite, which probably reflects the fact that the ΔV for carbonate dissolution represents the volume change for the reaction given in Equation (14).

The strong influence of melt composition on CO_2 solubility, as indicated by differences in $K°$ values can be seen in Figure 14A. Figure 14B illustrates the variable effect of T on CO_2 solubility, ranging from negligible for tholeiites to negative for the strongly alkaline leucitite composition. Similar differences can be seen in comparing CO_2 solubility in rhyolitic melts, which contain only molecular CO_2 with the basaltic melts which have only CO_3^{2-} (Fig. 15).

Calculation of CO_2 solubility in rock melts

The model parameters extracted from experimental studies of CO_2 solubility are given in Table 5. Weight percent concentration of CO_2 in a melt at a given P and T is calculated as follows:

1. Find K using Equation (13) and data ($V_{CO_2}^{o,m}$, $\Delta H_{CO_2}^o$, ln $K°$) from Table 5.

2. Find f_{CO_2} at P and T from an equation of state. It is important to use the same equation of state that was used to fit the parameters in Equation (13).

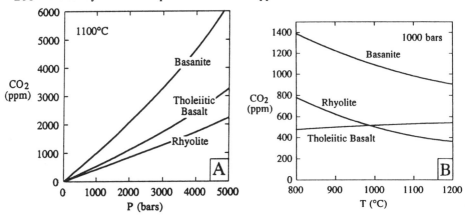

Figure 15. Calculated CO_2 solubility in basanite, tholeiitic basalt and rhyolite composition melts at low pressures using parameters listed in Table 5. (A) Solubility vs. pressure at 1100°C. (B) Solubility vs. temperature at P = 1000 bars.

3. Substitute Equation (15) into Equation (12) and solve for the mole fraction of carbonate:

$$X_{CO_3^{2-}}^{melt} = \frac{K_f}{1 - K_f} \tag{21}$$

where $K_f = K \cdot f_{CO_2}$ (22)

(In rhyolitic melt, with only molecular CO_2, Equation (21) is $X_{CO_2}^{melt} = K_f$)

4. Convert from mole fraction to weight percent using:

$$wt\%_{CO_2}^{melt} = \left[\frac{44.01 \cdot X_{CO_3^{2-}}^{melt}}{44.01 \cdot X_{CO_3^{2-}}^{melt} + (1 - X_{CO_3^{2-}}^{melt}) \cdot FW_{one}} \right] \tag{23}$$

where FW_{one} is the formula weight for one mole of volatile-free melt and 44.01 the formula weight of CO_2.

MIXED-VOLATILE SOLUBILITIES IN SILICATE MELTS

Water-carbon dioxide mixtures

Wyllie and Tuttle (1959) conducted the first experimental study of volatile solubilities in granitic melts equilibrated with a mixed CO_2-H_2O fluid. Work done by Kadik and his colleagues (e.g., Kadik and Lukanin, 1973) focused on rhyolitic, andesitic, and basaltic melts at pressures of 1 to 5 kbar. Their data appear to suggest ideal mixing between H_2O, CO_2, and the silicate melts (i.e., Henrian behavior), but the behavior of CO_2 is difficult to view in detail because of the use of the weight-loss technique for measurement of volatile contents. This procedure involves piercing a cold capsule to release CO_2 vapor at the end of an experiment (assuming H_2O remains as ice) and taking the difference in weight of CO_2 loaded into the capsule and the weight change before and after piercing as a measure dissolved CO_2. Because of the inaccuracies inherent in this technique, their data include analytical uncertainties larger than any of the apparent variation in CO_2 solubility. In contrast, their water solubility data show regular variations as a function of pressure and are in better agreement with results from other laboratories.

Low pressure experiments on both rhyolitic (Blank et al., 1993b) and basaltic (Dixon et al., in press) compositions have demonstrated that, within experimental uncertainty, H_2O has no effect on the solubility of CO_2 in the melt (Fig. 13A,B). The amount of CO_2 in the melt is lowered because the activity of CO_2, and hence f_{CO_2}, in the fluid phase, is lowered by dilution (due to the presence of H_2O); this effect is easily accounted for using Equation (22).

In contrast, experiments by Mysen and others in the mid-1970's were performed at higher pressure from 5 to 20 kbar (Mysen et al., 1976; Holloway, 1981). These workers examined volatile solubilities in natural and synthetic melts and concluded that the addition of small amounts of water to the fluid phase enhances the solubility of CO_2. In the highest pressure experiments there was up to ~40% increase in solubility relative to the that of CO_2 in a melt equilibrated with a pure CO_2 fluid. These observations do not necessarily disagree with the low pressure studies on basalt and rhyolitic melts in which the amounts of total dissolved H_2O and CO_2 are much lower. Enhancement of CO_2 solubility by H_2O may well occur at high pressure where both CO_2 and H_2O solubilities are large.

Degassing at low pressures . The degassing behavior of systems with two dissolved volatiles has been discussed by Holloway (1976), Anderson et al. (1989) and Dixon et al. (in press), among others, and can be described with two-volatile solubility diagrams such as shown in Figures 16 and 17. In these diagrams the CO_2 and H_2O contents of the melt can be noted from the coordinate axes, with the exact position on a given isobar being controlled by the bulk H_2O and CO_2 contents of the system. The dashed lines in Figures 16 and 17 thus show the paths that would be followed during closed system degassing of magma with different initial H_2O/CO_2. Ternary phase diagrams of melt-fluid equilibria are given by Eggler and Kadik (1979).

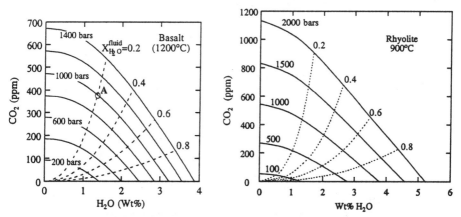

Figure 16 (left). Two-volatile solubility diagram for CO_2 and H_2O in tholeiitic basalt at 1200°C as a function of pressure and fluid composition. See text for discussion. Modified from Dixon et al. (in press).

Figure 17 (right). Two-volatile solubility diagram for CO_2 and H_2O in rhyolite at 850°C as a function of pressure and fluid composition. Pure-fluid CO_2 solubilities were calculated using parameters in Table 5, and H_2O solubilities using the Burnham method and parameters given in Table 4. Calculated using the two-volatile program in the Appendix.

Figures 16 and 17 were constructed by first selecting a temperature and pressure and calculating the solubility of CO_2 and H_2O in melts saturated with pure CO_2 and H_2O, respectively. The studies on basaltic and rhyolitic melts described above have shown that the amount of either volatile dissolved depends only on its pure gas fugacity weighted by its mole fraction in the fluid. This allows a chosen value of H_2O activity (between 0 and 1) to be equated to the mole fraction of H_2O in the fluid. Because the fluid in this calculation is a binary mixture, $X_{CO_2}{}^{fl} = 1 - X_{H_2O}{}^{fl}$, and thus the CO_2 and H_2O contents of the melt can be calculated using Equations (6) and (21). The data points from such calculations can be seen in Figure 17.

Degassing of a basaltic melt containing approximately 375 ppm CO_2 and 1.4 wt % H_2O will follow the dashed curve passing through point A in Figure 16. At a pressure of 1400 bars, a magma with this abundance of H_2O and CO_2 lies below the 1400 bar saturation curve in Figure 16, indicating that the melt is undersaturated in volatiles and so the system will not contain a fluid phase. Upon lowering of pressure, e.g., by magma ascent, the system will remain fluid-absent until the pressure reaches 1000 bars, whereupon the saturation curve crosses over point A, and the melt becomes saturated in a CO_2-H_2O fluid whose molar composition (given by the dashed line through A) is 0.2 mole fraction H_2O. The exact path that a degassing magma will follow on the CO_2-H_2O diagram will depend on the extent to which fluid is lost from the system (open system behavior) or remains with the melt (closed system behavior). In either case, the melt and fluid compositions can be calculated if degassing occurs under equilibrium conditions. Calculation procedures are given in Holloway (1976, with which the modern algorithms for CO_2 solubility given above should be used), and in Dixon et al. (in press).

CO-CO₂ fluids: Carbon species solubility under reducing conditions

The CO/CO_2 ratio in a C-O fluid is a linear function of f_{O_2} at constant P and T. At constant f_{O_2} the ratio is maximized at high T and low P. The maximum ratio is limited by the graphite saturation surface. The mole fraction of CO in graphite-saturated C–O fluids, and the relations between P, T, f_{O_2} and fluid composition are shown in Figure 3.

Experimental measurements suggest that CO has negligible solubility in basaltic compositions for pressures up to 10 kbar (Pawley et al., 1992). In these experiments all of the carbon present appears as carbonate in infrared spectra of the quenched glasses. These observations indicate that dissolved carbonate is directly proportional to the activity of CO_2 in the system, and to the mole fraction of CO_2 in the fluid phase if a fluid phase is present. The CO_2 solubility can be calculated as described in the previous section by calculating f_{CO_2} as $f_{co_2} = f_{co_2}^o \cdot X_{co_2}^{fl}$.

The variation in the dissolved CO_2 content of a MORB liquid in equilibrium with a CO-CO₂ fluid is shown as a function of f_{O_2} at 1200°C and 1 kbar in Figure 18 together with the mole fraction of CO_2 in the coexisting fluid. Note that at oxygen fugacities greater than NNO-1 (1 $\log_{10} f_{O_2}$ unit below the NNO buffer curve) the fluid is nearly pure CO_2 and so the amount of dissolved carbon in the melt is essentially the same as the case of a pure CO_2 fluid. At f_{O_2} values lower than NNO-1 there is a nearly linear relationship between the mole fraction of CO_2 in the fluid and ΔNNO; because the dissolved carbon content of the melt is proportional to $X_{co_2}^{fl}$, there is a linear relationship between dissolved CO_2 and $\log f_{O_2}$. The lower curve in Figure 18 is the fluid saturation surface in a melt containing only C-O volatiles. For the given P and T, any combination of f_{O_2} and carbon (as CO_2) falling below the curve indicates that the melt is fluid-undersaturated.

Results similar to those observed for tholeiitic basalt have been found for an olivine leucitite by Thibault and Holloway (1994) at pressures up to 10 kbar. However, their experimental results at 15 and 20 kbar suggest that the total amount of dissolved carbon

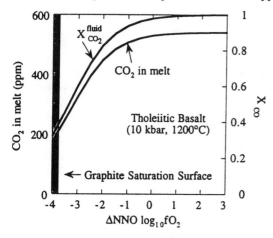

Figure 18. The variation of CO_2 content of a tholeiitic basalt melt saturated with CO-CO_2 fluid at 1200°C and 1 kbar as a function of f_{O_2} relative to NNO (O'Neill, 1987). The CO_2 content is directly proportional to the mole fraction of CO_2 in the fluid (also shown).

was greater than that which could be attributed to CO_2 alone. Earlier evidence of CO solubility at pressures above 20 kbar was found by Eggler et al. (1979).

For rhyolitic melts in equilibrium with CO-rich fluids there is no direct evidence of how the effect of CO on total C solubility. However, the observation that argon solubility is very high in rhyolitic, compared to basaltic compositions (White et al., 1989; Carroll and Stolper, 1993) suggests that CO should have a solubility roughly equal to CO_2 in rhyolitic compositions. If that is true there would be little or no effect of f_{O_2} on total dissolved carbon in rhyolitic systems.

STABLE ISOTOPE PARTITIONING BETWEEN VOLATILE SPECIES IN FLUID AND MELT

In the preceding sections we presented an overview of experimentally-determined equilibrium water and carbon dioxide solubilities in natural silicate melt compositions. We also outlined two ways to calculate, using the experimental data and thermodynamic models, solubilities of these volatiles as a function of P and T. Additional important information on the source(s) and behavior of volatiles in magmas can be obtained through stable isotope measurements. Below, we review (1) the terminology used to describe stable isotope fractionation between vapor and melt, (2) basic forms of two end-member models that describe the coevolution of the isotopic composition and concentration of a volatile phase that remains in the melt during degassing, and finally (3) the experimental work that has determined quantitatively the isotopic partitioning behavior between major volatile species and melt. Experimental studies pertaining to stable isotopic variations between coexisting magmatic volatiles and natural silicate melts have been summarized previously by Taylor (1986). While it is beyond the scope of this volume to present a comparably detailed review, new work in this field warrants a brief reexamination of the subject.

Fractionation of isotopes among gaseous, liquid, and solid compounds approaches zero at very high temperature (Urey, 1947) but may remain sufficiently large at magmatic temperatures to cause measurable isotopic variation resulting from gain or loss of volatiles. As a consequence, stable isotope studies of C-H-O fluids provide a valuable means of examining interactions between volatiles and magmas. Measurements of stable isotope ratios in magmatic gases, erupted glasses, and melt inclusions in phenocrysts can be used to interpret the volatile history of natural systems (e.g., Allard, 1983; Taylor et al., 1983; Sakai et al., 1984; Gerlach and Casadevall, 1986; Mathez et al., 1990; Sarda and Graham, 1990). However, all models incorporating isotopic variation ultimately depend on how well equilibrium isotopic fractionation between species is known. The magnitude of this fractionation is dependent on factors such as temperature, speciation, chemical environment, and reaction kinetics (cf. O'Neil, 1977). With respect to the principal magmatic volatiles, water and carbon dioxide, the isotopes of carbon and hydrogen have a greater potential for isotope variation than those of oxygen isotopes; oxygen isotope ratios are buffered by the large amount of oxygen present in silicate melts, and thus loss or gain of volatiles has little effect on the bulk oxygen isotope composition of the melt/magma. Stable isotope studies of magmatic volatiles therefore focus on the stable isotopes of carbon and hydrogen and variations in their ratios, $^{13}C/^{12}C$ and D/H.

Terminology

Mass spectrometers measure the difference in the isotopic ratio of a sample relative to a standard. Stable isotope data are reported using the δ notation (e.g., O'Neil, 1986), in permil (‰), such that for a sample, A, the isotopic composition is defined by:

$$\delta_A = (\frac{R_A - R_{std}}{R_{std}})\ 10^3 \tag{24}$$

where $R_A = (^{13}C/^{12}C)_A$, $(D/H)_A$, etc., and likewise R_{std} is the corresponding ratio in the standard. Typical reproducibility of measurements is 1 to 2 ‰ for δD and 0.1 ‰ for $\delta^{13}C$.

The *isotopic fractionation factor* between two substances, A and B, is defined as:

$$\alpha_{A-B} = \frac{R_A}{R_B} = \frac{\delta_A + 1000}{\delta_B + 1000} \tag{25}$$

and is related to the equilibrium constant for the isotope exchange reaction between them (Bigeleisen, 1955). The value of the fractionation is usually very close to unity and therefore is commonly reported in permil form as $1000 \ln\alpha$.

Another way to express the isotopic fractionation factor is as the difference in measured δ values of the two substances. This difference:

$$\Delta_{A-B} = \delta_A - \delta_B \tag{26}$$

is written conventionally so as to be a positive number (J.R. O'Neil, pers. comm., 1994). Δ can be related to the fractionation factor through the use of the approximation $10^3 \ln (1.00X) \approx X$, so that

$$\Delta_{A-B} \approx 1000 \ln \alpha_{A-B}. \tag{27}$$

The validity of this approximation diminishes for values of Δ greater than around 10 permil (cf. O'Neil, 1986; Table A1.), but for most fractionations occurring at magmatic temperatures the agreement is well within analytical uncertainties (O'Neil, 1977). Laboratory fractionation experiments can be used to determine values of Δ_{v-m} and α_{v-m}, which can then be incorporated into degassing models. With respect to fractionation between vapor and coexisting silicate melt this relation can be written as:

$$\Delta_{v-m} = \delta_v - \delta_m \approx 10^3 \ln \alpha_{v-m} \tag{28}$$

where δ_v and δ_m are the isotopic compositions of C or H in the vapor and melt, respectively.

Degassing models

Magmatic degassing occurs as a result of supersaturation of a volatile species in a magma, often in response to depressurization (with magma ascent) or crystallization (which causes a relative increase in the concentration of volatiles in the residual melt). Volatile loss can have a profound influence on the stable isotopic composition as well as the volatile content of the residual magma. For this reason, isotopic ratios of natural lavas and gases can be used to deduce information regarding the nature and amounts of degassing, provided the relevant isotopic fractionations are known. The isotopic shift accompanying vapor saturation (degassing) in magmas can be modeled by one, or a combination of, two end-member processes: closed-system degassing and open-system (Rayleigh) degassing (e.g., Rayleigh, 1896; Brown et al., 1985). These processes are analogous to equilibrium and fractional crystallization, respectively. The simple forms of these degassing models are presented below.

In ideal closed-system degassing, the exsolved vapor and the melt are in continuous communication. The bulk volatile content and isotopic ratio of the vapor-melt system remain constant, and the isotopic composition and concentration of the volatile in the melt and vapor phases respond to its partitioning between melt and vapor during degassing. The exsolved volatile phase is in equilibrium with the melt, and the isotopic composition of the melt will increase or decrease depending on the proportion of volatile species dissolved in the melt and the magnitude of the fractionation. This can be modeled using a mass balance approximation:

$$\delta_i \approx F\delta_f + (1-F)(\delta_f + 1000 \ln \alpha_{v-m}) \qquad (29)$$

where δ_f, δ_i, are the final and initial isotopic compositions of the element of interest in the melt, F is the mole fraction (normalized) of the volatile species remaining dissolved in the melt (1 at the beginning, 0 at the end), and α_{v-m} is the vapor-melt fraction factor. This reduces to:

$$\delta_f \approx \delta_i - (1-F) \, 10^3 \ln \alpha_{v-m}. \qquad (30)$$

In a closed-system degassing regime, variation in the isotopic composition of the melt with the concentration of volatile species dissolved in the melt is approximately linear, and if vapor-melt isotopic equilibrium is maintained continually, the isotopic shift does not depend on whether the vapor is evolved continuously, in a single step, or in a number of steps. Under closed-system degassing conditions, the maximum amount of isotopic shift in the melt is limited to the magnitude of the fractionation factor.

The other end-member model is open-system degassing, in which discrete parcels of exsolved volatile are removed systematically from the melt. Removal of vapor separated from the magma prohibits maintenance of continuous isotopic equilibrium between vapor and melt. Open-system degassing, also called Rayleigh fractionation, is described by:

$$\frac{R_f}{R_i} = F^{\alpha_{v-m} - 1} \qquad (31)$$

where R_i is the initial isotopic ratio in the melt, R_f is the isotopic ratio in the melt after a fraction (1-F) of the volatile has been removed via degassing, and F and α_{v-m} are as described above. Equation (31) can be rewritten using the δ-notation as

$$\delta_f = (\delta_i + 1000)(F^{\alpha_{v-m} - 1}) - 1000 \qquad (32)$$

where δ_f and δ_i are as defined above. An approximation based on this exact expression,

$$\delta_f \approx \delta_i + 10^3(F^{\alpha_{v-m} - 1} - 1)$$
(33)

has been shown to be valid for most cases by Broecker and Oversby (1971). This open-system degassing model requires continuous exchange and removal of infinitely small aliquots of fluid, each before the volatilization of the next. As a consequence, the potential magnitude for changes in isotopic composition of the partially-degassed melt and the sequentially evolved vapor is much more pronounced than in the case for closed-system degassing.

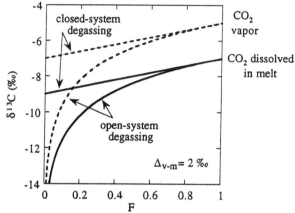

Figure 19. Comparison of closed- and open-system degassing models, illustrating the predicted relative change of the stable isotopic composition of the melt as the heavy isotope is partitioned preferentially into the vapor phase during degassing. Plotted are the $\delta^{13}C$ values of both the melt (solid curves) and co-existing CO_2 vapor in equilibrium with the melt (dashed curves) as a function of F, the fraction of CO_2 remaining dissolved in the melt. See text for specific degassing equations. The example here is for CO_2 degassing from tholeiitic magma, but a similar trend can be calculated for water degassing. Initial conditions for the model calculations are $\delta_i^{13}C$ (melt) = -7, $[CO_{2, melt}]$ = 1650 ppm. A value of Δ_{v-m} = 2‰ was used.

The different degrees of influence that these two end-member degassing processes exhibit on the isotopic composition of the melt are depicted in Figure 19. The example here is for CO_2 degassing from basaltic magma. Solid curves indicate the evolving isotopic composition of the melt as the initial CO_2 content decreases. Dashed curves indicate the isotopic composition of the corresponding vapor in equilibrium with the melt at each fraction, and are fixed according to Δ_{v-m}. At high fractions of the initial total CO_2 in the melt, the two models have quite similar results, but as degassing progresses the differences become more pronounced. The region in which these models diverge depends largely on the magnitude of the fractionation factor; values with greater deviations from unity will result in more pronounced changes in the $\delta^{13}C$ of the residual melt. At late stages of degassing, whereupon most of the volatile has partitioned into the vapor phase, open-system degassing can produce enormous changes in the isotopic composition of the residual melt. The closed-system degassing model thus serves as a minimum estimate of the shift in isotopic composition due to volatilization, whereas the open-system degassing model gives a maximum value. Each of these models assumes isotopic equilibrium at some point between the coexisting vapor phase and the melt. This is not necessarily achieved in nature; for example, non-equilibrium isotopic fractionation may occur (e.g., Anderson and Fink, 1989), or bubbles may remain in the same vicinity of origin in a magma until they are sufficiently large to segregate buoyantly. The end-member models are clearly more simplistic than degassing in dynamic, natural systems. Although natural systems differ from these end-member cases in a number of ways (cf. Gerlach and Taylor, 1990), simple degassing models serve to bracket many of the observed variations in volatile contents and isotopic compositions of associated lavas. These processes are also relevant to describe release of fluids through metamorphism (cf. review by Valley, 1986).

Experimentally-determined isotopic fractionations and application to natural systems

Experimentally-determined Δ_{v-m} values for water and carbon dioxide in equilibrium with volatile species dissolved in natural silicate melts are summarized in Table 6. Note that the fractionations are species-dependent. The dissolved *molecular* species appear to exhibit no fractionation and thus their exsolution will not affect the isotopic composition of the residual melt.

Theoretical calculations (Bottinga, 1969; Richet et al., 1977) predict that the equilibrium isotopic fractionations among minor C-O-H volatile species are significant and thus the presence of other C- and H-bearing gases would affect measured differences in isotopic composition between CO_2 or H_2O vapor and dissolved volatiles (see also, Taylor, 1986) if these minor species were not taken into account. There are currently no published experimentally-determined Δ_{v-m} values for these minor volatiles and silicate melts. Interaction of gases in a mixed-volatile vapor phase (e.g., H_2O+CO_2, Rosenbaum, 1993), will also affect the observed fractionations between melt and vapor species.

Table 6. Experimentally-determined equilibrium isotope fractionation factors for C and H in coexisting vapor and natural silicate melts.

Melt	Melt Species [a]	Δ_{v-m} (‰) [b]	T(°C)	Reference
CO_2				
basalt	CO_3^{2-}	4.3 (±0.4)	1120-1280	1
"		2.0 (±0.2)	1200-1400	2
rhyolite	CO_2	0.0 (±0.2)	800-1200	3
H_2O				
rhyolite	H_2O+OH^-	≈ 23.6	950	4
"	OH^-	51-40 (±1) [c]	530-850	5
"	OH^-	36 (±10)	850	6
"	H_2O	0 (±10)	850	6

[a] Volatile species dissolved in the melt determined using infrared spectroscopy; [b] Isotopic fractionation (Δ_{v-m}) defined as $\delta_{vapor} - \delta_{melt}$. Numbers in parenthesis indicate 2σ deviations from calculated mean value for Δ_{v-m} ^{13}C determinations or an estimated error for δD measurements. [c] Dobson et al. (1989) report Δ_{v-m} as a function of temperature, expressed as $1000\ln\alpha_{v-m}$ [a regression using their results for 530-750°C for the system water vapor-OH^- (in rhyolite) yields 10^3 ln $\alpha_{water-OH^-} \approx -16.89$ $(10^6/T^2)$ + 23.12, where T is in (K) and $r^2 = 0.98$]. References: [1] Javoy & co-workers (Javoy et al. 1978; Trull et al., 1991); [2] Mattey (1991); [3] Blank & Stolper (1993); [4] Taylor & Westrich (1985); [5] Dobson et al. (1989); [6] Ihinger (1991).

CO₂ *CO₂*. The isotopic fractionation between CO_2 fluid and dissolved carbon has been measured for natural melts of basaltic and rhyolitic compositions (Table 6). For silicate melts in which only one dissolved C-bearing species has been detected (i.e., carbonate in basalt, molecular CO_2 in rhyolite), the isotopic fractionation can be expressed using Equation (28). Silicate melts of intermediate composition have both dissolved CO_3^{2-} and molecular CO_2 present (e.g., Fogel and Rutherford, 1990), and thus a more complex relation involving the three carbon-bearing species (and their relative abundances) is required.

Application of stable isotope data to questions related to CO_2 degassing from natural melts is currently limited to basaltic lavas. The value of Δ_{v-m} for CO_2-basalt at magmatic temperature is controversial and is discussed in Chapter 5. Authors have used Δ_{v-m} of 2-4 ‰ in conjunction with the degassing models similar to those mentioned above to relate through degassing histories the observed variations in the $\delta^{13}C$ compositions of natural basaltic glasses (e.g., Pineau and Javoy, 1983; Blank et al., 1993; MacPherson and Mattey, 1994; Javoy and Pineau, 1994) and fumarolic gases (Gerlach and Taylor, 1990;); a similar approach has been used to estimate the initial carbon content of undegassed basaltic magma and calculate the flux of CO_2 from mid-ocean ridge volcanism to the atmosphere (e.g., Javoy et al., 1982; Des Marais, 1985; Gerlach, 1991).

The absence of a discernible isotopic fractionation under equilibrium conditions between CO_2 and rhyolitic melt suggests that variations in the $\delta^{13}C$ among natural suites of this lava type will reflect processes other than degassing. For example, assimilation, magma mixing, or late-stage contamination or entrainment of air could affect the measured $\delta^{13}C$. Kinetic isotopic fractionation associated with degassing may also occur. At present, there are no published measurements of the $\delta^{13}C$ composition of natural rhyolites, principally due to the very low abundance of dissolved CO_2 in subaerially-erupted lavas.

H₂O. Taylor and Westrich (1985) performed the first H_2O-rhyolite isotope fractionation experiments and determined a *bulk* $10^3 \ln \alpha_{v-m}$ fractionation value of ~23.6‰ for water-saturated experiments held at 950°C and 500 bars. These authors conducted pioneering work to document how the δD of rhyolitic lava domes and tephra obsidians (e.g., Taylor et al., 1983; see also, Taylor, 1988, for review) could be related by degassing. Subsequent studies (Newman et al., 1988; Dobson et al., 1989; Ihinger, 1991) have reexamined their model incorporating information about speciation of dissolved water in rhyolitic melts and glasses.

The hydrogen isotope fractionation between melt and aqueous fluid,

$$\Delta_{v-m} = \delta D_{vapor} - \delta D_{melt} \tag{33}$$

is complicated by the presence of water dissolved both as hydroxyl groups and as molecules of water. If hydrogen speciation in hydrous glasses reflects speciation in fluid-saturated melts, then we can use the approximation:

$$\delta D_{melt} \approx \delta D_{glass} \approx X_{H_2O,mol.} \, \delta D_{H_2O} + (1 - X_{H_2O,mol.}) \, \delta D_{OH} \tag{34}$$

where $X_{H_2O,mol.}$ is the fraction of total dissolved water present as water molecules in the melt. This expression can be corrected to account for speciation changes upon quenching; see Chapter 4.

Because the relative proportions of dissolved water species change as a function of the total amount of water remaining dissolved in the melt, resulting in an increased proportion of OH^- (which has a large D/H fractionation with water vapor) as degassing progresses, the bulk value of Δ_{v-m} (and $10^3 \ln \alpha_{v-m}$) will increase as a magma degasses. The greatest changes in isotopic composition of the melt resulting from degassing will occur at low total dissolved water contents. Dobson et al. (1989) applied their experimentally-determined value of $\alpha_{v\text{-}m}$ for $H_2O_{(vapor)}$-$OH^-_{(rhyolite)}$ to degassing models using numerical solutions to a system of coupled linear differential equations relating the speciation of water in the melt, the partitioning of hydrogen isotopes between the hydrous melt species and the water vapor, and the fraction of initial water remaining in the melt. Their models were able to predict closely the wide range of δD variations observed in several volcanic centers in the Western United States.

Degassing of natural systems. Isotopic shifts observed among natural eruption products are governed in part by the solubilities of the gaseous species in magmas as a function of pressure. If a magma is saturated with respect to a volatile phase while it is in its magma chamber, then it will immediately begin to exsolve a separate vapor phase upon ascent prior to eruption. The rate, distance, and nature of this rise to the surface influences the style and amount of degassing, and thus the magnitude of isotopic fractionation that occurs. With respect to natural systems, application of stable isotopes to degassing systematics has focused on basaltic lavas for $\delta^{13}C$ and rhyolitic lavas for δD. This is probably due to the fact that, at the pressures under which these lava types typically erupt—submarine environments at pressures of several hundred bars for basalts, subaerial environments for rhyolites—degassing is affecting the concentrations of CO_2 and H_2O, respectively. Samples from both settings have been shown to reflect the degassing processes described above.

Among suites of basaltic lavas, CO_2 contents and $\delta^{13}C$ appear to more closely resemble open system degassing trends (e.g., Blank et al., 1993; MacPherson and Mattey, 1994). Taylor (1988) notes that relative to the hydrogen isotope shifts, depressurization is generally likely to be associated with a more "open-system," while a magma which generally reaches and maintains vapor saturation primarily via crystallization and cooling may exhibit a hydrogen isotope fingerprint resembling vapor separation in a "closed-system."

With respect to closed-system degassing, continuous equilibrium fractionation requires rapid re-equilibration between magmas and separated vapor. This demands maintenance of physical contact between magma and separated vapor and melt, and homogenization of isotopes in the magma by convection of the magma and/or diffusion of the volatile species. Both of these conditions are unlikely to occur in small magma bodies because of loss of vapor, but might be approximated by homogeneous bubble nucleation in larger, deeper-seated magma bodies where continual, rapid loss of vapor is inhibited.

Used in conjunction with solubility measurements, knowledge of stable isotope fractionations between dissolved and gaseous volatiles can be used as sensitive probes of late-stage evolutionary histories of natural eruptive products. The recognition of isotopic shifts related to degassing can also be utilized in reconstructing the isotopic composition of an undegassed magma, which can be used to characterize volatile source regions.

ACKNOWLEDGMENTS

This chapter was written while JRH was supported by NSF grants OCE-9123453 and EAR-9312498, and while JGB was supported by the CROUS. We are grateful to Mike Carroll, Jackie E. Dixon, Patrick F. Dobson and Vivian Pan for timely and careful reviews. JGB thanks Claude Jaupart and the members of the Laboratoire de Systèmes Dynamiques et Géologiques.

REFERENCES

Allard P (1983) The origin of hydrogen, carbon, sulfur, nitrogen and rare gases in volcanic exhalations: Evidence from isotope geochemistry. In: H Tazieff and J-C Sabroux (eds) Forecasting Volcanic Events. Elsevier, Amsterdam, p 337-386

Anderson SW Fink JH (1989) Hydrogen-isotope evidence for extrusion mechanisms of the Mount St. Helens lava dome. Nature 341:521-523

Anderson AT Jr., Newman S, Williams SN, Druitt TH, Skirius C, Stolper E (1989) H$_2$O, CO$_2$, Cl, and gas in Plinial and ash-flow Bishop rhyolite. Geol 17:221-225

Anderson GM, Crerar DA (1993) Thermodynamics in Geochemistry. Oxford University Press, New York, 588 p

Baker MB, Grove TL (1985) Kinetic controls on pyroxene nucleation and metastable liquid lines of descent in a basaltic andesite. Am Mineral 70:279-287

Bigeleisen J (1955) Statistical mechanics of isotopic systems with small quantum corrections. I. General considerations and the rule of the geometric mean. J Chem Phys 23:2264-2267

Blank JG, Delaney JR, Des Marsais DJ (1993a) The concentration and isotopic composition of carbon in basaltic glasses from the Juan de Fuca Ridge. Geochim Cosmochim Acta

Blank JG, Stolper EM, Carroll MR (1993b) Solubilities of carbon dioxide and water in rhyolitic melt at 850°C and 750 bars. Earth Planet Sci Lett 119:27-36

Bottinga Y (1969) Calculated fractionation factors for carbon and hydrogen isotope exchange in the system calcite-carbon dioxide-graphite, methane-hydrogen-water vapor. Geochim Cosmochim Acta 33:49-64

Bottinga Y, Javoy M (1989) MORB degassing: Evolution of CO$_2$. Earth Planet Sci Lett 95:215-225

Brodholt JP, Wood BJ (1994) Measurements of the PVT properties of water to 25 kbars and 1600°C from synthetic fluid inclusions in corundum. Geochim Cosmochim Acta 58:2143-2148

Broecker WS Oversby VM (1971) Chemical Equilibria in the Earth. New York, McGraw-Hill

Brown PE, Bowman JR Kelly WC (1985) Petrologic and stable isotopic constraints on the source and evolution of skarn-forming fluids at Pine Creek, California. Econ Geol 80:72-95

Burnham CW, Davis NF (1971) The role of H$_2$O in silicate melts: I. P-V-T relations in the system NaAlSi$_3$O$_8$-H$_2$O to 10 kilobars and 1000°C. Am J Sci 270:54-79

Burnham CW, Davis NF (1974) The role of H$_2$O in silicate melts: II. Thermodynamic and phase relations in the system NaAlSi$_3$O$_8$-H$_2$O to 10 kilobars, 700°C -1100°C. Am J Sci 274:902-940

Burnham CW, Jahns RH (1962) A method for determining the solubility of water in silicate melts. Am J Sci 260:721-745

Burnham CW, Nekvasil H (1986) Equilibrium properties of granitic magmas. Am Mineral 71:239-263

Carmichael ISE (1991) The redox state of basic and silicic magmas. Contrib Mineral Petrol 106:129-141

Carroll M, Blank JG (1994) The solubility of water in phonolitic melts. EOS Trans Am Geophys Union Spring Meeting, p 353

Carroll MR, Stolper EM (1993) Noble gas solubilities in silicate melts and glasses: New experimental results for argon and the relationship between solubility and ionic porosity. Geochim Cosmochim Acta 57:5039-5051.

Chase MW Jr., Davies CA, Downey JR Jr., Frurip DJ, McDonald RA, Syverud AN (1985) JANAF Thermochemical Tables, 3rd Edn, J Phys Chem Ref Data, 1856

Cocheo PA, Holloway JR (1993) The solubility of water in basanitic melts at low pressures. EOS Trans Am Geophys Union 74:350

Dayhoff MO, Lippincott ER, Eck RV, Nagarajan G (1967) Thermodynamic equilibrium in prebiological atmospheres of C, H, O, N, P, S, and Cl. Clearinghouse for Federal Scientific and Technical Information, Washington, DC, 259 p

DesMarais DJ (1985) Carbon exchange between the mantle and the crust, and its effect upon the atmosphere. In: The Carbon Cycle and Atmospheric CO$_2$: Natural Variations Archean to Present, 602-611, American Geophysical Union, Washington, DC

Dingwell DB, Harris DM, Scarfe CM (1984) The solubility of H$_2$O in melts in the system SiO$_2$-Al$_2$O$_3$-Na$_2$O-K$_2$O at 1 to 2 kbar. J Geol 92:387-395

Dixon JE, Stolper EM, Holloway JR (in press) An experimental study of water and carbon dioxide solubilities in midoceanic ridge basaltc liquids. J Petrol

Dobson PF, Epstein S and Stolper EM (1989) Hydrogen isotope fractionation between coexisting vapor and silicate glasses and melts at low pressure. Geochim Cosmochim Acta 53:2723-2730

Eggler DH, Kadik AA (1979) The system NaAlSi$_3$O$_8$-H$_2$O-CO$_2$ to 20 kbar pressure: I. compositional and thermodynamic relations of liquids and vapors coexisting with albite. Am Mineral 64:1036-1048

Eggler DE, Mysen BO, Hoering TC and Holloway, JR (1979) The solubility of carbon monoxide in silicate melts at high pressures and its effect on silicate phase relations. Earth Planet Sci Lett 43:321-33

Eugster H, Skippen GB (1967) Igneous and metamorphic equilibria involving gas equilibria. Res Geochemistry 2:492-526

Fine GJ, Stolper EM (1986) Dissolved carbon dioxide in basaltic glasses: Concentrations and speciation. Earth Planet Sci Lett 76:263-278

Fogel RA, Rutherford MJ (1990) The solubility of carbon dioxide in rhyolitic melts: a quantative FTIR study. Am Mineral 75:1311-1326

French BM (1966) Some geological implications of equilibrium between graphite and a C-H-O gas phase at high temperatures and pressures. Rev Geophys 4:223-253

French BM, Eugster HP (1965) Experimental control of oxygen fugacities by graphite-gas equilibriums. J Geophys Res 70:1529-1539

Frost RB (1979) Mineral equilibria involving mixed-volatiles in a C-O-H fluid phase: the stabilities of graphite and siderite. Am J Sci 279:1033-1059

Gerlach TM (1991) Comment on "Mid-ocean ridge popping rocks: implications for degassing at ridge crests" by P. Sarda and D. Graham. Earth Planet Sci Lett 105:566-567.

Gerlach TM, Casadevall TJ (1986) Evaluation of gas data from high-temperature fumaroles at Mount St. Helens, 1980-1982. J. Volcanol Geotherm Res 28:107-140.

Gerlach TM, Taylor BE (1990) Carbon isotopic constraints on degassing of carbon dioxide from Kilauea Volcano. Geochim Cosmochim Acta 54:2051-2058

Gordon S, McBride BJ (1971) Computer Program for Calculation of Complex Chemical Equilibrium Compositions, Rocket Performance, Incident and Reflected Shocks, and Chapman-Jouguet Detonations, 177 p, National Technical Information Service, Springfield, VA

Hamilton DL (1987) Solubility of water in albite-melt determined by the weight loss method: A reply. J Geol 95:585

Hamilton DL, Burnham CW, Osborn EF (1964) The Solubility of water and effects of oxygen fugacity and water content on crystallization in mafic magmas. J Petrol 5:21-39

Holloway JR (1976) Fluids in the evolution of granitic magmas:Consequences of finite CO_2 solubility. Bull Geol Soc Am 87:1513-1518

Holloway JR (1981) Volatile interactions in magmas, Thermodynamics of Minerals and Melts, 273-293, Springer-Verlag, New York

Holloway JR (1984) Graphite-CH_4-H_2O-CO_2 equilibria at low-grade metamorphic conditions. Geol 12:455-458

Holloway JR (1987) Igneous fluids, Thermodynamic Modelling of Geological Materials:Minerals, Fluids and Melts. Rev Mineral 17:211-232

Holloway JR, Jakobsson S (1986) Volatile solubilities in magmas:Transport of volatiles from mantles to planet surfaces. J Geophys Res 91:D505-D508

Holloway JR, Pan V, Gudmundsson G (1992) Melting experiments in the presence of graphite:Oxygen fugacity, ferric/ferrous ratio and dissolved CO_2. Eur J Mineral 4:105-114

Holloway JR, Reese RL (1974) The generation of N_2-CO_2-H_2O fluids for use in hydrothermal experimentaion. I. Experimental method and equilibrium calculation in the C-O-H-N system. Am Mineral 59:589-597

Holtz F, Behrens H, Dingwell DB, Taylor RP (1992) Water solubility in aluminosilicate melts of haplogranite composition at 2 kbar. Chem Geol 96:289-302

Ihinger PD (1991) The interaction of water with granitic melt. PhD dissertation, California Institute of Technology, Pasadena, CA

Jakobsson S, Oskarsson N (1994) The system C-O in equilibrium with graphite at high pressure and temperature:An experimental study. Geochim Cosmochim Acta 58:9-17.

Javoy M, Pineau F (1994) Strong degassing at ridge crests—the behavior of dissolved carbon and water in basalt glasses at 14°N, Mid-Atlantic Ridge. Earth Planet Sci Lett 123:179-198

Javoy M, Pineau F, Allègre CJ (1982) Carbon geodynamic cycle. Nature 300:831.

Javoy M, Pineau F, Liyama I (1978) Experimental determination of the isotopic fractionation between gaseous CO_2 and carbon dissolved in tholeiitic magma. Contrib Mineral Petrol 67:35-39

Joyce DB, Holloway JR (1993) An experimental determination of the thermodynamic properties of H_2O-CO_2-NaCl fluids at high pressures and temperatures. Geochim Cosmochim Acta 57:733-746

Kadik AA, Lukanin OA (1973) The solubility-dependent behavior of water and carbon dioxide in magmatic processes. Geochem Int'l 10:115-129

Karsten JL, Holloway JR, Delaney JR (1982) Ion microprobe studies of water in silicate melts:Temperature-dependent water diffusion in obsidian. Earth Planet Sci Lett 59:420-428

Kerrick DM, Jacobs GK (1981) A modified Redlich-Kwong equation for H_2O, CO_2 and H_2O-CO_2 mixtures. Am J Sci 281:735-767

Khitarov NI, Lebedev EB, Rengarten EV, Arseneiva RV (1959) Comparative characteristics of the solubility of water in basaltic and granitic melts. Geokhimiya 5:479-492

MacPhearson C, Mattey D (1994) Carbon-isotope variations of CO_2 in central Lau Basin basalts and ferrobasalts. Earth Planet Sci Lett 121:263-276

Mader UK, Berman RG (1991) An equation of state for carbon dioxide to high pressure and temperature. Am Mineral 76:1547-1559

Mathez EA, Pineau F (1990) Carbon isotopes in xenoliths from the Hualalai Volcano,Hawaii, and the generation of isotopic variability. Geochim Cosmochim Acta 54:217-227

Mattey (1991) Carbon dioxide solubility and carbon isotope fractionation in basaltic melt. Geochim Cosmochim Acta 55:3467-3473

McMillan PF, Holloway JR (1987) Water solubility in aluminosilicate melts. Contrib Mineral Petrol 97:320-332

Mysen BO, Eggler DH, Seitz MG, Holloway JR (1976) Carbon dioxide in silicate melts and crystals. Part I. Solubility measurements. Am J Sci 276:455-479

Newman S, Epstein S, Stolper E (1988) Water, carbon dioxide, and hydrogen isotopes in glasses from the ca. 1340 A.D. eruption of the Mono Craters, California:Constraints on degassing phenomena and initial volatile content. J Volcanol Geotherm Res 35:75-96

O'Neil JR (1977) Stable isotopes in mineralogy. Phys Chem Minerals 2:105-123

O'Neil JR (1986) Theoretical and experimental aspects of isotopic fractionation. In: JW Valley, H.P. Taylor Jr, JR O'Neil (eds) Stable Isotopes in High Temperature Geological Processes. Rev Mineral 16:1-40

O'Neill HSC (1987) Free energies of formation of NiO, CoO, Ni_2SiO_4, and Co_2SiO_4. Am Mineral 72:280-291

Paillat O, Elphick SC, Brown WL (1992) Solubility of water in $NaAlSi_3O_8$ melts:a re-examination of Ab-H_2O phase relationships and critical behavior at high pressures. Contrib Mineral Petrol 112:490-500

Pan V, Holloway JR, Hervig RL (1991) The pressure and temperature dependence of carbon dioxide solubility in tholeiitic basalt melts. Geochim Cosmochim Acta 55:1587-1595.

Pan V, Holloway JR, Hervig RL (submitted) Experimental determination of carbon dioxide solubility in basanitic melts. Am Mineral

Pawley AR, Holloway JR, McMillan P (1992) The effect of oxygen fugacity on the solubility of carbon-oxygen fluids in basaltic melt at low pressure. Earth Planet Sci Lett 110:213-225

Pineau F and Javoy M (1983) Carbon isotopes and concentrations in midocean ridge basalts. Earth Planet Sci Lett 62:239-257

Prausnitz JM, Lichtenthaler RN, Axevedo EGd (1986) Molecular Dynamics of Fluid Phase Equilibria, 600pp, Prentice Hall, Englewood Cliffs, New Jersey

Rayleigh JWS (1896) Theoretical considerations respecting the separation of gases by diffusion and similar processes. Philosophical Magazine 42:493

Richet P, Bottinga Y, Javoy M (1977) A review of hydrogen, carbon, nitrogen, oxygen, sulphur, and chlorine stable isotope fractionation among gaseous molecules. Ann Rev Earth Planet Sci. 5:65-110

Rosenbaum JM (in press) Stable isotope fractionation between coexisting carbon dioxide and calcite at 900°C. Geochim Cosmochim Acta

Sakai H, Des Marais DJ, Ueda, A and Moore, JG (1984) Concentrations and isotopic ratios of C, N, and S in ocean-floor basalt. Geochim Cosmochim Acta 48:2433-2441

Sarda P and Graham D (1990) Mid-ocean ridge popping rocks:implications for degassing at ridge crests. Earth Planet Sci Lett 97:268-289

Saxena SK, Fei Y (1987) High pressure and high temperature fluid fugacities. Geochim Cosmochim Acta 51:783-791

Silver L, Stolper E (1989) Water in albitic glasses. J Petrol 30:667-709

Silver LA, Ihinger PD, Stolper EM (1990) The influence of bulk composition on the speciation of water in silicate glasses. Contrib Mineral Petrol 104:142-162

Silver LA, Stolper EM (1985) A thermodynamic model for hydrous silicate melts. J Geol 93:161-178

Spera FJ, Bergman SC (1980) Carbon dioxide in igneous petrogenesis:I. Aspects of the dissolution of CO_2 in silicate liquids. Contrib Mineral Petrol 74:55-66

Stolper E (1982a) The speciation of water in silicate melts. Geochim Cosmochim Acta 46:2609-2620

Stolper E (1989) Temperature dependence of the speciation of water in rhyolitic melts and glasses. Am Mineral 74:1247-1257

Stolper E, Holloway JR (1988) Experimental determination of the solubility of carbon dioxide in molten basalt at low pressure. Earth Planet Sci Lett 87:397-408

Stolper EM (1982b) Water in silicate glasses:An infrared spectroscopic study. Contrib Mineral Petrol 81:1-17

Stolper EM, Fine GJ, Johnson T, Newman S (1987) The solubility of carbon dioxide in albitic melt. Am Mineral 72:1071-1085

Taylor BE (1986) Magmatic volatiles:Isotopic variation on C, H, and S, Stable Isotopes in High Temperature Geological Processes, Rev Mineral 16:185-226

Taylor BE (1988) Degassing of rhyolitic magmas:hydrogen isotope evidence and implications for magmatic-hydrothermal ore deposits. In: RP Taylor, DF Strong (eds) Recent Advances in the Geology of Granite-Related Mineral Deposits. Special vol, Canadian Inst Mining Metallurgy, Ottawa, p 33-49

Taylor BE, Eichelberger JC, Westrich HR (1983) Hydrogen isotope evidence of rhyolitic magmas degassing during shallow intrusion and eruption. Nature 306:541-545

Taylor BE, Westrich HR (1985) Hydrogen isotope exchange and water solubility in experiments using natural rhyolite obsidian. EOS Trans Am Geophys Union 66:387.

Taylor WR, Green DH (1988) Measurement of reduced peridotite-C-O-H solidus and implications for redox melting of the mantle. Nature 332:349-352

Thibault Y, Holloway JR (1994) Solubility of CO_2 in a Ca-rich leucitite:effects of pressure, temperature and oxygen fugacity. Contrib Mineral Petrol 116:216-224

Trull T, Pineau F, Bottinga Y, Javoy M (1991) Experimental study of CO_2 bubble growth and $^{13}C/^{12}C$ isotopic fractionation in tholeiitic melt, 4th Silicate Melt Workshop:Program/Abstracts, 19-23 March, 1991. Abbrev. Journal 7

Ulmer P, Luth RW (1991) The graphite-COH fluid equilibrium in P T fO_2 space. An experimental determination to 30 kbar and 1600°C. Contrib Mineral Petrol 106:265-272

Urey HC (1947) The thermodynamic properties of isotopic substances. J Royal Chem Soc 1:562-581

Valley JW (1986) Stable isotopic geochemistry of metamorphic rocks. In: HP Taylor Jr, JW Valley, JR O'Neil (eds) Stable Isotopes in High Temperature Geological Processes, Rev Mineral 16:445-489

White BS, Brearley M, Montana A (1989) Solubility of argon in silicate liquids at high pressures. Am Mineral 74:513-529

Wyllie PJ, Tuttle OF (1959) Effect of carbon dioxide on the melting of granite and feldspars. Am J Sci 257:548-655

Zhang Y, Stolper EM, Wasserburg GJ (1991) Diffusion of water in rhyolitic glasses. Geochim Cosmochim Acta 55:441-456.

APPENDIX

This appendix contains the following programs:

(1) Spera-Stolper H2O version 1.0. A program to calculate H_2O solubilities in silicate melts using the method outlined in Spera and Bergman (1980) and Silver and Stolper (1985).

(2) CO2-H2O Degas version 1.0 A program to calculate two-volatile saturation as a function of pressure.

Each of these contains subroutines which can be lifted out and used separately to calculate MRK and Saxena and Fei fugacities and volumes, and Burnham solubilities.

Please reference this *Reviews in Mineralogy* chapter when referring to use of these programs or subroutines.

* * * * * * * * * *

(1) Spera-Stolper H2O version 1.0

```
'Spera-Stolper H2O version 1.0
' A program to calculate H2O solubilities in silicate melts
'with the Stolper model. (e.g. Silver and Stolper, 1989.
'Written 7/19/94 by John R. Holloway, Geology Dept.
'Arizona State University. The language is Microsoft
'Quickbasic 1.0 for the Macintosh.
'
'Gas constant in cm3 bar/K mol:
R = 83.14
'
'regular solution parameters:
aprime =   0.403
bprime = 15.333
cprime = 10.894
'
'molecular H2O parameters for Basalt from Dixon et al (in press):
'these parameters must be changed for other silicate compositions.
```

```
'molar volume of H2O in basaltic liquid.
vm = 12!
'mole fraction of H2O in one-oxygen melt at reference P and T.
xmref =  0.0000328
'one-oxygen formula weight of the silicate melt.
fw1 = 36.6
'

'Calculation temperature in degress celsius. Change for other
'Temperatures
'

TC = 1200!
'

PRINT "T (°C) = ";
PRINT USING "#####";TC
     TK=TC+273.15
'

PRINT
PRINT "At the next prompt, enter pressure in bars, followed by return."
PRINT "A zero for P ends program."
PRINT
PRINT "P bars   fH2O   XH2O,mol   XOH    XB     wt% H2O"
'

start:
INPUT;"",pbars
IF pbars<1 THEN GOTO done
'

'call subroutine to solve by iteration the reqular solution equation
GOSUB stolper
'

xb = xm + .5 * xoh
'

xr=xb/(1!-xb)
xr=18.015*xr/fw1
wth2om=xr*100!/(1!+xr)
'wth2om=(18.015*xb)/((18.015*xb) + (1!-xb)/fw1)

PRINT USING "  #####";fh2o;
PRINT USING "   #.###";xm;
PRINT USING "   #.###";xoh;
PRINT USING "   #.###";xb;
PRINT USING "   ###.##";wth2om
'

GOTO start
done:
STOP
END
'

stolper:
'routine to calculate molecular H2O in the melt.
'

'calculate H2O fugacity at reference P and T
'for rhyolite this is 1 bar and 850°C.
'for basalt it is 1 bar and 1200°C.
Pmrk= 1!
Tmrk = 1473.15  'T in Kelvins
'call MRK routine to calculate the fugacity coefficient of H2O
```

```
'at the reference Pand T specified.
GOSUB MRK
'calculate the reference H2O fugacity.
fref = Pmrk * FP
'

'calculate H2O fugacity at actual P and T
Pmrk = pbars
Tmrk = TK
GOSUB MRK
fh2o = pbars * FP
'

'calc volume term
exv = EXP(-vm*(pbars-1!)/(R*TK))
'

'calc mole fraction of molecular H2O
xm = xmref *(fh2o/fref)*exv
'

'The following is an iterative routine to solve the Silver et al. (1990)
'reqular solution model. The required input is xm, the one-oxygen mole
'fraction of molecular H2O (XH2O) and the three regular solution parameters.
'This routine solves for the mole fraction of OH in the melt (XOH).
'

'set iteration parameters
kount=0
tol = .00001
damper = xm*2!
sign = 1!
'

xoh = .5 * xm       'set the initial value of XOH
'

'call rutine to find the difference between the two sides of the
'regular solution equation.
GOSUB calcy
IF ydiff<0 THEN sign = -1!
'

nexty:
kount = kount + 1
yold = ydiff
GOSUB calcy
'

IF ABS(ydiff)>tol THEN
   IF ydiff*yold < 0 THEN
      damper = .6 * damper
      sign = -sign
   END IF
xold = xoh
xoh = xoh * (1! +sign*damper)
GOTO nexty
END IF
'

RETURN
'

calcy:
xln = -LOG(xoh^2/(xm*(1!-xoh-xm)))
prime = aprime + bprime * xoh + cprime*xm
ydiff = xln - prime
```

```
RETURN
'
MRK:
'MRK H2O routine for MSA short course 1994 J R Holloway
'Pressure in bars, temperature in deg K, volume in cm3

b=14.5
'
 RXT=R*Tmrk
 RT=R*Tmrk^1.5* .000001
'
'calculate the MRK a parameter for H2O using a fit to 1981
'equation from 600-1600 deg. C. June 9, 1994

ah2o = 115.98 - .0016295# * Tmrk - 1.4984E-05 * Tmrk^2
'
 A2B=ah2o/(b*RT)
 BP=Pmrk*b/RXT

'Call REDKW to find volume and fugacity
GOSUB REDKW
'calculate the molar volume for H2O. This quantity is given
'for reference only, it is not used in this program.
 VOL=Z*RXT/Pmrk
RETURN
'
REDKW:
'A ROUTINE TO CALC COMPRESSIBILITY FACTOR AND FUGACITY
'COEFFICIENT WITH THE REDLICK-KWONG EQUATION FOLLOWING
'EDMISTER(1968). THIS SOLUTION FOR SUPERCRITICAL FLUID.
'WRITTEN BY JOHN R. HOLLOWAY, 1973, REVISED APRIL, 1983
'RETURNS A VALUE OF 1 FOR FP IF ARGUMENTS OUT OF RANGE.
'TRANSLATED INTO BASIC FOR MAC NOV, 1985 BY JRH.
'
 TH =.333333
 IF A2B < 1E-10 THEN A2B=.001
 RR=-A2B*BP*BP
 QQ=BP*(A2B-BP-1!)
 XN=QQ*TH+RR-.074074
 xm=QQ-TH
 XNN=XN*XN/4!
 XMM=xm*xm*xm/27!
 ARG=XNN+XMM
 IF ARG > 0! THEN GOTO 5
 IF ARG < 0! THEN GOTO 15
 FP=1!
 Z=1!
 RETURN
15  COSPHI=SQR(-XNN/XMM)
 IF XN > 0! THEN COSPHI=-COSPHI
 TANPHI=SQR(1!-COSPHI*COSPHI)/COSPHI
 PHI=ATN(TANPHI)*TH
 FAC=2!*SQR(-xm*TH)
'SORT FOR LARGEST ROOT
 R1=COS(PHI)
 R2=COS(PHI+2.0944)
```

```
R3=COS(PHI+4.18879)
'SORT FOR LARGEST ROOT
RH=R2
IF R1 > R2 THEN RH=R1
IF R3 > RH THEN RH=R3
Z=RH*FAC + TH
GOTO 20
5  x=SQR(ARG)
F=1!
XN2=-XN/2!
XMM=XN2+x
IF XMM < 0! THEN F=-1!
XMM=F*((F*XMM)^TH)
F=1!
XNN=XN2-x
IF XNN < 0! THEN F=-1!
XNN=F*((F*XNN)^TH)
Z=XMM+XNN+TH
20  ZBP=Z-BP
IF ZBP < .000001 THEN ZBP=.000001
BPZ=1!+BP/Z
FP=Z-1!-LOG(ZBP)-A2B*LOG(BPZ)
IF FP < -37 OR FP > 37 THEN FP=.000001
FP=EXP(FP)
RETURN
```

* * * * * * * * * *

(2) CO_2-H_2O Degas version 1.0

```
'CO2-H2O Degas version 1.0
'A program to calculate two-volatile saturation
'curves as a function of temperature, pressure and fluid composition.
'Written 5/21/94 by John R. Holloway, Geology Department
'Arizona State University. Written in Microsoft Quickbasic 1.0
'for the Macintosh.
'
' Includes the Burnham H2O model and the Spera-Stolper CO2 model.
'
'this example version is for rhyolite at 900°C.
'
DEFSNG A-Z
DIM  xkh2o(6), xkfco2(6), ppmco2(6),xh2o(6),pstore(6)
tb$=CHR$(9)
'
'specify number of pressures
npr=5
'
'define thepressures in bars. These can be changed as desired.
DATA 2000., 1500., 1000., 500., 100.
'
FOR i=1 TO npr
READ pstore(i)
NEXT i
'
'Define parameters for rhyolitic melts
'Burnham H2O parameters
FW8=263!
Kfactor = .2
```

```
'Define CO2 parameters for rhyolitic melts
FWone = 36.6
kref=3.6259E-07
deltah=1293!
deltav=22.87
Tref=850!                    'T in Celsius
Pref=1!
'
'set  calculation temperature in degrees celsius
TC=900!
'
TK=TC+273!
'
'the following statements produce a tab delimited file with
'wt% H2O as the independent variable and ppm CO2 as the
'dependent variable along isobars and isopleths.
'
ffname$="CO2H2ORhyolite"
OPEN ffname$ FOR OUTPUT AS #3
'
'the following line makes labels for columns of constant pressure
PRINT#3, "wt%H2O";tb$;"2000";tb$;"1500";tb$;"1000";tb$;"500";tb$;"100";
'
'the following line makes labels for columns of constant
'mole fraction of H2O in the fluid
PRINT #3, tb$;"0.2X";tb$;"0.4X";tb$;"0.6X";tb$;"0.8X"
'
NewPTpt:
'begin pressure loop
FOR i=1 TO npr
pb=pstore(i)
'
'Burnham routine to calculate H2O solubilities in rock melts
'For tholeiitic basalt, andesite, dacite, rhyolite, albite, nepheline, quartz
'The eight-oxygen formula weights are, respectively:
'292 ,276,265,263,262,284!,240.4
'The K-factor values are:
'.236,.173,.153,.203,0!,-.47,.47
'
     LNP=LOG(pb)                 'natural log
'
'Calculate Burnham Henry's law analogue constant for albite composition.
LNK=5!+LNP*((.00000004481#)*TK^2-(.000151#)*TK-1.137)
LNK=LNK+ (LNP^2)*((.00000001831#)*TK^2-(.00004882#)*TK+(.04656#))
LNK=LNK+(LNP^3)*(.0078#)-(LNP^4)*(.0005012#)
LNK=LNK+ (.004754#)*TK-(.000001621#)*TK^2
'
'adjust Henry's law analogue for melt composition
lnkrock = LNK + Kfactor
xk = EXP(lnkrock)
'
'call subroutine to calculate CO2 fugacity
CALL fco2calc(pb,TK,fco2)
'
'calculate Henry's law constant for CO2
```

```
K = kref * EXP (-deltav*(pb-Pref)/(83.28*TK) - (deltah/1.987)*((1!/TK)-(1!/Tref)))
'
'store the pressure dependent constants
xkh2o(i)=xk
xkfco2(i)=K*fco2!
'
'start the fluid composition loop
'define activity of H2O (assume it is equal to mole fraction of H2O in fluid)
'
FOR ah2o=0 TO 1.1 STEP .1
'
XCO2fl = 1! - ah2o          'calc mole fraction CO2 in fluid assuming ideal mixing
'
'Calculate H2O in the melt
xh = SQR(ah2o/xkh2o(i))       'calc mole fraction H2O in melt using CWB model
    IF xh>.5 THEN
    xh= LOG(ah2o/(.25*xkh2o(i)))/(6.52-2667!/TK) + .5
    END IF
'
'calc wt % H2O
X=xh/(1!-xh)
Z=18.02*X/FW8
wh2om=Z*100!/(1!+Z)
'
'calc CO2 content of the melt
kf!=xkfco2(i)*XCO2fl
xco2m= kf!/(1! + kf!)
'calc wt % CO2 in melt
wco2 = 4401! * xco2m / (FWone + 44.01 * xco2m)
ppmco2m=wco2 * 10000!       'convert to ppm by weight
IF ppmco2m < .000001 THEN ppmco2m = 0!
'
PRINT #3, USING " ##.##";wh2om;
PRINT #3, tb$;
'
   FOR ip = 1 TO i-1
      PRINT #3, tb$;
   NEXT ip
'
PRINT #3,USING " #####";ppmco2m
'
NEXT ah2o
'end of fluid composition loop
NEXT i
'end of pressure loop
'
'Start constant fluid composition loop
FOR jh2o = 2 TO 8 STEP 2
ah2o=.1*jh2o
XCO2fl = 1! - ah2o                  'calc mole fraction CO2 in fluid assuming ideal mixing
'
FOR i = 1 TO npr
'Calculate H2O in the melt
xh = SQR(ah2o/xkh2o(i))       'calc mole fraction H2O in melt using CWB model
    IF xh>.5 THEN
    xh= LOG(ah2o/(.25*xkh2o(i)))/(6.52-2667!/TK) + .5
```

```
        END IF
'
'calc wt % H2O
X=xh/(1!-xh)
Z=18.02*X/FW8
wh2om=Z*100!/(1!+Z)
'
'calc CO2 content of the melt
kf!=xkfco2(i)*XCO2fl
xco2m= kf!/(1! + kf!)
'calc wt % CO2 in melt
wco2 = 4401! * xco2m / (FWone + 44.01 * xco2m)
ppmco2m=wco2 * 10000!        'convert to ppm by weight
IF ppmco2m < .000001 THEN ppmco2m = 0!
'
PRINT #3, USING " ##.##";wh2om;
PRINT #3, tb$;
'
    FOR ia = 1 TO ((jh2o/2)+npr-1)
        PRINT #3, tb$;
    NEXT ia
'
PRINT #3,USING " #####";ppmco2m
'
NEXT i
NEXT jh2o
'
INPUT "finished, hit return"; dum
'
CLOSE #3
STOP
END
'
SUB fco2calc(pb,TK,stdfco2) STATIC
'USES MRK AT P BELOW 4 KB AND SAXENA AT P ABOVE 4 KB
'CALCULATE FUGACITY COEFFICIENT FOR CO2.
'TK IS T IN KELVIN, pb IS PRESSURE IN BARS.
'
DEFSNG A-Z
 PO=4000!/1.013
 PA=pb/1.013
 IF (pb > 4000!) THEN
    'CALC FP AT T AND 4 KB.
    CALL RKCALC(TK,PO,PUREG)
    CALL SAXENA(TK,pb,XLNF)
        PUREG=PUREG+XLNF
 ELSE
    CALL RKCALC(TK,PA,PUREG)
 END IF
    'CONVERT FROM LN FUGACITY TO FUGACITY
        stdfco2=EXP(PUREG)
 END SUB
'
 SUB RKCALC(T,P,PUREG) STATIC
'LAST MODIFiED JUNE 13, 1987
'CALCULATION OF PURE GAS MRK PROPERTIES FOLLOWING
```

```
'HOLLOWAY 1981, 1987.
'P IS IN ATM, T IN KELVIN.  RETURNS LN FUGACITY.
'
DEFSNG A-Z
 R =82.05
 RR = 6732.2
 pb=1.013*P
 PBLN=LOG(pb)
 TCEL=T-273.15
 RXT=R*T
 RT=R*T^1.5*.000001
'
'CALCULATE T DEPENDENT MRK A PARAMETER CO2
 ACO2M=73.03-.0714*TCEL+2.157E-05*TCEL*TCEL
'
'DEFINE MRK B PARAMETER FOR CO2
 BSUM=29.7
 ASUM=ACO2M/(BSUM*RT)
 BSUM=P*BSUM/RXT
'CALL REDKW TO FIND PURE GAS FUGACITY COEFFICIENT.
 CALL REDKW((BSUM),(ASUM),XLNFP)
 ' CONVERT TO NATURAL LOG OF FUGACITY
 PUREG=XLNFP+PBLN
END SUB
'
SUB REDKW(BP,A2B,XLNFP) STATIC
'the RK routine
'A ROUTINE TO CALC COMPRESSIBILITY FACTOR AND FUGACITY
'COEFFICIENT WITH THE REDLICK-KWONG EQUATION FOLLOWING
'EDMISTER(1968). THIS SOLUTION FOR SUPERCRITICAL FLUID.
'WRITTEN BY JOHN R. HOLLOWAY, 1973, REVISED APRIL, 1983
'RETURNS A VALUE OF 1 FOR FP IF ARGUMENTS OUT OF RANGE.
'TRANSLATED INTO BASIC FOR MAC NOV, 1985 BY JRH.
 TH =.333333
 IF A2B < 1E-10 THEN A2B=.001
 RR=-A2B*BP*BP
 QQ=BP*(A2B-BP-1!)
 XN=QQ*TH+RR-.074074
 XM=QQ-TH
 XNN=XN*XN/4!
 XMM=XM*XM*XM/27!
 ARG=XNN+XMM
 IF ARG > 0! THEN GOTO 5
 IF ARG < 0! THEN GOTO 15
 FP=1!
 Z=1!
 EXIT SUB
15 COSPHI=SQR(-XNN/XMM)
 IF XN > 0! THEN COSPHI=-COSPHI
 TANPHI=SQR(1!-COSPHI*COSPHI)/COSPHI
 PHI=ATN(TANPHI)*TH
 FAC=2!*SQR(-XM*TH)
'SORT FOR LARGEST ROOT
 R1=COS(PHI)
 R2=COS(PHI+2.0944)
 R3=COS(PHI+4.18879)
```

```
'SORT FOR LARGEST ROOT
RH=R2
IF R1 > R2 THEN RH=R1
IF R3 > RH THEN RH=R3
Z=RH*FAC + TH
GOTO 20
5  X=SQR(ARG)
F=1!
XN2=-XN/2!
XMM=XN2+X
IF XMM < 0! THEN F=-1!
 XMM=F*((F*XMM)^TH)
F=1!
XNN=XN2-X
IF XNN < 0! THEN F=-1!
 XNN=F*((F*XNN)^TH)
Z=XMM+XNN+TH
20  ZBP=Z-BP
 IF ZBP < .000001 THEN ZBP=.000001
BPZ=1!+BP/Z
FP=Z-1!-LOG(ZBP)-A2B*LOG(BPZ)
IF FP < -37 OR FP > 37 THEN FP=.000001
XLNFP = FP
END SUB
'
 SUB SAXENA(TK,pb,XLNF) STATIC
'HIGH PRESSURE CORRESPONDING STATES ROUTINES FROM SAXENA & FEI
'(1987) GCA VOL. 51, 783-791
'RETURNS NATURAL LOG OF THE RATIO F(P)/F(4000 BAR) AS XLNF ARRAY
DEFSNG A-Z
DEFINT NFL, i, J, K
'
'Define integration limit
 PO = 4000!
'
'CRITICAL TEMPERATURES AND PRESSURES FOR CO2
    TR=TK/304.2
    PR=pb/73.9
    PC=73.9
'
'VIRIAL COEFFICIENTS
    A=2.0614 - 2.2351/TR^2 - .39411*LOG(TR)
    B=.055125/TR + .039344/TR^2
    C=-1.8935E-06/TR - 1.1092E-05/TR^2 - 2.1892E-05/TR^3
    D=5.0527E-11/TR - 6.3033E-21/TR^3
'
'CALCULATE MOLAR VOLUME
    Z=A + B*PR + C*PR^2 + D*PR^3
    V=Z*83.0117 * TK/pb
'
'INTEGRATE FROM P0 (4000 BARS) TO P TO CALCULATE ln FUGACITY
    LNF = A*LOG(pb/PO)+(B/PC)*(pb-PO)+(C/(2!*PC^2))*(pb^2-PO^2)
    LNF=LNF+(D/(3!*PC^3))*(pb^3-PO^3)
 XLNF= LNF
END SUB
```

Chapter 7

SOLUBILITIES OF SULFUR, NOBLE GASES, NITROGEN, CHLORINE, AND FLUORINE IN MAGMAS

Michael R. Carroll

Department of Geology
Wills Memorial Building
University of Bristol
Bristol, BS81RJ U.K.

James D. Webster

Department of Mineral Sciences
American Museum of Natural History
Central Park West at 79th Street
New York, New York 10024-5192 USA

INTRODUCTION

Water and carbon dioxide comprise the most abundant volatile species in most magmas, yet other volatile species are also of interest for a wide variety of reasons, ranging from their utility in studying the degassing behavior of mantle-derived magmas, to their importance in the formation of mineral deposits, to their effects on magma physical properties and phase relations. Vapor saturation of a magma occurs when the sum of the vapor pressures of individual volatile species exceeds the total pressure. Because magmas contain a number of volatile species, vapor saturation may occur even though the solubility of no single volatile species is exceeded. Once volatile saturation is reached, all volatile species will be partitioned between magma and vapor phase in proportion to their relative solubilities and abundances (e.g., Khitarov and Kadik, 1973; Holloway, 1976; Shilobreyeva et al., 1983; Gerlach, 1986; Newman et al., 1988; Anderson et al., 1989). Thus, in order to model and understand the degassing behavior of ascending magmas quantitatively, it is necessary to understand the interactions between different volatile species and the potential influence of all volatile species on magmatic processes.

In this chapter we examine what is known about the solubility behavior of sulfur, noble gases, nitrogen, and the halogens F and Cl. Studies of volcanic gases show that their sulfur dioxide abundances are generally only exceeded by those of H_2O and CO_2 (e.g., Symonds et al., 1994). Sulfur solubility in melts is of interest because many ore deposits are composed of metal sulfide minerals, and there has been considerable recent interest in the potential influence of volcanic sulfur emissions on the climate, as well as interest in use of sulfur emissions for prediction of volcanic eruptions, and investigation of magma oxidation state and elemental cycling in subduction zones. Fluorine and chlorine may occur at concentrations of several wt % in some enriched magma types, and their potential effects on melt physical properties, phase relations, and as complexing ligands for a variety of metal species in ore-forming fluids have prompted considerable study of the factors affecting halogen solubilities in melts, their partitioning between melts and aqueous fluids, and their influence on magmatic processes, especially in evolved, highly differentiated magma types. Volcanic emissions of F and Cl may influence the climate, because some F- and Cl-bearing compounds are greenhouse gases that also play a role in stratospheric ozone depletion and the generation of acid rain. The behaviors of F and Cl in magmatic systems are quite different, because F selectively partitions into the melt, while Cl favors aqueous fluids. Hence, Cl has a greater influence on properties of hydrothermal fluids and F has greatest effects on melt structures, melting temperatures, viscosities, phase equilibria, and diffusion (Wyllie and Tuttle, 1961; Manning, 1981; Dingwell, 1985; Baker and Watson, 1988). The noble gases, although they typically occur at sub-ppm concentration levels in magmas, have been the subject of extensive study because their isotopic composition and abundance patterns provide insights into the large scale degassing behavior of the earth/mantle system and the evolution of the atmosphere (e.g., Dymond and Hogan, 1973, 1978; Kurz et al., 1982a,b; Allégre et al., 1986; Fisher, 1986a,b; Staudacher et al., 1986, 1989; Sarda and Graham, 1990; see also Chapter 12). Their inert nature makes them ideal tracers of magmatic processes and their solubilities can be understood by considering the effects of atomic size on their solubility in melts. Nitrogen occurs in magmas at abundance levels similar to those of the noble gases but has been the subject of less study. Unlike the noble gases, nitrogen may occur in several different forms (e.g., N_2, ammonia, nitrates) and thus nitrogen speciation may have a large influence on its solubility.

SULFUR SOLUBILITY IN MELTS

General solubility behavior

Experiments on a variety of silicate melt compositions show that sulfur solubility behavior is closely linked with oxidation state and that sulfur dissolves primarily as sulfide (reduced) and sulfate (oxidized) species (e.g., Fincham and Richardson, 1954). The proportions of these species in a given melt are controlled by the equilibrium

$$S^{2-}_{melt} + 2O_{2gas} = SO_4^{2-} {}_{melt} \tag{1}$$

with the equilibrium constant

$$K_1 = \frac{[SO_4^{2-}]}{[S^{2-}]f_{O_2}^2} \tag{2}$$

where brackets refer to activities of the dissolved sulfur species in the melt and f_{O_2} is the fugacity of oxygen. As long as activity coefficients of the sulfur species do not vary greatly with sulfur content, the concentration ratio of sulfate/sulfide will be proportional to the square of the oxygen fugacity at fixed P and T. The data of Fincham and Richardson (1954) for simple synthetic melts at ~1500°C show dissolved sulfur occurs mainly as sulfide at oxygen fugacities below those corresponding to an extrapolation of the fayalite-magnetite-quartz buffer (FMQ). A similar conclusion may be drawn from results obtained from experiments on sulfur solubility in basaltic to rhyolitic melts over a range of T and $f(O_2)$; in general dissolved sulfate sulfur only becomes significant (>10% of total sulfur) at oxygen fugacities greater than ~1 log $f(O_2)$ unit above the FMQ buffer (e.g., Connolly and Haughton, 1972; Nagashima and Katsura, 1973; Katsura and Nagashima, 1974; Carroll and Rutherford, 1988).

Differences in the size and bonding behavior of oxidized and reduced sulfur species result in differences in solubility behavior. The solubilities of sulfide and sulfate in anhydrous systems may be described by two reactions, depending on prevailing oxidation state:

$$1/2\ S_2 + O^{2-} = 1/2\ O_2 + S^{2-} \tag{3}$$

and $$1/2\,S_2 + 3/2\,O_2 + O^{2-} = SO_4^{2-} \tag{4}$$

where S_2 and O_2 refer to gas species, S^{2-} and SO_4^{2-} refer, respectively, to sulfide and sulfate species dissolved in melt, and O^{2-} refers to oxygen anions in the melt (Fincham and Richardson, 1954). The sulfur solubility can thus be expected to depend on oxygen fugacity [$f(O_2)$], sulfur fugacity [$f(S_2)$], and the O^{2-} anion activity in the melt. Because the S^{2-} anion is significantly larger than the O^{2-} anion, it is unlikely that S can substitute for oxygens involved with tetrahedrally-bound cations (either bridging or non-bridging oxygens, which are those bound with 2 and 1 Si, respectively), and instead it is expected that sulfur will replace free oxygen species (those bound only to non-tetrahedral cations; Toop and Samis, 1962; Hess, 1977) Also, Fincham and Richardson (1954) observed that increased Al_2O_3 and SiO_2 lowered the sulfide carrying capacity of melts, further suggesting that the exchangeable oxygen in Equation (3) is associated with non tetrahedrally-bound cations.

For naturally-occurring basaltic magmas it appears that dissolution of sulfur as sulfide is strongly dependent on melt FeO content, as might be described by an equilibrium of the form

$$FeO_{melt} + 1/2\,S_2 = FeS_{melt} + 1/2\,O_2 \tag{5}$$

where S_2 and O_2 are gas species and FeS and FeO are species in the silicate melt. This equilibrium suggests a preference of S^{2-} for exchange with O^{2-} anions associated with ferrous iron in melts and is in accord with observations of strong correlation between dissolved sulfur content (at <FMQ) and melt FeO content in both experimental (Haughton et al., 1974; Katsura and Nagashima, 1974; Wendlandt, 1982; Carroll and Rutherford, 1985, 1987; Luhr, 1990) and natural systems (Mathez, 1976; Devine et al., 1984; Wallace and Carmichael, 1992).

In hydrous magmas, sulfide sulfur in the vapor phase will occur predominantly as H_2S when oxygen fugacities are below FMQ+1 and total pressures exceed a few hundred bars. The chemical similarity between H_2O and H_2S led Burnham (1979) to propose that H_2S dissolution in melts can be described by

$$H_2S_{\,gas} + O^{2-}{}_{\,melt} = HS^-{}_{\,melt} + OH^-{}_{\,melt} \tag{6}$$

This reaction does not take into account evidence for preferential association between Fe and S in both hydrous (Carroll and Rutherford, 1985) and anhydrous melts (Haughton et al., 1974; Buchanan and Nolan, 1979; Buchanan et al, 1983), and thus, in hydrous natural melts the solution mechanism for H_2S may involve reactions such as

$$H_2S_{\,gas} + FeO_{\,melt} + O^{2-}{}_{\,melt} = FeS_{\,melt} + 2\,OH^-{}_{\,melt} \tag{7}$$

or, $$H_2S_{\,gas} + FeO_{\,melt} = FeS_{\,melt} + H_2O_{\,gas} \tag{8}$$

both of which suggest a negative correlation between H_2O fugacity or melt water content and sulfur solubility, given constancy of other intensive variables ($f(S_2)$, $f(O_2)$, etc.). The available experimental data on sulfur solubility in hydrous, sulfide saturated melts are not sufficiently well constrained in terms of component activities in the melt and vapor phase to test these postulated solution mechanisms, but the solubility behavior of H_2S will clearly be of importance in hydrous magmas.

Under relatively oxidizing conditions, magmas may contain substantial amounts of sulfur as sulfate (e.g., Katsura and Nagashima, 1974; Carroll and Rutherford, 1985, 1987, 1988; Luhr, 1990). Although there is a suggestion of a positive correlation between melt Ca content and sulfate solubility (Katsura and Nagashima, 1974; Luhr, 1990), there are no solubility data with melt Ca content as the only variable. Sulfur-rich oxidized melts containing anhydrite ($CaSO_4$) as a stable, near-liquidus phase, suggest a preferential association between dissolved sulfate and Ca^{2+} in melts. Studies of sulfate solubility in simple synthetic melts also show a strong correlation between sulfate solubility and the concentration of alkali metal and alkaline earth components (e.g., Nagashima and Katsura, 1973; Popadopolous, 1973).

One major complication regarding sulfur behavior in natural systems is the saturation of magmatic liquids with a sulfur-rich condensed phase, most commonly an immiscible iron sulfide liquid (or pyrrhotite at lower T), and more rarely, crystalline $CaSO_4$ in oxidized magmas. For example, the majority of natural basaltic magmas contain trace amounts of immiscible iron sulfide, which is actually an Fe-S-O liquid containing up to ~10 wt % dissolved oxygen, the amount of which depends mainly on $f(O_2)$ and $f(S_2)$ of the system (e.g., Naldrett, 1969; Shima and Naldrett, 1975). For a given melt composition at fixed P and T, the development of an immiscible sulfide phase uniquely defines $f(O_2)$ and $f(S_2)$, as well as the activities of FeS and FeO in the coexisting sulfide and silicate melts. If the chemical potential of FeO in the silicate melt is less than that in a hypothetical sulfide liquid at given conditions of P, T, $f(O_2)$, and $f(S_2)$, then no immiscible sulfide will form and the silicate melt will remain undersaturated with respect to sulfur. The application of such ideas to sulfur behavior in magmas is discussed below in reference to experimental studies of sulfur solubility in both saturated and undersaturated silicate melts.

Experimental methods of sulfur solubility determination

Laboratory studies of sulfur solubility relations can be divided into (1) those done at atmospheric pressure, with gas fugacities controlled by exposing the samples to flowing gases of fixed composition (system open to S_2 and O_2), and (2) those done at elevated pressures, some with gas fugacities constrained by buffer assemblages, and some with gas fugacities more poorly constrained by the bulk composition of the experimental charge (generally closed or partly-closed systems). The methods used in the 1 atmosphere experiments are similar to those used for other gas-mixing experiments, except that care must be taken not to allow condensation of elemental sulfur to stop flow of gas out of the furnace. For higher pressure experiments, standard procedures are also used, but AgPd capsules must be avoided because they react to form silver sulfide. Further details of experimental techniques are given in papers dealing with experimental studies of sulfur solubility (e.g., Haughton et al., 1974; Katsura and Nagashima, 1974; Danckwerth et al., 1979; Mysen and Popp, 1980; Carroll and Rutherford, 1985, 1987; Luhr, 1990).

Almost all studies have used the electron microprobe for analysis of S in experimental run products, except for Nagashima and Katsura (1973) and Katsura and Nagashima (1974), who used wet-chemical analysis of S-undersaturated samples to determine both total S concentration and the speciation of sulfur as either sulfide or sulfate (discussed below). The sulfur contents of natural glasses have also been determined almost exclusively by electron microprobe analysis. With sufficiently long counting times (up to several minutes on peak and background) and beam currents of 25 to 40 nA it is possible to achieve lower levels of detection for sulfur of ~25 ppm, and precisions of better than 10% at the 1000 ppm concentration level.

An important distinction that must be made in discussing sulfur solubility is between solubility in melts saturated with a sulfur-rich condensed phase (e.g., pyrrhotite, FeS-rich, immiscible Fe-S-O melt, or $CaSO_4$), and in those in which sulfur is partitioned only between a sulfur-bearing vapor phase and coexisting silicate melt; we will refer to the latter as undersaturated, and the former as saturated. The response of melt sulfur content to changes in intensive and extensive variables may differ markedly in these two cases (sulfur saturated vs. undersaturated), and thus, in the following sections we will clearly differentiate between saturated and undersaturated systems in discussion of sulfur solubility behavior.

Experimental studies at atmospheric pressure

The solubility of sulfur in melts at atmospheric pressure has been studied in a variety of systems and these results help define the relative effects on solubility of $f(O_2)$, $f(S_2)$, T and melt composition (e.g., Fincham and Richardson, 1954; Richardson and Fincham, 1956; Abraham and Richardson, 1960; Abraham et al., 1960; Shima and Naldrett, 1975; Haughton et al., 1974; Katsura and Nagashima, 1974; Buchanan and Nolan, 1979. The experiments used gas mixtures of H_2-CO_2-SO_2 or CO-CO_2-SO_2, flowed through a sealed furnace tube, with the sample suspended in a constant temperature zone at the center of the furnace. The major sulfur species in the gas phase in such experiments include S_2, SO_2 and COS, with H_2S as an additional component in experiments with H_2 as one of the mixing gases. Experiments have generally been done over a range of $f(O_2)$, controlled by varying the CO/CO_2 or H_2/CO_2 ratio in a gas mixture containing a constant amount of SO_2. At high temperature and depending on the $f(O_2)$, the input SO_2 reacts with the other gas species to produce mainly S_2, COS, H_2S, and SO_3, the fugacities of which can be calculated from the mixing ratios of the inlet gases using standard thermodynamic data (e.g., *JANAF Thermochemical Tables;* Stull and Prophet, 1971). Figure 1 shows how the abundances of the different gas species vary with oxygen fugacity for the systems C-O-S (Fig. 1a), and H-C-O-S (Fig. 1b); here and elsewhere we give $f(O_2)$ relative to the FMQ buffer curve, or as ΔFMQ, which is in $\log_{10} f(O_2)$ units relative to FMQ (e.g., ΔFMQ = +1 means an $f(O_2)$ 1 log unit above FMQ). Both systems show similar behavior, with

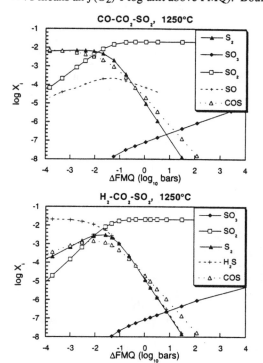

Figure 1. Calculated gas fugacities as a function of oxygen fugacity, at 1 bar total pressure and 1250°C. ΔFMQ refers to oxygen fugacity relative to the fayalite-magnetite-quartz (FMQ) equilibrium (from O'Neill, 1987); e.g., ΔFMQ = -1 corresponds to an $f(O_2)$ 1 $\log_{10}$ unit below FMQ. (a) species present in CO-CO_2-SO_2 gas mixture with constant 2.1 vol % SO_2 in the inlet gas mixture. (b) species present in H_2-CO_2-SO_2 gas mixture, 1250°C, with constant 2.1 vol % SO_2 in the inlet gas mixture.

reduced sulfur species (H_2S, S_2, COS) dominating at $f(O_2)$ below ~FMQ-1, giving way to dominantly SO_2 and minor SO_3 at oxygen fugacities above FMQ. In the C-O-S system, the fugacities of S_2 and COS are approximately constant at $f(O_2) <$ FMQ-1, but in the H-C-O-S system, H_2S is the species with nearly constant abundance at <FMQ-1, whereas S_2 and COS abundances continue to decrease as $f(O_2)$ decreases below FMQ-1.

In sulfur undersaturated melts equilibrated with a H_2-CO_2-SO_2 gas mixtures containing a constant amount of SO_2 in the inlet gas, sulfur solubility shows a strong dependence on both oxidation state and melt composition. Results from the study of Katsura and Nagashima (1974) (Fig. 2), document a minimum in S solubility between ~FMQ and FMQ+1, with sulfur solubility rising steeply towards more reducing conditions, and rising more gradually towards more oxidizing conditions. The position of the minimum in sulfur solubility at ~FMQ+1 shown in Figure 2 does not coincide with the change from H_2S-to SO_2-dominated speciation in the gas phase at ~FMQ-1.5 (Fig. 1), but rather is approximately coincident with the $f(O_2)$ where the SO_3 fugacity overtakes those of the reduced sulfur species (~FMQ+1). Decreasing Fe^{2+}/Fe^{3+} in the melt may partly explain this observation, but it may also indicate that H_2S, S_2, and possibly SO_3 are appreciably more soluble than SO_2; this highlights how sulfur speciation in the gas phase may influence the solubility in melts, because of differences in solubility behavior of different gaseous species, but it should be noted that the $f(O_2)$ at the solubility minimum will depend on the relative activity coefficients of dissolved sulfur species and need not coincide with change from H_2S to SO_2-dominated speciation in the vapor phase.

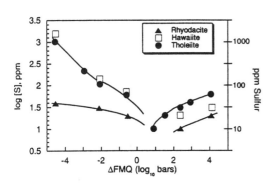

Figure 2. Sulfur solubility as a function of $f(O_2)$ in sulfur undersaturated, rhyolitic, hawaiitic, and tholeiitic melts (data from Katsura and Nagashima, 1974). All experiments done at 1250°C with constant 2.1 vol % SO_2 in the gas mixture. Melt FeO* (total iron as FeO), CaO, and SiO_2 contents are, respectively: Tholeiite: 11.75, 10.4, 50.4; Hawaiite: 13.1, 7.9, 45.8; Rhyodacite: ~ 3.0, 2.6, 66.8 (wt %). ΔFMQ as described in Figure 1. Sulfur contents at minimum (~FMQ to FMQ+1)were too low (<10 ppm S) for reliable analysis by the wet-chemical method used (Katsura and Nagashima, 1974); estimated errors for other samples are similar to symbol size.

The speciation of dissolved sulfur in quenched silicate melts as a function of $f(O_2)$ is shown in Figure 3. Katsura and Nagashima (1974) and Nagashima and Katsura (1973) determined the proportions of sulfate- and sulfide-sulfur by wet chemical analysis of experimental samples, while the other results are based on electron microprobe measurement of the wavelength shift of S K_α x-rays, which was calibrated as a function of $f(O_2)$ with experimental samples synthesized at known $f(O_2)$ by Carroll and Rutherford (1988). The data both from experimental studies (Katsura and Nagashima, 1974; Nagashima and Katsura, 1973; Carroll and Rutherford, 1988) and from natural samples (Nilsson and Peach, 1993; Wallace and Carmichael, 1994) indicate a large increase in the proportion of sulfur dissolved as sulfate as $f(O_2)$ increases from FMQ to FMQ+2; the data for simple sodium-trisilicate melt (Nagashima and Katsura, 1973) show a much greater proportion of sulfate at a given $f(O_2)$ than the data for natural melt compositions. The $f(O_2)$ of the natural samples was determined from measurement of Fe^{3+}/Fe^{2+}, which can be related to $f(O_2)$ via the experimental work of Kress and Carmichael (1991); there is

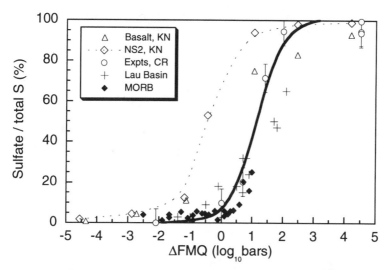

Figure 3. Percent sulfur present as sulfate as a function of oxygen fugacity, relative to the FMQ buffer. Open circles with error bars from experimental data of Carroll and Rutherford (1988), for hydrous andesitic to dacitic melts synthesized at 1 to 2.9 kbar, 920° to 1025°C, with sulfur speciation estimated from measured wavelength shift of S $K\alpha$ x-rays. Open triangles from wet chemical measurements of Katsura and Nagashima (1974) on tholeiitic basalt synthesized at 1 atm, 1250°C and known $f(O_2)$; points at high ΔFMQ have large associated errors (up to ±30%) and averages of possible range are shown. Open diamonds from wet chemical measurements of Nagashima and Katsura (1973) on $Na_2O \cdot 3SiO_2$ melts synthesized at 1 atm and controlled $f(O_2)$, $f(S_2)$; the higher sulfate content of these melts at given $f(O_2)$ is likely due to their alkali-rich composition. Crosses and filled diamonds from Nilsson and Peach (1993) and Wallace and Carmichael (1994), respectively, corresponding to submarine basaltic to andesitic glasses with $f(O_2)$ determined from measured Fe^{3+}/Fe^{2+} and sulfur speciation determined by S $K\alpha$ wavelength shifts.

overall good agreement between the natural sample data, with $f(O_2)$ determined independently, and the experimental data of Carroll and Rutherford (1988), with perhaps a slight tendency for the natural samples to indicate a smaller proportion of oxidized sulfur at a given $f(O_2)$. The data shown in Figure 3 comprise results for melt compositions ranging from dry MORB basalts to hydrous back-arc basalts and experimentally-synthesized dry basalt to hydrous andesitic to dacitic melts. This small variability in speciation over a large composition range in consistent with T, P, and magma water content being of secondary importance relative to oxidation state in determining the speciation of magmatic sulfur.

The effect of melt composition on total sulfur solubility, in contrast, is significant and as shown in Figure 2, lower solubilities are generally associated with more SiO_2-rich melt composition (compare rhyodacite, tholeiite in Fig. 2). Under reducing conditions S solubility (as sulfide) is positively correlated with melt FeO content, as illustrated for sulfide-saturated melts in Figure 4. The data in Figure 4 include sulfide-saturated basaltic melts from the experimental work of Haughton et al. (1974), as well as natural, sulfide-saturated submarine basalts analyzed by Mathez (1976) and Wallace and Carmichael (1992). The results for the experimental and natural samples form sub-parallel arrays with positive correlations between sulfur solubility and melt FeO content. The lack of exact correspondence between the experimental sulfur solubility values and those in natural melts may be a result of the fact that the experimental data are isothermal (1200°C), but polybaric in $f(O_2)$ and $f(S_2)$, while the natural samples are likely polythermal and polybaric in $f(O_2)$ and $f(S_2)$. The combined effects of these variables on sulfur solubility are complicated and must be considered if we wish to understand the observed variations in magma sulfur contents; further discussion is deferred to a later section.

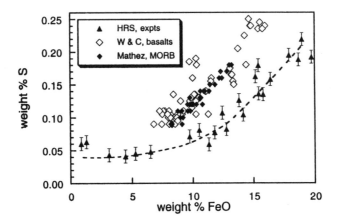

Figure 4. Sulfur versus total iron as FeO in sulfide-saturated basaltic to basaltic andesite melts. Triangles correspond to submarine basaltic glasses (from Mathez, 1976) and indicate an approximately linear relation between FeO* and S content. More recent data for MORB glasses (diamonds – Wallace and Carmichael, 1992) are essentially coincident with the Mathez (1976) data, with a slightly greater degree of scatter. Triangles with error bars (±100 ppm S – also applicable to natural sample data) are from 1 atm, 1200°C, sulfide saturated experiments of Haughton et al. (1974) on a basaltic starting composition equilibrated with CO-CO$_2$-SO$_2$ gas mixtures at ΔFMQ = -0.5 to -3.6 and log $f(S_2)$ = -0.82 to -2.64. Changes in melt S and FeO* contents result from changes in $f(O_2)$ and $f(S_2)$, which affect compositions of the coexisting sulfide and silicate melts (discussed in text).

The effect of temperature on S-solubility in undersaturated tholeiitic melt is shown in Figure 5. At a given ΔFMQ value, the S-solubility is approximately independent of T. At constant absolute value of $f(O_2)$, however, the solubility will increase with increasing T under reduced conditions (<FMQ) and decrease with increasing T under oxidizing conditions (because of the positive $\Delta f(O_2)/\Delta T$ slope of the FMQ buffer reaction). For example, if log $f(O_2)$ = -8.5 (ΔFMQ = -0.09) at 1200°C, raising T to 1400°C is analogous to lowering $f(O_2)$ relative to FMQ (to ΔFMQ = -2.11) and thus S solubility increases (as shown in Fig. 2). Figure 6 shows how S solubility in undersaturated melts decreases as the fugacities of sulfur species in the gas phase decrease, as would be expected from consideration of a simple exchange reaction between S in the gas phase and dissolved S. These data illustrate the major factors affecting S solubility in undersaturated melts but they give few insights into the behavior of S in saturated melts, which appear to be the norm in nature (e.g., Mathez, 1976; Czamanske and Moore, 1977; Luhr et al., 1984; Whitney, 1985; Wallace and Carmichael, 1992).

Changes in sulfur solubility in S-saturated melts with changes in $f(O_2)$ and $f(S_2)$ are complex but the experimental data can be easily understood if it is remembered that the experiments are open systems with respect to sulfur, and consequently changes in $f(S_2)$ may cause changes in melt FeO content, due to formation of immiscible sulfide melt. For example, Figure 7 shows the experimental results of Danckwerth et al. (1979) on sulfur solubility in a lunar basaltic composition as a function of $f(S_2)$ at constant temperature (1250°C) and log $f(O_2)$ = FMQ-3.7±0.2. The FeO content of the experimental melts, shown next to the symbols, are variable because of precipitation of varying proportions of

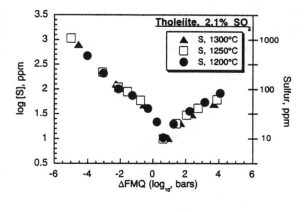

Figure 5. Sulfur solubility (log scale) in sulfur-undersaturated tholeiite melt, 1200° to 1300°C vs. $f(O_2)$ relative to the FMQ buffer (Katsura and Nagashima, 1974). Plotted against absolute $f(O_2)$, data show that at constant $f(O_2)$, the S solubility increases with increasing T under reduced conditions (<FMQ), while under oxidized conditions the solubility decreases with increasing temperature (>FMQ). When referenced relative to the FMQ buffer this distinction disappears, except perhaps for a very slight negative temperature-dependence under oxidizing conditions; errors are not clearly described in original work but are believed to be similar to size of symbols.

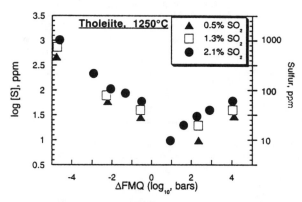

Figure 6. Sulfur solubility (log scale) in S-undersaturated tholeiitic basalt at 1250°C as a function of $f(O_2)$ for different amounts of SO_2 in the gas mixture let into the furnace (data from Katsura and Nagashima, 1974). Higher S contents with higher amounts of SO_2 in the inlet gas reflect higher fugacities of S-species (S_2, COS, H_2S, SO_2) in the high temperature gas phase with which the sample equilibrated.

immiscible sulfide with different FeO content, depending on $f(S_2)$; some variability in melt FeO also results from use of Pt-Fe alloy wire loops (sample holders) with different compositions. Sulfide saturated melts (square symbols) define the sulfide saturation surface shown by a heavy line with negative $\Delta S/\Delta f(S_2)$ slope. The dashed lines with positive slope in Figure 7 are approximate isopleths for melt FeO content and are constructed using the results of sulfur undersaturated experiments (oval symbols). The major influence of increasing $f(S_2)$ is to induce saturation with an immiscible Fe-S-O melt, which, if the $f(S_2)$ is sufficiently high, can cause depletion of FeO in the melt. The negative slope of the sulfide saturation surface implies that increasing $f(S_2)$ will lead to a decrease in sulfur solubility, associated with decreased melt FeO content. Consider a melt with 15 wt % FeO at $f(S_2) = -3$; as $f(S_2)$ is increased with $f(O_2)$ held constant the S content of the melt will increase, following the schematic 15% FeO* isopleth in Figure 7. At approximately log $f(S_2) = -2.4$ and 3200 ppm dissolved sulfur the melt will become saturated with an immiscible iron sulfide phase containing a small amount of dissolved oxygen. Increasing $f(S_2)$ further will decrease the FeO activity in the sulfide phase, and will lead to removal of FeO from the silicate melt to form more sulfide (e.g., see Naldrett, 1969; Shima and Naldrett, 1975). The removal of FeO from the silicate results in lower sulfur solubility, and thus, sulfur contents of sulfide saturated melts decrease with increasing $f(S_2)$ if all other intensive parameters are held constant. In natural systems the $f(S_2)$ is not likely to be externally controlled and there is a limited amount of sulfur available, thus significant changes in melt FeO content should not occur due to sulfide segregation.

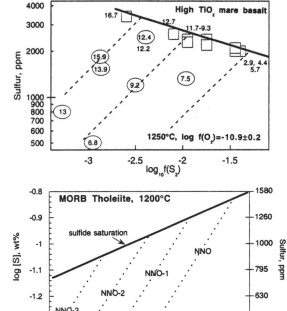

Figure 7. Sulfur solubility (log scale) in basaltic melts vs. sulfur fugacity at fixed $f(O_2)$ and temperature (data from Danckwerth et al., 1979). Boxes indicate sulfide saturated melts, ellipses indicate sulfur undersaturated samples; numbers near/inside symbols give total Fe as FeO (wt % FeO*) in the experimental melts. Heavy line shows sulfide saturation surface, and dashed lines are schematic isopleths of melt FeO* (5, 10, 15 wt%); see text for discussion

Figure 8. Sulfur solubility vs. $f(S_2)$ for basaltic melt of constant composition (9.55 wt % FeO*) at 1200°C (redrawn after Wallace and Carmichael, 1992, Fig. 6); NNO = FMQ+0.68 at this temperature. Heavy solid line is sulfide saturation surface, dashed lines are $f(O_2)$ isobars. Construction of diagram is described in Wallace and Carmichael (1992) and is based on their multiple regression of experimental sulfur solubility data, to yield sulfur solubility as a function of T, $f(O_2)$, $f(S_2)$ and melt composition.

The sulfide saturation surface is constrained by the equilibrium between sulfide melt and silicate melt (i.e., 'FeS' activity is ~1 at saturation) and is calculated from thermodynamic data (Robie et al., 1979) for the reaction $1/2 \, S_{2 \, gas} + FeO_{silicate \, melt} = FeS_{sulfide \, melt} + 1/2 \, O_{2 \, gas}$ and the model of Snyder and Carmichael (1992) for FeO activity in silicate melt.

An alternative way of looking at the interdependence of sulfur solubility, $f(O_2)$, and $f(S_2)$ is to consider the case of invariant melt composition (except for Fe^{3+}/Fe^{2+}) and examine how the sulfide saturation surface depends on $f(O_2)$ and $f(S_2)$ (e.g., Shima and Naldrett, 1975; Wallace and Carmichael, 1992). Figure 8 illustrates the important relationships between these variables for a basaltic melt (9.55 wt % FeO*) at 1200°C. The sulfide saturation surface in this case has a positive slope, and log $f(O_2)$ isobars are shown as dashed lines, also with positive slope. Increasing $f(S_2)$ for a sulfur undersaturated melt will cause the sulfur content to increase until saturation is achieved, while oxidation at constant $f(S_2)$ will cause S contents to decrease. Melt Fe^{3+}/Fe^{2+} is likely to provide the major influence on $f(O_2)$ in natural systems, and thus oxidation of sulfide-saturated magma at constant T and melt composition will result in increased $f(S_2)$ and some immiscible sulfide will dissolve in order to increase the sulfur content of the silicate melt. In natural systems the situation is likely more complex because fractionation during cooling may change melt composition and Fe^{3+}/Fe^{2+}, which in turn may influence $f(S_2)$ in sulfur saturated magmas; additional details of such processes are discussed in Wallace and Carmichael (1992, 1994).

High pressure sulfur solubility experiments

Experimental determinations of sulfur solubility at pressures greater than 1 bar provide information on the effects of P on sulfur solubility in magmas. The fugacities of gaseous species are not as easily controlled in such experiments, but the results provide useful information concerning sulfur solubility in magmas at depth in the earth.

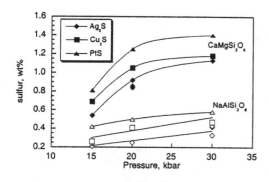

Figure 9. Sulfur solubility in diopside (filled symbols) and albite (open symbols) composition melts as a function of pressure with three different sulfur fugacity buffers (from Mysen and Popp, 1980). Diamonds = Ag-Ag$_2$S, squares = Cu-Cu$_2$S, triangles = Pt-PtS; sulfur fugacities increase in the order Ag<Cu<Pt. All experiments at 1650°C, except for two on Alb, 30 kbar, Ag-Ag$_2$S buffer which were at 1450°C and 1550°C —these show ~0.1 wt % decrease in sulfur solubility from 1450° to 1650°C. Typical 1σ errors on melt sulfur contents are ±0.01 to ±0.05 wt% (similar to symbol size).

Mysen and Popp (1980) determined the solubility of sulfur in melts of CaMgSi$_2$O$_6$ (Di) and NaAlSi$_3$O$_8$ (Alb) compositions at 15 to 30 kbar, 1650°C in the presence of graphite and a C-O-S fluid phase, with f(S$_2$) buffered by Cu-Cu$_2$S, Ag-Ag$_2$S, or Pt-PtS. Their results, shown in Figure 9, indicate that sulfur solubility increases with increasing P, and that sulfur solubility in Di melt is significantly higher than in Alb melt. This suggests that sulfur reacts preferentially with non-bridging oxygens in these Fe-free melts. Most of the sulfur should have been present as sulfide in these experiments, but speciation was not determined directly. The exact relationship between solubility and partial pressure of sulfur-bearing gas species can not be determined because the effect of P on the f(S$_2$) of the buffers used in the experiments is not known (and some of the buffers would have been liquid at the conditions of the experiments).

The high P solubility of sulfur in several sulfide-saturated natural melts has been determined by Wendlandt (1982). These experiments, also done in graphite capsules, involved equilibrating samples consisting of 60% basaltic to andesitic melt and 40 wt % sulfide with a C-O-S fluid phase at 10 to 30 kbar, 1300° to 1450°C. The experiments showed that the sulfur content of sulfide saturated melts decreased with increasing P (Fig. 10 a) and increased with increasing T (Fig 10b). The reason for the difference in P effect on S solubility in these experiments compared with those of Mysen and Popp (1980) is not clear. The relative f(O$_2$) of the C-CO-CO$_2$ equilibrium will increase with increasing P (FMQ-1.96 at 10 kbar, to FMQ-1.48 at 30 kbar; Ulmer and Luth, 1991) but it is less easy to estimate how much the sulfur fugacity varies in these experiments. Similarly, the positive correlation between S-solubility and T shown in Figure 10b could be in part a result of the T dependence of the C-CO-CO$_2$ equilibrium, which results in a reduction relative to FMQ when T increases (at 20 kbar, from FMQ-1.4 at 1300°C to FMQ-1.7 at 1460°C); how sulfur fugacity varies with T in these experiments is not well constrained. These uncertainties in the relative changes in f(S$_2$) with P and T make application of high P experimental data to natural systems difficult and further work is needed on this topic.

Several recent studies have investigated sulfur solubility in hydrous andesitic to dacitic melts in equilibrium with an H-O-S composition fluid phase (Carroll and Rutherford, 1985, 1987; Luhr, 1990). The fugacities of the different sulfur species in these experiments are difficult to define with high accuracy but the data are useful for defining the limits of sulfur solubility in hydrous, sulfur-saturated magmas at crustal pressures. In an H-O-S vapor phase, the H$_2$S/SO$_2$ ratio increases with increasing P at fixed ΔFMQ, thus high pressure magmatic gases associated with hydrous magmas will be more H$_2$S-rich than low P gases at similar T and f(O$_2$) (see Chapter 1). A significant result of these studies is that the mineral anhydrite (CaSO$_4$) may crystallize from S-rich magmas at f(O$_2$) above

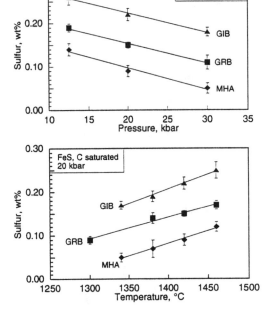

Figure 10. Effects of (a) pressure, and (b) temperature on sulfur solubility in three sulfide-saturated silicate melt compositions (from Wendlandt, 1982). All experiments run in graphite capsules, sealed in Pt, and thus equilibrated with a C-O-S gas phase with $f(O_2)$ near the graphite surface. Starting compositions include MHA = Mt. Hood Andesite (60.8 wt % SiO$_2$, 5.4 wt % FeO*), GIB = Goose Island basalt (45.9 wt % SiO$_2$, 14 wt % FeO*), GRB = Grande Ronde basalt (54.4 wt % SiO$_2$, 11.1 wt % FeO*). All experiments have less total Fe in the silicate melt than was present in the glass starting material because of Fe transfer from silicate to coexisting sulfide melt; see text for discussion.

~FMQ+2.2; the anhydrite may either coexist with or replace Fe-sulfide as the major sulfur-bearing condensed phase. The exact conditions for stabilizing anhydrite are only known to within ~1 log $f(O_2)$ unit because of the lack of suitable $f(O_2)$ buffers for use at oxidation states above FMQ+1. The experimental data on anhydrite stability are in good agreement, however, with values of $f(O_2)$ estimated from natural anhydrite-bearing magmas erupted at El Chichon (Luhr et al., 1984), Mount Pinatubo (Rutherford, 1991) and Lascar, Chile (Mathews et al., 1994). Figure 11 shows how sulfur solubility in the El Chichon trachyandesite varies with T at constant P in Fe$_3$O$_4$-Fe$_2$O$_3$ (MH; ≈FMQ+4.5) and Ni-NiO (NNO; ≈FMQ+0.7) buffered experiments; the MH buffered melts are anhydrite-saturated, and the NNO-buffered samples are saturated with liquid or crystalline Fe-sulfide. At near-liquidus temperatures, the sulfur solubility in anhydrite-saturated melts is up to 5-fold greater than in the sulfide-saturated experiments, but this difference decreases with decreasing T, along with the magnitude of the S solubilities, until at 800° to 850°C the solu-

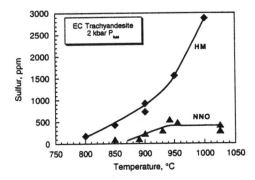

Figure 11. Sulfur solubility versus T (°C) at constant P of 2 kbar for MH- and NNO-buffered experiments on El Chichon trachyandesite starting composition (data from Luhr, 1990; Carroll and Rutherford, 1987). All samples coexist with H-O-S composition fluid phase but MH-buffered experiments are anhydrite-saturated while NNO-buffered experiments contain either pyrrhotite or an Fe-S-O immiscible melt. Only the highest T experiments are above the liquidus; melts at 800° to 850°C are rhyodacitic.

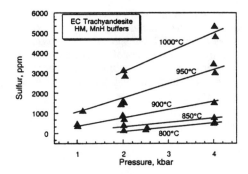

Figure 12. Sulfur contents as functions of pressure for anhydrite-saturated experiments on El Chichon trachyandesite starting material (from Luhr, 1990; Carroll and Rutherford, 1987). Experiments with Fe_3O_4-Fe_2O_3 (MH) and MnO-Mn_3O_4 (MnH) buffers are not distinguished, but for highest T pairs of points (2 and 4 kbar, 1000°C and 950°C), the slightly higher S-contents occur in the MnH-buffered experiments.

bilities are almost indistinguishable at a few 100 ppm S. Figure 12 shows the variation in sulfur solubility with P for various isotherms in anhydrite-saturated melts (El Chichon trachyandesite) equilibrated with an H-O-S fluid phase at 1 to 4 kbar total pressure. In contrast with the sulfide-saturated melts (NNO, Fig 11) which show little variation in S-solubility with T, the anhydrite saturated melts (MH, Fig. 11, Fig. 12) show a strong positive correlation of S-solubility with T, and also with P at T ≥ 900°C (Fig. 12). In general the sulfur solubilities in hydrous, sulfide saturated melts differ little from the results of Haughton et al. (1974) and melt FeO content appears to be the most important influence on sulfur solubility (Carroll and Rutherford, 1985, 1987; Luhr, 1990). There appears to be a more pronounced positive effect of P on S-solubility in anhydrite saturated melts, especially at near-liquidus T, but the exact relations between fugacities of S-species and S-solubility are unknown because the gas fugacities are insufficiently constrained (Carroll and Rutherford, 1985, 1987; Luhr, 1990). Because melt composition is also changing with T in these experiments it is impossible to be more quantitative concerning T effects on S solubility, although the trends observed should be applicable to natural magmas.

Sulfur behavior in natural systems

The behavior of sulfur in natural systems can be conveniently divided into (1) those systems in which the predominant form of sulfur is sulfide, which includes mainly water-poor basaltic to ultramafic magmas, and (2) more oxidized systems in which oxidized sulfur comprises a significant portion of the total sulfur, such as many subduction-related and back-arc magmatic systems. Excellent overviews of sulfur behavior in sulfide-dominated systems have been recently presented by Poulson and Ohmoto (1990) and Wallace and Carmichael (1992). The most important factors affecting sulfur solubility in such systems are melt FeO content and oxidation state, with the highest sulfur solubilities observed in the most FeO-rich melt compositions. The sulfur contents of primitive mid-ocean ridge magmas do not appear to vary greatly (~800 to 1000 ppm S) and most are sulfide saturated. Differentiation leading to Fe-enrichment is associated with increased sulfur concentrations in ferrobasalts (up to ~2500 ppm S), until melt Fe contents start to decrease following magnetite saturation in evolved ferrobasaltic compositions (and sulfur decreases in parallel with melt FeO content). Less is known regarding sulfur behavior in subduction-related magmas but oxidized (≥FMQ+1) back-arc basalts may have a significant proportion of their sulfur as sulfate and the most oxidized and evolved samples can be very sulfur-poor (lacking sulfide and some with <200 ppm S in melt), presumably due to transfer of sulfur to a magmatic vapor phase, the sulfur content of which will increase with increased oxidation state (Carroll and Rutherford, 1987; Nilsson and Peach, 1993; Wallace and Carmichael, 1994). The recent eruptions of anhydrite-bearing magmas at El Chichon, Mount Pinatubo, and Lascar all involved relatively S-rich magmas (0.1 to

>1wt % sulfur in whole-rock samples) and all were accompanied by large emissions of SO_2 gas. The glass phase in the eruptive products, and in subaerial magmas in general, is typically highly degassed (<100 ppm S). The high sulfur solubilities in oxidized magmas, especially at T > 900°C, makes them potentially efficient transporters of sulfur from mantle or lower-crustal depths to near-surface volcanic environments. The nature of the S-bearing gas phase associated with such magmas is discussed in Chapter 1, and pre-eruptive magmatic sulfur contents are discussed further in Chapter 8.

NOBLE GAS SOLUBILITIES IN SILICATE MELTS

Experimental measurements

Two major approaches have been taken to determine noble gas solubilities in melts. For atmospheric pressure experiments, samples are equilibrated at high T with a gas phase of known composition and the quenched samples are analyzed by mass spectrometry (e.g., Kirsten, 1968; Hayatsu and Waboso, 1985; Jambon et al., 1986; Lux, 1987; Broadhurst et al., 1990, 1992). Results from such experiments are available for He through Xe in a variety of melt compositions and quoted analytical precisions are typically ±10 to ±20% relative or better. For higher pressure experiments, samples may be equilibrated with the gas of interest, used as the pressure medium or contained within a sealed capsule, and the run products can be analyzed by microbeam or bulk analysis methods (e.g., Faile and Roy, 1966; Shelby, 1976; White et al., 1989; Carroll and Stolper, 1991, 1993; Roselieb et al., 1992; Carroll et al., 1993). Accuracy of these results is also typically within ±10 to ±20% relative and there are published data for Ar in a wide range of melt compositions, and a smaller number of results for He, Ne, Kr, and Xe in simple synthetic and natural melt compositions.

General solubility behavior

The available data show that inert gas solubility in melts and glasses is a strong function of the melt composition and the size of the gas atom, an approximately linear function of P between ~10^{-3} and 10^3 bar gas pressure (i.e., Henry's Law behavior), and almost independent of T (exceptions are SiO_2 glass at low T (e.g., Shelby, 1974; Carroll and Stolper, 1991), and Ar in $K_2Si_4O_9$ melt at high P (White et al., 1989)). At gas fugacities greater than 500-1000 bars few data exist, but reported solubilities are less than would be predicted by Henrian extrapolation of low P data (Shelby, 1976; White et al., 1989; Carroll and Stolper, 1991, 1993; Roselieb et al., 1992). Solubilities may vary 10- to 100-fold between SiO_2 glass and natural basaltic compositions (e.g., Lux, 1987; White et al., 1989; Carroll and Stolper, 1991, 1993) and the smaller gas atoms show the highest solubilities in a given silicate composition (Kirsten, 1968; Hayatsu and Waboso, 1985; Jambon et al., 1986; Lux, 1987; Broadhurst et al., 1990, 1992; Roselieb et al., 1992).

Melt composition and noble gas solubility

Noble gas solubilities are generally higher in more silica-rich melts, and the solubility decreases with increasing size of the gas atom in a given melt (Doremus, 1966, 1973; Kirsten, 1968; Studt et al., 1970; Perkins and Begeal, 1971; Shackelford et al., 1972; Shelby, 1976; Hayatsu and Waboso, 1985; Jambon et al., 1986; Lux, 1987; Broadhurst et al., 1992; Roselieb et al., 1992). Solubilities of He through Xe in rhyolitic, basaltic, and komatiitic melt compositions, calculated from melt ionic porosities as described below, are shown as a function of gas atom radius in Figure 13. The calculated Xe solubilities differ by almost 100-fold between komatiitic and rhyolitic melt, while He solubilities vary approximately three-fold over the same compositional range. Such large solubility variations with melt composition have lead to treatment of noble gas solubility from a mainly physical point of view in which the size of the gas atom and the openness of the silicate melt structure are the major influences on solubility. Gas dissolution is considered to involve an interstitial mechanism with gas atoms occurring, for example, in the centers of rings of interconnected silicate tetrahedra (Doremus, 1973; Shelby, 1974; Shackelford, 1982). Such interstitial locations for dissolved gas atoms are referred to as "holes" or "solubility sites". Lower solubility of the larger gas atoms is consistent with the existence of a continuum of hole sizes, with a greater number of small holes in which atoms like He and Ne may be accommodated (Doremus, 1973; Shelby, 1974; Shackelford, 1982). Increasing content of network-forming silica or alumina tends to increase the openness of the melt structure, and thus increase the concentration of holes that can be occupied by gas atoms. Incorporation of network-modifying cations in highly polymerized compositions leads to a contraction of the open framework structure, and occupation of space in the structure that might otherwise be populated by a gas atom; both effects reduce the number of potential sites for inert gas atoms and reduce noble gas solubility.

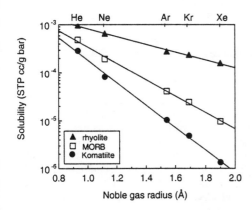

Figure 13. Solubilities of He through Xe in vapor saturated melts (1200° to 1400°C, 1 bar gas pressure) as a function of gas atom radius (Pauling, 1927); calculated from ionic porosity relations shown in Figure 15, using regression parameters given in Table 1. Relation between ionic porosity and noble gas solubility are discussed in text; see also Carroll and Stolper, 1993 for calculation of ionic porosity).

The large variations in noble gas solubility with melt composition (Fig. 13) require a means of quantifying the compositional dependence of solubility over the wide range of melt compositions encountered in nature. Previous studies of inert gas solubility in different melt compositions have shown that solubility variations may be correlated with melt density (Lux, 1987; White et al., 1989) and with melt molar volume (Broadhurst et al., 1990, 1992). Recently, Carroll and Stolper (1993) have shown that noble gas solubility also correlates well with the melt ionic porosity (IP; see also Dowty, 1980; Fortier and Giletti, 1989), defined as

$$IP = 100*[1 - (V_{ca}/V_m)] \qquad (9)$$

where V_{ca} is the total volume of spherical cations plus anions in one gram of melt, calculated using Shannon and Prewitt (1969) ionic radii, and V_m is the melt specific volume (cm^3/g), calculated using the oxide partial molar volumes of Lange and Carmichael (1987). The small to negligible T dependence of noble gas solubility (discussed below) allows T effects on solubility to be ignored and we have calculated melt volumes at 1200° to 1400°C, with most at 1300°C (mainly because this is near the T of many experimental solubility determinations). If the noble gases dissolve by an interstitial mechanism in amorphous silicates then it is plausible to suppose that the solubility of each of the noble gases would be related to the availability of space within a given melt composition; the ionic porosity is one measure of such space and it has previously been shown to be useful as a predictor of diffusivities in minerals (Dowty, 1980; Fortier and Giletti, 1989).

Table 1.

Linear least-squares regressions of noble gas-solubility vs. ionic porosity

for: $ln \ [X_i] = m*IP + b$

	slope (m)	intercept (b)
He	0.344 (±0.046)	-29.719 (±2.222)
Ne	0.608 (±0.126)	-43.118 (±5.988)
Ar	0.796 (±0.025)	-53.352 (±1.235)
Kr	0.855 (±0.071)	-56.573 (±3.437)
Xe	1.125 (±0.059)	-70.303 (±2.838)

Fits to data shown in Figure 15. Correlation coefficients (r^2) are:
  He (0.948), Ne (0.796), Ar (0.980), Kr (0.954), Xe (0.989)

X_i are mole fractions dissolved gas, calculated using oxide components for the melts. Ionic porosities calculated as described in and by Carroll and Stolper (1993).

Figure 14 shows variations in the logarithm of the 1 bar Ar solubilities for a wide range of silicate melt compositions with ionic porosity, melt density, and molar volume (see legend for data sources, calculation methods). Figure 15 extends this comparison of ionic porosity and rare gas solubility to the other stable rare gases, and best-fit slope and intercept values for the IP-solubility correlations are given in Table 1. Most published measurements of noble gas solubilities in multicomponent melts are included in Figures 14 and 15; excluded results, mainly for simple $CaO-MgO-Al_2O_3-SiO_2$ (CMAS) melts, are noted in the figure legend. We do not understand why the results for CMAS melts do not agree with the other data, but this does not diminish the utility of the correlations shown in Figure 15 for predicting the effects of melt composition on noble gas solubilities in natural systems. The results in Figure 15 show an increase in slope (Δsolubility/ΔIP) from the lighter, smaller rare gases to the heavier, larger ones, implying that the sensitivity of rare gas solubility to melt composition increases as the size of the gas atom increases. This suggests that the space available for the larger rare gases (as measured by ionic porosity) decreases more rapidly with overall void space in the melt than does the space available for the smaller rare gases (Carroll and Stolper, 1993). These results support treatment of noble gas solubility behavior from a largely physical point of view, characterized by systematic relations between the gas atom size and solubility. Average differences between measured and calculated solubilities are small (He, ±3%; Ne, ±11%; Ar, ±9%; Kr, ±15%; and Xe, ±6%), supporting the utility of the ionic porosity for modelling noble gas solubility variations in natural melt compositions.

Alternative treatment of melt composition effects on solubility. The compositional dependence of helium solubility in a wide range of silicate melts and glasses has been successfully modelled by Jambon (1987) and Chennaoui-Aoudjehane and Jambon (1990) using an empirical approach which treats the logarithm of the gas solubility as a linear combination of the mole fractions of the oxide components, i.e., $ln \ X_{He} = \Sigma d_i X_i$, where X_{He} is the solubility (molar) of He at 1 bar gas pressure, the X_i are the mole

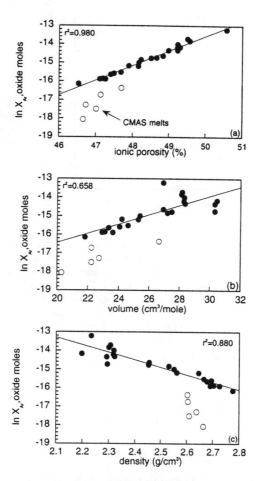

Figure 14. Comparison of correlations between the natural logarithm of mole fraction (oxide basis) dissolved Ar at one bar gas pressure (1000° to 1400°C) and (a) ionic porosity, (b) melt molar volume, and (c) melt density. All trends are approximately linear, but ionic porosity model yields closest corespondence between measured and calculated solubilities, especially for more silica-rich compositions. Ionic porosity calculations, described in detail by Carroll and Stolper (1993), utilize melt densities and molar volumes (at 1200° to 1400°C, depending on T range of solubility data) calculated using data of Lange and Carmichael (1987), and ionic volumes caluated assuming spherical geometries and ionic radii from Shannon and Prewitt (1969). The ionic porosity, expressed as a percentage, corresponds to the difference between the bulk melt specific volume (V_m, in cm^3/g), and that which would be calculated from the ionic volumes of the constituent anions and cations (V_{ca}, in cm^3/g); when these quantities are equal the ionic porosity is zero, but in general the macroscopic specific volume is greater than the summed volumes of constituent species, and this difference is taken to be the ionic porosity. Results for simple $CaO-MgO-Al_2O_3-SiO_2$ melts (two from White et al., 1989; three from Broadhurst et al., 1990), shown by open symbols, are excluded from plotted linear regressions. Poor results for these samples may indicate an inability to calculate their densities and ionic porosities correctly, experimental or analytical errors, or limitations to the hypothesis that ionic porosity is a good predictor of noble gas solubility. Data sources are Hayatsu and Waboso (1985), Jambon et al. (1986), Lux (1987), White et al. (1989), Broadhurst et al. (1990), Carroll and Stolper (1991, 1993), and Roselieb et al. (1992). Thermodynamic model described in Carroll and Stolper (1991) was used to extract 1 bar solubilities from high pressure data of White et al. (1989), Carroll and Stolper (1991, 1993), and Roselieb et al. (1992).

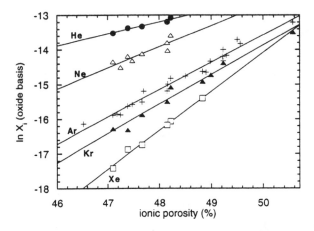

Figure 15. Noble gas solubilities, expressed as mole fraction bar^{-1} (mole fractions on oxide basis), as a function of melt ionic porosity in melts ranging from ~40 to 100 wt % SiO_2 at 1000° to 1400°C. Linear regression lines are shown for each gas species; linear regression parameters are in Table 1, data sources the same as Figure 14.

fractions of individual oxide components in the melt, and the d_i are constant coefficients for each oxide, determined by multiple linear regression of experimental solubility data. The functional form of this relationship derives from consideration of Henry's law constants in multicomponent solvents (Prausnitz, 1969; p. 372-375) assuming the multicomponent solvent is an ideal mixture. This approach works well for prediction of He and Ar solubilities (including CMAS melts; Carroll, 1993, unpublished results) because there are many experimental data for a wide range of melt and glass compositions, but there are only limited data available for Ne, Kr, and Xe solubilities, and thus it is difficult to calculate well-constrained d_i values for these gases.

Pressure and temperature dependence of noble gas solubility

With data over a sufficient range of P and T, the solubility of a given noble gas may be modelled by fitting measured solubilities to the following simple thermodynamic expression:

$$ln\,(a_i/f_i) = ln\,(a_0/f_0) - V°_i\,(P - P_0)/RT - \Delta H° \,(1/T - 1/T_0)/R \qquad (10)$$

where a_i is the activity of gas species i in vapor saturated melt at pressure P (bars) and temperature T (Kelvins), f_i is the fugacity of gas i at P and T, a_0 and f_0 refer to the activity and fugacity of gas i at reference conditions P_0 and T_0, R is the gas constant, with appropriate units, $V°_i$ is the partial molar volume of gas i in the melt in its standard state, and $\Delta H°$ is the molar enthalpy of solution of gas i in the melt at low P (assuming constant $V°_i$ and $\Delta H°$). This approach has been used to model water and carbon dioxide solubilities in melts (e.g., Fine and Stolper, 1985; Silver and Stolper, 1989; Fogel and Rutherford, 1990; Silver et al., 1990; Pan et al., 1991) as well as noble gas solubilities (e.g., White et al., 1989; Carroll and Stolper, 1991, 1993; Carroll et al., 1993); the reader is referred to those references for more detailed descriptions (see also Chapter 6).

In melts, the effect of temperature on noble gas solubility is small and often near the resolution of experimental data (e.g., Jambon et al., 1986; Lux, 1987; White et al., 1989), and thus such T effects are of limited importance for magmas. In contrast, in glasses, there is experimental evidence that solubility and temperature are negatively correlated and large solubility variations have been observed at temperatures of ~100° to 700°C (e.g., Shelby, 1974; Shackelford et al., 1972; Jambon and Shelby, 1980; Carroll and Stolper, 1991, 1993).

The pressure dependence of noble gas solubility is controlled by the $V°_i$ term in Equation (10). At low pressure, or small values of $V°_i$, Equation (10) reduces to Henry's law, which predicts a linear proportionality between fugacity and mole fraction dissolved gas; i.e.,

$$f_i = K_i \cdot X_i ,\qquad(11)$$

where f_i is the fugacity of gas i, X_i is the mole fraction of dissolved gas, and K_i is the Henry's Law constant (which equals $1/a_0$ in Eqn. (10) when $f_0 = 1$ bar). All of the available noble gas solubility data approach Henry's law behavior at P < ~1000 bars, but solubilities at higher gas pressures are lower than would be predicted by a Henrian extrapolation of low P results. The noble gas abundances observed in mantle-derived magmas, however, are sufficiently low (from <<1 to ~30 ppm, by weight) that even at total pressures of 10 to 30 kbar the fugacities would be less than a few bars. Even if the mantle was an order of magnitude richer in He and Ar (the most abundant noble gases), the fugacities would be on the order of tens to possibly hundreds of bars, which is still well within the range of Henrian solubility behavior.

Melt-vapor partitioning of noble gases in natural systems

During ascent of magma towards the surface, the pressure will decrease and at some point a gas phase may form, the composition of which will depend on the magma's initial volatile content and the relative solubilities of the different volatile species present. For example, in submarine basaltic lavas erupted at depths greater than ~500 m, the exsolved gas is almost pure CO_2, whereas in shallower submarine or subaerial eruptions, exsolved water or S-bearing species become more abundant (Moore, 1970; Moore and Schilling, 1973; Gerlach and Graeber, 1985; Gerlach, 1986). These observations can be explained by the lower solubility of CO_2 relative to H_2O and sulfur-gases (S_2, H_2S, SO_2). How the noble gas abundances in a magma will respond to magma degassing processes will thus depend on the abundance of all magmatic volatiles and on their relative solubilities. Given knowledge of the variation in gas solubilities with pressure and some estimation of initial magma volatile contents, it is possible to calculate how the composition of a magmatic vapor phase will change during magma ascent to lower pressures (e.g., Holloway, 1976; Newman et al., 1988; Anderson et al., 1989; Bottinga and Javoy, 1990a, 1990b).

Solubilities of noble gases, H_2O, and CO_2 in basaltic and rhyolitic melts are shown as a function of pressure in Figure 16. Water solubility exceeds those of CO_2 and the noble gases by a large amount (>5-fold), but the molar solubilities of He, H_2O and CO_2 do not differ greatly between basalt and rhyolite. In contrast, the heavier noble gases (Ar, Kr, Xe) are much less soluble in basaltic melts than in rhyolitic melts. We do not distinguish between different species of water (molecular H_2O, OH^- groups) or carbon dioxide (molecular CO_2, CO_3^{2-} groups) in this comparison but rather consider only total solubilities. Activity-composition relationships for noble gases in H_2O-CO_2 fluids are not known so we must assume ideal mixing in the vapor phase. Because of the low solubilities of the noble gases and CO_2 relative to H_2O, the noble gases will follow CO_2 in being preferentially partitioned into the early-formed vapor phase in an ascending magma. The extent to which the noble gases follow CO_2 into the vapor phase will depend on their relative solubilities and abundances. The less soluble noble gases (Xe, Kr, possibly Ar, depending on melt composition) will be enriched in the vapor phase relative to the melt, in inverse proportion to their solubility (i.e., with Xe most strongly partitioned into the vapor, and He the most "compatible" in the melt). As an example of the effects of vesiculation on noble gas abundances, Figure 17 shows how the noble gases are partitioned between melt and vesicles for magma vesicle fractions between 10^{-4} and 1; the stippled region along the vesicularity axis covers the range of vesicularities found in submarine basalts (from Moore,

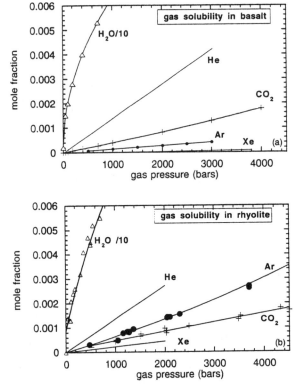

Figure 16. Solubility (molar-oxide basis) versus gas pressure for H_2O, CO_2 and He-Xe in (a) basalt, and (b) rhyolite. Data sources for rhyolite are: H_2O, Silver et al. (1990); CO_2, Fogel and Rutherford (1990; corrected using IR molar absorptivity for molecular CO_2 reported in Blank et al., 1993); Ar, Carroll and Stolper (1993), He, Xe, calculated from ionic porosity (Table 1). Data sources for basalt: H_2O and CO_2, from calculations based on data and model of Dixon et al. (1994; identical to data of Hamilton et al., 1964, but contains results at lower pressures); Ar, calculated from data of Carroll and Stolper (1993); He and Xe, as for rhyolite.1979).

The curve for CO_2 would plot between those for Ar and Ne. This simple mass balance calculation (following Jambon et al., 1986) shows that a major proportion of magmatic noble gases (and CO_2) will reside in bubbles when vesicularity exceeds approximately 1 vol %. Such behavior is in accord with He and Ar abundances in basaltic glasses which have been crushed to release vesicle gas, then heated to high temperature to release gas dissolved in glass (e.g., Kurz and Jenkins, 1981; Jambon et al., 1985; Marty and Ozima, 1986; Staudacher et al., 1989; Javoy and Pineau, 1991). For the heavier rare gases such comparisons are more difficult to make because the low amount of gas present in the glass is typically close to the limits of analysis.

Figure 17. Calculated fraction of total gas in basaltic magma which is present in vesicles as a function of vesicularity (i.e., amount of degassing). Calculation done following Jambon et al. (1986; p. 406). Typical range of vesicularities in submarine oceanic basalts is indicated on horizontal axis (from Moore, 1979).

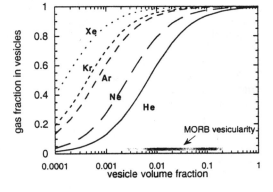

Abundance fractionations and magma degassing

Whether solubility differences can produce fractionations in noble gas abundances during magma degassing will depend on how much degassing occurs between magma generation and eventual eruption, and whether degassing occurs under open or closed system conditions. Closed system degassing refers to the situation where bubbles formed by exsolved gas are not separated from the magma. In open system degassing, vapor bubbles are segregated from the magma, as a result of their low density, thus reducing the magma's bulk volatile content. For closed system degassing, the relative abundances of noble gases measured in a sample will be unchanged from their original values, provided that gas in both vesicles and glass (± crystals) is analyzed. If vesicle gas is not included in the analysis the magma will appear to be enriched in the more soluble light rare gases relative to the less soluble heavy rare gases. If vapor and magma are separated, the noble gas abundances in the residual melt will vary as a function of the amount of gas loss and more extreme fractionations of elemental abundance ratios may be produced (relative to the closed system case).

Taking He and Ar as an example, the theoretical maximum fractionation in the melt phase under closed system conditions is simply the initial He/Ar ratio of the magma, multiplied by the ratio of their relative solubilities, i.e.,

$$[He]_{glass} / [Ar]_{glass} = [He]_0 / [Ar]_0 \cdot S_{He}/S_{Ar} \tag{12}$$

where brackets denote concentration in glass or original magma ($[i]_0$), and S refers to solubility of the subscripted species (see also Jambon et al., 1985, 1986). The sigmoidal curves in Figure 18 illustrate how the He/Ar ratio in a basaltic melt and coexisting vesicle gas would vary as a function of magma vesicularity during closed system degassing. The closed system example shown in Figure 18 corresponds to a magma with initial He/Ar = 3 and S_{He}/S_{Ar} of ~10; as required by Equation (12), He/Ar in the melt approaches a plateau value of ~30 at vesicle fractions greater than a few percent. Partial degassing and enrichment of magma in He relative to Ar provides a reasonable explanation for the observation that many analyses of MORB show He/Ar ratios greater than 3, which is approximately the radiogenic production ratio for the mantle (based on estimated K and U abundances). Elevated He/Ar ratios in MORB could also be produced by open system degassing, involving removal of bubbles from the volume of magma in which they formed. The steeply rising curves in Figure 18 (labeled open system) show the evolution of melt and vesicle He/Ar for melts that have had vesicles removed in 1×10^{-4} volume

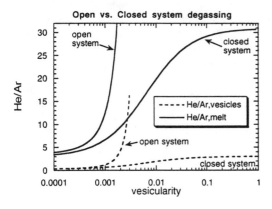

Figure 18. Calculated variation in He/Ar ratio of basaltic melt (solid curves) and coexisting gas (dashed curves) for open system and closed system degassing. Starting point is basaltic melt with He/Ar = 3. Closed system degassing is simply related to relative solubilities of He and Ar, as described in text (see also Jambon et al., 1986). Open system degassing was simulated by removing gas in 0.01vol% increments and using the residual melt He/Ar ratio as the starting point for the next increment of gas removal. In this case the vesicularity scale corresponds to the integrated vesicularity as each 0.01% increment is considered to be removed from the system (see text for discussion).

fraction increments. This was done from the same starting conditions as in the closed system case, but gas was removed stepwise, and the melt He/Ar ratio after each step of gas removal (which causes He/Ar of melt to increase) was used as the starting point for the subsequent step. The vesicularity scale for the open system curves thus corresponds to the integrated vesicularity of the samples and not the actual vesicle volume fraction that would be observed (10^{-4}). This simple example illustrates the power of open system degassing to produce magmas with relatively small (preserved) vesicle fractions yet having highly fractionated noble gas abundances.

NITROGEN SOLUBILITY IN MAGMAS

The solubility of nitrogen in melts of geological interest is poorly known. In studies related to industrial processes the solubility behavior of nitrogen has been characterized as either 'physical' or 'chemical', depending on the form of nitrogen (molecular N_2, or reactive N-H, N-H-C, N-O species) and whether it is interpreted to react with the silicate melt (chemical solubility) or whether it dissolves by an interstitial, substitutional reaction (physical solubility, like the noble gases) (Mülfinger and Meyer, 1963; Müllfinger, 1966; Kelen and Mülfinger,1968; Davies and Meherali, 1971; Mülfinger et al., 1972; Frischat and Schrimpf, 1980). Under highly reducing conditions, nitrogen solubility in simple silicate and oxynitride glasses may approach several weight percent, predominantly in the form of nitrides, CN, HCN, and ammonia species, depending on the composition of the gas atmosphere used (e.g., Mulfinger, 1966; Loehman, 1979; Frischat and Schrimpf, 1980). In general the chemical solubility of nitrogen is 10 to 10^5 times greater than the physical solubility, obtained by equilibrating silicate melts with N_2 gas (Mülfinger, 1966).

Published work on nitrogen solubility in silicate melts under conditions appropriate to magmatic processes is limited to experiments on $NaAlSi_3O_8$ melt composition at 1 to 4.6 kbar P_{total} (Kesson and Holloway, 1974; Shilobreyeva et al., 1994), and several abstracts reporting nitrogen solubility in basaltic melts (Javoy, 1984; Miyazaki et al., 1992; Shilobreyeva et al., 1994). Kesson and Holloway (1974) studied nitrogen solubility in hydrous albitic melt, using experiments at 960°C and up to 4.6 kbar P_{total} in which the N_2 partial pressure (P_{N_2}) was ~0.5 P_{total}; they found nitrogen solubility to be an approximately linear function of P_{N_2} and nitrogen solubility increased by ~200 ppm N per kilobar N_2 pressure. Shilobreyeva et al. (1994) report lower N solubilities in anhydrous albitic melt at 1250°C, ranging from 58 to 74 ppm N at 3 kbar, with the higher solubility observed in more oxidizing experiments (the $f(O_2)$ varied from IW to FMQ+0.7). Nitrogen solubility in basaltic melt is reported by Javoy (1984) to increase by 100 ppm/kbar P_{N_2} but no details are given regarding experimental methods. Shilobreyeva et al. (1994) studied nitrogen solubility in basaltic melt at 1 to 3 kbar P_{N_2} and they report higher solubilities under more reducing conditions; e.g., 80 ppm at 1 kbar to 230 ppm N at 3 kbar P_{N_2}, IW buffer, and 38 ppm N at 1 kbar, FMQ+0.7. Experiments on N solubility in basaltic melt at atmospheric pressure and $f(O_2)$ = FMQ+0.7 (Miyazaki et al., 1992) yield a Henry's law constant which predicts a N solubility at 1 kbar of 540 ppm, but this result may include some contamination of the sample by organic nitrogen (Miyazaki et al., 1992). If the solubility of molecular nitrogen can be considered to be mainly a physical process then the similar sizes of N_2 and Ar support a molecular N_2 solubility on the order of 75 to 150 ppm/kbar, depending on melt composition as discussed with reference to noble gas solubilities. It is impossible to comment further on nitrogen solubilities in magmas until additional experimental solubility data are available.

EXPERIMENTAL DETERMINATION OF HALOGEN SOLUBILITIES

Analytical techniques for F and Cl

Fluorine and chlorine concentrations in igneous rocks and natural and synthetic silicate glasses are determined by a variety of methods. Bulk analysis techniques for fluorine include ion-selective electrodes (Christiansen and Lee, 1986; Haselton et al., 1988; Holtz et al., 1993), neutron activation analysis (Dingwell, 1984; Mosbah et al., 1991), solution-based ICP-AES (Holtz et al., 1993), and spectrophotometry (Kogarko et al., 1968; Kovalenko, 1977), with typical detection levels of ~0.01 wt % F and uncertainties of 5 to 10 % relative. Bulk analysis techniques for Cl include ion-selective electrodes (Haynes and Clark, 1972), neutron activation (Johansen and Steinnes, 1967), X-ray fluorescence (Wahlberg, 1976; Christiansen and Lee, 1986), beta-track mapping (Watson, 1991), colorimetry (Shinohara et al., 1989), and atomic absorption (Iwasaki et al., 1952; Kilinc and Burnham, 1972), with detection levels of 0.001 to 0.01 wt % Cl and typical uncertainties of 5 to 15 % relative. Attempts have also been made to determine Cl concentrations in glasses based on changes in glass refractive index with Cl content (Koster van Groos and Wyllie, 1969; Barker, 1976). There are a variety of problems associated with bulk analytical methods, the most serious being that they are generally destructive and they are sensitive to inclusions of other phases in the material being analyzed.

Glasses from experimental run products and natural samples may contain other phases (phenocrysts, quench crystals, quenched melts, fluid inclusions, etc.) and hence, microbeam techniques are preferred because of the high spatial resolution they offer. Useful microbeam techniques for determination of Cl and F concentrations in silicate glasses include the electron microprobe, ion microprobe (SIMS), and the nuclear microprobe (PIGME). Although laser ablation ICP-MS offers the potential to analyze a large number of constituents in silicate materials simultaneously, this technique suffers from high detection limits for Cl and F, due to high ionization potentials and resultant low ionization yields for these elements (David Lambert, pers. comm.).

The electron microprobe has frequently been used to determine Cl and F in experimental samples (e.g., Manning et al., 1980; Dingwell, 1984; Luth, 1988a; Webster, 1990; Webster, 1992a;b; Metrich and Rutherford, 1992; Holtz et al., 1993; Bai and Koster van Groos, 1994) and in natural glasses and glass inclusions in phenocrysts (Anderson, 1974; Lowenstern and Mahood, 1991; Webster and Duffield, 1991; Dunbar and Kyle, 1992; Hervig and Dunbar, 1992; many others). F in hydrous glasses may behave in a volatile manner during electron microprobe analysis, but reliable analyses can be obtained using low accelerating potentials, low beam currents, and a defocused electron beam (≥5µm in diameter) (Webster, 1990). Count rates for F are not particularly good when using crystals such as TAP, but newer synthetic crystals (e.g., OVONICS-60) can give analytical precisions on the order of 5% for total F concentrations of 0.1 wt % (e.g., ARL-SEMQ, at 15 keV, 10nA gives ~50 cps for granitic glass with 0.1 wt % F). Cl count rates are very stable during electron microprobe analysis of water-saturated glasses when using a slightly defocused beam at the same probe conditions (Webster and Holloway, 1988; Webster 1992a;b); replicate analyses conducted on a single glass spot demonstrate that Cl counts do not vary with time. Furthermore, the count rates with an ARL-SEMQ at these conditions are good; with a PET crystal and glass sample containing 0.1 wt % Cl, the relative precision for a 60 second analysis is roughly ±5%. Microprobes newer than the ARL-SEMQ generally use lower take-off angles, which result in lower count rates, and in the case of light elements such as F sensitivities may be diminished by as much as 20% relative (John J. Donovan, pers. comm.).

Secondary ion mass spectrometry (SIMS) has also been used to determine F and Cl contents of silicate glasses (Kovalenko et al., 1988; others), because it offers high spatial resolution and relatively low limits of detection. This technique, however, requires standards with compositions similar to those of the unknowns. A detailed description of the analytical methodology for determining F and Cl contents of glasses by SIMS is given in Chapter 2.

In a recent study, F concentrations of natural silicate glasses were determined with a 30 μm diameter proton beam by proton microprobe (PIGME; Mosbah et al., 1991). The limit of detection is 30 ppm for samples having a thickness ≥60 μm; the analytical precision for glasses containing 0.01 wt % F is 10% relative. This technique may also be used to determine concentrations of other volatiles, e.g., Cl, CO_2, S, B, and N, in silicate glasses and these applications are currently under investigation (Mosbah et al., 1991).

Chlorine solubilities: general approach

Chlorine solubilities are determined experimentally as the concentration of Cl in a silicate glass (quenched melt) that was equilibrated with a Cl-bearing aqueous fluid or an immiscible chloride melt; the Cl content of quenched glass is typically assumed to equal that of melt prior to quench. The quench products of experiments involving coexisting silicate and alkali chloride melts are silicate glass with small dark spherical aggregates of alkali chloride crystals, interpreted to have been alkali chloride melt at high T (Koster van Groos and Wyllie, 1969); alkali chlorides do not quench to glass in most experimental apparatus. The maximum dissolved Cl content in a molten silicate, i.e., the solubility limit, has been investigated in a variety of systems, but most work has been confined to felsic melts in equilibrium with molten NaCl.

Chlorine solubility experiments involving fluid/melt partitioning of Cl present a variety of potential difficulties. With fluid-present experiments one cannot directly determine the Cl content of fluids containing ≥4.5 molal Cl at run conditions, because alkali chloride minerals precipitate on quench; this necessitates collection and analysis of all fluid precipitates in addition to analyzing the fluid. This problem is exacerbated with mixed H_2O-CO_2 fluids because of their inherently lower Cl solubilities, and in response, some investigators have computed the abundance of Cl in fluids by mass balance (Webster and Holloway, 1988; Webster, 1992a; 1992b; Metrich and Rutherford, 1992). Furthermore, if Cl is added as a solution enriched in HCl, then the fluid/melt ratio must be low or the fluid will selectively dissolve alkali metals from melt, decreasing the melt (Na+K)/Al ratio (Webster and Holloway, 1988). Conversely, experiments involving alkali chloride-bearing starting fluids may add alkali metals to the melt, leading to peralkaline melts and changes in melt Na/(Na+K). These latter difficulties can be minimized by using small (fluid/melt) ratios. Alternatively, one can add Cl for O in the starting glasses via the exchange operator Cl_2O_{-1} by substituting two moles of NaCl or KCl for each mole of Na_2O or K_2O, respectively.

Chlorine solubilities

Maximum Cl solubilities in anhydrous melts of granitic to basaltic composition equilibrated with chloride melts at 1 bar range from one-half to several wt %, while felsic silicate melts equilibrated with Cl-bearing aqueous fluids may contain maximum Cl of one-third to over 1 wt % Cl. The effects of P, T, and system composition on Cl solubility are not well-constrained, but in the following sections we review the available experimental data.

Water- and silica-poor silicate melts. The solubility of Cl in basaltic melts saturated with various chloride melts has been established for 20 kbar (Zharikov et al., 1985, discussed in Malinin et al., 1989), but in general, Cl solubilities in anhydrous mafic melts are not well-known. Cl solubility in tholeiitic melt saturated with different molten chlorides varies from 2.5 wt % Cl ($CaCl_2$ or $FeCl_3$ saturation), to 2.2 wt % ($AlCl_3$ saturation), to ~1.25 wt % (NaCl or KCl saturation). The minimum Cl content of ~0.1 wt % occurs in basaltic melts equilibrated with molten $MgCl_2$. The strong influence of composition on Cl solubility in basaltic melts is also supported by low P experiments in which silicate melts were equilibrated with 1 bar P_{HCl} at 1200° C; these experiments yield Cl solubilities ranging from ≤1.45 wt % Cl in basaltic melt, to ≤0.4 wt % Cl in andesitic melt (Iwasaki and Katsura, 1967). The effect of P is undetermined, but increasing T from 1200° to 1290°C yields an ~30 % decrease in HCl solubility in basaltic melt (Iwasaki and Katsura, 1967).

Water-poor felsic melts. Chlorine solubility in anhydrous, broadly granitic melts, equilibrated with anhydrous alkali chloride melts ranges from ~0.5 to 1.5 wt % at 1 to 2000 bar and 1100° to 1200°C (Fig. 19). The effects of P and T on Cl solubilities in such melts are poorly known, but with subaluminous and metaluminous compositions at ≤2 kbar the solubility of Cl appears to be independent of pressure. The only data on the T dependence is for a rhyolitic (liparite) melt at 1bar P_{HCl} in which Cl concentration decreases by approximately 20% relative (from 0.05 to 0.04 wt % Cl) as T increases from 1200° to 1290°C (Iwasaki and Katsura, 1967).

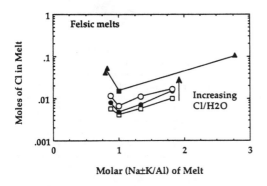

Figure 19. Solubility of Cl in "granitic" melts equilibrated with Cl-bearing aqueous fluids (3 lowermost curves) or alkali chloride melts with no water added (uppermost curve with triangles and solid square). Chlorine solubility in felsic silicate melts is smallest for melts with molar (Na±K)/Al of one, and Cl solubility increases as molar (Na±K)/Al varies from one. Chlorine solubility in silicate melt increases as Cl/H_2O increases; see text for discussion. Data from experiments of Ryabchikov (1963) in the system: SiO_2-Al_2O_3-Li_2O-Na_2O-K_2O-Rb_2O-Cs_2O-Cl_2O_{-1} conducted at 1 atm and 1100°C (triangles); molar [total alkalis+Al]/Si ranges from 0.1 to 0.4. Experiments of Webster (1992b) in the system: SiO_2-Al_2O_3-Na_2O-K_2O-Cl_2O_{-1}±H_2O were conducted at 2 kbar and 1200°C (solid squares) and 800°C; molar [Na+Al+K]/Si ranges from 0.22 to 0.25. For 800°C experiments, computed Cl contents of aqueous fluids are 40 wt % (open circle), 20 wt % (solid circle), and 10 wt % (open squares).

Chlorine solubility in natural and synthetic anhydrous, felsic melts also depends strongly on melt composition, especially the (total alkalis + Al_2O_3)/SiO_2 and total alkalis/Al_2O_3 (Figs. 19 through 21; Table 2). For example, molten silica and molten NaCl are almost entirely immiscible (Kotlova et al., 1960), which implies that the solubility of Cl in SiO_2 melt is extremely low. The solubility limit of Cl increases as Na_2O is added, and up to 4.4 wt % Cl dissolves in Na-rich, Si-poor melts at 1 bar and 1100° C (Ryabchikov, 1963). Experiments on peraluminous felsic melts also show enhanced Cl solubilities, and in general, the lowest solubilities occur in metaluminous and subaluminous granitic melts because Cl solubility increases with increasing concentrations of network-modifying cations. Na-enriched haplogranite melt in equilibrium with NaCl-KCl melt at 2.0 kbar and

Table 2.
Chlorine concentrations of anhydrous silicate glasses[†] equilibrated with molten salt(s)

Study[††]	Cl in glass (wt %)	Temperature (°C)	Pressure (bars)	System composition
Glasses with Molar (Na±K)/Al >1				
1	1.42	1400°	1	75% SiO_2-9.2% CaO-15.8%Na_2O[†††] -NaCl
2	1.5–1.8	1100° -1200°	1	$NaAlSi_3O_8$-NaCl
2	1.33	1550°	600	$NaAlSi_3O_8$-NaCl
3	1.17	900°	2000	natural pantellerite-NaCl
Glasses with Molar (Na±K)/Al <1				
4	1.46	1100°	1	SiO_2-$NaAlSi_3O_8$-NaCl
Glasses with Molar (Na±K)/Al ≈1				
5	0.52	1200°	1	natural obsidian-NaCl
6	0.54	1180°	2000	haplogranite-NaCl

[†] Water concentrations of most glasses not determined.
[††] References: (1) Bateson and Turner (1939), (2) Kogarko and Ryabchikov (1969), (3) Lowenstern (1994), (4) Ryabchikov (1963), (5) Zünckel and Hempel (1917), (6) Webster (1992a).
[†††]Composition of starting material before NaCl was added.

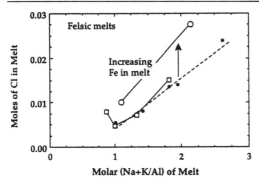

Figure 20. Solubility of Cl in granite melts equilibrated with Cl-bearing aqueous fluids; data at 1 kbar and ~840°C (circles) from Metrich and Rutherford (1992) and with 20 wt % Cl in fluid at 2 kbar and 800°C (squares; molar Si/Al ranges from 8 to 11; Webster, 1992b). Cl solubility increases in metaluminous and peralkaline, felsic silicate melts as FeO content of melt increases and as P decreases; see text for discussion. Open circles represent experiments with natural pantellerite and phonolite melts in which FeO concentration increases from 0.6 to 7.9 wt % as molar (Na+K)/Al increases, SiO_2 inmelt of 54 to 74 wt %, and Cl in fluid is variable but unreported. Small filled circles represent experiments with synthetic haplogranites (molar Si/Al: 10 to 17).

Figure 21. Solubility of Cl in granite melts equilibrated with Cl-bearing aqueous fluids; data at 1 kbar and ~840°C (circles) from Metrich and Rutherford (1992), with 20 wt % Cl in fluid at 2 kbar and 800°C (squares; molar Si/Al from 8 to 11; Webster, 1992b). Cl solubility increases in felsic silicate melts as SiO_2 in melt and P decrease; see text for discussion. Small open circles represent data for synthetic haplogranites (no FeO, and molar Si/Al is 9.3). Small filled circles represent experiments with synthetic haplogranites (molar Si/Al of 10 to 17); Cl content of fluid is variable but unreported.

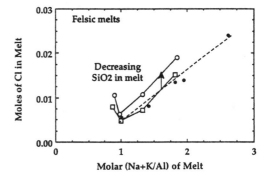

1180°C (Table 2), for example, contains 0.54 wt % Cl (Webster, 1992b), and a NaCl- and KCl-saturated rhyolitic melt at 1 bar contains 0.52 wt % Cl at 1200° C (Zünckel and Hempel, 1917).

Cl partitioning between hydrous felsic melts and fluids. Experiments involving hydrous, felsic melts equilibrated with aqueous solutions define the solubility of Cl in granitic melt relative to that in the coexisting fluid phase(s); the Cl concentration in melt is often expressed as the fluid/melt distribution coefficient for Cl (D_{Cl}). It should be noted that Cl solubility in silicate melts is a strong function of phase relations in aqueous fluid-, liquid-, and vapor-bearing systems (Shinohara et al., 1989). The solubility limit for Cl in all hydrous felsic melts studied to date is ≤1 wt %.

Investigations of Cl solubility with mixed volatiles have been limited to systems comprised of $H_2O–Cl±CO_2±F$, and they show that Cl solubility is a complex function of P, T, and Cl concentration. Cl solubility varies little with P and T in granitic melts containing less than 0.05 wt % Cl, but larger variations are observed in more Cl-rich melts. In Cl-rich systems (Kilinc and Burnham, 1972; Webster and Holloway, 1988; Shinohara et al., 1989; Malinin et al., 1989) Cl solubility in melt decreases with increasing P and decreasing T, and the rate of change is composition dependent (for P ≤ 8 kbar and T of 750° to 1000°C) (Fig. 22a). Cl solubilities in granitic melts at low magmatic temperatures are virtually unconstrained, and more experimental data are needed because many mineralizing magmas exsolve Cl-rich hydrothermal fluids at these conditions. The solubility of Cl in subaluminous and peralkaline haplogranite melts at 2 kbar increases strongly with increasing T (Webster and Holloway, 1988), in contrast to the T effect on Cl solubility in HCl-bearing basaltic and rhyolitic melts at 1 bar (Iwasaki and Katsura, 1967).

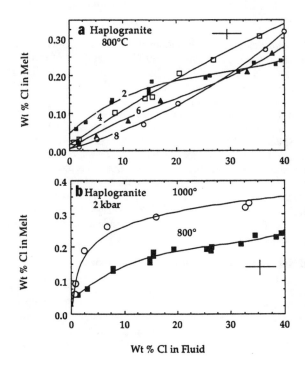

Figure 22. Solubility of Cl in subaluminous haplogranite melts equilibrated with alkali chloride-bearing aqueous fluids as a function of P at 800°C (a), and T at 2 kbar (b). Cl contents of fluid computed by mass balance (Webster, 1992a). With ≤30 wt % Cl in fluid, Cl solubility in aluminosilicate melt decreases as P increases from 2 (solid squares) to 4 (open squares) to 6 (triangles) to 8 kbar (circles) as noted on figure; at higher concentrations of Cl in system the behavior of Cl is erratic. The solubility of Cl in melt increases strongly with increasing T, from 800°C (squares) to 1000°C (circles) as noted on figure. One-sigma error bars are shown on right.

The solubility of Cl in broadly "granitic" melts also depends strongly on system composition. Concentrations of Cl in hydrous granitic melts increase, relative to that in coexisting aqueous fluid(s), with increasing abundances of F (Webster and Holloway, 1990), CO_2 (Webster and Holloway, 1988), Cl (Webster and Holloway, 1988; 1990; Webster, 1992a; 1992b), and FeO (Metrich and Rutherford, 1992) (Fig. 20) and increasing Na/(Na+K) (Malinin et al., 1989; Webster, 1992b). The ratio of Na+K to Al has a large effect on Cl solubility in hydrous granitic melts (Fig. 21), with a solubility minimum at molar (Na+K)/Al = 1, slightly higher solubilities in peraluminous melts, and much higher solubilities in peralkaline melts (Metrich and Rutherford, 1992; Webster, 1992b). Fluid-saturated peralkaline haplogranite and pantellerite melts, and melts in the system albite-nepheline-H_2O-NaCl, for example, contain 0.75 to 0.92 wt % Cl at 0.5 to 4 kbar and 800° to 940° C (Binsted, 1981; Webster, 1992b; Metrich and Rutherford, 1992), whereas subaluminous haplogranites contain $\leq$ 0.35 wt % Cl at similar conditions. Cl solubility increases as the activities, a_i, of network modifying cations (Al^{VI}+Na+K+Fe+Ca) in melt increase relative to a_{SiO2} (Fig. 21). Silicate melts, in fluid-saturated granitic systems that are enriched in CO_2, contain little CO_2 because of its low inherent solubility (see Chapters 5 and 6); thus, it is unlikely that a small quantity of CO_2 in melt could directly influence Cl solubility in melt. We speculate that introduction of CO_2 increases the CO_2 content of the coexisting aqueous fluid and decreases the fluid's mean dielectric constant, the activity of water in fluid and melt, and the relative solubility of Cl in fluid.

Summary of Cl solubilities. In fluid saturated igneous systems under conditions of geological interest, Cl solubilities in most aluminosilicate melts range from a few thousand ppm to ~2 wt %. Cl typically partitions in favor of aqueous fluids rather than melts, which explains why alkali chloride contents of exsolved magmatic fluids may exceed 70 wt % (Roedder, 1984), while Cl contents of many silicate melts may be as low as several hundred parts-per-million.

The solubility of HCl in aluminosilicate melts is generally similar to those of CO_2 and H_2S, and the solubilities of each of these species are significantly less that those of H_2O or HF at equivalent P and T in equivalent melt compositions. These differences in solubilities can be understood by comparing the sizes, the intermolecular attractive forces, and the tendencies to ionize or cause ionization for each volatile. For HCl, the Redlich-Kwong *b* parameter, an indication of molecular size (Holloway, 1981; see also Chapter 6), is roughly twice that of HF or H_2O, and the ionic radius of Cl^- (1.81 Å) is greater than those of F^- (1.33 Å) or OH^- (1.37 Å) (Shannon, 1976); the *b* parameter for HCl is more similar to those of CO_2 or H_2S (Holloway, 1981). Moreover, the Redlich-Kwong *a* parameter (which indicates the total intermolecular attractive force) for HCl corresponds more closely with those of CO_2 and H_2S (Holloway, 1981), and the dipole moments (which predict the tendency to ionize or cause ionization) of HCl and H_2S are very similar (Holloway, 1981). Hence, HCl should exhibit solubilities similar to those of CO_2 and H_2S, because of the relative ease with which HCl fits into the silicate melt structure, its tendency to ionize, and its ability to interact chemically with the "solvent".

Solubility mechanisms and speciation of Cl in silicate magmas

The speciation of Cl in silicate melts of geologic interest is not well known. Solubility studies, however, provide indirect information on melt complexes, and thus help to constrain bonding and local structures in Cl-bearing silicate melts. Although the solubility of Cl is generally low in silicate melts, its speciation is of interest if we wish to develop any predictive models of Cl solubility as a function of melt composition, P, and T; such information has applications ranging from understanding the differentiation trends of Cl-rich peralkaline melts (Ryabchikov, 1963; Barker, 1976), to studying the importance of Cl as a complexing ligand in the formation of mineral deposits associated with evolved felsic magmas (Holland, 1972; Burnham, 1967). Solubility data for compositionally-simple, anhydrous felsic melts suggest that although Cl does not complex with Si in melts (Kotlova et al., 1960), Cl may complex with network modifying Na ions (Ryabchikov, 1963; Kogarko, 1974). Moreover, the minimum in Cl solubility as a function of melt Na+K/Al (Figs. 19, 20, 21), and the high solubilities in peralkaline melts support a preferential assocaition of Cl with network modifying cations, especially the alkali-metals, Na and K. In experiments involving melts of $NaAlSi_3O_8$-SiO_2-NaCl-H_2O and $KAlSi_3O_8$-SiO_2-KCl-H_2O (Webster, 1992b), Cl solubility is higher in sodic melts, and as a result, Cl may preferentially pair with Na. Complexes of Fe and Cl may also be important, because Cl solubility is greater in natural, iron-enriched pantellerite melts than in haplogranitic equivalents at similar conditions.

For hydrous granitic systems, the dissolution of HCl in silicate melt has been described as:

$$HCl_{fluid} + O^{2-}_{melt} = Cl^-_{melt} + OH^-_{melt} \qquad (13)$$

(Kogarko, 19764; Burnham, 1979), but HCl should not significantly influence the bulk polymerization of melts given its low solubility. Alternatively, Cl may dissolve:

$$Me^{+1}_2O_{melt} + 2\ Me^{+2}Cl_{2melt} = 2\ Me^{+1}Cl_{melt} + Me^{+2}_2O_{melt} \qquad (14)$$

where Me^{+i} refers to a metal cation of charge i.. We do not know the identity of the various melt species (Cl^-_{melt}, $Me^{+2}Cl_{2melt}$, or $Me^{+1}Cl_{melt}$), but, based on experimental solubility studies, a significant fraction of the Cl in silicate melts appears to complex with network modifying Na, K, Ca, Fe, and to a lesser extent with Al and Mg. Furthermore, $HCl°$ apparently makes an insignificant contribution to the total Cl dissolved in melts at low P because of the low overall solubility of Cl in HCl-saturated melts. It is unclear why F affects Cl solubility, but it is likely a result of the depolymerizing effect of F on melt structure, rather than a direct result of complexing between F and Cl.

F solubilities: general approach

Solubility limits for F have been determined for a limited number of aluminosilicate melts of geologic interest. Few experimental studies have involved F-bearing systems similar in composition to natural magmas, few studies have specifically dealt with the effects of P and T on F solubility or partitioning between melt and vapor, and few studies have established solubility limits for F in silicate melts saturated in fluoride-minerals and/or fluoride-melts (Tables 3 and 4). In the following discussion it should be noted that F solubilities determined from references published prior to 1960 were extracted via chemographic analysis of published phase equilibrium data, and as a result the solubilities may include unquantified, significant errors.

Experiments on F-bearing melts and aqueous fluids involve a variety of challenges. If F is added to the system as a comparatively large quantity of HF solution, the melt composition may change as a result of melt constituents dissolving into the fluid. Furthermore, experiments involving alkali fluoride-bearing fluids may alter the silicate melt composition, yielding more alkaline (or peralkaline) melts. These problems may be minimized by adding F to starting glasses via the exchange operator F_2O_{-1} (substituting 2 moles of NaF for each mole of Na_2O, for example) and through the use of small fluid/melt (mass ratios of ≤ 0.1) in preparing experimental charges.

Table 3. Concentrations of F (expressed as wt.% alkaline earth-fluoride) in silicate glasses from experiments comprised of: silicate melt ± fluoride melt.

Study[†]	Maximum MeF$_2$ in silicate glass (mole%)	Temperature (°C)[††]	Pressure (bars)	System composition
50 mole% SiO$_2$ in glass				
1	10[◊]	≥ 1450°	1	MgF$_2$-MgO-SiO$_2$
1	34[◊]	≥ 1480°	1	SrF$_2$-SrO-SiO$_2$
1	37[◊◊]	≥ 1565°	1	BaF$_2$-BaO-SiO$_2$
2	40[◊]	≥ 1420°	1	MgF$_2$-Al$_2$O$_3$-SiO$_2$
2	42[◊]	≥ 1410°	1	CaF$_2$-Al$_2$O$_3$-SiO$_2$
2	43[◊◊]	≥ 1410°	1	SrF$_2$-Al$_2$O$_3$-SiO$_2$
2	46[◊◊]	≥ 1410°	1	BaF$_2$-Al$_2$O$_3$-SiO$_2$
75 mole% SiO$_2$ in glass				
2	16[◊◊]	≥ 1420°	1	MgF$_2$-Al$_2$O$_3$-SiO$_2$
2	17[◊◊]	≥ 1410°	1	CaF$_2$-Al$_2$O$_3$-SiO$_2$
2	19[◊◊]	≥ 1410°	1	SrF$_2$-Al$_2$O$_3$-SiO$_2$
2	21[◊◊]	≥ 1410°	1	BaF$_2$-Al$_2$O$_3$-SiO$_2$

[†] References: (1) Ershova and Olshanskii (1957) and (2) Ershova (1957).
[††] Temperatures for melts with these silica contents are only approximately known.
[◊] MeF$_2$ content extracted from interpolations of experimental results summarized in phase diagrams and is approximate only; MeF$_2$ concentration is maximum allowed for silicate melt in the one liquid stability field.
[◊◊] MeF$_2$ content extracted from extrapolations of experimental results summarized in phase diagrams and may include significant errors; MeF$_2$ concentration is maximum allowed for silicate melt in the one liquid stability field.

Table 4. Fluorine concentrations of silicate glasses from experiments comprised of silicate melt ± aqueous or aqueous-carbonic fluid ± fluoride melt ± silicate minerals.

Study[†]	F in silicate glass (wt %)	Temperature (°C)	Pressure (bars)	System composition
1	≤ 1[◊]	630° - 800°	980	granite-LiF-H$_2$O
2	≤ 2[◊]	600° - 770°	980	natural granite-NaF-H$_2$O
3	≤ 2[◊]	810° - 820°	980	natural granite-HF-H$_2$O
4	≤ 3.3[◊]	800°	980	natural granite-HF-H$_2$O
5	< 1 - 8.5	775° - 1000°	500-5000	topaz rhyolite-H$_2$O±NaF±AlF$_3$
6	< 1 - 6.5	795°	2000	haplogranite-H$_2$O±NaF±AlF$_3$
7	18.9	1400°	15000	NaAlSi$_3$O$_8$-F$_2$O$_{-1}$
8	25[††]	690° - 850°	1000	NaAlSi$_3$O$_8$-NaF-(25%)H$_2$O
8	25[††]	600° - 660°	4000	NaAlSi$_3$O$_8$-NaF-(50%)H$_2$O

[†] References: (1) Glyuk et al. (1980), (2) Anfiligov et al. (1973), (3) Glyuk and Anfiligov (1973), (4) Kovalenko (1977), (5) Webster (1990), (6) Webster and Holloway (1990), (7) Dingwell (1987), (8) Koster van Groos and Wyllie (1967).
[◊] System *apparently* consisted of silicate melt, fluoride melt, and aqueous fluid ± silicate minerals; the experiments were conducted with very large starting ratios of (fluid/solid) in the starting materials. The F content of the silicate glass was extracted from interpolations of experimental results summarized in phase diagrams and is approximate only.
[††] This is the maximum F concentration in melt, as experimentally constrained; however, system exhibits almost complete miscibility between hydrous albite melt and hydrous villiaumite melt; see text for discussion.

Water- and silica-poor silicate melts. Studies of phase relations with immiscible fluoride and silicate melts at P-T-X conditions where F-bearing minerals and aqueous fluids are unstable provide information on F solubilities in anhydrous silica-poor systems. The solubility of alkaline earth fluorides in such melts is inversely related to silica activity and dependent on the identities and activities of other cations. With MeF_2-MeO-SiO_2, where Me includes Mg, Sr, and Ba, the width of the fluoride melt—silicate melt miscibility gap decreases with increasing cation radius, implying a positive correlation between alkaline-earth cation radius and fluoride solubility in silicate melt. Thus, F may be more likely to bond with the larger alkaline earth metals in silicate melts because of the lesser tendency to unmix with these cations. At 1450° to 1565° C and 1 bar (Fig. 23), alkaline earth-rich fluoride melts coexist with silicate melts containing approximately 50 mole % SiO_2 and 10 to 15 wt % F (Ershova and Olshanskii, 1957). These phase relations are compatible with the results of Baak and Ölander (1955), Mukerji (1965), and Ershova (1957) on CaO-CaF_2-SiO_2 at 1 bar, which indicate stable liquid immiscibility at 1450°C, with the silicate melt containing ~44 wt % SiO_2 and ≤16 wt % F. Luth (1988a) found one melt containing 44 wt % SiO_2 and 3.8 to 8.4 wt % F at 1550° to 1600°C; whereas, Susaki et al. (1990) found no evidence of silicate-fluoride immiscibility in experiments on CaO-CaF_2-SiO_2 with trace quantities of silver and ≤ 0.6 wt % S_2 at 1200° to 1350° C.

Fluorine solubilities inferred from 1 bar phase relationships in MeF_2-Al_2O_3-SiO_2 systems, where Me includes Mg, Ca, Sr, and Ba, also show systematic variations with cation size, width of miscibility gap, and MeF_2 solubility (Table 3, Fig. 24). As was observed with Cl solubilities, F solubilities vary inversely with a_{SiO2}. In these simple aluminosilicate melts, F solubility is greater than in the Al-free systems (16 to 23 wt % F, vs. 8 to 16 wt % F), but a similar trend of increasing solubility with increasing radius of the alkaline earth cation is observed (Ershova, 1957). The higher F solubilities in these Al-bearing melts are consistent with experience in the glass industry, showing that addition of Al_2O_3 reduces the liberation of F during glass-making (Kogarko and Ryabchikov, 1961b).

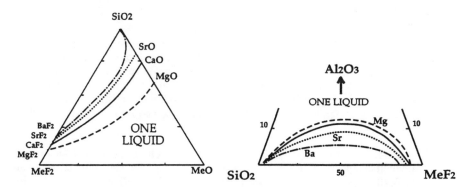

Figure 23 (left). Schematic curves (dashed for MgO/MgF_2; solid for CaO-CaF_2; dotted for SrO/SrF_2; dotted-dashed for BaO/BaF_2) after Ershova and Olshanskii (1957) showing phase equilibria (2-liquid fields) in systems MeF_2-MeO-SiO_2 at 1 bar and 1400° to 1670°C. Systems consist of single melt to lower right of each curve and immiscible silicate and fluoride melts to upper left of curves. Locations of curves and concentrations (mole %) are approximate.

Figure 24 (right). Schematic curves for MgF_2 (dashed), CaF_2 (solid), SrF_2 (dotted), and BaF_2 (dotted-dashed) after Ershova (1957) showing phase melt-melt phase equilibria in the alumina-poor region of the systems MeF_2-Al_2O_3-SiO_2 at 1 bar and 1360° to 1640°C; *note vertical exaggeration*. Field above each curve system consists of single melt, and in fields below each curve systems consist of immiscible silicate and fluoride melts. Locations of curves and concentrations (mole %) are approximate.

Fluorine-rich experimental melts also exsolve from anhydrous, silica-poor melts at elevated P. Immiscibility occurs in the system $KAlSiO_4$-Mg_2SiO_4-SiO_2 at 28 kbar and ~1500° C with 10 wt % F (bulk silicate composition of 44% kalsilite, 39% forsterite, 17% silica, with 10% F substituting for O; Foley et al., 1986). Conversely, only one melt is stable in this system with 4 wt % F, suggesting F solubility is >4 but <10 wt % (Foley et al., 1986). Application of these results to natural mafic systems leads us to speculate that the exsolution of F-rich immiscible melts from mafic and ultramafic magmas appears unlikely because the F contents of such magmas are typically an order of magnitude or more below the ~10 to >20 wt % F required to stabilize immiscible melts.

Water-poor felsic melts. F solubility in dry, felsic aluminosilicate melts is a strong function of T and composition, although available experimental data are limited. The system NaF-$NaAlSi_3O_8$ at 1 bar shows a eutectic where the villiaumite (NaF) saturated melt contains approximately 7 wt % F, and has molar Na/Al of ~2 (Koster van Groos and Wyllie, 1967). With increasing T, the NaF-saturated liquids become more F- and Na-rich, until at superliquidus conditions ($\geq$1050° C, 1 bar), there appears to be either complete miscibility in this system (Koster van Groos and Wyllie, 1967), or a poorly-defined region of silicate-fluoride liquid immiscibility (Kogarko, 1967; Bailey, 1977).

The effect of pressure on F solubility is poorly known, but decreasing P does not reduce F solubilities to the same extent that it reduces H_2O solubilities. For example, only ~0.1 wt % H_2O dissolves in granitic melts at 1 bar $P_{(H_2O)}$ whereas silicate melts in the systems AlF_3-SiO_2, NaF-SiO_2, $Na_2O\pm Al_2O_3\pm SiO_2\pm F_2O_{-1}$, and CaO-CaF_2-SiO_2 dissolve up to   10 wt % F in substitution for O at similar conditions (Mysen and Virgo, 1985a; 1985b; Luth, 1988a).

The P-T-X conditions that have been examined in studies of water-poor, silica-rich melts are not particularly relevant to natural felsic magmas. Nonetheless, the existing data for simple MeF_2-Al_2O_3-SiO_2 systems indicate F solubilities of 7 to 22 wt % F and recent studies show that such melts may dissolve from 8 to >13 wt % F (Mysen and Virgo, 1985b; Webster, 1990).

Hydrous felsic melts. The F contents of felsic melts saturated with respect to F-enriched aqueous fluid have been the focus of many experimental studies (Table 4), mainly because peraluminous rhyolites and granites are associated with a large variety of hydrothermal, F-enriched granophile mineral deposits. The F concentration in melt is often expressed as the fluid/melt distribution coefficient for F (D_F); for most melts of geologic interest, F dissolves preferentially in the silicate melt phase and D_F is typically less than unity (Webster, 1990).

Fluorine solubility in fluid-saturated aluminosilicate melts and values of D_F depend strongly on melt composition. F solubility increases with the (alkali/Al) ratio (Koster van Groos and Wyllie, 1967; Anfiligov et al., 1973; Glyuk et al., 1980), and F is marginally more soluble in rhyolitic melt as small quantities of CO_2 are added to the system at 1 to 5 kbar and <1000° C (Webster, 1990). The effects of other compositional variables, such as H_2O content and the total alkali/Al and Na/(Na+K) ratios of melt, have not been established. The effect of the melt Na content on F solubility for NaF-$NaAlSi_3O_8$-H_2O is akin to that observed in the anhydrous system (Koster van Groos and Wyllie, 1967), and the results are consistent with Na-F complexing in hydrous melts.

The effects of T and P on F solubility are less well known. For example, the F contents of topaz rhyolite melts coexisting with $H_2O\pm CO_2$ fluids remain constant although D_F increases as T increases from 800° to 993° C at 2 kbar, and consequently, F solubility may be independent of T for granitic melts with ≤1.2 wt % F (Webster, 1990; Fig. 25). In contrast, Kovalenko (1977) observed increasing F solubility with increasing T for granite melts in equilibrium with HF solutions; however, the melt (Al+Na+K)/Si changed simultaneously with T, so the solubility variations are not solely attributable to T. Interpretation of the inadequate data base also suggests that F solubility is independent of P from 2 to 5 kbar for fluids containing $H_2O\pm CO_2$ at 800° to 950° C (Webster, 1990), but more data are needed to verify this.

Fluorine solubility is limited in some systems by the exsolution of fluoride melt from silicate melt (Grigorev and Iskyul, 1937); however, the occurrence of equilibrium liquid immiscibility in hydrous, felsic systems is controversial. Studies of fluid-saturated systems with large ratios of fluid/solid in the starting materials (Glyuk and Anfiligov, 1973; Anfiligov et al., 1973; Kovalenko, 1977; Glyuk et al., 1980) report textural evidence of liquid immiscibility with as little as 2 wt % F in silicate melt. Other studies conducted at similar conditions (0.5 to 5 kbar and 680° to 1000°C) but with smaller fluid/melt find no evidence of silicate-fluoride liquid immiscibility (Wyllie and Tuttle, 1961; Burnham, 1967; Hards, 1976; Manning, 1981; Dingwell, 1984, Webster and Holloway, 1987; London et al., 1988; Webster and Holloway, 1990; Webster, 1990; Keppler, 1993; and others), and in some of the latter experiments, the hydrous silicate glasses contain up to 6.9 wt % F. Although it has been suggested that the presence of two melts in the former studies may be a result of metastable or quench liquation (Manning, 1981), it is also likely that coexisting melts were stabilized in some of the former experiments using large fluid/melt ratios due to selective loss of some melt components to the aqueous fluid.

Summary of F solubilities.   Compared to most other volatiles, F is highly soluble in silicate melts, with many "granitic" melts dissolving more than 10 wt % F. In aqueous fluid-saturated igneous systems under conditions of geological interest, F dissolves preferentially into aluminosilicate melts rather than in fluids, and alkali fluoride contents of exsolved magmatic fluids may be quite low (Webster, 1990). See Figure 25. Fluorine in granitic melt must exceed 8 wt % before D_F exceeds unity and F partitions in favor of aqueous fluids. Although liquid immiscibility limits F solubilities in some anhydrous silicate systems (Ershova and Olshanskii, 1957; Ershova, 1957; Kogarko et al., 1968), the exsolution of F-rich melts from water-enriched and/or fluid saturated granites and its geologic relevance are less well understood.

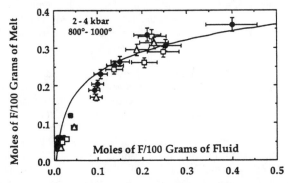

Figure 25.   Solubility of F in peraluminous haplogranite and topaz rhyolite melts at 2 kbar, 800°C (circles), 2 kbar, 1000°C (squares), and 4 kbar, 800°C (triangles) relative to computed concentrations of F in aqueous fluids. Solubility of F in melts is greater than in aqueous fluids and independent of P and T over the noted ranges. Molar Al / (Ca+Na+K) of melts ranges from 1.07 to 1.18. Coexisting silicate and fluoride melts were not observed in the run products from any of these experiments; data from Webster (1990) and J.D. Webster (unpublished data).

HF solubilities in silicate melts should be similar to those of H_2O because of similarities in their dipole moments and Redlich-Kwong a and b parameters (Holloway, 1981). In fact, the solubility of F in natural silicate melts at pressures of 1 to 4 kbar is $\geq$ than that of H_2O and is significantly greater than those of Cl, H_2S, H_2SO_4, or CO_2 (for similar P-T-X conditions). These relative solubilities can be understood in a qualitative sense by comparing dipole moments and Redlich-Kwong a and b parameters for HF and the other volatiles (Holloway, 1981) assuming that the HF Redlich Kwong parameters are analogous to those of the predominant fluoride species in silicate melts. For example, the dipole moment of HF is most similar to that of H_2O and significantly greater than those of HCl, H_2S, and CO_2; thus, the relative tendency of HF to ionize or induce ionization is most like that of H_2O. Moreover, the Redlich-Kwong b parameter for HF is also most similar to that of H_2O, so the tendency of each molecule to fit into the melt structure should be similar.

Solubility mechanisms and speciation of F in silicate magmas

Fluorine speciation and solubility mechanisms in highly polymerized aluminosilicate melts are strong functions of composition, in accord with the large effects of melt composition on F solubility described previously. A variety of simple F-bearing silicate melts prepared from Na_2O-$F_2O_{-1}$$\pm$$Al_2O_3$$\pm$$SiO_2$ have been investigated by Raman and FTIR spectroscopy, for example, and although these methods provide only indirect information on F speciation (Schaller et al., 1992), the data support a variety of metal-fluorine compexes, such as Al+F, Na+F, and Si+F (reviewed in Dingwell, 1985). It has been proposed that F should replace O in bridging SiO_4 tetrahedra of felsic glasses and melts, because of similarities in the anionic radii of F^- (1.33 Å) and O^{2-} (1.32 Å) (Dietzel, 1941; Buerger, 1948). Vibrational spectra have been used to suggest that $SiO_{4-x}F_x$ complexes occur in MeF-SiO_2 glasses, where Me is K, Rb, or Cs, (Takusagawa, 1980), SiO_3F^{3-} complexes occur in silica (Dumas et al., 1982), and $SiO_2F_2^{2-}$ (Duncan et al., 1986) and SiF_6^{2-} (Nakamoto, 1978; Mysen and Virgo, 1985b) complexes may also occur in compositionally simple felsic melts. Conversely, a recent study of Na_2O-SiO_2-NaF-$Al_2O_3$$\pm$$H_2O$ glasses (with molar Na/Al = 1) by NMR spectroscopy (MAS and CP/MAS methods) concluded that Si-F bonds in glasses are unlikely, although they could not be ruled out entirely (Schaller et al., 1992). Thus, it appears that if Si-F-O complexes occur in felsic glasses and melts, they are not particularly abundant (Mysen and Virgo, 1985a), and this is compatible with the observation that F does not depolymerize molten SiO_2 by complexing with Si. Rather, F increases the activity of highly-polymerized crystalline phases such as quartz in aluminosilicate melts (Kogarko and Krigman, 1973; Manning, 1981).

Species involving preferential associations of Al, Na, and F are important in aluminosilicate melts of geologic interest (Mysen and Virgo, 1985a; Schaller et al., 1992), but there are conflicting interpretations of spectroscopic data concerning the relative stabilities of the different fluoride species. It has been suggested that $NaF°$ complexes may occur in NaF-SiO_2 melts (Dingwell, 1985; Mysen and Virgo, 1985a), that $AlF_3°$ complexes may occur in AlF_3-SiO_2 melts(Mysen and Virgo, 1985a), and that $NaF°$ and $AlF_3°$ predominate over other F-bearing complexes in melts of albite-F_2O_{-1} and albite$_{50}$silica$_{50}$-F_2O_{-1} (Mysen and Virgo, 1985c). The preferential association of Na and F would explain the strong influence of Na on F solubility in Na-rich felsic melts (Range and Willgallis, 1964; Koster van Groos and Wyllie, 1967), and the observation that melts of NaF-Na_2O or $NaAlSi_3O_8$-NaF are completely miscible at superliquidus temperatures and 1 bar. The NMR spectroscopic data on F-bearing aluminosilicate melts also led investigators to suggest that F preferentially bonds with Al in AlF_6^{3-} complexes (Kohn et al., 1991;

Schaller et al., 1992), and possibly also in AlF_5^{2-} and AlF_4^- complexes (Kohn et al., 1991) in $Na_2O-SiO_2-NaF-Al_2O_3\pm H_2O$ glasses. However, Schaller et al. (1992) observed little evidence to support Na-F complexing in these glasses, and as a result, they proposed that F dissolution may be described by:

$$3\ NaAlSi_3O_{8\ melt} + 3\ F_2O_{-1} = Na_3AlF_{6\ melt} + Al_2O_{3\ melt} + 9\ SiO_{2\ melt} \quad (15)$$

which is equivalent to the reaction proposed by Manning et al. (1981) for hydrous haplogranite melts at 1 kbar:

$$AlSi_3O_8^-{}_{melt} + 6\ F^-{}_{melt} = AlF_6^{3-}{}_{melt} + 2\ O^{2-}{}_{melt} + 3\ SiO_{2\ melt} \quad (16)$$

Equation (16) may explain why the quartz-alkali feldspar cotectic and the minimum melt compositions shift away from the quartz apex in the haplogranite system as the F content of the melt increases (Manning et al., 1981). As a consequence of Equation (16), cations such as Na^+ that were initially held in a charge-balancing role with tetrahedrally-coordinated aluminum ions may be released to act as network modifiers, leading to melt depolymerization. With regard to these conflicting results, it should also be kept in mind that the relative importance of Na-F versus Al-F complexing apparently depends on composition, i.e., on the relative abundances of Na versus Al in melt (Dingwell, 1985).

Vibrational spectroscopy has also been applied to F- and alkaline earth-bearing silicate melts. However, these methods do not directly identify metal-fluorine bonds involving Na, Al, Ca, and Mg, because the Raman intensities are low owing to the highly ionic nature of some bonds (Luth, 1988a). Nevertheless, F appears to complex with alkaline earth metals in melts, for example:

$$CaSiO_{3\ melt} + F_2O_{-1} = CaF_{2\ melt} + SiO_{2\ melt}. \quad (17)$$

If the atomic (Ca/Si) in melt is greater than 1, then F is coordinated with excess Ca ions and it will not break Si-O bonds; conversely, with ratios less than 1, it has been suggested that F may complex with Si and Ca (Tsunawaki 1980). With more complicated melt compositions, other F-bearing complexes may occur; Foley et al. (1986) speculate that complexes of Mg-F and K-F may be stabilized:

$$Mg_2SiO_{4\ melt} + 2\ HF = MgSiO_{3\ melt} + MgF_{2\ melt} + H_2O \quad (18)$$

$$KAl^{IV}O_{2\ melt} + 2\ K^+{}_{melt} + 4\ HF = Al^{IV}F_{3\ melt} + KF_{melt} + 2\ H_2O, \quad (19)$$

and with $NaAlSiO_4-CaMgSi_2O_6-SiO_2-F_2O_{-1}$, F may dissolve (Luth, 1988a;b) as:

$$CaMgSi_2O_{6melt} + F_2O_{-1} = CaMgSi_2O_{6melt} + MgF_{2melt} + CaF_{2\ melt} + SiO_{2melt}. \quad (20)$$

In Equation (18), network modifying Ca and Mg ions are removed from the silicate framework and the silicate melt becomes increasingly polymerized as F is added to the system (Luth, 1988a). When melt polymerization increases because of F addition, this effect indicates that Si-F complexes are unimportant because Si-F bonding should lead to depolymerization of aluminosilicate melts. A detailed discussion of the influence of F on melt polymerization is beyond the scope of this chapter; interested readers should consult Dingwell (1985), Foley et al. (1986), Luth (1988a,b), and Chapter 9 of this volume.

At present, there are few published data that constrain the stability and relative importance of other fluoride complexes in aluminosilicate melts. For instance, the relative stabilities of NaF° and KF° in melt are undetermined, and even though HF° is an important species in greisenizing magmatic-hydrothermal fluids derived from mineralizing granitic magmas (Burt, 1981), the significance of H- and F-bearing complexes in granite magma is poorly known. It has been observed that the behavior of F in anhydrous albitic, jadeitic, and nephelinitic glasses is similar to that in hydrous haplogranites, and furthermore, that undissociated HF° is not a major species in the haplogranite glasses or presumably in the melts (Schaller et al., 1992). Although F⁻ (1.33 Å) substitutes for OH⁻ (1.37 Å) in some minerals (Shannon, 1976), it behaves differently in melts. Based on MAS-NMR evidence, the solution mechanism for F is dissimilar from that of H_2O in jadeite + NaF and jadeite + cryolite (Na_3AlF_6) glasses prepared at 3 to 3.5 kbar (Kohn et al., 1991). Hence, HF° may dissolve via a two-step process in silicate melts: (1) by dissociation and (2) through the formation of new complexes with other cations in the melt, perhaps in a manner similar to those of Equations (18) and (19). Given the potential importance of HF° in shallow, mineralizing magmatic systems, further work is needed to establish the speciation of F in natural, hydrous aluminosilicate melts at low P.

HALOGENS IN Cl- AND F-ENRICHED MAGMAS

Research on F and Cl in magmas and magmatic-hydrothermal systems, in light of experimental solubility data, helps constrain when and if halogen saturation occurs in natural systems. In this section we compare the halogen contents observed in experimental studies with those of natural systems. Most basaltic rocks contain from 0.001 to 0.1 wt % Cl (Correns, 1956; Brehler and Fuge, 1974; Michael and Schilling, 1989) and 0.01 to 0.1 wt % F (Correns, 1956; Allman and Koritnig, 1974), whereas most granites and rhyolites contain from 0.01 to 0.2 wt % Cl and F (Brehler and Fuge, 1974; Christiansen and Lee, 1986). Alkaline varieties of mafic to felsic igneous rocks may contain significantly greater halogen abundances, which is consistent with the observed effects of Na and K on Cl solubilities in melts (this is also due to the incompatible nature of F and Cl and their enrichment in evolved compositions). Some pantellerite glasses, for example, contain up to 1.3 wt % Cl and 1.5 wt % F (Kovalenko et al., 1988; Lowenstern and Mahood, 1991). However, most igneous whole rocks are not chemically representative of the magmas from which they were derived because Cl, F, and other volatiles degas from magma and hot rocks during and after emplacement/eruption (Fig. 26). Accordingly, we focus mainly on

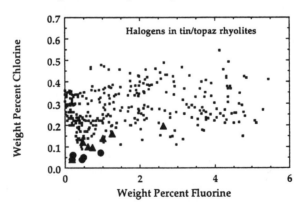

Figure 26. Abundances of Cl and F in felsitic topaz rhyolites (circles), vitrophyric topaz rhyolites (triangles), and >300 glass inclusions from tin and topaz rhyolites (small squares) of the western United States and central Mexico. Halogen abundances of glass inclusions are generally higher than those of vitrophyres which exceed those of felsitic rhyolites and are among the highest concentrations reported for metaluminous to peraluminous rhyolites. Analyses of vitrophyres and felsites are from Christiansen et al. (1986) and Congdon and Nash (1988); glass inclusion data are from Gavigan et al. (1989), Payette and Martin (1990), Webster and Duffield (1991; 1994), Webster and Burt (unpublished data).

data from glass inclusions trapped within crystals. Recent studies (Anderson et al., 1989; Hervig and Dunbar, 1992; Webster and Duffield, 1994) have shown that compositions of rhyolite glass inclusions are not significantly affected by boundary layer effects or post entrapment crystallization of the host phenocryst, and thus these glasses can provide useful estimates of pre-eruptive magmatic volatile contents (Roedder, 1984; Anderson et al., 1989; many others—see Chapter 8).

Chlorine in magmas

There are few constraints on the solubility and behavior of Cl in natural mafic igneous systems, even though some are of considerable economic importance. Research on volatiles in Pt-group element (PGE)-bearing mafic layered intrusions, for instance, suggests that the PGE's may move via Cl-rich magmatic fluids (Boudreau et al., 1986; Barnes and Campbell, 1988; Mathez et al., 1989), but the initial Cl contents of magma and the partitioning of Cl between mafic melts and fluids are unknown. Data on Cl in mafic magmas are also limited and not internally consistent; glass inclusions in basalts from Etna contain only 0.2 wt % Cl (Metrich and Clocchiatti, 1989), whereas other studies imply that Cl contents of some basalts may approach experimentally-determined saturation limits. For example, holocrystalline basaltic dikes crosscutting salt beds in New Mexico contain up to 1.15 wt % Cl (Calzia and Hiss, 1978; Roedder, 1992), which is roughly similar to the Cl solubility in molten tholeiite basalt equilibrated with molten NaCl at 1200° C, 20 kbar (1 to 2.5 wt %; Malinin et al., 1989), or with HCl vapor at 1 bar ($\leq$1.45 wt % Cl; Iwasaki and Katsura, 1967). Although glass inclusions from some mafic and intermediate volcanic rocks (Anderson, 1974; Metrich and Clocchiatti, 1989) contain significantly greater abundances of Cl and/or F than their whole-rock equivalents, we will now concentrate on Cl- and F-rich samples from highly evolved, halogen-enriched felsic magmas, mainly because it is in these systems that the halogens are most abundant.

Glass inclusions from a variety of chemically-evolved, silicic magmas have Cl and F abundances far exceeding those in glassy and holocrystalline whole-rock samples (Fig. 26). Quartz phenocrysts from mineralized tin and topaz rhyolites of the western United States and central Mexico contain glass inclusions with up to 0.5 wt % Cl and 5.5 wt % F, suggesting that these and other geochemically similar magmas (i.e., tin granites, tin greisen deposits, Climax-type molybdenum-mineralized porphyry systems) have higher halogen contents than would be inferred from analysis of degassed extrusive samples of similar whole-rock composition (Webster and Duffield, 1994). Because these high pre-eruptive halogen abundances are equivalent to experimentally determined Cl solubilities for felsic melts (Webster, 1993), it is likely that some highly evolved, mineralizing felsic magmas, or perhaps some volatile-enriched portions of these magmas, may saturate in alkali chloride melts. Other igneous systems show similar relationships; Cl contents of ~0.9 to 1.3 wt % in glass inclusions and matrix glasses from Pantelleria, Italy, (Kovalenko et al., 1988) are very similar to those observed in anhydrous Na-rich pantellerite glasses (quenched melts) equilibrated with molten NaCl at 2 kbar and 900° C (Lowenstern, 1994). Furthermore, glass inclusions in dacites erupted from Augustine volcano in 1976 contain up to 0.5 wt % Cl (Johnston, 1980) and rhyodacitic matrix glasses from mid-ocean ridge volcanics may contain up to 0.8 wt % Cl (Michael and Schilling, 1989). The available observations thus suggest that many evolved "granitic" magmas contain high pre-eruptive Cl abundances that are similar to those in experimental studies of coexisting silicate and chloride melts, and as a result, hydrous saline melts may exsolve from some felsic magmas.

Highly saline fluid inclusions with or without coexisting Cl-rich melt inclusions occur in phenocrysts from a variety of felsic igneous environments, and it has been suggested

(Roedder, 1992) that they are the direct result of liquid immiscibility and not a consequence of boiling aqueous fluids. For example, phenocrysts from Ascension Island contain inclusions of glass ± coexisting alkali chloride minerals and aqueous fluids (Roedder and Coombs, 1967), interpreted to indicate that an alkali chloride-rich melt or "salt melt" had exsolved from granitic magma. These metaluminous to mildly peralkaline glass inclusions have recently been determined to be volatile-rich, containing 0.25 to 0.54 wt % Cl, 3.5 to ~5.5 wt % H_2O, and 0.3 to 1.5 wt % F (J.D. Webster and E. Roedder, unpublished data). Alkali chloride-rich aqueous fluids have also apparently exsolved from granite magmas associated with porphyry copper deposits (Quan et al., 1987), Sn-W greisens (Roedder, 1984), and from non-mineralized alkaline granites (Frost and Touret, 1989; Roedder, 1992). Roedder (1992) has described 14 other igneous systems involving coeval felsic silicate and saline melts, thus suggesting that saline melts may be a common feature of highly differentiated felsic magmas, and in such systems, the activity of Cl in silicate melt may be buffered by coexisting salt melt and/or Cl-rich aqueous fluid phase(s).

Fluorine in magmas

Compositions of glass inclusions from tin and topaz rhyolites indicate that some fractions of highly evolved magmas contain ≥5 wt % F (Fig. 26), and this agrees with F abundances estimated by other methods. For example, Carten (1987) and Carten et al. (1988) observed up to 7 wt % F in whole-rock samples from the Mo-mineralized Seriate stock at Henderson, Colorado, and mass balance constraints suggest that magma in the apex of this stock carried at least 5 wt % F. Likewise, matrix glasses from topaz rhyolites contain up to 3 wt % F (Congdon and Nash, 1988), and Mongolian ongonites have been reported to contain up to 4.9 wt % F (Kovalenko, 1973). Experiments (Table 4, above) demonstrate that anhydrous, metaluminous to mildly peraluminous "granitic" melts may dissolve as much as 8 wt % F at 0.5 to 5 kbar and 775° to 1000°C, and albitic melts may hold up to roughly 19 wt % at 1400°C and 15 kbar without exsolving a separate F-rich melt phase. Thus, F contents of peraluminous to peralkaline granitic magmas may reach many wt % without-causing fluoride-silicate immiscibility. Water-rich felsic melts may achieve similarly high F concentrations, but the F contents of melts saturated in an aqueous fluid will be moderated by partitioning of F between melt and fluid (Webster, 1990). Overall, F concentrations in most natural, highly-evolved felsic magmas are far lower than experimentally determined solubilities, and hence, the fluorine activity (a_F) in most natural magmas will not be buffered by exsolving fluoride melts; more likely, the magma a_F will be buffered by the crystallization of F-rich minerals or the degassing of F into coexisting (F-poor) aqueous fluids. We have not addressed the behavior of F in less common silicate melts such as nepheline syenites or carbonatites, however, and it is possible that F behavior may be quite different in such systems. For example, the activities of fluoride species in these and other alkaline magmas may be buffered by the precipitation of magmatic villiaumite or other alkali-fluoride and/or alkaline earth-fluoride minerals (Stormer and Carmichael, 1970; Kogarko, 1976).

Volcanic degassing of F and Cl

Volcanic halogen emission influences the degradation of the ozone layer in the atmosphere (Johnston, 1980; Albritton, 1989) and contributes to production of acid rain through acid halide species. Consequently, to constrain volcanic degassing of Cl and F and to identify those magma types which may be associated with large halogen emissions during eruption, we must better determine halogen solubilities in magmas at shallow crustal conditions. Although few published studies specifically address degassing of F and Cl from hydrous magmas during decompression and ascent toward the surface, data interpretation suggests that anhydrous melts at 1 bar may retain significant quantities of F during and after eruption and this has important consequences for magma viscosities during and after eruption (e.g., Dingwell et al., 1985). Lower melt viscosities also influence vapor bubble growth and ascent, lava flow rates, and under special circumstances, high F contents facilitate fountain-fed flows of rhyolite lava (Webster and Duffield, 1994). Exsolution of acid halides (HF and HCl) also affects volcanic processes such as hydrothermal alteration and associated mineralization. This is of particular importance for Cl, because of its inherently low solubility in aluminosilicate melts and its preference to dissolve in aqueous fluids.

The effects of F and Cl on volcanic eruption processes and on Earth's atmosphere and climate are a direct function of the predominant fluoride and chloride species that degass. Unfortunately, little is known about halide speciation in vapors and melts. Thermodynamic modelling based on volcanic gas and condensates from felsic magma indicates that the predominant chloride complexes in volcanic gas involve $HCl°$ and up to 10 metal-chloride species (see Chapter 1) and that the predominant fluoride complexes include $HF°$ and various Si-F and Al-F species (Symonds et al., 1992). Applicable solubility data are extremely limited, but the concentration of $HCl°$ in melt apparently increases relative to those of other chloride species ($NaCl°$, $KCl°$, Fe-Cl complexes, and others) as P decreases (Shinohara et al., 1989). The effects of P and T on fluoride species in melts are virtually unknown.

Thermodynamic constraints on Cl and F solubilities

Theoretical treatments of F and Cl dissolution in silicate and aluminosilicate melts have, for the most part, been limited to computing stabilities of halide species (based on differences in electronegativity) and fugacities of acid halide species. Although activities of silicate species in F-bearing aluminosilicate melts have been addressed to some extent (Kogarko and Krigman, 1973; Luth, 1988b), the activities of F-bearing species are poorly known. Equilibrium constants of simple exchange reactions such as:

$$MeX_{melt} + 1/2\ H_2O_{fluid} = 1/2\ Me_2O_{melt} + HX_{fluid}, \quad (21)$$

where Me is a metal cation and X = Cl or F, favor the degassing of HF and HCl from melts (melt) with increasing cation electronegativity (Kogarko, 1974). Because the electronegativity of Si exceeds that of other major cations in the order K<Na<Ca<Mg<Al<Fe$\leq$Si (Pauling, 1960), halogen solubilities should decrease as the melt a_{SiO_2} increases, and consequently, alkali halide complexes should be more stable than complexes involving halogens and Fe, Al, or alkaline earth metals. This is consistent with experimental observations, spectroscopic data, and with speciation and solubility mechanisms discussed previously.

The fugacities of HF, f_{HF}, and HCl, f_{HCl}, and relative fugacities $f_{HCl}\ /\ f_{H_2O}$ and f_{HF}/f_{H_2O} may be calculated using data for mineral assemblages containing hydrous phases such as micas, amphiboles, apatite, and topaz or other minerals containing F or Cl as an essential constituent (e.g., villiaumite or marialite) (many references summarized in Munoz, 1984; Förster, 1990; Zhu and Sverjensky, 1991). For example, f_{HCl}/f_{H_2O} and f_{HF}/f_{H_2O} may be constrained by exchange equilibria such as

$$mica(OH) + HX = mica(X) + H_2O \quad (23)$$

where X is either Cl⁻ or F⁻ (detailed in Munoz, 1984). In this case the ratio of H_2O to HCl fugacities can be obtained from mica compositions by using the relation

$$\log\left[\frac{f_{H_2O}}{f_{HCl}}\right] = \frac{5151}{T} - 5.01 - 1.93X_{Mg} - \log\left[\frac{X_{Cl}}{X_{OH}}\right] \qquad (22)$$

which is based on experimental calibration. Many studies of micas or amphiboles have constrained absolute and relative fugacities of HF and HCl by using these equilibria and by accounting for the effects of T, the cation population of the octahedral sheet, and the effects of secondary fluid-rock interactions on halogen concentrations. The validity of this approach was recently verified by the derivation of standard-state thermodynamic properties for F and Cl endmember phases (Zhu and Sverjensky, 1991; 1992); their results exhibit internal consistency and are also compatible with other thermodynamic constraints. Variations in F and Cl concentrations of apatites have also been used to constrain relative halogen fugacities in volcanic and plutonic magmas (Piccoli and Candela, 1994). Vapor saturated and vapor undersaturated magmas have been thermodynamically modeled by assuming ideal mixing of the quasichemical components Cl, H_2O, O°, OH in the anion matrix of Cl-bearing melts and of F, H_2O, O°, OH in the anion matrix F-bearing melts (Candela, 1986) using reactions like:

$$4Ca_5(PO_4)_3Cl_{(Apatite)} = 18CaO_{(melt)} + 6P_2O_{5(melt)} + 4Ca_{0.5}Cl_{(melt)} \qquad (24)$$

This technique can also be used to estimate initial halogen concentrations of granitic magmas. Studies such as these have shown that micas and apatites from a large variety of mineralized and barren granites and rhyolites exhibit log (f_{HCl}/f_{H_2O}) ranges of -1.5 to -5.0 and log (f_{HF}/f_{H_2O}) of -1.9 to -5.5 (Förster et al., 1989; Förster, 1990) at magmatic equilibration temperatures (i.e., ≥550° C). Given geologically reasonable ranges in water fugacities, it follows that values of f_{HF} normally range from 0.01 to 1 bar and rarely exceed 10 bar; f_{HCl} is normally ≤1 bar (Förster, 1990).

The apparent halogen concentrations of natural magmas and associated experimental results, discussed previously, permit us to speculate about the mixing behavior of F and Cl in aluminosilicate melts. Ideal mixing assumes that dissolved ions or molecules do not interact preferentially with other melt constituents; however, aluminosilicate melts at their solubility limits exhibit comparatively low solubilities for Cl and exsolve alkali halide-rich melts. Thus, small increases in Cl concentration will apparently have a significant effect on increasing the activity coefficient of Cl, γ_{Cl}, in such melts (Webster, 1992a), and as a consequence, it is not appropriate to assume ideal mixing for many Cl-enriched magmas. On the other hand, solubility limits for F in aluminosilicate melts far exceed the apparent F contents of most mafic to felsic magmas, and as a result it may be safe to assume that F mixes ideally in many of such magmas.

CONCLUSIONS

In this brief overview of solubilities of the less abundant magmatic volatile species we have attempted to outline the current state of knowledge concerning solubilities as a function of intensive and extensive variables, mainly as derived from experimental studies, augmented by data from natural samples where appropriate. The behavior of volatile species in natural systems, and applications of experimental solubility data are discussed further in other chapters of this volume. Our general findings and suggestions for future research can be summarized as follows:

(1) *Sulfur*— solubilities under reducing conditions where S^{2-} predominates are reasonably well known but few data exist concerning more oxidizing conditions where sulfate sulfur may play an important role in determining solubility behavior. Sulfur speciation in the gas phase (e.g., as H_2S, S_2, SO_2, SO_3) may have a

significant influence on S solubility in magmas but details of the relative solubilities of different sulfur species are not well known. Oxidized sulfur species may be particularly important in subduction zone magmas, where oxidation states are often higher than those typical of MOR settings. Also poorly understood are potential redox interactions between iron and sulfur, and the factors that control variations in sulfur and oxygen fugacities in cooling, differentiating magmas.

(2) *Noble gases* — solubilities are relatively well known at low pressures, almost independent of T, and almost linearly dependent on gas fugacity below ~1000 bars. The effects of melt composition on solubility can be quantified by considering systematic relationships between gas atom size and the ionic porosity or molar volume (<70% SiO_2 melts) of the silicate matrix. The effects of melt compressibility on high pressure solubilities are not known but such effects could be important. A major uncertainty in describing noble gas behavior in natural systems is how they are partitioned between minerals and melts and further experimental work is required on this topic.

(3) *Nitrogen* — little is known regarding nitrogen solubility in magmas, either from experimental studies or from analyses of natural samples. Molecular N_2 solubility appears to be similar to that of Ar, which is similar in size, but further work is needed on N-H and N-O species that may be important for some conditions within the earth.

(4) *Chlorine* — partitioning behavior between melts and aqueous fluids has been studied for a limited range of compositions and conditions, and available data show relatively low Cl solubilities in silicate melts, with a preference for Cl to concentrate in aqueous fluids. Solubilities appear to be higher in more mafic melts, and in water-rich aqueous fluids. However, the speciation of Cl in silicate melts, the effects of melt and fluid composition, and P and T on melt–vapor partitioning are still not well known for the range of compositions encountered in nature.

(5) *Fluorine* — solubility in silicate melts is high and, thus, F is strongly concentrated in magmas relative to any coexisting vapor phase. As is true for Cl, the effects of melt and fluid composition, and P and T on F solubility in melts are still not well known for the range of compositions encountered in nature. The high F abundances in some evolved felsic magmas undoubtedly have a large influence on melt physical properties and phase equilibrium relations, and further experimental work is needed before such effects can be fully quantified.

ACKNOWLEDGEMENTS

This chapter was written while MRC was supported by NERC Grants GR9/884 and GR3/7776, and JDW was supported by NSF Grant EAR-9315683. We thank Paul Wallace, Bob Luth, and Don Dingwell for their thoughtful reviews.

REFERENCES

Abraham KP, Richardson FD (1960) Sulphide capacities of silicate melts Part II. J Iron Steel Inst 196:313-317
Abraham KP, Davis MW, Richardson FD (1960) Sulphide capacities of silicate melts. J Iron Steel Inst 196:309-312
Albritton DL (1989) Stratospheric ozone depletion: global processes. In: Ozone Depletion, Greenhouse Gases, and Climate Change. National Research Council, Washington, DC, p 10-18

Allegre CJ, Staudacher T, Sarda P (1986) Rare gas systematics: formation of the atmosphere, evolution and structure of the Earth's mantle. Earth Planet Sci Lett 81:127-150

Allman R, Koritnig S (1974) Fluorine. In Wedepohl KH (ed), Handbook of Geochemistry. Springer-Verlag, Berlin, 9-A-9-O

Anderson, AT (1974) Chlorine, sulfur, and water in magmas and oceans. Geol Soc Am Bull 85:1485-1492

Anderson AT Jr, Newman S, Williams SN, Druitt TH, Skirius C, Stolper E (1989) H_2O, CO_2, Cl, and gas in plinian and ash-flow Bishop rhyolite. Geology 17:221-225

Anfiligov VN, Glyuk DS, Trufanova LG (1973) Phase relations in interactions between granite and NaF at water vapour pressure of 1000 kg/cm^2. Geochem Int'l 15:30-32

Baak T, Ölander A (1955) The system $CaSiO_3$-CaF_2. Acta Chemica Scandia 9:1350

Bai TB, Koster van Groos A (1994) Diffusion of chlorine in granitic melts. Geochim Cosmochim Acta 58:113-124

Bailey JC (1977) Fluorine in granitic rocks and melts: A review. Chem Geol 19:1-42

Baker DR, Watson EB (1988) Diffusion of major and trace elements in compositionally complex Cl- and F-bearing silicate melts. J Non-Cryst Solids 102:62-70

Barker DS (1976) Phase relations in the system $NaAlSiO_4$-SiO_2-NaCl-H_2O at 400°-800° C and 1 kilobar, and petrologic implications. J Geol 84:97-106

Barnes SJ, Campbell IH (1988) Role of late magmatic fluids in Merensky-type platinum deposits: A discussion. Geology 16:488-491

Binsted N (1981) The system Ab-Ne-NaCl-H_2O. NERC Prog Exp Petrol D-18:34-36

Blank JG, Stolper EM, Carroll MR (1993) Solubilities of carbon dioxide and water in rhyolitic melts. Earth Planet Sci Lett 119:27-36

Bottinga Y, Javoy M (1990a) Mid-ocean ridge basalt degassing: Bubble nucleation. J Geophys Res 95:5125-5131

Bottinga Y, Javoy M (1990b) MORB Degassing: Bubble growth and ascent. Chem Geol 81:255-270

Boudreau AE, Mathez EA, McCallum IS (1986) Halogen geochemistry of the Stillwater and Bushveld Complexes: Evidence for transport of the platinum-group elements by Cl-rich fluids. J Petrol 27:967-986

Brehler B, Fuge R (1974) Chlorine. In Wedepohl KH (ed) Handbook of Geochemistry. Springer-Verlag, Berlin, 17-A-17-O

Broadhurst CL, Drake MJ, Hagee BE, Bernatowicz TJ (1990) Solubility and partitioning of Ar in anorthite, diopside, forsterite, spinel, and synthetic basaltic liquids. Geochim Cosmochim Acta 54:299-309

Broadhurst CL, Drake MJ, Hagee BE, Bernatowicz, TJ (1992) Solubility and partitioning of Ne, Ar, Kr, and Xe in minerals and synthetic basalt melts. Geochim Cosmochim Acta 56:709-723

Buchanan DL, Nolan J (1979) Solubility of sulfur and sulfide immiscibility in synthetic tholeiitic melts and their relevance to Bushveld-complex rocks. Can Mineral 17:483-494

Buchanan DL, Nolan J, Wilkinson N, DeVilliers JPR (1983) An experimental investigation of sulfur solubility as a function of temperature in synthetic silicate melts. Spec Pub Geol Soc S Africa 7:383-391

Buerger MJ (1948) The structural nature of the mineralizer action of fluorine and hydroxyl. Am Mineral 33:11-12

Burnham CW (1967) Hydrothermal fluids at the magmatic stage. In Barnes HL (ed) Geochemistry of Hydrothermal OreDeposits. John Wiley & Sons, New York, 34-76

Burnham CW (1979) Magmas and hydrothermal fluids. In: Barnes HL (ed) Geochemistry of Hydrothermal Ore Deposits, John Wiley & Sons 2:71-136

Burt DM (1981) Acidity-salinity diagrams: Application to greisen and porphyry deposits. Econ Geol 76:832-843

Calzia JP, Hiss WL (1978) Igneous rocks in northern Delaware Basin, New Mexico and Texas. New Mexico Bur Mines Mineral Res Circ 159:39-45

Candela PA (1986) Toward a thermodynamic model for the halogens in magmatic systems: An application to melt-vapor-apatite equilibria. Chem Geol 57:289-301

Carroll MR, Rutherford MJ (1985) Sulfide and sulfate saturation in hydrous silicate melts. J Geophys Res 90:C601-C612

Carroll MR, Rutherford MJ (1987), The stability of igneous anhydrite: experimental results and implications for sulfur behavior in the 1982 El Chichon trachyandesite and other evolved magmas. J Petrol 28:781-801

Carroll MR, Rutherford MJ (1988) Sulfur speciation in hydrous experimental glasses of varying oxidation state: results from measured wavelength shifts of sulfur X-rays. Am Mineral 73:845-849

Carroll MR, Stolper EM (1991) Argon solubility and diffusion in silica glass: Implications for the solution behavior of molecular gases. Geochim Cosmochim Acta 55:211-225

Carroll MR, Stolper EM (1993) Noble gas solubilities in silicate melts and glasses: new experimental results for Ar and the relationship between solubility and ionic porosity. Geochim Cosmochim Acta 57:5039-5051.

Carroll MR, Sutton S, Rivers M, Woolum D (1993) Krypton solubility and diffusion in silicic melts. Chem Geol 109:9-28

Carten RB (1987) Evolution of immiscible Cl- and F-rich liquids from ore magmas, Henderson porphyry molybdenum deposit, Colorado. Geol Soc Am Absts Prog 19:613

Carten RB, Geraghty EP, Walker BM, Shannon JR (1988) Cyclic development of igneous features and their relationship to high-temperature hydrothermal features in the Henderson porphyry molybdenum deposit, Colorado. Econ Geol 83:266-296

Chennaoui-Aoudjehane H, Jambon A (1990) He solubility in silicate glasses at 250°C: a model for calculation. Eur J Mineral 2:539-545

Christiansen EH, Lee DE (1986) Fluorine and chlorine in granitoids from the Basin and Range province, western United States. Econ Geol 81:1484-1494

Christiansen EH, Sheridan MF, Burt DM (1986) The geology and geochemistry of Cenozoic topaz rhyolites from the western United States. Geol Soc Am Spec Paper 205:82

Congdon RD, Nash WP (1988) High fluorine rhyolite: An eruptive pegmatite magma at the Honeycomb Hills. Geol 16:1018-1021

Connolly JWD, Haughton DR (1972) The valence of sulfur in glass of basaltic composition formed under low oxidation potential. Am Mineral 57:1515-1517

Correns CW (1956) The geochemistry of the halogens. In Ahrens LH, Rankama K, Runcorn SK (eds) Physics and Chemistry of the Earth. Pergamon Press, London 181-233

Czamanske GK, Moore JG (1977) Composition and phase chemistry of sulfide globules in basalt from the Mid-Atlantic Ridge rift valley near 37°N lat. Geol Soc Am Bull 88:587-599

Danckwerth PA, Hess PC, Rutherford MJ (1979) The solubility of sulfur in high TiO_2 mare basalts. Proc Lunar Planet Sci Conf 10:517-530

Davies MW, Meherali, SG (1971) The equilibrium solubility of nitrogen in aluminosilicate melts. Metal Trans 2:2729-2733

Devine JD, Sigurdsson H, Davis AN, Self S (1984) Estimates of sulfur and chlorine yields to the atmosphere from volcanic eruptions and potential climatic effects J Geophys Res 89:6309-6325

Dietzel A (1941) Strukturchemie des glasses. Naturwis 29:537

Dingwell DB (1984) Investigations of the role of fluorine in silicate melts: Implications for igneous petrogenesis. Unpub. PhD dissertation, University of Alberta, Edmonton, 149 p

Dingwell DB (1985) The structure and properties of fluorine-rich magmas: a review of experimental studies. In: Taylor RP, Strong DF (eds) Recent Advances in the Geology of Granite-Related Mineral Deposits. Can Inst Mining Metal 39:1-12

Dingwell DB (1986) Volatile solubilities. Mineral Assoc Canada Short Course on Silicate Melts

Dingwell DB (1987) Melt viscosities in the system $NaAlSi_3O_8-H_2O-F_2O_{-1}$. In Mysen BO (ed) Magmatic Processes: Physicochemical Principles, The Geochem Soc 1:423-432

Dingwell DB, Scarfe CM, Cronin DJ (1985) The effect of fluorine on viscosities in the system $Na_2O-Al_2O_3-SiO_2$: Implications for phonolites, trachytes and rhyolites. Am Mineral 70:80-87

Dixon JE, Stolper EM, Holloway JR (1994) An experimental study of water and carbon dioxide solubility in midoceanic ridge basalt. J Petrol, submitted.

Doremus RH (1966) Physical solubility of gases in fused silica. J Am Ceram Soc 49:461-462.

Doremus RH (1973) Glass Science. Wiley, New York, 21-145

Dowty E (1980) Crystal-chemical factors affecting the mobility of ions in minerals. Am Mineral 65: 174-182

Dumas P, Corset J, Carvalho W, Levy Y, Neuman Y (1982) Fluorine-doped vitreous silica analysis of fiber optic performs by vibrational spectroscopy. J Non-Cryst Solids 70:239-241

Dunbar NW, Kyle PR (1992) Volatile contents of obsidian clasts in tephra from the Taupo Volcanic Zone, New Zealand: Implications to eruptive processes. J Volcanol Geotherm Res 49:127-145

Dymond J, Hogan L (1973) Noble gas abundance pattern in deep-sea basalts - primordial gases from the mantle. Earth Planet Sci Lett 20:131-139

Dymond J, Hogan L (1978) Factors controlling the noble gas abundance patterns of deep-sea basalts. Earth Planet Sci Lett 38:117-128

Ershova ZP (1957) Equilibrium of immiscible liquids in the systems of the $MeF_2-Al_2O_3-SiO_2$ type. Geochem 4:350-358

Ershova ZP, Olshanskii YI (1957) Equilibrium of immiscible liquids in the systems of the $MeF_2-MeO-SiO_2$ type. Geochem 3:257-266

Faile SP, Roy DM (1966) Solubilities of Ar, N_2, CO_2, and He in glasses at pressures to 10 kbars. J Am Ceram Soc 49:638-643

Fincham CJB, Richardson FD (1954) The behavior of sulfur in silicate and aluminate melts. Phil Trans Roy Soc London A223:40-62

Fine G, Stolper E (1985) The speciation of carbon dioxide in sodium aluminosilicate glasses. Contrib Mineral Petrol 91:105-121

Fisher DE (1986a) Radiogenic rare gases and the evolutionary history of the depleted mantle. J Geophys Res 90:1801-1807

Fisher DE (1986b) Rare gas abundances in MORB. Geochim Cosmochim Acta 50:2531-2541

Fogel RA, Rutherford MJ (1990) The solubility of carbon dioxide in rhyolitic melts: a quantitative FTIR study. Am Mineral 75:1311-1326

Foley SF, Taylor WR, Green DH (1986) The effect of fluorine on phase relationships in the system $KAlSiO_4$-Mg_2SiO_4-SiO_2 at 28 kbar and the solution mechanism of fluorine in silicate melts. Contrib Mineral Petrol 93:46-55

Förster H-J (1990) Halogen fugacities (HF, HCl) in melts and fluids: A survey of published data. Z. Geol. Wiss. 18:255-266

Förster H-J, Thomas R, Tischendorf G (1989) Physicochemical conditions as controlling factors on magmatism and metallogenesis. In Tischendorf G (ed) Silicic Magmatism and Metallogenesis of the Erzgebirge. ICL Comm Pub, Potsdam, 0171:221-243

Fortier SM, Giletti BJ (1989) An empirical model for predicting diffusion coefficients in silicate minerals. Science 245:1481-1484

Frischat GH, Schrimpf C (1980) Preparation of nitrogen-containing Na_2O-CaO-SiO_2 glasses. J Am Ceram Soc 63:714-715

Frost BR, Touret JLR (1989) Magmatic CO_2 and saline melts from the Sybille Monzosyenite, Laramie Anorthosite complex, Wyoming. Contrib Mineral Petrol 103:178-186

Gavigan TH, Nash WP, Webster JD (1989) Preeruptive gradients in silicic magma as recorded by melt inclusions. Trans Am Geophys Union EOS 70:1417

Gerlach TM (1986) Exsolution of H_2O, CO_2, and S during eruptive episodes at Kilauea Volcano, Hawaii. J Geophys Res 91:12177-12185

Gerlach TM, Graeber EJ (1985) Volatile budget of Kilauea volcano. Nature 313:273-277

Glyuk DS, Anfiligov (1973) Phase equilibria in the system granite-H_2O-HF at a pressure of 1000 kg/cm^2. Geokhim 3: 434-438

Glyuk DS, Trufanova LG (1980) Phase relations in the granite-H_2O-LiF system at 1000 kg/cm^2. Geokhim 9:1327-1342

Grigorev DP, Iskyul EV (1937) Differentsiatsiya Neokotorykh Silikatnykh Rasplavov Kak Rezul'tat Obrazovaniya Dvukh Nesmeshivayuschchikhsys Zhidkostei (Differentiation in certain silicate melts as the result of formation of immiscible liquids). Izvestiya 1, Izd. Akademii Nauk SSSR:77

Hamilton DL, Burnham CW, Osborn EF (1964) The solubility of water and effects of oxygen fugacity and water content on crystallization in mafic magmas. J Petrol 5:1-39

Hards NJ (1976) Distribution of elements between the fluid phase and silicate melt phase of granites and nepheline syenites. NERC Rpt Prog Exp Petrol 3:88-90

Haselton HT, Cygan GL, D'Angelo WM (1988) Chemistry of aqueous solutions coexisting with fluoride buffers in the system K_2O-Al_2O_3-SiO_2-H_2O-F_2O_{-1} (1 kbar, 400°-700°C). Econ Geol 83:163-173

Haughton DR, Roedder PL, Skinner BJ (1974) Solubility of sulfur in mafic magmas. Econ Geol 69:451-467

Hayatsu A, Waboso CE (1985) The solubility of rare gases in silicate melts and implications for K-Ar dating. Chem Geol 52:97-102

Haynes SJ, Clark AH (1972) A rapid method for the determination of chlorine in silicate rocks using ion-selective electrodes. Econ Geol 67:378-382

Hervig R, Dunbar NW (1992) Cause of chemical zoning in the Bishop (California) and Bandelier (New Mexico) magma chambers. Earth Planet Sci Let 111:97-108

Hess PC (1977) Structure of silicate melts. Can Mineral 15:162-178

Holland HD (1972) Granites, solutions, and base metal deposits. Econ Geol 67:281-301

Holloway JR (1976) Fluids in the evolution of granitic magmas: Consequences of finite CO_2 solubility. Geol Soc Am Bull 87:1513-1518

Holloway JR (1981) Volatile interactions in magmas. In: Newton RC, Navrotsky A, Wood BJ (eds) Thermodynamics of Minerals and Melts. 1 Springer-Verlag, New York:273-294

Holloway JR, Reese RL (1974) The generation of N_2-CO_2-H_2O fluids for use in hydrothermal experimentation I: Experimental method and equilibrium calculation in the C-O-H-N system. Am Mineral 59:587-597

Holtz F, Dingwell DB, Behrens H (1993) Effects of F, B_2O_3, and P_2O_5 on the solubility of water in haplogranite melts compared to natural silicate melts. Contrib Mineral Petrol 113:492-501

Iwasaki B, Katsura T (1967) The solubility of hydrogen chloride in volcanic rock melts at a total pressure of one atmosphere and at temperatures of 1200° C and 1290° C under anydrous conditions. Bull Chem Soc Jap 40:554-561

Iwasaki I, Utsumi S, Ozawa T (1952) New colorimetric determination of chloride using mercuric thiocyanate and ferric iron. Bull Chem Soc Japan 25:226

Jambon A (1987) He solubility in silicate melts: a tentative model of calculation. Chem Geol 62:131-136

Jambon A, Shelby JE (1980) Helium diffusion and solubility in obsidians and basaltic glass in the range 200°-300°C. Earth Planet Sci Lett 51:206-214

Jambon A, Weber HW, Begemann F (1985) Helium and argon from an Atlantic MORB glass: concentration, distribution and isotopic composition. Earth Planet Sci Lett 73:255-267

Jambon A, Weber H, Braun O (1986) Solubility of He, Ne, Ar, Kr, and Xe in a basalt melt in the range 1250-1600°C: Geochemical implications. Geochim Cosmochim Acta 50:401-408

Javoy M (1984) Solutions des carbone et d'azote dans les magmas tholeitiques: aspects chemiques et isotopiques. Proceedings CNRS Colloque Prevision et Surveillance des Eruptions Volcaniques, Clermont-Ferrand, France

Javoy M, Pineau F (1991) The volatiles record of a "popping" rock from the Mid-Atlantic Ridge at 14°N: chemical and isotopic composition of gas trapped in the vesicles. Earth Planet Sci Lett 107:598-611

Johansen O, Steinnes E (1967) Determination of chlorine in U.S.G.S. standard rocks by neutron activation analysis. Geochim Cosmochim Acta 31:1107-1109

Johnston D (1980) Volcanic contribution of chlorine to the stratosphere: More significant to ozone that previously estimated? Science 209:491-492

Katsura T, Nagashima S (1974) Solubility of sulfur in some magmas at 1 atmosphere. Geochim Cosmochim Acta 38:517-531

Kelen T, Mulfinger HO (1968) Mechanism of chemical solution of nitrogen in glass melts. Glastech Ber 41:230-242

Keppler H (1993) Influence of fluorine on the enrichment of high field strength trace elements in granitic rocks. Contrib Mineral Petrol 114:479-488

Keppler H, Wyllie PJ (1991) Partitioning of Cu, Sn, Mo, W, U, and Th between melt and aqueous fluid in the systems haplogranite-H_2O-HCl and haplogranite-H_2O-HF. Contrib Mineral Petrol 109:139-150

Kesson SE, Holloway JR (1974) The generation of N_2-CO_2-H_2O fluids for use in hydrothermal experimentation II. Melting of albite in a multispecies fluid. Am Mineral 59:598-603.

Khitarov NI, Kadik AA (1973) Water and carbon dioxide in magmatic melts and peculiarities of the melting process. Contrib Mineral Petrol 41:205-215

Kilinc IA, Burnham CW (1972) Partitioning of chloride between a silicate melt and coexisting aqueous phase from 2 to 8 kilobars. Econ Geol 67:231-235

Kirsten T (1968) Incorporation of rare gases in solidifying enstatite melts. J Geophys Res 73:2807-2810

Kogarko LN (1967) Immiscibility region in the melts containing Si, Al, Na, O, and F. Dokl AN SSSR 176:221-225

Kogarko LN (1974) Role of volatiles. In Sorensen H (ed) The Alkaline Rocks. John Wiley & Sons, p 474-487

Kogarko VI, Krigman LD (1973) Structural position of fluorine in silicate melts (according to melting curves). Geokhim 1:49-56

Kogarko LN, Ryabchikov ID (1961a) Differentiation of alkali magmas rich in volatiles. Geokhim 12: 1439-1450

Kogarko LN, Ryabchikov ID (1961b) Dependence of the content of halogen compounds in the gaseous phase on the chemistry of the magma. Geochem 12:1192-1201

Kogarko LN, Krigman LD, Sharudilo NS (1968) Experimental investigations of the effect of alkalinity of silicate melts on the separation of fluorine into the gas phase. Geokhim 8:948-956

Kohn SC, Dupree R, Mortuza MG, Henderson CMB (1991) NMR evidence for five- and six-coordinated aluminum fluoride complexes in F-bearing aluminosilicate glasses. Am Mineral 76:309-312

Koster van Groos AFK, Wyllie PJ (1967) Melting relationships in the system $NaAlSi_3O_8$-NaF-H_2O to 4 kb pressure. J Geol 76:50-70

Koster van Groos AFK, Wyllie PJ (1969) Melting relationships in the system $NaAlSi_3O_8$-NaCl-H_2O at one kilobar pressure, with petrological applications. J Geol 77:581-605

Kotlova AG, Ol'shanskii YI, Tsvetkov AI (1960) Some trends in immiscibility effects in binary silicate and borate systems. Mineral i Geokhim Tr Inst Geol Rudn Mest AN SSSR 42:3

Kovalenko VI (1977) The reactions between granite and aqueous hydrofluoric acid in relation to the origin of fluorine-bearing granites. Geochem Int'l 14:108-118

Kovalenko VI (1973) Distribution of fluorine in a topaz-bearing quartz keratophyre dike (ongonite) and solubility of fluorine in granitic melts. Geokhim 1:57-66

Kovalenko VI, Hervig RL, Sheridan MF (1988) Ion microprobe analyses of trace elements in anorthoclase, hedenbergite, aenigmatite, quartz, apatite and glass in pantellerite: Evidence for high H_2O contents in pantellerite melt. Am Mineral 73:1038-1045

Kress VC, Carmichael ISE (1991) The compressibility of silicate liquids containing Fe_2O_3 and the effect of composition, temperature, oxygen fugacity and pressure on their redox states. Contrib Mineral Petrol 108:82-92

Kurz MD, Jenkins WJ (1981) The distribution of He in oceanic basalt glasses. Earth Planet Sci Lett 53: 41-54

Kurz MD, Jenkins WJ, Hart SR (1982a) Helium isotope systematics of oceanic islands and mantle heterogeneity. Nature 297:43-47

Kurz MD, Jenkins WJ, Schilling JG, Hart SR (1982b) Helium isotopic variations in the mantle beneath the central North Atlantic Ocean. Earth Planet Sci Lett 58:1-14

Lange RM, Carmichael ISE (1987) Densities of $Na_2O-K_2O-CaO-MgO-FeO-Fe_2O_3-Al_2O_3-TiO_2-SiO_2$ liquids: new measurements and derived partial molar properties. Geochim Cosmochim Acta 51: 2931-2946

Loehman RE (1979) Preparation and properties of Yttrium-silicon-aluminum oxynitride glasses. J Am Ceram Soc 62:491-494

London D, Hervig RL, Morgan GB (1988) Melt-vapor solubilities and element partitioning in peraluminous granite-pegmatite systems: Experimental results with Macusani glass at 200 MPa. Contrib Mineral Petrol 99:360-373

Lowenstern JB (1994) Chlorine, fluid immiscibility and degassing in peralkaline magmas from Pantelleria, Italy. Am Mineral 79:353-369

Lowenstern JB, Mahood GA (1991) New data on magmatic H_2O contents of pantellerites, with implications for petrogenesis and eruptive dynamics at Pantelleria. Bull Volcanol 54:78-83

Luhr JF (1990) experimental phase relations of water and sulfur-saturated arc magmas and the 1982 eruptions of El Chichon volcano. J Petrol 31:1071-1114

Luhr JF, Carmichael ISE, Varekamp JC (1984) The 1982 eruptions of El Chichon volcano, Chiapas, Mexico: mineralogy and petrology of the anhydrite-bearing pumices. J Volc Geotherm Res 23:69-108

Luth RW (1988a) Raman spectroscopic study of the solubility mechanisms of F in glasses in the system $CaO-CaF_2-SiO_2$. Am Mineral 73:297-305

Luth RW (1988b) Effects of F on phase equilibria and liquid structure in the system $NaAlSiO_4-CaMgSi_2O_6-SiO_2$. Am Mineral 73:306-312

Lux G (1987) The behavior of noble gases in silicate liquids: solution, diffusion, bubbles, and surface effects, with applications to natural samples. Geochim Cosmochim Acta 51:1549-1560

Malinin SD, Kravchuk IF, Delbove F (1989) Chloride distribution between phases in hydrated and dry chloride-aluminosilicate melt systems as a function of phase composition. Geochem Int'l 26:32-38

Manning DAC (1981) The effect of fluorine on liquidus phase relationships in the system Qz-Ab-Or with excess water at 1 kbar. Contrib Mineral Petrol 102:1-17

Manning DAC, Hamilton DL, Henderson CMB, Dempsey MJ (1980) The probable occurrence of interstitial Al in hydrous, F-bearing and F-free aluminosilicate melts. Contrib Mineral Petrol 75:257-262

Marty B, Ozima M (1986) Noble gas distribution in oceanic basalt glasses. Geochim Cosmochim Acta 50:1093-1097

Mathews SJ, Jones AP, Gardeweg MC (1994) Lascar Volcano, Northern Chile: Evidence for steady-state disequilibrium. J Petrol 35: 401-432

Mathez E A (1976) Sulfur solubility and magmatic sulfides in submarine basaltic glass. J Geophys Res 81:4269-4275

Mathez EA, Dietrich VJ, Holloway JR, Boudreau AE (1989) Carbon distribution in the Stillwater Complex and evolution of vapor during crystallization of Stillwater and Bushveld magmas. J Petrol 30:153-173

Metrich N, Clocchiatti R (1989) Melt inclusion investigation of the volatile behaviour in historic alkali basalt magmas of Etna. Bull Volcanol 51:185-198

Metrich N, Rutherford MJ (1992) Experimental study of chlorine behavior in hydrous silicic melts. Geochim Cosmochim Acta, 56:607-616

Michael PJ, Schilling J-G (1989) Chlorine in mid-ocean ridge magmas: Evidence for assimilation of seawater-influenced components. Geochim Cosmochim Acta 53:3131-3143

Miyazaki A, Hiyagon H, Sugiura N, Hashizume K (1992) Solubility of nitrogen and noble gases in silicate melt. Terra Nova (Abstr Sup) 4:30

Moore JG (1970) Water content of basalt erupted on the ocean floor. Contrib Mineral Petrol 28:272-279

Moore JG (1979) Vesicularity and CO_2 in mid-ocean ridge basalt. Nature 282:250-253

Moore JG, Schilling J-G (1973) Vesicles, water and sulfur in Reykjanes Ridge basalts. Contrib Mineral Petrol 41:105-118

Mosbah M, Metrich N, Massiot P (1991) PIGME fluorine determination using a nuclear microprobe with application to glass inclusions. Nuc Instr Meth Phys Res B58:227-231

Mukerji J (1965) Phase equilibrium diagram $CaO-CaF_2-2CaO-SiO_2$. J Am Ceram Soc 48:210-213

Mulfinger HO (1966) Physical and chemical solubility of nitrogen in glass melts. J Am Ceram Soc 49:462-467

Mulfinger HO, Meyer H (1963) Physical and chemical solubility of nitrogen in glass melts. Glastech Ber 36:481-483

Mulfinger HO, Deitzel A, Navarro JMF (1972) Physical solubility of helium, neon, and nitrogen in glass melts. Glastech Ber 45,:389-396

Munoz JL (1984) F-OH and Cl-OH exchange in micas with applications to hydrothermal ore deposits. In: Bailey SW (ed) Micas. Rev Mineral 13:469-494

Mysen BO, Popp RK (1980) Solubility of sulfur in $CaMgSi_2O_6$ and $NaAlSi_3O_8$ melts at high pressure and temperature with controlled $f(O_2)$ and $f(S_2)$. Am J Sci 280:78-92

Mysen BO, Virgo D (1985a) Interaction between fluorine and silica in quenched melts on the joins SiO_2-AlF_3 and SiO_2-NaF determined by raman spectroscopy. Phys Chem Minerals 12:77-85

Mysen BO, Virgo D (1985b) Structure and properties of fluorine-bearing aluminosilicate melts: The system $Na_2O-Al_2O_3$- SiO_2-F at 1 atm. Contrib Mineral Petrol 91:205-220

Nagashima S, Katsura T (1973) The solubility of sulfur in Na_2O-SiO_2 melts under various oxygen partial pressures at 1200, 1250 and 1300°C. Bull Chem Soc Japan 46:3099-3103

Nakamoto K (1978) Infrared and Raman Spectra of Inorganic and Coordination Compounds (3rd edn). John Wiley & Sons, New York, 182 p

Naldrett AJ (1969) A portion of the system Fe-S-O between 900 and 1080°C and its application to sulfide ore magmas. J Petrol 10:174-201

Newman S, Epstein S, Stolper EM (1988) Water, carbon dioxide, and hydrogen isotopes in glasses from the ca 1340 AD eruption of Mono Craters California: constraints on degassing phenomena and initial volatile content. J Volc Geotherm Res 35:75-96

Nilsson K, Peach CL (1993) Sulfur speciation, oxidation state, and sulfur concentration in backarc magmas. Geochim Cosmochim Acta 57:3807-3813

O'Neill H St C (1987) The quartz-fayalite-iron and quartz-fayalite-magnetite equilibria and the free energies of formation of fayalite (Fe_2SiO_4) and magnetite (Fe_3O_4). Am Mineral 72:67-75

Pan V, Holloway JR, Hervig RL (1991) The pressure and temperature dependence of carbon dioxide solubility in tholeiitic basalt melts. Geochim Cosmochim Acta 55:1587-1595

Pauling L (1927) The sizes of ions and the structure of ionic crystals. J Am Chem Soc 49:765-790

Pauling L (1960) The Nature of the Chemical Bond. Third Edition, Cornell University Press, 644

Payette C, Martin RF (1990) Melt inclusions in the quartz phenocrysts of rhyolites from Topaz and Keg Mountains, Thomas Range, Utah. Geol Soc Am Spec Paper 246:89-102

Perkins WG, Begeal DR (1971) Diffusion and permeation of He, Ne, Ar, Kr, and D_2 through silicon oxide thin films. J Chem Phys 54:1683-1694

Piccoli P, Candela P (1994) Apatite in felsic rocks: A model for the estimation of initial halogen concentrations in the Bishop Tuff (Long Valley) and Tuolumne intrusive suite (Sierra Nevada batholith) magmas. Am J Sci 294:92-135

Popadopolous K (1973) The solubility of SO_3 in soda-lime-silicate melts. Phys Chem Glasses 14:60-65

Poulson SR, Ohmoto H (1990) An evaluation of the solubility of sulfide sulfur in silicate melts from experimental data and natural samples. Chem Geol 85:57-75

Prausnitz JM (1969) Molecular Thermodynamics of Fluid Phase Equilibria. Prentice Hall, Englewood Cliffs, New Jersey

Quan RA, Cloke PL, Kesler SE (1987) Chemical analyses of halite trend inclusions from the Granisle porphyry copper deposit, British Columbia. Econ Geol 82:1912-1930

Range KY, Willgallis A (1964) Über reaktionen in schmelzen des systems $NaF-SiO_2-Na_2O-(H_2O)$. Radex-Rundschau 2:75

Richardson FD, Fincham CJB (1956) Sulphur in silicate and aluminate slags. J Iron Steel Inst 178:4-15

Robie RA, Hemingway BS, Fisher JR (1979) Thermodynamic properties of minerals and related substances at 298.15K and 1 bar (10^5 Pascals) pressure and higher temperatures. U S Geol Surv Bull 1452

Roedder E (1984) Fluid Inclusions. Rev Mineral 12, 644 p

Roedder E (1992) Fluid inclusions evidence for immiscibility in magmatic differentiation. Geochim Cosmochim Acta 56:5-20

Roedder E, Coombs DS (1967) Immiscibility in granitic melts, indicated by fluid inclusions in ejected granitic blocks from Ascension Island. J Petrol 8:417-451

Roselieb K, Rammensee W, Büttner H, Rosenhauer M (1992) Solubility and diffusion of noble gases in vitreous albite. Chem Geol 96:241-266

Rutherford MJ (1991) Pre-eruption conditions and volatiles in the 1991 Pinatubo magma. Trans Am Geophys Union EOS 72:62

Ryabchikov ID (1963) Experimental study of the distribution of alkali elements between immiscible silicate and chloride melts. Dokl Akad Nauk SSSR 149:190-192

Sarda P, Graham D (1990) Mid-ocean ridge popping rocks: implications for degassing at ridge crests. Earth Planet Sci Lett 97:268-289

Schaller T, Dingwell, DB, Keppler H, Knöller W, Merwin L, Sebald A (1992) Fluorine in silicate glasses: A multinuclear nuclear magnetic resonance study. Geochim Cosmochim Acta 56:701-707

Shackelford JF (1982) A gas probe analysis of structure in bulk and surface layers of vitreous silica. J Non-Cryst Solids 49:299-307

Shackelford JF, Studt PL, Fulrath RM (1972) Solubility of gases in glass II. He, Ne, and H_2 in fused silica. J Appl Phys 43:1619-1626

Shannon RD (1976) Revised effective ionic radii and systematic studies of interatomic distances in halides and chalcogenides. Acta Cryst A32:751-767

Shannon RD, Prewitt CT (1969) Effective ionic radii in oxides and fluorides. Acta Cryst B 25:1046-1048

Shelby JE (1974) Helium diffusion and solubility in K_2O-SiO_2 glasses. J Am Ceram Soc 57:236-263

Shelby JE (1976) Pressure dependence of helium and neon solubility in vitreous silica. J Appl Phys 47:135-139

Shilobreyeva S, Kadik A, Matveev S, Chapyzhnicov B (1994) Solubility of N_2 in basalt and albite melt at pressures of 1-3 kbar and temperature 1250°C. In: Matsuda JI (ed) Proc Int'l Workshop on Noble Gas Geochemistry and Cosmochemistry, Kyoto, Japan (Osaka Univ), 21-22

Shilobreyeva SN, Kadik AA, Lukanin OA (1983) Outgassing of ocean-floor magmas as a reflection of volatile conditions in the magma generation region. Geokhim 9:1257-1274

Shima H, Naldrett AJ (1975) Solubility of sulfur in an ultramafic melt and relevance of the system Fe-S-O. Econ Geol 70:960-967

Shinohara H, Iiyama JT, Matsuo S (1989) Partition of chlorine compounds between silicate melt and hydrothermal solutions. Geochim Cosmochim Acta 53:2617-2630

Silver LA, Ihinger PD, Stolper E (1990) The influence of bulk composition on the speciation of water in silicate glasses. Contrib Mineral Petrol 104:142-162

Silver LA, Stolper E (1989) Water in albitic glasses. J Petrol 30:667-710

Snyder DA Carmichael ISE (1992) Olivine-liquid equilibria and the chemical activities of FeO, NiO, Fe_2O_3 and MgO in natural basic melts. Geochim Cosmochim Acta 56:303-318

Staudacher T, Kurz M, Allegre CJ (1986) New noble gas data on glass samples from Loihi seamount and Haulalai and on dunite samples from Loihi and Reunion island. Chem Geol 56:193-205

Staudacher T, Sarda P, Richardson SH, Allegre CJ, Sagna I, Dmirtriev LV (1989) Noble gases in basalt glasses from a Mid-Atlantic Ridge topographic high at 14°N: geodynamic consequences. Earth Planet Sci Lett 96:119-13326

Stormer JC, Carmichael ISE (1970) Villiaumite and the occurrence of fluoride minerals in igneous rocks. Am Mineral 55:126-134

Studt PL, Shackelford JF, Fulrath RM (1970) Solubility of gases in glass--a monatomic model. Jour Appl Phys 41:2777-2780

Stull DR, Prophet H (eds) (1971) JANAF Thermochemical Tables U S Bur Standards, National Standard Reference Data Series NSRDS-NBS 37:1141 p

Susaki K, Maeda M, Sano N (1990) Sulfide capacity of CaO-CaF_2-SiO_2 slags. Metall Transa B 21B:121-129

Symonds RB, Reed MH, Rose WI (1992) Origin, speciation, and fluxes of trace-element gases at Augustine volcano, Alaska: Insights into magma degassing and fumarolic processes. Geochim Cosmochim Acta 56:633-657

Takusagawa N (1980) Infrared absorption spectra and structure of fluorine-containing alkali silicate glasses. J Non-Crystal Sol 42:35-40

Toop GW, Samis CS (1962) Some new ionic concepts of silicate slags. Can Metal Q 1:129-152

Ulmer P, Luth RW (1991) The graphite-COH fluid equilibrium in P, T, $f(O_2)$ space: An experimental determination to 30 kbar and 1600°C. Contrib Mineral Petrol 106:265-272

Wahlberg JS (1976) Analysis of rocks and soils by X-ray fluorescence. U S Geol Surv Prof Paper 954-A:A11-A12

Wallace P, Carmichael ISE (1992) Sulfur in basaltic magmas. Geochim Cosmochim Acta 56:1863-1874

Wallace P, Carmichael ISE (1994) S speciation in submarine basaltic glasses as determined by measurements of SKa X-ray wavelength shifts. Am Mineral 79:161-167

Watson EB (1991) Diffusion of dissolved CO_2 and Cl in hydrous silicic to intermediate magmas. Geochim. Cosmochim. Acta 55:1897-1902

Webster JD (1990) Partitioning of F between H_2O and CO_2 fluids and topaz rhyolite melt: Implications for mineralizing magmatic-hydrothermal fluids in F-rich granitic systems. Contrib Mineral Petrol 104:424-438

Webster JD (1992a) Water solubility and chlorine partitioning in Cl-rich granitic systems: Effects of melt composition at 2 kbar and 800° C. Geochim Cosmochim Acta 56:679-687

Webster JD (1992b) Fluid-melt interactions involving Cl-rich granites: Experimental study from 2 to 8 kbar. Geochim Cosmochim Acta 56:659-678

Webster JD (1993) Exsolution of F- and Cl-rich melts from felsic silicate melts: Constraints from experiments and glass inclusions. Geol Soc Am Abstr Prog 25:A372

Webster JD, Duffield WA (1991) Volatiles and lithophile elements in Taylor Creek Rhyolite: Constraints from glass inclusion analysis. Am Mineral 76:1628-1645

Webster JD, Duffield WA (1994) Extreme halogen abundances in tin rhyolite magma at Taylor Creek, New Mexico. Econ Geol 89:134-144

Webster JD, Holloway JR (1987) Partitioning of halogens between topaz rhyolite melt and aqueous and aqueous-carbonic fluids. Geol Soc Am Abstr Prog 19:884

Webster JD, Holloway JR (1988) Experimental constraints on the partitioning of Cl between topaz rhyolite melt and H_2O and H_2O + CO_2 fluids: New implications for granitic differentiation and ore deposition. Geochim Cosmochim Acta 52:2091-2105

Webster JD, Holloway JR (1990) Partitioning of F and Cl between magmatic hydrothermal fluids and highly evolved granitic magmas. Geol Soc Am Spec Paper 246:21-34

Webster JD, Holloway JR, Hervig RL (1987) Phase equilibria of a Be, U and F-enriched vitophyre from Spor Mountain, Utah. Geochim Cosmochim Acta 51:389-402

Wendlandt RF (1982) Sulfide saturation of basalt and andesite melts at high pressure and temperature. Am Mineral 67:877-885

White BS, Brearley M, Montana A (1989) The solubility of argon in silicate liquids at high pressures. Am Mineral 74:513-529

Whitney JA (1985) Fugacities of sulfurous gases in pyrrhotite-bearing silicic magmas. Am Mineral 69:69-78

Wyllie PJ, Tuttle OF (1961) Experimental investigations of silicate systems containing two volatile components. Am J Sci 259:128-143

Zharikov VA, Ishbulatov RA, Kosyakov AV (1985) Magma i magmaticheskiye flyuidy. Tez dokl Izd IEM AN SSSR:69

Zhu C, Sverjensky DA (1991) Partitioning of F-Cl-OH between minerals and hydrothermal fluids. Geochim Cosmochim Acta 55:1837-1858

Zhu C, Sverjensky DA (1992) F-Cl-OH partitioning between biotite and apatite. Geochim Cosmochim Acta 56:3435-3467

Zunckel R, Hempel W (1917) Die Losung von NaCl im Obsidian Schmelzen. Z Vulkanol, p 256

Chapter 8

PRE-ERUPTIVE VOLATILE CONTENTS OF MAGMAS

Marie C. Johnson
Lamont-Doherty Earth Observatory
Columbia University
Palisades, NY 10964 USA

Alfred T. Anderson, Jr.
Department of the Geophysical Sciences
The University of Chicago
Chicago, IL 60637 USA

Malcolm J. Rutherford
Department of Geological Sciences
Brown University
Providence, RI 01292 USA

INTRODUCTION

Magmas transfer volatiles from the Earth's interior to its oceans and atmosphere. Hence, understanding planetary outgassing and oceanic and atmospheric evolution requires knowledge of the flux of magmatic volatiles. Within the framework of the solid Earth, magmas influence the partitioning of volatiles between crust and mantle. Mantle-derived magmas such as mid-ocean ridge basalts transfer volatiles from mantle to crust. When this basaltic crust is subsequently subducted in trenches, some volatiles are recycled to the crust via arc petrogenesis. Other volatiles may survive dehydration and partial melting and be returned to the deep mantle along with the slab residue. Thus, the complementary fluxes at ridges and trenches affect the development of volatile reservoirs within the solid Earth. On the scale of an individual lava, volatiles affect nearly all aspects of its physical and chemical behavior. For example, volatiles affect solidus temperature, extent of melting, melt viscosity and thermodynamic properties, and eruption style.

Dissolved volatiles are present in nearly all terrestrial magmas. Many magmas also evolve a separate fluid phase at some stage during their evolution. The total volatile content of a magma includes the volatiles dissolved in the melt as well as those in the fluid and crystalline phases. As magma migrates within the Earth, its total volatile content may fluctuate. If the activity of a volatile species in the wall rock is greater than in the magma, volatiles may be gained from surrounding wall rocks. Opportunities for volatile exchange without magma/rock mixing are limited, however, by barriers such as quenched magma. Magmatic volatile contents are generally assumed to decrease as magmas ascend because the solubilities of the dominant species (H_2O and CO_2) decrease markedly with decreasing pressure. This decrease in solubility may cause nucleation and growth of gas bubbles that may enlarge and escape with continued ascent. Magma degassing is significant because it effects such processes as hydrothermal veining and ore formation as well as the nature of volcanic eruptions. This chapter is primarily concerned with assessing magmatic volatile contents prior to and during eruptions, a task which involves determining the total volatile content of a magma prior to its emergence at the surface.

Assessing the pre-eruptive volatile content of magma is difficult, but not impossible. One necessary step toward accomplishing this task is to determine the concentration of volatiles dissolved in the melt. The concentrations of dissolved volatiles have been estimated by various techniques with significant agreement in specific cases. One technique is to analyze glass inclusions, samples of melt trapped at depth within crystallizing phenocrysts. These melt inclusions are quenched to glass by rapid cooling on eruption. Strong host phenocrysts may serve as pressure containers and allow volatiles dissolved in the melt at depth to remain dissolved in the glass after eruption. The ongoing development of micro analytical techniques such as electron microprobe, Fourier transform infra-red (FTIR) spectroscopy, and ion probe, allow volatiles dissolved in these trapped melts to be analyzed quantitatively.

While glass inclusion studies can tightly constrain the species and abundance of pre-eruptive dissolved volatiles, one must further assess whether a separate fluid phase existed prior to eruption. If a fluid did exist, one must determine its composition to define the complete volatile budget. One approach to addressing these questions is experimental petrology. Experimental petrology allows parameters such as pressure, temperature, and water fugacity to be varied independently under controlled conditions. By systematically varying these parameters, their influence on mineral stabilities and compositions can be calibrated. Once these influences are calibrated, the natural phase assemblage can be compared with laboratory data to deduce the physical conditions including the possible existence and composition of a fluid just prior to eruption. Here, we focus on inferring pre-eruptive volatile contents by combining accurate determination of dissolved volatiles in natural samples with phase equilibria constraints. This combined approach offers several possibilities for assessing total pre-eruptive magmatic volatiles including those present in any associated magmatic gas.

In this chapter, we first present an overview of the general utility of glass inclusion data and experimental phase equilibria. We then review existing data on the volatile contents of some basaltic and silicic magmas. Kilauea Volcano, Hawaii has been the subject of many such studies, and we present a detailed analysis of volatiles in this volcanic system. Next, we summarize data for basalts erupted along mid-ocean ridges, in arcs associated with subduction zones, and in back-arc basins. The following section concentrates on volatiles associated with silicic magmas. Particular attention is paid to three well-studied examples: Mt. St. Helens, Mt. Pinatubo, and the Bishop Tuff, Long Valley, CA. We integrate these data in a discussion that focuses first on the current state of knowledge of the Kilauean magmatic system and then on the origin of high-Al_2O_3 arc basalts. The final sections consider the questions of gas saturation in subvolcanic calc-alkaline magma bodies, the degassing of obsidians, and the relationship between volatiles and magmatic oxidation state. We conclude with some speculations on the significance of volatiles in evolved plutonic vs. volcanic magmas.

CONSTRAINTS ON PRE-ERUPTIVE VOLATILES FROM NATURAL GLASSES

Magmatic volatiles are routinely estimated by measuring dissolved volatiles in natural glasses such as submarine basalt glasses, obsidians, and glass inclusions trapped within phenocrysts. Volatiles such as S, F, and Cl can be analyzed directly using an electron microprobe. Lighter elements such as H and C cannot be measured directly, but the difference between 100% and the electron microprobe total for all analyzable elements is often assigned to dissolved H_2O and CO_2. This "difference method" can produce reliable volatile contents if care is taken during microprobe analyses. FTIR spectroscopy allows H_2O and CO_2 concentrations and speciations to be measured directly (Stolper, 1982; Fine and Stolper, 1986; Newman et al., 1986), and thus provides more complete and potentially accurate information than provided by the difference method. Spectroscopic results, however, depend on calibrating molar absorptivity by analyzing volatiles with other techniques. Ion probe technology also allows H_2O concentrations to be analyzed directly. An advantage of these microbeam techniques is that specific target areas can be chosen, and heterogeneous, bubble or crystal-rich areas can be avoided. A disadvantage is that gas in included bubbles is difficult to analyze. Alternatively, both gas in bubbles and volatiles dissolved in melt can be analyzed by heating hand-picked glass chips or phenocrysts containing glass inclusions. Any released volatiles are collected, and analyzed by manometry or quadrupole mass spectrometry. This technique provides an averaged volatile content, and cannot reveal the detailed spatial information potentially available with microbeam techniques; further details of volatile analysis are given in Chapter 2.

Basaltic glasses are common on the sea-floor. A glassy outer rim is produced by rapid quenching when hot magma contacts sea-water. The glass preserves the melt composition, including any volatiles in solution at the moment of quenching, because the melt is quenched at a pressure corresponding to sea-water depth (1000 m = 100 bar). The quenched glassy rims of sea-floor basalts make them ideal samples for measuring pre-eruptive volatiles. At the other end of the compositional spectrum, obsidians are non-vesicular rhyolitic glasses. Most young, fresh obsidians from lava flows and domes have ~0.1 wt % H_2O, roughly the solubility of water at 1 bar (Hamilton et al., 1964; Dingwell et al., 1984; Silver et al., 1989). Many apparently fresh glasses (pitchstones) have H_2O concentrations of 4 to 6 wt %. These obsidians are commonly judged to have undergone post-eruptive hydration because their D/H ratios suggest H_2O addition at low temperatures (Friedman and Smith, 1958; Garlick and Dymond, 1970).

The hydration of some rhyolitic glasses highlights a danger implicit in examining whole rock volatile data. Whole rock (glass) volatile analyses may not be representative of volatile contents at depth. After eruption or solidification, magmatic volatile contents may be enhanced by secondary alteration and hydration as commonly observed for obsidian glasses. Volatiles may also be lost during degassing in sub-surface storage chambers and upon eruption. This eruptive volatile loss can not be evaluated simply by analyzing compositionally equivalent intrusive rocks. Although intrusive rocks may have solidified at pressures greater than 1 bar, their high crystallization temperatures cause dominantly volatile-free minerals to crystallize forcing the residual liquid to evolve towards gas saturation, and this gas may escape.

Glass inclusions trapped in phenocrysts provide an independent method of assessing magmatic volatiles at depth prior to eruption. Glass inclusions as large as 0.1 mm in diameter rarely survive adiabatic decompression of more than ~1 kbar (Tait, 1992). Because many glass inclusions do survive, decompression is either rapid or not adiabatic. With cooling, pressure on inclusions inside a crystal decreases because, in general, the melt contracts more than the crystal, and pressure within the inclusion may become negative if no gas bubble forms. Small gas bubbles comprising less than a few volume percent of the inclusion commonly contain partial vacuums (Roedder, 1979). This observation reflects contraction at low temperatures where bubble growth rates exceed diffusion rates yielding disequilibrium between melt and gas.

Interpreting glass inclusion analyses requires allowance for several potential problems. For example, trapped melt may differ compositionally from the liquid that existed at depth. Before entrapment, the melt may have been locally modified by growth of the host crystal depleting elements essential to the host mineral and enriching other incompatible elements. This effect has not been demonstrated for naturally cooled materials, however (Anderson, 1974b), possibly because crystal growth in silicate melts is limited by thermodynamic equilibrium and SiO_2 concentration. Silica has a low diffusivity in melts, and silica diffusion limits crystal growth rates so that concentration gradients in other diffusing species are negligible. Compositions of quenched glasses support this view (Bottinga et al., 1966).

The trapped melt composition may also be modified by post-entrapment processes such as trapped-mineral crystallization on inclusion walls. Alternatively, melt trapped in inclusions may equilibrate with external melt during decompression. Inclusions in Fe-rich minerals such as olivine may undergo extensive redox reactions and diffusive transfer of hydrogen species (Pasteris and Wanamaker, 1988). In quartz phenocrysts, the beta to alpha quartz inversion at 573°C (Ghiorso et al., 1979) causes an ~1 vol. % contraction of the quartz host. This contraction increases the pressure on inclusions in quartz and may help to keep gas bubbles from forming. Diffusive volatile loss may be promoted, however, by dislocations and twin boundaries which accompany the stresses that develop around inclusions during this inversion.

Glass inclusion analyses often yield a large range of volatile concentrations, even within a given sample. Although variations may reflect the accuracy of different analytical methods, some variations may reflect different approaches to sample selection. Three approaches have been used in presenting and interpreting volatiles in glass inclusions: the average approach, the high approach, and the one-by-one approach. In the average approach, a group of inclusions is analyzed and the results are averaged. Interpretations are based on trends in the averages for various samples. In the high approach, attention is focused on those few inclusions that have high volatile concentrations. The high concentrations are assumed to represent most accurately inclusions unaffected by leakage. In the one-by-one approach, selection criteria are applied to individual inclusions on a case-by-case basis. Inclusions passing the selection criteria are considered neither to have leaked nor to have gained secondary volatiles. All of these approaches have disadvantages. Experience has revealed that the averaging approach can easily ignore real relationships. The high approach, though more justifiable than the average approach, has the peril of being subject to concentrations enhanced by weathering and alteration. The one-by-one approach, though tedious, is generally the approach of choice. Despite the potential problems, important insights into the volatile budgets of magmas have resulted from glass inclusion studies.

CONSTRAINTS ON PRE-ERUPTIVE VOLATILES FROM PHASE EQUILIBRIA

Mineral-melt phase equilibria can provide complementary information to that obtained from natural glasses and can help to define pre-eruptive conditions and volatile concentrations in magmas at depth. The natural phenocryst-melt equilibria are fingerprints of the ambient intensive parameters prior to eruption or solidification. The variables affecting phase equilibria are bulk composition, temperature, total pressure, water fugacity (f_{H_2O}), oxygen fugacity (f_{O_2}), and potentially carbon dioxide and sulfur dioxide fugacities (f_{CO_2}, f_{SO_2}). The challenge is to quantify the relationship between these intensive parameters and the resulting phase equilibria. With these types of experimental data, it is increasingly possible to determine the conditions that existed in a magma at depth by comparing natural mineral and melt (glass) compositions with data from experimental samples.

The effects of dissolved volatiles were first investigated in silicic systems in part because their low liquidus temperatures make them experimentally more tractable than basaltic systems. As improvements in furnace design and temperature control were implemented, the phase equilibria of volatile-rich basaltic systems were also studied. Tuttle and Bowen (1958) and coworkers pioneered studies of the influence of dissolved H_2O on the phase equilibria and crystallization temperatures of silicic magmas. More recently, these phase equilibria have been studied experimentally by Luth et al. (1964), and theoretically by Nekvasil and Burnham (1987). In a classic paper concerning the influence of volatiles on basalt phase equilibria, Yoder and Tilley (1962) further demonstrated the profound effect of H_2O in decreasing anhydrous mineral crystallization temperatures. Yoder and Tilley (1962) noted that the magnitude of the temperature decrease differed for different minerals. They observed that silicate crystallization temperatures were decreased much more than Fe-Ti oxide crystallization temperatures. These observations can be rationalized by recalling that water dissolution involves breaking bridging oxygen bonds consequently depolymerizing the melt. Complex framework and chain sub-species in the melt are destroyed. The destruction of these complex melt subspecies suppresses nucleation of complex crystalline silicates like plagioclase and clinopyroxene while nucleation of simple crystalline structures like Fe-Ti oxides is relatively enhanced (e.g., Sekine and Wyllie, 1983). Water dissolution alters melt structure, and may change the order of mineral crystallization. Because mineral crystallization controls residual melt composition, dissolved water may alter melt evolution paths.

At a given pressure, a magma may not be in equilibrium with a gas phase. If a gas phase does exist, the magma may be water-saturated (in equilibrium with a H_2O-rich gas), or it may be water-undersaturated (in equilibrium with a H_2O-poor gas). Recognizing this, experimentalists (e.g., Maaloe and Wyllie, 1975; Whitney, 1975; Naney and Swanson, 1980) determined the phase equilibria for a range of silicic magmas with fixed amounts of H_2O up to and beyond that required for melt saturation at the experimental pressure. These results have proven generally useful for interpreting the phase equilibria of granites and rhyolites, particularly the quartzo-feldspathic phases. For example, Whitney and Stormer (1985) used the Whitney (1975) and Naney and Swanson (1980) data to confirm their estimates of P, T, and P_{H_2O} in the dacitic Fish Canyon magma system prior to its eruption.

The above method of determining phase equilibria, however, did not allow oxygen fugacity to be controlled, and therefore information on the stability and reactions of ferromagnesian minerals is inadequate for understanding natural systems. Eugster (1957) and his group (Eugster and Wones, 1962; Ernst, 1962) showed that f_{O_2} affects the stability and composition of minerals such as biotite and hornblende crystallizing in a silicic magma. Osborn (1959) and Buddington and Lindsley (1964) noted the importance of f_{O_2} in determining iron oxide compositions (Fig. 1). These studies demonstrated that f_{O_2} buffered experiments are required to gain maximum information from a phase equilibria study. Recently, the phase equilibria of several silicic magmas have been determined using redox buffering techniques and mixed CO_2-H_2O fluids at various pressures (Clemens and Wall, 1981; Merzbacher and Eggler, 1984; Rutherford et al., 1985; Clemens and Wall, 1988; Johnson and Rutherford, 1989a). These methods were developed originally by Eggler and Burnham (1973) and Holloway and Burnham (1972).

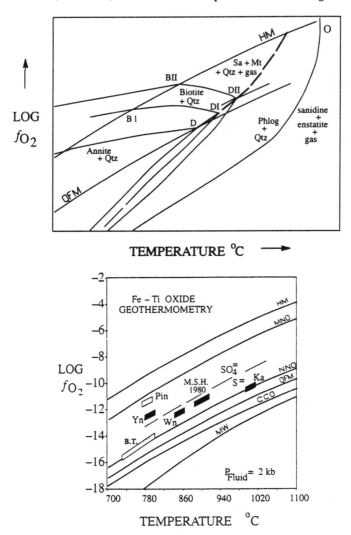

Figure 1. (a) Schematic f_{O_2}-temperature relationships for biotite + quartz at moderate pressure under H_2O-saturated conditions (modified after Wones and Eugster, 1965). Curves Annite + Quartz, BI, and BII represent increasingly Mg-rich biotite coexisting with magnetite, hematite and quartz; D, DI, and DII represent the maximum thermal stability of a given biotite and quartz, and curve O represents the maximum stability of phlogopite + quartz. Curves QFM and HM are the quartz-fayalite-magnetite and hematite-magnetite oxygen buffering equilibria respectively. (b) Log f_{O_2}-T diagram showing the conditions in pre-eruptive magmas at Long Valley, California (B.T.), Mount St. Helens (1980, Yn, Wn, Ka) and Mount Pinatubo (Pin). The long dashed curve shows the change from sulfide to sulfate sulfur dissolved in silicate melt.

In the case of basaltic phase equilibria, Helz (1973, 1976) further improved the early work of Yoder and Tilley (1962) by controlling oxygen fugacity in experiments with an H_2O-rich fluid at 2 and 5 kbar. Holloway and Burnham (1972) and Spulber and Rutherford (1983) first investigated basalt crystallization when $P_{H_2O} < P_{total}$. Spulber and Rutherford (1983) showed that basalt crystallization in the presence of an H_2O-

bearing fluid could reproduce the Thingmuli, Iceland magmatic sequence described by Carmichael (1964). The SiO_2 enrichment required to produce the evolved magmas is caused by early Fe-Ti oxide crystallization at low pressures and by hornblende crystallization at higher pressures. More recently, Devine (in press) used the Spulber and Rutherford (1983) data set to show that crystal fractionation can explain the range of evolved fine-grained basalts found in the southern Lesser Antilles (Fig. 2).

Following their investigations of f_{O_2} effects on biotite stability, Wones and Eugster (1965) used the relationship between f_{O_2}, f_{H_2}, and f_{H_2O} in hydrothermal systems to develop an expression relating f_{H_2O} to the compositions of coexisting biotite, alkali feldspar, and magnetite. Although this expression generally yields H_2O estimates subject to large uncertainties (e.g., 2±1 kbar), it has great potential if the activity-composition relation for biotite can be better determined. Another promising method of estimating the f_{H_2O} of a silicic magma uses a solution model fit to the composition data of coexisting plagioclase and hydrous silicate melt (Kudo and Weill, 1970; Housh and Luhr, 1991). Many phase equilibria studies have illustrated the sensitivity of this equilibrium to the water content of the coexisting melt, but a reliable calibration has been prevented by difficulty in obtaining feldspar equilibrium in hydrothermal experiments.

VOLATILES IN BASALTIC AND ANDESITIC MAGMAS

Concentrations of dissolved pre-eruptive volatiles in basic magmas have been estimated by a variety of methods with significant agreement. Estimating the total volatile content of volatile-rich magma has proven more controversial. Conflicting estimates of the bulk volatile budget have relied on combinations of volcanic gas monitoring, eruption dynamics modelling, and dissolved pre-eruptive volatile determinations. The various approaches and the sources of disagreement are illustrated by the case of Kilauea volcano, Hawaii, which has been a focus of volcanic gas research for more than half a century. Accordingly, we review first the pre-eruptive volatile content of tholeiitic basalt from Kilauea. Kilauea may be regarded to be broadly representative of other ocean island basalt volcanoes such as Reunion and Galapagos that are dominated by abundant tholeiitic basalts.

Volatiles in magmas from Kilauea Volcano, Hawaii

Early studies of Kilauean magmatic volatiles involved analyzing both rocks (Chamberlin, 1908; Shepherd, 1938) and gas (Jaggar, 1940). Shepherd (1938) emphasized the difficulty of interpreting rock analyses; volatiles are lost during crystallization and eruption and subsequently adsorbed from air. Subsequent to these pioneering efforts, workers have assessed volatile abundance in Kilauean basalts by (1) analyzing glassy basalts and basaltic glasses, (2) monitoring volcanic gas, and (3) modelling eruption dynamics.

H_2O in Kilauean magmas. Deep sea dredging has recovered Kilauean basalts quenched at hundreds of bars. Moore (1965) showed that the most rapidly quenched, glassy, Kilauean basalts contain roughly constant (~0.5 wt %) H_2O concentrations for samples quenched at pressures greater than ~50 bars. Moore interpreted this constancy to signify that those basalts erupted at pressures greater than 50 bars were H_2O-undersaturated. Dixon et al. (1991) analyzed the same and additional submarine basaltic glasses from Kilauea and found higher H_2O concentrations (mostly 0.4 to 0.9 wt %) consistent with Moore's (1965) work after correcting for the phenocryst content of the basalts. H_2O concentrations have recently been determined for glassy fragments of

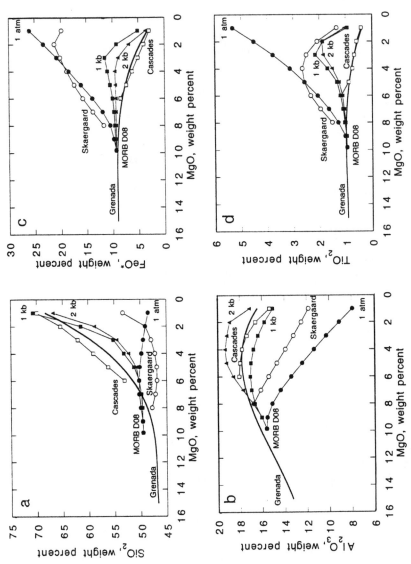

Figure 2. Melt evolution diagrams for (a) SiO$_2$, (b) Al$_2$O$_3$, (c) FeO and (d) TiO$_2$ vs. MgO for the Spulber and Rutherford (1983) set of experimental melts produced by basalt crystallization (MORB DO8) with increasing water pressures (modified after Devine, in press). The X_{H_2O} in coexisting fluid equals 0.67 in both the 1 and 2 kbar experiments; the 1 bar experiments were dry. The oxygen fugacity was below FMQ. The trend of the experimental melt (glass) compositions is compared with the trends defined by the Skaergaard residual melt, the Cascades volcanic rocks, and fine-grained volcanic rocks on Grenada in the Lesser Antilles.

submarine Kilauean basalt (Clague et al., in press); some of these fragments have ~15 wt % MgO, and are considered close in composition to Kilauean parental magma. Such glasses contain ~0.4 wt % H_2O (Fig. 3) which, if crystallized in a closed system free of gas, would yield basaltic liquids with 7 wt % MgO and ~0.55 wt % H_2O, similar to the results of Moore (1965) and Dixon et al. (1991).

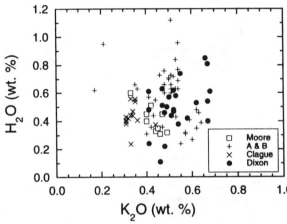

Figure 3. H_2O and K_2O contents of Kilauean basaltic glasses and glassy basalts. The data plotted include glass on rims of submarine basalts (circles; Dixon et al., 1991), grains of glass from turbidite (crosses; Clague et al., in press), glassy submarine basalts (squares; Moore, 1965), and glass inclusions in olivine phenocrysts from the 1959 Kilauea Iki eruption (+ ; Anderson and Brown, 1993).

Glass inclusions in olivine phenocrysts from the 1959 Kilauean eruption of picritic basalt have been analyzed using four different methods. (1) Anderson and Brown (1993) using FTIR spectroscopy reported a range from 0.24 to 1.1 wt % H_2O. (2) Anderson's (1974 a, b) summation deficit for an inclusion from this eruption is consistent with less than ~0.5 wt % H_2O, a result at the detection limit for this technique. (3) Harris and Anderson (1983) using a tedious but simple manometric approach reported 0.2 wt % H_2O. This result is at the low end of the spectroscopic data (Anderson and Brown, 1993) and may reflect analytical bias combined with H_2O loss during pre-analysis drying in vacuum as observed by Moore and Schilling (1973). (4) Muenow et al. (1979) reported a maximum of 0.004 wt % H_2O by pyrolizing polycrystalline olivine samples and assuming that certain gas surges resulted from bursting of glass inclusions.

The H_2O concentrations reported by Anderson and Brown (1993) correlate with eruption date. Early erupted inclusions comprise a high-H_2O group (mostly 0.70±0.10 wt % H_2O). Most inclusions that erupted later have less than 0.40 wt % H_2O. In all, 21 inclusions have less than 0.5 wt % H_2O, and three inclusions have more than 0.9 wt % H_2O. Thus, lower than average H_2O concentrations are more common than high H_2O concentrations. H_2O does not correlate with K_2O (Fig. 3). Based on the total dissolved volatiles, some inclusions have formation pressures up to ~2000 bars indicating that basaltic magma with ~0.7 wt % H_2O exists at considerable depths beneath Kilauea.

Gerlach and Graeber (1985) and Greenland et al. (1985) independently estimated the pre-eruptive H_2O contents of Kilauean basaltic magma by combining gas monitoring data with published glass inclusion and submarine glass analyses. Gerlach and Graeber's (1985) estimate (0.30 wt % H_2O) relies mostly on eruptive gas sample compositions collected by Shepherd and Jaggar from the active Halemaumau lava lake in 1917 and 1918 and a sulfur normalization procedure. This procedure entails calculating sulfur loss from included and residual (degassed) glass and combining this information with the gas

H_2O/SO_2 ratio to yield the amount of H_2O lost by exsolution. Adding the residual H_2O dissolved in erupted scoriaceous glass to the amount lost yields the pre-eruptive H_2O. The estimate of Greenland et al. (1985) is based on eruptive and non-eruptive volcanic gas emission rates during the 1980's and involves a CO_2 normalization procedure. CO_2 emission derived from the volume flux of lava is used to estimate the CO_2 lost per unit mass of magma. This information together with the H_2O/CO_2 ratio of coeval eruptive gas samples yields a concentration of 0.32 wt % H_2O in the pre-eruptive magma.

Eruption dynamics provide an independent approach to estimating pre-eruptive magmatic volatile contents. Wilson and Head (1981) applied their eruption dynamics model to the 1959 Kilauea eruption and noted that the observed lava fountain heights and mass eruption rates calculated from lava lake filling were consistent with 0.19 to 0.27 wt % H_2O exsolved. These numbers assume a crystal-free system and should be adjusted to 0.24 to 0.34 wt % in view of up to 20 wt % olivine phenocrysts. They should be further increased by ~0.06 (Harris, 1981) to 0.09 (Friedman, 1967; Stolper, 1982) wt % to account for H_2O remaining in the glass to yield an estimate of 0.31 to 0.43 wt % pre-eruptive H_2O. In addition, the 1959 glasses are unusually MgO-rich (~10 wt % for early erupted scoria; Helz, 1987). For comparison, a further increase to 0.34 to 0.47 wt % H_2O is appropriate because ~10 wt % olivine crystallization is required to yield melt with 6.2 wt % MgO similar to Kilauean basaltic glasses (Moore, 1965; Dixon et al., 1991) and summit lavas extruded in 1917-1918. Although the gas erupted in 1959 was not analyzed, significant SO_2 and CO_2 were probably present in the eruptive gas (Gerlach, 1986) implying that the adjusted Wilson and Head (1981) estimate is a maximum.

Wilson and Head (1981) assumed that melt degassing is arrested at the pressure where foam disrupts into spray (assumed to occur at ~70 vol % gas). Their model requires a disruption pressure of ~9 bars to form a 300 m lava fountain. The scoria glass has, however, degassed to pressures less than ~1.2 bars (Dixon, 1992). About 0.2 wt % H_2O would exsolve between 9 and 1.2 bars according to the H_2O solubility data of Dixon (1992). Whether the Wilson and Head (1981) estimate should be increased by this H_2O content depends upon whether the additional 0.2 wt % H_2O contributes to the kinetic energy of the erupting lava and thus lava fountain height. After magma disrupts, gas may escape to the surface independently of molten spray. Such non-eruptive gas escape is suggested by active gas emission from non-incandescent gas vents proximal to eruptive vents where lava is extruding (Greenland, 1987). If the post-disruption exsolved H_2O is assumed not to contribute to lava fountain height, then the implied pre-eruptive H_2O concentration must be increased by 0.2 wt % to 0.54 to 0.67 wt %. The similarity between this result and the H_2O concentrations measured in Kilauean glass inclusions and submarine basaltic glasses may be real, but it must be recalled that the effect on eruption dynamics of any CO_2 and SO_2 present in coeruptive gas has been ignored.

CO_2 in Kilauean magmas. Pre-eruptive CO_2 concentrations in Kilauean basaltic magma are constrained by five different methods which yield different concentration estimates. For example, some non-Kilauean Hawaiian olivines contain inclusions of dense CO_2-rich gas requiring ~4 kbar entrapment pressures at 1200°C (Roedder, 1965). A basaltic melt in equilibrium with 4 kbar of CO_2 pressure (CO_2 fugacity = 10,600 bar) would dissolve ~0.53 wt % CO_2 according to a linear extrapolation of the Pawley et al. (1992) solubility relation. Dissolved CO_2 (as CO_3^{2-}; Fine and Stolper, 1986) concentrations are substantial in both submarine basaltic glass and glass inclusions in olivine phenocrysts from Kilauea (Fig. 4). The highest measured CO_2 concentration from a glass inclusion in a Kilauean olivine phenocryst is 758 ppm (Anderson and Brown, 1993). Much CO_2, however, may be stored in the accompanying gas bubble.

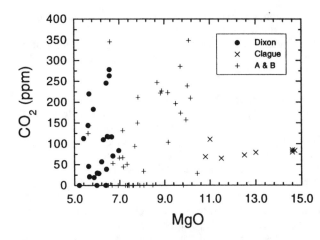

Figure 4. CO_2 and MgO contents of Kilauean basaltic glasses. Data plotted include glass on rims of submarine basalts (● ; Dixon et al., 1991), grains of glass from turbidite (crosses; Clague et al., in press), and glass inclusions in olivine phenocrysts from the 1959 Kilauea Iki eruption (+ ; Anderson and Brown, 1993; only includes raw CO_2 in the glass; does not include CO_2 that may reside in a gas bubble; one inclusion with 758 ppm CO_2 plots off the diagram). Note the several glass grains from turbidite that have ~15 wt % MgO.

Estimating the total CO_2 in the inclusion (melt plus gas) gives ~0.2 wt % CO_2 for this relatively differentiated basaltic glass.

Ground deformation at Kilauea depends partly on the volume of compressible gas in Kilauea's stored magma (Johnson, 1992). Johnson's (1992) assessment of the compressibility of the summit reservoir suggests ~0.3 wt % gas is present. Volcanic gas monitoring data yield estimates of 0.32 (Greenland et al., 1985) and 0.65 (Gerlach and Graeber, 1985) wt % CO_2 in parental Kilauean basaltic magma. Greenland et al. (1985) assumed that the CO_2 escaping non-eruptively from the summit was derived from magma erupting almost simultaneously from the east rift zone. Gerlach and Graeber's (1985) estimate is partly based on analyses of CO_2-rich gas collected in 1917-1918 from the active summit lava lake. They argued that the gas resulted from a single stage of degassing, and that the magma retained its original volatile complement until the moment of gas release from the lava lake. This closed-system view of parental magma ascent is difficult to reconcile with views that Kilauean parental magma is picritic (Wright, 1984; Clague et al., 1991) because the 1917-1918 lavas are basalts with ~7.8 wt % MgO (Wright, 1971). Bottinga and Javoy (1991) pointed out that the CO_2-rich gas composition would result in ascent rates too great to allow magma and gas to remain in equilibrium. The gas composition then reflects quenching from a greater pressure where bubbles are small and CO_2-rich. Another possibility is that rising gas-rich magma loses H_2O by convectively mixing with previously degassed melt that convects downwards. Estimates of pre-eruptive magmatic CO_2 derived by volcanological approaches are greater than solubility will allow within Kilauea's summit magma storage reservoir and imply the presence of excess gas in parental magma as it enters the magma storage reservoir.

Kilauean gas and glass carbon isotopic compositions suggest appreciable degassing of CO_2 (Gerlach and Taylor, 1990). The $^{13}C/^{12}C$ ratio of Kilauean gases and glasses decreases with degassing, and Gerlach and Taylor (1990) interpret this trend to result from batch degassing in two stages; non-eruptive release of CO_2-rich gas from Kilauea's summit area followed by release of the remaining magmatic CO_2 during rift zone eruptions.

In summary, the CO_2 content of parental Kilauean magma is uncertain and may have varied secularly. Volcanic gas data and ground deformation studies suggest that magma stored in Kilauea's summit reservoir contains excess CO_2-rich gas. Glass inclusion data reflect only dissolved CO_2 and suggest a lower limit of ~0.2 wt % CO_2. These data are consistent with the assessment of Greenland et al. (1985) of 0.3 wt % CO_2. Factor of 2 uncertainties exist, however, and Gerlach and Graeber's (1985) estimate of ~0.6 wt % cannot be discounted.

Sulfur in Kilauean magmas. Pre-eruptive sulfur concentrations in Kilauean basaltic magma have been assessed based on analyses of submarine basaltic glass (Moore and Fabbi, 1971; Dixon et al., 1991) and glass inclusions in phenocrysts (Anderson and Wright, 1972; Anderson, 1974a; Harris and Anderson, 1983; Anderson and Brown, 1993). Most analyses have 0.08 to 0.15 wt % S, although concentrations as low as 0.02 and as high as 0.19 wt % S have been reported (usual microprobe precision is about ±0.01 wt % S absolute). A strong positive correlation exists between S and H_2O in the submarine Kilauean glasses (Dixon et al., 1991), but not in glass inclusions (Anderson and Brown, 1993). The trend in the submarine glasses is interpreted to indicate that this basalt, or a component of it, has been affected by degassing which lowed both water and sulfur abundances. Greenland et al. (1985) estimated eruptive degassing of 0.06 wt % S which together with 0.02 wt % S in residual erupted basaltic glass (Moore and Fabbi, 1971) yields a pre-eruptive magmatic S content of 0.08 wt %. This volcanological estimate is at the low end of the results from inclusion analyses, and this discrepancy is not well explained. The abundance of sulfur in Kilauean basalts is less than that required for sulfide saturation, and immiscible sulfides are not present in the glasses, but rare sulfide-rich olivine phenocrysts suggest that the picritic parent magma may be sulfide saturated (Helz, 1987). Any breakdown of this immiscible sulfide, however, would tend to make the volcanological sulfur estimates higher than dissolved gas analyses. Several possible explanations exist for the range of sulfur abundances observed in the glass inclusions. Sulfur solubility is enhanced by magma oxidation (Carroll and Rutherford, 1985; 1987; Wallace and Carmichael, 1992). Perhaps different inclusions had different oxidation histories, although no evidence exists that these basalts underwent such oxidation state changes. Another possibility is that some glass inclusions may contain relict high temperature super-saturated sulfur concentrations. Finally, it is possible that some pulses of these magmas have suffered degassing of a CO_2 and S-rich volatile phase at depth, and the olivines have grown in variably degassed melt.

Pre-eruptive volatiles in basaltic magmas from other tectonic settings

Mid-ocean ridge basalts. Following his early studies of Kilauean submarine basaltic glasses, Moore and coworkers pioneered the study of dissolved volatiles in other submarine basalts (Moore, 1970; Moore and Schilling, 1973; Moore, 1979). Since then, basaltic glasses from many different sea-floor settings have been analyzed by various techniques. Although all techniques yield similar dissolved H_2O contents, CO_2 measurements appear technique-dependent.

Reported analyses of dissolved water in submarine basaltic glasses from seamounts and fast and slow spreading mid-ocean ridges are presented in Figure 5 where total magmatic H_2O (OH$^-$ and H_2O molecular in spectroscopic data, H_2O^+ in Penfield analyses) is plotted against sample recovery depth. The experimentally determined solubility of H_2O in basaltic melt as a function of pressure is also shown (Hamilton et al., 1964). Total H_2O correlates poorly with depth and is generally less than the amount of H_2O required for saturation indicating that most melts were not saturated with an H_2O-rich fluid when quenched to glass. Virtually all deep sea basaltic glasses contain vesicles,

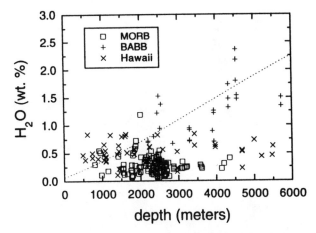

Figure 5. Published H_2O contents of glassy submarine basalts versus depth of recovered dredge for mid-ocean ridges, back-arc basins, and Hawaiian tholeiites. Dashed line represents experimentally determined solubility of water as a function of pressure (Hamilton et al., 1964). Most MORB glasses are H_2O-undersaturated. (Data for MORB from: Moore, 1970; Delaney et al., 1978; Harris, 1981; Byers et al., 1983; 1984; 1986; Dixon et al., 1988; Michael, 1988; Jambon and Zimmerman, 1990; data for BABB from: Garcia et al., 1979; Muenow et al., 1980; Volpe et al., 1987; Aggrey et al., 1988b; Jambon and Zimmerman, 1990; Danyushevsky et al., 1993; data for Hawaii from: Muenow et al., 1979; Harris, 1981; Byers et al., 1985; Jambon and Zimmerman, 1990; Dixon et al., 1991.)

however, indicating gas saturation. These observations may be reconciled by concluding that the ambient gas was H_2O-poor.

Water and K_2O abundances in mid-ocean ridge basalt (MORB) glasses are positively correlated (Fig. 6), with H_2O/K_2O ratios ranging from 1:4 to 1:1. Both H_2O and K_2O behave incompatibly during partial melting of MORB-type mantle (Michael, 1988; Hess, 1992), and H_2O/K_2O ratio variations mainly reflect compositional variations in the source. The limited range of H_2O/K_2O ratios implies that these elements are linked during source enrichment processes. Since the trend of the data intersects the

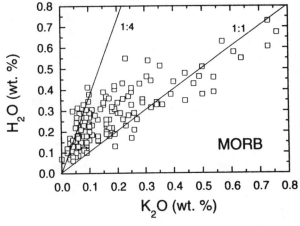

Figure 6. Compilation of literature data showing measured H_2O content of submarine MORB glasses as a function of K_2O. Only glasses with less than 0.8 wt % K_2O are shown. K_2O/H_2O ratios of 1:1 and 1:4 are indicated; most MORB have intermediate K_2O/H_2O ratios. (Data compiled from: Moore, 1970; Harris, 1981; Byers et al., 1983; 1984; 1986; Delaney et al., 1978; Aggrey et al., 1988a; Dixon et al., 1988; Michael, 1988; Jambon and Zimmerman, 1990.)

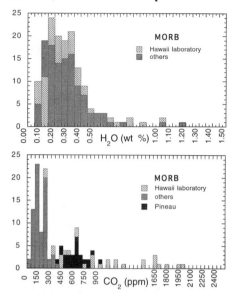

Figure 7. (a) Histogram of water contents for mid-ocean ridge basaltic glasses. Striped bars are data from the Hawaii-laboratory (Delaney et al., 1978; Byers et al., 1983; 1984; 1986; Aggrey et al., 1988a), gray bars are other available data (Moore, 1970; Harris, 1981; Brett et al., 1987; Dixon et al., 1988; Michael, 1988; Jambon and Zimmerman, 1990). (b) Histogram of CO_2 contents for mid-ocean ridge basaltic glasses. Striped bars are data from the Hawaii-laboratory (Delaney et al., 1978; Byers et al., 1983; 1984; 1986), black bars are from Pineau et al., 1983, gray bars are other available data (Moore, 1970; Harris, 1981; Fine and Stolper, 1986; Brett et al., 1987; Dixon et al., 1988). Although similar water contents are reported by all groups, the CO_2 contents reported by the Hawaii-laboratory are uniformly higher than most other analyses.

positive H_2O axis rather than the origin, H_2O is inferred to behave slightly more compatibly than K_2O during mantle melting (Michael, 1988).

Measurements of CO_2 in submarine MORB glasses are controversial. Most studies report CO_2 contents of less than 400 ppm (Harris, 1981; Fine and Stolper, 1986; Brett et al., 1987; Dixon et al., 1988). Data collected using vacuum fusion and quadrupole mass spectrometry (Delaney et al., 1978; Byers et al., 1983; 1984; 1986) yield results a factor of 2-10 greater (Fig. 7). One study employing vacuum fusion and manometry reports a mean dissolved CO_2 content of 615±30 ppm (Pineau and Javoy, 1983). Measured CO_2 contents are plotted as a function of sample recovery depth in Figure 8. The experimentally determined variation in CO_2 solubility as a function of pressure is also shown (Dixon et al., in press, a). This work includes new measurements made using an improved background subtraction scheme on CO_2-saturated glasses produced in an earlier experimental study (Stolper and Holloway, 1988). These new results suggest that at a given pressure, CO_2 saturation occurs with ~20% less dissolved CO_2 than previously reported by Pawley et al. (1992).

Little correlation exists between CO_2 content and recovery depth for most MORB glasses (Fig. 8, but see Harris, 1981). Relative to the experimental solubility data of Dixon et al. (in press, a), most analyzed MORB glasses plot above the equilibrium saturation line indicating that these glasses are CO_2 supersaturated. If the Pawley et al. (1992) solubility curve is considered, then some glasses plot below equilibrium

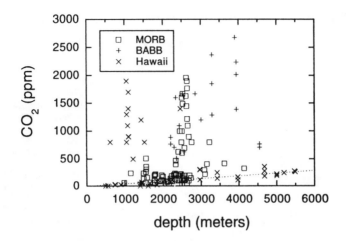

Figure 8. Compilation of literature data showing measured CO_2 contents in submarine basaltic glasses versus depth of recovered dredge for mid-ocean ridges, back-arc basins, and Hawaiian tholeiites. Dashed line represents experimentally determined solubility of CO_2 as a function of pressure (Dixon et al., in press, a). Most MORB glasses are super-saturated in CO_2. (Data for MORB from: Moore, 1970; Delaney et al., 1978; Harris, 1981; Byers et al., 1983; 1984; 1986; Dixon et al., 1988; data for BABB from: Garcia et al., 1979; Muenow et al., 1980; Volpe et al., 1987; Aggrey et al., 1988b; data for Hawaii from: Muenow et al., 1979; Harris, 1981; Byers et al., 1985; Dixon et al., 1991.)

saturation. Samples interpreted to be CO_2 super-saturated must have equilibrated at higher pressures. Apparently, ascent rates were rapid enough to prevent bubbles from nucleating and growing, CO_2 remained dissolved in the melt, and melt/fluid equilibrium was not maintained.

A dissolved CO_2 content cannot be translated directly into an equilibration depth unless the coexisting gas composition is known. In general, MORB gas compositions are CO_2-rich (Moore et al., 1977). If the coexisting gas is pure CO_2, then the experimental data allow equilibration pressures to be calculated directly. At oxygen fugacities greater than 1 log unit below NNO, however, substantial CO will be present in the gas phase reducing P_{CO_2} to significantly less than P_{total}. In this case, the measured CO_2 contents can not be directly equated with equilibration pressures unless CO_2 fugacity is known (Pawley et al., 1992).

In contrast to H_2O, CO_2 is not correlated with melt K_2O (Fig. 9). Yet, CO_2 must behave as an incompatible element during MORB formation as no CO_2-rich minerals are thought to be present in the mantle in significant quantities at the pressures where MORB melting begins. The CO_2 found in MORB is probably derived from carbon present in the MORB mantle as graphite, diamond, or amorphous carbon depending on pressure and oxygen fugacity. The lack of correlation between K_2O and CO_2 suggests that melt CO_2 contents are affected by a heterogeneous equilibrium between melt and a CO_2-rich vapor. This relationship may not be simple. Reduced C decorating fractures in phenocrysts and vesicle walls occurs in MORB samples (Mathez and Delaney, 1981). This reduced carbon is not detectable by all analytical methods, and its origin is problematical.

The basalt CO_2 contents reported by Byers et al. (1983; 1984; 1986) and Delaney et al. (1978) are uniformly higher (up to an order of magnitude) than those observed by

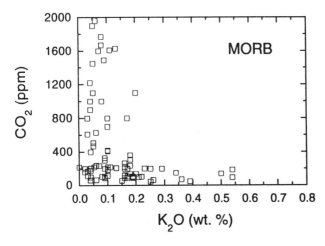

Figure 9. Compilation of literature data showing lack of correlation between measured CO_2 content of submarine MORB glasses and melt K_2O content. Only glasses with less than 0.8 wt % K_2O are shown. (Data compiled from: Moore, 1970; Delaney et al., 1978; Harris, 1981; Byers et al., 1983; 1984; 1986; Dixon et al., 1988.)

most other workers who have analyzed similar rocks (Fig. 7). One explanation is that different techniques analyze different phases. FTIR spectroscopy measures only IR-active dissolved volatiles. Data reported by the above authors were collected by heating carefully selected glass fragments to high-temperature and releasing volatiles from both the glass and any trapped bubbles. The bubbles in submarine basalt glasses are CO_2-rich and H_2O-poor (Moore et al., 1977), and they represent a possible CO_2 source not analyzed by FTIR spectroscopy. If these bubbles were included in mass spectrometric analyses but not in FTIR analyses, then the mass spectrometric results would be expected to have higher CO_2 contents and equivalent H_2O contents as observed (Fig. 7). The ideal gas law suggests that at equilibrium at 1200°C and 200 bars, a magma containing 1.5 vol % of CO_2 bubbles would have ~300 ppm CO_2 in the bubbles. Indeed, Pineau et al. (1983) report dissolved CO_2 contents from 370-900 ppm (Fig. 7), and note that these CO_2 contents may be contaminated by CO_2 from fluid inclusions which did not burst during earlier laser heating.

Another difference between the Byers and coworkers data sets and most other data sets is that reduced species of C (CO and CH_4) are commonly reported, yet these species are not reported in other studies (e.g., Dixon et al., 1988; Brett et al., 1987; Moore, 1970). Perhaps, the CO and CH_4 are derived from rare amorphous carbon from vesicles and cracks (Mathez and Delaney, 1981). Amorphous carbon may be oxidized during sample heating forming CO as well as CO_2. This C was probably present in samples analyzed by other groups using pyrolitic methods, although no reduced C-species are reported, and measured CO_2 contents are generally lower than those of Byers and coworkers. These differences may simply be a reflection of the fact that carbon analyses at these low concentrations are extremely difficult, and contamination is a potentially serious problem. According to DesMarais and Moore (1984) various forms of carbon are adsorbed onto basalt at room temperature and contribute to carbon species emitted during pyrolysis.

Arc basalts. Most submarine basalts from mid-ocean ridges and Hawaii are volatile-poor. Early studies of pillow basalts from the Marianas arc and trough (Garcia et al., 1979) suggested that arc basalts also had low volatile contents (<1 wt %). A volatile-poor nature of arc basalts became accepted by some (Perfit et al., 1980). Compared to sea-floor basalt studies, estimates of dissolved volatiles in subaerially erupted arc magmas are hampered by two complications. First, arc basalts rarely quench to glass upon subaerial eruption. Second, arc basalts commonly erupt at near atmospheric pressures, and their magmatic volatile contents are strongly affected by exsolution. Even if a whole rock analysis represents a liquid, the primary dissolved volatile content is unlikely to be preserved as volatiles can be lost during degassing or gained during secondary alteration.

One approach to minimizing these problems is to analyze glass inclusions. Harris and Anderson (1984) using vacuum fusion manometry and electron microprobe techniques measured H_2O, CO_2, and Cl abundances in melts trapped in olivine phenocrysts from the 1974 eruption of Fuego volcano, Guatemala. They discovered that the least evolved melt had an H_2O content of 1.6 wt %. Sisson and Layne (1993) reanalyzed Harris and Anderson's glass inclusions plus additional inclusions from phenocrysts in the Cascade magmatic arc in the western USA (Mount Shasta and Black Crater) and report up to 6.2 wt % water for Fuego and 0.2 wt % for Black Crater. These results are consistent with tectonic setting as Fuego is part of a frontal arc and Black Crater (near the Medicine Lake Highlands) represents a back-arc environment. Intermediate values (3.3 wt %) were reported for volcanoes north of Mt. Shasta.

Back-arc basin basalts. The compositions of back-arc basin basalts are in many respects intermediate between MORB and arc basalt compositions. This characteristic probably reflects the tectonic setting of back-arcs, small spreading centers located behind mature arcs. FTIR studies of back-arc basin basalt glasses from the Southwest Pacific (Danyushevsky et al., 1993) reveal that their intermediate geochemical signature extends to their dissolved water contents (1-3 wt %) that are intermediate between typical MORB and arc values. These water contents are higher than those reported for Lau Basin glasses (~0.2 wt % H_2O; Jambon and Zimmerman, 1990) and Woodlark Basin glasses (0.2-0.6 wt % H_2O; Muenow et al., 1991). An independent study of Lau Basin and North Fiji Basin glasses found H_2O concentrations slightly higher than these reports (0.4-1.8 wt %; Aggrey et al., 1988b). High water contents have also been reported for submarine basalt glasses from the East Scotia Sea back-arc basin (~1 wt % H_2O; Muenow et al., 1980) and from the Susimu Rift (> 1 wt % H_2O; Hochstaedter et al., 1990).

In basaltic glasses from the Mariana back-arc basin H_2O and K_2O are positively correlated suggesting common sources and incompatible behavior of both H_2O and K_2O during magmatic evolution (Stolper and Newman, 1994). Their study also emphasizes the significant effect of small amounts of H_2O on the percentage melt produced.

Phase equilibria constraints on volatiles in basalts

The volatile contents of most mid-ocean ridge and back-arc basin basalts are generally too low to cause crystallization of hydrous or other volatile-bearing minerals as the magma crystallizes. Volatile concentrations are much higher in arc basalts, and their resulting influence on phase equilibria is correspondingly more dramatic. For example, arc basalts commonly contain volatile-bearing minerals such as hornblende. Other effects, more subtle than hydrous mineral crystallization, are noted from arc basalt phase equilibria studies.

The most abundant basaltic lavas erupted at arcs are low-MgO, high-Al_2O_3 basalts (HAB). Because of their ubiquity, HAB have been a principle focus of experimental study which has centered on investigating melting and crystallization under nominally anhydrous conditions (Johnston, 1986; Gust and Perfit, 1987; Bartels et al., 1991; Johnston and Draper, 1992). Recently, however, experiments to investigate the role of water in high alumina basalt evolution have been conducted (Beard and Lofgren, 1992; Sisson and Grove, 1993a and b). Sisson and Grove examined the effect of water on liquid evolution using natural aphyric HAB and intrusive equivalents as starting materials. Under dry conditions, most HAB have olivine accompanied by (or closely followed by) plagioclase on their liquidus up to 11 kbar. Under water-saturated conditions at 2 kbar, Cr-spinel or magnetite becomes the liquidus mineral (Sisson and Grove, 1993a), and plagioclase crystallization is suppressed. This change in liquidus mineralogy reflects the depolymerization of the silicate melt structure by dissolved water.

Mineral indicators. Beyond altering the sequence of mineral appearance by depolymerizing the melt, water also affects equilibrium mineral compositions. Plagioclase is the most well known example of this effect. Yoder et al. (1957) showed that high water pressure causes plagioclase to begin to crystallize at lower temperatures than in dry systems, but does not significantly change the shape of the plagioclase-melt binary loop. Detailed studies by Johannes (1978), however, suggest that the loop may change shape as water pressure increases. The plagioclase commonly found in arc basalts (An_{90-95}) is more anorthite-rich than the plagioclase found in MORB. This difference in An-content may be attributed to a higher water fugacity in arc basalts than in ridge basalts. Sisson and Grove (1993a) defined a plagioclase-melt exchange K_d

$$K_d = \frac{(Ca/Na)_{plag}}{(Ca/Na)_{liquid}} . \tag{1}$$

For dry melts at pressures below 20 kbar, this K_d is always less than 2.5. K_d increases with increasing water in the melt, and reaches 5.5 for water-saturated liquids at 2 kbar (Sisson and Grove, 1993a). These results quantify the observation that at a given temperature and bulk composition, plagioclase coexisting with hydrous melts is more An-rich than plagioclase coexisting with dry melts.

Housh and Luhr (1991) also investigated the change in plagioclase composition that results from increasing water pressure by regressing 54 new plagioclase-melt pairs from experimental studies on basalts and andesites at 1 to 4 kbar water pressure together with water-saturated experimental data from the literature. They calibrated two expressions, one for albite and one for anorthite, that allow melt water contents to be calculated if magmatic temperature is known independently. Sisson and Grove (1993b) note that the derived expressions were calibrated using a data set containing no truly basaltic liquids ($SiO_2 \leq 52$ wt %, anhydrous) and report that although the albite expression consistently overestimates their experimental melt water contents, the anorthite expression recovers melt water content with an average error of 0.35 wt % H_2O.

The effect of water on clinopyroxene composition has also been explored (Gaetani et al., 1993). Pyroxene compositions produced in dry experiments at 1 bar and 8 to 15 kbar are plotted in Figure 10 (Gaetani et al., 1993). In these anhydrous experiments, increasing pressure decreases the wollastonite (Wo) component and increases dissolved Al_2O_3 in the clinopyroxenes. The increase in dissolved Al_2O_3 is seen as a shift towards the $Ca(Al,Cr)SiAlO_6$ apex with increasing pressure. The clinopyroxenes coexisting with hydrous melts are clearly different from the clinopyroxenes crystallized in the dry experiments; the wollastonite component is dramatically increased without a concomitant increase in dissolved Al_2O_3 creating a "wet field" ($P_{H_2O} < 2$ kbar; Fig. 10). Clinopyroxene compositions from olivine-clinopyroxene-bearing arc lavas plot in the wet field suggesting that they crystallized at low total pressures but under relatively high water pressures (Gaetani et al., 1993; Fig. 10). Finally, clinopyroxenes in cumulates from subduction-related settings have wollastonite contents distinctly higher than clino-

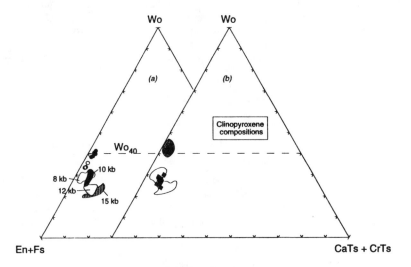

Figure 10.   The Wo (CaSiO$_3$) – En+Fs (MgSiO$_3$+FeSiO$_3$) – CaTs+CrTs (CaAlSiAlO$_6$+CaCrSiAlO$_6$) ternary diagram showing fields of experimentally produced clinopyroxenes.   (a) Filled circles are clinopyroxenes produced in 2-kbar water-saturated experiments. Open circles are 1 bar anhydrous clinopyroxene compositions. Other fields (8 to 15 kbar) are labeled according to anhydrous pressure. (b) Open circle is anhydrous high pressure data from (a).   Clinopyroxenes from wehrlite and clinopyroxenite xenoliths from non-subduction related settings (diamonds) and subduction related settings (gray field) are shown. Data from and figure after Gaetani et al. (1993).

pyroxene cumulates from non-subduction-related settings (Gaetani et al., 1993; Fig. 10). This difference is so dramatic that Gaetani et al. argue that simply by studying accumulated crystal compositions, associated magmatic water contents may be inferred even when direct liquid samples no longer exist.   As an example, Gaetani et al. plot clinopyroxenes from ophiolite bodies.   These cumulates fall into the field defined by clinopyroxenes crystallized at modest water pressures suggesting that at least some ophiolites are related to convergent rather than divergent margins (see also, Muenow et al., 1990).

Hornblende phenocrysts, together with anorthite-rich plagioclase and wollastonite-rich clinopyroxene, are classic phase equilibria indicators of high water pressure. Merzbacher and Eggler (1984) showed that 4 to 5 wt % dissolved water is necessary to stabilize hydrous hornblende in basaltic to dacitic melts.   Hence, the occurrence of hydrous hornblende phenocrysts in natural rocks requires that at least several weight percent water must have been dissolved in the coexisting melt.   Sisson and Grove (1993a), expanding on earlier work (Cawthorne, 1976; Cawthorne and O'Hara, 1976), have added a caveat to this observation.   They found that hornblende crystallization is a strong function of melt Na$_2$O content.   Hornblende has its maximum thermal stability in alkali basalt compositions.   This observation is expected as the bulk composition of hornblende projects into the alkali basalt region of the basalt tetrahedron.   Hornblende thermal stability in low-Ca pyroxene normative compositions is lower than in alkali basalt compositions.   Hornblende crystallization temperature, however, does not appear to increase steadily with increasing Na$_2$O in the pyroxene normative field.

Mineral-melt indicators. Experimental phase equilibria data can be used to develop geothermometers and hydrometers. For example, Sisson and Grove (1993b) calibrated several equations that relate natural glass and phenocryst compositions to magmatic temperature and dissolved water content. For liquids saturated with plagioclase, high-Ca pyroxene or hornblende, ± olivine, they present an empirical relationship between temperature and melt water content.

$$T(^{\circ}C) = 969 - 33.1 * H_2O + 0.0052 * (P-1) + 742.7 * Al^{\#} - 138 * NaK^{\#} + 125.3 * Mg^{\#} \qquad (2)$$

where P is in bars, $Al^{\#}$ = wt % $Al_2O_3/(SiO_2+Al_2O_3)$,

$NaK^{\#}$ = wt % $(Na_2O+K_2O)/(Na_2O+K_2O+CaO)$, and

$Mg^{\#}$ = molar $Mg/(Mg+Fe)$.

Their tests demonstrate that model temperatures are correct to within 7°C for both dry and wet systems. They develop a second independent method of estimating temperature by calibrating Mg distribution between olivine and liquid for liquids saturated with olivine and plagioclase but not orthopyroxene (Sisson and Grove, 1993b).

$$\log_{10}(K^{Mg}) = 4129 * (1/T\ K) - 2.082 + 0.0146 * (P-1)/(T\ K) \qquad (3)$$

where $K^{Mg} = (X^{Mg}_{ol})/((X^{Mg}_{L})(X^{Si}_{L})^{0.5})$ and X^{Mg}_{ol} and X^{Mg}_{L} are the mole fractions of MgO in olivine and liquid, respectively. For this formulation, temperature for wet and dry experiments is accurate to within an average of ~10°C.

Equations (2) and (3) contain three unknowns: pressure, temperature, and melt water content. Simultaneous solution of (2), (3), and the anorthite expression from Housh and Luhr (1991) will yield pressure, temperature, and melt water content. Sisson and Grove (1993b) point out, however, that pressure is incorporated into Housh and Luhr's model only to the extent that the orthoclase activity of plagioclase is affected. Since most HAB are K_2O-poor, the plagioclase expression is relatively insensitive to pressure. For HAB, Sisson and Grove (1993b) recommend specifying pressure based on other knowledge, and solving for temperature and melt water content. The three equations can be combined in three different pairs to test for consistency.

In summary, natural glass and experimental phase equilibria data suggest that basalts erupted from divergent margins have low dissolved volatile contents, and are characterized by CO_2-rich fluids. Basalts that erupt in subduction zones have 3 to 6 wt % dissolved water contents, and are characterized by H_2O-rich fluids. Basalts erupted from intraplate settings have characteristics intermediate between these two end-members perhaps reflecting an average of these two geological settings.

Phase equilibria constraints on volatiles in andesites

The phase equilibria of andesites are not as well studied as basaltic and rhyodacitic compositions, mostly because of the difficulty in experimenting with H_2O-rich compositions without H_2 loss by diffusion or iron loss to sample containers. The experiments of Eggler (1972) and Eggler and Burnham (1973) probably changed composition (and oxidation state) through iron loss (Sisson and Grove, 1993a), but their phase equilibria are likely to be correct qualitatively (Fig. 11); many of the water-rich experiments were at sufficiently low temperature to minimize iron-loss. An effort was made to control the f_{O_2} at the QFM oxygen buffer. At low P_{H_2O}, pyroxenes and/or olivine + plagioclase crystallize early and the liquidus temperature is about 1200°C. With increasing H_2O in the melt at 3 to 5 kbar, the liquidus temperature decreases from

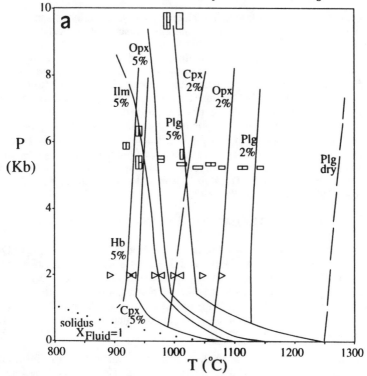

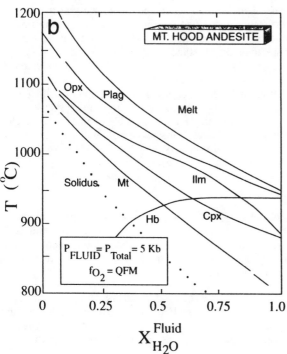

Figure 11. Phase equilibria of Mount Hood Andesite after Eggler and Burnham (1973). (a) P-T phase relations at H_2O-undersaturated conditions. High temperature liquidi are drawn for 0, 2 and 5 wt % H_2O in the melt. (b) T-X_{H_2O} plot of the same data for a constant pressure of 5 kbar. All experiments are nominally at the QFM oxygen buffer.

>1200°C to approximately 1000°C and plagioclase crystallization is suppressed relative to the pyroxenes and Fe-Ti oxides. Amphibole crystallization occurs at 950-1000°C depending on melt composition, providing H_2O in the melt is greater than ~4 wt % (see also Merzbacher and Eggler, 1984; Rutherford and Hill, 1993). Mineral and glass compositions were not obtained from these early experiments. Some phase compositions are available from basaltic composition experiments (Spulber and Rutherford, 1983; Sisson and Grove, 1993a and b) because these experiments produced andesitic and more evolved melts. Numerous andesite phase equilibria questions remain unanswered, however. What reactions control olivine resorption in an andesitic magma, and how do these reactions change with P, T and P_{H_2O}? Is amphibole involved in these reactions? What is the source of variable H_2O, K_2O, and Na_2O in andesitic magmas, and how do these variations affect phase equilibria (Sisson and Layne, 1993)? What is the oxidation state of andesitic magmas? Does oxidation state correlate with H_2O abundance (Carmichael, 1991; Rutherford et al., 1993)?

VOLATILES IN DACITIC AND RHYOLITIC MAGMAS

Silicic magmas may have substantial C-O-H-S volatile concentrations. Qualitative evidence includes the presence of H_2O- and S-bearing phenocrysts, dissolved volatiles in glass inclusions in phenocrysts, violently explosive eruptions, and the vesicular texture of tephra and lavas. Giant crystals in granitic pegmatites, and the common association of granites with various volatile-bearing ore deposits provide evidence from plutonic settings. It is much more difficult, however, to determine the quantitative abundances of these volatiles in magmas at depth.

Clemens (1984) reviewed various lines of evidence and resulting estimates of H_2O in silicic and intermediate plutonic and volcanic magmas. He concluded that such magmas have H_2O contents ranging from <1 wt % to ~7 wt % and suggested that these water contents are related to particular source rocks and melting processes. At the time of Clemens' review various methods had been applied to different systems, and it was difficult and rare to apply more than one method to the same sample or system. Dissolved H_2O estimates varied greatly as summarized by Clemens. Subsequent to Clemens review, spectroscopic methods pioneered by Stolper and coworkers have enabled the rapid and non-destructive analysis of CO_2 and H_2O in samples of silicic glass as small as twenty microns in diameter. Many glass inclusions in phenocrysts are of this size. The new, spectroscopic analytical capability made it possible to analyze these glass inclusions and thereby compare indirect estimates of dissolved H_2O with direct analyses. With the recent improvement of experimental solubility relations for CO_2 as well as H_2O in silicic melts (see Chapter 6), directly measured CO_2 and H_2O concentrations could be used to estimate minimum crystallization pressures for comparison with estimates derived from other approaches.

In view of these developments, our review focuses primarily on studies of magmatic systems which include spectroscopic analyses of glasses for both CO_2 and H_2O, especially those systems for which estimates by other methods have been made of dissolved H_2O or crystallization pressure or both. We compare several results and discuss possible factors affecting their agreement.

Methods used to make quantitative estimates of volatile (mainly H_2O) concentrations in silicic magmas have been based on geological inference (for example, textural evidence for gas saturation combined with estimates of burial depth), eruption dynamics, mineral composition and sequence, and glass inclusions. SO_2 erupted to the atmosphere is also helpful. Whereas mineral composition and crystallization sequence are applicable to both plutonic and volcanic systems, eruption dynamics and glass inclusions are limited mainly to volcanic systems. Eruption dynamics reflect the total (exsolved as well as dissolved) volatile content, whereas mineral composition and crystallization sequence reflect dissolved volatiles. These limitations lead to our emphasis on volcanic systems. We do not intend to imply that the history of volcanic rocks accurately reflects or represents plutonic magmas.

Rhyolitic glass inclusions

Dissolved H_2O and CO_2 in rhyolitic glass inclusions from a variety of tectonic settings range up to 8 wt % and from 10 to 1000 ppm, respectively (Anderson et al., 1989; Johnson and Rutherford, 1989a; Lowenstern, in press; Wallace and Gerlach, in press). Corresponding pressures of gas saturation range up to 4 kbar. Anderson et al. (1989) interpreted a negative correlation between H_2O and CO_2 in glass inclusions from the early and late-erupted parts of the Bishop Tuff in terms of gas saturation and crystallization induced differentiation. Additional work (Lu, pers. comm.) has shown that the CO_2 and H_2O variation, although consistent with crystallization, is opposite to the expected sequence; inclusions of more differentiated melt formed first in the late-erupted material and were succeeded by inclusions of less differentiated, CO_2-rich melt. Similarly, Hervig and Dunbar (1992) concluded that glass inclusions from the Bishop and Bandelier tuffs reflect a history of magma mixing. These results reveal that textural documentation is important to interpret correctly variations in pre-eruptive volatiles as preserved in glass inclusions. In a study of only the early-erupted part of the Bishop Tuff, glass inclusions show a range of gas saturation pressures consistent with a slight overall increase in gas saturation pressure with continued eruption (Wallace, pers. comm.). This type of pattern is expected for a gas-saturated magma because early-erupted magma tends to tap the upper, low-pressure part of the magma body (e.g., Smith and Bailey, 1966; but see Gardner et al., 1991). Similarly, Wallace and Gerlach (in press) show that quartz-bearing phenocrysts in the Pinatubo magma (dacite with rhyolitic melt) contained dissolved CO_2 (278-416 ppm) and H_2O (6.1-6.6 wt %) suggesting gas saturation at ~2.6 kbar, slightly higher than the pressure indicated by phase equilibria studies, but within the error limits on that estimate (Rutherford and Devine, in press). The porphyry molybdenum-associated rhyolitic pumice of Pine Grove, Utah contains glass inclusions in quartz that have 6-8 wt % dissolved H_2O and 60-960 ppm CO_2 (Lowenstern, pers. comm.). Lowenstern interprets the large CO_2 variation at a roughly constant H_2O concentration to reflect formation of inclusions during decompressive ascent of magma from 16 km (4.3 kbar) to 9 km (2.5 kbar) depth.

It should be noted that both H_2O and CO_2 may be lost by inclusions after initial formation. Inclusions with cracks, unusually large gas bubbles, or unusually free of devitrification products are especially suspect. In some cases, gas may have leaked from inclusions that were cracked. Leakage appears to be the case for particularly explosive eruption deposits, e.g., the June 15, 1991 Pinatubo eruption, as indicated by the fragmented nature of phenocrysts, and the fact that inclusions in hornblende contain less water than is required to stabilize hornblende (Westrich and Gerlach, 1992). Loss of only 1 wt % H_2O from an inclusion will generate either a 3 vol. % bubble or a large negative pressure within the inclusion if the host crystal is rigid. Many glass inclusions in these explosive eruption products contain anomalously large vapor bubbles. Alternatively, the low H_2O concentrations may reflect diffusive H_2O loss after initial entrapment and before extrusion as suggested by Qin et al. (1992); this process would require a change in the H_2O-content of the melt outside the crystal. Some inclusions, however, appear uncracked and contain significantly less than 4 wt % H_2O (Johnson and Rutherford, 1989a; Webster, pers. comm.), possibly reflecting very slow diffusive loss of H_2 or H_2O over time.

Post-formation CO_2 loss is particularly likely for slightly devitrified inclusions. Experimentally revitrified inclusions from the Bishop Tuff have low CO_2, but occur in units that also contain pristine inclusions with much higher CO_2 concentrations (Skirius et al., 1990; Wallace, unpublished data). The low-CO_2, revitrified inclusions have subtle speckles in the quartz near inclusion corners that could represent healed cracks. In addition, much CO_2 in inclusions may be localized in small gas bubbles within the inclusions. Qin (1994) analyzed glass inclusions before and after laboratory heating and found that the CO_2 dissolved in glass increased after heating homogenized the inclusion by dissolving the bubble. Qin (1994) also found that with prolonged laboratory heating, most inclusions in quartz phenocrysts lost CO_2 but not H_2O.

Phase equilibria constraints

In the following sections, we present the determination of magmatic conditions in several well-studied volcanic complexes using mixed volatile and oxygen buffered experiments. Although the data generated are specific to the bulk compositions studied, the trends revealed by the experimental data have general applicability.

Mount St. Helens mafic dacites. The recent (1980-1986) eruptions at Mount St. Helens produced pumice deposits (May 18 through October 16, 1980) and lava domes of compositionally similar dacitic magma (62 to 64 wt % SiO_2). The dacites contain plagioclase, low-Ca pyroxene, hornblende, magnetite, ilmenite, and rare Ca-rich pyroxene phenocrysts. Late erupted magma is characterized by a microlite-rich glass matrix and reaction rims mantling hornblendes (Rutherford and Hill, 1993). Other phenocrysts, including the iron-titanium oxides show little compositional variation through the eruption (Cashman, 1992; Rutherford and Hill, 1993). Recalculated Fe-Ti oxide phenocryst compositions (Stormer, 1983) used with the Anderson and Lindsley (1988) algorithm to calculate pre-eruptive temperatures yield $900\pm20°C$ and an oxygen fugacity of NNO+1 (1.0 log unit above the NNO buffer; Rutherford and Hill, 1993). One exception is the October 1986 lava dome eruption that produced magma last equilibrated at 860°C. The phase equilibria of the 1980 dacite composition have been investigated at NNO+1 oxygen fugacity as a function of P, T, and X_{H_2O} in the coexisting gas (Rutherford et al., 1985; Rutherford and Devine, 1988). Water undersaturated conditions were achieved by equilibrating a mixed H_2O/CO_2 fluid phase of known composition with the phenocryst-melt assemblage.

Figures 12 and 13 show the phase equilibria of Mount St. Helens dacites. At 900°C, the natural 1980-86 phenocryst phase assemblage is stable at pressures above 1.6 kbar under water-saturated conditions, but the plagioclase is more anorthite-rich than the natural plagioclase at the higher pressures. If the magma was not water-saturated, the range of conditions that could produce the observed phenocryst assemblage is expanded; at 2.2 kbar (Fig. 13) the natural assemblage is stable in melts coexisting with a fluid phase ranging from $X_{H_2O} = 1$ to $X_{H_2O} = 0.67$. At 3.2 kbar total pressure, the required phase assemblage is stable over a range of fluid compositions from X_{H_2O} equal to 0.5 to 0.7 (Rutherford et al., 1985), but the plagioclase in this range is not as evolved as the natural phase (An_{49}). In addition, the experimental residual melts do not match the 1980 matrix glass. As a result, the magma phase assemblage prior to eruption was determined to be at 920°C at 2.2 kbar total pressure (equivalent to 8.5 km depth; Williams et al.,

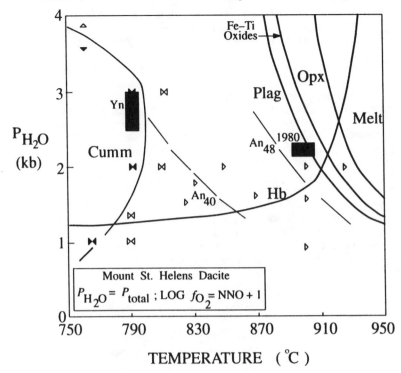

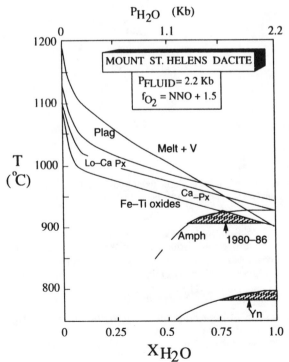

Figure 12 (above). Water-saturated phase equilibria of dacite composition erupted recently (1980-86) at Mount St. Helens (modified after Rutherford et al., 1985; Rutherford and Devine, 1988; Geschwind and Rutherford, 1992). Shaded boxes show the conditions in the pre-eruption 1980 and the 3500 ybp Yn magma systems.

Figure 13 (left). T-X_{H_2O} phase relations for the Mount St. Helens dacite described in Figure 12. Symbols as in Figure 12 caption.

1987). At this P and T, the residual melt composition as indicated by the matrix glass of the May 18, 1980 pumice is best approximated by experimental melt in equilibrium with a fluid of $X_{H_2O} = 0.5$ to 0.7 composition (Rutherford and Devine, 1988).

This phase assemblage result is consistent with experimental data on plagioclase compositions and amphibole stability, and with the abundance of H_2O determined in glass inclusions trapped in natural phenocrysts. Figure 14 shows plagioclase compositions in the dacitic magma as functions of X_{H_2O} in the coexisting fluid at 920°C and 2.2 kbar. The experimental plagioclase composition equals the composition in equilibrium with the natural melt when X_{H_2O} in the coexisting fluid is 0.67. At this point, the melt contains 4.6 wt % dissolved water according to analyses of the experimental glasses (Rutherford and Devine, 1988) and available solubility data (Silver et al., 1989). Analyses of glass inclusions trapped in natural plagioclase and hornblende phenocrysts yield water contents of 4.6±1.0 wt % by the difference method (Rutherford et al., 1985), in agreement with the water content predicted by the plagioclase composition. Note also that hornblende is stable in melts at this P, T, and X_{H_2O}, but is unstable in melts with lower H_2O abundances. The water content of a magma containing both hornblende and Ca-rich pyroxene would be buffered if both phases were present, assuming that hornblende forms at the expense of Ca-rich pyroxene in these magmas.

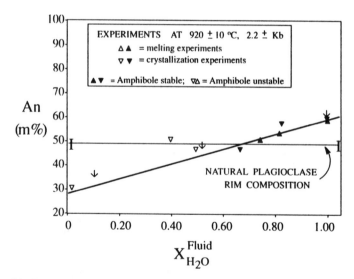

Figure 14. Composition of plagioclase in experiments on dacite (after Rutherford and Devine, 1988) vs. the X_{H_2O} of the coexisting H_2O-CO_2 fluid phase compared to plagioclase rim compositions in the 1980 Mount St. Helens' magmas.

The phase equilibria, P, T, and volatile content of the Mount St. Helens magma system have also been estimated for conditions just prior to six other explosive eruptions that occurred over the past 4000 years. In these studies, Geschwind and Rutherford (1992) and Gardner et al. (in press) determined the dissolved H_2O in the pre-eruptive melt by ion probe analyses of glass inclusions. These data, together with temperature estimates from Fe-Ti oxides, the dacite phase equilibria, and additional data on plagioclase compositions yield estimates of both P_{H_2O} and P_{total} or the magma storage zone depths for these eruptions. The results indicate that the zone has moved with time from 11 to 12 km (3 kbar) ~3500 years ago to 6 to 7 km (1.6 kbar) ~200 years ago

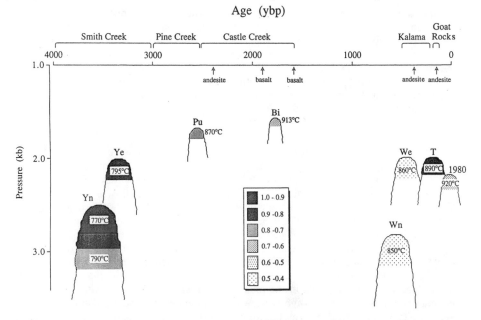

Figure 15. Schematic diagram showing variations in total pressure, water pressure and temperature as a function of age for the dacitic magma reservoir at Mount St. Helens over the last 4000 years (after Gardner et al., in press). Shading indicates the X_{H_2O} in the coexisting fluid.

(Fig. 15). Additionally, the magma system was occasionally water saturated, particularly when the temperature was low, i.e., prior to the cummingtonite-bearing 790°C eruption at ~3.5 ka. The results of the H_2O estimates from these studies compare well with those obtained from the Housh and Luhr (1992) expression relating plagioclase and coexisting melt compositions although the difference between the albite (AB) and the anorthite (AN) expressions suggest this model could be further refined.

Mount Pinatubo dacite. The main phase of the June 1991 eruption of Mount Pinatubo (Philippines) tapped a more silica-rich dacite (65 wt % SiO_2) than the recent events at Mount St. Helens (Pallister et al., in press). More significantly, the Pinatubo dacite contains a diverse phenocryst assemblage including plagioclase (An_{41} rims), hornblende, biotite, cummingtonite (usually as rims on hornblende), quartz, magnetite, ilmenite, anhydrite, apatite, and rare alkali feldspar. This diversity allows recently developed expressions relating mineral and melt compositions to be used to estimate pre-eruptive magmatic conditions. This multiple-technique approach offers a check on the validity of different phase equilibria methods. The initial eruptions of Pinatubo (June 7-12, 1991) produced an andesite that is a mixture of phenocryst-rich dacite and basalt, and presents an opportunity to study magma mixing (Pallister et al., 1992; Rutherford et al., 1993). The large volume, main phase of the eruption produced only dacite, and we focus primarily on this magma. Approximately 5 to 10 km^3 of this dacite was erupted from a 40 to 90 km^3 magma reservoir (estimated from seismic evidence; Mori et al., 1993).

Iron-titanium oxides in the 1991 Pinatubo dacites indicate that the pre-eruptive magma equilibrated at 780±10°C under very oxidizing conditions, 3 log units above the NNO oxygen buffer (Rutherford and Devine, 1991; in press). The oxidation state and temperature are confirmed by other phase equilibria. The presence of ubiquitous (<1 vol %) anhydrite phenocrysts evenly distributed in the dacite and occasionally included in other phenocrysts requires an oxygen fugacity above ~NNO+1.5 (Carroll and Rutherford, 1987). Glass inclusions in plagioclase and hornblende phenocrysts contain 65±20 ppm S present entirely as sulfate according to the S X-ray peak position (Carroll and Rutherford, 1988). In addition to being oxidized, this S is identical in abundance to that of an anhydrite saturated silicic melt at 800°C (Rutherford and Devine, in press); if the pre-eruptive magma temperature had been higher, the glass inclusions would have contained higher S abundances to maintain equilibrium.

The pressure on the pre-eruptive Pinatubo dacitic magma has been estimated to be 2.2±0.5 kbar using the aluminum-in-hornblende geobarometer of Johnson and Rutherford (1989b) which was calibrated at the Pinatubo dacite magma temperature. The large error on this estimate reflects the hornblende (rim) compositional variability in this dacite. The Al-in-hornblende pressure estimate agrees with the cummingtonite + quartz assemblage stability limits as defined by experiments on Pinatubo dacite (Rutherford and Devine, in press). The stable water-saturated phase assemblage is shown as a function of P and T in Figure 16. The shaded bar in Figures 16 and 17 represents the P and T derived from Fe-Ti oxide geothermometry and Al-in-hornblende geobarometry. If the pressure was more than 0.5 kbar different or if the temperature was 20°C hotter, the cummingtonite + quartz assemblage would be unstable in the presence of a water-rich fluid. Alternatively, the magma could only have been slightly water-undersaturated, but the coexisting gas had to contain >75% water to stabilize cummingtonite (Fig. 17). Thus, the phase equilibria require the pre-eruptive Pinatubo dacite to be water-rich at 2.2±0.5 kbar and 780±10°C. Recognizing that the experimental and natural melt (glass) compositions depend on the nature and extent of crystallization, Rutherford and Devine (in press) compare the melts (glasses) in the various experiments with the natural matrix glass. They found that the experiments that best match the matrix glass are water-saturated at 2.0 to 2.2 kbar.

Estimates of pre-eruptive volatiles in the Pinatubo dacitic magma are also available from analyses of volatile species (H_2O, CO_2, SO_2) present in glass inclusions trapped in phenocrysts. Many glass inclusions are identical in major element composition to the matrix glass and contain ~6.4 wt % H_2O, 60 to 70 ppm S, and <20 ppm CO_2 (Westrich and Gerlach, 1992; Rutherford and Devine, in press). Some glass inclusions with low H_2O probably leaked. The 6.4 wt % H_2O determined by ion probe analysis as well as the electron microprobe difference method, indicates P_{H_2O} = 2 kbar in the pre-eruptive magma; the CO_2 and SO_2 contents in the glass inclusions indicate a P_{CO_2} < 250 bars and a P_{SO_2} ~ 110 bars (Rutherford and Devine, in press). An estimate of the pre-eruptive volatiles in this magma by Wallace and Gerlach (in press) suggests a similar P_{H_2O}, but a significant P_{CO_2} (~400 bars) based on FTIR analyses of CO_2 (348 ppm) and H_2O (6.3 wt %) in bubble-poor, slightly devitrified glass inclusions in quartz phenocrysts. Wallace and Gerlach's volatile concentrations yield a total pressure of 2.6 kbar which is within the limits indicated by the phase equilibria. The higher CO_2 determination may mean that the clear glass inclusions common in plagioclase all leaked CO_2, or that melt in the resorbed looking quartz phenocrysts represents an earlier period in the Pinatubo magmatic history.

The pre-eruptive P_{H_2O} of the Pinatubo dacitic magma can also be estimated from the plagioclase-melt model of Housh and Luhr (1991). Pinatubo matrix glass and plagioclase rim compositions yield H_2O contents of 4.4 and 6.4 wt % from the AN and AB models, respectively. Comparing these H_2O estimates with ion probe analyses of the glass

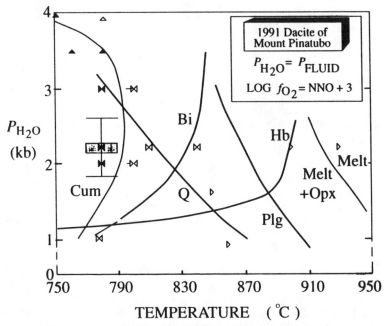

Figure 16. P-T diagram for the Pinatubo dacite composition under water-saturated and very oxidized (NNO+3) conditions (modified after Rutherford and Devine, in press). Triangles show location of experiments and point in the direction of approach to the final conditions. Curved phase boundaries indicate appearance of a phase in this dacite composition. Symbols at 780°C and 2.2 kbar show the conditions in the pre-eruptive magma from Fe-Ti-oxide geothermometry and Al-in-hornblende geobarometry. Hb, hornblende; Opx, orthopyroxene; Plg, plagioclase; Bi, biotite; Q, quartz; Cum, cummingtonite. Anhydrite is stable up to 850°C and possibly higher depending on the wt % sulfur in the magma.

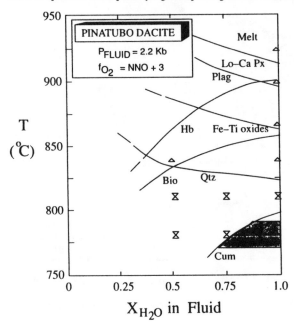

Figure 17. T-X_{H_2O} for the Pinatubo dacite compositions at 2.2 kbar and NNO+3 (Re+ReO$_2$ buffer) oxygen fugacity. Symbols as in the Figure 16 caption. Shaded area indicates where the Pinatubo phenocryst phase assemblage is stable at 780°C.

inclusions (6.4 wt %) suggests that the AB solution model better estimates water solubility in silicic melts.

Bishop Tuff rhyolite. The Bishop Tuff is a high-silica rhyolite significantly more evolved than the magma erupted at Pinatubo. Results of numerous studies, summarized by Bailey et al. (1976), Hildreth (1979) and recently Anderson et al. (1989), indicate that this magma tapped a large chemically and thermally zoned magma body. The early part of the eruption produced an airfall pumice and contemporaneous ash-flows (Wilson and Hildreth, 1991) containing phenocrysts (5 to 10 vol. %) of quartz, sanidine, plagioclase, biotite, magnetite, ilmenite, zircon, and allanite. The iron-titanium oxide geothermometer suggests a pre-eruptive magmatic temperature for the airfall unit of 743°C based on recalculations of the Hildreth (1979) analyses. Later pyroclastic flows (Hildreth, 1979) contain higher temperature, more phenocryst-rich magma. Both low-Ca and high-Ca pyroxene phenocrysts appear at temperatures above ~760°C in this series of eruptions. The temperature of the last erupted flow is now estimated to have been 820°C based on Hildreth's analyses. Because the later samples cooled as part of a pyroclastic flow, the Fe-Ti oxide temperatures may have been reset during cooling (Lindsley et al., 1991), thus we restrict our discussion to the early unwelded units.

Based on six glass inclusions from the early Bishop Tuff, Anderson et al. (1989) conclude that the magma that produced the airfall pumice was volatile-rich and coexisted with a water-rich, CO_2-bearing gas. The sum of the gas partial pressures was determined to be 1.6 to 2.6 kbar. Phase equilibria as a function of temperature and coexisting H_2O-CO_2 fluid composition at 2 kbar total pressure (Fig. 18) have been determined for the Bishop Tuff based on this early pressure estimate. The experiments indicate that the plagioclase-sanidine-quartz-biotite-magnetite-ilmenite phenocryst phase assemblage would be stable over a wide range of fluid pressure conditions at 2 kbar gas-saturated. The shaded bar shows the X_{H_2O} in the fluid deduced from the glass inclusions and Fe-Ti

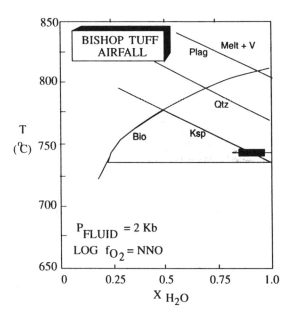

Figure 18. T-X_{H_2O} for the Bishop Tuff airfall composition at 2.0 kbar and NNO+1 oxygen fugacity (M.J. Rutherford, unpublished data). The small shaded bar shows the conditions deduced for the airfall unit based on glass inclusion data (Anderson et al., 1989; see text).

oxide temperature estimate for the airfall unit. Thus, the experimental phase equilibria do not additionally constrain volatile estimates from glass inclusions. Once determined, the plagioclase composition in equilibrium with various 2 kbar melts may well tighten the phase equilibria constraints. The recent melt inclusion data of Skirius et al. (1990) also suggest total pressures as high as 3 kbar; the phase equilibria of this biotite rhyolite need to be determined for this higher pressure.

Experiments to determine the phase equilibria of natural rhyolites and dacites have not focused on the crystallization of minor phases such as apatite and zircon. Studies exist, however, which show that crystallization of these phases depends on magmatic water content or f_{H_2O} and for apatite on f_{Cl}, f_F, f_{CO_2}, and f_{SO_2} (Watson, 1979; Piccolli and Candella, 1994). The apatite phase equilibria are particularly interesting because this phase may potentially yield information about all the above fugacities for the melt from which it crystallized (Tacker and Stormer, 1989). Piccolli and Candella (1994) attempt to decipher this information for the Bishop Tuff airfall sample and find the magma began to crystallize apatite at 860°C when it contained 700 to 960 ppm Cl and 160 to 330 ppm F. These estimates compare well with those based on glass inclusion analyses, indicating that H_2O-rich gas coexisting with the magma contained 3.1 to 4.2 wt % Cl (Piccolli and Candella, 1994).

DISCUSSION

Kilauean basaltic magma system

Results cited earlier suggest that the pre-eruptive H_2O content of Hawaiian tholeiitic basalt ranges from ~0.2 to 0.7 wt %. The cause of this variation and even the source of the H_2O is not well established. The range in H_2O contents may be explained by gain or loss of H_2O. Variable H_2O and relatively constant K_2O in glass inclusions (Anderson and Brown, 1993) and submarine basalt glass (Dixon et al., 1991) signify that H_2O was added or lost independently of K_2O, consistent with preferential mobility of H_2O in an H_2O-rich gas. The CO_2 contents of most glass inclusions analyzed by Anderson and Brown (1993) imply partial pressures of CO_2 that are greater than 200 bars whereas the H_2O contents imply H_2O partial pressures of less than ~80 bars; and gas that would be in equilibrium with the melt is correspondingly CO_2-rich and H_2O-poor. Loss or gain of significant H_2O by gas transfer at inclusion formation pressures would require more than three times as much gas as the mass of H_2O transferred. Since inclusions with low H_2O are more common than inclusions with high H_2O, H_2O may be more commonly lost than gained.

Gain of H_2O from ground-water could explain high H_2O concentrations. Invasion of ground water accompanies episodic magma withdrawal and unusual explosive eruptions (Decker and Christiansen, 1984; Holcomb, 1987). Refilling of Kilauea's shallow magma reservoir (Ryan, 1987) apparently began in 1952 following a 1924 withdrawal episode. Invading magma may have surrounded pillars and septa of ground-water containing rock. Uptake of ground-water derived from sea-water is consistent with hydrogen isotopic data (Kyser and O'Neil, 1984). The depth of the transition between meteoric and marine ground-water beneath Kilauea is unknown, but currently is probably lower than sea level and probably the depth of shallow magma storage. Following an episode of magma withdrawal, hydrostatic pressure could allow upward intrusion of marine ground water into regions formerly filled with magma or meteoric-water-saturated rock. Some submarine basalts may have formed while the edifice was submerged.

A range of pre-eruptive CO_2 concentrations (~0.2 to 0.65 wt % CO_2) is consistent with the available data, but not yet proven for parental Kilauean basaltic magma. This variation represents either different analytical approaches or secular variations of the pre-eruptive CO_2 content. The pressure at which Kilauean parental magmas become gas saturated is uncertain. It is likely to be greater than ~3 kbar because most estimates of pre-eruptive CO_2 are at least 0.3 wt %. Densities of gas inclusions in olivine phenocrysts, measured CO_2 concentrations in glass inclusions, and geophysical data combine to suggest that magma in Kilauea's shallow magma storage reservoir is gas-saturated and contains significant exsolved gas.

The pre-eruptive S content of most Kilauean basaltic magmas is between ~0.02 and 0.19 wt %. The range of S concentrations could reflect pre-eruptive degassing (Dixon et al., 1991) and mixing of degassed magma with less degassed magma. The 1959 magma was a mixture of hot, deeply derived magma and cooler volatile-poor magma that was stored in Kilauea's summit magma storage reservoir (Wright, 1973; Helz, 1987; Schwindinger and Anderson, 1989). The stored magma probably had partially degassed and was volatile-poor (Greenland et al., 1985; Gerlach and Graeber, 1985; Gerlach, 1986). Degassed magma from the lava lake drained back into the vent following the later 1959 eruptive episodes, and glass inclusions crystallized from such outgassed magma may account for the preferential occurrence of low-H_2O inclusions in the late-erupted 1959 magma. The lack of a positive correlation between H_2O and S in Kilauean glass inclusions (Anderson and Brown, 1993) is inconsistent with loss of gas rich in both H_2O and S, however.

The correlation between H_2O and K_2O in basaltic glasses from the wholly submarine Loihi volcano (Harris, 1982) compared to the independent behavior of H_2O and K_2O in basaltic glasses from Kilauea may reflect pressures in the Loihi magma reservoir too high for significant H_2O to enter the gas phase. Alternatively, the contrast may reflect episodic withdrawal of magma from Kilauea's shallow magma storage reservoir. It is sobering to realize that the extensive knowledge that decades of Kilauean studies have provided compel the conclusion that yet further studies are required to establish the relations between Kilauea's pre-eruptive, parental, and primary magmatic volatiles.

Volatile constraints on high-alumina basalt magma genesis

High-alumina basalts (HAB) are common in subduction zone environments, and have compositions suitable to being parental magmas for the commonly associated andesites, dacites, and rhyolites. The origin of HAB is controversial. Low-MgO HAB are not thought to be direct melts of the mantle because they have low-MgO contents, low Mg$^{\#}$'s, and low Ni contents. Instead, they may be daughter melts of rare primitive high-MgO basaltic magmas. Such potential parental magmas may be primitive as they are multiply saturated with mantle minerals at relevant pressures (~12 kbar; Johnston and Draper, 1993). This model has been tested by crystallizing MgO-rich basalts in the laboratory, and monitoring daughter liquid compositions. Because arc basalts were assumed to be water-poor by analogy with sea-floor basalts, most experimental studies were conducted under nominally anhydrous conditions (Gust and Perfit, 1987; Bartels et al., 1991). These studies have documented that high pressure crystallization drives these transitional compositions towards nepheline-normative liquids away from natural low-MgO HAB (Bartels et al., 1991). Low pressure crystallization causes plagioclase to crystallize extensively producing Al_2O_3-poor residual liquids rather than the observed high Al_2O_3 compositions (Grove et al., 1982; Bartels et al., 1991).

The lack of success in experimentally reproducing low-MgO, high-Al_2O_3 basalts has led some workers to conclude that they do not represent liquid compositions, but instead result from substantial plagioclase accumulation (Crawford et al., 1987). Trapped melts in olivine phenocrysts from arc basalts have extremely high Al_2O_3 contents, however, even allowing for minor host mineral crystallization that slightly increases melt Al_2O_3 (Sisson and Layne, 1993). The lack of mid-ocean ridge HAB and the reproducible association of HAB with arcs must reflect fundamental differences in MORB and arc basalt petrogenesis.

One intrinsic difference may be that HAB petrogenesis involves melting under H_2O-rich conditions (Baker and Eggler, 1983; Beard and Lofgren, 1992; Sisson and Grove, 1993a and b). Recent discoveries of substantial dissolved water in arc melts contradict the assumption that mafic arc magmas are volatile-poor. Thus, previous experimental work conducted under anhydrous conditions may not be germane to HAB petrogenesis. As noted by Holloway and Burnham (1972) and recently by Sisson and Grove (1993a), water stabilizes an early appearing FeTi-oxide phase. FeTi-oxide crystallization results in FeO depletion and SiO_2 enrichment in the evolving melt. This trend is the signature of the classic calc-alkaline differentiation trend of arc volcanoes. Previous experimental efforts to reproduce this trend under anhydrous conditions invoked geologically unreasonably high oxygen fugacities to generate the required early oxide crystallization.

Besides stabilizing an oxide as a liquidus phase, dissolved water expands the primary phase volumes of olivine and clinopyroxene relative to plagioclase (Fig. 19; Yoder, 1965; Sisson and Grove, 1993a; Gaetani et al., 1993). Olivine and clinopyroxene crystallize for an extended interval under hydrous conditions before plagioclase becomes a liquidus phase. The interval of Fe-Mg mineral crystallization and delay in plagioclase nucleation produces derivative liquids low in MgO and high in Al_2O_3, exactly the attributes of low-MgO HAB.

Under anhydrous low-pressure conditions, most basalts crystallize olivine and plagioclase, while at high pressure olivine and clinopyroxene become the first liquidus phases. This observation has led workers to conclude that olivine and clinopyroxene cumulates are a signature of high pressure crystallization. Gaetani et al. (1993) report that a Lau Basin basalt crystallized along the NNO oxygen buffer at 2 kbar water-saturated produced a long interval of olivine and clinopyroxene crystallization even though olivine and plagioclase are the 1 bar liquidus phases. This result emphasizes that dissolved water delays the crystallization of plagioclase relative to olivine and clinopyroxene. Thus, olivine and clinopyroxene cumulates may not reflect high pressure crystallization but rather modest water pressure. When global arc compositions are compiled, they reflect long intervals of olivine-clinopyroxene crystallization rather than olivine-plagioclase crystallization (Fig. 20; Gaetani et al., 1993). The key to creating this sequence of mineral crystallization is modest water pressure. Global arc compositions are thus consistent with glass inclusion analyses, and require that at depth arc melts had significant dissolved water contents.

Sisson and Grove (1993b) applied their magmatic geothermometers and hydrometers (Eqns. 2 and 3) together with Housh and Luhr's anorthite model to natural arc compositions. They found that low-MgO high-Al_2O_3 basalts commonly contain 4 to 6 wt % dissolved water. These water contents suggest that most low-MgO HAB reach saturation with an H_2O-rich gas at low pressure (1 to 2 kbar). Gas saturation and concomitant volatile exsolution then drive explosive volcanic eruptions. As the melt exsolves volatiles, the liquidus temperature increases dramatically forcing rapid

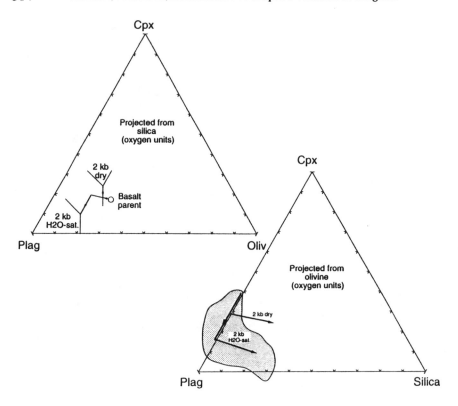

Figure 19 (upper left). Projection from silica onto clinopyroxene–plagioclase–olivine ternary. Mineral components are calculated in oxygen units. The experimentally determined 2 kbar anhydrous and water-saturated pseudo-eutectics are shown. A typical basalt composition is shown by the open circle, and its crystallization under dry or wet conditions is shown by the dark arrows. Note that the plagioclase primary phase volume decreases with increasing water pressure. Data from and figure after Gaetani et al. (1993).

Figure 20 (lower right). Projection from olivine onto clinopyroxene – plagioclase – silica ternary. Mineral components are calculated in oxygen units. Dark arrows show crystallization path under 2 kbar anhydrous and 2 kbar water-saturated conditions. Field of arc lavas is shown in gray. Arc lavas compositions appear to follow the hydrous crystallization path. Data from and figure after Gaetani et al. (1993).

crystallization. The crystallization caused by volatile exsolution may explain the near absence of quenched glasses from arc environments. Sisson and Grove (1993b) also estimated temperatures and water contents for high-MgO arc basalts which may be parental to low-MgO HAB. These parental magmas are estimated to contain 1 to 2 wt % dissolved water, higher than the water contents measured in sea-floor basalts.

The above phase equilibria demonstrate that for primitive high-MgO basalt, a water pressure of 2 kbar is sufficient to delay plagioclase crystallization and enhance Fe-Ti-oxide crystallization, so that low-MgO HAB basalts are produced as daughter products. Sisson and Grove (1993a and b) also demonstrated that the crystallizing hydrous liquids follow the classic calc-alkaline differentiation trend as shown previously by Spulber and Rutherford (1983). These phase equilibria data combined with glass inclusion evidence for significant dissolved water contents support the conclusion that arc basalt magmas have high dissolved water contents prior to eruption.

Volatile saturation in magmas

Evidence for an excess gas phase in basalts erupting at a center such as Kilauea suggests volatile saturation initially with a CO_2-rich gas at depth which evolves through a continuum of compositions to become more H_2O-rich close to the surface (Gerlach and Graeber, 1985; Gerlach, 1986). The proportion of S and Cl in these fluids is not well constrained, however. Experimental data on the distribution of S and Cl between a basaltic melt and a CO_2-rich fluid is needed. Additional data is also required on the C content of mantle-derived basaltic melts. Together with experimental data on CO_2 solubility (Pan et al., 1991; Pawley et al., 1992), these data would allow a more rigorous assessment of when an ascending magma could first generate an excess CO_2-rich gas, and what its compositions would be. Answers to these questions are important for models of basaltic magma ascent and degassing. They are also very important in understanding processes in more silicic magma bodies, many of which are underplated or intruded by basalt and may have their volatile budget altered by basalt degassing (Hildreth, 1981; Pallister et al., 1993).

Volatile saturation of magma prior to eruption is suggested for each of the silicic magma systems described above although the evidence is sometimes equivocal. Phase equilibria and glass inclusion data for the 1980-86 Mount St. Helens magma system indicate that it was water-undersaturated, but whether a mixed volatile gas (CO_2 + SO_2 + H_2O) was present or whether the system was fluid-undersaturated is unknown. To distinguish between these alternatives, glass inclusion analyses for CO_2 and S are required. Melt inclusions indicate the sulfur abundance is ~100 ppm which corresponds to P_{SO_2} <100 bars (Carroll and Rutherford, 1985), but glass inclusions large enough for FTIR analysis of CO_2 have not been found. Since the P_{H_2O}, P_{total}, and T for the reservoir magma remained constant over a 5 year period (Rutherford and Hill, 1993), however, the magma may have been buffered by a CO_2-H_2O gas phase in which X_{H_2O} was 0.67 (Fig. 14). This conclusion is supported by the distribution and abundance of sulfur in this magma. Sulfide is present only in the cores of Fe-Ti oxide grains. In addition, melt inclusions trapped in phenocrysts generally contain less sulfur (~100 ppm) than would be present at sulfide saturation in a 900°C melt at this f_{O_2} (330 to 500 ppm; Carroll and Rutherford, 1987). These observations suggest that sulfide saturation did occur early in the phenocryst growth stage of this magma but was followed by sulfur oxidation, generating a gas phase (no other sulfur-bearing phase is stable) and the sulfide-undersaturated melt found in all melt inclusions.

Evidence for an excess gas phase in the 1991 Pinatubo dacite is strong in that estimates of total pressure equal the sum of the gas partial pressures (Rutherford and Devine, in press; Wallace and Gerlach, in press). The questions that need to be addressed are those concerning the composition and abundance of the excess gas. Westrich and Gerlach (1992), Gerlach et al. (in press), and Wallace and Gerlach (in press) are convinced that ~5 vol. % of an excess gas phase was present in the 5-10 km^3 of dacite erupted at Pinatubo on June 15, 1991. This fluid is suggested as the source of the huge SO_2 mass injected into the atmosphere (Bluth et al., 1992). It is not yet possible to demonstrate that excess fluid was this abundant because at least some of the excess sulfur may have been derived from anhydrite breakdown during magma ascent (Devine et al., 1984; Baker and Rutherford, 1992). Further work on this problem will be interesting because the development of a significant fluid volume in such a magma system has important consequences for the physical properties of the magma and, therefore, the explosivity of the eruptions. It is important to know whether and under what conditions this volatile-rich magma could exist at depth. An answer may come from studying chlorine behavior in the magma. No phenocryst phase exists to buffer Cl abundance in the melt as is the case for S, and it should partition strongly into the vapor (Webster and Holloway, 1988). The chlorine concentration in the pre-eruptive melt was ~1200 ppm, well below chloride saturation in high-SiO_2 melt (Webster and Holloway, 1988; Metrich and Rutherford, 1992). A constant or decreasing melt chlorine content during magma crystallization would suggest the generation of an excess fluid phase. Evidence for this chlorine decrease may exist in the glass inclusions trapped in early formed An-rich plagioclase, or in the chlorine content of phenocrysts such as hornblende, biotite and apatite. Experimental study of the fluid compositions coexisting with anhydrite saturated melts (Baker and Rutherford, 1992) could also constrain the amount of excess fluid required to form the excess atmospheric sulfur by better defining the fluid composition in equilibrium with anhydrite.

An excess gas phase in the pre-eruptive Bishop Tuff magma seems to be indicated by the melt inclusion studies, although the result is tentative; neither geologic considerations nor phase equilibria place very tight constraints on the total magmatic pressure. Using his experimental data for CO_2 solubility in rhyolites (Fogel and Rutherford, 1990) along with the Anderson et al. (1989) glass inclusion analyses, Fogel (1989) calculated P_{fluid} to have been 1.94 kbar with P_{CO_2} = 132 bars. A new, large data set comprising 50 glass inclusions was recently compiled by Skirius (1990); it included analyses of reheated and rehomogenized inclusions (Skirius et al., 1990) from early ash-flow as well as plinian pumice clasts. The gas saturation pressures for these early Bishop glass inclusions range from 1.3 to 3.0 kbar with no significant difference between ash-flow and plinian inclusions. Dunbar and Hervig (1992), however, reported a significant gradient in H_2O for the early Bishop magma (from 6 to 3.5 wt %) based on ion-microprobe analyses of melt inclusions. By comparison, only one of Skirius' early inclusions has significantly less than 5.0 wt % H_2O (4.5 wt % H_2O). Stratigraphic non-equivalence does not seem to explain these analytical differences; it may be that different sample preparation procedures led to the analysis of leaked inclusions. Earlier estimates of H_2O in the Bishop magma by Hildreth (1979) were based on biotite stability and yielded values around 5 wt % for the early-erupted Bishop magma. Stormer and Whitney (1985) estimated a total pressure of 4 to 6 kbar for the early Bishop magma based on coexisting feldspars and oxides, but this method appears to produce anomalously high pressure estimates (Johnson and Rutherford, 1989a). More recently, Frost and Lindsley (1992) estimated a range of total pressures of crystallization of later-erupted Bishop magma from 2.1 to 3.3 kbar based on oxide and pyroxene equilibria. These total pressures are thus consistent with the gas-saturation pressures, but Frost and Lindsley (1992) point out problematic equilibrium relations between oxides and pyroxenes in the late-erupted Bishop magma that may reflect magma mixing. The sum of all this work seems to be that the early Bishop magma was likely, although not certainly, saturated with gas over a significant pressure range.

Although we do not know if crystallizing silicic magmas are commonly gas-saturated, the above data on explosive silicic eruptions argues for the existence of several silicic magmas that were saturated with a gas containing CO_2 at pressures as great as 4 kbar. This evidence, together with recent large estimates of excess gas production during some eruptions (Gerlach et al., in press), suggests that the source regions of some granitic magmas may be very gas-rich and may contain an excess gas phase. The physical properties of such magmas would be different from gas-undersaturated magmas, and these differences should be reflected in their migration and accumulation.

Obsidians: the Mono Craters obsidian clasts and domes

It is interesting and relevant to consider the significance of pre-eruptive volatiles for the origin and eruption of obsidian flows and domes. Are obsidian-producing magmas hot and dry compared to ash-producing magmas or do obsidians result from different eruptive and degassing processes?

The Mono Craters rhyolites in NE California contain unaltered, recently-extruded clasts of both obsidian and pumice in early-erupted tephra deposits as well as slightly later-erupted obsidian domes (Sieh and Bursik, 1986). The tephra are phenocryst-poor. Concentrations in the obsidians of CO_2 (5-38 ppm in pyroclasts and ≤ 2 ppm in domes), H_2O (0.5-0.7 wt % in pyroclasts and 0.1-0.3 wt % in domes) and D/H ratios (-51 to -76 per mil in pyroclasts and -102 to -127 in dome obsidians) were measured by Newman et al. (1988) who noted that the obsidians with the highest H_2O concentrations were erupted first. Equilibration temperatures of magnetite and ilmenite phenocrysts in several dome obsidians range from 790 to 860°C (Carmichael, 1967). Such temperatures, if magmatic, correspond to minimum (P = P_{H_2O}) H_2O concentrations between ~3 and 4 wt % (Wyllie et al., 1976). This estimate assumes that the Fe-Ti oxides temperature is the temperature of the pre-eruptive magma, and that no heat loss or oxide re-equilibration occurred during magma ascent or immediately following extrusion. Experiments have shown that Fe-Ti oxides will completely re-equilibrate to a new temperature (by recrystallization) in as little as four days at 850°C (Gardner et al., in press). The assumption of essentially adiabatic upwelling of this magma may be justified, however; Rutherford and Hill (1993) found strong evidence that the pre-eruptive magma temperature and the extrusion temperature were equal and essentially constant during extrusion of the 1980-1985 Mount St. Helens lava domes. Newman et al. (1988) interpreted their volatile and isotopic measurements as best explained by closed system decompression of tephra-yielding magma which contained a total (exsolved + dissolved) of 5.0 wt % H_2O and 1.2 wt % CO_2, followed by open system degassing at pressures less than ~50 bars to yield domes. An important implication of the Mono Craters work is that the pre-eruptive magmas contained much more CO_2 (mostly exsolved) and significantly more H_2O than remains dissolved in the obsidians.

The data on the Mono Craters obsidians place helpful constraints on eruption and degassing processes, topics that continue to be debated (see Fink et al., 1992). Dissolved H_2O concentrations in other dome obsidians range up to ~1.0 wt %, but most young fresh obsidians contain ~0.1 wt % H_2O (Macdonald et al., 1992) suggesting that most dome obsidians have degassed to ~1 bar (0.10 wt % is the 1 bar concentration derived by extrapolating the low-pressure work of Silver et al., 1990). The process whereby obsidians attain such low H_2O concentrations is the subject of much recent interest. Eichelberger and Westrich (1981) and Eichelberger et al. (1986) offer an explanation descriptively denoted the permeable foam model. According to this model, silicic magma foam is the primary extrusive product. During the final stages of upwelling, during extrusion and immediately following extrusion, bubbles in the foam disappear to form dense obsidian. Eichelberger and coworkers argue that low gas pressures within the foam allow loading caused by thickening of the flow to resorb bubbles into the glass at ~10 m depth. High temperatures during foam deflation promote obliteration of bubble walls (Westrich et al., 1988) and thus shard textures, such as observed in welded tuffs, are destroyed. Fink et al. (1992) emphasize, however, that some structures and textures require post-extrusion vesicle growth rather than shrinkage.

The above numbers imply the escape of ~90% of the pre-eruptive volatiles through the magma and the conduit walls during ascent and decompression. The Mono Craters data (Newman et al., 1988) are consistent with late-stage, low-pressure, open-system loss of gas. Geological relations indicate that eruption may take only a few months (Sieh and Bursik, 1986), insufficient time for volatiles to diffuse through meters of ascending foam. Therefore, the open-system gas loss implies cracks or other highly permeable channelways in magma for gas escape. The physical mechanism for open-system gas loss is unknown, but foam collapse (e.g., Vergniolle and Jaupart, 1990) was suggested by Newman et al. (1988).

The Mono Craters obsidians also present some problems to these developing ideas of eruptive degassing (see also Chapter 11). Seven of the Mono Craters pyroclastic obsidians have significantly less than 100 bars worth of dissolved H_2O and CO_2 (Newman et al., 1988, their Fig. 5), and both the D/H and CO_2 vs. H_2O for these samples are consistent with closed system degassing. Gas bubbles would comprise more than 85 vol. % of the magma at 100 bars (about 89 vol. % at 70 bars) and 800°C, however, if the magma had exsolved 4.0 wt % H_2O and 1.2 wt % CO_2, as implied above. This volume is close to the maximum porosity of the most vesicular pumices (Whitham and Sparks, 1985), and greater than the 77 vol. % at which magmatic foam is typically inferred to disrupt into spray (Heiken, 1972). The Newman et al. (1988) data and interpretations imply that early erupted rhyolitic tephra evolved under closed conditions in foams with vesicle porosities as great as 89%. How did obsidian clasts found in the tephra lose volatiles and become bubble-free while the enclosing tephra vesiculated in a closed system?

The origin of the tephra-forming rhyolite and the dome-forming rhyolite can apparently be explained by invoking an eruptive process similar to that proposed by Eichelberger and Westrich (1981) and Newman et al. (1988) and modified by Fink et al. (1992). The tephra-producing magma retained all or nearly all of its gas until the moment of eruption, and upon eruption negligible additional exsolution occurred. The dome-producing magma evolved from remnants of the tephra-producing magma that continued to decompress, exsolve and lose gas (by partial collapse of subvolcanic foam during non-eruptive degassing) down to pressures as low as a few bars. Many extruded rhyolites have less than 50 vol. % vesicle porosity and less than 0.3 wt % H_2O, seeming to require the loss of gas from magmatic foams at low pressures (a few bars) but considerable depths; i.e., conduit pressures considerably less than lithostatic. These pressures might result from slow viscous movement of rhyolite in the conduit such that transitory connection of pipe-like vesicle trains is achieved, or from tensional stresses in a carapace inflating from below.

Another alternative is plausible for the Mono Craters tephra. The H_2O versus CO_2 concentrations may reflect buffering by a large amount of gas with ~15 mole % CO_2. In this interpretation, tephra-producing melts would be buffered to the observed CO_2 concentrations by abundant gas. The range of D/H ratios found for the obsidian tephra (-51 to -76 per mil; Newman et al., 1988) is broadly consistent with equilibration between melt and a gas of constant D/H ratio because of concentration-dependent dissolved species and the isotopic fractionations (Dobson et al., 1989). If the Mono Craters obsidian tephra formed in a gas-buffered environment, then the initial bulk volatile content of the rhyolitic magma is constrained only to exceed the observed maximum dissolved concentrations of ~2.6 wt % H_2O and 40 ppm CO_2 (Newman et al., 1988) and the closed-system vesicle porosity need not exceed 70% at 100 bars. The dome obsidians might still evolve by open system degassing, or their D/H ratios might reflect some interaction with H_2O of meteoric origin.

Seismic studies suggest that the roof of a magma body lies about 8 to 10 km beneath the Mono Craters (Achauer et al., 1986). The tephra-producing magma with 5 wt % H_2O and 1.2 wt % CO_2 would contain several vol % gas at such depths. The seismic velocity contrast for the anomalous region (magma) is only 6%, and, as Achauer et al. (1986) note, may be explained by a small melt fraction. As the Mono Craters rhyolites are crystal-poor, the possibility should be considered that the anomalous region contains mainly a small amount of gas and little or no melt.

In summary, the 5 wt % H_2O and 1.2 wt % CO_2 inferred by Newman et al. (1988) for the Mono Craters rhyolitic magma is plausible and consistent with the crystallization temperatures of Carmichael (1967) and seismic anomalies located 10 km below. It appears possible to reconcile the data with a much lower CO_2 concentration (~50 ppm) and less H_2O (about 3 wt %), however.

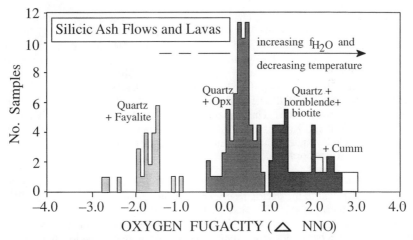

Figure 21. Plot (modified after Carmichael, 1991) showing abundance and oxidation state of quartz-bearing volcanic rocks. Note the trend to increasingly oxidized and H_2O-rich magmas at lower temperatures.

Volatiles and the oxidation state of magmas

The oxidation states of terrestrial magmas vary over eight orders of magnitude from NNO-3 to NNO+5 (Fig. 21; Carmichael, 1991; see also Chapter 6). The oxidation state of a magma may reflect the oxidation state of the source region (Carmichael, 1991). Other models suggest that magmatic oxidation states may be significantly altered by volatile loss or gain. For example, SO_2 loss from basalt would tend to reduce the magma because sulfur is primarily present as S^{2-} in basalts (Katsura and Nagashima, 1974; Hattori, 1993). Degassing of SO_2 was interpreted to have reduced basalts erupted at Hawaii (Moore and Fabbi, 1971; Anderson and Wright, 1972). Carbon occurs in basalts either as elemental carbon or dissolved as CO_3^{2-} (Fine and Stolper, 1986; Fogel, 1989); elemental carbon degassing as $CO_2 + CO$ would cause reduction of the residual magma, dissolved CO_3^{2-} degassing would have a weak oxidizing influence with the effect depending on the oxidation state (Fe^{3+} content) of the magma prior to degassing. An oxidation process involving H_2 loss from hydrous basalts has also been proposed by Sato and Wright (1966) and Sato (1978). For a basalt, the oxidation-reduction potential of SO_2 and CO_2 degassing is controlled by the amounts and oxidation states of S, C, and Fe in the magma. Similarly, the potential oxidation effect of H_2 loss is limited in basalts by low dissolved H_2O abundances and low H_2/H_2O ratios. Observations of Kilauean summit and rift gases fail to show any relative change in oxygen fugacity (Gerlach, 1993). The effect can be significant, however, in more silica-rich magmas where the iron content is lower and the H_2O content is greater (Candela, 1986).

The role of oxidation-reduction in both basaltic and silicic magmas is illustrated by the recently erupted rocks of Mount Pinatubo. At NNO+3, this dacite is one of the most oxidized magmas yet reported. The dacite was also water-rich (6.4 wt % in the melt; hornblende, biotite and cummingtonite are present as phenocryst phases) and sulfur-rich. The 0.4 wt % sulfur in the dacite (Bernard et al., 1991; Pallister et al., in press) is present primarily as anhydrite phenocrysts. An additional 350 to 1000 ppm sulfur may have been present as a coexisting gas phase in the pre-eruptive magma (Gerlach and Westrich, 1992; Wallace and Gerlach, in press). Thus, this magma is noteworthy because it is highly oxidized, water-, and sulfur-rich, characteristics that are potentially related. Equally water-rich magmas occur at Long Valley, CA (Anderson et al., 1989), Fish Canyon, CO (Johnson and Rutherford, 1989a) and Mount St. Helens (Geschwind and Rutherford, 1992), but these magmas are less oxidized (NNO, NNO+1, and NNO+2 respectively). They also contain less than 100 to 200 ppm dissolved sulfur. Basalt was erupted at Mount Pinatubo and occurs as pillow-like masses in andesite. The andesite, clearly a dacite and basalt mixture (Pallister et al., 1992), dominated the early dome and pumice deposits of the 1991 eruption. The oxidation state of the basalt suggested by the S^{2-} to SO_4^{2-} ratio (Carroll and Rutherford, 1988) in melt trapped in Fo_{87} olivine was ~NNO. The andesite oxidation state is intermediate between the basalt and dacite at NNO+1 and 900° to 1000°C (Rutherford et al., 1993).

The sulfur-rich character of the Pinatubo dacite may be related to basalt emplacement adjacent to the dacitic reservoir and subsequent volatile exchange (Pallister et al., 1992; Matthews et al., 1992; Hattori, 1993). Pallister et al. (1993) note that basalt incursion has occurred several times in the ~30,000 year eruption history of Mount Pinatubo. Does this basalt degassing model explain the unique oxidation state and H_2O-rich character as well as the sulfur content of the dacite? If basalt were emplaced adjacent to the cool (780°C) dacite, cooling of basalt across a boundary layer would induce crystallization eventually yielding gas saturation in the basalt. Evidence from deep (35 km) long period earthquakes (White et al., in press) indicates that the magma at depth in this system was gas saturated before reaching the dacite storage region at 7 to 11 km. Thus, gas saturation of the basalt is likely before it encountered the dacite. Based on available data concerning CO_2, H_2O and S solubility in basaltic melt, the early fluid available for transfer from the basalt to the dacite at depth would be rich in CO_2 and S with a low H_2O fraction because the basalt was far from H_2O saturation (Rutherford et al., 1993 find 1000 ppm sulfur and ~1 to 2 wt % H_2O in olivine glass inclusions). For the maximum conceivable oxidation potential we may assume that the sulfur in this fluid occurred exclusively as SO_2; gases evolved from basalt at low pressure have large SO_2 to H_2S ratios (Gerlach, 1986). Transfer of such a fluid phase from basalt to dacite could oxidize the dacite, given that the sulfur in the basaltic magma is primarily S^{2-} (Katsura and Nagashima, 1974; Caroll and Rutherford, 1988). The fluid could also serve as a mechanism for increasing total sulfur in the dacite, as the S-carrying capacity of the C-S fluid would tend to be decreased by reaction with water-rich dacite producing sulfide or sulfate crystallization in the dacite. It is not obvious, however, that this SO_2-bearing fluid has the potential to oxidize the dacite magma beyond the sulfide-sulfate boundary (NNO+1.5; Carroll and Rutherford, 1988) because $CaSO_4$ formation requires $1/2O_2$ in addition to SO_2 and CaO. We conclude that the Pinatubo dacite probably underwent an additional process such as H_2-loss to achieve the NNO+3 oxidation state.

Eruption dynamics

One of the earliest estimates of the pre-eruptive H_2O content of a silicic magma was 4 wt % H_2O based on the 1955 Bezymianni eruption dynamics (Markhinen, 1962). Wilson et al. (1980) developed a comprehensive model of eruptive column dynamics by combining observations and theory. Their Figure 6 shows how column height and erupted clast size can be used to infer pre-eruptive magmatic volatile content. They note that the inferred volatile content increases with increasing conduit wall strength and the difference between lithostatic and magmatic pressure. Wilson notes (written communication, 1992) that implicit in his models of basaltic lava fountains is the assumption that volatile exsolution is quenched at the moment that magmatic foam disrupts into spray. This assumption is justified by rapid expansion and acceleration of the two-phase system, causing chilling within a few tens of seconds (see also Chapter 11). The same assumption is implicit for plinian eruptions of rhyolitic magma. Magma disruption pressures calculated by Wilson et al. (1980) are in the range of hundreds of bars, and the implied concentrations of residual H_2O remaining dissolved in quenched glass (pumice) are generally 1 to 2 wt %. Magmatic H_2O remaining in fresh glassy pumice is generally thought to be close to the 1 bar solubility of ~0.1 wt %; few analyses exist of H_2O in historic age rhyolitic pumice where adsorbed H_2O in vesicles etc. has been evaluated, however. If residual magmatic H_2O in pumice is ~0.1 wt %, then pre-eruptive H_2O estimates based on eruption dynamics could be ~2 wt % low.

Eruption models of Wilson et al. (1980) and co-workers relate mass eruption rate (proxied by eruptive column height and tephra dispersal patterns) and vent exit velocity (proxied by maximum clast size) to exsolved H_2O content. Woods (1988), in reviewing various eruption models, emphasized that eruption column height is sensitive also to eruption temperature. Further work integrating petrological and observational data on erupting magma with theoretical models should yield firmer estimates of eruptive and pre-eruptive volatile concentrations for specific, well-documented eruptions.

At present the most stringent constraints on pre-eruptive volatiles come from observational measurements of eruptive SO_2 interpreted in light of compositions of glass inclusions and residual glasses in pumice (e.g., Allard et al., 1994; Gerlach et al., in press, a and b). Combining published estimates based on satellite data of the amount of SO_2 erupted by Pinatubo on June 15, 1991 with estimates of the pre-eruptive magmatic gas composition based on melt inclusion data, Wallace and Gerlach (in press) argue that the pre-eruptive Pinatubo magma contained 0.7 to 1.3 wt % exsolved gas at a total pressure of ~2.8 kbar, yielding a total (exsolved + dissolved) volatile concentration of ~3.7 to 4.0 wt % H_2O, 0.12 to 0.24 % CO_2 and 0.04 to 0.08 wt % exsolved SO_2 in magma with 45 wt % of volatile-free crystals and about 0.4 wt % of anhydrite. Several other possible sources of the erupted sulfur have been ruled minor by Gerlach et al. (in press). It remains uncertain, however, how much of the atmospheric sulfur was derived from high-pressure magmatic gas and how much was derived from syneruptive breakdown of anhydrite (Rutherford and Devine, in press).

The significance of volatiles in silicic magmas: some speculations

On Earth, H_2O is the most abundant volatile species and important for lowering rock melting temperatures. Removal of H_2O from the lower crust renders the crust less susceptible to melting and thereby tends to increases its strength and thickness. A weaker crust would be thinner and possibly submarine rather than continental. If silicic magmas were poor in H_2O, they could not dehydrate and strengthen the crust. The effectiveness of silicic magmas in transferring volatiles upwards is demonstrated by their H_2O contents.

Rhyolitic magmas ranging widely in alkalinity and erupted in different tectonic environments usually contain at least ~4 wt % H_2O and up to ~1000 ppm dissolved CO_2 prior to their eruption according to recent studies described above. Some of these magmas contained excess gas during crystallization. The observed H_2O concentrations in silicic magmas may reflect some critical threshold H_2O concentration required to form a large enough volume fraction of partial melt (at crustal temperatures) to be readily erupted. Alternatively, H_2O concentrations in silicic magmas may reflect the effect of H_2O on magma properties whereby partial melts with more than ~4 wt % H_2O are prone to coalesce and form uniform magma bodies.

Are granites simply failed rhyolites and vice versa, or do important differences exist between different silicic magmas that promote or inhibit eruption? Burnham (1979) has argued that the effect of dissolved H_2O on crystallization temperatures impedes ascent and eruption of H_2O-rich magmas. Yet, eruption is not generally an equilibrium process, and the potential energy for eruption increases with magmatic volatile content, other factors being equal. Silicic parental magmas with <4 wt % H_2O may have erupted and formed rhyolites or dacites, but presently available data do not confirm this supposition. Establishing the pre-eruptive H_2O contents of granitic magmas is more difficult, but some appear to have contained less than 4 wt % dissolved H_2O in their early stages of crystallization (e.g., Maaloe and Wyllie, 1975; Wyllie et al., 1976).

Volcanic silicic magmas may contain more H_2O than magmas that form most granites, despite similar crystallization pressures. This contrast, if true, may be explained by the tendency of volatiles to concentrate in residual melts that ascend and accumulate in the upper parts of solidifying magma bodies. Eruptions may be expected mainly to tap the upper parts of magma bodies. Most large ash-flow tuffs are compositionally and texturally stratified suggesting sequential tapping of progressively less differentiated and more crystal-rich magma. These features are displayed by the Bishop Tuff where late-erupted silicic magmas, presumably erupted from greater depths, have lower dissolved H_2O concentrations.

CO_2 adds an important conceptual wrinkle to the problem of granites vs. rhyolites. In the late-erupted ash-flows of the Bishop Tuff, glass inclusions not only have lower H_2O but greater CO_2. The resulting range of gas-saturation pressures is similar for both the early-erupted and late-erupted magmas. Possibly, the late-erupted magmas were not gas-saturated, and were derived from a greater depth. It is equally possible that the two systems occupied different lateral positions separated by country rocks. In either case, the differences between the magmas represented by the early- and late-erupted ash-flows seem analogous to the inferred differences between volcanic and plutonic magmas. The suggestion is that the volcanic magma can not be contained as an intrusion, and the plutonic magma can not be erupted. Perhaps, in general, volcanic and plutonic silicic magmas are complimentary. We expect cases exist where a volatile-rich silicic magma cannot erupt, possibly for either size or tectonic reasons, and crystallizes as a water-rich granite or pegmatite.

It is interesting to contrast two statements published in the mid 1970s that reflect views derived from experimental studies of plutonic rocks on the one hand and from primarily analytical studies of volcanic rocks on the other: "The H_2O content of large granitic bodies is less than 1.5 wt %." (Wyllie et al., 1976, p. 1007). "It seems probable that volcanic rocks are generally volumetrically small portions of subsurface reservoirs which may be vapor-saturated." (Anderson, 1975, p. 37). Twenty years later it still seems that plutonic magmas are drier than volcanic magmas.

REFERENCES

Achauer U, Greene L, Evans JR, Iyer HM (1986) Nature of the magma chamber underlying the Mono Craters area, eastern California, as determined from teleseismic travel time residuals. J Geophys Res 91:13,873-13,891

Aggrey KE, Muenow DW, Batiza R (1988a) Volatile abundances in basaltic glasses from seamounts flanking the East Pacific Rise at 21°N and 12-14 °N. Geochim Cosmochim Acta 52::2115-2119

Aggrey KE, Muenow DW, Sinton JM (1988b) Volatile abundances in submarine glasses from the North Fiji and Lau back-arc basins. Geochim Cosmochim Acta 52:2501-2506

Allard P, Carbonnelle J, Metrich N, Loyer N, Zettwoog P (1994) Sulphur output and magma degassing budget of Stromboli volcano. Nature 368:326-330

Andersen DJ, Lindsley DH (1988) Internally consistent solution models for Fe-Mg-Mn-Ti oxides. Am Mineral 73:57-71

Anderson AT Jr. (1974a) Chlorine, sulfur and water in magmas and oceans. Geol Soc Am Bull 85:1485-1492

Anderson AT Jr. (1974b) Evidence for a picritic, volatile-rich magma beneath Mt. Shasta, California. J Petrol 15:243-267

Anderson AT Jr. (1975) Some basaltic and andesitic gases. Rev Geophys Space Physics 13:37-55

Anderson AT Jr., Brown GG (1993) CO_2 and formation pressures of some Kilauean melt inclusions. Am Mineral 78:794-803

Anderson AT Jr., Newman S, Williams SN, Druitt TH, Skirius C, Stolper E (1989) H_2O, CO_2, Cl and gas in Plinian and ash-flow Bishop rhyolite. Geology 17:221-225

Anderson AT Jr., Wright TL (1972) Phenocrysts and glass inclusions and their bearing on oxidation and mixing of basaltic magmas, Kilauea Volcano, Hawaii. Am Mineral 57:188-216

Bailey RA, Dalrymple GB, Lanphere MA (1976) Volcanism, structure, and geochronology of Long Valley Caldera, Mono County, California. J Geophys Res 81:725-744

Baker DR, Eggler DH (1983) Fractionation paths of Atka (Aleutians) high-alumina basalts: constraints from phase relations. J Volc Geotherm Res 18:387-404

Baker L, Rutherford MJ (1992) Anhydrite breakdown as a possible source of excess sulfur in the 1991 Mount Pinatubo eruption. EOS Trans Am Geophys Union 73:625

Bartels KS, Kinzler RJ, Grove TL (1991) High pressure phase relations of primitive high-alumina basalts from Medicine Lake volcano, northern California. Contrib Mineral Petrol 108:253-270

Beard JS, Lofgren GE (1992) An experiment-based model for the petrogenesis of high-alumina basalts. Science 258:112-115

Bernard A, Demaiffe D, Mattielli N, Punongbayan RS (1991) Anhydrite-bearing pumices from Mount Pinatubo: Further evidence for the existence of sulfur-rich silicic magmas. Nature 354:139-140

Bluth GJS, Doiron SD, Schnetzler CC, Krueger AJ, Walter LS (1992) Global tracking of the SO_2 clouds from the June 1991 Mount Pinatubo eruptions. Geophys Res Lett 19:151-154

Bottinga Y, Javoy M (1991) The degassing of Hawaiian tholeiite. Bull Volcanol 53:73-85

Bottinga Y, Kudo A, Weill D (1966) Some observations on oscillatory zoning and crystallization of magmatic plagioclase. Am Mineral 51:792-806

Brett R, Evan HT Jr., Gibson EK Jr., Hedenquist JW, Wandless M-V, Sommer MA (1987) Mineralogical studies of sulfide samples and volatile concentrations of basalt glasses from the southern Juan de Fuca Ridge. J Geophys Res 92:11,373-11,379

Buddington AF, Lindsley DH (1964) Iron-titanium oxide minerals and synthetic equivalents. J Petrology 5:310-357

Burnham CW (1979) The importance of volatile constituents. In:Yoder HS (ed) The evolution of the igneous rocks: fiftieth anniversary perspectives. Princeton University Press, 439-482

Byers CD, Christie DM, Muenow DW, Sinton JM (1984) Volatile contents and ferric-ferrous ratios of basalt, ferrobasalt, andesite and rhyodacite glasses from the Galapagos 95.5°W propagating rift. Geochim Cosmochim Acta 48:2239-2245

Byers CD, Garcia MO, Muenow DW (1985) Volatiles in pillow rim glasses from Loihi and Kilauea volcanoes, Hawaii. Geochim Cosmochim Acta 49:1887-1896

Byers CD, Garcia MO, Muenow DW (1986) Volatiles in basaltic glasses from the East Pacific Rise at 21 °N: implications for MORB sources and submarine lava flow morphology. Earth Planet Sci Lett 79: 9-20

Byers CD, Muenow DW, Garcia MO (1983) Volatiles in basalts and andesites from the Galapagos Spreading Center, 85° to 86 °W. Geochim Cosmochim Acta 47:1551-1558

Candela PA (1986) The evolution of aqueous vapor from silicate melts: effect on oxygen fugacity. Geochim Cosmochim Acta 50:1205-1211

Carmichael ISE (1964) The petrology of Thingmuli, a Tertiary volcano in Eastern Iceland. J Petrology 5:435-460

Carmichael ISE (1967) The iron-titanium oxides of salic volcanic rocks and their associated ferromagnesian silicates. Contrib Mineral Petrol 14:36-64

Carmichael ISE (1991) The redox states of basic and silicic magmas: A reflection of their source regions. Contrib Mineral Petrol 106:129-141

Carroll MR, Rutherford MJ (1985) Sulfide and sulfate saturation in hydrous silicate melts. J Geophys Res 90:C601-C612

Carroll MR, Rutherford MJ (1987) The stability of igneous anhydrite: Experimental results and implications for sulfur behavior in the 1982 El Chichon trachyandesite and other evolved magmas. J Petrology 28:781-801

Carroll MR, Rutherford MJ (1988) Sulfur speciation in hydrous experimental glasses of varying oxidation state: Results from measured wave-length shifts of sulfur X-rays. Am Mineral 73:845-849

Cashman KV (1992) Groundmass crystallization of Mount St. Helens dacite, 1980-1986 a tool for interpreting shallow magmatic processes. Contrib Mineral Petrol 109:431-449

Cawthorn RG, O'Hara MJ (1976) Amphibole fractionation in calc-alkaline magma series. Am J Sci 276: 309-329

Cawthorne RG (1976) Melting relations in part of the system $CaO-MgO-Al_2O_3-SiO_2-Na_2O-H_2O$ under 5kb pressure. J Petrol 17:44-72

Chamberlin RT (1908) The gases in rocks. Carnegie Inst Wash Publ 106:1-80

Clague DA, Moore JG, Dixon JE, Friesen WB (in press) Petrology of submarine lavas from Kilauea's Puna Ridge, Hawaii. J Petrol

Clague DA, Weber WS, Dixon JE (1991) Picritic glasses from Hawaii. Nature 353:553-556

Clemens JD (1984) Water contents of silicic to intermediate magmas. Lithos 17:273-287

Clemens JD, Wall VJ (1981) Origin and crystallization of some peraluminous (S-type) granitic magmas. Can Mineral 19:111-131

Clemens JD, Wall VJ (1988) Controls on the mineralogy of S-type volcanic and plutonic rocks. Lithos 21:53-66

Crawford AJ, Falloon TJ, Eggins S (1987) The origin of island-arc high-alumina basalts. Contrib Mineral Petrol 97:417-430

Danyushevsky LV, Falloon TJ, Sobolev AV, Crawford AJ, Carroll MR, Price RC (1993) The H_2O content of basalt glasses from Southwest Pacific back-arc basins. Earth Planet Sci Lett 117:347-362

Decker RW, Cristiansen RL (1984) Explosive eruptions of Kilauea volcano, Hawaii. In: Explosvie volcanism: inception, evolution, and hazards. National Academy Press, Washington, DC, 122-132

Delaney JR, Muenow DW, Graham DG (1978) Abundance and distribution of water, carbon and sulfur in the glassy rims of submarine pillow basalts. Geochim Cosmochim Acta 42:581-594

DesMarais DJ, Moore JG (1984) Carbon and its isotopes in mid-oceanic basaltic glasses. Earth Planet Sci Lett 69:43-57

Devine JD (in press) Petrogenesis of the basalt-andesite-dacite association of Grenada, Lesser Antilles island arc, revisited. J Volcanol Geothermal Res

Devine JD, Sigurdsson H, Davis AN (1984) Estimates of sulfur and chlorine yield to the atmosphere from volcanic eruptions and potential climatic effects. J Geophys Res 89:6309-6325

Dingwell D, Harris DM, Scarfe CM (1984) The solubility of H_2O in melts in the system $SiO_2-Al_2O_3-Na_2O-K_2O$ at 1 to 2 kbars. J Geol 92:387-395

Dixon JE, Clague DA, Stolper EM (1991) Degassing history of water, sulfur, and carbon in submarine lavas from Kilauea volcano, Hawaii. J Geol 99:371-394

Dixon JE, Stolper E, Delaney JR (1988) Infrared spectroscopic measurments of CO_2 and H_2O in Juan de Fuca Ridge basaltic glasses. Earth Planet Sci Lett 90:87-104

Dixon JE, Stolper EM, Holloway JR (in press, a) An experimental study of water and carbon dioxide solubilities in mid-ocean ridge basaltic liquids Part A: Speciation and solubility results. J Petrol

Dixon JE (1992) Water and carbon dioxide in basaltic magmas. PhD dissertation. California Institute of Technology, Pasadena, CA

Dobson PF, Epstein S, Stolper EM (1989) Hydrogen isotope fractionation between coexisting vapor and silicate glasses and melts at low pressure. Geochim Cosmochim Acta 53:2723-2730

Dunbar NW, Hervig RL (1992) Petrogenesis and volatile stratigraphy of the Bishop Tuff: evidence from melt inclusion analysis. J Geophys Res 97:15,129-15,150

Eggler DH (1972) Water saturated and undersaturated melting relations in a Paricutin andesite and an estimate of H_2O content in the natural magma. Contrib Mineral Petrol 34:261-271

Eggler DH, Burnham CW (1973) Crystallization and fractionation trends in the system andesite-H_2O-CO_2-O_2 at pressures to 10 kb. Geol Soc Am Bull 84:2517-2532

Eichelberger JC, Carrigan CR, Westrich HR, Price RH (1986) Non-explosive silicic volcanism. Nature 323:598-602

Eichelberger JC, Westrich HR (1981) Magmatic volatiles in explosive rhyolitic eruptions. Geophys Res Lett 8:757-760

Ernst WG (1962) Synthesis, stability relations and occurrence of reibeckite and reibeckite-arfvesdsonite solid solutions. J Geol 70:689-736

Eugster H, Wones DR (1962) Stability relations of the ferruginous biotite, annite. J Petrol 3:81-125

Eugster HP (1957) Heterogeneous reactions involving oxidation and reduction at high pressures and tempera-tures. J Chem Phys 26:1760

Fine GJ, Stolper E (1986) Dissolved carbon dioxide in basaltic glasses: concentrations and speciation. Earth Planet Sci Lett 76:263-278

Fink J, Anderson SW, Manley CR (1992) Textural constraints on effusive silicic volcanism: Beyond the permeable foam model. J Geophys Res 97:9073-9083

Fogel RA (1989) The role of C-O-H-Si volatiles in planetary igneous and metamophic processes: Experimental and theoretical studies. PhD dissertation. Brown University, Providence, Rhode Island

Fogel RA, Rutherford MJ (1990) The solubility of carbon dioxide in rhyolitic melts: A quantitative FTIR study. Am Mineral 75:1311-1326

Friedman I (1967) Water and deuterium in pumice from the 1959-60 eruption of Kilauea Volcano, Hawaii. U S Geol Surv Prof Paper 575B:B120-B127

Friedman I, Smith RL (1958) The deuterium content of water in some volcanic glasses. Geochim Cosmochim Acta 15:218-228

Frost BR, Lindsley DH (1992) Equilibria among Fe-Ti oxides, pyroxenes, olivine and quartz: Part II. Application. Am Mineral 77:1004-1020

Gaetani GA, Grove TL, Bryan WB (1993) The influence of water on the petrogenesis of subduction-related igneous rocks. Nature 365:332-334

Garcia MO, Liu NW, Muenow DW (1979) Volatiles in submarine volcanic rocks from the Mariana Island arc and trough. Geochim Cosmochim Acta 43:305-312

Gardner JE, Rutherford MJ, Carey S, Sigurdsson H (in press) Experimental constraints on pre-eruptive water contents and changing magma storage prior to explosive eruptions of Mount St. Helens volcano. Bull Volc

Gardner JE, Sigurdsson H, Carey SN (1991) Eruption dynamics and magma withdrawal during the plinian phase of the Bishop Tuff eruption, Long Valley, CA. J Geophys Res 96:8069-8080

Garlick GD, Dymond JR (1970) Oxygen isotope exchange between volcanic materials and ocean water. Geol Soc Am Bull 81:2137-2142

Gerlach TM (1986) Exsolution of H_2O, CO_2 and S during eruptive episodes at Kilauea volcano, Hawaii. J Geophys Res 91:177-12

Gerlach TM (1993) Oxygen buffering of Kilauea volcanic gases and the oxygen fugacity of Kilauea basalt. Geochim Cosmochim Acta 57:795-814

Gerlach TM, Graeber EJ (1985) Volatile budget of Kilauea volcano. Nature 313:273-277

Gerlach TM, Taylor BE (1990) Carbon isotope constaints on degassing of carbon dioxide from Kilauea Volcano. Geochim Cosmochim Acta 54:2051-2058

Gerlach TM, Westrich HR, Casadevall TJ, Finnegan DL (in press, a) Vapor saturation and accumulation in magmas of the 1989-1990 eruption of Redoubt volcano. J Volc Geotherm Res

Gerlach TM, Westrich HR, Symonds RB (in press, b) A gas-saturated magma source for the stratospheric SO_2 cloud from the June 15, 1991 eruption of Mount Pinatubo. In: Punongbayan RS, Newhall CG (eds) The 1991-1992 Eruptions of Mt. Pinatubo, Phillipines. U S Geol Surv Spec Publication.

Geschwind CHG, Rutherford MJ (1992) Cummingtonite and the evolution of the Mount St. Helens magma system: An experimental study. Geology 20:1011-1014

Ghiorso MS, Carmichael ISE, Moret LK (1979) Inverted high-temperature quartz. Unit cell parameters and properties of the α-ß inversion. Contrib Mineral Petrol 68:307-323

Greenland LP (1987) Hawaiian eruptive gases. U S Geol Survey Prof Paper 1350:759-770

Greenland LP, Rose WI, Stokes JB (1985) An estimate of gas emissions and magmatic gas content from Kilauea Volcano. Geochim Cosmochim Acta 49:125-129

Grove TL, Gerlach DC, Sando TW (1982) Origin of calc-alkaline series lavas at Medicine Lake volcanoe by fractionation, assimilation and mixing. Contrib Mineral Petrol 80:160-182

Gust DA, Perfit MR (1987) Phase relations of a high-Mg basalt from the Aleutian Island Arc: Implications for primary island arc basalts and high-Al basalts. Contrib Mineral Petrol 97:7-18

Hamilton DL, Burnham CW, Osborn EF (1964) The solubility of water and effects of oxygen fugacity and water content on crystallization in mafic magmas. J Petrol 5:21-39

Harris DM (1981) The concentration of CO_2 in submarine tholeiitic basalts. J Geol 89:689-701

Harris DM (1982) H_2O, K_2O, P_2O_5, and Cl in basaltic glasses from Kilauea Volcano and Loihi Seamount, Hawaii. EOS Trans Am Geophys Union 63:1138

Harris DM, Anderson AT Jr. (1983) Concentrations, sources and losses of H_2O, CO_2 and S in Kilauean basalt. Geochim Cosmochim Acta 47:1139-1150

Harris DM, Anderson AT Jr. (1984) Volatiles H_2O, CO_2, and Cl in a subduction related basalt. Contrib Mineral Petrol 87:120-128

Hattori K (1993) High sulfur magma, a product of fluid discharge from underlying mafic magma: Evidence from Mount Pinatubo, Philippines. Geology 21:1083-1086

Heiken G (1972) Morphology and petrography of volcanic ashes. Geol Soc Am Bull 83:1961-1987

Helz RT (1973) Phase relations of basalt in their melting ranges at PH_2O = 5kb as a function of oxygen fugacity. J Petrol 14:249-302

Helz RT (1976) Phase relations of basalt in their melting rnages at PH_2O = 5kb. Part II: Melt compositions. J Petrol 17:139-193

Helz RT (1987) Diverse olivine types in lava of the 1959 eruption of Kilauea volcano and their bearing on eruption dynamics. U S Geol Survey Prof Paper 1350:691-722

Hervig RL, Dunbar NW (1992) Causes of chemical zoning in the Bishop (California) and Bandelier (New Mexico) magma chambers. Earth Planet Sci Lett 111:97-108

Hess PC (1992) Phase equilibria constraints on the origin of ocean floor basalts. In: JP Morgan, DK Blackman, JM Sinton (eds) Mantle Flow and Melt Generation at Mid-Ocean Ridges. Geophysical Monogr 71:67-102, American Geophysical Union, Washington, DC

Hildreth W (1979) The Bishop Tuff: Evidence for the origin of compositional zonation in silicic magma chambers. Geol Soc Am Special Paper 180:43-73

Hildreth W (1981) Gradients in silicic magma chambers: implications for lithospheric magmatism. J Geophys Res 86:10,153-10,192

Hochstaedter AG, Gill JB, Kusakabe M, Newman S, Pringle M, Taylor B, Fryer P (1990) Volcanism in the Sumisu Rift, I. Major element, volatile, and stable isotope geochemistry. Earth Planet Sci Lett 100: 179-194

Holcomb RT (1987) Eruptive history and long-term behavior of Kilauea volcano. U S Geol Survey Prof Paper 1350:261-350

Holloway JR, Burnham CW (1972) Melting relations of basalt with equilibrium water pressure less than total pressure. J Petrol 13:1-29

Housh TB, Luhr JF (1991) Plagioclase-melt equilibria in hydrous systems. Am Mineral 76:477-492

Jaggar TA (1940) Magmatic gases. Am J Sci 238:313-353

Jambon A, Zimmermann JL (1990) Water in oceanic basalts: evidence for dehydration of recycled crust. Earth Planet Sci Lett 101:323-331

Johannes W (1978) Melting of plagioclase in the system Ab-An-H_2O and Qz-Ab-An-H_2O at PH_2O = 5 kbar, an equilibrium problem. Contrib Mineral Petrol 66:295-303

Johnson DJ (1992) Dynamics of magma storage in the summit reservoir of Kilauea Volcano, Hawaii. J Geophys Res 97:1807-1820

Johnson MC, Rutherford MJ (1989a) Experimentally determined conditions in the Fish Canyon Tuff, Colorado, magma chamber. J Petrology 30:711-737

Johnson MC, Rutherford MJ (1989b) Experimental calibration of the Al-in-hornblende geobarometer with application to Long Valley Caldera, Calif. volcanic rocks. Geology 17:837-841

Johnston AD (1986) Anhydrous P-T phase relations of near-primary high-alumina basalt from the South Sandwich Islands. Contrib Mineral Petrol 92:368-382

Johnston AD, Draper DS (1992) Near-liquidus phase relations of an anhydrous high-magnesia basalt from the Aleutian Islands: Implications for arc magma genesis and ascent. J Volcan Geotherm Res 52:27-41

Katsura T, Nagashima S (1974) Solubility of sulfur in some magmas at 1 atm pressure. Geochim Cosmochim Acta 38:517-531

Kudo AM, Weill DF (1970) An igneous plagioclase thermometer. Contrib Mineral Petrol 25:52-65

Kyser TK, O'Neil JR (1984) Hydrogen isotope systematics of submarine basalts. Geochim Cosmochim Acta 48:2123-2133

Lindsley DH, Frost RB, Ghiorso MS, Sack RO (1991) Oxides lie: The Bishop Tuff did not erupt from a thermally zoned magma body. EOS Trans Am Geophys Union 72:312

Lowenstern JB (in press) Dissolved volatile concentrations in an ore-forming magma. Geology

Luth WC, Jahns RH, Tuttle OF (1964) The granite system at pressures of 4 to 10 kb. J Geophys Res 69:759-73

Maaloe S, Wyllie PJ (1975) Water content of a granite magma deduced from the sequence of crystallization determined experimentally under water-undersaturated conditions. Contrib Mineral Petrol 52:175-191

Macdonald R, Smith RL, Thomas JE (1992) Chemistry of the Subalkalic Silicic Obsidians. U S Geol Surv Prof Paper 1523

Markhinen EK (1962) On the possibility of estimating the amount of juvenile water participating in volcanic explosion. Bull Volcanol 24:187-191

Mathez EA, Delaney JR (1981) The nature and distribution of carbon in submarine basalts and peridotite nodules. Earth Planet Sci Lett 56:217-232

Matthews SJ, Jones AP, Bristow CS (1992) A simple magma-mixing model for sulfur behaviour in calc-alkaline volcanic rocks: Mineralogical evidence from Mount Pinatubo 1991 eruption. J Geol Soc London 149:863-866

Merzbacher C, Eggler DH (1984) A magmatic geohygrometer: Application to Mount St. Helens and other dacitic magmas. Geology 12:587-590

Metrich N, Rutherford MJ (1992) Experimental study of chlorine behavior in hydrous silicic melts. Geochim Cosmochim Acta 56:607-616

Meunow DW, Graham DG, Liu NWK, Delaney JR (1979) The abundance of volatiles in Hawaiian tholeiitic submarine basalts. Earth Planet Sci Lett 42:71-76

Michael PJ (1988) The concentration, behavior and storage of H_2O in the suboceanic upper mantle: Implications for mantle metasomatism. Geochim Cosmochim Acta 52:555-566

Moore JG (1965) Petrology of deep-sea basalt near Hawaii. Am J Sci 263:40-52

Moore JG (1970) Water content of basalt erupted on the ocean floor. Contrib Mineral Petrol 28:272-279

Moore JG (1979) Vesicularity and CO_2 in mid-ocean ridge basalt. Nature 282:250-253

Moore JG, Batchelder JN, Cunningham CG (1977) CO_2-filled vesicles in mid-ocean basalt. J Volcan Geotherm Res 2:309-327

Moore JG, Fabbi BP (1971) An estimate of the juvenile sulfur content of basalt. Contrib Mineral Petrol 33:118-127

Moore JG, Schilling J-G (1973) Vesicles, water, and sulfur in Reykjanes Ridge basalts. Contrib Mineral Petrol 41:105-118

Mori J, Eberhart-Phillips D, Harlow D (1993) Three dimensional velocity structure at Mount Pinatubo, Philippines: resolving magma bodies and earthquake hypocenters. EOS Trans Am Geophys Union 74:667

Muenow DW, Garcia MO, Aggrey KE, Bednarz U, Schmincke HU (1990) Volatiles in submarine glasses as a discriminant of tectonic origin: application to the Troodos ophiolite. Nature 343:159-161

Muenow DW, Graham DG, Liu NKW (1979) The abundance of volatiles in Hawaiian tholeiitic submarine basalts. Earth Planet Sci Lett 42:71-76

Muenow DW, Liu NWK, Garcia MO, Saunders AD (1980) Volatiles in submarine volcanic rocks from the spreading axis of the East Scotia Sea back-arc basin. Earth Planet Sci Lett 47:272-278

Muenow DW, Perfit MR, Aggrey KE (1991) Abundances of volatiles and genetic relationships among submarine basalts from the Woodlark Basin, southwest Pacific. Geochim Cosmochim Acta 55:2231-2239

Naney MT, Swanson SE (1980) The effect of Fe and Mg on crystallization in granitic systems. Am Mineral 65:639-53

Nekvasil H, Burnham CW (1987) The calculated individual effects of pressure and H_2O content on phase equilibria in the granitic system. Geochem Soc Special Pub 1:433-446

Newman S, Epstein S, Stolper E (1988) Water, carbon dioxide, and hydrogen isotopes in glasses from the ca. 1340 A.D. eruption of the Mono Craters, California: Constraints on degassing phenomena and initial volatile content. J Volc Geotherm Res 35:75-96

Newman S, Stolper EM, Epstein S (1986) Measurement of water in rhyolitic glasses: Calibration of an infrared spectroscopic technique. Am Mineral 71:1527-1541

Osborn EF (1959) Role of oxygen pressure in the crystallization and differentiation of basaltic magma. Am J Sci 257:609-47

Pallister J, Meeker GP, Newhall CG, Noblill RP, Martinez M (1993) 30,000 years of the "same old stuff" at Pinatubo. EOS Trans Am Geophys Union 74:667

Pallister JS, Hoblim RP, Meeker GP, Knight RJ, Siems DF (in press) Magma mixing at Mount Pinatubo: Petrographic and chemical evidence from the 1991 deposits. In: RS Punongbayan, CG Newhall (eds) The 1991-1992 Eruptions of Mt. Pinantubo, Phillipines. U S Geol Surv Special Publ

Pallister JS, Hoblit RP, Reyes AG (1992) A basalt trigger for the 1991 eruptions of Pinatubo volcano. Nature 356:426-428

Pan V, Holloway JR, Hervig RL (1991) The pressure and temperature dependence of carbon dioxide solubility in tholeiitic basalt melts. Geochim Cosmochim Acta 55:1587-1595

Pasteris JD, Wanamaker BJ (1988) Laser microprobe analysis of experimentally re-equilibrated fluid inclusions in olivine: some implications for mantle fluids. Am Mineral 73:1074-1088

Pawley AR, Holloway JR, McMillan PF (1992) The effect of oxygen fugacity on the solubility of carbon-oxygen fluids in basaltic melt. Earth Planet Sci Lett 110:213-225

Perfit MR, Gust DA, Bence AE, Arculus RJ, Taylor SR (1980) Chemical characteristics of island-arc basalts: implications for mantle sources. Chem Geol 30:227-256

Piccoli P, Candela P (1994) Apatite in felsic rocks; a model for the estimation of initial hologen concentrations in the Bishop Tuff (Long Valley) and Tuolumne Intrusive Suite (Sierra Nevada Batholith) magmas. Am   J Sci 294:92-135

Pineau F, Javoy M (1983) Carbon isotopes and concentrations in mid-oceanic ridge basalts. Earth Planet Sci Lett 62:239-257

Qin Z (1994) Melting and diffusive equilibration in igneous processes. PhD dissertation, Univ Chicago, Chicago, Illinois

Qin Z, Lu F, Anderson AT Jr. (1992) Diffusive reequilibration of melt and fluid inclusions. Am Mineral 77:565-576

Roedder E (1965) Liquid CO_2 inclusions in olivine-bearing nodules and phenocrysts from basalts. Am Mineral 50:1746-1782

Roedder E (1979) Origin and significance of magmatic inclusions. Bull Mineral 102:487-510

Rutherford MJ, Baker L, Pallister JS (1993) Petrologic constraints on timing of magmatic processes in the 1991 Mount Pinatubo volcanic system. EOS Trans Am Geophys Union 74:671

Rutherford MJ, Devine J (1991) Pre-eruption conditions and volatiles in the 1991 Pinatubo magma. EOS Trans Am Geophys Union 72:62

Rutherford MJ, Devine JD (1988) The May 1, 1980 eruption of Mount St. Helens: 3, Stability and chemistry of amphibole in the magma chamber. J Geophys Res 93:949-959

Rutherford MJ, Devine JD (in press) Pre-eruption P-T conditions and volatiles in the 1991 Pinatubo magma. In: RS Punongbayan, CG Newhall (eds) The 1991-1992 Eruptions of Mt. Pinatubo Volcano, Phillipines.   U S Geol Surv Special Publ

Rutherford MJ, Hill PM (1993) Magma ascent rates from amphibole breakdown: An experimental study applied to the 1980-86 Mount St. Helens eruptions. J Geophys Res 98:19,667-19,686

Rutherford MJ, Sigurdsson H, Carey S (1985) The May 18, 1980 eruption of Mount St. Helens, 1. Melt compositions and experimental phase equilibria. J Geophys Res 90:2929-2947

Ryan MP (1987) Elasticity and contractancy of Hawaiian olivine tholeiite and its role in the stability and structural evolution of subcaldera magma reservoirs and rift systems. U S Geol Surv Prof Paper 1395-1447

Sato M (1978) Oxygen fugacity of basaltic magmas and the role of gas-forming elements. Geophy Res Lett 5:447-449

Sato M, Wright TL (1966) Oxygen fugacities directly measured in magmatic gases. Science 153:1103-1105

Schwindinger KR, Anderson AT Jr. (1989) Synneusis of olivine and crystal settling beneath Kilauea. Contrib Mineral Petrol 103:187-198

Sekine T, Wyllie PJ (1983) Effect of H_2O on liquidus relationships in $MgO-Al_2O_3-SiO_2$ at 30 kilobars. J Geol 91:195-210

Shepherd ES (1938) The gases in rocks and some related problems. Am J Sci 35A:311-351

Sieh K, Bursik M (1986) Most recent eruption of the Mono Craters, eastern central California. J Geophys Res 91:12,539-12,571

Silver LA, Ihinger PD, Stolper E (1989) The influence of bulk composition on the speciation of water in silicate glasses. Contrib Mineral Petrol 104:142-162

Sisson TW, Grove TL (1993a) Experimental investigations of the role of H_2O in calc-alkaline differentiation and subduction zone magmatism. Contrib Mineral Petrol 113:143-166

Sisson TW, Grove TL (1993b) Temperatures and H_2O contents of low-MgO high-alumina basalts. Contrib Mineral Petrol 113:167-184

Sisson TW, Layne GD (1993) H_2O in basalt and basaltic andesite glass inclusions from four subduction-related volcanoes. Earth Planet Sci Lett 117:619-635

Skirius CM (1990) Pre-eruptive H_2O and CO_2 content of the plinian and ash-flow Bishop Tuff magma. PhD dissertation. University of Chicago, Chicago, IL

Skirius CM, Peterson JW, Anderson AT Jr. (1990) Homogenizing silicate glass inclusions. Am Mineral 75:1381-1398

Smith RL, Bailey RA (1966) The Bandelier Tuff: A study of ash-flow eruption cycles from zoned magma chambers. Bull Volcanol 29:83-103

Spulber S, Rutherford MJ (1983) The origin of rhyolite and plagiogranite in oceanic crust: an experimental study. J Petrol 24:1-25

Stolper E (1982) Water in silicate glasses: an infrared spectroscopic study. Contrib Mineral Petrol 81:1-17

Stolper E, Newman S (1994) The role of water in the petrogenesis of Mariana trough magmas. Earth Planet Sci Lett 121:293-325

Stolper EM, Holloway JR (1988) Experimental determination of the solubility of carbon dioxide in molten basalt at low pressure. Earth Planet Sci Lett 87:397-408

Stormer JC (1983) The effects of recalculation on estimates of temperature and oxygen fugacity from analyses of multi-component iron-titanium oxides. Am Mineral 68:586-594

Stormer JC Jr., Whitney JA (1985) Two feldspar and iron-titanium oxide equilibria in silicic magmas and the depth of origin of large volume ash-flow tuffs. Am Mineral 70:52-64

Tacker RC, Stormer JC Jr. (1989) A thermodynamic model for apatite solid solutions, applicable to high-temperature geologic problems. Am Mineral 74:877-888

Tait S (1992) Selective preservation of melt inclusions in igneous phenocrysts. Am Mineral 77:146-155

Tuttle OF, Bowen NL (1958) Origin of granite in the light of experimental studies in the system $NaAlSi_3O_8$-$KAlSi_3O_8$-SiO_2-H_2O. Geol Soc Am Memoir 74:154

Vergniolle S, Jaupart C (1990) Dynamics of degassing at Kilauea volcano, Hawaii. J Geophys Res 95:2793-2809

Volpe AM, Macdougall JD, Hawkins JW (1987) Mariana trough basalts (MTB): Trace element and Sr-Nd isotopic evidence for mixing between MORB like- and arc like- melts. Earth Planet Sci Lett 82:241-254

Wallace DJ, Gerlach TM (in press) Magmatic vapor source for SO_2 released during volcanic eruptions: New evidence from Mount Pinatubo. Science

Wallace P, Carmichael ISE (1992) Sulfur in basaltic magmas. Geochim Cosmochim Acta 56:1863-1874

Watson EB (1979) Apatite saturation in basic to intermediate magmas. Geophys Res Lett 6:937-940

Webster JD, Holloway JR (1988) Experimental constraints on the partitioning of Cl between topaz rhyolite melt and H_2O and H_2O + CO_2 fluids: New implications for granitic differentiation and ore deposition. Geochim Cosmochim Acta 52:2091-2105

Westrich HR, Gerlach TM (1992) Magmatic gas source for the stratospheric SO_2 cloud from the June 15, 1991 eruption of Mount Pinatubo. Geology 20:867-870

Westrich HR, Stockman HW, Eichelberger JC (1988) Degassing of rhyolitic magma during ascent and emplacement. J Geophys Res 93:6503-6511

White RA, Harlow DH, Chouet BA (in press) Precursorary deep long period earthquakes at Mt. Pinatubo: Spatio-temporal link to a basalt trigger. In: RS Punongbayan, CG Newhall (eds) The 1991-1992 Eruptions of Mt. Pinatubo Volcano, Phillipines. U S Geol Surv Special Publ

Whitham AG, Sparks RSJ (1986) Pumice. Bull Volcanol 48:209-223

Whitney JA (1975) The effects of pressure, temperature, and $X(H_2O)$ on phase assemblage in four synthetic rock compositions. J Geol 83:1-31

Whitney JA, Stormer JC Jr. (1985) Mineralogy, petrology, and magmatic conditions from the Fish Canyon Tuff, Central San Juan Volcanic Field, Colorado. J Petrology 26:726-762

Williams DL, Abrams G, Finn C, Dzurisin D, Johnson DJ, Denlinger R (1987) Evidence from gravity data for an intrusive complex beneath Mount St. Helens. J Geophys Res 92:10,207-10,222

Wilson CJN, Hildreth W (1991) Bishop Tuff revisited: new insights on eruption timing. Geol Soc Am Abstracts Prog 23:109

Wilson L, Head JW (1981) Ascent and eruption of basaltic magma on the Earth and Moon. J Geophys Res 86:2971-3001

Wilson L, Sparks RSJ, Walker GPL (1980) Explosive volcanic eruptions - IV. The control of magma properties and conduit geometry on eruption column behavior. Geophys J Roy Astron Soc 63:117-148

Wones DR, Eugster HP (1965) Stability of biotite: experiment, theory and application. Am Mineral 50:1228-1272

Woods AW (1988) The fluid dynamics and thermodynamics of eruption columns. Bull Volcanol 50:169-193

Wright TL (1971) Chemistry of Kilauea and Mauna Loa in space and time. U S Geol Surv Prof Paper 735, 40 p

Wright TL (1973) Magma mixing as illustrated by the 1959 eruption of Kilauea Volcano, Hawaii. Geol Soc Am Bull 84:849-858

Wright TL (1984) Origin of Hawaiian tholeiite: a metasomatic model. J Geophys Res 89:3233-3252

Wyllie PJ, Huang W-L, Stern CR, Maaloe S (1976) Granitic magmas: possible and impossible sources, water contents, and crystallization sequences. Can J Earth Sci 13:1007-1019

Yoder HS (1965) Diopside-anorthite-water at five and ten kilobars and its bearing on explosive volcanism. Carnegie Inst Wash Year Book 64:82-89

Yoder HS, Stewart DB, Smith JR (1957) Ternary feldspars. Carnegie Inst Wash Year Book 56:206-214

Yoder HS, Tilley CE (1962) Origin of basalt magmas; an experimental study of natural and synthetic rock systems. J Petrology 3:342-532

Chapter 9

THE EFFECT OF H_2O, CO_2 AND F ON THE DENSITY AND VISCOSITY OF SILICATE MELTS

Rebecca A. Lange

Department of Geological Sciences
The University of Michigan
Ann Arbor, Michigan 48109 USA

INTRODUCTION

The physical properties of silicate liquids are of fundamental importance to our understanding of the dynamic behavior and chemical differentiation of magmas. The dramatic role that dissolved volatiles play in modifying the transport properties of silicate melts has long been recognized (e.g., Leonteva, 1940), yet few data constraining their effect are available. In this chapter, attention is focused primarily on the effects of H_2O, CO_2, and F on silicate melt density and viscosity. Although these components typically make up less than 5 wt % of most magmatic liquids, their low molecular weights translate their concentrations into high mole fractions; incorporation of these components in viscosity and density models is imperative in order to be applicable to natural liquids. Other volatile components such as Cl and SO_3 are not discussed in this chapter simply because the requisite data do not exist. No measurements of the effect of dissolved sulfate on either melt density or viscosity have been obtained, even though recent eruptions of Mt. Pinatubo and El Chichón clearly demonstrate that SO_3 can be an important component in melts from the arc environment (Chapter 7).

This chapter begins with a discussion of silicate melt densities. Because the molar volume of *natural* silicate liquids can be described as a linear function of composition, temperature, and pressure (Lange and Carmichael, 1990), model equations are readily extended to include the volatile components. However, most of our information on the partial molar volume of H_2O and CO_2 is not direct, but is derived from the pressure dependence of their respective solubilities in silicate melts. Uncertainties in what values should be applied to natural liquids are significant, and a comprehensive experimental program to directly measure the densities of volatile-bearing liquids is imperative. In the meantime, the data that are available are used in this chapter to demonstrate the dramatic effect that H_2O, CO_2, and F have on melt densities, and hence, the dynamic behavior of magmas.

Next, attention is turned to silicate melt viscosity. Because its variation with temperature, pressure, and composition is complex, it is necessary to first outline the limitations of current viscosity models, new insights into the microscopic mechanisms of flow obtained in recent years from high temperature NMR (nuclear magnetic resonance) experiments, and the successful application of the configurational entropy theory applied to silicate melt viscosities. It is within the context of these three issues that the effect of H_2O, CO_2, and F are discussed and the framework for a comprehensive model for silicate melt viscosities proposed.

SILICATE MELT DENSITIES

The effect of temperature, pressure, and composition

An equation of state or pressure-temperature-density relation for multicomponent silicate liquids is essential for calculations of magmatic phase relations at pressure and for determining the direction and velocity of magma transport in the Earth. The experimental goal, therefore, is to obtain a sufficient data base that allows a model equation to be formulated for silicate liquid volumes as a function of temperature, pressure and composition, which includes the volatile components. Before discussing the effect of H_2O, CO_2, and F on the density of silicate liquids, and how a general model could be formulated to incorporate these volatile components, it is useful to first outline the development of a density model equation that includes the anhydrous compositional components, as well as temperature and pressure.

Composition and temperature at 1 bar. Model equations describing the one bar volume of *volatile-free* silicate liquids as a function of composition and temperature have been available in the literature for almost twenty-five years (Bottinga and Weill, 1970); however, the precision of such models has increased by almost an order of magnitude in recent years (Lange and Carmichael, 1987), reflecting an increase in the quality and quantity of density data since 1970. Moreover, the great success in modeling the volume (and hence, density) of silicate liquids at one bar is largely dependent on the fact that the volume and thermal expansion of silicate melts can be described as a linear function of composition, where:

$$V_{liq}(T) = \sum X_i \left[\overline{V}_{i,T_{ref}} + \frac{d\overline{V}_i}{dT}\left(T - T_{ref}\right) \right] \tag{1}$$

and $V_{liq}(T)$ is the molar volume of the melt at temperature T, X_i is the mole fraction of each oxide component, $\overline{V}_i$ is the partial molar volume of each oxide component, $d\overline{V}_i/dT$ is the temperature derivative of $\overline{V}_i$, and T_{ref} is a reference temperature (representing the midpoint in the experimental temperature range). Conversion from the molar volume to the density of a silicate melt is achieved using the following relationship:

$$\rho_{liq}(T) = \frac{\sum X_i \ (M.W.)_i}{V_{liq}(T)} \tag{2}$$

where $(M.W.)_i$ is the molecular weight of oxide component i.

Implicit in Equation (1) is the assumption that the partial molar volume of each oxide component is independent of composition. Lange and Carmichel (1987) found that the measured volumes of 36 metaluminous Na_2O-K_2O-CaO-MgO-Al_2O_3-SiO_2 liquids could be modeled by Equation (1) with a relative standard error of 0.2%. This is within the experimental uncertainty of the double-bob Archimedean technique (0.2 to 0.3%) that is used to measure density (Bockris et al., 1956). When liquids containing FeO and Fe_2O_3 are added to the data base, the errors on the volume equation increase to 0.4%, which may be attributed to errors in estimating the ferric-ferrous ratio in experimental liquids (Lange and Carmichael, 1987, 1989) and/or composition-induced coordination change of Fe^{3+} (Dingwell and Brearley, 1988). The addition of TiO_2 as a component also complicates matters, for there is clear evidence that $\overline{V}_{TiO_2}$ is strongly dependent on composition in alkali silicate melts (Lange and Carmichael, 1987; Dingwell, 1992; Lange, 1993), possibly reflecting a change in Ti^{4+} coordination. There is growing evidence, however, that Ti^{4+} is predominantly six-fold coordinated in natural liquids, and that $\overline{V}_{TiO_2}$ is independent of composition over the range of natural liquids; a detailed review of these issues can be found in Lange and Carmichael (1990). The question for petrologists, however, is what parameters for $\overline{V}_i$ and $d\overline{V}_i/dT$ should be employed to calculate the density of natural liquids using Equation (1). A set of parameters from Lange and Carmichael (1990) that are recommended for application to metaluminous and peralkaline igneous liquids is presented in Table 1.

Table 1. Partial molar volumes, thermal expansions and compressibilities of oxide components.

$$V_{liq}(T,P,X_i) = \Sigma\, X_i\, [V_{i,1673K} + dV_i/dT\,(T\text{-}1673\ K) + dV_i/dP\,(P\text{-}1bar)]$$

	$V_{i,1673K}$ (cc/mole)	$(dV_i/dT)_{1bar}$ $(10^{-3}$ cc/mole-K)	$(dV_i/dP)_{1673K}$ $(10^{-4}$ cc/mole-bar)	$[(dV_i/dP)]/dT$ $(10^{-7}$ cc/mole-bar-K)
SiO₂	26.90 ± .06	0.00 ± 0.50	-1.89 ± .02	1.3 ± 0.1
TiO₂	23.16 ± .26	7.24 ± 0.46	-2.31 ± .06	-----
Al₂O₃	37.11 ± .18	2.62 ± 0.17	-2.26 ± .09	2.7 ± 0.5
Fe₂O₃	42.13 ± .28	9.09 ± 3.49	-2.53 ± .09	3.1 ± 0.5
FeO	13.65 ± .15	2.92 ± 1.62	-0.45 ± .03	-1.8 ± 0.3
MgO	11.45 ± .13	2.62 ± 0.61	0.27 ± .07*	-1.3 ± 0.4
CaO	16.57 ± .09	2.92 ± 0.58	0.34 ± .05*	-2.9 ± 0.3
Na₂O	28.78 ± .10	7.41 ± 0.58	-2.40 ± .05	-6.6 ± 0.4
K₂O	45.84 ± .17	11.91 ± 0.89	-6.75 ± .14	-14.5 ± 1.5
Li₂O	16.85 ± .15	5.25 ± 0.81	-1.02 ± .06	-4.6 ± 0.4
Na₂O-Al₂O₃	-----	-----	10.18 ± .50	----

Volume and thermal expansion data from Lange and Carmichael (1987).
Compressibility data from Kress and Carmichael (1991).
**Fitted parameters should not be applied to pure MgO and CaO liquids.*

Pressure. Successful extension of Equation (1) to pressures in excess of one bar is based primarily on sound speed measurements on silicate liquids to obtain compressibility (Rivers and Carmichael, 1987; Manghnani et al., 1986; Kress et al., 1988; Kress and Carmichael, 1991). The first comprehensive model equation describing silicate liquid compressibilities as a systematic function of composition was presented by Rivers and Carmichael (1987). They measured the sound speeds (c) in 25 multicomponent silicate liquids to derive isothermal compressibilities (β_T) using the relation:

$$\beta_T = \frac{V_T}{c^2} + \frac{V_T T\alpha^2}{C_p} \tag{3}$$

where V_T is molar volume at temperature T, α is the coefficient of thermal expansion, C_p is heat capacity, and

$$V_T\beta_T = -\left(\frac{\partial V}{\partial P}\right)_T \tag{4}$$

where $\left(\partial V/\partial P\right)_T$ is the change in volume with pressure at constant temperature. Rivers and Carmichael (1987) used their calculated values of $\left(\partial V/\partial P\right)_T$ for each experimental liquid, combined with the analyzed compositions, to calibrate a model that describes the change in volume of a silicate melt with pressure as a function of composition:

$$\left(\frac{\partial V}{\partial P}\right)_T = \sum X_{i,T}\left(\frac{\partial \overline{V}_i}{\partial P}\right)_T \tag{5}$$

where $\left(\partial \overline{V}_i/\partial P\right)_T$ is the change in the partial molar volume of each oxide component i with pressure at temperature T. The most recent calibration of Equation (5) has been performed by Kress and Carmichael (1991) and their recommended values of $\partial \overline{V}_i/\partial P$ are presented in Table 1 for use in the comprehensive model:

$$V_{liq}(X_i,T,P) = \sum\left[\overline{V}_i\left(T_{ref},1\ bar\right) + \frac{\partial \overline{V}_i}{\partial T}\left(T - T_{ref}\right) + \frac{\partial \overline{V}_i}{\partial P}\left(P - 1\ bar\right)\right]X_i \tag{6}$$

This equation is recommended for use over the interval 0 to 20 kbar; at higher pressures, changes in the compressibility of silicate melts with pressure become important, and information on $d\beta/dP$ is required. For a detailed discussion of sink/float and shock wave experiments that provide constraints on this parameter, the reader is referred to Agee and Walker (1993), Rigden et al. (1989), and Miller et al. (1991). For the purposes of this chapter, discussion is restricted to the interval 0 to 20 kbar, and Equation (6) will be used as the framework for incorporation of H_2O, CO_2, and F into a model that describes the density of volatile-bearing igneous liquids.

The importance of water

There is little doubt that an equation of state must first incorporate the effect of water before it can be applied to the entire range of igneous liquids generated in the upper mantle and crust. Routine determination of water concentrations in glass inclusions trapped in phenocryst phases using Fourier transform infrared (FTIR) spectroscopy and the ion microprobe has established that substantial amounts of water can occur in both silicic and mafic magmas (see Chapter 8). For example, several recent studies have shown that rhyolitic and dacitic magmas commonly contain up to 4 to 6 wt % H_2O (e.g., Newman et al., 1988; Anderson et al., 1989; Hervig et al., 1989; Webster and Duffield, 1991; Geschwind and Rutherford, 1992). Similarly, pre-eruptive water concentrations in peralkaline rhyolites (previously assumed to be relatively anhydrous; Nicholls and Carmichael, 1969) have been shown to range up to 4.3 wt % H_2O (Kovalenko et al., 1988, 1989; Lowenstern and Mahood, 1991). Such high water concentrations are not restricted to silicic magmas; values up to and exceeding 6 wt % have also been measured in arc-related basaltic and basaltic andesite glass inclusions from the 1974 Fuego eruption in Guatemala (Harris and Anderson, 1984; Sisson and Layne, 1993). In addition, basaltic andesite glass inclusions from Goosenest Volcano, Oregon contain up to 3.3 wt % H_2O (Sisson and Layne, 1993). Even back-arc basaltic magmas from the Mariana trough (Stolper and Newman, 1994) and the Lau, North Fiji, and Woodlark basins of the southwest Pacific (Danyushevsky et al., 1993) have up to 1.6 to 1.9 wt % H_2O. This growing body of data on pre-eruptive water contents in natural samples is consistent with recent experimental data provided by Sisson and Grove (1993) that indicate that high alumina basalts, which can be the dominant volcanic product in some arcs (e.g., Aleutians), could have contained between 4 to 6 wt % water at crustal depths.

The effect of water concentrations of 2 to 6 wt % on the density of multicomponent silicate liquids is significant because of the low molecular weight of H_2O. For example, since the molecular weight of H_2O is 18.016 g/mole, whereas that of SiO_2, Al_2O_3, and Na_2O are 60.085, 101.961, and 61.979 g/mole respectively, 5 wt % H_2O translates into ~15 mole % H_2O water in albite melt. In addition, if the partial molar volume of water in multicomponent silicate liquids is at all similar to the molar volume of ice (19.6 cc/mole at 0°C and 1 bar), this translates into an extremely low density component (~ 0.9 g/cc) relative to most silicate melts (~2.3 for a rhyolitic melt and ~2.7 for a basaltic melt). Given the potentially large effect of water on the density of magmatic liquids, it is essential to quantify its partial molar volume, thermal expansivity, and compressibility.

Direct P-V-T measurements

Currently, very few direct measurements are available on the density of hydrous silicate liquids. These are confined to the P-V-T determinations made by Burnham and Davis (1971) on H_2O-$NaAlSi_3O_8$ liquids and to a falling sphere measurement of Kushiro (1978) on a hydrous calc-alkaline andesite. The volumetric measurements of Burnham and Davis (1971) were performed with the same apparatus as that used for the determination of the specific volume of water by Burnham et al. (1969), to which the reader is referred for a more detailed description. Burnham and Davis (1971) fitted their volumetric data on hydrous albitic melts (with 5.86, 8.25 and 10.9 wt % dissolved water) to a third-order polynomial in P and T. The reported water concentrations reflect those weighed into the sample capsules and are not direct analyses of quenched charges. The albite starting material was a natural specimen from a granitic pegmatite near Quadeville, Ontario; it contained ~8% each of the orthoclase and anorthite components. Their derived values for $\overline{V}^m_{H_2O,total}$ are therefore for water dissolved in a plagioclase liquid solution rather than in pure molten albite. Calculated values of $\overline{V}^m_{H_2O,total}$ vary from 14.2 to 20.5 cc/mole between 750 and 950°C and 3.5 to 8.5 kbar. The overall uncertainty in the measured volumes of the hydrous albite melts was reported to be ±0.4%, although the error in the partial molar volume of water was estimated to be significantly larger, namely ±3.2% at pressures above 2.5 kbar. The uncertainty at 1 kbar was estimated to be about twice as large (~±6%). At 20 cc/mole this is approximately ±1.2 cc/mole, approximately an order of magnitude larger than the uncertainties on $\overline{V}_i$ for the anhydrous components.

Derivation of $\overline{V}^m_{H_2O,total}$ from solubility curves

An alternative method for determining the partial molar volume of total water in a silicate melt is to examine the solubility of water as a function of pressure. Consider the following expression for the equilbrium relationship between H_2O in a vapor and a coexisting silicate melt:

$$H_2O(vapor) = H_2O^{total}(melt) \qquad (7)$$

Regardless of how water is speciated in the melt, the equilibrium reaction in Equation (7) under isothermal conditions can be described as follows:

$$\Delta G_T(P) = 0 = \Delta G^{\circ}_{T,1\ bar} + RT \ln \frac{a^m_{H_2O,total}}{f^{\circ}_{H_2O}} + \int_{1}^{P} \overline{V}^m_{H_2O,total}\ dP \qquad (8)$$

where $\Delta G_T(P)$ is the change in free energy of the reaction under equilibrium conditions as a function of pressure and at a constant temperature T, $\Delta G^{\circ}_{T,1\ bar}$ is the standard state change in free energy at one bar and temperature T, R is the gas constant, $a^m_{H_2O,total}$ is the activity of total water in the melt, $f^{\circ}_{H_2O}$ is the fugacity of pure water vapor, and $\overline{V}^m_{H_2O,total}$ is the partial molar volume of the total water component dissolved in the silicate melt. This equation can be rearranged:

$$\ln \frac{f^{\circ}_{H_2O}}{X^m_{H_2O,total}} = \frac{\Delta G^{\circ}_{T,1\ bar}}{RT} + \ln \gamma^m_{H_2O,total} + \int_{1}^{P} \frac{\overline{V}^m_{H_2O,total}}{RT} \qquad (9)$$

where $\gamma^m_{H_2O,total}$ is the activity coefficient for the total H_2O component dissolved in the silicate melt. If the activity coefficient is assumed to be independent of pressure, then under isothermal conditions this parameter is a constant. If it is further assumed that the partial molar volume of total water is also independent of pressure, then Equation (9) essentially describes a straight line:

$$\ln\frac{f^{\circ}_{H_2O}}{X^m_{H_2O,total}} = C' + \overline{V}^m_{H_2O,total} * \left(\frac{P-1}{RT}\right) \qquad (10)$$

where

$$C' = \frac{\Delta G^{\circ}_{T,1\ bar}}{RT} + \ln\gamma^m_{H_2O} \qquad (11)$$

and is a constant at constant temperature. Therefore, a plot of $\ln\dfrac{f^{\circ}_{H_2O}}{X^m_{H_2O,total}}$ versus $\dfrac{P-1}{RT}$ obtained from a series of water solubility experiments over a range of pressure but at constant temperature will result in a line which has a slope equal to $\overline{V}^m_{H_2O,total}$.

This approach can be applied to the recent water solubility measurements of Paillat et al. (1992) on molten albite. As discussed in detail by Ihinger (1991), Paillat et al. (1992) and McMillan (chapter 4, this volume), considerable care must be taken when evaluating water solubility data on molten silicates. For example, the data of early studies may well represent minimum values if water concentrations in the quenched charges were determined by weight loss measurements. More recent techniques for analyzing water contents in silicate glasses are based on infrared spectroscopy, the ion microprobe, and manometry and are considerably more reliable (see Ihinger et al., Chapter 2). However, Newman et al. (1986) emphasized that several precautions must be taken in order to obtain an accurate H_2O yield from a silicate glass during manometry. In addition, water analyses on silicate glasses are particularly troublesome when bubbles are present. If the bubbles represent water that was dissolved in the melt at elevated P and T and has subsequently exsolved during the quench, then ignoring the water in the bubbles will lead to an underestimation of water solubility. Alternatively, if the bubbles represent trapped vapor in a viscous melt, then incorporation of the bubbles in a water analysis will lead to an overestimation of water solubility.

For the purpose of extracting the partial molar volume of total water in a silicate melt from solubility data, the data of Paillat et al. (1992) on molten albite were chosen because: (1) the experiments were conducted in an internally heated pressure vessel with precise control of pressure (±100 bars); (2) several experiments at constant temperature (1400°C) and variable pressure (2.0-4.5 kbar) were performed; and (3) water concentrations in the quenched charges were analyzed using an ion microprobe. Experiments at pressures between 5.0-7.3 kbar and 1400°C were not used in this analysis because of the presence of bubbles in the quenched glasses. Values of $X^m_{H_2O,total}$ for each experiment were calculated from the weight percent values reported by Paillat et al. (1992).

The fugacity of pure water vapor was calculated using the modified Redlich-Kwong equation of state from Brodholt and Wood (1993). A plot of $\ln\dfrac{f^{\circ}_{H_2O}}{X^m_{H_2O,total}}$ versus $\dfrac{P-1}{RT}$ is presented in Figure 1 and the derived value for $\overline{V}^m_{H_2O,total}$ is 14.8±4.2 cc/mole. This value represents the average value for $\overline{V}^m_{H_2O,total}$ in albite melt between 2.0-4.5 kbar at 1400°C. It is also considerably less than that calculated by extrapolation of the model of Burnham and Davis (1971; 26.6 cc/mole at 1400°C and 4.5 kbar). This may indicate that the model of Burnham and Davis (1971) should not be extrapolated beyond the temperature and pressure range of their experiments.

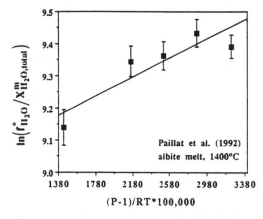

Figure 1. Plot of $\ln(f^{o}_{H_2O}/X^{m}_{H_2O,total})$ as a function of $(P-1)/RT \cdot 100,000$ from the total water solubility data of Paillat et al. (1992) on molten albite. Error bars are based on an uncertainty in P and $f^{o}_{H_2O}$ of ±100 bars and in $X^{m}_{H_2O,total}$ of ± 0.002. The slope of the line is $\overline{V}^{m}_{H_2O,total}$ = 14.8±4.2 cc/mole.

What about the speciation of water?

While the measurements of Burnham and Davis (1971) and Paillat et al. (1992) provide no distinction between molecular water and hydroxyl groups, it is well established from spectroscopic investigations that water dissolves in silicate glasses and melts in these two forms (Orlova, 1962; Ostrovsky et al., 1964; Bartholomew et al., 1980; Stolper, 1982a,b; McMillan et al., 1983; Eckert et al., 1987; 1988; Silver and Stolper, 1989). The two hydrous species are postulated to interact through the following homogeneous equilibrium:

$$H_2O_{molecular} \text{ (melt)} + O^{2-} \text{ (melt)} = 2 \text{ OH}^- \text{ (melt)}, \tag{12}$$

where O^{2-} = an oxygen atom, OH⁻ = a hydroxyl group and H₂O molecular = a water molecule in the melt (Stolper, 1982b; Silver and Stolper, 1989; Stolper, 1989). If the partial molar volume of molecular water is different from the partial molar volume of an hydroxyl group, then there will be a pressure dependence to the equilibrium reaction in Equation (12). Nogami and Tomozawa (1984) inferred from their study on the effect of stress on water diffusion in silica glass that ΔV for the reaction in Equation (12) is *negative*. In contrast, examination of the pressure dependence on the speciation of water in two natural rhyolite glasses (~0.8 wt % total water at 550°C and 5, 10, 20 and 25 kbar) by Zhang (1993) indicates a small but *positive* ΔV for the reaction in Equation (12) of ~0.7 cc/mole between 1 bar and 10 kbar, but which changes sign and becomes larger and more negative (~-3 cc/mole) between 20 and 25 kbar. However, these data apply to silicate *glasses*. The data for silicate *liquids* are more difficult to interpret because of problems with quenching the high temperature water speciation in a liquid to a glass (Zhang et al., 1989, 1991; Ihinger, 1991; Zhang, pers. comm.).

Derivation of $\overline{V}_{H_2O,molecular}$ from solubility curves

Estimates of the partial molar volume of *molecular* water in silicate liquids can be derived from water solubility curves in a manner analogous to that for total water. The reaction to consider is that between pure water vapor and molecular water dissolved in a coexisting silicate melt:

$$H_2O(vapor) = H_2O^{molecular}(melt) \tag{13}$$

The equilibrium reaction in Equation (13) under isothermal conditions can then be described as:

$$\Delta G_T(P) = 0 = \Delta G^{\circ}_{T,1\ bar} + RT \ln \frac{a^m_{H_2O,molecular}}{f^{\circ}_{H_2O}} + \int_1^P \overline{V}^m_{H_2O,molecular}\ dP \qquad (14)$$

where $a^m_{H_2O,molecular}$ is the activity of molecular water, and $\overline{V}^m_{H_2O,molecular}$ is the partial molar volume of molecular water dissolved in the silicate melt. Several studies have demonstrated that the activity of molecular water in silicate melts can be described by Henry's Law (e.g., Silver et al., 1990; Holloway and Blank, Chapter 6, this volume); i.e., $a^m_{H_2O,molecular} = K\ X^m_{H_2O,molecular}$, where K is a constant that is independent of T and P. Equation (14) is therefore readily rearranged:

$$\ln \frac{f^{\circ}_{H_2O}}{X^m_{H_2O,molecular}} = C" + \overline{V}^m_{H_2O,molecular} * \left(\frac{P-1}{RT} \right) \qquad (15)$$

where C" contains the standard state free energy of the reaction in Equation (13) and the Henrian constant K. Again, a plot of $\ln \dfrac{f^{\circ}_{H_2O}}{X^m_{H_2O,molecular}}$ versus $\dfrac{P-1}{RT}$ obtained from a series of molecular water solubility experiments over a range of pressure but at constant temperature will result in a line which has a slope equal to $\overline{V}^m_{H_2O,molecular}$.

In order to apply this method, direct determinations must be made of the molecular water concentration in silicate glasses quenched from melts equilibrated with pure water vapor. However, because of the difficulty of quenching the equilibrium water speciation in silicate melts at elevated P and T, the results of such an analysis must be viewed with caution (Zhang, personal communication). Nevertheless, one of the best studies to employ for the purpose of extracting $\overline{V}^m_{H_2O,molecular}$ is that of Ihinger (1991) on a rhyolite melt because: (1) water solubility experiments were performed at a relatively low temperature (850°C) in a rapid-quench cold-seal apparatus thus minimizing problems of quenching the equilibrium speciation; (2) both manometric and infrared measurements were made thus fully characterizing the speciation and total water contents of the quenched products; and (3) twenty-six individual experiments were performed between 202-1590 bars thus providing several replicate experiments, which is typically lacking in other studies. Values of $X^m_{H_2O,molecular}$ for each experiment were obtained from Table 1.7 in Ihinger (1991). The fugacity of pure water vapor was calculated using the modified Redlich-Kwong equation of state of Brodholt and Wood (1993); values for $f^{\circ}_{H_2O}$ are within 10 bars of those calculated based on the equation of state from Holloway (1977). A plot of $\ln \dfrac{f^{\circ}_{H_2O}}{X^m_{H_2O,molecular}}$ versus $\dfrac{P-1}{RT}$ is presented in Figure 2a. Error bars assigned to $\ln \dfrac{f^{\circ}_{H_2O}}{X^m_{H_2O,molecular}}$ reflect the propagated errors associated with $f^{\circ}_{H_2O}$ ($\pm$ 10 bars) and $X^m_{H_2O,molecular}$ (± 0.0015). Because the uncertainties in these two terms represent much larger proportions of their values at low pressures, the errors on $\ln \dfrac{f^{\circ}_{H_2O}}{X^m_{H_2O,molecular}}$ tend to increase with decreasing pressure. In fact, these errors may be underestimated based on the scatter observed for 12 experiments performed between 486 and 546 bars (Fig. 2). It should be emphasized, however, that the reason this scatter can be evaluated is because of the care taken to replicate so many experiments. If the outlier in Figure 2a (at the lowest pressure) is ignored, then the slope of a straight line fitted through the data leads to a value for $\overline{V}^m_{H_2O,molecular}$ of -4.0±4.7 cc/mole, or essentially *zero*. This value is very similar to those presented by Silver et al. (1990) and Blank et al. (1993) for $\overline{V}^m_{H_2O,molecular}$ in molten rhyolite (-3.3±2.8 and -2.5±3.0 cc/mole respectively) derived from molecular water solubility data.

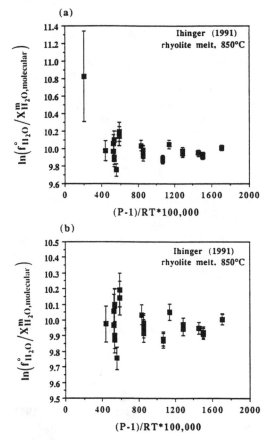

(a)

(b)

Figure 2. $ln(f^{\circ}_{H_2O}/X^m_{H_2O,molecular})$ as a function of $(P-1)/RT \bullet 100,000$ from the moecular water solubility data of Ihinger (1991) on molten rhyolite. Error bars are based on an uncertainty in P and f_{H_2O} of ±10 bars and in $X^m_{H_2O,total}$ of ± 0.0015. B. Same plot as in A without the outlier at the lowest pressure. The slope of the line is $\overline{V}^m_{H_2O,molecular}$ = -4.0±4.7 cc/mole.

An obvious question is how real is this derived value of zero for $\overline{V}^m_{H_2O,molecular}$ in rhyolite liquids? As a fitted parameter in a thermodynamic model to calculate water solubilites in rhyolite melts (e.g., Silver et al., 1990; Holloway and Blank, this volume), it is perfectly appropriate. However, the scatter evident in Figure 2 leads this author to doubt that the partial molar volume of molecular water is necessarily zero. This point of view is substantiated when values for $\overline{V}^m_{H_2O,molecular}$ derived from water solubity data on albite melt (~22 cc/mole at 1000°C; Silver et al., 1990), orthoclase melt (~25 cc/mole; Silver et al., 1990), a calcium aluminosilicate melt (~16 cc/mole; Silver et al., 1990), and a tholeiite melt (~12 cc/mole; Dixon et al., 1994) are all considered. It is not clear why $\overline{V}^m_{H_2O,molecular}$ should vary so much (0-25 cc/mole) as a function of composition. Moreover, it is difficult to reconcile $\overline{V}^{rhyolite\ melt}_{H_2O,molecular}$ = 0 with a value for $\overline{V}^{rhyolite\ glass}_{H_2O,total}$ of 12.0 cc/mole at 25°C and 1 bar. The latter value is based on the density of several hydrous rhyolite glasses (with up to

75% of their total water as molecular water), measured at room temperature and 1 bar by Silver et al. (1990). The value for $\overline{V}^{rhyolite\ glass}_{H_2O,total}$ is actually much closer to the value estimated for $\overline{V}^{albite\ melt}_{H_2O,total}$ (~14±4 cc/mole) from the solubility data of Paillat et al. (1992). Nor is it likely that $\overline{V}^m_{H_2O,molecular} = 0$ when $\overline{V}^m_{H_2O,total} = \sim 14$ cc/mole, since that would imply a very large change in volume associated with the reaction in Equation (12). This in turn would lead to a very strong pressure dependence to the speciation of water, which is not observed (Zhang, 1993).

What value of $\overline{V}^m_{H_2O,total}$ should be applied to natural liquids?

Although estimates of $\overline{V}^m_{H_2O,molecular}$ and $\overline{V}^m_{H_2O,total}$ from thermodynamic analyses of solubility curves are useful, considerable uncertainty still surrounds what value for $\overline{V}^m_{H_2O,total}$ should be applied to natural liquids. Currently, the best estimate that can be broadly applied to a wide range of melt compositions at variable temperature and pressure is approximately 17±5 cc/mole (Table 2). Although there is also the nagging possibility that it may be zero in molten rhyolite at low pressures. Unfortunately, this much uncertainty in $\overline{V}^m_{H_2O,total}$ translates into enormous uncertainties in the densities of natural liquids containing substantial amounts of water. This can best be seen by calculating the density of a hydrous basaltic andesite liquid represented by one of the Fuego 1974 glass inclusions (5-1 from sample 178-2) analyzed for its water content by Sisson and Layne (1993). This basaltic andesite glass inclusion contains 5.5 wt % H_2O; its calculated liquid density at 1100°C, 2 kbar, and an fO_2 equivalent to that of the Ni-NiO buffer (using the volume data of Lange and Carmichael, 1987 and Kress and Carmichael, 1991) is 2.55, 2.46 and 2.38 g/cc assuming $\overline{V}^m_{H_2O,total} = 12$, 17, and 22 cc/mole, respectively. In other words, an uncertainty in $\overline{V}^m_{H_2O,total}$ of ±5 cc/mole, leads to an uncertainty in the liquid density of ±3.4%. This much variation in the density of the basaltic andesite (±0.085 g/cc) could also be achieved with a ±640 degree temperature change.

Table 2. Current estimates of the partial molar volume of H_2O in silicate liquids

Composition	$\overline{V}^m_{H_2O,total}$ (cc/mole)	P range (kbar)	T range (°C)	Reference
$NaAlSi_3O_8$	14-20	3.5-8.5	750-950	Burnham &Davis (1971)
$NaAlSi_3O_8$	~14	2.0-4.5	1400	Paillat et al. (1992)
$CaMgSi_2O_6$	~17	20	1240	Hodges (1974)

Composition	$\overline{V}^m_{H_2O,molecular}$ (cc/mole)	P range (kbar)	T range (°C)	Reference
$NaAlSi_3O_8$	~22	1-8	1000	Silver et al. (1990)
$KAlSi_3O_8$	~25	1-7	900-1340	Silver et al. (1990)
Ca-Al-silicate	~16	1-5	1180	Silver et al. (1990)
Basalt	~12	1-20	1200	Dixon et al. (1994)
Rhyolite	~0	≤ 1.5	850	Silver et al. (1990)
Rhyolite	~0	≤ 1.5	850	Ihinger (1991)

The effect of the large uncertainty in $\overline{V}^m_{H_2O,total}$ on the calculated density of the Bishop Tuff rhyolitic magma is equally large. For example, Skirius et al. (1990) reported an average value for the total concentration of water in homogenized glass inclusions from the Plinian deposit of the Bishop Tuff of ~5.5 wt %. The calculated density of the Bishop Tuff magma at 750°C and 2 kbar is 2.31, 2.23, and 2.16 g/cc assuming $\overline{V}^m_{H_2O,total}$ = 12, 17, and 22 cc/mole, respectively. In this case, this much variation in the density of the rhyolite liquid (±0.075 g/cc) could also be achieved with a ±1200 degree temperature change.

Application to magma dynamics

It has been suggested from a number of laboratory and theoretical investigations (e.g, Sparks et al., 1984) that rather small changes in melt density, of the order of ~0.5%, can exert a profound influence on the dynamic evolution of magma chambers and can lead to a variety of important convective phenomena. Martin et al. (1987) argued that thermal and compositional density gradients can contribute equally to convection velocities in the interior of magma chambers. They pointed out that even under conditions of very low cooling rates, the thermal Rayleigh number Ra is almost always large enough for both mafic and granitic magmas to convect. An interesting question is how much of a water gradient is required to offset the effect of a thermal density gradient in a magma?

Recent isotopic studies of the chemically zoned high-silica Bishop Tuff rhyolite (Halliday et al., 1989; Christensen and DePaulo, 1993) have led to the conclusion that convection could not have involved the *entire* magma body since that would have wiped out observed Rb/Sr isochrons. Yet, the temperature gradient inferred for the Bishop Tuff magma of ~70 degrees (720 to 763°C for the early-erupted units and 760 to 790°C for the later units; Hildreth, 1981) is sufficient to drive convection. Viscosity is not likely to be a significant hindrance due to the high pre-eruptive water contents of 4 to 6 wt % determined by FTIR on trapped glass inclusions (Anderson et al., 1989; Skirius et al., 1990). However, one aspect of this puzzle originally addressed by Hildreth (1977) is the effect of a water gradient on the magma density profile. If the partial molar volume of water is either 12, 17, or 22 cc/mole, then a difference of less than 0.4, 0.3, or 0.2 wt % H_2O, respectively, between the lowest and highest temperature units (720 and 790°C, respectively) is all that is required to offset the thermal density gradient. However, if $\overline{V}^m_{H_2O,total}$ is *zero* in the Bishop Tuff magma, which would be consistent with the data of Silver et al. (1990) and Ihinger (1991), then the thermal density gradient would remain a strong driving force for convection. This example serves to emphasize how important the volumetric properties of dissolved water are to magmatic processes. There is clearly a pressing need to establish a solid experimental basis for calculations of the density, thermal expansion and compression of hydrous silicate melts.

The importance of CO_2

At the present time, there are not any direct density measurements on CO_2-bearing silicate liquids that have been published (to the author's knowledge), unlike the case for hydrous silicate liquids. Yet the presence of CO_2 in the upper mantle and its enhanced solubility in undersaturated (silica-poor and alkali-rich) magmas have been well documented (Mysen et al., 1975; Brey and Green, 1975; Eggler, 1978; Wendlandt and Mysen, 1978). The intimate association of carbon dioxide with the alkaline and kimberlite rock clan has long been recognized (Heinrich, 1966; Tuttle and Gittins, 1966; Wyllie, 1967). As a consequence, information on the partial molar volume of CO_2 in these magmatic liquids is central to understanding both their transport to the Earth's surface as well as their equilibrium phase relationships at elevated pressures.

Derivation of $\overline{V}^m_{CO_2,total}$ from solubility curves

Despite the lack of any direct volume measurements on CO_2-bearing liquids, the partial molar volume of dissolved CO_2 in silicate melts can be extracted from CO_2 solubility curves as outlined above for water. Spera and Bergman (1980) were the first to analyze the available data on the pressure dependence of CO_2 solubility in silicate melts in order to derive a value for the partial molar volume of the total CO_2 dissolved in silicate melts. Regardless of the precise form in which CO_2 enters silicate melts, the reaction between CO_2 in a fluid and a coexisting silicate melt may be expressed as follows:

$$CO_2(vapor) = CO_2^{total}(melt) \tag{16}$$

The equilibrium reaction in Equation (16) at constant temperature can then be described as:

$$\Delta G_T(P) = 0 = \Delta G^{\circ}_{T,1\ bar} + RT \ln \frac{a^m_{CO_2,total}}{f^{\circ}_{CO_2}} + \int_1^P \overline{V}^m_{CO_2,total}\ dP \tag{17}$$

where $a^m_{CO_2,total}$ is the activity of total water in the melt, $f^{\circ}_{CO_2}$ is the fugacity of pure carbon dioxide vapor, and $\overline{V}^m_{CO_2,total}$ is the partial molar volume of the total CO_2 component dissolved in the silicate melt. This equation can be rearranged:

$$\ln \frac{f^{\circ}_{CO_2}}{X^m_{CO_2,total}} = \frac{\Delta G^{\circ}_{T,1\ bar}}{RT} + \ln \gamma^m_{CO_2,total} + \int_1^P \frac{\overline{V}^m_{CO_2,total}}{RT} \tag{18}$$

where $\gamma^m_{CO_2,total}$ is the activity coefficient for the total CO_2 component dissolved in the silicate melt. If the activity coefficient is assumed to be independent of pressure, then under isothermal conditions this parameter is a constant. If it is further assumed that the partial molar volume of total CO_2 is also independent of pressure, then Equation (18) essentially describes a straight line:

$$\ln \frac{f^{\circ}_{CO_2}}{X^m_{CO_2,total}} = C''' + \overline{V}^m_{CO_2,total} * \left(\frac{P-1}{RT} \right) \tag{19}$$

where

$$C''' = \frac{\Delta G^{\circ}_{T,1\ bar}}{RT} + \ln \gamma^m_{CO_2} \tag{20}$$

and is a constant at constant temperature. Therefore, a plot of $\ln \dfrac{f^{\circ}_{CO_2}}{X^m_{CO_2,total}}$ versus $\dfrac{P-1}{RT}$

obtained from a series of CO_2 solubility experiments over a range of pressures but at constant temperature will result in a line which has a slope equal to $\overline{V}^m_{CO_2,total}$.

Spera and Bergman (1980) extracted values for the partial molar volume of total dissolved CO_2 in a wide range of melt compositions (albite, jadeite, and nepheline as well as andesite, tholeiite and olivine melilite) for which CO_2 solubility data were available. The average of their derived values for $\overline{V}^m_{CO_2,total}$ of 33.4±0.4 cc/mole is remarkably well constrained considering the wide range of compositions explored.

What about the speciation of CO_2?

The essentially constant value for $\overline{V}^m_{CO_2,total}$ derived by Spera and Bergman (1980) is significant in light of the spectroscopic measurements of Fine and Stolper (1985, 1986) that demonstrate that CO_2 can dissolve in silicate liquids as both carbonate (CO_3^{2-}) and molecular CO_2. The two species are postulated by Fine and Stolper (1985) to interact through the following homogeneous equilibrium:

$$CO_{2, \text{molecular}} (\text{melt}) + O^{2-} (\text{melt}) = CO_3^{2-} (\text{melt}). \qquad (21)$$

Fine and Stolper (1985) also observed that the proportion of dissolved molecular CO_2 relative to carbonate is strongly dependent on the activity of silica in the melt, with molecular CO_2 contents highest in the most silicic liquids. For example, most of the dissolved CO_2 in albitic liquid is molecular CO_2, whereas most in nephelinitic melt is expected to be carbonate (Fine and Stolper, 1985). The fact that Spera and Bergman (1980) found $\overline{V}_{CO_2,total}^m$ to be essentially identical in both of these melt compositions, despite the difference in their speciation of CO_2, suggests that the volume change associated with the reaction in Equation (9) is small.

The partial molar volume of molecular CO_2 in molten albite. The results of Spera and Bergman (1980) on albitic melt are based on the CO_2 solubility measurements of Mysen et al. (1976). However, the more recent CO_2 solubility and speciation results of Stolper et al. (1987) are significantly different from those of Mysen et al. (1976) and are probably more reliable because they did not use the older, imprecise KBr pellet technique for infrared determinations of carbonate concentrations that was used in the earlier study. From a thermodynamic analysis of the solubility and speciation of CO_2 in albite melt, Stolper et al. (1987) derived a value of 28.6 cc/mole for $\overline{V}_{CO_2,molecular}^m$ and found that the volume change associated with the reaction in Equation (21) is approximately -3.9±0.3 cc/mole in albite melt, reflecting the subtle and gradual increase in dissolved carbonate to molecular CO_2 with increasing pressure.

The partial molar volume of dissolved carbonate in molten basalt. In a subsequent study, Stolper and Holloway (1988) found that virtually all of the CO_2 dissolved in basaltic liquids is carbonate; from a thermodynamic analysis of CO_2 solubility based on the following reaction:

$$CO_2 (\text{vapor}) + O^{2-} (\text{melt}) = CO_3^{2-} (\text{melt}) \qquad (22)$$

they derived a value of 33.0±0.5 cc/mole for $\overline{V}_{CO_3^{2-}}^m - \overline{V}_{O^{2-}}^m$, which is the effective partial molar volume of total CO_2 dissolved as carbonate in the melt. This value suggests that there may be a compositional dependence to the partial molar volume of carbonate dissolved in silicate melts, since the inferred value for $\overline{V}_{CO_3^{2-}}^m - \overline{V}_{O^{2-}}^m$ in albite melt is ~24.7 cc/mole (Stolper et al., 1987).

The importance of $f_{CO_2}^\circ$ to derived values of $\overline{V}_{CO_2,total}^m$

An additional point to emphasize is that the values for $\overline{V}_{CO_2,molecular}^m$ and $\overline{V}_{CO_3^{2-}}^m - \overline{V}_{O^{2-}}^m$ derived from thermodynamic analyses of solubility curves are heavily dependent on the equation of state employed to calculate $f_{CO_2}^\circ$. Spera and Bergman (1980) used a modified Redlich-Kwong equation of state similar to that employed by Stolper et al. (1987) and Stolper and Holloway (1988). More recently, Pan et al. (1991) provided a thermodynamic analysis of their CO_2 solubility measurements in a tholeiitic basalt to 20 kbar. In that study, they employed both a modified Redlich-Kwong equation of state at low pressures (< 4 kbar) and the model of Saxena and Fei (1987) for CO_2 fluid at high pressures (> 4 kbar) to derive a value for $\overline{V}_{CO_3^{2-}}^m - \overline{V}_{O^{2-}}^m$ of 23±1 cc/mole. They further showed that if the Saxena and Fei (1987) equation of state were employed by Stolper and

Holloway (1988), their derived value for $\overline{V}^m_{CO_3^{2-}} - \overline{V}^m_{O^{2-}}$ would shift from 33.0 to 27.7 cc/mole. The study of Pan et al. (1991) clearly demonstrates that uncertainties in derived estimates of $\overline{V}^m_{CO_2}$ and $\overline{V}^m_{CO_3^{2-}} - \overline{V}^m_{O^{2-}}$ are derived, in part, from uncertainties in $f^{\circ}_{CO_2}$.

The effect of CO₂ on the density of magmatic liquids

Currently, the best estimates for $\overline{V}^m_{CO_2,total}$ fall between 21 and 28 cc/mole, and depends upon whether molecular CO₂ or carbonatite ion is the dominant species. In rhyolitic glasses, the only observed carbon species is molecular CO₂; therefore, the appropriate partial molar volume to apply to rhyolitic liquids is that derived for $\overline{V}^m_{CO_2,molecular}$ (~28 cc/mole; Stolper et al., 1987; Blank et al., 1993; Table 3). However, solubilities of CO₂ are so low in rhyolitic liquids, that the effect on density is negligible. In contrast, silica-undersaturated liquids (e.g., leucitites; Thibault and Holloway, 1984) can contain substantial concentrations of dissolved CO₂, primarily as the carbonate ion. In this case, the appropriate partial molar volume to use is that for $\overline{V}^m_{CO_3^{2-}} - \overline{V}^m_{O^{2-}} = \overline{V}^m_{CO_2,total}$

(= 21 to 28 cc/mole; Table 3). An average value of ~24 cc/mole can be used to examine how the liquid density of an alkali olivine basalt will vary as a function of total CO₂ content. In Figure 3, the density of this lava type is plotted as a function of pressure under both dry and CO₂ = 3 wt % conditions at 1400°C. In order to directly compare the effect of carbon dioxide relative to water on liquid densities, the density of the alkali olivine basalt liquid containing 3 wt % H₂O is also shown, assuming a value for $\overline{V}^m_{H_2O,total}$ of 17 cc/mole. The data indicate that the effect of adding 3 wt % CO₂ (4.1 mol %) is to decrease the liquid density by ~3.3%; this much variation in density could also be achieved by a change in pressure of ~6 kbar. In contrast, the effect of adding 3 wt % H₂O (9.5 mol %) is to decrease the density by ~4.8%, which could also be achieved by reducing the pressure by more than 9 kbar. Although the effect of water on decreasing the density of the liquid is more pronounced (primarily because of its higher mole fraction), the effect of ~3 wt % CO₂ is significant. Its high solubility in undersaturated magmas undoubtedly plays a role in the rapid transport of CO₂-bearing alkalic lavas to the Earth's surface.

Table 3. Best estimates of the partial molar volume of CO₂ in silicate liquids.

Composition	$\overline{V}^m_{CO_2,molecular}$ (cc/mole)	Reference
NaAlSi₃O₈	28.6	Stolper et al. (1987)
Rhyolite	28.3	Blank et al. (1993)

Composition	† $\overline{V}^m_{carbonate}$ (cc/mole)	Reference
Tholeiite	27.7	Stolper and Holloway (1988)
Tholeiite	23.1	Pan et al. (1991)
Leucitite	21.5	Thibault and Holloway (1994)
NaAlSi₃O₈	24.7	Stolper et al. (1987)

† $\overline{V}^m_{CO_3^{2-}} - \overline{V}^m_{O^{2-}}$

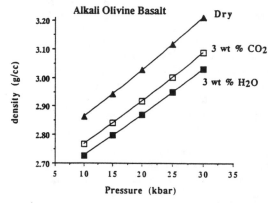

Figure 3. The density of an alkali olivine basalt liquid (from the Honolulu volcanic series, Oahu; Clague and Frey, 1982) at 1400°C as a function of pressure under dry, $CO_2 = 3$ wt %, and $H_2O = 3$ wt % conditions. Values of $\overline{V}^m_{H_2O,total;} = 17$ cc/mole and $\overline{V}^m_{CO_2,total;} = 24$ cc/mole were assumed (see Tables 2 and 3). The effect of adding 3 wt % CO_2 vs. 3 wt % H_2O is equivalent to decreasing pressure by ~6 vs. ~9 kbar, respectively.

The effect of F on the density of silicate melts

Although fluorine is generally of low concentration in common basaltic rocks (< 500 ppm; Schilling et al., 1980; Aoki et al., 1981; see also Chapter 7), it can reach significant levels in mantle-derived ultrapotassic melts and has long been recognized as an important component in late-stage granitic systems (Bailey, 1977). Moreover, concentrations in excess of 5 wt % F have been reported in some topaz-bearing rhyolites (Congdon and Nash, 1988). There is, therefore, considerable interest in incorporating this component into a general volume equation for igneous liquids.

To date there have been three sets of density measurements reported in the literature on fluorine-bearing silicate melts. The first was performed by Kogarko and Krigman (1981) on a series of melts in the Na_2O-SiO_2-NaF system at 1250°C using the single-bob Archimedean method. As discussed by Lange and Carmichael (1987), the single-bob versus double-bob Archimedean method does not allow for correction of surface tension acting on the bob's stem during a density measurement; therefore, density data obtained by the single-bob method tend to be systematically too high. Nonetheless, the partial molar volume for the NaF component derived from the data of Kogarko and Krigman (1981) is ~25.3 cc/mole, which (if cast as $\overline{V}_{Na_2F_2} = $~50.6 cc/mole) is significantly larger than the derived value for $\overline{V}_{Na_2O}$ of 27.7 at 1250°C (calculated from Table 1). The implication is that substitution of two moles of fluorine per mole of oxgyen (F_2O_{-1}) leads to a positive volume change of approximately 22 cc/mole.

More recently, density measurements on F-bearing aluminosilicate liquids have been reported by Grjotheim et al. (1990) and Dingwell et al. (1993). These studies are probably more relevant for igneous melts since recent NMR studies indicate that F is preferentially complexed to aluminum rather than sodium in $Na_2O-Al_2O_3-SiO_2$ glasses (Kohn et al., 1991; Schaller et al., 1992). Therefore, a more appropriate F-bearing component to incorporate into Equation (6) may be AlF_3 rather than NaF. The single-bob density measurements of Grjotheim et al. (1990) were performed on five melts along the Na_3AlF_6-$KAlSi_3O_8$ join, whereas those of Dingwell et al. (1993) were made on two supercooled haplogranite melts doped with 2.86 and 4.55 wt % F respectively.

The density determinations of Dingwell et al. (1993) are based on a method developed by Webb et al. (1992) for determining the thermal expansion of silicate supercooled liquids using both differential scanning calorimetry and glass dilatometry. This approach is based on the assumption that there is an equivalence of the relaxation parameters of volume and enthalpy through the glass transition. In other words, a comparison is made between calorimetric measurements of dH/dT (heat capacity) and dilatometric measurements of dV/dT (thermal expansion) on a glassy sample as it is heated through the glass transition into the supercooled liquid region. Although the dV/dT measurements are meaningless above the glass transition because of the viscous body forces acting on the cylindrical sample in this configuration, the heat capacity measurements are not. Therefore, by normalizing the the thermal expansion curve to the heat capacity data, a measure of dV/dT is obtained for the supercooled liquid. From a measurement of the density of the glass at room temperature, the density of supercooled liquid can then be calculated using the thermal expansion data.

As pointed out by Dingwell et al. (1993), it is not possible to derive a precise value for the partial molar volume of the fluorine component in silicate melts when it is present in such low concentrations (2.86 and 4.55 wt % F) in only two melts. Nonetheless, they derive a value for $\overline{V}_{F_2O_{-1}}$ of 14.2±1.3 cc/mole at 750°C, which is the change in volume associated with the substitution of two moles of fluorine per mole of oxygen. Although the error on $\overline{V}_{F_2O_{-1}}$ is an order of magnitude larger than the errors associated with $\overline{V}_i$ for the non-volatile components in Table 1, the available data clearly indicate that the effect of adding fluorine to a silicate melt is to significantly reduce its density.

An example of how to calculate the density of a fluorine-bearing melt (when the concentration of all components are given in wt %) is provided in Appendix 1. As an example, the measured density for Dingwell et al.'s F-free haplogranite melt (2.295 g/cc) is within 1.1% of that calculated from the model of Lange and Carmichael (2.321 g/cc; Table 1). A discrepancy of ~1% is not surprising given the large extrapolation in temperature between the density data used to calibrate the model in Table 1 (most measurements were made between 1200 and 1600°C) and the temperature of the measurements made by Dingwell et al. (750°C). Incorporation of a value for $\overline{V}_{F_2O_{-1}}$ of 14.2 cc/mole at 750°C to the model of Lange and Carmichael (Table 1) leads to calculated densities for Dingwell et al.'s two fluorine-bearing melts that are within 1.5 and 1.2% of the measured values (see Appendix 1). In order to reduce the uncertainty in $\overline{V}_{F_2O_{-1}}$, further experiments are required on aluminosilicate melts containing much higher concentrations of fluorine (~10 to 20 mol %).

An additional point raised by Dingwell et al. (1993) is that their extrapolated value for $\overline{V}_{F_2O_{-1}}$ of ~19 cc/mole at 1250°C is significantly lower than that derived (~22 cc/mole) from the density measurements of Kogarko and Krigman (1981) on Na_2O-SiO_2-NaF liquids. This discrepancy may be accounted for if the speciation of fluorine in alumino-silicate liquids is considered. Recent nuclear magnetic resonance spectroscopy studies on Na_2O-Al_2O_3-SiO_2 glasses indicate that F is preferentially coordinated with Al rather than Na to form AlF_5^{2-} and AlF_6^{3-} complexes (Kohn et al., 1991; Schaller et al., 1992). The occurrence of [5]Al and [6]Al is in marked contrast to its predominant tetrahedral coordination to oxygen in flourine-free, non-peraluminous silicate liquids at one bar (McKeown et al., 1984; White and Minser, 1984; Stebbins and Sykes, 1990). As a consequence, the smaller value for $\overline{V}_{F_2O_{-1}}$ derived from density measurements on aluminosilicate liquids may reflect the smaller volume of 5- and 6-fold coordinated aluminum rather than any change in the partial molar volume of the fluorine component.

SILICATE MELT VISCOSITIES

Virtually every magmatic process is strongly dependent on silicate melt viscosity;

examples include melt segregation and migration in source regions, magma re-charge, mixing, and convection in crustal chambers, and mode of eruption. Attempts to model these dynamic processes is made difficult by the particularly strong and complex variations in the viscosity of silicate melts with composition, temperature, and pressure. Unlike molar volume, the viscosity of molten silicates cannot be modeled as a simple function of these parameters, as seen in Equation (6). In order to develop a theoretical framework for an appropriate model equation, information on the microscopic mechanisms of viscous flow is critical. New insights are beginning to be shed by *in-situ*, high temperature NMR (nuclear magnetic resonance) spectroscopy measurements on molten silicates (e.g., Stebbins et al., 1992), and can be related to the configurational entropy model of viscosity (described below).

It is remarkable that as numerical models of dynamic magmatic processes become increasingly sophisticated, calculations of silicate melt viscosity continue to be based on the landmark models of either Bottinga and Weill (1972) or Shaw (1972). The reason that both models continue to be so heavily employed has less to do with their reliability over a wide range of magmatic conditions and more to do with the extreme difficulty in calibrating something better. Neither model incorporates the effect of pressure; this must be included before the effect of dissolved H$_2$O and CO$_2$ can be explored, since the solubility of these volatile components is strongly dependent on pressure. However, development of a theoretical framework for the incorporation of pressure and dissolved volatiles into a comprehensive model equation for silicate melt viscosities first requires that the effects of temperature and composition be understood. Before this is done, it is useful to briefly discuss Newtonian and non-Newtonian viscosity behavior.

Newtonian and non-Newtonian rheology

Numerous experimental studies of silicate melt viscosities demonstrate that most silicate melts behave as Newtonian liquids; that is, the measured shear rate ($\dot{\gamma}$) is linearly proportional to the applied stress (σ) or shear force per unit area:

$$\sigma = \eta \, \dot{\gamma} \qquad (23)$$

where η is the viscosity of the liquid. Small departures from Newtonian behavior (pseudoplastic or dilatant) can be described with a power law model:

$$\eta = \frac{\sigma}{\dot{\gamma}} = m\dot{\gamma}^{n-1} \qquad (24)$$

In cases where the power law parameter exceeds unity, the liquid can be considered dilatant; the shear rate decreases as the applied stress increases. This rheology is observed in foamed (bubble porosities > 25%) rhyolite liquids (Bagdassarov and Dingwell, 1993), where the viscosity behavior is dilatant because of viscous deformation of the bubbles. At low flow rates, the bubbles exert little influence on magmatic flow, but at high flow rates the viscosity increases. The opposite behavior is termed pseudoplastic or shear-thinning, where viscosity decreases with increasing shear rate and the power law parameter in Equation (24) is < 1. Stein and Spera (1993) observed very small departures from Newtonian behavior in a few silica-rich sodium aluminsilicate liquids between 1125 and 1200°C. In all cases, however, the departure remained small and the power law parameter was ≤ unity, indicating pseudoplastic behavior. Shear-thinning is also observed in viscous melts with a dilute suspension of bubbles (< 5 vol %; Stein and Spera, 1992). Thus the combined studies of Bagdassarov and Dingwell (1993) and Stein and Spera (1992) demonstrate that rhyolite magma viscosity can either increase or decrease with bubble content depending upon the strain rate during viscous flow. For thorough reviews of this topic, the reader is referred to Kraynik (1988), Bagdassarov and Dingwell (1992, 1993),

Stein and Spera (1992), and Dingwell et al. (1993).

Viscoelastic behavior applies to materials that partially return to their original shape after an applied stress is removed. The transition from viscous to viscoelastic behavior is one mechanical definition of the transition between silicate glass and supercooled liquid. The onset of viscoelastic behavior in silicate melts as a function of strain rate has been investigated by Webb and Dingwell (1990). For all of the measurements that will be discussed in this chapter, however, Newtonian viscous behavior is dominant.

The effect of composition and temperature at 1 bar

Qualitatively, it is well known how composition and temperature affect silicate melt viscosity; namely that the viscosities of anhydrous rhyolite melts at 700 to 900°C ($\sim 10^9$ to 10^{12} poise) are several orders of magnitude greater than those of anhydrous basalts at 1100 to 1200°C ($\sim 10^2$ to 10^4 poise). Abundant super-liquidus viscosity measurements on binary, ternary, and natural silicate melts further demonstrate that changes in temperature of a few hundred degrees can lead to orders of magnitude change in melt viscosity, typically following an Arrhenius relationship. Incorporation of low-temperature, supercooled liquid viscosity measurements, immediately above the glass transition, complicates the picture. For some compositions, like pure SiO_2 and $NaAlSi_3O_8$ liquids, the temperature dependence is strictly Arrhenius over the entire range of viscosity measurements (10^{13} to 10^2 poise); for others, like molten anorthite ($CaAl_2Si_2O_8$) and diopside ($CaMgSi_2O_6$), the relationship is far more complex (Fig. 4).

It should be emphasized that measurements of silicate liquid viscosity are typically constrained to two distinct temperature ranges: (1) ~300 degree interval immediately above the glass transition and (2) super-liquidus conditions. For many silicates, there is an intermediate temperature range where viscosity measurements on the pure liquid are not possible due to problems of crystallization. However, for many purposes (e.g., modeling the transport of liquid in a crystal mush), this is exactly the temperature range over which the viscosity data are to be applied.

The models of Bottinga and Weill (1972) and Shaw (1972) are strictly Arrhenius and are calibrated on viscosity data obtained at super-liquidus temperatures on binary, ternary, quaternary, and (in Shaw's model) natural silicate liquids. For the diopside example,

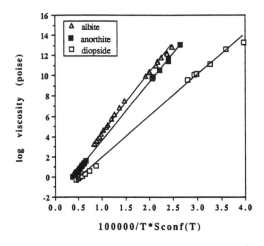

Figure 4. The log viscosity of albite, anorthite, and diopside liquids as a function of inverse temperature (10^4/T). Note that only the viscosity data for albite display an Arrhenius behavior. The high temperature viscosity data for all three liquids is from Urbain et al. (1982). The low temperature viscosity data for albite, anorthite, and diopside liquids is from Hummel and Arndt (1985) and Taniguchi (1992).

extrapolation to lower temperatures (1100 to 1200°C; the range over which basaltic magmas crystallize), leads to an underestimation of melt viscosity. Similarly, melt viscosities obtained only at low temperatures, just above the glass transition, also exhibit an Arrhenius behavior; extrapolation of these measurements to higher temperatures (1100 to 1200°C) will also lead to estimates that are too low by ~2 log units.

Additional problems with both models are found in how the compositional dependence to melt viscosity is formulated. Bottinga and Weill (1970) employed a strictly linear compositional model:

$$\ln \eta(T) = \sum_i D_i(T)X_i \qquad (25)$$

where $\eta(T)$ is the viscosity in poise at temperature T, X_i is the mole fraction of component i, and $D_i(T)$ is a fitted parameter for each component i over a specified silica mole fraction range. The $D_i(T)$ values were obtained by least squares fitting of Equation (25) to binary, ternary and quaternary silicate liquids exhibiting strictly Arrhenius behavior. Shaw (1972) also used the Arrhenius equation as a starting point:

$$\ln \eta = A + \frac{B}{T} \qquad (26)$$

and derived an arithmetic method to calculate the constants A and B in Equation (26), using the D_i values of Bottinga and Weill (1972) as well as some additional viscosity data on molten rocks. The most notable difference between Shaw's (1972) algorithm and that of Bottinga and Weill's (1972) is the incorporation of water as a compositional component.

Although an additive compositional dependence to silicate melt viscosity works reasonably well at high temperatures (>1300°C), it breaks down at lower temperatures. This is best demonstrated by the recent low-temperature viscosity measurements of Neuville and Richet (1991) on molten pyroxenes. They observed that molten $Ca_2Si_2O_6$ and $Mg_2Si_2O_6$ have nearly identical viscosities at 800°C (~$10^{10.5}$ poise), whereas at the same temperature, molten $CaMgSi_2O_6$ has a viscosity that is more than one and a half orders of magnitude lower (Fig. 5). A strictly additive model predicts that diopside would have a similar viscosity to that of the two end-member melts, which is clearly not the case. The complex temperature and compositional dependence to melt viscosity, particularly at temperatures <1300°C, can be understood using the concept of configurational entropy.

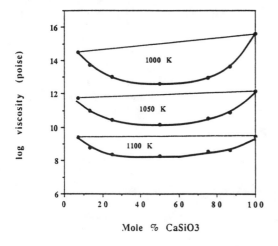

Figure 5. Low-temperature viscosity measurements of liquids along the join $CaSiO_3$-$MgSiO_3$ from Neuville and Richet (1991). Deviations from an additive compositional model are largest at low temperatures and reflect a mixing contribution to the configurational entropy of the liquids (see Eqn. 29 in text).

Configurational entropy theory of viscosity

Considerable progress has been made in recent years in developing a theoretical framework for explaining the compositional and temperature dependence of silicate melt viscosities over the entire range of magmatic temperatures (Richet, 1984; Hummel and Arndt, 1985; Richet et al., 1986; Neuville and Richet, 1991; Neuville et al., 1993). It is based upon the Adam and Gibbs (1965) configurational entropy theory of relaxation, where viscous flow is assumed to occur through the cooperative rearrangement of configurational states in the liquid. The more configurational states that are available to the liquid, the more readily viscous flow occurs. Richet (1984) and Richet et al. (1986) demonstrated that this theory of relaxation can be quantitatively applied to changes in viscosity with temperature for many silicate and aluminosilicate liquids through the simple relationship:

$$\ln \eta = A_e + \frac{B_e}{T \, S^{conf}(T)} \tag{27}$$

where A_e is a pre-exponential term, B_e is a potential energy barrier hindering the structural rearrangement in the liquid, and $S^{conf}(T)$ is related to the configurational states available to the liquid:

$$S^{conf}(T) = S^{conf}(T_g) + \int_{T_g}^{T} \frac{C_p^{conf}(T)}{T} \, dT \tag{28}$$

where $S^{conf}(T)$ is the change in configurational entropy of the liquid with temperature and $S^{conf}(T_g)$ is the residual configurational entropy in the glass at the glass transition temperature. $C_p^{conf}(T)$ is the configurational contribution to liquid heat capacity, and is defined as:

$$C_p^{conf}(T) = C_p^{liq}(T) - C_p^{glass}(T_g) \tag{29}$$

where $C_p^{liq}(T)$ is the heat capacity of the liquid at temperature T, and $C_p^{glass}(T_g)$ is the heat capacity of the glass at the glass transition temperature (T_g). Use of Equation (29) is based upon the fact that at temperatures below T_g, the heat capacity of the glass is essentially vibrational. It is generally assumed that the abrupt increase in heat capacity observed upon heating through T_g is due to the sudden achievement of configurational rearrangements in the liquid. In fact, Stebbins et al. (1984) pointed out that constant volume heat capacities of silicate liquids can be as much as 1.5 times greater than the ideal harmonic vibrational limit, thus providing strong evidence for a major configurational contribution to silicate liquid heat capacities.

Application of the configurational entropy theory of viscosity to diopside, anorthite, and albite melts by Richet (1984; Fig. 6) demonstrates the advantage of Equation (26) over a strictly Arrhenius model. For those compositions, values of $C_p^{conf}(T)$ are readily obtained from heat capacity measurements of the glasses and liquids. In addition, $S^{conf}(T_g)$ is also derived from calorimetric measurements as seen below:

$$S^{conf}(T_g) = \int_0^{T_{fusion}} \left[\frac{C_p^{crystal}(T)}{T} \right] dT + \Delta S_{fusion} +$$

$$\int_{T_{fusion}}^{T_g} \left[\frac{C_p^{liquid}(T)}{T} \right] dT + \int_{T_g}^{0} \left[\frac{C_p^{glass}(T)}{T} \right] dT. \tag{30}$$

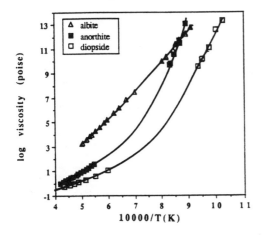

Figure 6. The log viscosity of albite, anorthite, and diopside liquids plotted as a function of inverse temperature (1/T) times the inverse configurational entropy of the liquid ($1/S^{conf}\{T\}$). The configurational entropy values for each liquid were derived from calorimetric measurements presented by Richet (1984) and Richet et al. (1986).

Calorimetric determinations of $S^{conf}\left(T_g\right)$ require that heat capacity data are available for both the crystalline and glassy forms of the mineral down to 0 K. In addition, the entropy of fusion and heat capacity of the liquid are also required. Clearly, $S^{conf}(T_g)$ can only be *calorimetrically* determined for those liquid compositions that correspond to a single crystalline compound.

Application of Equations (27) and (28) to the viscosity-temperature behavior of natural liquids, which do not crystallize to single, stoichiometric compounds, requires that $S^{conf}\left(T_g\right)$ be an additional fit parameter (along with A_e and B_e) when modeling liquid viscosity measurements over a wide temperature range. Fortunately, this is not a serious problem; Richet et al. (1986) demonstrated that the value of $S^{conf}\left(T_g\right)$ for diopside glass could be readily obtained by fitting diopside liquid viscosity measurements over a wide temperature range to Equations (27) and (28). Moreover, their derived value (26.4 J/mol-K) is the same within error to that obtained from Equation (14) using the laborious calorimetric measurements (24.3±3 J/mol-K).

Natural liquids. Extension of the configurational entropy theory of viscosity to natural liquids requires that both high and low temperature viscosity data be available in order to obtain fitted values for A_e, B_e, and $S^{conf}\left(T_g\right)$. Unfortunately, most viscosity measurements on natural silicate liquids have been restricted to super-liquidus temperatures. Recently, both low and high temperature viscosity measurements on a molten rhyolite and a molten andesite were obtained by Neuville et al. (1993); they fitted their data to Equations (27) and (28) to obtain values for A_e, B_e, and $S^{conf}\left(T_g\right)$ for each sample. A comparison of their modelled viscosity-temperature curves versus those derived by only fitting the high temperature data to an Arrhenius model is shown in Figure 7. This diagram illustrates that the viscosity of the rhyolite and andesite will be under-estimated by ~2 log units at 1000°C if only the high temperature data were used to calibrate their respective viscosity-temperature behavior.

Figure 7 also demonstrates that the slope of log viscosity with temperature becomes increasingly steep at low temperatures, indicating that $S^{conf}(T)$ exerts a stronger influence at low temperatures than at high temperatures. The reason for this can be seen in Figure 8, where the functions $1/S^{conf}(T)$, $1/T$, and $1/TS^{conf}(T)$ are each plotted as a function of temperature. The parameter, $1/TS^{conf}(T)$, is simply the product of $1/S^{conf}(T)$ and $1/T$. As temperature increases, the $1/S^{conf}(T)$ term is thus "weighted" by increasingly smaller values

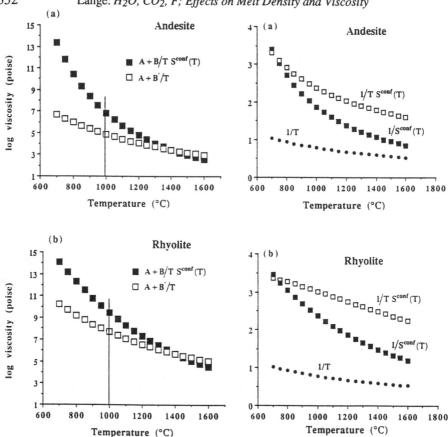

Figure 7(left). Comparisons of measured values of log viscosity for (a) an andesite melt and (b) a rhyolite melt obtained by Neuville et al. (1993) fitted to a configurational entropy model (solid boxes) versus values of log viscosity derived by only fitting the high temperature data to an Arrhenius model (open boxes). At 1000°C, an Arrhenius model calibrated only on superliquidus viscosity data (e.g., Bottinga et al., 1972; Shaw, 1972) underestimates the viscosity of the rhyolite and andesite melts by ~2 log units.

Figure 8 (right). Plots of 1/T, 1/S^{conf}(T), and the product of these two functions, 1/T*S^{conf}(T), for the andesite and rhyolite melts studied by Neuville et al. (1993). The effect of decreasing temperature is to "weight" the configurational entropy paramter, 1/S^{conf}(T), by increasing values of 1/T. The compositional dependence to the configurational entropy of a silicate liquid affects liquid viscosity more at low versus high temperatures. This is why additive compositional models for silicate melt viscosities break down at low temperatures (<1100°C).

of 1/T. In contrast, as temperature decreases, the configurational entropy term is "weighted" by increasingly higher values of 1/T. As a consequence of this trend, if there is a compositional dependence to S^{conf}, then it follows that deviations from an additive compositional dependence to silicate melt viscosity will be more evident at low temperatures than at high ones.

A compositional dependence to S^{conf} is clearly demonstrated by the low temperature viscosity measurements on molten pyroxenes by Neuville and Richet (1991; Fig. 5). This compositional dependence can be expressed as a configurational entropy of mixing:

$$S^{conf} = \sum_i X_i S_i^{conf} - nR \sum_i X_i \ln X_i, \tag{31}$$

where S_i^{conf} is the configurational entropy of end-member i (Neuville and Richet, 1991). The reason why the viscosity of the rhyolite liquid is lower than that of albite liquid at ~800°C may be the result of a mixing contribution (from the orthoclase and albite components in the molten rhyolite) to its configurational entropy (Neuville et al., 1993).

Although application of the configurational entropy theory of viscosity to natural silicate liquids is only in its infancy, the success that Richet (1984) and Richet et al. (1986) demonstrated in modeling the viscosity-temperature behavior of diopside, anorthite and albite liquid, using independently obtained calorimetric determinations of $S^{conf}(T_g)$, confirms that the configurational entropy theory provides a useful theoretical framework for a comprehensive model equation for silicate melt viscosity; moreover, it may prove useful when extending the model to pressures in excess of one bar.

Microscopic mechanisms of viscous flow: NMR constraints

A detailed view of the microscopic mechanisms of viscous flow is highly desirable in order to develop a theoretical framework for modeling silicate melts viscosities, particularly at elevated pressures. Information on the microscopic structure of silicate liquids is based primarily on spectroscopic data. Most techniques (Raman, infrared, Mössbauer) provide a "snapshot" view of a silicate liquid structure because the time scale of the technique is on the order of lattice vibrations. For example, high-temperature Raman spectra obtained on silicate liquids (Seifert et al., 1981; Mysen, 1990; McMillan et al., 1992) indicate that Q^n species (where Q^n refers to the number of bridging oxygens (n) bonded to silicon) retain their identity on the vibrational time scale. In other words, silicate tetrahedra remain intact for periods of time that are on the average greater than ~100 fs at temperatures >1000°C.

In contrast, nuclear magnetic resonance (NMR) spectroscopic studies on molten silicates (e.g., Stebbins et al., 1985; Liu et al., 1988; Stebbins et al., 1992) can probe much lower frequency motions and hence, longer time scales that are applicable to viscous flow. NMR spectra obtained on melts under *in-situ*, high temperature conditions (Farnan and Stebbins, 1990; Stebbins and Farnan, 1992) demonstrate that the lifetime of a silicate tetrahedron is relatively short on the NMR time scale (seconds to microseconds) because only one line is observed in the spectra of samples where, by stoichiometry, more than one silicon environment must exist. This is true for samples of greatly differing silica contents and shows that even when a silicate system contains a large proportion of weak bonds, the Si-O bonds in the system are breaking and re-forming rapidly. For example, in a high-temperature NMR study of ^{29}Si line shapes observed for alkali silicate liquids, Liu et al. (1988) showed that the exchange of oxygens between bridging and non-bridging sites takes place at the microsecond to nanosecond time scale at liquidus temperatures. They concluded that this process must be a fundamental step in viscous flow and configurational entropy generation.

Line shape analysis by Farnan and Stebbins (1990) indicates that the rate of the Si-O bond-breaking process in $K_2Si_4O_9$ melt at ~1000°C is approximately 2 μs. The observed Q^n species exchange process obtained from the spectra allows a temperature-dependent exchange time, τ, to be derived. Because this exchange involves the breaking and re-forming of Si-O bonds, it is probably related to the basic mechanism of viscous flow and relaxation. Farnan and Stebbins (1990) demonstrated this by calculating the viscosity of the $K_2Si_4O_9$ liquid using a simple Eyring model of viscosity:

$$\eta = \frac{kT\tau}{d^3} \tag{32}$$

assuming jump frequencies (τ) equal to the exchange frequencies obtained from the NMR results at each T, and a jump distance (d) taken as the Si-Si distance of 0.3 nm, which is typical of many crystalline silicates.

The calculated viscosity of molten $K_2Si_4O_9$ from Farnan and Stebbins (1990) using Equation (17) compares extremely well with that measured for this composition by Shartsiss et al. (1952) at high temperatures, but deviates from the low temperature measurements of Webb and Dingwell (1990). Farnan and Stebbins (1990) pointed out, however, that this could be accounted for by recognizing two processes which affect motional narrowing of the ^{29}Si NMR spectrum. The first process is termed a "non-exchange event", where an intermediate state is formed but relaxes back to its original configuration. The second process is termed a "complete exchange event", where the intermediate state relaxes to a new configuration. At high temperatures, the probability of the complete exchange process is close to 1 after the intermediate state is formed, whereas at low temperatures, the probability decreases. Because the "non-exchange" process contributes to NMR line-shape narrowing, the NMR derived values of τ underestimate the viscosity. However, Farnan and Stebbins (1990) were able to use a two-dimensional NMR experiment at a temperature just above T_g where the two Q species exchange processes are decoupled, and the "complete" exchange process can be observed independently of the "non-exchange" process. In that experiment, an exchange time scale on the order of 0.5 seconds was observed and leads to viscosities calculated from Equation (32) that match the values measured at low temperature.

What is the relationship between the parameters in Equation (32) and those in the configurational entropy model of Equation (27)? From the original work of Adam and Gibbs (1965), the relaxation time scale of a silicate melt is expected to be inversely proportional to the configurational states available to the system. Therefore, the parameter τ in Equation (32) should be inversely proportional to $S^{conf}(T)$. The next question is what is the nature of the intermediate state inferred from the NMR results of Farnan and Stebbins (1990), and can it be related to the parameter B_e in Equation (27), which is considered to represent an energy barrier to a change in configurational states. The NMR results of Liu et al. (1988), and Stebbins et al. (1992) directly address this issue and provide constraints on the nature of the intermediate step during viscous flow. This is based on the more general NMR approach to studying the dynamics in both liquids and solids and involves measurements of the spin-lattice relaxation times, T_1. It is beyond the scope of this paper to explain this parameter, and the reader is referred to Stebbins et al. (1992) for a discussion. The principal point is that the time scale of T_1 is often just that needed to explore diffusion, and it is observed that a fundamental change in the relaxation mechanism for ^{29}Si takes place at or near the glass transition. Liu et al. (1988) and Stebbins et al. (1992) suggested that ^{29}Si spin-lattice relaxation may be caused by the formation of short-lived distorted sites that are intermediates in the Q^n species exchange reactions. They proposed a mechanism for Q^n species exchange that is inspired by the molecular dynamic simulations of Soules (1979), Angell et al. (1983), Brawer (1985) and Kubicki and Lasaga (1988). A possible scenario is that when a non-bridging oxygen moves close to another SiO_4 tetrahedron, a distorted five-coordinated Si site is created. Some of the time, the newly arrived non-bridging oxygen "jumps" away and the initial configuration is maintained. At other times, one of the original bridging oxygens "jumps" away from the distorted five-coordinated site, and a new configuration is achieved. When this happens, both local viscous flow and Q^n speciation has occurred.

Although it not possible for NMR experiments to test this model, NMR spectra obtained by Stebbins (1991) and Xue et al. (1991) have detected measurable abundances of the proposed intermediate state, namely five-coordinated Si, in glasses quenched from melts equilibrated both a 1 bar and elevated pressures. Abundances up to 0.10% have been observed in $K_2Si_4O_9$ glasses quenched from liquids equilibrated at one bar. Not surprisingly, abundances of both Si^V and Si^{VI} increase in $K_2Si_4O_9$ glasses quenched from liquids equilibrated at increasing pressures; at 6 GPa, quenched abundances of Si^V and Si^{VI} approach 3.9 and 1.5% respectively. Clearly, a link between pressure-induced coordination change of Si and the role of Si^V as an intermediate step in viscous flow is compelling. Such a relationship may explain the complex pressure dependence to melt viscosity.

The effect of pressure on silicate melt viscosity

A thorough review of the pressure dependence of silicate melt viscosities can be found in Scarfe (1986) and Scarfe et al. (1987). Currently, most of the published data on the viscosity of molten silicates at elevated pressures have been obtained by the falling-sphere method (Shaw, 1963; Kushiro, 1978) in a piston-cylinder apparatus. The technique employs Stoke's relation:

$$\eta = \frac{2g\left(\rho^{liq} - \rho^{sphere}\right)r^2}{9v} \tag{33}$$

where ρ^{liq} and ρ^{sphere} are the density of the liquid and falling sphere respectively, g is the gravitational constant, r is the radius of the sphere, and v is the velocity at which the sphere falls through the silicate melt. This relation must be combined with the Faxen correction to correct for drag along the walls of the crucible (Shaw, 1963). Most of the error in derived values of η is derived from the uncertainty in the velocity at which the sphere falls; this is obtained from several experiments quenched at different time intervals to document the distance traversed by the sphere per unit time. Reported uncertainties in viscosity range from 15 to 25%. Since viscosity varies by several orders of magnitude as a function of composition, temperature, and pressure, this is not a large relative error.

The results of high-pressure falling sphere viscometry applied to a wide variety of synthetic and natural liquid compositions have been summarized by Scarfe et al. (1987). Although some silicate melts (e.g, $CaMgSi_2O_6$) show a viscosity increase with increasing pressure, consistent with other inorganic liquids, most silicate and aluminosilicate liquids ($NaAlSi_3O_8$, $NaAlSi_2O_6$, $K_2O-MgO-5SiO_2$, an andesite, and two tholeiitic basalts) show a decrease in viscosity with applied pressure. In virtually all cases, the variation in viscosity over a 20 kbar interval is $\leq$ 1 log unit. An exception to this is the viscosity decrease of more than 2 log units between 0 and –25 kbar for molten $K_2Si_4O_9$ (Dickinson et al., 1990), which has the highest concentration of five-coordinated Si in quenched liquids from 0 to 6 GPa (Stebbins and McMillan, 1989; Xue et al., 1991).

One of the limitations of the falling-sphere technique is that it cannot be applied to extremely fluid liquids. Uncertainties become unacceptably large when viscosities are $\leq$ 10 poise because the velocity of the falling sphere is too rapid for accurate measurement. As a consequence, alternative methods are required in order to obtain viscosity data for silicate melts at pressure > 5 GPa, which are also at high temperatures. To overcome this problem, Kanzaki et al. (1987) developed an *in-situ* technique for measuring the velocity of Pt spheres in silicate melts at pressures $\leq$ 10 GPa using synchrotron radiation to produce X-ray shadowgraphs. The most serious limitation of this method, however, are the severe temperature gradients that exist (Kanzaki, 1987).

Another approach for estimating the viscosities of fluid liquids at elevated temperatures and pressures is based on the Erying relationship applied to O diffusivity data (e.g., Rubie et al., 1993). From Equation (16), diffusivity can be related to viscosity as follows:

$$D = kT/\eta\lambda \qquad (34)$$

where D is O diffusivity, k is the Boltzmann constant, T is absolute temperature, η is viscosity and λ is the diffusive jump distance. This relationship can only be applied to those melts where a similar mechanism for oxygen diffusion and viscous flow is operative. The successful application of Equation (34) to molten $NaAlSi_2O_6$ (Shimizu and Kushiro, 1984) and $Na_2Si_4O_9$ (Rubie et al., 1993) implies that Si-O and Al-O bond breaking is a fundamental feature of both viscous flow and oxygen diffusion in these two polymerized melts.

The results of Rubie et al. (1993) indicate that the viscosity of molten $Na_2Si_4O_9$ at 1825°C decreases with increasing pressure, but only by ~0.8 log units over a 100 kbar interval. Perhaps more importantly, Rubie et al. found that the activation energy for O self-diffusion at 4 GPa is considerably lower than the value obtained from 1 bar viscosity measurements (55±17 versus 174 to –180 kJ/mol). The implication is that the energy barrier to a change in configurational sates (the parameter B_e in Equation 27) has been reduced and the intermediate step for viscous flow is more readily accessible in molten $Na_2Si_4O_9$ at pressures approaching 4 GPa.

It is tempting to speculate that the general trend of decreasing viscosity with increasing pressure in polymerized melts is related to an increase in the abundance of five-coordinated silicon (or aluminum), which may be the intermediate step in viscous flow. The observation that the degree of melt polymerization appears to control whether viscosity will increase or decrease with pressure may be related to the compositional control on the formation of five- and six-coordinated silicon inferred from the NMR results of Xue et al. (1991). In that study, it was concluded that the degree of polymerization of a silicate melt plays a major role in controlling the formation of high-coordinated [5]Si and [6]Si species with increasing pressure. Xue et al. (1991) proposed a model in which five- and six-coordinated Si is formed by conversion of non-bridging O atoms to bridging O atoms. The mechanism is expressed in the following simple reactions:

$$Q^3 + Q^4 = Q^{4*} + {}^{[5]}Si \qquad (35)$$

$$2Q^3 + Q^4 = 2Q^{4*} + {}^{[6]}Si \qquad (36)$$

where Q^3 refers to a silicon atom with three bridging oxygens and one non-bridging oxygen, Q^4 refers to a silicon atom with four bridging oxygens, and Q^{4*} is a SiO_4 species with three [4]Si and one [6]Si or [5]Si neighbors. This model implies that the formation of high-coordinated Si is favored in those melts where *both* Q^3 and Q^4 species are maximized. In other words, if the abundance of Q^4 species is diminished, by an increase in alkali *or water concentration*, then the reactions in Equations (19) and (20) will not be driven toward the creation of high-coordinated Si. Stebbins et al. (1992) observed a strong correlation between Q^4 abundance and the activity of silica in the melt; it follows therefore that the most silica undersaturated magmas will form less five-coordinated silicon with increasing pressure, and therefore will not have melt viscosities that decrease as a function of pressure. This has been observed, in fact, for an andesite melt with 2.9 wt % (9.5 mole%) H_2O (Kushiro, 1978). Not only does the presence of water reduce the viscosity of the andesite melt by a factor of 20, but it also changes the pressure dependence of viscosity.

There are, in addition, a variety of silicate liquids that demonstrate a complex pressure dependence to their viscosities; for example, Scarfe et al. (1987) observed that the viscosity of molten $Na_2Si_2O_5$ initially increases by ~0.6 log units between 1 and 15 kbar, and then proceeds to decrease by ~1.0 log units between 15 and 25 kbar. In contrast, Brearley et al. (1986) found that the viscosity of molten $Ab_{25}Di_{75}$ initially decreases by less than 0.5 log units between 0-5 kbar, and then increases by ~0.5 log units between 5 and 25 kbar. The bottom line for petrologists, however, is that most natural silicate liquids (rhyolites through basalts) have anhydrous melt viscosities that decrease with pressure, often by less than 1 log unit over a 20 kbar interval. A much more significant reduction in melt viscosity is observed with the addition of water.

The effect of H_2O on silicate melt viscosities

Published measurements of the viscosity of hydrous silicate melts are currently restricted to four siliceous compositions: rhyolite (Sabatier, 1956; Saucier, 1952; Shaw, 1963; Friedman et al., 1963; Persikov et al., 1990), andesite (Kushiro, 1978b; Persikov et al., 1990), albite (Dingwell and Mysen, 1985; Dingwell, 1987; Persikov et al., 1990), and sanidine (White and Montana, 1990). All experiments are based on the falling sphere technique described above and demonstrate that the addition of ~2 wt % H_2O can dramatically reduce melt viscosity by up to ~3 log units. For comparison, a temperature increase of ~200 degrees is required to achieve the same viscosity reduction in albite and rhyolite melts (Urbain et al., 1982; Neuville et al., 1993). There have been no reported viscosity measurements on hydrous basaltic or otherwise depolymerized (e.g., diopside) melt compositions despite the evidence cited earlier that both silicic and mafic magmas can contain significant amounts of dissolved water. The primary reason for this paucity of data is the difficulty of performing falling sphere viscometry to extremely fluid liquids.

The effect of increasing water content on the viscosity of rhyolite, albite, and sanidine liquids is shown in Figure 9. Although the data cover a wide range of pressures (2 to 15 kbar) and temperatures (850 to 1500°C), a general trend is observed where there is an initial, rapid decrease in viscosity with the first 3 wt % of total dissolved water concentration, which begins to level off at ~5 wt % total water. As originally discussed by Stolper (1982a), this trend is consistent with the water speciation results of Stolper (1982b), Silver and Stolper (1989), and Silver et al. (1991) on rhyolite, albite, and sandine liquids; their data demonstrate that the hydroxyl concentration increases rapidly as water is first added to the melt. As total water concentrations increase beyond ~5 wt %, most of the additional water is incorporated as molecular H_2O. The strong correlation between the reduction in melt viscosity and the increase in hydroxyl concentration with total water content

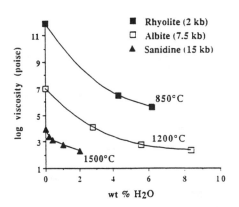

Figure 9. A plot of log viscosity for molten albite (Dingwell, 1987), sanidine (White and Montana, 1990) and rhyolite (Shaw, 1963) as a function of total dissolved water. The effect of ~2 wt % total H_2O is more pronounced at 850°C (~3 log unit reduction) than at 1500°C (~1.7 log unit reduction).

was noted by Stolper (1982b) and is illustrated in Figure 10. The striking correlation reflects the role of OH groups in depolymerizing the highly siliceous melts, $NaAlSi_3O_8$, $KAlSi_3O_8$ and rhyolite, that under anhydrous conditions are nearly fully polymerized with virtually all of their oxygens as bridging oxygens. The speciation of dissolved water according the reaction in Equation (12) requires that for every two hydroxyl groups formed, one bridging oxygen is consumed, which both depolymerizes the melt and increases the number of available configurational states. This, in turn, according to Equation (27), leads to a decrease in melt viscosity, as observed. It appears that dissolved molecular water has little effect on melt viscosity.

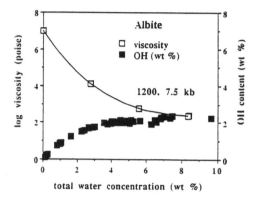

Figure 10. A plot of log viscosity for albite liquid (Dingwell, 1987; right axis) and concentration of hydroxyl groups dissolved in albite liquid (Silver and Stolper, 1989; left axis) as a function of total dissolved water content. The correlation between vis-cosity and hydroxyl concentration indicates that dissolved molecular water has little effect on silicate melt viscosity.

The reduction in silicate melt viscosity caused by dissolved hydroxyl groups is strongly dependent on temperature. This is illustrated by the data presented in Figure 9 and Table 4, where the effect of ~2 wt % total water on melt viscosity is more pronounced at lower than higher temperatures. For example, at 850°C (and 2 kbar), the effect of 2 wt % H_2O is to reduce the melt viscosity of rhyolite by ~3.1 log units. In contrast, at 1200°C (and 7.5 kbar), the viscosity of molten albite is reduced by ~2.3 log units, whereas at 1500°C (and 20 kbar), the viscosity of molten sanidine is reduced by only ~1.7 log units. This does not seem to be related to pressure, since the viscosity of molten albite with 2 wt % H_2O is virtually identical ($10^{2.3}$ vs $10^{2.4}$ poise) at different pressures (7.5 vs 15 kbar) but at the same temperature (1200°C). In contrast, at constant pressure (15 kbar), but at different temperatures (1200 vs 1400°C), the effect of ~2 wt % H_2O in albite melt leads to

Table 4. Variable effect of water on silicate melt viscosity

Composition	T (°C)	P (kbars)	log η dry (poise)	log η 2 wt % H_2O (poise)	Δ log η (poise)	Reference
Rhyolite	850	2.0	11.9	8.8	3.1	Shaw (1963)
Albite	1200	7.5	7.0	4.7	2.3	Dingwell (1987)
Albite	1200	15.0	6.6	4.2	2.4	Dingwell (1987)
Albite	1400	15.0	4.5	2.9	1.6	Dingwell (1987)
Sanidine	1500	20.0	3.9	2.2	1.7	White and Montana (1990)

different values of viscosity ($10^{2.4}$ versus $10^{1.6}$ poise). Nor can this variable effect of water on melt viscosity as a function of temperature be related to an increase in hydroxyl groups at low temperatures. To the contrary, Stolper (1989) and Zhang (1993) have shown that at P < 20 kbar, the effect of increasing temperature on the reaction in Equation (12) is to increase the production of OH groups. The most likely reason for the increased effect of water on melt viscosities at lower temperatures is that the configurational entropy term in Equation (27) is "weighted" more heavily at lower versus higher temperatures as seen in Figure 8.

The effect of water on melt viscosity is further enhanced when F (and possibly other volatile components) are also dissolved in the melt. This is caused by the entropy of mixing contribution (Eqn. 31) to $S^{conf}(T)$, rendering an additive compositional model for viscosity (e.g., Bottinga and Weill, 1972; Shaw, 1972) incorrect. Evidence for this is seen in the results of Dingwell (1987) on the combined effect of fluorine and H_2O on albite melt viscosity. At 1400°C, deviations from an additive model are slight, and a maximum discrepancy of ~0.2 log units is observed at equal molar proportions of F and H_2O (5.8 wt % F and 2.79 wt % H_2O). At 1200°C, the maximum discrepancy increases to ~0.4 log units, whereas at 1000°C, an additive model overestimates the viscosity by ~0.6 log units. Once again the configurational entropy of mixing term becomes increasingly important at lower temperatures.

The combined effect of both CO_2 and H_2O on silicate melt viscosities has not been studied as yet, although some data may exist inadvertently due to the extreme difficulty in keeping any sample completely anhydrous in a piston-cylinder apparatus (Fine and Stolper, 1985; Stolper et al., 1987). This effect will be discussed below.

The effect of CO₂ on silicate melt viscosities

Currently, data on the effect of CO_2 on silicate melt viscosities are fairly limited; only three melt compositions have been explored ($KAlSi_3O_8$, $NaAlSi_3O_8$, and $NaCaAlSi_2O_7$) in two published studies (Brearley and Montana, 1989; White and Montana, 1990). White and Montana (1990) measured the viscosity of sanidine liquid with 0.25, 0.50, 1.00, and 1.50 wt % CO_2 between 15 and 25 kbar and 1500 to 1600°C. The results (Fig. 11) are somewhat surprising in that increasing concentrations of dissolved CO_2 have virtually no effect on the melt viscosity (within error) at 20 kbar. In sharp contrast, however, the effect of just 0.5 wt % CO_2 appears to significantly reduce the melt viscosity at 25 kbar. This contrasting effect of dissolved CO_2 on sanidine melt viscosity between 20 and 25 kbar may be due to the somewhat anomalous viscosity value for sanidine melt obtained under dry conditions at 1500°C and 25 kbar (White and Montana, 1990). In Figure 12, a comparison of measured sanidine and albite melt viscosities as a function of pressure (Brearley and Montana, 1989; White and Montana, 1990) indicates that the value for sanidine at 1500°C and 25 kbar is surprisingly high. This value is especially unusual because it indicates that the viscosity of sanidine melt actually increases between 20 and 25 kbar; such an increase in viscosity with pressure is usually observed for only depolymerized melts (e.g., diopside) and not for such highly polymerized liquids as sanidine. Moreover, this behavior is not seen at 1600°C, where the sanidine melt viscosity decreases between 20 and 25 kbar (White and Montana, 1990), similar to the trend observed for albite melt (Brearley and Montana, 1989). White and Montana (1990) pointed out that 1500°C corresponds to the high-temperature bracket of the sanidine liquidus at 25 kbar (Boettcher et al., 1984); therefore, the high viscosity value may reflect incipient crystallization.

Another possible explanation for the contrasting effect of CO_2 on sanidine melt

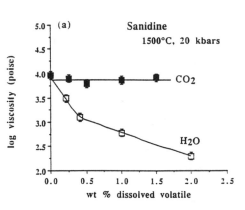

Figure 11. A comparison of the effect of H_2O versus CO_2 on the viscosity of sanidine melt from White and Montana (1990). At 20 kbars and 1500°C, the effect of 0.5 to 2.0 wt % total dis-solved CO_2 has virtually no effect on the viscosity of sanidine melt within error (±15 % in viscosity and ±0.08 in log viscosity). In contrast, at 25 kbars and 1500°C, the effect of 0.5 and 1.0 wt % CO_2 is to reduce the viscosity of sanidine melt by more than 0.5 log units. These results should be interpreted with caution, however, since Fine and Stolper (1985) and Stolper et al. (1987) have shown that it is virtually impossible to keep water out of nominally anhydrous CO_2-bearing samples in a piston-cylinder apparatus. Less than 0.3 wt % H_2O is all that is required to reduce the viscosity of nominally anhydrous CO_2-bearing sanidine melts at 25 kbars.

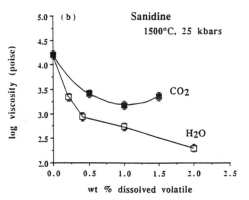

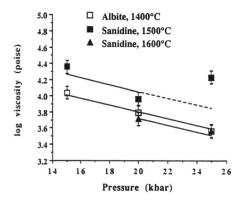

Figure 12. A plot of log viscosity of albite melt at 1400° (Brearley and Montana, 1989) and sanidine melt at 1500 and 1600°C (White and Montana, 1990) as a function of pressure.

viscosity at 20 versus 25 kbar (at 1500°C) is variable water contamination. White and Montana (1990) acknowledged that it is virtually impossible to completely eliminate the presence of H_2O in their piston-cylinder experiments, which are characterized by run durations of 3 to 45 minutes. Fine and Stolper (1985) found that in their CO_2-speciation studies of nominally anhydrous sodium aluminosilicate liquids, piston-cylinder run durations of only 30 minutes led to total water concentrations in the samples that ranged from 0.10 to 0.65 wt %. At these low concentrations, virtually all of the water is dissolved as hydroxyl groups and would exert a strong influence on measured melt viscosities. From Figure 11b, in which the effect of dissolved H_2O and CO_2 on sanidine melt viscosity are compared, it can be seen that less than 0.3 wt % H_2O is all that is required to reduce the viscosity of the CO_2-bearing sanidine melt to the values obtained at 25 kbar. As a consequence, it is not possible to attribute the reduction in sanidine melt viscosity at 25 kbar to the presence of dissolved CO_2 alone until the water concentration of each experimental charge has been determined by spectroscopic methods. Moreover, if there is a small concentration of water in the CO_2-bearing sanidine melts, then the presence of both dissolved water and CO_2 will contribute an entropy of mixing term to the configurational entropy of the melt, which should further reduce melt viscosity.

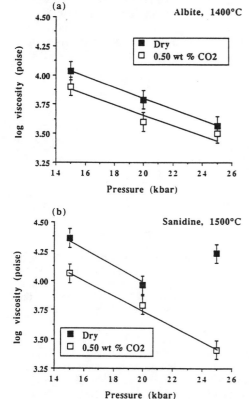

Figure 13. A comparison of the effect of 0.5 wt % CO_2 on the viscosity of albite and sanidine melts (Brearley and Montana, 1989; White and Montana, 1990).

A comparison of the effect of dissolved CO_2 on the viscosity of albite versus sandine liquid is shown in Figure 13 as a function of pressure. The measurements of Brearley and Montana (1989; Fig. 13a) indicate that the effect of 0.5 wt % CO_2 is to reduce the viscosity of albite liquid by ~0.15 log units at 1400°C. In contrast, the measurements of White and Montana (1990; Fig. 13b) indicate that 0.5 wt % CO_2 reduces the viscosity of sanidine liquid by ~0.30 log units at 1500°C. Although the effect of CO_2 is more pronounced in sanidine versus albite melt, it is significantly less than the effect of dissolved water on the viscosity of sanidine melt as seen in Figure 11, particularly at 20 kbar.

From the CO_2 speciation measurements of Fine and Stolper (1985) and Stolper et al. (1987) on albite liquid, it is well established that CO_2 dissolves both as molecular CO_2 and as a carbonate species. At a concentration of ~0.50 wt % total CO_2, 20 kbar pressure and 1450°C, the concentration of molecular CO_2 is ~0.39 wt % and that of CO_3^{2-} is ~0.14 wt % in albite melt (Stolper et al., 1987). Given a total dissolved CO_2 concentration of 0.5 wt %, if molecular CO_2 exerts little influence on the structure and hence viscosity of albite liquid, then it is not surprising that the effect of ~0.14 wt % dissolved carbonate is to reduce the viscosity by only ~0.15 log units. The fact that 0.5 wt % CO_2 appears to have a more pronounced effect on the viscosity of sanidine melt may indicate a higher proportion of total CO_2 dissolved as carbonate in sanidine versus albite melt under otherwise similar conditions.

The effect of 2 wt % CO_2 on the viscosity of a depolymerized melt composition, $NaCaAlSi_2O_7$, was also explored by Brearley and Montana (1989). From the results of Fine and Stolper (1985), Stolper and Holloway (1988), and Pan et al. (1991), it is anticipated that virtually all of the CO_2 is dissolved as carbonate. From the viscosity data plotted in Figure 14 it is clear that 2 wt % CO_2 has no measurable effect on the viscosity of sodium melilite liquid. Nor is there any resolvable pressure effect between 10 and 20 kbar, either dry or with 2 wt % CO_2.

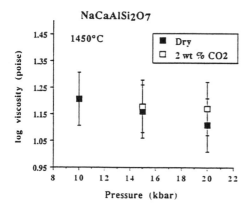

Figure 14. The effect of 2 wt % CO_2 on the viscosity of sodium melilite ($NaCaAlSi_2O_7$) liquid as a function of pressure. Errors on log viscosity are ±0.1 for this extremely fluid composition (Brearely and Montana, 1989).

The general conclusion that can be drawn from these initial studies is that regardless of its speciation, dissolved CO_2 appears to have a minimal effect on silicate melt viscosity. In those experiments where the measured viscosities of CO_2-bearing melts are significantly reduced, it is unclear how much can be attributed to small concentrations (≤0.3 wt %) of dissolved water. The importance of analyzing each experimental charge for both H_2O and CO_2 cannot be overemphasized, in order to fully resolve the influence of dissolved CO_2 on silicate melt viscosity.

The effect of F on silicate melt viscosities

The effectiveness of F as an agent to reduce the viscosity of silicate melts has been utilized for centuries in the glass and ceramics industry. Experimental work conducted by these two industries over the past fifty years has been primarily applied to Al-poor slag melts at one bar (e.g., Owens-Illinois Glass Company, General Research Laboratory, 1944; Hirayama and Camp, 1969). An exception is the work of Bills (1963) on CaO-Al₂O₃-SiO₂-CaF₂ liquids, also at one bar. As discussed below, the effect of F on the viscosity of Al-free vs. Al-bearing silicate melts is significantly different, and only the latter liquids are relevant to the study of magmas. In the recent geologic literature, viscosity measurements on F-bearing silicate melts have been performed both at one bar by concentric cylinder viscometry (five sodium aluminosilicates and diopside; Dingwell et al., 1985; Dingwell, 1989) and at elevated pressure (2.5 to 22.5 kbar) on molten albite by falling sphere viscometry (Dingwell and Mysen, 1985; Dingwell, 1987). Of all the volatile components considered in this chapter, fluorine is by far the most soluble in silicate melts at one bar, thus allowing viscosity measurements to be performed on F-rich liquids at atmospheric pressure.

A comparison of the effect of fluorine on the viscosity of molten albite (0 versus 5.8 wt % F) and molten diopside (0 versus 4.7 wt % F) is shown in Figure 15 as a function of inverse temperature at one bar. Most obvious is the pronounced effect that F has on reducing the viscosity of fully polymerized, Al-bearing melts such as albite relative to depolymerized, Al-free melts such as diopside. Although some ultrapotassic, mantle-derived melts (i.e., lamproites) can contain up to 2 wt % F (Aoki et al., 1981), the measurements of Dingwell (1989) indicate that this amount of F in a lamproite melt will reduce viscosity by less than 0.2 log units. In sharp contrast, at low temperatures (< 1100°C) the effect of almost 6 wt % F is to reduce the viscosity of albite melt by more than 3 log units. It is quite clear that fluorine will have a marked effect on the magma chamber dynamics and eruptive style of high-silica rhyolites. This is confirmed by field studies of F-rich topaz rhyolites that erupted to form unusually long lava flows with little pyroclastic activity (Christiansen et al., 1983).

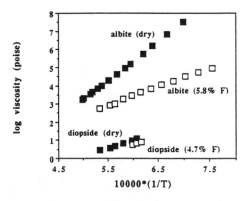

Figure 15. A comparison of the effect of fluorine on albite vs. diopside liquid vis-cosity. Data are from the 1 bar concentric cylinder viscosity measurements of Urbain et al. (1982) for dry albite, Scarfe et al. (1983) for dry diopside, Dingwell et al. (1985) for albite with 5.8 % F, and Dingwell (1989) for diopside with 4.7 % F.

As observed for H₂O, the reduction in aluminosilicate melt viscosity caused by dissolution of fluorine is strongly dependent on temperature. In the case for albite, the effect of 5.8 wt % F is to reduce melt viscosity by more than 3 log units at 1100°C and less than 1 log unit at 1600°C. The cause for this strong temperature dependence is related to the marked difference in the respective slopes of log viscosity versus 1/T curves for the F-bearing and F-free albite melts. The slope of each curve is proportional to the parameter B_e in Equation (27), which is the energy barrier to a change in configurational states and is considerably lower in the F-bearing albite melt. The implication is that the intermediate step for viscous flow is more readily accessible in the F-bearing than the F-free albite melt. As discussed above, recent NMR studies have led to suggestions that five- and six-fold coordinated Si and Al may serve as intermediates in Q^n species exchange and viscous flow (Stebbins et al., 1992). This interpretation is consistent with NMR evidence for five- and six-fold coordinated aluminum fluoride complexes (Kohn et al., 1991; Schaller et al., 1992) in sodium aluminosilicate glasses. Moreover, the removal of Al from the tetrahedral network by complexing with F will cause depolymerization of the melt and a consequential reduction in viscosity. Although both water and fluorine significantly reduce alumino-silicate melt viscosities, the solution mechanism of F appears to be entirely different from that of H_2O (Kohn et al., 1989).

A comparison of the effect of 5.8 wt % F versus 2.8 wt % H_2O on albite melt viscosity as a function of pressure is shown in Figure 16 (Dingwell and Mysen; 1985). Although dissolved water appears to be far more effective than dissolved fluorine on a weight per cent basis, concentrations of 5.8 wt % F and 2.8 wt % H_2O (in albite) are similar in terms of the molar quantity $X/(X+O)$ where $X = F, OH$. The effect of the two volatiles on melt viscosity is therefore nearly equivalent on a molar basis; dissolved hydroxyl is somewhat more effective in reducing melt viscosity (~0.3 log units difference between the two components at 15 kbar and 1400°C). The most effective reduction in melt viscosity is acheived, however, when both hydroxyl and fluorine ions are present. The configurational mixing contribution to the viscosity is substantial at 1000°C (0.6 log units) as seen in the data of Dingwell (1987) and discussed above in the section on the effect of H_2O on melt viscosity.

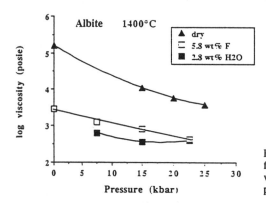

Figure 16. A comparison of the effect of fluorine (5.8 wt %) vs. water (2.8%) on the viscosity of molten albite as a function of pressure.

Toward a comprehensive model equation for silicate melt viscosity

Many more measurements of silicate melt viscosity containing a wide variety of volatile components, including Cl and SO_3, are required before a general model equation applicable to a wide variety of magmatic liquids becomes available. The experiments that are sparse are those at low temperatures, where the effect of variable composition on melt viscosity is particularly strong. Of all the volatile components, H_2O and F are by far the most important to silicate melt viscosity models. What has not yet been taken into account, however, in current viscosity models (e.g., Shaw, 1972) is the strong temperature dependence to the effect of water on melt viscosity. Because this can be understood using the theory of configurational entropy, successful development of a general model should be based on this concept.

ACKNOWLEDGMENTS

Support from the National Science Foundation (EAR-9219070) is acknowledged. This manuscript was significantly improved by comments from N. Bagdassarov, M. Carroll, E. Essene, W.G. Ernst, P. Richet, Y. Zhang, and one anonymous reviewer.

REFERENCES

Adam G, Gibbs JH (1965) On the temperature dependence of cooperative relaxation properties in glass-forming liquids. J Chem Phys 43:139-146

Agee CB, Walker D (1993) Olivine flotation in mantle melt. Earth Planet Sci Lett 114:315-324

Anderson AT, Newman S, Williams SN, Druitt TH, Skirius C, Stolper E (1989) H_2O, CO_2, Cl, and gas in Plinian and ash-flow Bishop rhyolite. Geology 17:221-225

Angell CA, Cheeseman PA, Tamaddon S (1983) Water-like transport property anomalies in liquid silicates investigated at high T and P by computer simulation techniques. Bull Minéral 106:87-97

Aoki K, Ishikawa K, Kanisawa S (1981) Fluorine geochemistry of basaltic rocks from continental and oceanic regions and petrogenetic application. Contrib Mineral Petrol 76:53-59

Bagdassarov NS, Dingwell DB (1992) A rheological investigation of vesicular rhyolite. J Volcanol Geotherm Res 50:307-322

Bagdassarov NS, Dingwell DB (1993) Frequency dependent rheology of vesicular rhyolite. J Geophys Res 98:6477-6487

Bailey JC (1977) Fluorine in granitic rocks and melts: a review. Chem Geol 19:2-42

Bartholomew RF, Butler BL, Hoover HL, Wu CK (1980) Infrared spectra of water-containing glass. J Am Ceram Soc 63:481-485.

Bills PM (1963) Viscosities in silicate slag systems. Journal of the Iron and Steel Institute 201:133-140

Blank JG, Stolper EM, Carroll MR (1993) Solubilities of carbon dioxide and water in rhyolite melt at 850°C and 750 bars. Earth Planet Sci Lett 119:27-36

Bockris JO'M, Tomlinson JW, White JL (1956) The structure of the liquid silicates: partial molar volumes and expansivities. Faraday Soc Trans 52:299-310

Bottinga Y, Weill D (1970) Densities of liquid silicate systems calculated from partial molar volumes of oxide components. Am J Sci 269:169-182

Bottinga Y, Weill DF (1972) The viscosity of magmatic silicate liquids: A model for calculation. Am J Sci 272:438-475

Brawer S (1985) Relaxation in Viscous Liquids and Glasses. American Chemical Society, Columbus, Ohio

Brearley M, Montana A (1989) The effect of CO_2 on the viscosity of silicate liquids at high pressure. Geochim Cosmochim Acta 53:2609-2616

Brearley M, Dickinson JE, Scarfe CM (1986) Pressure dependence of melt viscosities on the join diopside-albite. Geochim Cosmochim Acta 50:2563-2570

Brey G, Green DH (1975) The role of CO_2 in the genesis of olivine melilitite. Contrib Mineral Petrol 49:93-103

Brodholt JP, Wood BJ (1993) Simulations of the structure and thermodynamic properties of water at high pressures and temperatures J Geophys Res 98:519-536

Burnham CW, Davis NF (1971) The role of H_2O in silicate melts. 1. P-V-T relations in the system $NaAlSi_3O_8-H_2O$ to 10 kilobars and 1000 °C. Am J Sci 270:54-79

Burnham CW, Holloway JR, Davis NF (1969) The specific volume of water in the range 1000 to 8900 bars, 20 to 900 °C. Am J Sci 267-A:70-95

Christensen JN, DePaulo DJ (1993) Time scales of large volume silicic magma systems: Sr isotopic systematics of phenocrysts and glass from the Bishop Tuff, Long Valley, California. Contrib Mineral Petrol, 113:100-114

Christiansen EH, Sheridan MF, Burt DM (1983) The geology and geochemistry of Cenozoic topaz rhyolites from the western United States. Geol Soc Am Special Paper 205, 82 p

Clague DA, Frey FA (1982) Petrology and trace elemeny geochemistry of the Honolulu volcanics, Oahu: implications for the ocean mantle below Hawaii. J Petrol 23:447-504

Congdon RD, Nash WP (1988) High-fluorine rhyolite: an eruptive pegmatite magma at the Honeycomb Hills, Utah. Geology 16:1018-1021

Danyushevsky LV, Falloon TJ, Sobolev AV, Crawford, AJ, Carroll M, Price RC (1993) The H_2O content of basalt glasses from Southwest Pacific back-arc basins. Earth Planet Science Lett 117:347-362

Dickinson JE, Scarfe CM, McMillan P (1990) Physical properties and structure of $K_2Si_4O_9$ melt quenched from pressures up to 2.4 GPa. J Geophys Res 95:15675-15682

Dingwell DB (1987) Melt viscosities in the system $NaAlSi_3O_8-H_2O-F_2O_{-1}$. In: Mysen BO (ed) Magmatic Processes: Physiochemical Principles. The Geochemical Society, Spec Pub 1:423-431

Dingwell DB (1989) Effect of fluorine on the viscosity of diopside liquid. Am Min 74:333-338

Dingwell DB (1992) Density of some titanium-bearing silicate liquids and the compositional dependence of the partial molar volume of TiO_2. Geochim Cosmochim Acta 56:3403-3408

Dingwell DB, Bagdassarov NS, Bussod GY, Webb SL (1993) Magma rheology. In: Luth RW (ed) Experiments at High Pressure and their Applications to the Earth, Short Course Handbook. Mineralogical Association of Canada 21:131-196.

Dingwell DB, Brearley M (1988) Melt densities in the $CaO-FeO-Fe_2O_3-SiO_2$ system and the compositional dependence of the partial molar volume of ferric iron in silicate melts. Geochim Cosmochim Acta 52:2815-2825

Dingwell DB, Knoche R, Webb SL (1993) The effect of F on the density of haplogranite melt. Am Min 78:325-330

Dingwell DB, Mysen BO (1985) Effects of water and fluorine on the viscosity of albite melt at high pressure: a preliminary investigation. Earth Planet Sci Lett 74:266-274

Dingwell DB, Scarfe CM, Cronin DJ (1985) The effect of fluorine on viscosities in the system Na_2O-Al_2O_3-SiO_2: implications for phonolites, trachytes and rhyolites. Am Min 70:80-87

Dingwell DB, Webb SL (1990) The onset of non-newtonian rheology of silicate melts: A fiber elongation study. Phys Chem. Minerals 17:125-132

Dixon JE, Stolper EM, Holloway JR (1994) An experimental study of water and carbon dioxide solubilities in mid-oceanic ridge basaltic liquids. J Petrol (in press)

Eckert H, Yesinowski JP, Stolper EM, Stanton TR, Holloway, JR. (1987) The state of water in rhyolitic glasses: a deuterium NMR study. J Non-Cryst Solids 93:93-114.

Eckert H, Yesinowski JP, Silver LA, Stolper EM (1988) Water in silicate glasses: quantitation and structural studies by ^{1}H solid echo and MAS-NMR methods. J Chem Phys 92:2055-2064

Eggler DH (1978) The effect of CO_2 upon partial melting of peridotite in the system Na_2O-CaO-Al_2O_3-MgO-SiO_2-CO_2 to 35 kb, with an analysis of melting in a peridotite-H_2O-CO_2 system. Am J Sci 278:305-343

Farnan I, Stebbins JF (1990) A high-temperature ^{29}Si NMR investigation of solid and molten silicates. J Am Chem Soc 112:32-39

Fine G, Stolper E (1985) Dissolved carbon dioxide in basaltic glasses: Concentrations and speciation. Earth Planet Sci Lett 76:263-278

Fine G, Stolper E (1986) The speciation of carbon dioxide in sodium aluminosilicate glasses. Contrib Mineral Petrol 91:105-121

Friedman I, Long W, Smith RL (1963) Viscosity and water content of rhyolite glass. J Geophys Res 68:6523-6535

Geschwind CH, Rutherford MJ (1992) Cummingtonite and the evolution of the Mount St. Helens (Washington) magma system: an experimental study. Geology 20:1011-1014.

Grjotheim K, Matiasovsky K, Danek V, Stubergh JR (1990) Electrochemical deposition of metals; alloys and oxygen from natural minerals - I. Physicochemical properties of molten cryolite-potassium feldspar mixtures. Canad Metall Quart 29:39-42

Halliday AN, Mahood GA, Holden P, Metz JM, Dempster TJ, Davidson JP (1989) Evidence for long residence times of rhyolitic magma in the Long Valley magmatic system: the isotopic record in precaldera lavas of Glass Mountain. Earth Planet Sci Lett 94:274-290

Harris DM, Anderson AT (1984) Volatiles H_2O, CO_2, and Cl in a subduction related basalt. Contrib Mineral Petrol 87:120-128

Heinrich EW (1966) The Geology of Carbonates. Rand McNally, New York.

Hervig RV, Dunbar N, Westrich HR, Kyle PR (1989) Pre-eruptive water content of rhyolitic magmas as determined by ion microprobe analyses of melt inclusion in phenocrysts. J Volcanol. Geotherm. Res 36:293-302

Hildreth WE (1977) The magma chamber of the Bishop Tuff: gradients in temperature, pressure, and composition. Ph.D. thesis, University of California, Berkeley.

Hirayama C, Camp E (1969) The effect of fluorine and chlorine substitution on the viscosity and fining of soda-lime and potassium-barium silicate glass. Glass Tech 10:123-127

Hodges FW (1974) The solubility of H_2O in silicate melts. Carnegie Inst. Wash. Year Book 73:251-255

Holloway JR (1977) Fugacity and activity of molecular species in supercritical fluids. In: Fraser DG (ed) Thermodynamics in Geology. D Reidel, Dordrectht, Holland, p 161-181.

Hummel W, Arndt J (1985) Variation of viscosity with temperature and composition in the plagioclase system. Contrib Mineral Petrol 90:83-92

Ihinger PD (1991) An experimental study of the interaction of water with granitic melt. Ph.D. thesis, California Institute of Technology

Kanzaki M (1987) Physical properties of silicate melts at high pressures Ph.D. thesis, Geophysical Institute, University of Tokyo, Tokyo, Japan

Kanzaki M, Kurita K, Fujii T, Kato T, Shimomura O, Akimoto S (1987) A new technique to measure the viscosity and density of silicate melts at high pressure. In: Manghnani MH and Syono Y (eds) High-pressure research in mineral physics. Geophysical Monograph 39:195-200. Am Geophys Union, Washington, D.C.

Kogarko LN, Krigman LD (1981) Fluorine in Silicate Melts and Magmas. 159 p Nauka, Moscow.

Kohn SC, Dupree R, Smith ME (1989) A multinuclear magnetic resonance study of the structure of hydrous albite glasses. Geochim Cosmochim Acta 53:2925-2935

Kohn SC, Dupree R, Mortuza MG, Henderson CMB (1991) NMR evidence for five- and six-coordinated aluminum fluoride complexes in F-bearing aluminosilicate glasses. Am Min 76:309-312

Kovalenko VI, Hervig RL, Sheridan MF (1988) Ion microprobe analyses of trace elements in anorthoclase, hedenbergite, aenigmatite, quartz, apatite and glass in pantellerite. Am Min 73:1038-1045

Kovalenko VI, Hervig RL, Schauer S (1989) Volatile contents of pantellerites. New Mex Bur Mines Mineral Resource Bull 131:156

Kraynik AM (1988) Foam flows. Ann Rev Fluid Mech 20:325-357

Kress VC, Williams Q, Carmichael ISE (1988) Ultrasonic investigation of melts in the system Na_2O-Al_2O_3-SiO_2. Geochim Cosmochim Acta 52:283-293

Kress VC, Carmichael ISE (1991) The compressibility of silicate liquids containing Fe_2O_3 and the effect of composition, temperature, oxygen fugacity and pressure on their redox states. Contrib Mineral Petrol 108:82-92

Kubicki JD, Lasaga AC (1988) Molecular dynamic simulations of SiO_2 melt and glass: ionic and covalent models. Am Min 73:941-955

Kushiro I (1978) Density and viscosity of hydrous calc-alkaline andesite magma at high pressure. Res Carnegie Inst Wash Yearbk 77:675-677

Lange RA (1993) Densities of TiO_2-bearing silicate melts: evidence for composition-induced coordination change of Ti^{4+}. Geol Soc Am Abastr Programs 25:A212

Lange RA and Carmichael ISE. (1987) Densities of Na_2O-K_2O-CaO-MgO-FeO-Fe_2O_3-Al_2O_3-TiO_2-SiO_2 liquids: new measurements and derived partial molar properties. Geochim Cosmochim Acta 53:2195-2204

Lange RA, Carmichael ISE (1989) Ferric-ferrous equilibria in Na_2O-FeO-Fe_2O_3-SiO_2 melts: effects of analytical techniques on derived partial molar volumes. Geochim Cosmochim Acta 53:2195-2204

Lange RA, Carmichael ISE (1990) Thermodynamic properties of silicate liquids with emphasis on density, thermal expansion and compressibility. In: Nicholls J and Russell JK (eds) Modern Methods of Igneous Petrology. Rev Mineral 24:25-64

Leonteva A (1940) Measurements of the viscosity of obsidian and of hydrated glasses (in Russian), Izv Akad Nauk SSSR, Ser Geol 2:44-54

Liu S-B, Stebbins JF, Schneider E, Pines A (1988) Diffusive motion in alkali silicate melts: an NMR study at high temperature. Geochim Cosmochim Acta 52:527-538

Lowenstern JB, Mahood GA (1991) New data on magmatic H_2O contents of pantellerites with implications for petrogenesis and eruptive dynamics at Pantelleria. Bull Volcanol 54:78-83

Manghnani MH, Sato H, Rai CS (1986) Ultrasonic velocity and attenuation measurements on basalt melts to 1500°C: role of composition and structure in the viscoelastic properties. J Geophys Res 91:9333-9342

Martin D, Griffiths RW, Campbell IH (1987) Compositional and thermal convection in magma chambers. Contrib Mineral Petrol 96:465-475

McKeown DA, Galeener FL, Brown GE 91984) Raman studies of Al coordination in silica-rich sodium aluminosilicate glasses and some related minerals. J Non-Cryst Solids 68:361-378

McMillan PW, Jakobsson S, Holloway JR, Silver LA (1983) A note on the Raman spectra of water-bearing albite glass. Geochim Cosmochim Acta 47:1937-1943

McMillan PW, Wolf GH, Poe BT (1992) Vibrational spectroscopy of silicate liquids and glasses. Chem Geol 96:351-366

Miller GH, Stolper EM, Ahrens TJ (1991) The equation of state of a molten komatiite. 1. Shock wave compression to 36 GPa. J Geophys Res 96:11,831-11,848

Mysen BO (1990) Effects of pressure, temperature, and bulk composition on the structure and species distribution in depolymerized alkali aluminosilicate melts and quenched melts. J Geophys Res 95:15733-15744

Mysen BO, Arculus RJ, Eggler DH (1975) Solubility of carbon dioxide in natural nephelinite, tholeiite and andesite melts to 30 kbar pressure. Contrib Mineral Petrol 53:239

Mysen BO, Eggler DH, Setiz MG, Holloway JR (1976) Carbon dioxide in silicate melts and crystals. Part I. Solubility measurements. Am J Sci 276:455-479

Neuville D, Richet P (1991) Viscosity and mixing in molten (Ca, Mg) pyroxenes and garnets. Geochim Cosmochim Acta 55:1011-1019

Neuville D, Courtial P, Dingwell DB, Richet P (1993) Thermodynamic and rheological properties of rhyolite and andesite melts. Contrib Mineral Petrol 113:572-581

Newman S, Stolper EM, Epstein S (1986) Measurement of water in rhyolitic glasses: Calibration of an infrared spectroscopic technique. Am Min 86:1527-1541

Newman S, Epstein S, Stolper EM (1988) Water, carbon dioxide, and hydrogen isotopes in glasses from the ca. 1340 A.D. eruption of the Mono Craters, California: constraints on degassing phenomena and initial volatile content. J Volcanol Geotherm Res 35:75-96

Nicholls J, Carmichael ISE (1969) Peralkaline acid liquids: a petrological study. Contrib Mineral Petrol 20:268-294

Nogami M, Tomozawa M (1984) Effect of stress on water diffusion in silica glass. J Am Ceram Soc 67:151-154

Orlova GP (1962) The solubility of water in albite melts. Int Geol Rev 6:254-258

Ostrovskiy IA, Orlova GP, Rudnitskaya YS (1964) Stoichiometry in the solution of water in alkali-aluminosilciate melts. Doklady Akad Nauk SSR 157:149-151

Owens-Illinois Glass Company, General Research Laboratory (1944) Effect of fluorine and phosphorous pentoxide on properties of soda-dolomite lime-silica glass. J Am Ceram Soc 27:369-372

Paillat O, Elphick EC, Brown WL (1992) The solubility of water in NaAlSi3O8 melts: a re-examination of Ab-H2O phase relationships and critical behavior at high pressures Contrib Mineral Petrol 112:490-500

Pan V, Holloway JR, Hervig RL (1991) The pressure and temperature dependence of carbon dioxide solubility in tholeiitic basaltic melts. Geochim Cosmochim Acta 55:1587-1596

Persikov ES, Zharikov VA, Bukhtiyarov PG, Pol'skoy SF (1990) The effect of volatiles on the properties of magmatic melts. Eur J Mineral 2:621-642

Richet P (1984) Viscosity and configurational entropy of silicate melts. Geochim Cosmochim Acta 48:471-483

Richet P, Robie RA, Hemingway BS (1986) Low-temperature heat capacity of diopside glass (CaMgSi2O6): a calorimetric test of the configurational entropy theory applied to the viscosity of liquid silicates. Geochim Cosmochim Acta 50:1521-1533

Rigden SM, Ahrens TJ, Stolper EM (1989) High pressure equation of state of molten anorthite and diopside. J Geophys Res 94:9508-9522

Rivers ML, Carmichael ISE (1987) Ultrasonic studies of silicate melts. J Geophys Res 92:9247-9270

Rubie DC, Ross CR, Carroll MR, Elphick SC (1993) Oxygen self-diffusion in Na2Si4O9 liquid up to 10 GPa and estimation of high-pressure melt viscosities. Am Min 78:574-582

Sabatier G (1956) Influence de la teneur en eau sur la viscosité d'une rétinite, verre ayant la composition chimique d'un granite. C R Aca Sci 242:1340-1342

Saucier PH (1952) Quelques expérience sur la viscosité à haute température de verres ayant la composition d'un granite. Influence de la vapeur d'eau sous pression. Bull Soc fr Min Cristallogr 75:1-45

Saxena SK, Fei Y (1987) High pressure and high temperature fluid fugacities. Geochim Cosmochim Acta 51:783-791.

Scamehorn CA, Angell CA (1991) Viscosity-temperature relations in fully polymerized aluminosilicate melts from ion dynamics simulations. Geochim Cosmochim Acta 55:721-730.

Scarfe CM (1986) Viscosity and density of silicate melts. In: Scarfe CM (ed) Silicate Melts: Their Properties and Structure Applied to Problems in Geochemistry, Petrology, Economic Geology, and Planetology, Short Course Handbook. Mineral Assoc'n Canada, 36-56

Scarfe CM, Mysen BO, Virgo D (1987) Pressure dependence of the viscosity of silicate melts. In: Mysen BO (ed) Magmatic Processes: Physiochemical Principles. Geochemical Society Spec Pub 1:59-67

Schaller T, Dingwell DB, Keppler H Knöller W, Merwin L, Sebald A (1992) Fluorine in silicate glasses: A multinuclear NMR study. Geochim Cosmochim Acta 56:701-707

Schilling J-G, Bergeron MB, Evans R (1980) Halogens in the mantle beneath the North Atlantic. Phil Trans Roy Soc Lond ser A 297:147-178

Seifert FA, Mysen BO, Virgo D (1981) Structural similarity of glasses and melts relevant to petrological processes. Geochim Cosmochim Acta 45:1879-1884

Shaw HR (1963) Obsidian-H2O viscosities at 1000 and 2000 bars in the temperature range 700-900°C. J Geophys Res 68:6337-6343

Shaw HR (1972) Viscosities of magmatic silicate liquids: an empirical method of prediction. Am J Sci 272:870-893

Shimizu N and Kushiro I (1984) Diffusivity of oxygen in jadeite and diopside melts at high pressures Geochim Cosmochim Acta 48:1295-1303

Silver LA, Ihinger PD and Stolper E (1990) The influence of bulk composition on the speciation of water in silicate glasses. Contrib Mineral Petrol 104:142-162

Silver L and Stolper E (1989) Water in albitic glasses. Journal of Petrology, 30:667-709

Sisson TW, Grove TL (1993) Experimental investigations of the role of H2O in calc-alkaline differentiation and subduction zone magmatism. Contrib Mineral Petrol 113:143-166

Sisson TW, Layne GD (1993) H2O in basalt and basaltic andesite glass inclusions from four subduction-related volcanoes. Earth Planet Science Lett 117:619-635

Skirius CM, Peterson JW, Anderson AT (1990) Homogenizing rhyolitic glass inclusions from the Bishop Tuff. Am Min 75:1381-1398

Sparks RJS, Huppert HE, Turner JS (1984) The fluid dynamics of evolving magma chambers. Royal Soc Lond Phil Trans Ser A 310:511-534

Spera F,, Bergman SC (1980) Carbon dioxide in igneous petrogenesis: I. Aspects of the dissolution of CO2 in silicate liquids. Contrib Mineral Petrol 74:55-66

Stebbins JF (1991) NMR evidence for five-coordinated silicon in a silicate glass at atmospheric pressure. Nature 351:638-639

Stebbins JF, Farnan I (1992) Effects of high temperature on silicate liquid structure: a multinuclear NMR study. Science 255:586-589

Stebbins JF, McMillan P (1989) Five- and six-coordinated Si in $K_2Si_4O_9$ liquid at 1.9 GPa and 1200°C. Am Min 74:965-968

Stebbins JF, Sykes D (1990) The structure of $NaAlSi_3O_8$ liquid at high pressure: new constraints from NMR spectroscopy. Am Min 75:943-946

Stebbins JF, Carmichael ISE, Moret LK (1984) Heat capacities and entropies of silicate liquids and glasses. Contrib Mineral Petrol 86:131-148

Stebbins JF, Farnan I, Xue X (1992) The structure and dynamics of alkali silicate liquids: A view from NMR spectroscopy. Chem Geol 96:371-385

Stebbins JF, Murdoch JB, Schneider E, Carmichael ISE, Pines A (1985) A high temperature nuclear magnetic resonance study of ^{27}Al, ^{23}Na, and ^{29}Si in molten silicates. Nature 314:250-252

Stein DJ, Spera FJ (1992) Rheology and microstructure of magmatic emulsions: theory and experiments. J Volcanol Geotherm Res 49:157-174

Stein DJ, Spera FJ (1993) Experimental rheometry of melts and supercooled liquids in the system $NaAlSiO_4$-SiO_2: implications for structure and dynamics. Am Min 78:710-723

Stolper EM (1982a) Water in silicate glasses: an infrared spectroscopy study. Contrib Mineral Petrol 94:178-182

Stolper EM (1982b) The speciation of water in silicate melts. Geochim Cosmochim Acta 46:2609-2620

Stolper E (1989) Temperature dependence of the speciation of water in rhyolitic melts and glasses. Am Min 74:1247-1257

Stolper E, Fine G, Johnson T, Newman S (1987) The solubility of carbon dioxide in albitic melt. Am Min 72:1071-1085

Stolper E, Holloway JR (1988) Experimental determination of the solubility of carbon dioxide in molten basalt at low pressure. Earth Planet Sci Lett 87:397-408

Stolper E, Newman S (1994) The role of water in the petrogenesis of Mariana trough magmas. Earth Planet Sci Lett 121:293-325

Taniguchi H (1992) Entropy-dependence of viscosity and the glass-transition temperature of melts in the system diopside-anorthite. Contrib Mineral Petrol 109:295-303

Thibault Y, Holloway JR (1994) Solubility of CO_2 in Ca-rich leucitite: effects of pressure, temperature and oxygen fugacity. Contrib Mineral Petrol 116:216-224

Tuttle OF, Gittens J (eds) (1966) Carbonatites. Interscience 591 p

Urbain G, Bottinga Y and Richet P (1982) Viscosity of liquid silica, silicates and aluminosilicates. Geochim Cosmochim Acta 46:1061-1072

Webb SL, Dingwell DB (1990) The onset of non-Newtonian rheology of silicate melts. Phys Chem Min 17:125-132

Webb SL, Knoche R, Dingwell DB (1992) Determination of silicate liquid thermal expansivity using dilatometry and calorimetry. Eur J Mineral 4:95-104

Webster JD, Duffield WA (1991) Pre-eruptive concentrations of volatiles and lithophile elements in Taylor Creek Rhyolite: analysis of glass inclusions in quartz phenocrysts. Am Min 76:1628-1645

Wendlandt RF, Mysen BO (1978) Melting phase relations of natural peridotite and CO_2 as a function of degree of partial melting at 15 and 30 kbar. Carnegie Inst Wash Year Book 77:756-761

White BS, Montana A (1990) The effect of H_2O and CO_2 on the viscosity of sanidine liquid at high pressures J Geophys Res 95:15,683-15,693

White WB, Minser DG (1984) Raman spectra and structure of natural glasses. J Non-Cryst Solids 67:45-49

Wilding MC, Macdonald R, Davies JE, Fallick AE (1993) Volatile characteristics of peralkaline rhyolites from Kenya: an ion microprobe, infrared spectroscopic and hydrogen isotope study. Contrib Mineral Petrol 114:264-275

Wyllie P (1967) Ultramafic and Related Rocks. Wiley, New York

Xue X, Stebbins JF, Kanzaki M, McMillan P, Poe B (1991) Pressure-induced silicon coordination and tetrahedral structural changes in alkali oxide-silica melts up to 12 GPa: NMR, Raman, and infrared study. Am Min 76:8-26

Zhang Y (1993) Pressure dependence of the speciation of water in rhyolitic glasses. Trans. Am Geophys Union 74:631

Zhang Y, Stolper EM, Ihinger PD (1989) Reaction kinetics of $H_2O + O = 2 OH$ and its equilibrium, revisited. V.M. Goldschmidt Conference Abstr, p 94

Zhang Y, Stolper EM, Wasserburg GJ (1991) Diffusion of water in rhyolitic glasses. Geochim Cosmochim Acta 55:441-456

Chapter 10

DIFFUSION IN VOLATILE-BEARING MAGMAS

E. Bruce Watson

Department of Earth and Environmental Sciences
Rensselaer Polytechnic Institute
Troy, New York 12180 USA

INTRODUCTION

The role of volatiles in determining the phase equilibria, physical properties and eruption dynamics of magmas has been an area of intensive investigation for several decades. Indeed, this "Reviews" volume is a testimonial not only to the perceived importance of magmatic volatiles, but also to the magnitude of the intellectual effort put toward their understanding.

In the context of the long history of research into the nature and effects of magmatic volatiles, investigations focusing specifically upon diffusion came on the scene relatively recently, beginning with a ground-breaking effort by H.R. Shaw (1974) published just twenty years ago. This late attention to diffusion can probably be attributed in part to the former tendency among experimentalists to concentrate upon strictly equilibrium phenomena, and perhaps also to the limited availability of analytical instruments suitable for high-resolution, in-situ characterization of dissolved volatile concentrations. Despite a general paucity of volatile-related diffusion data prior to the late 1970s, theorists and experimentalists alike had developed a strong suspicion that dissolved volatiles (water in particular) have a strong enhancing effect on diffusion—theorists because they appreciated the inverse correlation between viscosity and diffusion embodied in the Stokes-Einstein equation, and experimentalists because they routinely observed that crystals grew bigger and faster in hydrous melts than in dry melts of the same composition at the same temperature.

The purpose of this chapter is to summarize the results of experimental studies that have contributed to our present understanding of diffusion in magmas containing dissolved volatiles. Two kinds of investigation are recognized here and treated in separate sections: those focusing on diffusion of volatiles themselves, and those addressing the effects of dissolved volatiles upon diffusion of other magmatic components. In order to keep the scope of the chapter manageable, only passing references are made to diffusion of volatiles in solid glasses, though a sizable literature on this topic does exist (especially for water in SiO_2). To assist readers who may be unfamiliar with basic diffusion theory or with experimental techniques used in characterizing diffusion in melts containing dissolved volatiles, a brief, practical overview of these subjects is provided in the sections immediately following.

OVERVIEW OF BASIC DIFFUSION THEORY

Nature and theoretical conceptualizations of diffusion

Definitions. In this chapter and in most correct usages of the word, "diffusion" refers to the translational motion of atoms or molecules dissolved in a phase; in general, the reference frame for gauging motion is the lattice or structure of the phase of interest.

Relatively inexplicit volatile-related magmatic processes such as "vapor-phase transport" and "volatile streaming" fall outside the realm of what can be called strictly diffusion, and so are not discussed in this chapter.

Diffusion theory is typically put forth in separate but complementary approaches: *atomistics* (deductions based upon consideration of the random motion of individual atoms) and *phenomenology* (presentation of the empirical equations governing diffusion in a continuum). Available space in this chapter (and expertise of the writer!) precludes comprehensive treatment of these approaches, but a brief and selective overview will expedite the clear presentation of data and concepts later in the chapter. For a more in-depth review of diffusion theory, the reader is referred to Manning (1968), Shewmon (1989) and Cussler (1984). This last volume is an especially readable presentation of theory and applications of diffusion to mass transport in liquid and/or gas systems; it is laden with interesting examples and solved problems on topics ranging from ion-exchange columns to artificial kidneys (no geology, unfortunately).

Atomistics. Classic atomistic treatments of diffusion are based upon a crystalline lattice characterized by a limited number of site types (usually just one). This framework is not very realistic for magmatic liquids, in which long-range structure is lacking and site characteristics are determined at least in part by the identity of the occupying atom. Nevertheless, certain aspects of atomistic diffusion theory are readily transferable to molten silicates. It is still appropriate, for example, to consider a dissolved ion oscillating in a (momentarily) fixed position, at frequency υ, as having the potential in any given oscillation cycle to leave its original position and "relocate" a small distance away. This displacement would constitute a diffusive "jump", and would amount to a net motion of the ion in the melt. Even in liquids, the fraction of oscillations that result in jumps is small, being given by $\exp(-E/kT)$, where E is the energy barrier that must be surmounted to achieve a jump (contributions to E include the breaking of bonds at the original site, as well as the transient, close approach to neighboring ions that accompanies a jump). In general, then, the jump frequency of a dissolved ion (in sec^{-1}) from a given position or site is given by

$$\Gamma = \upsilon \, \exp \, (-E/kT) \qquad\qquad (1)$$

where υ is the oscillation frequency (typically $\sim 10^{14}$ Hz), k is Boltzmann's constant and T is temperature in kelvins.

Equation (1) applies to the totally random translational motion of an atom, ion or molecule in its host phase. If one could tag a specific atom and follow its progress, that atom would make periodic jumps in random directions, possibly even revisiting sites it occupied previously. This process is sometimes referred to as "random walk" diffusion, and it occurs in all substances regardless of the degree of initial heterogeneity (i.e., no concentration gradient is required). The values of E and υ, of course, are affected not only by the identity of the diffusing species of interest, but also by its immediate atomic environment and the general structural (lattice) characteristics of the host phase. Recognizing the existence of random-walk diffusion is critical to appreciating the meanings and distinctions among the terms *chemical diffusion*, *self diffusion*, *isotopic diffusion* and *tracer diffusion* (see section on "Kinds" of diffusion).

In crystalline solids, the random atomic-jump process described above may involve one of several possible mechanisms, the most important of which are probably the *vacancy* (where the species of interest jumps into a near-neighbor vacant site), the *interstial* (where the diffusant is small enough to occupy a position between or among normal lattice sites) and the *ring mechanism* (in which simultaneous rotation of a cluster of atoms results in net motion of all of them). In cases where the vacancy mechanism operates, it is clear that the otherwise random-directional aspect of any single diffusive jump can be heavily biased by the presence of a neighboring vacancy. It is also clear that the frequency of jumps (hence the rate of diffusion) may be strongly dependent upon the concentration of vacancies. In magmas and other complex liquids, the conceptualization of diffusion as mechanism-specific may be a little tricky to defend; in fact, it seems likely that for network-modifying species in magmatic liquids (which most dissolved volatiles are), a variety of atomic migration mechanisms can operate, because there is some local flexibility in the way in which atoms fill space. In silicate melts, the main determinant of available diffusion mechanisms may well be the requirement that local charge balance be recovered once a jump has occurred.

Atomistic models of diffusion take on directionality when a concentration gradient is introduced. Consider a series of closely-spaced planes (in a crystal, these would be lattice planes) perpendicular to a one-dimensional gradient (along x) in the concentration of B atoms (solid circles in Fig. 1). The flux of atoms from plane 1 to plane 2 can be described in terms of the jump frequency, Γ, introduced previously. Recalling that the units of flux are atoms·area^{-1}·time^{-1} (the area being perpendicular to the flux direction), and avoiding reference to any diffusion mechanism, the *net* flux, J, from (1) to (2) is simply the number of atoms jumping from $1 \to 2$ minus the number jumping back $(2 \to 1)$:

$$J_{1 \to 2} = n_1(\Gamma / N) - n_2(\Gamma / N) \qquad (2)$$

where $n_{1,2}$ represent the densities of diffusant atoms on the specified plane and N is the number of possible jump directions (i.e., only one jump in N would move an atom from $1 \to 2$ or vice versa; in a metal, N would be six—the coordination number). Converting atom densities to concentrations ($c = n/\alpha$, where α is the distance between planes), we have

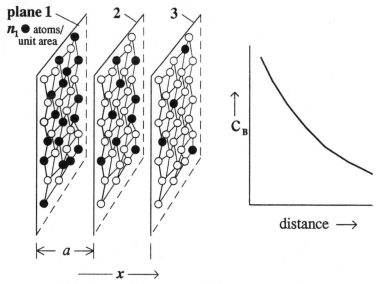

Figure 1. Atomistic representation of diffusion in the presence of a concentration gradient. Planes 1, 2, and 3 are separated by one atomic jump distance, a (in a crystal, this would be the lattice spacing), and are oriented perpendicular to a 1-dimensional gradient in the concentration of B atoms (l), which decreases in the x direction (i.e., from left to right). Because the medium of interest is a melt, the atoms in each plane exhibit no long-range order. See text for discussion.

$$J_{1\rightarrow 2}=\frac{\Gamma\alpha}{N}(c_1-c_2) \tag{3}$$

Assuming that concentration is a continuous function of x, so that $c_2 = c_1 + \alpha(\partial c/\partial x)$, we arrive at an equation relating the flux to the concentration gradient:

$$J=-\left[\frac{\Gamma\alpha^2}{N}\right]\left(\frac{\partial c}{\partial x}\right) \tag{4}$$

where the term in brackets is a collection of constants characteristic of the diffusant of interest and the diffusion medium. In general, we would expect this term, $\Gamma\alpha^2/N$, to be constant for diffusion of a specific species in a specific medium; its actual value, of course, would be impossible to estimate a priori for a substance as complex as a silicate melt.

In the interest of saving space, the above analysis will not be extended to address the question of how a concentration gradient would evolve over time as a consequence of "random-jump" diffusion. The extension is straightforward, though, and is presented in a lucid manner by Shewmon (1989).

The conclusions that emerge from the foregoing atomistic analysis are that: (1) reduced to its simplest terms, diffusion involves inherently random motions of atoms, ions or molecules in their host media; and (2) diffusion manifests itself as a net flux only when there is a pre-existing concentration gradient in the species of interest. Armed with these conclusions, we can be appropriately skeptical about the conventional wisdom that "a gradient in concentration (or chemical potential) is required for diffusion." In reality, diffusion is an unavoidable consequence of atomic vibrations in any substance at high temperature, and does not require a concentration gradient to occur. Under most circumstances, however, we can observe the results of diffusion only when a preexisting concentration gradient is present.

Phenomenology: diffusion equations. It can be shown by experiment (and predicted on the basis of intuition) that an initially inhomogeneous phase, if annealed at sufficiently high temperature, will eventually reach a state of homogeneity. This state is achieved by migration of atoms from high- to low-concentration regions. In 1855 Adolph Fick recognized this behavior (see Cussler, 1984, p. 18) and noted the mathematical parallels between diffusion of matter and the conduction of heat (Fourier's law) and electricity (Ohm's law). As aids in developing intuition about diffusion, these parallels are still worth emphasizing today, supplemented by the additional analogy with Darcy's law for fluid flow. Fick stated (but did not actually prove) that the one-dimensional flux of a component (i) is proportional to the gradient in its concentration

$$J_i=-D_i\frac{\partial c_i}{\partial x} \tag{5}$$

through a constant, D_i, which we now recognize as the *diffusion coefficient* or *diffusivity* (units = m^2/s). Equation (5) has the expected property that, as the concentration gradient goes to zero, so does the flux. This equation has come to be referred to as Fick's first law—or the steady-state diffusion equation—and its equivalence to Equation (4) is clear. Fick went on to develop a more general mass-conservation equation, now written (for one-dimensional diffusion) as

$$\left(\frac{\partial c_i}{\partial t}\right)_x = \frac{\partial}{\partial x}\left[D_i\frac{\partial c_i}{\partial x}\right]_t \tag{6}$$

or, if D_i is not a function of x (i.e., of concentration)

$$\left(\frac{\partial c_i}{\partial t}\right)_x = D_i \frac{\partial^2 c_i}{\partial x^2} \qquad (7)$$

This second-order partial differential equation describes the time evolution of concentration, and so is referred to as the nonsteady-state diffusion equation (and also as Fick's second law). Solutions to this equation have been obtained for many different boundary conditions; a summary can be found in the classic reference work by Crank (1979). As we will show shortly, experimental studies of diffusion in volatile-bearing magmas usually incorporate one of three relatively simple solutions to Equation (7), because only a small number of sample geometries and initial concentration distributions are experimentally tractable.

Complications. This discussion would be incomplete to the point of negligence without a few additional observations concerning the diffusion equations above. First, the diffusivity, D_i, cannot generally be considered independent of c_i over large changes in concentration. As we shall see, dissolved water provides a good example of concentration-dependent diffusion. Dependence of D_i upon concentration can be due to effects intrinsic to the diffusant of interest, or it may be attributable to changing influences of other components (or both; see section on "kinds" of diffusion). Diffusion of dissolved CO_2, sulfur and chlorine, as well as most cations, is strongly dependent upon the water content of a magma.

A second and not unrelated observation is that, contrary to the simple formulation embodied in Equation (5)—which could be stated in words as "a concentration gradient will lead to an observable flux"—chemical thermodynamics suggests a more correct statement: "a gradient in *chemical potential* (μ_i) will lead to an observable flux." Since μ_i is not independent of the concentrations of other system components, the implications of this alternative phrasing are profound for systems characterized by strong gradients in most or all components, and the complexity of rigorous treatment rapidly escalates beyond the scope of this practical primer (but see discussion of chemical diffusion in the next section).

A final note concerning the diffusion equations is that concentration gradients in natural magmas rarely exist in one dimension only. The more general statements of Equations (5) and (7), for D_i independent of position, are:

$$\mathbf{J}_i = -D_i \nabla c_i \qquad \text{(steady-state)} \qquad (8)$$

where $\mathbf{J}_i$ is a vector quantity, and

$$(\partial c_i / \partial t) = D_i \nabla^2 c_i \qquad \text{(nonsteady-state)} \qquad (9)$$

For ease in extracting diffusivities from laboratory samples, experiments are usually designed so that diffusion occurs in one dimension only. In melts, of course, because of their isotropic structure, the resulting diffusivities are appropriate for modeling more complicated diffusion processes in magmas.

"Kinds" of diffusion

The preceding discussion should provide a good basis for appreciating the significance of the various adjectives used to modify the word "diffusion"—e.g., "self", "isotope", "tracer", "chemical", and "multicomponent", among others. The distinctions among these can be very significant to geologists concerned with volatiles in magmas, who routinely consider transport processes involving major elements, trace elements and isotopes. Definitions of the kinds of diffusion seem to vary somewhat among authorities and among disciplines, but we can probably write down interpretations that would be generally agreed upon. Inasmuch as this chapter deals with diffusion specifically in melts, we can omit discussion of terms such as "grain-boundary", "surface", "intrinsic", "extrinsic" and "defect" diffusion, all of which are significant only in crystals or polycrystals.

Strictly speaking, *self diffusion* refers to diffusion of the constituent atoms of a pure or nearly-pure substance; however, the term has also been applied to diffusion of components at any concentration level (e.g., major or trace elements), as long as it occurs in the absence of a concentration gradient. Characterizing self diffusion necessitates distinguishing among different atoms of the same element, so an isotopic tracer must be used, and the assumption made that the mass difference among isotopes is unimportant (which is valid for high-Z elements). The random-walk process described earlier represents self diffusion. Geochemists tend to use the term "isotopic diffusion" for the same phenomenon, in reference to a process that involves the exchange of isotopes with an implied absence of a gradient in chemical potential of the element of interest. This usage is reasonable, but in geological systems it is probably rare (and definitely fortuitous) that disequilibrium or heterogeneity in the isotopic composition of an element could exist in the absence of disequilibrium in total abundance of that element. In the context of volatiles in magmas, examples of self- or isotopic diffusion include: homogenization of the isotopes of carbon or sulfur in dissolved CO_2 or SO_2; and the effects of dissolved water upon homogenization of radiogenic isotope ratios (e.g., for Pb, Sr, Nd, or Hf).

Because isotopic tracers are needed to characterize self diffusion, the term *"tracer diffusion"* is often equated with self diffusion. There is little risk in doing this, but it is worth noting that virtually all tracer diffusion studies on geologic melts are done by introducing an isotope (usually radioactive) at one end of an experimental sample— technically, this introduces a small chemical concentration gradient in the species of interest even if the sample already contains a high concentration of other isotopes of that species. However, because very small quantities of tracer are used, the chemical concentration gradient is diminishingly small, and the resulting diffusivities are indistinguishable from self-diffusion coefficients.

Chemical diffusion refers to diffusion in a chemical potential gradient, for which conditions are rife in magmas. The chemical or interdiffusion coefficient for a binary system is often defined simply in terms of Fick's first law: $\tilde{D} = -J / (\partial c / \partial x)$. This interdiffusion coefficient generally varies with concentration (hence with distance across the interdiffusion zone), but a solution to Fick's second law exists that will accommodate variation in $\tilde{D}$ with x:

$$\tilde{D}_{c'} = \frac{-1}{2t} \frac{dx}{dc} \int_{c_{-\infty}}^{c'} x\,dc \qquad (10)$$

where dx/dc is evaluated at c' (a concentration of interest along the interdiffusion profile) and x is the distance from the zero-net-flux plane at $x=0$ (this plane is referred to as the Matano interface). Implementation of Equation 10 is referred to as Boltzmann-Matano analysis (see Shewmon, 1989, p. 34-37); formerly done graphically, it can now be accomplished more easily by fitting a functional form to c vs. x data, and performing the differentiation and integration at various concentrations along the curve.

The relationship between the two diffusivities just discussed—i.e., the *tracer* diffusivity (D^*) and the *chemical interdiffusion* coefficient ($\tilde{D}$)—is relatively simple for binary metallic systems, as shown by Darken (1948):

$$\tilde{D} = [X_2D_1^* + X_1D_2^*](1 + dln\gamma_1/dlnX_1) \tag{11}$$

Here X represents the mole fraction of the component indicated by the subscript and γ is the activity coefficient. Although it strictly applies only to diffusion in two-component systems in which charge-balance considerations do not arise (i.e., not to magmas), the Darken equation is useful here in making the point that in ideal (or dilute ideal) binary solutions the chemical diffusion coefficient is given simply by the weighted sum of the endmember tracer diffusivities. Equations analogous to no. 11 have been derived for simple ionic materials as well (see Manning, 1968, p. 21; Brady, 1975, p. 980). For complex systems, some researchers have resorted to use of *effective binary diffusion coefficients* (EBDCs; see Cooper 1968) for cases in which one component is imagined as interdiffusing with a mixture of other components—the system is treated as binary, even though one "endmember" is itself really a mixture. For some studies described later involving diffusion of highly-soluble volatile components in magmas (e.g., H_2O), the diffusivities extracted are actually EBDCs. The same is true of diffusivities for major or minor components obtained by analysis of concentration gradients produced upon dissolution of a simple mineral into a melt. The only disadvantage of these values is that they are valid only for the specific system, conditions, and range of concentrations directly investigated. Lesher (1994) applied the Darken equation (Eqn. 11) to volatile-free magmas using EBDCs in combination with γ-X relationships extracted from Soret diffusion experiments. To the author's knowledge, no similar attempt has been made for volatile-bearing magmas, probably because γ-X data are generally lacking.

Rigorous treatment of multicomponent (chemical) diffusion in systems of more than two components begins with Onsager's (1945) extended version of Fick's first law:

$$J_i = -\sum_{j=1}^{n-1} D_{ij} (\partial c_i / \partial x) \tag{12}$$

where n is the number of components in the system (n-1 of the diffusivities are independent). Not only are the D_{ij} values composition dependent, the flux of one component also depends upon the concentration gradients in all other components, leading to the possibility of "uphill" diffusion—i.e., migration of a component *up* a gradient in its concentration. Obtaining D_{ij}s for a complex system requires precise and accurate analysis of diffusion profiles for all components followed by rigorous statistical treatment of data; no such analysis has been attempted for magmas or even simple analog melts containing dissolved volatiles. For in-depth discussion of multicomponent diffusion, with applications to simple melts, the reader is referred to Zhang (1993), Richter (1993), Kress and Ghiorso (1993; 1994), Lesher (1993) and Trial and Spera (1994).

Temperature and pressure dependence of diffusion

Effect of temperature. From the glimpse into the atomistics of diffusion provided earlier in this chapter, it is expected that diffusion is a temperature-dependent process. Empirically, D depends on temperature according to

$$D = D_o exp(-Q/RT) \tag{13}$$

where D_o and Q are temperature-independent constants characteristic of the diffusant and the medium, and R is the gas constant. The so-called activation energy for diffusion, Q, can be readily obtained from an Arrhenius plot (*ln*D vs. 1/T), on which the slope is -Q/R. There are examples in the literature involving diffusion in crystalline materials where the expected linear relation between *ln*D and 1/T fails (usually at low temperature), but no such cases have been documented for diffusion in magmatic liquids. The possibility may exist, in the case of dissolved volatiles whose speciation in the melt depends upon temperature, for apparent deviations from Arrhenius behavior. If a composite diffusivity is measured by a technique that does not distinguish different dissolved species (e.g., molecular CO_2 or CO_3^{2-}), the measured value conceivably could reflect changes in speciation with temperature. This topic will surface later in reference to water diffusion in melts.

Pressure dependence of D. Theories addressing the pressure dependence of D are well developed for simple crystals and van der Waals liquids (see Watson, 1979a for some classic references). In general, it has been shown that

$$D_{T,P} = D_T exp(-PV_a/RT) \qquad (14)$$

where $D_{T,P}$ is the diffusivity at some temperature and pressure, D_T is the "zero-pressure" diffusivity at the same temperature (given by Eqn. 13) and V_a is a constant characteristic of the diffusant and the medium, having the units of volume. The atomistic significance of V_a (called the activation volume) depends upon the operative diffusion mechanism. For a vacancy mechanism, it has been equated with the sum of the partial molar volumes of vacancies and activated complexes (see Shewmon, 1989). The significance of V_a in complex melts is open to debate. For alkali diffusion in alkali silicate *glasses*, it has been shown to approach the molar volume of the diffusant (which implies that the effects of pressure on diffusion will be greatest for large ions; see Hamann, 1965). Existing interpretations of the activation volume imply that it will be positive in sign, which is consistent with the intuitively-reasonable prediction that compressing a material should restrict interatomic space (perhaps "squeezing out" vacancies) and thus slow diffusion. There do exist cases involving diffusion in molten silicates where V_a is close to zero (e.g., Watson and Bender, 1980; Dingwell and Scarfe, 1984) or demonstrably negative (e.g., Baker, 1990). A reasonable interpretation of a negative activation volume is that the melt structure itself—and also, perhaps, the diffusion mechanism—is pressure dependent. Baker (1990) hypothesized the formation of transient, high-coordination complexes (e.g., Si^V or Si^{VI}) during the diffusive jump as an alternative explanation of the negative V_a. As we shall see, data concerning pressure effects on diffusion in volatile-bearing magmas are relatively sparse.

EXPERIMENTAL AND ANALYTICAL APPROACHES

General considerations

There are basically two constraints upon the design of experiments involving diffusion in volatile-bearing magmas: (1) usually (though not always) the experiments must be done at high pressures, because most volatiles of petrologic interest are either too insoluble in melts or too fugitive at atmospheric pressure; and (2) the experiments must produce diffusion gradients that can be accurately characterized and readily manipulated mathematically to extract a diffusivity.

High-pressure diffusion experiments on liquids presented new problems to experimentalists because of the need to worry about not only the usual petrologic variables (P, T, and possibly f_{O_2}), but also sample geometry (which must remain constant over the course of an experiment). The first experiments on diffusion in volatile-bearing magmas (Jambon et al., 1978; Watson, 1979b) were performed in fluid-medium pressure devices, which have the advantage of maintaining hydrostatic pressure throughout a diffusion anneal, thus minimizing sample deformation. In most laboratories, however, pressures achievable in fluid-medium pressure devices are limited to less than about 0.6 GPa, so for higher-pressure studies, solid-media (piston-cylinder) techniques were eventually developed. Strategies to combat the sample-deforming tendency of the piston-cylinder have included surrounding the sample with a weak solid incapable of supporting shear stresses (Watson, 1979a), or using the opposite approach of encapsulating the melt in a container strong enough to resist deformation during pressurization (Watson, 1991).

In most cases, diffusion in melts is sufficiently fast to generate diffusion gradients on the order of at least 100 μm (sometimes much more) in a practical laboratory time frame of hours to weeks. This means that the spatial resolution of the analysis method rarely needs to be better than ~10 μm, which opens up the possibility of applying a wide range of analytical techniques, including: Fourier transform infrared spectroscopy (FTIR; for H_2O and CO_2); serial sectioning for radiotracer counting; ß-track mapping; and electron-, proton-, *x*-ray, or ion-microprobe (in traversing mode).

The thin-source geometry

Experimental technique. A technique popular for its simplicity is that involving the deposition of a thin, planar layer of diffusant, usually a radiotracer, on one end of the experimental sample. In typical studies of diffusion in volatile-bearing melts, a prefabricated glass cylinder is polished at one end and the radiotracer added as an aqueous solution (subsequently dried off) or by vapor deposition (see Fig. 2). Prevention of the escape of volatiles during the diffusion anneal necessitates an impermeable container, usually a noble metal tube. Rather than retrofit a container to a glass specimen, it may be more convenient to draw a melt into a platinum tube at atmospheric conditions and slice up the resulting glass-filled tube into lengths suitable for high-pressure diffusion experiments (e.g., Hofmann and Magaritz, 1977; Watson, 1979a; Watson et al., 1982). Alternatively, some workers have found it easier to pre-fabricate glass specimens at high pressure, especially if initially uniform dissolved volatile contents are desired (e.g., Jambon et al., 1978; Watson, 1979b; Watson, 1991; Watson et al., 1993). The pre-synthesis step is followed by polishing the quenched glass sample at one end, addition of tracer, and returning the sample to the high-pressure apparatus for the diffusion anneal.

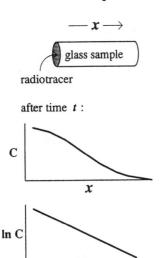

Figure 2. Schematic representation of a typical thin-source experiment. A cylindrical sample is depicted because melts generally are contained in noble metal tubes (not shown), with the tracer introduced at one end. Annealing for time *t* produces a gradient in radiotracer activity (*c*) described by Equation (15), which is linearized by plotting ln*c* vs. x^2 (Eqn. 16).

diffusion couple

at $t = 0$:

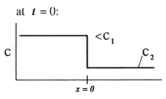

after time t:

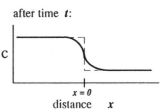

Figure 3. Schematic representation of a diffusion-couple experiment. See text for discussion and Equation (17) for the mathematical description of c vs. x for a given annealing time.

Extraction of diffusivity. The unique aspect of the thin-source boundary condition is that the diffusant tracer introduced on one surface of the specimen dissolves completely at the start of the diffusion anneal. With increasing time, the tracer penetrates into the specimen, so the concentration at the surface drops. The solution to the nonsteady-state diffusion equation for the circumstance just described, given infinite x in the direction of the diffusive flux, is

$$c_{x,t} = \frac{M}{(\pi Dt)^{1/2}} \exp(-x^2/4Dt) \tag{15}$$

where $c_{x,t}$ is the concentration of the tracer at time t and some distance x into the sample. M is the total amount of tracer initially deposited on the surface (at $x = 0$), and D is the diffusivity (see Crank, 1975). In practice, knowledge of M is generally unnecessary for calculating D. Because the pre-exponential quantity, $M/(\pi Dt)^{1/2}$, is a constant for any given experiment, Equation (16) can be re-written

$$\ln c_{x,t} = -x^2/(4Dt) + \text{const} \tag{16}$$

so a simple plot of $\ln c_{x,t}$ vs. x^2 should yield a straight line having a slope of $-(4Dt)^{-1}$, making extraction of D straightforward. Use of the thin-source solution assumes, of course, that D is independent of concentration, which is reasonable in tracer experiments (see previous section of "kinds" of diffusion).

The diffusion couple and related approaches

Experimental configuration. In the context of diffusion studies on volatile-bearing magmas, diffusion couples have been popular mainly (but not exclusively) in the determination of cation diffusivities in melts containing dissolved water. In this case, the strategy is to juxtapose, along a planar interface, two melt reservoirs having different concentrations of the element(s) of interest (see Fig. 3). As with thin-source experiments, cylindrical samples are generally used for convenience in fabrication and encapsulation. Two cylinders are polished at one end, the polished faces placed in contact to form a diffusion couple, and the assembly annealed at the conditions of interest, causing a diffusive flux across the original interface.

Extraction of diffusivity. Given a composition-independent D and a diffusion couple that is effectively infinite in length, the flux resulting from the configuration described above leads to a distribution of diffusant described by (Crank, 1975)

$$c_{x,t} - c_2 = 0.5(c_1 - c_2)\mathrm{erfc}\{x/(4Dt)^{1/2}\} \qquad (17)$$

where x is the distance from the original interface at $x = 0$, c_1 is the initial concentration at negative x, c_2 is the initial concentration at positive x, and erfc denotes the error function complement. The diffusivity is extracted either from a nonlinear fit of $c_{x,t}$ vs. x data to Equation (17) or by inverting the data through the error function in a plot of x vs. $\mathrm{erfc}^{-1}\{2(c_{x,t}-c_2)/(c_1-c_2)\}$; in the ideal case, this plot is linear, with slope $(4Dt)^{-1/2}$. Instances of a composition-dependent D are revealed by a non-linear plot, which might necessitate an alternative treatment to that represented by Equation (17) (see earlier section dealing with chemical diffusion).

A useful characteristic of the "diffusion-couple" solution discussed above is that the concentration at $x = 0$ is constant [at $c_2 + 1/2(c_1 - c_2)$] for all $t > 0$, so essentially the same solution to the nonsteady-state diffusion equation applies to cases in which diffusion occurs from a surface held at constant concentration into a medium that is infinite in just one direction (i.e., semi-infinite). The relation describing concentration as a function of distance for this "constant surface" boundary condition is

$$(c_{x,t} - c_0)/(c_i - c_0) = \mathrm{erf}\{x/(4Dt)^{1/2}\} \qquad (18)$$

where c_i is the initial concentration of diffusant in the sample of interest and c_0 is the surface concentration. This solution has been implemented for diffusion in volatile-bearing melts in several ways, including: (1) exposing an initially volatile-free specimen to an effectively infinite reservoir of fluid, thus generating uptake profiles in which the surface concentration of the dissolved volatile is equal to its solubility (and therefore constant over time); (2) annealing samples initially containing dissolved volatiles at conditions conducive to diffusive *loss*; and (3) partially dissolving minerals into volatile-bearing melts to evaluate the effect of dissolved volatiles on diffusion of the components of the dissolving mineral.

In the last case above, the diffusant concentration is held constant at the surface of the dissolving mineral ($x = 0$) as dictated by the solubility of that mineral in the melt (see Fig. 4). The process of mineral dissolution actually results in a *moving* mineral/melt interface, thus invalidating Equation (18) as a rigorous description of $c_{x,t}$ vs. x in the melt. For highly insoluble minerals (e.g., apatite and zircon in silicic melts), the interface displacements involved in dissolution may have an insignificant effect on the diffusion gradient in the contacting melt, so the solution above is still serviceable for extracting diffusivities. (Strictly speaking, the volatile-uptake and -loss approaches also involve a moving interface if a change in volume accompanies dissolution or volatilization, but such effects are likely to be negligibly small). In the context of moving boundaries, it should be noted that analytical solutions to the nonsteady-state diffusion equation do exist for some moving-boundary situations (see Crank, 1975), and almost any degree of complication can be modeled numerically using finite-difference techniques. In most situations, of course, it is the objective of the experimenter to design his or her experiments around the simplest possible data acquisition and reduction, keeping the number of variables to a minimum.

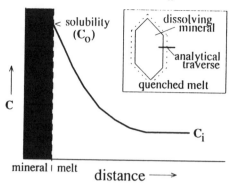

Figure 4. Schematic representation of chemical diffusion in a melt resulting from partial dissolution of a simple mineral. See text and Equation (18).

Diffusion in more than one dimension

For certain problems in volatile diffusion, it is convenient to use what are best described as bulk-gain or bulk-loss experiments—i.e., ones in which a volatile component diffuses into or out of a sample over its entire surface area. Prior to the advent of in situ analytical techniques, this approach was necessary (as in Shaw's pioneering work on H_2O diffusion in granitic melt) because, even though concentration profiles could not be determined, bulk loss or gain of a component could be gauged gravimetrically. Appropriate solutions to Equation (9) exist for dealing with situations in which diffusion occurs, for example, radial to the axis of a cylinder or the center of a sphere, and the only quantity that can be measured is the total amount of material diffused in or out (see Crank, 1975, pp. 90*ff*). In general, the diffusivity is assumed to be independent of concentration in applying such solutions, but characterization of the shape of sorption (or desorption) vs. (time)$^{1/2}$ curves can permit an assessment of the existence of any concentration dependence (e.g., Shaw, 1974; Jambon et al., 1992). Depending upon the uncertainty in measuring gain or loss of volatiles, and also upon the degree to which the sample approaches an ideal shape (sphere, cylinder, etc.), D values resulting from bulk sorption or desorption can have relatively high uncertainties.

Even if an in situ analytical technique is available for a component of interest, it may still be convenient to use a spherical sample geometry—e.g., a melt droplet in contact with an atmosphere that serves as a source or sink for the volatile of interest. Such a case involves radial diffusion, and the resulting concentration profiles differ from those caused by diffusion in one dimension only. The experimenter is faced with the choice of fitting measured profiles to a slightly more complex equation for spherical diffusion (Crank, 1975, p. 91), or ignoring the radial aspect and invoking Equation (18). The latter approach is perfectly acceptable when the total length of the diffusion profile is a small fraction of the sample radius.

DIFFUSION OF DISSOLVED VOLATILES IN MAGMATIC LIQUIDS

Introduction

The strategy in this section is to review, in chronological order of publication (as far as is practical), experimental studies of dissolved volatile diffusion in magmas. The various magmatic volatiles are discussed independently, more or less in order of abundance, beginning with water and progressing to CO_2, sulfur species, halogens, and eventually to noble gases. An attempt is made to present enough details of the experimental techniques used in each study for the reader to appreciate the general quality and range of applicability of the data. For clarity in presentation and ease of interstudy comparison, available data for a given volatile are presented on a single Arrhenius diagram, omitting actual data points in most cases. A comparative summary diagram of available data on diffusion of all dissolved volatiles appears at the end of the section (Fig. 13, below). For ease in comparison among various diffusants, Arrhenius equations taken from the numerous sources of data are standardized to SI units, and lines appearing on the summary figures are keyed to equations in the text. Many readers are probably more familiar with the now unfashionable units of calories and centimeters, so the following conversion table may be helpful.

units used in this chapter		conversion factors		
parameter	*units*	*to convert from*	*to*	*multiply by*
D	m^2/s	cm^2/s	m^2/s	10^{-4}
D_o	m^2/s			
Q	J/mole	calories	joules	4.184
R	J/deg·mole (8.3143)			

Water and related species

Silicic melt compositions. As noted previously, the first study of H_2O diffusion in a magmatic liquid was that of Shaw (1974), who investigated the rate of H_2O uptake by cylinders of granitic obsidian. Although much more is known today about the details and mechanisms of dissolved H_2O transport in melts, Shaw brought to light the general systematics of the diffusion process, and his data and interpretations, as well as the applications he discussed, still stand as fundamental contributions.

Shaw's bulk-uptake approach was simple and appropriate to the general unavailability (at the time) of in situ analytical techniques for hydrogen and/or oxygen. Cylinders of granitic obsidian (Valles Caldera, NM, USA; 77% SiO_2) were milled to ~2.5 mm in diameter and sealed in gold capsules with an amount of water just sufficient for saturation of the obsidian at run conditions. Using cold-seal pressure vessels, the capsules were subjected to water pressures of 10 to 200 MPa at temperatures of 750° to 850°C for durations of 1 to 150 hours. At any given pressure and temperature, water diffused into the obsidian cylinders to an extent determined by the run duration. Upon quenching, the total water uptake was gauged both by the weight gained during the high P-T hydration and the weight lost upon subsequent complete dehydration at 1000°C in air.

Shaw recognized the possibility that water diffusivity would depend upon water content, but his bulk-uptake technique limited his ability to quantitatively determine D_{water}[#] as a function of X_{H_2O}. By inspection of the form of weight-gain vs. hydration-time curves for specific conditions, and by analogy with concentration-dependent diffusion in other materials, he was able to deduce the general sense of D_{water} variation with concentration, which is reproduced in Figure 5 in comparison with results from later studies. The diffusivity ranges from ~10^{-13} m^2/s at vanishingly small H_2O content to ~10^{-11} m^2/s at 6 wt % dissolved H_2O. Shaw's best estimate of the activation energy, Q, for diffusion was ~60±20 kJ/mole (see lines 19 and 20 in Fig. 6), which, he noted, compared favorably with previously-determined values for hydrogen and water diffusion in silica

---

[#] The subscript "water" is used to denote an apparent diffusivity that incorporates the migration and interaction of all water-related species. The importance of this distinction will become clear later.

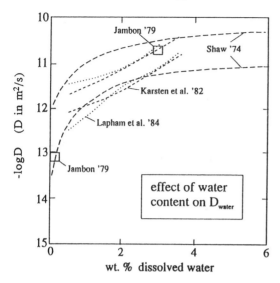

Figure 5. Dependence of apparent total water diffusivity (D_{water}) upon H_2O concentration, according to several research groups. All information applies to 800°C except for the data of Jambon (1979), which pertain to 850°C. Note that Jambon's value plotted at 3% water actually represents melts containing 1 to 5% water—an average water content is shown here for clarity. Note also that his value at ~0.1% water is shown as corrected by the later work of Jambon et al. (1992). The dashed or dotted lines show envelopes of uncertainty proposed by the authors of each study, except for those of Karsten et al. (1982), which were estimated by Lapham et al. (1984).

Figure 6. Summary of data on diffusion of water-related species in melts and glasses. The numbers associated with specific Arrhenius lines refer to numbered equations in the text; other data are identified in the legend. The indicated percentages denote water content in wt %. Most information pertains to granitic compositions except for the lines labeled 24 and 25, which refer to hydrogen diffusion in dry albite glass and total water diffusion in basalt melt, respectively. Line 23 represents diffusion specifically of molecular H_2O; all other lines depict total water diffusivities (except #24). See text for details and discussion.

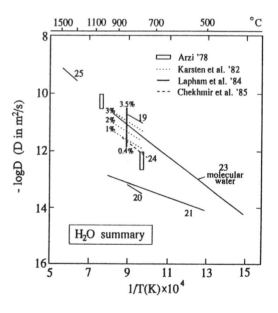

glass. Arrhenius equations for D_{water} at high and low water contents are given approximately by:

$$D_{water} \sim 1.8 \cdot 10^{-8} \exp(-65,000/RT) \qquad \textit{(6 wt \% H}_2\textit{O)} \qquad (19)$$

$$D_{water} \sim 1.6 \cdot 10^{-10} \exp(-74,000/RT) \qquad \textit{(nominally dry)} \qquad (20)$$

where the diffusivity and pre-exponential term are in m²/s and R is 8.3143 J/deg·mole (the activation energy therefore being in J/mole).

An interesting aspect of Shaw's study, and one that is shared with most subsequent investigations of diffusion in H_2O-bearing granitic compositions, is that the diffusion medium varied from *glass* containing a small amount of dissolved water to hydrous granitic *melt*. At 750°- 850°C and 200 MPa, nominally dry obsidian is metastable (possibly even below the glass transition temperature, T_g), whereas water-saturated obsidian is a stable, superliquidus melt. Perhaps the change in D_{water} with increasing water content (Fig. 5) occurs as the diffusion medium varies from a rigid glass to a low-viscosity melt. Clearly, this is a circumstance in which the diffusant of interest has a profound effect upon the stability and physical properties of the diffusion medium. Indeed, it seems to have been Shaw's own earlier work on obsidian-H_2O viscosities (Shaw, 1963) that stimulated his interest in diffusion.

Shortly after the Shaw's paper appeared, a study of diffusion in obsidian at low temperatures (~100° to 250°C) was published by Friedman and Long (1976). Because this review concerns diffusion specifically in magmas, this study of glasses is mentioned only in passing; it is worth noting, however, that the up-temperature extension of the hydration-rate data acquired by Friedman and Long passes in the general vicinity of Shaw's inferred values for diffusion in hydrous obsidian melt. As we shall see, this coincidence presaged a later conclusion: diffusion of water in amorphous silicates is insensitive to the rigidity of the medium.

Following that of Shaw, the next study to yield data on water diffusion in magmas was that of Arzi (1978), who examined the melting kinetics of water-bearing rocks, with emphasis on the Westerly granite. In an internally-heated, gas-medium pressure vessel, Arzi exposed large (~1-cm) samples of natural rock to water pressures of 0.2 GPa and temperatures up to 1060°C; this treatment resulted in the migration of a melting front from the exposed surface of the sample inward. He hypothesized that the migration rate of this front was rate-limited by diffusion of H_2O through the melt, and on this basis deduced information on H_2O diffusion by measuring melt-front position as a function of time in a series of five experiments at ~1020°C (actually 980° to 1060°C). Arzi also applied his technique for extracting D_{water} to an earlier set of experimental data on melt-front penetration in granulite at 760° and 0.2 GPa (Mehnert et al., 1973). Combined with his own value for 1020°, the resulting lower-temperature diffusivity enabled him to estimate the activation energy for H_2O diffusion at ~140 kJ/mole. Although his data are qualitatively similar to those of Shaw (1974; see Fig. 6), it seems clear, in the light of more recent data, that Arzi's indirectly calculated activation energy is almost certainly too high. Moreover, his additional conclusion that D_{water} is independent of H_2O concentration has been called into question by later work (see below), in which Shaw's strongly concentration-dependent D is confirmed.

Jambon (1979) performed dehydration and hydration experiments on a natural obsidian (Valles Caldera, NM, U.S.A.; 76.9% SiO_2) to determine D_{water} at low (<0.4 wt %) and high (1 to 5 wt %) water concentrations, respectively. The dehydration runs involved simple heating of obsidian slabs in air at 500° to 980°C, and the diffusivities were computed from bulk weight-loss measurements. The hydration runs, like those of Shaw (1974), were made in cold-seal pressure vessels (0.1 GPa; 800° to 900°C), but Jambon was able to characterize actual diffusive uptake profiles (rather than just bulk uptake), by making weight-loss measurements on microsectioned wafers. Another innovation in Jambon's study was the addition of obsidian powder to the water reservoir in the hydration experiments. Rapid cation exchange between the water and the glass powder established chemical equilibrium between the reservoir and the diffusion sample for all components except H_2O. The results of the study reveal a 3-order-of-magnitude difference in D_{water}

between obsidians with low and high water contents. The "water rich" values resemble those of Shaw (~10^{-11} m²/s; see Fig. 5), and the "water-poor" data define an Arrhenius line giving lower diffusivities:

$$D_{water} = 1.3 \cdot 10^{-11} \exp(-47,100/RT)^*$$ (21)

The activation energy for H_2O diffusion in the water-poor samples (~47 kJ/mole) is within the overall range of 60±20 kJ/mole suggested by Shaw (cf. line 21 with lines 19 and 20 in Fig. 6).

Delaney and Karsten (1981) were the first to employ a truly in situ analytical technique to determine D_{water}. They used an ion microprobe to characterize the concentration gradients in H_2O across axial sections of some of Shaw's original obsidian samples that had been partially hydrated at 850°C and 70 MPa water pressure. (The ion microprobe actually measures H concentration in the glass, which was assumed to represent dissolved H_2O.) Armed with "H_2O" concentration vs. distance curves, Delaney and Karsten could calculate D_{water} more precisely than Shaw was able to do from his bulk weight-gain data. The shapes of the uptake profiles necessitated a concentration-dependent diffusivity to achieve good fits; in the final analysis, however, Shaw's original estimates for D_{water}—and even its dependence upon H_2O concentration—held up remarkably well (see Figs. 5 and 6). Delaney and Karsten concluded that D_{water} increases exponentially with H_2O content in the 0.2 to 3.7 wt % range represented in their study.

Karsten et al. (1982) later extended the work just described to cover a 300° range in temperature (650° to 950°C), with the aim of accurately determining the activation energy for water diffusion. They ran a new series of experiments on the same obsidian originally used by Shaw, but modified the experimental technique so that H_2O diffused only into the exposed, polished ends of 2.5 mm-diameter, platinum-jacketed cylinders encapsulated in $Ag_{70}Pd_{30}$ tubes. The diffusion anneals were done in an argon-medium, internally-heated pressure vessel at 70 MPa pressure, f_{O_2} between FMQ and NNO, for durations of 1 to 6 hours. The resulting diffusivities confirmed both the marked dependence of D_{water} upon water content (see Fig. 5) and the activation energy for diffusion originally suggested by Shaw (1974): at a constant water content of 2 wt %, the diffusivity ranged from ~10^{-12} m²/s at 650°C to ~10^{-11}m²/s at 950°C, which translates into an activation energy of ~80 kJ/mole (see Fig. 6).

The first study aimed at determining the effect of pressure on diffusion of water in a geologic melt was that of Lapham et al. (1984). In preference to the hydration technique used in earlier studies, these workers chose a diffusion-couple configuration for their experiments so they could vary pressure without also causing major changes in the concentration of dissolved water (water solubility in melts is strongly pressure-dependent; see Chapter 3, 4, 6). In a gas-medium pressure vessel, they prepared uniformly hydrated cylinders (~3.7 wt % H_2O) of the same obsidian originally used by Shaw, and subsequently juxtaposed these with nominally dry cylinders of the natural obsidian. The couples were run at 850°C and pressures of 70 to 500 MPa for 3 or 6 hours. The resulting "H_2O" diffusion profiles were characterized by ion microprobe analysis for H, and diffusivities were extracted using Boltzmann-Matano analysis (see Eqn. 10). Concerning the main effect they were investigating—that of pressure—Lapham et al. concluded that it is too small to see within the uncertainty of their data, but placed the maximum allowable activation volume at ~4 cm³/mole. Their overall data set is in excellent agreement with earlier results (see Fig. 6), with D_{water} again showing a strong dependence upon

---

* This is a corrected version of the relation originally published by Jambon (1979), which follows from a more accurate estimate of the initial water content of the obsidian.

concentration (Fig. 5). A useful additional piece of information provided in the study is that the alkali concentrations remain constant across the diffusion couples even in the face of extensive water diffusion, thus eliminating the possibility that the diffusion process involves exchange of H^+ with Na^+ and/or K^+ (and perhaps suggesting that water diffuses as a neutral species...?).

Lapham et al. also included in their study some experiments involving D_2O diffusion. These were done using the techniques employed earlier by Karsten et al. (1982), except that, because the source reservoir was 99.9% D_2O, gold rather than AgPd was used for the outer container (gold is less permeable to H_2 and D_2). Deuterium-uptake profiles were characterized with the ion microprobe. The systematics of D_2O diffusion were found to be very similar to those for water, the Arrhenius parameters being the same as those determined by Karsten et al. (1982) except for the pre-exponential constant, D_0, which was a factor of two lower for D_2O. The pilot study of Lapham et al. (1984) on D_2O diffusion was followed by additional data from the ASU group (Stanton et al., 1985) obtained from experiments on the same obsidian using a similar D_2O-uptake approach. The main improvement seems to have been in analytical procedure: Stanton et al. obtained previously unavailable standards of known D_2O content for use in the ion microprobe analysis. The resulting diffusivities differed somewhat from the preliminary results of Lapham et al., but were indistinguishable from the earlier D_{water} values obtained by Karsten et al. (1982). The coincidence in diffusion behavior of water and heavy water was cited by Stanton et al. as a demonstration that water diffuses as something other than free protons or H_2 molecules [the factor-of-two difference in mass between H and D would have resulted in a significantly lower pre-exponential term (by a factor of $\sqrt{2}$) for diffusion of D_2O relative to H_2O if protons and deuterons (or H_2 and D_2) were the diffusing species].

Several years elapsed before further advances were made in our understanding of water diffusion in magmas. The crucial foundation for further progress was laid by Stolper and co-workers in the early 1980s, who investigated the speciation of dissolved water in a variety of silicate melts and glasses using infrared (IR) spectroscopy. Briefly summarized, the principal conclusion of these studies was that H_2O dissolves as both hydroxyl groups (OH^-) and molecular water, the former dominating at low total H_2O content (~0 to 2 wt %), the latter becoming increasing important at higher concentrations (see, e.g., Stolper 1982; see also Chapters 3, 4, 6). Recognizing the potential significance of speciation to diffusion, and applying a mathematical formulation for diffusion of a multi-species component, Zhang et al. (1991) published a definitive study of water diffusion in rhyolitic glass. They worked with polished slabs of natural glass from Mono Craters, California, initially containing 0.2 to 1.7 wt % H_2O. These slabs were partially dehydrated by heating in N_2 at 403° to 550°C, and water-loss profiles were characterized by FTIR spectroscopy in terms of both OH^- and molecular H_2O. The resulting profiles were used to model diffusion governed by

$$\frac{\partial[\text{water}]}{\partial t} = \frac{\partial}{\partial x}\Big[D_{H_2O}\frac{\partial[H_2O]}{\partial x} + D_{OH}\frac{\partial[OH]/2}{\partial x}\Big] \qquad (22)$$

which the reader will recognize as the one-dimensional, nonsteady-state diffusion equation (cf. Eqn. 6) describing the change in concentration of "bulk" water at some point in the sample as a consequence of the mobilities of *two* dissolved species (OH^- and H_2O). Zhang et al. (1991) were able to show that at water contents exceeding ~1 wt % (depending somewhat on temperature and whether the experiment involved hydration or dehydration), speciation equilibrium was achieved at all points along the H_2O diffusion gradients. They used both a forward finite difference model and inversion of the measured profiles (a Boltzmann-Matano type of analysis) to show convincingly that D_{OH^-}/D_{H2O} is close to zero.

In other words, diffusion of dissolved water occurs by migration of the molecular species, the hydroxyls being effectively immobile. This means that the concentration of OH^- groups along a diffusion gradient is not attributable to diffusion but to local conversion from H_2O molecules—a finding consistent with earlier models for simple glasses (e.g., Doremus, 1969). An important conclusion of the Zhang et al. study is that earlier data on water diffusion in obsidian (see above) pertain to an apparent diffusivity that results from rapid diffusion of H_2O and local $H_2O + O = 2\ OH^-$ equilibrium (they refer to this diffusivity as the "apparent total water diffusivity"). Using calculated equilibrium values for $[H_2O]/[OH^-]$, Zhang et. al. were able to extract D_{H_2O} from the published apparent values and show that a down-temperature extrapolation of previous data (i.e., those of Karsten et al., 1982 and Lapham et al., 1984) is entirely consistent with the new lower-temperature data (see Fig. 6). Interestingly, as noted by Zhang et al., this means that the transition from hydrous *melt* at ~950°C to water-bearing *glass* at 400°C is transparent to diffusion of water. The Arrhenius relation proposed by Zhang et al. to describe diffusion of molecular water in rhyolitic melts or glasses containing 1 to 3 wt % H_2O is

$$D_{H_2O} = 4.6^{+18}_{-3.3} \cdot 10^{-7} \exp(-103,000 \pm 5,000)/RT \qquad (23)$$

where the units are as indicated for Equations (19) and (20). The authors noted that the activation energy for diffusion of H_2O molecules falls on a monotonic trend of Q vs. atomic (molecular) radius of neutral species (see Fig. 11, below).

In a re-visitation of the earlier work by Jambon (1979), Jambon et al. (1992) extended the data base on D_{H_2O} in obsidian to low, natural water contents of ~0.114 wt % [this concentration represents a re-determination of the 0.38% originally used by Jambon (1979)]. As noted previously, the experiments consisted of simple dehydration of wafers of obsidian from Valles Caldera (76.9 wt % SiO_2). Because the 1979 run products were lost, the 1992 re-examination involved no new analytical determinations. Rather, the original, bulk-loss data were re-interpreted using the newly-determined initial water content (0.114 wt %) and the "mobile molecular water" model described above. The results were found to be consistent with extrapolation of existing data for wetter samples.

A final paper concerning transport of water-related species in silicic melts is that of Chekhmir et al. (1985) on hydrogen (H_2) diffusion in dry albitic glass ($NaAlSi_3O_8$) and water-saturated albitic melt. This study is discussed outside of the chronological progression just completed because it addresses geologically unusual circumstances of extremely reducing conditions.

The authors used a technique quite different from those implemented in the studies described above: basically, they used redox color indicators to estimate the extent of hydrogen penetration into their experimental samples. Albite glasses were prepared in a methane-oxygen flame, which resulted in dissolved CO_2 concentrations of ~0.05 wt %. Monitoring diffusion of hydrogen is made possible by the fact that dissolved CO_2 imparts no color to the specimens, but C produced by reaction with diffusing hydrogen causes the sample to turn grey. Diffusion experiments were performed by placing the pre-melted glass specimens in Pt containers open to an atmosphere of pure H_2 at a pressure of 180 MPa. The extent of hydrogen penetration was assessed by measuring the thickness of the darkened outer zone; the diffusivity could be calculated directly from this measurement because the concentration of CO_2 was not high enough to consume a significant amount of the inward diffusing hydrogen. Over the temperature range 750° to 880°C, Chekhmir et al. obtained the following Arrhenius relation for hydrogen diffusion in (initially) dry albite glass

$$D_{H_2} = 3.7 \cdot 10^{-7} \exp(-113,000/RT) \qquad (24)$$

which has been included on Figure 6 for comparison with water diffusion systematics in obsidian.

Chekhmir et al. (1985) used a similar technique for additional experiments on hydrous, molten albite. In this case, the starting specimens for the diffusion experiments were prepared at P_{H_2O} = 50 MPa, T = 1000°C (giving X_{H_2O} ~ 1.7 wt %) in a gas-medium pressure vessel. Included in the albite was 7 wt % Mn_2O_7 to serve as a hydrogen trap and redox indicator. The hydrogen diffusion experiments were performed by returning the Mn oxide-doped glasses to the pressure vessel in open Pt tubes, so at run conditions they were exposed to a buffered atmosphere with P_{H_2} ~ 500 kPa. Inward-diffusing hydrogen reduced the Mn to the +2 state, causing a color change and allowing the depth of H_2 penetration to be determined. Taking into account the consumption of mobile hydrogen by the reduction reaction, the authors estimated D_{H_2} in hydrous albite melt at $8 \cdot 10^{-8}$ m²/s, which is much higher than any diffusivity in a silicate melt measured before or since. Assuming no melt-composition effects on hydrogen transport and extrapolating to basalt liquidus temperatures, Chekhmir et al. suggested a value of $1 \cdot 10^{-6}$ m²/s for D_{H_2} in molten basalt, which is comparable with diffusion of heat.

Basaltic melt. The only published data on diffusion of water-related species in basaltic melt are those of Zhang and Stolper (1991). Their starting material was a basaltic glass from the Juan de Fuca Ridge containing 50.6 wt % SiO_2 and having a natural water content of ~0.4 wt %. Pieces of this glass were dehydrated at 1 atmosphere pressure, ground into disks, and juxtaposed in diffusion couples against similar disks of pristine, water-bearing glass. The couples were run in graphite-lined Pt containers at 1 GPa and 1300° to 1500°C using a piston-cylinder apparatus. Upon quenching and depressurization, the samples were cut into wafers and the concentration profiles of OH^- groups and molecular H_2O were determined by FTIR spectroscopy, stepping across the region perpendicular to the interface of each couple. Diffusivities were determined by Boltzmann-Matano analysis and by modeling the gradients as resulting from diffusion of molecular H_2O only, with the additional assumption that local equilibrium is maintained between H_2O and OH^- groups. As in the case of the much more silicic melts discussed earlier, the apparent total water diffusivity varies with water content, and is significantly higher than the up-temperature extrapolation from previous data for more silicic melts (see Fig. 6). The governing Arrhenius equation is:

$$D_{water} = 3.8^{+35}_{-3.4} \cdot 10^{-6} \exp(-126,000 \pm 32,000/RT) \qquad (25)$$

Carbon dioxide

The first study of dissolved CO_2 diffusion in silicate melts was that of Watson et al. (1982), who acquired data for two melts: (1) a simple, sodium aluminosilicate melt (60 wt % SiO_2), for which diffusion was characterized over a broad temperature-pressure range (800° to 1500°C; 0.05 to 1.8 GPa); and (2) an iron-free, haplobasaltic melt (~53 wt % SiO_2), investigated at 1.5 GPa and 1350° to 1500°C. All experiments were made using a thin-source, radiotracer technique in which ^{14}C (in the form of sodium carbonate) was deposited on one end of cylindrical glass samples. Runs above 0.2 GPa were made in a piston-cylinder apparatus; those at 0.2 GPa and below were made in either cold-seal pressure vessels or an internally-heated, argon-medium apparatus.

The diffusion systematics for the two melts investigated by Watson et al. are very similar—indistinguishable, in fact, because of the relatively large uncertainty in the haplobasalt Arrhenius parameters resulting from the small temperature range over which

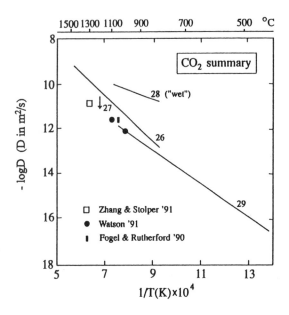

Figure 7. Summary of data on diffusion of dissolved CO_2 in various melts and glasses. The numbers identify equations in the text. Lines 28 and 29, along with the data of both Watson (1991) and Fogel and Rutherford (1990), are for granitic compositions. Lines 26 and 27 refer to a sodium aluminosilicate melt (60% SiO_2), and the datum of Zhang and Stolper (1991) is for molten basalt. See text for details and discussion.

that melt could be studied without encountering the liquidus. The temperature dependence of CO_2 diffusion in the sodium aluminosilicate melt at 50 MPa is well constrained by:

$$D_{CO_2} = 3.5 \cdot 10^{-4} \exp(-195,000/RT) \tag{26}$$

(see Fig. 7), and the pressure dependence—characterized at 1200°C only— is given by:

$$D_{CO_2@P} = 4.2 \cdot 10^{-11} \exp[-P (1.1 \cdot 10^{-5})/RT] \tag{27}$$

where the pre-exponential term is the diffusivity (m²/s) at 1200 and "zero" pressure, and the activation volume (cf. V_a in Eqn. 14) is in m³/mole [these units require that pressure be expressed in dyne/m² (Pa) and R in dyne·m/deg·mole (J/deg·mole)]. The pressure effect amounts to roughly an order-of-magnitude decrease in D_{CO_2} over a 2.5-GPa increase in pressure.

In the course of a general study of CO_2 solubility in a rhyolitic melt, Fogel and Rutherford (1990) obtained the first value for D_{CO_2} in a highly silicic composition (76.5 wt % SiO_2). Their technique involved sealing of polished obsidian slabs in Ag-Pd tubes along with sufficient $Ag_2C_2O_4$ to saturate the melt in CO_2 at run conditions (50 to 660 MPa; 950° to 1150°C). Using FTIR spectroscopy, they characterized CO_2 uptake profiles in some of the samples run at 1050°C, and were able to convert these into diffusivities by assuming a constant surface concentration of CO_2 and implementing a solution for diffusion in a cylindrical sample. The resulting mean diffusivity of $2.4(\pm0.5)\cdot10^{-12}$ m²/s differs by only about a factor of two from the value given by the Arrhenius relation of Watson et al. (1982; see Fig. 7), foretelling an insensitivity of D_{CO_2} to melt composition that would be borne out by future studies. [This insensitivity to large variations in melt composition is now a well-established characteristic of CO_2 diffusion; it is also somewhat enigmatic, because CO_2 dissolves primarily as carbonate anions in basic melts and as CO_2 molecules in polymerized, silicic melts (Fine and Stolper 1985; 1986)]

The topic of CO_2 diffusion was revisited by Watson (1991) with the specific objective of determining diffusivities in more silicic melts and characterizing the effect of dissolved water on CO_2 diffusion. The melts chosen for study were a metaluminous, rhyolitic obsidian from Lake county, Oregon (76.4 wt % SiO_2), and a natural dacite from the Aleutians (63.5 wt % SiO_2). A radiotracer technique similar to that employed earlier by Watson et al. (1982) was used, but the melts were contained in thick-walled Ni capsules lined with Au or Ag-Pd sleeves. All experiments were run at 1 GPa, over a temperature range of 800° to 1100°C. The main findings of the study are that: (1) there is little melt composition dependence of CO_2 diffusion, as indicated by the general similarity among Arrhenius parameters for dry melts of widely varying composition, including not only the sodium aluminosilicate and haplobasaltic compositions studied earlier, but also the new values for granitic and dacitic melts; and (2) dissolved water has a major effect on CO_2 diffusion, causing a three-order-of-magnitude increase in D_{CO_2} as water content is increased from 0 to 8 wt % at 800 (the effect is smaller at higher temperatures). For a granitic melt containing 8 wt % dissolved H_2O, diffusion of dissolved carbonate is described by:

$$D_{CO_2} \text{ (8\% } H_2O) = 6.5 \cdot 10^{-8} \exp(-75,000 \pm 21,000/RT) \tag{28}$$

In general, the effect of dissolved H_2O on CO_2 diffusion is to decrease both the activation energy and the pre-exponential constant (cf. Eqns. 28 and 26), which results in an increase in diffusivity at magmatically realistic temperatures (see Fig. 7 and also later section on the effects of dissolved water on cation diffusion). Interestingly, this pattern is typical of the effects of dissolved water on diffusion of cations, as we will show later in this chapter.

As a "by-product" of their study of water diffusion in basaltic melt, Zhang and Stolper (1991) also obtained one value for diffusion of CO_2 at 1300°C. The volatile-free half of each of their diffusion couples experienced uptake of CO_2 produced by oxidation of the contacting graphite container. The authors were able to characterize one of these uptake profiles by FTIR spectroscopy, and obtained a diffusivity almost identical to that determined earlier by Watson et al. (1982) for their simpler melt compositions. This result was encouraging not only in that it provided additional confirmation of the insensitivity of D_{CO_2} to melt composition, but also because it demonstrated that different labs using different experimental and analytical approaches (ß-track radiography vs. FTIR spectroscopy) could produce results in substantial agreement (see Fig. 7).

Stolper's group at Caltech followed up their initial foray into CO_2 diffusion with an in-depth study of diffusion in a granitic obsidian (Blank et al., 1991; Blank, 1993), using a technique similar to that of Fogel and Rutherford (1990). Cylinders or rectangular prisms of natural, water-poor obsidian were exposed to elevated pressures of CO_2 (50 to 105 MPa) over a broad range in temperature (450° to 1050°C) for periods ranging from ~1 day to 6 weeks. Following these diffusive-uptake experiments, wafers were cut from the samples and polished in preparation for analytical traverses with a focusing FTIR. The measured concentration profiles of dissolved CO_2 (all in molecular form) were inverted through error-function solutions for diffusion in a cylinder or slab to obtain diffusivities. Remarkably, the resulting data fall on a single Arrhenius line over the entire temperature range examined (see Fig. 7):

$$D_{CO_2} = 6.2^{+4.3}_{-2.6} \cdot 10^{-7} \exp(-144,600 \pm 4100/RT) \tag{29}$$

This line applies to CO_2 diffusion in a material of granitic composition spanning the range from solid glass on the low-temperature end to viscous melt at 1050°C, and passes close to the previous data of both Fogel and Rutherford (1990) and Watson (1991). The small range in pressure covered by Blank's experiments precluded calculation of a well-constrained activation volume, but a tentative value of at least 25 cm^3/mole was suggested. If V_a really is that large [in contrast with the value of ~11 cm^3/mole determined earlier by Watson et al. (1982; see above) for a more basic melt], then the coincidence of Blank's "low-pressure" Arrhenius line with the earlier 1-GPa data of Watson (1991) must be fortuitous—an activation volume of 25 cm^3/mole translates into a nearly 10-fold decrease in D_{CO_2} between 0.1 and 1 GPa. Either V_a is smaller than proposed by Blank (1993) or the agreement between data sets is not quite as good as one would conclude at first glance. [A strong dependence of diffusivity upon pressure is by no means unreasonable, and has been observed in studies of volatile-free melts—e.g., ferrous iron in sodium aluminosilicate melts (Dunn and Ratliffe, 1990).]

Sulfur-related species

We now turn from what are generally accepted as the two most abundant magmatic volatiles (H_2O and CO_2) to one that is relatively specific in (volumetric) importance to the late-stage evolution of calcalkaline magmas: i.e., SO_2 and its more reduced counterparts. Diffusion studies of sulfur-related species have been undertaken so recently that the only published data are those found in two abstracts (Watson et al., 1993; Baker and Rutherford, 1994); accordingly, this part of the present *review* is really more of a *preview* of coming attractions.

Like H_2O, sulfur dissolves in magmas in more than one form. In contrast to the case of H_2O, however, sulfur speciation is determined primarily by the oxidation state of the magma, with sulfide dominant at low f_{O_2} (<FMQ, roughly speaking) and sulfate at high f_{O_2} (>2 log f_{O_2} units above FMQ) [Carroll and Rutherford, 1988; see also Chapter 7]. Thus, the transition from sulfide- to sulfate-dominated melts occurs over a relatively small range in f_{O_2} that encompasses typical values for calcalkaline magmas. As we shall see, the studies described below do not answer definitively the question of speciation-dependence of D_{sulfur}, but they do provide a preliminary data set indicative of a relatively small f_{O_2} effect.

Watson et al. (1993) performed three types of experiments on a variety of melt compositions, including natural obsidian (76 wt % SiO_2) and dacite (68 % SiO_2) as well as synthetic (Fe-free) dacite, synthetic low-K andesite (61% SiO_2) and a lunar ultramafic composition (44 % SiO_2). The "lunar" melt was doped with 3000 ppm S and loaded on Pt wire loops for low-pressure, sulfur-loss experiments at 1300°C and log f_{O_2} = IW-0.7. Sulfur loss from the roughly spherical melt droplets (~1 mm dia.) was characterized by electron microprobe traverses on the quenched glasses, and D_{sulfur} was estimated by matching the measured concentration profiles with those predicted for diffusion in a sphere governed by various assumed values of D. Diffusion in the other melts was investigated by techniques more similar to those described earlier in this chapter for other volatiles. Specifically, two types of samples were run in a piston-cylinder apparatus at 1 GPa: (1) diffusion couples consisting of presynthesized, S-doped and S-free halves; and (2) thin-source, radiotracer experiments in which ^{35}S was deposited (as aqueous Na_2SO_4) at the end of prefabricated glass cylinders. Diffusion gradients in the quenched couples were characterized by electron microprobe traverses along axial sections, and chemical diffusivities were extracted by fitting the profiles to Equation (17). Tracer diffusivities were obtained from the thin-source experiments by making ß-track maps of the sectioned specimens, scanning these with a photodensitometer, and fitting the absorption vs. distance curves to Equation (16) (see Watson, 1991). The total range in temperature of the experiments was 1100° to 1500°C, although no single melt has been investigated over so

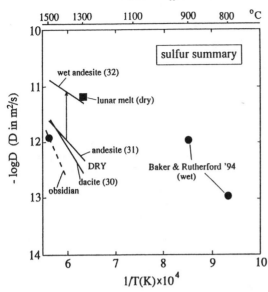

Figure 8. Summary of data on diffusion of sulfur in magmatic melts. The numbers identify Arrhenius equations in the text. Data on the left (high-temperature) side of the diagram are from Watson et al. (1993) and generally represent experiments performed under reducing conditions at 1 GPa ("wet" andesite contains ~5.5 wt % H_2O). The lunar melt was run under reducing conditions at near-atmospheric conditions. The values of Baker and Rutherford (1994) represent oxidizing conditions at 200 MPa water pressure (The values are switched in their published abstract; they appear correctly in this figure). See text for details.

large a range. The effect of up to 8.5 wt % dissolved H_2O on sulfur diffusion in the obsidian, dacite and andesite melts was assessed using the radiotracer technique.

Two low-pressure experiments on the lunar melt composition at 1300°C, of 40 and 180 minutes duration, returned the same diffusivity of $6.6 \cdot 10^{-12}$ m²/s. At the time of this writing, data at other temperatures are not available, so the temperature dependence of D_{sulfur} in ultramafic melts is unconstrained (but note that the activation energy for diffusion appears to decrease with decreasing SiO_2 content for the other melts; Fig. 8). It can probably be assumed that dissolved S is present as sulfide in the lunar melt (Carroll and Rutherford, 1988), so the diffusivity reported above might be considered an endmember value for very reduced magmas.

Sporadic coverage of T-f_{O_2} space has been achieved for the more silicic terrestrial melt compositions. Most of the diffusion-couple and radiotracer experiments run to date were made in Mo capsules, which impose very reducing conditions (near IW) on the contained melts. The body of data for sulfur diffusion in these reduced melts points to several general conclusions: (1) there is a systematic melt composition effect manifested as lower D_{sulfur} values and higher activation energies for diffusion as the SiO_2 content of the melt increases (the one value for the ultramafic lunar melt is consistent with this trend); (2) the effect of dissolved water is to significantly enhance sulfur diffusion, especially in highly silicic melts; and (3) D_{sulfur} in all but the lunar melt is significantly lower than the D_{CO_2} values summarized in the previous section, and markedly lower than D_{H_2O}, especially in silicic melts at typical liquidus temperatures. These general characteristics can be seen in the summary diagram presented as Figure 8, which include actual data points because the values are not yet published elsewhere. For dry Fe-free dacite and low-K andesite, Arrhenius equations based on 3 data points each (spanning 1300° to 1500°C) are as follows:

dry, reduced dacite: $\qquad D_{sulfur} \sim 1.4 \cdot 10^{-4} \exp(-263{,}000/RT)$ \hfill (30)

dry, reduced andesite: $\qquad D_{sulfur} \sim 1.0 \cdot 10^{-6} \exp(-191{,}000/RT)$ \hfill (31)

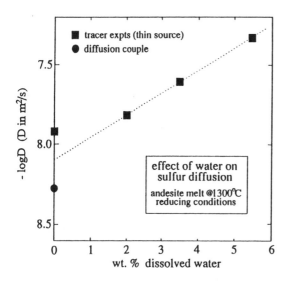

Figure 9. Effect of dissolved H_2O on sulfur diffusion in molten andesite at 1300°C (previously unpublished data of Watson, D.A. Wark and J. W. Delano). The apparent high quality of the log-linear fit (except for the "dry" values) is considered fortuitous in view of the estimated ~factor-of-2 uncertainty in each thin-source data point; however, the overall of influence of dissolved H_2O is thought to be accurately represented.

It is emphasized that these equations are preliminary, with uncertainties of at least 10% in Q and much greater in D_0. There are too few data at present to compute Arrhenius parameters for diffusion in dry obsidian melt, but it is clear that the trend represented by Equations (30) and (31)—toward lower D and higher Q with increasing SiO_2 content—continues up to the high-SiO_2 obsidian (see Fig. 8).

Effect of Water. The influence of water on D_{sulfur} has been characterized most thoroughly for the low-K andesite: As water content increases from 0 to 5.5 wt %, the enhancement in diffusivity exceeds an order of magnitude at temperatures below ~1100°C (see Fig. 8). On the basis of two data points (at 1300° and 1500°C), diffusion of sulfur in molten andesite containing ~5.5 wt % H_2O is described approximately by:

$$D_{sulfur} \sim 3.2 \cdot 10^{-8} \exp(-115,000/RT) \tag{32}$$

At 1300, where the data are most complete, D_{sulfur} appears to increase roughly linearly with added H_2O (see Fig. 9). The effect of dissolved H_2O appears to be even more pronounced in the obsidian melt, where an increase in D_{sulfur} by more than an order of magnitude occurs just in the interval 3.1 to 8.5 wt % H_2O at 1100°C (a reliable value for dry obsidian at this temperature is not yet available). The general systematics of the effect of H_2O on D_{sulfur} are similar to its effect on D_{CO_2} (and cation diffusion in general; see last section of this chapter)—specifically, added H_2O results in higher D at any given temperature and lower Q overall.

Role of oxygen fugacity. Although the picture concerning the influence of S oxidation state on D_{sulfur} is not fully developed, it can be tentatively concluded that the effect is significant but not large. Two sets of information point to this conclusion. First, Baker and Rutherford (1994) extracted D_{sulfur} values for rhyolite melt at 800° and 900°C (200 MPa water pressure) from anhydrite dissolution experiments made under highly oxidizing conditions (MnO-Mn_3O_4); their values appear to be a factor of 5 to 10 higher than the down-temperature extrapolation of the Watson et al. data for reduced rhyolite melt of similar H_2O content (see Fig. 8; note that the two diffusivities published by Baker and

Rutherford are switched in their abstract). However, it is important to bear in mind that at least some of the difference is attributable to the 0.8-GPa difference in pressure between the two sets of data. The other indication of a relatively small f_{O_2} effect on D_{sulfur} comes from additional data acquired by Watson et al. since the publication of their abstract. Several radiotracer experiments were made in which 5% hematite powder was added to the starting glasses as an internal f_{O_2} control, and the runs were made in Pt rather than Mo capsules. After both the presynthesis step and the actual diffusion anneal, the glasses contained small, dispersed euhedra of Fe-oxide—in some cases both hematite and magnetite. The diffusivities resulting from these experiments appear to be only slightly higher than D_{sulfur} values for reduced melts at similar conditions (see Fig. 8).

More experiments on S diffusion systematically incorporating f_{O_2} as a variable should eventually clarify the role of changes in S speciation; for now, however, it appears that the effect of oxidation is to enhance D_{sulfur} somewhat but not dramatically. A relatively small effect may imply that the diffusing species and/or diffusion mechanism is relatively independent of f_{O_2}.

Halogens

Relatively little information exists on diffusion of dissolved halogens in magmas, and much of the available data base pertains to dry, simple-system analogs over restricted ranges in temperature. To the author's knowledge, only fluorine and chlorine have been studied to date.

Fluorine. The first study of F diffusion in melts was that of Johnston et al. (1974), who used a radiotracer technique to measure D_F in a basic slag composition in the system $CaO-Al_2O_3-SiO_2$ (~40 wt % each of CaO and SiO_2) at ~1300° to 1600°C. Neglecting for the moment the questionable suitability of this depolymerized melt as a magma analog, the results of this study were intriguing in the respect that F was found to diffuse significantly faster—and with a lower activation energy—than any other melt component, including oxygen: D_F(tracer) ranged from ~$1 \cdot 10^{-9}$ to ~$4 \cdot 10^{-9}$ m^2/s over the temperature range investigated, which translates into an activation energy of ~80 kJ/mole (Fig. 10).

Using diffusion couples run in a piston-cylinder at 1200° to 1400°C and 1.0 to 1.5 GPa, Dingwell and Scarfe (1984) examined chemical diffusion of fluorine in a more geologically-relevant melt of jadeite ($NaAlSi_2O_6$) composition. One of the juxtaposed melts was of stoichiometric jadeite composition, the other jadeite in which some of the oxygen was replaced by fluorine (6 wt %). Diffusion couples were prepared by tightly packing glass powders of the two compositions into platinum tubes and sealing the tubes with welds. The couples were run with the denser, fluorine-free melt on the bottom, usually for 60 minutes. The extent of fluorine-oxygen interdiffusion was characterized by electron microprobe traverses on the quenched and sectioned samples. The authors were able to draw several important conclusions from their data: (1) the chemical diffusivity of fluorine (D_{F-O}) in a relatively polymerized melt (~60 wt % SiO_2) is significantly lower— and the activation energy for diffusion significantly higher—than the tracer values reported earlier for basic slags; (2) over the range of F/O ratios studied, there is no dependence of D_{F-O} upon concentration of F (i.e., the interdiffusion gradients conformed well to Eqn. 17); and (3) any effect of pressure on F-O interdiffusion is small (calculated activation volumes at 1200, 1300 and 1400°C are not statistically different from zero). Averaged over 1.0 to 1.5 GPa, fluorine-oxygen interdiffusion in jadeitic melt (see Fig.10).

$$D_{F-O} = 5.6 \cdot 10^{-6} \exp(-159,000/RT) \qquad (33)$$

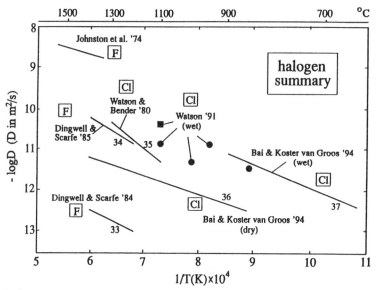

Figure 10. Summary of halogen (F and Cl) diffusivities in molten silicates. The "wet" Cl data of Watson (1991) are radiotracer values for molten obsidian (l) or dacite (n) containing ~8 wt % dissolved H_2O. Other diffusivities and diffusion media represented are: Cl tracer diffusion in soda-lime aluminosilicate melt with 56% SiO_2 (Watson and Bender 1980); F tracer diffusion in basic slag with 40% SiO_2 (Johnston et al. 1974); F-O interdiffusion in albite melt (Dingwell and Scarfe 1985); F-O interdiffusion in jadeite melt (Dingwell and Scarfe 1984); and Cl chemical diffusion in wet haplogranitic melt and in dry obsidian (Bai and Koster van Groos 1994). The numbered lines refer to Arrhenius equations in the text.

Dingwell and Scarfe followed up their initial paper on F diffusion in jadeite melts at high pressures with a study of other melts in the $Na_2O-Al_2O_3-SiO_2$ system at near-atmospheric pressure (Dingwell and Scarfe, 1985). These studies involved suspension of F-enriched glass spheres of albite, jadeite, and peraluminous haplogranite composition (75 wt % SiO_2) from Pt loops in an oxygen-flow furnace. At 1200° to 1400°C, these spheres exchanged fluorine for ambient oxygen at a rate determined by the F-O interdiffusion coefficient. As in the previous high-pressure study of F diffusion in jadeite melt, F-O interdiffusion was again found to be independent of F concentration, with D_{F-O} increasing in the order albite < peraluminous melt < jadeite. For the range in temperature covered, all diffusivities fall between 10^{-11} and 10^{-13} m²/s, and activation energies between ~120 and 180 kJ/mole. For albite melt, F-O interdiffusion conforms to

$$D_{F-O} = 2.3 \cdot 10^{-9} \exp(-124,000/RT) \tag{34}$$

Chlorine. The data base on chlorine diffusion is comparable with that on fluorine, being represented by just one abstract and two published papers. Watson and Bender (1980) measured tracer diffusivities for a variety of elements—including Cl—in a soda-lime aluminosilicate melt (56 wt % SiO_2) at 1100° to 1300°C and 0.6 to 1.8 GPa. As in radiotracer studies described earlier (e.g., of CO_2), the thin-source technique was used. Glass rods were prepared at atmospheric pressure by drawing the melt into a Pt capillary, which was subsequently cut into 6 to 8 mm lengths; one end of each length was polished and ^{36}Cl was introduced as aqueous NaCl. After drying off the solution, the sample was inserted in a graphite or boron nitride sleeve, which was plugged at the ends and run in a piston-cylinder apparatus. Diffusivities were calculated from photodensitometer scans of track maps made from axial sections of the samples. At the lowest pressure studied (0.6 GPa), the following Arrhenius relation was determined for Cl tracer diffusion in the Na_2O-CaO-Al_2O_3-SiO_2 melt:

$$D_{Cl}(tracer) = 3.4 \cdot 10^{-4} \exp(-207{,}000/RT) \qquad (35)$$

As in the case of F-O interdiffusion studied by Dingwell and Scarfe (1984), D_{Cl} is unaffected by changes in pressure: Watson and Bender reported an activation volume of ~1 cm^3/mole, which is not statistically different from zero.

The subject of chlorine diffusion was revisited by Watson (1991), who sought to characterize the effects of dissolved H_2O on D_{Cl} in rhyolitic and dacitic melts using techniques identical to those described for CO_2 in the same paper (see above). The part of the study devoted to Cl diffusion was not entirely successful in that reliable data could not be obtained for melts containing less than 8 wt % dissolved H_2O—in drier melts, the radiotracer tended to penetrate down the melt/capsule interface. One value was reported for the dacite (1100°C) and four for the obsidian (800° to 1100°C), all pertaining to melts with 8 wt % dissolved H_2O. Scatter in the data precludes an estimate of the Arrhenius parameters, but it is clear that D_{Cl} is lower than D_{CO_2} in melts of similar composition and water content (see Fig. 10). Beta-track maps of the failed experiments (in which ^{36}Cl penetrated along the capsule wall) indicated, furthermore, that D_{Cl} decreases sharply in melts containing less than 8% water.

Bai and Koster van Groos (1994) made a comprehensive study of Cl chemical diffusion in a haplogranitic melt (74.6 wt % SiO_2) and a natural obsidian (77.9 wt % SiO_2) that addressed the effects not only of temperature (650° to 1400°C) but also of pressure (to 0.46 GPa) and water content (0~7 wt %). The study incorporated three different experimental approaches: Nominally dry, high-temperature (850° to 1400°C) experiments were performed at atmospheric pressure by immersing spheres of the rhyolitic obsidian in molten NaCl sealed in Pt capsules. Higher pressure, hydrous runs in which NaCl-rich brines served as the Cl source were made in cold-seal pressure vessels at 850°C and 20 to 200 MPa, again using the natural glass as starting material. A final series of experiments was made at 0.1 to 0.46 GPa and 650° to 900°C on the haplogranitic melt in internally-heated or cold-seal pressure vessels. Chlorine uptake profiles, typically hundreds of microns long, were characterized by electron microprobe, and D_{Cl} values were calculated using the standard approach of inverting the c-x data through the error function.

For their low-pressure series of experiments at 850° to 1400°C, Bai and Koster van Groos (1994) report the following well-constrained Arrhenius relation for chemical diffusion of Cl in dry obsidian:

$$D_{Cl}(chemical) = 3.16 \cdot 10^{-9} \exp(-86{,}190/RT) \qquad (36)$$

The temperature effect on D_{Cl} in hydrous haplogranitic melts (coexisting with NaCl-rich brine at 0.2 GPa) is given by

$$D_{Cl}(chemical) = 6.46 \cdot 10^{-7} \exp(-110{,}900/RT) \qquad (37)$$

Comparison of Equations (36) and (37) indicates a substantial effect of dissolved water upon Cl diffusion (see also Fig. 15a, below). Note that Equation (37) is consistent with the isolated tracer diffusivity values reported by Watson (1991), but the newer data are of higher quality and provide a much better indication of the temperature dependence of D_{Cl} in hydrous silicic melts (see Fig. 10).

Bai and Koster van Groos were able to estimate activation volumes for Cl diffusion in both dry and hydrous granitic melts as 27±10 and 22±7 cm^3/mole, respectively. These differ significantly from the near-zero V_a value reported by Watson and Bender (1980), but the melt compositions and pressure regimes of the two studies are also markedly different.

Noble gases

Most studies of noble gas diffusion in amorphous silicates have been conducted on solid glasses [see Doremus (1994) for an up-to-date review; also Carroll and Stolper, 1991; 1993; and Carroll et al., 1993], so there exists only a tenuous basis upon which to draw conclusions about magmas. The one study devoted (in part) to diffusion in molten silicate is that of Lux (1987), who presented preliminary data for diffusion of He, Ne, Ar, Kr, and Xe in tholeiitic basalt (49.7 wt % SiO_2) at 1350°C and near-atmospheric pressure.

The technique used by Lux involved initial purging of natural tholeiite melt of its noble gases, loading the resulting glass fragments into a Pt tube (~3 cm long, 0.5 cm dia.), and re-heating the tube to 1350°C in a noble-gas atmosphere for 8 hours. The bulk uptake of each gas was determined by isotope-dilution mass spectrometry, and diffusivities were calculated by applying a solution for diffusion in a semi-infinite medium in which the surface concentration is constant over time (given by the gas solubility at the pressure of the experiment). Lux considered the resulting diffusivities to be preliminary because the experiments were not repeated, but the data are very systematic in showing a monotonic (though modest) decrease in diffusivity with increasing atomic radius, from $D_{He} \sim 5 \pm 1 \cdot 10^{-9}$ m^2/s, to $D_{Xe} \sim 2.5 \pm 1 \cdot 10^{-10}$ m^2/s. This relationship is good evidence that the results were not affected by convection in the melt—the main pitfall to which experiments of this type are prone.

Lux noted that the calculated diffusivities are higher than would be expected from up-temperature extrapolation of values measured on basaltic glasses, and also that the 20-fold change in D from the smallest (He) to the largest (Xe) rare gas is much smaller than the 8-order-of-magnitude difference reported by Hiyagon (1984) for diffusion in basaltic *glass* at 900°C. The apparent major differences between noble gas diffusion in basaltic glasses and melts led Lux to conclude that D values cannot be extrapolated through the glass transition temperature. This conclusion is quite different from that reached by Zhang et al. (1991) and Jambon et al. (1992) for diffusion of molecular H_2O—and by Blank (1994) for diffusion of CO_2—in obsidian, which obey single Arrhenius relations over a large range in temperature spanning the glass transition. Note also that the Arrhenius relation for He diffusion determined by Jambon and Shelby (1980) for basaltic glass at 200° to 300°C does extrapolate to a value generally consistent with Lux' data for diffusion of other noble gases in basalt melt at 1350°C (see Fig. 12, below). These divergent views concerning glass/melt similarity with respect to diffusion of neutral species are especially interesting in view of the fact that molecular H_2O, rare gases and CO_2 all fall on a smooth curve relating activation energy for diffusion in obsidian (glass) to molecular or atomic radius (Zhang et al., 1991; Blank, 1994; see Fig. 11). In obsidian, neutral H_2O and CO_2 molecules (and, by inference, rare gas atoms) seem to show consistent diffusion behavior through the glass transition; it is not yet clear, however, whether the same can be said of basalt. From studies of rare gas solubilities in natural melts, Lux concluded that solubility is negatively correlated with melt density, so perhaps the difference in diffusion properties of the two materials is due simply to structural differences between obsidian and basaltic glasses and melts.

A reasonable data base exists for rare gas diffusion in various solid glasses, including obsidian and basalt. However, because of the uncertain relevance of the information to magmas—and because other summaries exist—a comprehensive review is not attempted

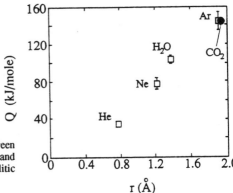

Figure 11. Diagram showing the correlation between atomic or molecular radius of neutral diffusants and the activation energy (Q) for diffusion in rhyolitic glass.

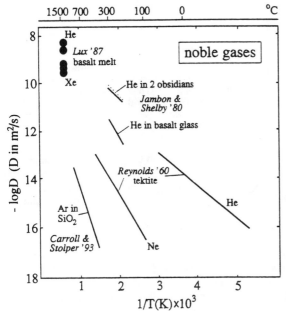

Figure 12. Summary of selected data on noble gas diffusion in various glasses, in comparison with the data of Lux (1987) for molten basalt. The relevance of the glass data to magmas is unclear, but it does appear that the low-temperature data for He diffusion in basalt glass (Jambon and Shelby 1980) extrapolate to values consistent with those for diffusion of other rare gases in basaltic melt. See text for discussion.

here. Selected noble-gas diffusivities for glasses do appear in Figure 12 for comparison with Lux's values for basaltic melts. For information on geologically-relevant glasses, the reader is referred to Reynolds (1960), Jambon and Shelby (1980), Carroll (1991), and Carroll et al. (1993). Noble-gas diffusivities for basaltic melts only (Lux, 1987) appear in Figure 13, which is a comprehensive summary diagram for diffusion of dissolved volatiles in magmatic liquids.

EFFECTS OF DISSOLVED VOLATILES (H_2O) ON CATION DIFFUSION

Introduction

The last topic taken up in this review concerns the effect of dissolved volatiles upon diffusion of other, non-volatile melt components. In almost all cases "dissolved volatiles" really means "water" here—only isolated data are available for the effects of other volatiles

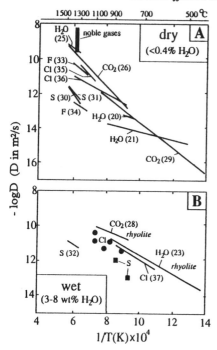

Figure 13. Overall summary/comparison diagrams for diffusion of various dissolved volatiles in magmatic liquids.(A) nominally dry melts (<0.4 wt % H_2O); (B) melts with 3 to 8 wt % dissolved H_2O. Numbers in parentheses refer to Arrhenius equations in the text.

(and these are restricted to F and Cl). Having seen the changes in D_{sulfur} and D_{CO_2} brought about by adding water to a melt (see above), the reader is in a good position to anticipate the consequences of dissolved water for diffusion in general. Interestingly, even before any diffusion measurements on hydrous melts were available, there existed a widespread expectation (e.g., Fenn, 1977) that water might have profound effects on diffusion, since a dramatic effect on melt viscosity had already been documented (Shaw, 1963).

Because H_2O is regarded as a common or even necessary component of most crustal magmas, numerous studies have been made of cation diffusion in water-bearing melts. In relatively few cases, however, has diffusion been examined specifically as a function of dissolved water content. For the sake of brevity, this review will focus on those studies in which the effects of water are specifically targeted, though other information will be brought in where relevant. The vast majority of data apply to melts of granitic composition.

Tracer diffusion

The first published results concerning the effects of water on diffusion were those of Jambon et al. (1978) for Cs in granitic melt (~73 wt % SiO_2, anhydrous). This study involved thin-source experiments on obsidian cylinders that had been pre-hydrated in open-ended Pt tubes at 800°C and a water pressure of 200 MPa (resulting in 5.5 wt % dissolved H_2O). After deposition of [134]Cs radiotracer at one end, a sample was returned to the pressure vessel (still in an open tube), pressurized with Ar to 300 MPa, and annealed at ~750° to 900°C. During this thermal treatment, the Cs radiotracer diffused into the hydrated glass, and, simultaneously, some water was lost to the dry Ar atmosphere. Profiles of tracer activity vs. penetration depth were determined by serial "sectioning" (with HF) and gamma-ray counting. Cesium diffusivities were computed for different water contents from the combined knowledge of ^{134}Cs-uptake and H$_2$O-loss profiles. The resulting values revealed a dramatic enhancement of Cs diffusion by the presence of dissolved water, amounting to a 4-order-of-magnitude increase in D between 0 and 2 wt % H$_2$O at 750°C (see Fig. 14a). Over this same water content interval, the activation energy for Cs diffusion was observed to drop from ~290 to ~70 kJ/mole.

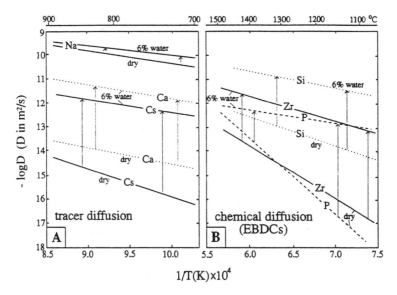

Figure 14. Comparison of diffusion in dry granitic melt (or glass) with diffusion in the same melt containing 6 wt % H$_2$O. (A) tracer diffusivities for Cs, Ca, and Na at relatively low temperatures (700° to 900°C); (B) effective binary diffusion coefficients (EBDCs) for Zr, P, and Si at relatively high temperatures (1100° to 1500°C). In general, Arrhenius lines are shown for the temperature range actually investigated. Note that if the Zr, P, and Si lines were extended down-temperature to geologically-plausible temperatures (for granitic melts), the effect of H$_2$O on diffusion would be much greater for these highly-charged diffusants than for the alkalies and alkaline earths. This is consistent with the broad observation that the slowest-moving species in dry melts are "accelerated" most by the addition of H$_2$O to the system.

Shortly after publication of the study by Jambon et al. (1978), Watson (1979b) reported qualitatively similar results for a nearly-identical system. Watson's thin-source technique was slightly different in that, after depositing radiotracer (^{134}Cs) at one end of a pre-hydrated obsidian cylinder (76 wt % SiO$_2$, anhydrous), he resealed the container and annealed the sample in a cold-seal pressure vessel at 200 MPa. Diffusivities were extracted not by serial sectioning but from densitometer scans of ß-track maps of the axially-sectioned samples. Watson referenced his results for melts containing ~6 wt % dissolved H$_2$O to values for dry obsidian reported earlier by Jambon and Carron (1973). The observed difference in D$_{Cs}$ was about 3.5 orders of magnitude at 700°C, and the activation energy for diffusion in the "wet" melt—at ~82 kJ/mole—was 2.5 smaller than the "dry" value of Jambon and Carron (see Figs. 14a and 15a). Thus, there is reasonable qualitative agreement with the Jambon et al. study, but significant differences in the details: relative to dry obsidian, Watson observed about the same enhancement in D$_{Cs}$ for ~6% added water that Jambon et al. produced with only 2%. This qualitative discrepancy may be attributed to the possible "open-system" nature of the experiments by Jambon et al.—their samples may have lost water to the Ar pressure medium. However, the consequences of water loss would seem more likely to cause an *under*estimation of the effect of water.

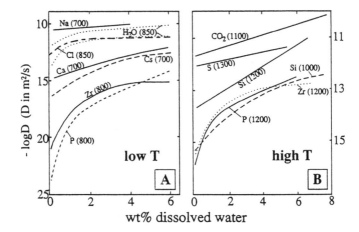

Figure 15. Diagram illustrating the dependence of diffusivities upon dissolved water content of granitic melt at low temperatures (A) 700° to 850°C, and high temperatures (B) 1000° to 1300°C. The 800°C curves for Zr and P were estimated by down-temperature extension of the Arrhenius lines for these elements shown in Figure 14 (see also Eqns. 38-41). Note the difference in the logD scales for (A) and (B), and that the influence of dissolved H_2O is much greater at low temperature. Note also that for highly-charged diffusants in particular, most of the increase in logD occurs over the first 2 to 3% of dissolved H_2O. Differences in pressure among the various studies represented in this diagram are ignored: In (A), all curves represent data acquired at 0.1 to 0.2 GPa, except for those marked Zr and P, which pertain to 0.8 GPa. In (B), all curves represent data acquired at 0.8 to 1.0 GPa. See text for further discussion.

Watson (1981) followed up his initial study of Cs diffusion in granitic melt containing ~6 wt % water with a more detailed investigation of Cs, combined with a first look at the effects of H_2O on Ca and Na tracer diffusion in the same melt. The results of [45]Ca radiotracer experiments on granitic obsidian containing ~0, 3.5 and 6.2 wt % H_2O, run at 200 MPa and 700° to 900°C, revealed strong enhancement of calcium diffusion by dissolved water, amounting to an increase in D_{Ca} by 2.5 orders of magnitude over the ~6% increase in water content. Although qualitatively similar, the effects of water on D_{Cs} and on D_{Ca} were noticeably different in several respects: (1) the overall enhancement of calcium diffusion was less pronounced; (2) within the uncertainties in the data, the activation energy for Ca diffusion (~110 to 120 kJ/mole) was unaffected by melt H_2O content; and (3) the effect of H_2O on $logD_{Ca}$ is close to linear, in contrast with the exponential effect on $logD_{Cs}$ (see Fig. 15a).

Sodium diffusion in granitic obsidian was examined by Watson (1981) for ~0 and 3.5 wt % H_2O using techniques identical to those employed for Cs and Ca, except that diffusion profiles were characterized by serial sectioning and counting the γ-activity of [22]Na radiotracer. At the data-reduction stage in this study, the results for Na were anticipated with some excitement, because enhancement in D_{Na} to the same degree as observed for Cs and Ca would lead to unprecedentedly high diffusivities (since Na diffusion is fast—at ~10^{-10} m²/s—even in dry obsidian at 700° to 900°C). The results were disappointing in the respect that H_2O had little influence on D_{Na}: relative to the dry case, 3.5% H_2O resulted in an increase in diffusivity by only about 10% at 700, and at higher temperatures the effect was even smaller (see Figs. 14a and 15a). Dissolution of H_2O does cause a significant decrease in the activation energy for Na diffusion, from ~96 to ~68 kJ/mole.

Although only 3 elements were examined in the study by Watson (1981), the results provided the first glimpse of what would become a recognizable characteristic of the effects of H_2O on diffusion in granitic melts; namely, the lower the diffusivity in the dry melt (or glass), the greater the enhancement in diffusivity—and reduction in activation energy—by the addition of water. This amounts to saying that the effect of H_2O on diffusion in granitic melts is to cause Arrhenius lines of highly variable slope (radiating from some high-temperature point?; see Hart, 1981) to collapse into a relatively narrow band near the high end of the "dry" range (see Fig. 14).

Chemical diffusion

Most studies aimed at characterizing the effects of changes in H_2O content on chemical diffusion have involved the "mineral dissolution" technique described previously in the section devoted to experimental approaches. If chemically simple minerals can be used, partial dissolution is a convenient way to obtain effective binary diffusion coefficients of the mineral components in the contacting melt (see Fig. 4). The technique works especially well for minerals that are relatively insoluble in the melt, because the changes in absolute abundance of melt components in the diffusion field surrounding the dissolving mineral will be small. Low solubility also means that the developing diffusion field can be modeled simply and accurately as a stationary-boundary process.

High field-strength elements. Information concerning the effects of dissolved H_2O on diffusion of high field-strength elements in granitic melts (~76.4 wt % SiO_2, anhydrous) was obtained by Watson and co-workers in studies of accessory mineral dissolution kinetics. By characterizing Zr concentration gradients against partially-dissolved zircons, Harrison and Watson (1983) characterized Zr diffusion in granitic melts containing ~0.1 to 6 wt % H_2O at 1020° to 1500°C and 0.8 GPa (over this temperature range, the solubility of zircon varies from ~1200 ppm to ~15,000 ppm Zr). Arrhenius lines were obtained for ~0.1% and 6% H_2O, and a systematic study of the effect of water was made at 1200°C, incorporating experiments with 0.1, 1.0, 2.3, 4.2, 6.0 and 6.3 wt % H_2O. Zirconium diffusion is slow in nominally dry granitic melts, and is enhanced considerably by addition of H_2O (see Fig. 14b):

$$D_{Zr} \, (dry) = 0.098^{+.14}_{-.06} \exp(-409,000 \pm 12,000/RT) \tag{38}$$

$$D_{Zr} \, (6\% \, H_2O) = 3.0^{+3.0}_{-1.5} \cdot 10^{-6} \exp(-198,000 \pm 7,950/RT) \tag{39}$$

As in the case of Cs, $logD_{Zr}$ varies with water content in a decidedly non-linear fashion: most of the "acceleration" of Zr occurs over the first 2% or so of dissolved H_2O (see Fig. 15b).

Harrison and Watson (1984) used the same partial-dissolution technique (this time with apatite) to obtain information on diffusion of phosphorus in granitic melts at 1100°-1500°C. Again, for a change in water content from ~0.1 to ~6 wt % H_2O, dramatic changes in P diffusivity are observed (see Fig. 14b):

$$D_P \, (dry) = 2.2^{+2.9}_{-1.3} \cdot 10^5 \exp(-601,000 \pm 12,000/RT) \tag{40}$$

$$D_P \, (6\% \, H_2O) \sim 10^{-9} \exp(-105,000/RT) \tag{41}$$

(Eqn. 41 is approximate because it is based upon only two points—at 1100° and 1300°C). To the author's knowledge, Equation (40) contains the highest activation energy (~600 kJ/mole) that has ever been reported for diffusion in a melt (Remarkably, it is also higher than most values for lattice diffusion in crystals!). The high Q and D_o combine to make P the slowest-diffusing melt component thus far characterized at temperatures below about 1300°C (Zr is slower above this temperature). In keeping with the tendency for the slowest-moving species to be most affected by dissolved H_2O, D_P increases by many orders of magnitude (depending upon temperature) for even modest additions of H_2O (see Figs. 14b and 15b). Significantly, parallel increases in the diffusivity of Ca—which, like phosphorus, is "released" during apatite dissolution—were not observed: because of its inherently higher diffusivity, Ca far outdistanced P in all experiments, but enhancement of D_{Ca} by added H_2O did not approach the effect observed for P.

In a study paralleling the earlier efforts by Harrison and Watson (1983; 1984), Rapp and Watson (1986) examined the dissolution kinetics of monazite ($[LREE]PO_4$) in granitic melts containing 1 to 6 wt % H_2O. Chemical diffusion of both P and REE was shown to be significantly enhanced by the presence of H_2O. In contrast to the case of Ca and P diffusing away from dissolving apatite, however, REE and P exhibited parallel diffusion gradients along which the stoichiometry of monazite was maintained at all concentration levels. The authors took this to mean that diffusion of REE and P is strongly coupled, and so did not associate the effect of H_2O specifically with either REE or P. It seems likely, however, that P was the rate-limiting diffusant, inasmuch as D_P from the monazite dissolution experiments was not greatly different from the values reported by Harrison and Watson (1984) in their study of apatite.

Silica. Chekhmir (1984) also made extensive use of the mineral dissolution technique for characterizing chemical diffusion, and incorporated H_2O content as a variable in some of his experiments (see Chekhmir and Epel'baum, 1991 for a summary in English). Of particular significance was his study (at 1000°C and 1 atm to 0.6 GPa pressure) of quartz dissolution in albite melts containing variable amounts of H_2O, in which he showed that $logD_{Si}$ increases exponentially with increasing H_2O content, with most of the increase occurring over the first ~4% H_2O (see Fig. 15b).

Additional information concerning the effect of H_2O on chemical diffusion of Si was provided by Baker (1991), who examined interdiffusion of dacite and rhyolite melts at both 3 and 6 wt % dissolved H_2O. The study covered a significant temperature range (1100° to 1400°C; all at 1 GPa), so the Arrhenius parameters for SiO_2 effective binary diffusion in silicic melts could be estimated for two different water contents:

$$D_{Si\ (3\%\ H_2O)} = 2.58 \cdot 10^8 \exp(126,500/RT) \tag{42}$$

$$D_{Si\ (6\%\ H_2O)} = 2.69 \cdot 10^{-7} \exp(-131,400/RT) \tag{43}$$

Baker emphasized that the uncertainties in D_0 and Q are large for the 6% H_2O case (80% in Q), but that the activation energy for SiO_2 diffusion in hydrous melts is clearly about half that for dry melts (see Fig. 14b). The dependence of $logD_{Si}$ upon water content appears to be roughly linear at 1200°C (see Fig. 15b).

Effects of H_2O: Recap and final observations

With rare exceptions, the effect of dissolved H_2O on both tracer and chemical diffusion in magmas is large—sometimes amounting to several orders of magnitude at a given temperature—and it is always positive. Two systematic trends seem to emerge: (1) diffusants that are particularly sluggish in dry melts or glasses exhibit the greatest enhancement by added water (e.g., Cs, Zr, P, Si); and (2) the addition of water to a melt lowers both D_0 and Q for all diffusants. Graphically, this latter trend means that a family of Arrhenius lines for diffusion in melts of differing H_2O content will be fan-shaped in overall appearance, with all lines intersecting at some high-temperature point (see discussion of diffusion "compensation" by Hart, 1981). Of the elements studied to date, diffusion of Na is least affected by addition of H_2O. The reader should bear in mind, however, that this conclusion can be strictly applied only to tracer diffusion. In chemical diffusion, the behavior of Na is known to be strongly coupled to that of SiO_2 (e.g., Watson, 1982; Baker, 1991), so during interdiffusion of dissimilar melts, any effect of H_2O on SiO_2 diffusion will also impact on Na.

At a given temperature, the increase in logD with increasing water content may be approximately linear (as in the cases of CO_2, S and SiO_2 in natural melts at 1200°C) or clearly exponential (as for Cs, Zr, P and SiO_2 in albite melt at 1000°C). However, a cautionary note is probably warranted here, also: In those cases where the "H_2O effect" appears to be log-linear, the uncertainties in the data are sufficiently large that some other functional form is also permitted. The "fanning-Arrhenius" character of the H_2O effect (see above) means that changes in logD with H_2O content are smaller at high temperatures. It may not be a coincidence that most of the apparently linear relationships between logD and dissolved H_2O content (see Fig. 15) apply to relatively high temperatures, where the inherent uncertainties in the measured D values can obscure the functional form of a relatively weak dependence. The manner in which D varies with melt H_2O content is of interest not only for predicting D values beyond the range of actual measurement, but also because of the obvious significance to melt structure and physical properties (see Chapter 9). In granitic melts at least, viscosity is known to depend strongly upon water content, and several authors have made a connection between the changes in diffusivity and in viscosity brought about by added water (e.g., Chekhmir and Epel'baum, 1991; Baker, 1991). The form of logD vs. H_2O content is also interesting from the standpoint of water speciation in the melt: a rapid initial increase in diffusivity with small amounts of added H_2O (dissolved mainly as OH^-), followed by a more asymptotic increase in D at higher H_2O content (where molecular H_2O becomes important), may indicate that it is dissolution of OH^-—and concomitant disruption of Si-O-Si bonds—that enhances diffusion.

Effects of other volatiles: Preliminary information on fluorine and chlorine

In an attempt to paint a general picture of chemical diffusion in silicic melts containing geologically-realistic levels of halogens, Baker and Watson (1988) executed 15 diffusion-couple experiments on halogenated natural and synthetic glasses. A fluorine-doped granitic obsidian (0.68 wt % F, 73.6% SiO_2) was juxtaposed against either a Cl-rich obsidian (0.53 wt % Cl; 71.6% SiO_2) or a synthetic Ca-Na-K-aluminosilicate glass (0.68 wt % F; 68.3% SiO_2). The couples were run in gas- or solid-medium pressure vessels at 0.01 to 1.0 GPa and 900° to 1400°C, and the resulting chemical diffusion profiles were characterized using an electron microprobe for the major elements and Synchrotron XRF for the minor and trace elements. Because systematic variations in halogen content were not included in the study, the effects of F and Cl could be assessed mainly by comparison with previous studies of diffusion in similar but halogen-free materials. The most direct comparison possible was between D_{Zr} in the halogenated melts and D_{Zr} values reported earlier by Harrison and Watson (1983) for the same melt lacking F and Cl. Baker and Watson (1988) noted a ~100-fold enhancement of Zr diffusion in the halogenated melt and a reduction in activation energy by ~36%—effects qualitatively similar to that of H_2O. The observed increase in diffusivity is consistent with the decrease in viscosity of albite melt caused by addition of fluorine (Dingwell and Mysen 1985).

Baker (1993) followed up the exploratory study by Baker and Watson (1988) with a more detailed investigation of the effects of F and Cl (at the ~1 wt % level) upon chemical diffusion of major elements in peralkaline rhyolitic and dacitic melts. As in previous studies of chemical diffusion in volatile-free magmas, the non-alkali elements (Al, Ti, Fe, Mg, and Ca) were found to diffuse at essentially the same rate as Si, so Baker focused his effort upon characterization of Si diffusion as a function of temperature and SiO_2 content of the melts. Interestingly, he found only minor differences—basically within experimental uncertainty—between D_{Si} in halogen-bearing and volatile-free melts. The data also suggested, however, that F causes a slight reduction in the activation energy for diffusion (relative to the volatile-absent case), whereas Cl causes a slight increase. The overall effects of halogens on major-element diffusion in peralkaline melts appear to be minor.

Baker and Bossànyi (1994) examined the combined effects of F and H_2O on Si diffusion in the same melts investigated by Baker (1993). They showed that in melts containing ~3 wt % dissolved H_2O, the addition of 1.3 wt % F further enhances diffusion by a factor of ~5 to 10 (depending on temperature) and causes a decrease in activation energy by ~30%. This finding indicates that, at least under some circumstances, the effects of H_2O and F are additive. In more hydrous melts (6 wt % H_2O), the effect of added F is small, however, the only perceptible change being a ~12% reduction in activation energy.

No other studies of the effects of halogens on diffusion in melts are known to the author, but this area looks very promising for future research. Although halogens rarely, if ever, achieve the abundance levels sometimes reached by water (but see Webster and Duffield, 1994; see also Chapter 7), it is clear from the initial work by Baker and colleagues that the effects may be important nevertheless. In terms of the significance to magmatic processes and to deciphering the mechanism of diffusion enhancement by halogens and H_2O, a further exploration of the additive nature of their effects would be especially interesting.

SIGNIFICANCE AND APPLICATIONS

The interrelationships among volatile abundance, transport properties and phase equilibria of magmatic melts lead to several key areas of application for data on diffusion. Many of these have been at least hinted at in the preceding review of experimental studies, because most experimenters have in mind specific uses of their data or specific hypotheses to test. The studies of Zr and P diffusion by Harrison and Watson (1983; 1984), for example—which were discussed above in the context of the effects of H_2O on high field-strength element diffusion—were actually conceived to test the survivability of accessory minerals under various circumstances of crustal melting. Thus it is already clear that the kinetics of crystal/liquid processes is one broad area to which diffusion in volatile-bearing magmas is highly germane. In the paragraphs below, this and other key areas of relevance are briefly discussed. On the grounds of impracticality, an exhaustive treatment is not attempted: the subject of diffusion in volatile-bearing magmas bears, in fact, on any and all magmatic processes that involve diffusion in the melt, and discussing all of these would require a chapter in itself.

Crystal/liquid processes

Growth and dissolution rates. Most experimentalists engaged in igneous phase equilibrium studies subscribe to a general view that crystals grow better in hydrous melts than in dry ones. Actual proof of a correlation between growth rate and melt H_2O content at a single temperature is elusive, but in light of what is now known about diffusion in H_2O-bearing melts, the experimentalists' preconception is almost certainly accurate: H_2O (and halogens?) can enhance the supply rate of components required for growth to a crystal surface, thus accelerating the growth process. As shown by Harrison and Watson (1983;

1984), the reverse of crystal growth—dissolution—can also be strongly influenced by volatile-induced changes to diffusion in the melt. A general understanding of the role of volatiles in diffusion is thus crucial to addressing a host of problems related to interaction between crystals and contacting melt.

Boundary-layer processes. A class of processes sometimes accompanying crystal growth might be best described as "boundary-layer phenomena." In general, these are interdependent with the process of growth just described. If a crystal grows rapidly (e.g., due to undercooling), incompatible components of the growth medium can accumulate ahead of the advancing crystal interface. The extent of accumulation depends upon the efficiency of diffusion in the melt at dispersing the component in question, so the extent of "pile-up" is related to the competition between the growth rate of the crystal (governed by diffusion of crystal-forming components in the melt) and diffusion of the incompatible component in the melt. It is easy to imagine that rapid growth of a crystal could also result in a boundary layer *depleted* in some slow-diffusing component, if that component were highly compatible in the crystal. For constant crystal growth rates, the mathematics of these phenomena are well understood, having been worked out, mainly in the 1950s, by researchers interested in the growth of semiconductor crystals (e.g., Smith et al. 1955). Complications may arise if boundary-layer depletion or accumulation of a particular component has a feedback effect on the growth rate or stability of the crystal. In such a case the complexity of the problem escalates rapidly (see, for example, Lasaga 1982). For the present purposes, it will suffice to point out simply that any changes in diffusivities of melt components—either absolute or relative—could markedly affect the behavior of the system. As we have seen, variations in H_2O content can have profound effects on diffusion of many components.

Even a brief discussion of boundary-layer phenomena would be incomplete without the observation that dissolved volatiles themselves may be subject to accumulation in advance of a growing crystal. Watson et al. (1982) called attention to this possibility in the context of H_2O/CO_2 values in glass inclusions. Because glass (melt) inclusions are thought to result from rapid crystal growth, it is conceivable that the abundances and ratios of volatile components in them may not reflect those of the bulk melt. In retrospect, and with the present advantage of considerably more information on diffusion of dissolved volatiles, it now appears that any effect on H_2O/CO_2 values in glass inclusions is likely to be minor unless crystal growth rates are extremely high. The case is by no means as clear for dissolved sulfur species, because these are now known to diffuse slowly in relation to other dissolved volatiles. It is even possible in this case that local, boundary-layer saturation in sulfide or sulfate could be induced by relatively rapid growth of a major mineral phase. These and related topics will be taken up in detail by Watson et al. (in prep.) as the data base for sulfur diffusion nears completion.

Bubble growth

The kinetics of bubble growth in magmas is a very promising area of application for data pertaining specifically to diffusion of dissolved volatiles. Once a bubble has nucleated in a magma (due to oversaturation resulting from decompression or crystallization or both), subsequent growth occurs by two mechanisms: (1) diffusion of volatile species into the bubble from the surrounding melt; and (2) expansion. In most cases, these operate simultaneously because both are inevitable responses to further decompression. Modeling bubble growth by expansion of a constant mass of vapor is relatively straightforward, requiring knowledge only of the P-V-T relations of the vapor of interest and the viscosity of the melt. Growth resulting from addition of mass to a bubble, on the other hand, requires detailed knowledge of the diffusivities in the melt of all volatiles contributing to the vapor—including any interplay among the various diffusivities (as in the case of H_2O affecting D_{CO_2} and D_{sulfur}). One of the first papers to incorporate measured diffusivities of dissolved volatiles into a bubble growth model is that of Sparks (1978), who applied Shaw's original data for D_{H_2O} to make general predictions of growth rates. Watson et al. (1982) used their newly-acquired data on D_{CO_2} to model growth of CO_2 bubbles in basaltic magmas, invoking the "bubble stability" equation derived much earlier by Epstein and Plesset (1950). Watson et al. also pointed out that disequilibrium vapor compositions (i.e., failure to achieve partitioning equilibrium between melt and vapor) could result if the diffusivities of contributing volatiles are markedly different. An important recent paper addressing growth of CO_2-rich bubbles in MORB is that of Bottinga and Javoy (1990), who solved numerically the transport equations for multicomponent bubble growth involving diffusion of CO_2, H_2O, He and Ar under both static and dynamic magma conditions. Even in its thoroughness, this last paper does not exhaust the problems to which volatile diffusivities can be applied, because it addresses a specific magma type and decompression path. Additional data for diffusion of H_2O, CO_2, and sulfur that have become available since 1990 also increase the level of detail with which behavior of volatiles during bubble growth can be modeled. For a broad review and perspective of magma degassing, the reader is referred to Chapter 11 of this volume.

Larger-scale magmatic processes

N.L. Bowen (1928, p. 22) justifiably concluded that diffusion is unimportant to magmatic processes occurring on a scale greater than meters, simply because heat conduction is orders of magnitude faster than diffusion, so a magma would freeze before diffusion could accomplish anything. Despite the pronounced effect of dissolved H_2O on diffusion—of which Bowen was unaware—his assessment is still largely accurate. What, then, is left to say about the role of diffusion in large-scale processes occurring in volatile-bearing magmas?

One of H.R. Shaw's principal interests in undertaking his study of H_2O diffusion in granitic magmas was to provide the data needed to assess the efficacy of diffusive transport on the scale of magma chambers. He was concerned, in particular, with such questions as the migration rates of hydration fronts and the influence of chemical and thermal boundary layers. Shaw did the first simple calculation to show convincingly that water cannot get deep into a magma body solely by diffusion from a wall-rock source (This was Bowen's point, and it later proved to be true of all other volatiles as well—the governing diffusivities are simply too low). It now appears that the only way a magma can take up H_2O and/or CO_2 from the crust is by assimilation of dispersed, volatile-bearing xenoliths (containing, for example, mica or carbonate).

More open to debate and future research are questions concerning the interplay between convection and diffusion in magma bodies. Shaw (1974) recognized that a marginal influx of water into a granitic magma body could generate a low-viscosity, low-density boundary layer that would be prone to buoyant rise, despite the influence of the broader thermal (cooler) boundary layer tending to sink. We now know, in addition, that diffusion rates of melt components in a wet boundary layer could be significantly accelerated relative to those in drier parts of the magma body. Since 1974, a large and diverse effort has been devoted to studies of convection/diffusion processes in magmas. To a great degree these seem to have been stimulated by the field and analytical work of Hildreth (e.g., 1979), whose reconstruction of a chemically-zoned magma chamber focused the attention of petrologists, geochemists, fluid dynamicists and numerical modelers on the plausibility and efficacy of various transport processes. Suffice it to say here that situations can arise where diffusion becomes important in large-scale processes despite its general ineffectiveness relative to magma advection and heat conduction (e.g., in cases where intra-layer convection in stably stratified bodies persists long enough for diffusive mass exchange between layers to be effective). We are now in a reasonably good position to include in models of such processes both the diffusion of dissolved volatiles themselves and their effects on diffusion of other magma components.

ACKNOWLEDGMENTS

The initial manuscript leading to this paper benefited from unofficial reviews by Dave Wark, Daniele Cherniak, John Hanchar, Dan Moore, Tobi Cohen and Sumit Chakraborty. Formal reviews by Don Baker, Todd Dunn and Youxue Zhang were extremely helpful in producing a more thorough and accurate final draft, and are greatly appreciated. Over the years, research on diffusion in melts at Rensselaer has been supported by the National Science Foundation, initially under grants EAR78-12980 and EAR80-25887, and most recently by EAR-9105055.

REFERENCES

Arzi AA (1978) Fusion kinetics, water pressure, water diffusion and electrical conductivity in melting rock, interrelated. J Petrol 19:153-159

Bai TB, Koster van Groos A (1994) Diffusion of chlorine in granitic melts. Geochim Cosmochim Acta 58:113-123

Baker DR (1990) Chemical interdiffusion of dacite and rhyolite: anhydrous measurements at 1 atm and 10 kbar, application of transition state theory, and diffusion in zoned magma chambers. Contrib Mineral Petrol 104:407-423

Baker DR (1991) Interdiffusion of hydrous dacitic and rhyolitic melts and the efficacy of rhyolite contamination of dacitic enclaves. Contrib Mineral Petrol 106:462-473

Baker DR (1993) The effect of F and Cl on the interdiffusion of peralkaline intermediate and silicic melts. Am Mineral 78:316-324

Baker DR, Bossànyi H (1994) The combined effect of F and H_2O on interdiffusion between peralkaline dacitic and rhyolitic melts. Contrib Mineral Petrol (in press)

Baker DR, Watson EB (1988) Diffusion of major and trace elements in compositionally complex Cl- and F-bearing silicate melts. J Non-Cryst Solids 102:62-70

Baker L, Rutherford MJ (1994) Diffusion rates of sulfur (S^{6+}) in hydrous silicic melts. Trans Am Geophys Union 75:353

Blank, JG (1993) An experimental investigation of the behavior of carbon dioxide in rhyolitic melt. PhD dissertation, California Institute of Technology, Pasadena, CA

Blank JG, Stolper EM, Zhang Y (1991) Diffusion of CO_2 in rhyolitic melt. Trans Am Geophys Union 72:312

Bottinga Y, Javoy M (1990) MORB degassing: Bubble growth and ascent. Chem Geol 81:255-270

Bowen NL (1928) The Evolution of the Igneous Rocks. Dover Publications, New York

Brady JB (1975) Reference frames and diffusion coefficients. Am J Sci 275:954-983

Carroll MR (1991) Diffusion of Ar in rhyolite, albite, and orthoclase composition glasses. Earth Planet Sci Lett 103:156-168

Carroll MR, Rutherford MJ (1988) Sulfur speciation in hydrous experimental glasses of varying oxidation state: Results from measured wavelength shifts of sulfur X-rays. Am Mineral 73:845-849

Carroll MR, Stolper EM (1991) Argon solubility and diffusion in silica glass: Implications for the solution behavior of molecular gases. Geochim Cosmochim Acta 55:211-225

Carroll MR, Stolper EM (1993) Noble gas solubilities in silicate melts and glasses: New experimental results for argon and the relationship between solubility and ionic porosity. Geochim Cosmochim Acta 57:5039-5051

Carroll MR, Sutton SR, Rivers ML, Woolum DS (1993) An experimental study of krypton diffusion and solubility in silicic glasses. Chem Geol 109:9-28

Chekhmir AS (1984) Experimental study of diffusion processes in magmatic melts. PhD dissertation, Vernadskii Inst of Geochemistry and Analytical Chemistry, Moscow, USSR.

Chekhmir AS, Epel'baum MB (1991) Diffusion in magmatic melts: New study. In: Physical Chemistry of

Magmas (Advances in Physical Geochemistry, v 9; LL Perchuk, I Kushiro, eds), Springer-Verlag, New York

Chekhmir AS, Persikov ES, Epel'baum MB, Bukhtiyarov PG (1985) Hydrogen transport through a model magma. Geokhimiya No. 5:594-598

Cooper AR (1968) The use and limitation of the concept of an effective binary diffusion coefficient for multi-component diffusion. In: Mass Transport in Oxides (JB Wachtman, AD Franklin, eds) National Bureau of Standards Publication 296:79-84

Crank J (1975) The Mathematics of Diffusion (2nd edn). Oxford University Press, London

Cussler EL (1984) Diffusion: Mass Transfer in Fluid Systems. Cambridge University Press, New York

Darken LS (1948) Diffusion, mobility and their interrelation through free energy in binary metallic systems. Trans AIME 175:184-201

Delaney JR, Karsten JL (1981) Ion microprobe studies of water in silicate melts: concentration-dependent diffusion in obsidian. Earth Planet Sci Lett 52:191-202

Dingwell DB, Mysen BO (1985) Effects of water and fluorine on the viscosity of albite melt at high pressures. Earth Planet Sci Lett 74:266-274

Dingwell DB, Scarfe CM (1984) Chemical diffusion of fluorine in jadeite melt at high pressure Geochim Cosmochim Acta 48:2517-2525

Dingwell DB, Scarfe CM (1985) Chemical diffusion of fluorine in melts in the system $Na_2O-Al_2O_3-SiO_2$. Earth Planet Sci Lett 73:377-384

Doremus RH (1969) The diffusion of water in fused silica. In: Reactivity of Solids: Proc 6th Int'l Symp (JW Mitchell, ed), p 667-673, Wiley Interscience, New York

Doremus RH (1994) Glass Science (2nd edn) Wiley Interscience, New York

Dunn JT, Ratliffe WA (1990) Chemical diffusion of ferrous iron in a peraluminous sodium aluminosilicate melt. J Geophys Res 95:15,665-15,673

Epstein PS, Plesset MS (1950) On the stability of gas bubbles in liquid-gas solutions. J Chem Phys 18:1505-1509

Fenn PM (1977) The nucleation and growth of alkali feldspars from hydrous melts. Can Mineral 15: 135-161

Fine G, Stolper EM (1985) The speciation of carbon dioxide in sodium aluminosilicate glasses. Contrib Mineral Petrol 91:105-201

Fine G, Stolper EM (1986) Dissolved carbon dioxide in basaltic glasses: concentration and speciation. Earth Planet Sci Lett 76:263-278

Fogel RA, Rutherford MJ (1990) The solubility of carbon dioxide in rhyolitic melts: a quantitative FTIR study. Am Mineral 75:1311-1326

Friedman I, Long W (1976) Hydration rate of obsidian. Science 191:347-352

Hamann SD (1965) The influence of pressure on electrolytic conduction in alkali silicate glasses. Austral J Chem 18:2-8

Harrison TM, Watson EB (1983) Kinetics of zircon dissolution and zirconium diffusion in granitic melts of variable water content. Contrib Mineral Petrol 84:66-72

Harrison TM, Watson EB (1984) The behavior of apatite during crustal anatexis: equilibrium and kinetic considerations. Geochim Cosmochim Acta 48:1467-477

Hart SR (1981) Diffusion compensation in natural silicates. Geochim Cosmochim Acta 45:279-291

Hildreth W (1979) The Bishop Tuff: Evidence for the origin of compositional zonation in silicic magma chambers. Geol Soc Am Spec Paper 180:43-76

Hiyagon H (1984) Experimental studies on noble gas diffusion in basalt glass and partition between basalt melt and olivine. PhD dissertation, University of Tokyo

Hofmann AW, Magaritz M (1977) Diffusion of Ca, Sr, Ba, and Co in a basalt melt: Implications for the geochemistry of the mantle. J Geophys Res 82:5432-5440

Jambon A (1979) Diffusion of water in a granitic melt: An experimental study. Carnegie Inst Wash Yearbook 78:352-355

Jambon A, Carron J-P, Delbove F (1978) Données préliminaires sur la diffusion dans les magmas hydratés: le césium dans un liquide granitique à 3 kbar. Compt Rendu Paris D 287:403-406

Jambon A, Carron JP (1973) Etude expérimentale de la diffusion des éléments alcalins K, Rb, Cs, dans une obsidienne granitique. Compt Rendu Paris D 276:3069-3072

Jambon A, Shelby JE (1980) Helium diffusion and solubility in obsidians and basaltic glass in the range 200°-300°C. Earth Planet Sci Lett 51:206-214

Jambon A, Zhang Y, Stolper EM (1992) Experimental dehydration of natural obsidian and estimation of D_{H_2O} at low water contents. Geochim Cosmochim Acta 56:2931-2935

Johnston RF, Stark RA, Taylor J (1974) Diffusion in liquid slags. Ironmaking and Steelmaking 1:220-227

Karsten JL, Holloway JR, Delaney JR (1982) Ion microprobe studies of water in silicate melts: Temperature-dependent water diffusion in obsidian. Earth Planet Sci Lett 59:420-428

Kress VC, Ghiorso MS (1993) Multicomponent diffusion in $MgO-Al_2O_3-SiO_2$ and $CaO-MgO-Al_2O_3-SiO_2$ melts. Geochim Cosmochim Acta 57:4453-4466

Kress VC, Ghiorso MS (1994) Multicomponent diffusion in basaltic melts. Geochim Cosmochim Acta (in press)

Lapham KE, Holloway JR, Delaney JR (1984) Diffusion of H_2O and D_2O in obsidian at elevated temperatures and pressures. J Non-Cryst Solids 67:179-191

Lasaga AC (1982) Toward a master equation in crystal growth. Am J Sci 282:1264-1288

Lesher CE (1994) Kinetics of Sr and Nd exchange in silicate liquids: Theory, experiments, and applications to uphill diffusion, isotopic equilibration, and irreversible mixing of magmas. J Geophys Res 99: 9585-9604

Lux G (1987) The behavior of noble gases in silicate liquids: Solution, diffusion, bubbles and surface effects, with applications to natural samples. Geochim Cosmochim Acta 51:1549-1560

Manning JR (1968) Diffusion Kinetics for Atoms in Crystals. Van Nostrand Co, Princeton, New Jersey

Mehnert KR, Büsch W, Schneider G (1973) Initial melting at grain boundaries of quartz and feldspar in gneisses and granulites. Neues Jb Mineral Mh 4:165-183

Onsager L (1945) Theories and problems of liquid diffusion. Ann NY Acad Sci 46:241-265

Rapp RP, Watson EB (1986) Monazite solubility and dissolution kinetics: Implications for the thorium and light rare earth chemistry of felsic magmas. Contrib Mineral Petrol 94:304-316

Reynolds JH (1960) Rare gases in tektites. Geochim Cosmochim Acta 20:101-114

Richter FM (1993) A method for determining activity-composition relations using chemical diffusion in silicate melts. Geochim Cosmochim Acta 57:2019-2032

Shaw HR (1963) Obsidian-H_2O viscosities at 1000 and 2000 bars in the temperature range 700 to 900°C. J Geophys Res 68:6337-6343

Shaw HR (1974) Diffusion of H_2O in granitic liquids: Part I. Experimental data. In: Geochemical Transport and Kinetics (AW Hofmann et al., eds), Carnegie Inst Wash Publ 634:139-154

Shewmon PG (1989) Diffusion in Solids (2nd edn). The Minerals, Metals and Materials Society, Warrendale, Pennsylvania

Smith VG, Tiller WA, Rutter JW (1955) A mathematical analysis of solute redistribution during solidification. Can J Phys 33:723-744

Sparks RSJ (1978) The dynamics of bubble formation and growth in magmas: A review and analysis. J Volcanol Geotherm Res 3:1-37

Stanton TR, Holloway JR, Hervig RL, Stolper, EM (1985) Isotopic effect on water diffusivity in silicic melts. Trans Am Geophys Union 66:1131

Stolper EM (1982) Speciation of water in silicate melts. Geochim Cosmochim Acta 46:2609-2620

Trial AF, Spera FJ (1994) Measuring the multicomponent diffusion matrix: Experimental design and data analysis for silicate melts. Geochim Cosmochim Acta (in press)

Watson EB (1979a) Calcium diffusion in a simple silicate melt to 30 kbar. Geochim Cosmochim Acta 43:313-322

Watson EB (1979b) Diffusion of cesium ions in H_2O-saturated granitic melt. Science 205:1259-1260

Watson EB (1981) Diffusion in magmas at depth in the Earth: the effects of pressure and dissolved H_2O. Earth Planet Sci Lett 52:291-301

Watson EB (1982) Basalt contamination by continental crust: some experiments and models. Contrib Mineral Petrol 80:73-87

Watson EB (1991) Diffusion of dissolved CO_2 and Cl in hydrous silicic to intermediate magmas. Geochim Cosmochim Acta 55:1897-1902

Watson EB, Bender JF (1980) diffusion of cesium, samarium, strontium, and chlorine in molten silicate at high temperatures and pressures. Geol Soc Am Abstr Progr 12:545

Watson EB, Sneeringer MA, Ross A (1982) Diffusion of dissolved carbonate in magmas: Experimental results and applications. Earth Planet Sci Lett 61:346-358

Watson EB, Wark DA, Delano JW (1993) Initial report on sulfur diffusion in magmas. EOS Trans Am Geophys Union 74:620

Webster JD, Duffield WA (1994) Extreme halogen abundances in tin-rich magma of the Taylor Creek rhyolite, New Mexico. Econ Geol (in press)

Zhang Y (1993) A modified effective binary diffusion model. J Geophys Res 98:11901-11920

Zhang Y, Stolper EM (1991) Water diffusion in basaltic melt. Nature 351:306-309

Zhang Y, Stolper EM, Wasserburg GJ (1991) Diffusion of water in rhyolitic glasses. Geochim Cosmochim Acta 55:441-456

Chapter 11a

PHYSICAL ASPECTS OF MAGMA DEGASSING I.

Experimental and theoretical constraints on vesiculation

R.S.J. Sparks[1], J. Barclay[1], C. Jaupart[3],

H.M. Mader[2] and J.C. Phillips[1]

[1] Department of Geology
University of Bristol
Wills Memorial Building, Queens Road
Bristol BS8 1RJ, U.K.

[2] Institute of Environmental and Biological Sciences
Lancaster University
Lancaster LA1 4YQ, U.K.

[3] Laboratoire de Dynamique des Systèmes Géologiques
Institut de Physique du Globe, 4 place Jussieu
75252 Paris Cedex 05, France

INTRODUCTION

One of the central phenomena of volcanology concerns the degassing of magma. Violent degassing results in explosive eruptions and formation of pyroclastic deposits. Slow degassing allows the separation of exsolving gases from the host magma and the eruption of degassed lava. Separated gases can contribute significant components to hydrothermal systems and cause fumarolic emissions. Segregation of gas from magma can also lead to chemical differentiation and can induce significant changes in magma physical properties such as crystal content, density and viscosity. This chapter reviews current understanding of the physical processes involved in degassing. The focus of the review is on the violent degassing that occurs in explosive eruptions, but we also consider briefly the slower degassing of magmas that erupt as lavas, since the former process can only be understood fully if the latter process is also well understood. The review concentrates on the degassing of high viscosity magmas, since degassing of basaltic magmas is comparatively well understood. As will emerge, degassing processes are complex and current understanding is far from complete. However recent advances in experimental simulations of degassing and theoretical modelling of bubble formation and flows in volcanic conduits allow a working hypothesis for explosive degassing to be proposed which departs significantly from previous ideas.

OBSERVATIONAL CONSTRAINTS

In this section we review what constraints can be placed on degassing processes from known physico-chemical properties of magmas and from observations of eruptions and their products. More detailed discussions of many of the issues can be found in other chapters of this volume and so our discussion will be brief. Key parameters in degassing are eruption rate, volatile solubility, volatile diffusivity, and magma composition and temperature, with their effects on viscosity and diffusivity. Surface tension effects are also important, but these are discussed thoroughly by Cashman and Mangan (Chapter 11b). These parameters are not independent of one another but their distinction helps to focus discussion of the information available.

Volatile solubility

For the purpose of discussing physical mechanisms of degassing the three volatiles of interest are H_2O, CO_2 and S as these are generally the principal volatile components. In most volcanic systems water is usually dominant, but there are examples of CO_2 rich systems such as kimberlites and carbonatites, and sulphur-rich magmas as in some calcalkaline systems. Water contents vary quite widely in magmas with contents typically in the range of 2 to 6 wt % H_2O in intermediate and silicic magmas. Basaltic magmas generally have significantly lower water contents with a broad dependence on tectonic environment. MORB typically contain between 0.1 and 0.75 wt % H_2O (e.g., Dixon et al., 1988) while alkali basalts and island arc basalts tend to have higher contents, perhaps up to 2 wt %.

The solubility of water is now known for a variety of magma types (see Chapters 3, 4, and 6), and as a first approximation water solubility in H_2O-saturated melts increases as the square root of pressure, P

$$C_S = k\ P^{0.5} \tag{1}$$

where C_S is the saturated concentration and k is the solubility constant. On a wt % basis, water is somewhat more soluble in rhyolite than in basalt, with the value of k being about a factor of 1.6 greater in rhyolite. The dissolution of water involves both OH^- and molecular H_2O (Stolper, 1982, 1989) with the concentration of these species being dependent on both temperature and pressure. Water solubility decreases slightly as temperature increases (Stolper, 1989).

From the perspective of volcanic processes the important consequence is that ascending magmas become supersaturated at shallow crustal depths. A rhyolite magma with 6 wt % dissolved H_2O will become saturated at approximately 200 MPa, equivalent to only about 6 km depth. Because of the square root dependence of solubility on pressure, and the strong pressure dependence of gas volume, the confining pressures at which substantial gas exsolution occurs are low and typically equivalent to a few hundred metres depth in a magma column. At this stage we are particularly careful to distinguish between confining pressure and depth. The common assumption that magma pressure will be close to the lithostatic or magmastatic (hydrostatic) pressure during volcanic eruptions will be shown below to be a poor assumption. Internal pressures can be greater or much less than the depth-equivalent lithostatic pressure in a volcanic event. There is also the caveat that the pressure in Equation (1) is the partial pressure of H_2O and this can be much lower than the confining pressure due to the presence of other volatile components.

CO_2 is very poorly soluble in common silicate magmas at crustal depths, with solubility being an almost linear function of pressure (Brey, 1976; Fine and Stolper, 1986; Stolper and Holloway, 1988; Chapters 5 and 6). Thus, most crustal magma chambers would be expected to be CO_2-saturated. Continuous release of CO_2 would therefore be expected during fractional crystallisation in crustal chambers and during magma ascent through the crust. Although H_2O is typically much more abundant than CO_2, the latter volatile may have an important role in the degassing behavior of H_2O and its distribution in a magma system. Some water partitions into the CO_2 gas phase (Holloway, 1976) even at depths where the water is nominally undersaturated. Thus, continual release of CO_2 and rise of bubbles through a magma can redistribute or progressively degas H_2O, and CO_2 fluxing can provide a mechanism of concentrating water at the top of a magma chamber (see Chapter 8).

The presence of sulphur in magmas can also influence water solubility. Under oxidizing conditions a sulphur-rich magma can result in greatly reduced partial pressures of water, due to high partial pressures of SO_2 (Carroll and Rutherford, 1985, 1987). Thus, a magma chamber at the same depth with the same water content may be characterized by a variety of water activities, depending on the activities of other volatile species such as H_2S and SO_2.

Basaltic (low viscosity) systems

Basalts have high temperatures, low viscosities and high gas diffusivities. They also tend to have more modest volatile contents than more evolved magmas. Many, although not all, aspects of degassing behavior of basaltic magmas are reasonably well understood. Many eruptions have been observed at close quarters, providing observations which cannot be easily made in more explosive high viscosity systems which tend to erupt much less frequently.

Low melt viscosity allows gas to segregate easily in basaltic magmas during ascent and in the vent environment. Typically activity varies from slow bubbling from a lava lake to intense fire-fountaining and strombolian explosive eruptions. The high gas diffusivities and ease of bubble coalescence allow large bubbles to form (Sparks, 1978; Sahagian, 1985; Head and Wilson 1987; Sahagian et al., 1989; Jaupart and Vergniolle, 1989). Bubbles in basaltic scoria are typically hundreds of microns to several millimetres in diameter (Mangan et al., 1993; Cashman and Mangan, this volume) and these sizes are consistent with the time-constraints for growth expected in basaltic eruptions (Sparks, 1978; Cashman and Mangan, this volume). Bubbles can coalesce easily because of the low melt viscosity (Sahagian et al., 1989) and this mechanism can account for the larger bubbles with diameters ranging from several centimetres to more than a metre which have been observed in ejecta and in large bursting bubbles in lava lakes and vents (Vergniolle and Jaupart, 1986). The physical mechanisms of gas segregation during fire-fountaining and strombolian activity have been well characterised (Vergniolle and Jaupart, 1986; Head and Wilson, 1987; Vergniolle and Jaupart, 1990).

Typical basaltic explosive eruptions produce rather coarse ejecta (e.g., Walker and Croasdale, 1972; Self et al., 1974). The larger clasts can be still hot and fluid and land to form spatter. Intense fire-fountaining can generate clastogenic lavas directly from the fall-out which cool negligibly and coalesce on landing (Swanson and Fabbi, 1973). Many Hawaiian lavas are directly fed from fire-fountains. Extreme degassing products can be produced such as pelées hair and reticulite, which demonstrates the ability of these magmas to expand to very high porosities due to gas exsolution without disruption (Cashman and Mangan, this volume; Thomas et al., 1994).

Silicic (high viscosity) systems

Degassing of rhyolitic and dacitic magmas with high viscosities is poorly understood and is the focus of this chapter. There are however a number of well-established facts and constraints on degassing from known physico-chemical properties and observations.

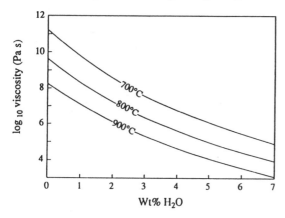

Figure 1. The variation of viscosity with water content for a rhyolitic composition reported by Dunbar and Kyle (1993) from the Taupo Volcanic Zone, New Zealand. Variations are shown for three different temperatures using the method of Shaw (1972).

Viscosity and diffusivity. Silicic magmas are highly polymerised systems (Hess, 1980) and a key factor in degassing is the profound influence of small amounts of dissolved water on viscosity and volatile diffusivity. Dissolution of water depolymerises silicic melts by reaction between H_2O and aluminosilicate chains and complexes (Burnham, 1975; Stolper, 1982). This is manifested at a fixed temperature by an increase in viscosity of several orders of magnitude (Fig. 1) when a magma containing a few percent water degasses completely to one atmosphere. This property, which is central to understanding the degassing of high viscosity magmas, poses several problems. Eruption of fully degassed basaltic lava is easily explained by the ease of segregation of the gas phase in low viscosity magmas. However the mechanisms in operation in basaltic magmas are not generally applicable to high viscosity silicic systems. Bubbles can ascend at speeds faster than or comparable to speeds of ascent of basaltic magma. However, speeds of bubble ascent in much more viscous magmas can be orders of magnitude less than ascent rates (Sparks, 1978) making simple bubbling away an improbable mechanism for degassing silicic magmas. Various models involving degassing through permeable foams have been proposed in order to explain eruption of fully degassed silicic lava which once contained high volatile contents (Eichelberger 1989; Eichelberger et al., 1986; Jaupart and Allegre, 1991). These models are not yet wholly convincing, because mechanisms of permeability development in ascending vesiculating magma are not well understood (Chapter 11b). Another problem posed by the very strong dependence of viscosity on gas content is how bubbles can expand in such very high viscosity liquids in explosive eruptions on apparently very short time-scales. This problem is addressed in a later section.

The diffusion of H_2O in silicic magmas is substantially slower than in basaltic magmas, as a consequence of lower temperature and melt structure (Shaw, 1974; Watson, 1981; Delaney and Karsten, 1981; Karsten et al., 1982). One consequence of this is that bubble sizes in silicic magmas are typically much smaller on the characteristic time-scales of volcanic eruptions and in broad terms this is born out by calculations (Sparks, 1978; Proussevitch et al., 1993) and observation (Whitham and Sparks, 1986; Sparks and Brazier; 1982). Water diffusivity is also strongly concentration-dependant (Shaw, 1975; Karstens and Delaney, 1981) which can be explained by the different speciations of water (Zhang et al., 1991).

Eruption styles. Silicic magmas display a wide range of phenomena during their eruptions. The two most important types of explosive style are plinian (fallout) eruptions and pyroclastic flow generation by fountain collapse. The controls on these two styles are primarily related to vent conditions, such as vent dimensions, magma gas content and magma discharge rate, and to interaction of the erupting jet with the atmosphere (Sparks and Wilson, 1976; Wilson et al., 1980; Neri and Dobran, 1994). Total exsolved gas content is an important factor in controlling fountain collapse, with low gas contents favoring collapse and high gas contents favoring plinian columns (Wilson et al., 1980). Models of conduit flow for these eruptions have assumed that the gas-melt system is always close to equilibrium at the local pressure conditions (e.g., Wilson et al., 1980; Dobran, 1992). In the models of Wilson et al. (1980) pressure in the ascending magma is lithostatic although this assumption is relaxed in the more general models of Dobran (1992) and Neri and Dobran (1994). The assumption of pressure close to lithostatic is shown below to be improbable. There may also be important non-equilibrium phenomena due to the kinetics of degassing which have not yet been recognised. Although these effects are expected to be second-order, a full evaluation of their influence on column stability must await a better understanding of the kinetic effects.

Both plinian and ignimbrite eruptions place useful constraints on time-scales for degassing. Discharge rates during plinian eruptions are high, typically ranging from 10^3 to 10^5 m^3s^{-1} (Walker, 1981; Sparks, 1986). They have been estimated to attain even higher values during major pyroclastic flow eruptions, e.g., 10^8 m^3s^{-1} in the Taupo ignimbrite eruption, New Zealand (Wilson, 1985). The time-scales for magma ascent and degassing are thus very short and can be illustrated by the Mount St. Helens plinian eruption of May 18th, 1980, where 0.2 km^3 of dacitic magma was erupted in 9 hours at an average rate of 10^4 m^3s^{-1} from a chamber at 6 to 7 km depth (Criswell, 1987; Carey et al., 1990). The magma contained 4.5 % H_2O and would therefore saturate at a pressure of 110 MPa. Assuming a conduit 15 m in radius the average ascent velocity of the magma would have been of order 15 ms^{-1}. This estimate indicates about 5 to 10 minutes as the total time for more or less complete degassing of the magma from the chamber to deposition as pumice clasts. Although the time could be increased somewhat by assuming a different geometry for the conduit the time-scale would still be of the same order. In fact the time-scale for the main degassing must be much shorter. The solubility of water and large expansion of the gas phase at low pressure requires that most gas exsolves and expands in the uppermost parts of the magma column and the degassing time is therefore more likely to be tens of seconds or even less, rather than minutes. This estimate is given further justification later in this chapter.

Ejecta are observed or deduced to travel at typical speeds of 200 to 500 ms^{-1} when they emerge from vents in explosive eruptions (Wilson et al., 1980; Wilson, 1980; Dobran, 1992). These velocities require a dramatic acceleration of the vesiculating magma. This acceleration must be associated with rapid expansion of the gas phase. Key issues here are how the gas expands explosively when surrounded by highly viscous melt and how this process relates to fragmentation.

Another major style of eruption is the extrusion of viscous lava domes which is sometimes associated with sudden violent vulcanian explosions. Lava dome eruptions pose some intriguing problems. Those domes that erupt quietly are often petrologically indistinguishable from associated pyroclastic ejecta (usually from earlier phases of activity). In some cases petrological observations indicate very similar pre-eruptive volatile contents for the batches of magma that erupt as lava, and those that erupted explosively. The problem is how such viscous magmas lost their gas during ascent.

Paradoxically perhaps, some lava domes can be destroyed by sudden onset of explosive activity, suggesting either the development of gas-rich magma below them or development of high internal gas pressures within the dome itself (Sato et al., 1992). Several months after the May 18th, 1980 eruption of Mount St. Helens, several cycles of dome growth and explosive eruption were observed. Lascar Volcano in Northern Chile provides a good example of an explosively unstable dome system. In the last few years a silicic andesite lava dome has been developing in the crater (Oppenheimer et al., 1993). Its development is associated with vigorous and continuous high temperature fumaroles with a substantial magmatic gas component (Andres et al., 1991). On several occasions since 1984, however, there have been sudden and very violent vulcanian eruptions (the latest on July 26th, 1994), and a short-lived but very intense plinian eruption in April 1993. Curiously these intense explosive events are preceded by subsidence of the lava dome. Such explosive activity is particularly dangerous due to its suddenness. Perhaps all that can be said at this stage is that the Lascar dome evolves to criticality as a consequence of the degassing process in the conduit system. How exactly this happens is not understood.

Eruption products. The products of explosive silicic eruptions are pumice and ash. The very commonness of these materials tends to hide the fact that there are some very fundamental problems with their interpretation. Indeed there is remarkably little quantitive data on their basic physical and textural features. There are extensive and excellent descriptions in the atlas by Heiken and Wohletz (1985), some recent systematic studies of density (Houghton and Wilson, 1989; Thomas et al., 1994) and a few studies of vesicle size distributions (Whitham and Sparks, 1986; Toramaru, 1990; see also Chapter 11b).

Despite the lack of data some statements can be made.

(1) Many pumices from plinian deposits and ignimbrites have mean vesicularities between 70 and 80% with quite narrow variations (Table 1; Thomas et al., 1994). However Thomas et al. (1994) have recognised when vesicularity is expressed as a voidage it tends to disguise important variations. For example they found for the Hatepe plinian pumice (New Zealand) that an observed vesicularity range from 68% to 78% represents a range of gas/liquid volume ratio of 2.1 to 3.5 respectively. This vesicularity range has been described as the fragmentation threshold (Sparks, 1978) implying that magmas disrupt when they achieve these amounts of expansion. The physical significance of this observation is however far from clear and is discussed more fully later. Thomas et al. (1994) in an analysis of several deposits (Table 1) found a correlation between mean vesicularity of pumice and inferred magma viscosity with the vesicularity increasing in lower viscosity magmas. However they found that there was no correlation between discharge rate and mean pumice vesicularity.

Table 1

Data on vesicularities of silicic pumices from Thomas et al. (1994) with estimated viscosities with dissolved water contents at level of fragmentation.

Eruption	Melt viscosity (Pas)	Mean vesicularity (%)	Range (%)
Bishop (Plinian)	2×10^8	71	64-78
Hatepe (Plinian)	1×10^7	73	68-77.9
Taupo (Plinian)	2×10^7	74	70.2-78.3
Taupo (ignimbrite)	2×10^7	73	70.6-77
Minoan (Plinian)	9×10^6	78	73.1-81.4
Minoan (ignimbrite)	9×10^6	77	75.2-79.8

(2) There is evidence for bimodal bubble size distributions (Sparks and Brazier, 1982; Whitham and Sparks, 1986) in some pumices. The physical explanation might be related to different nucleation events in the eruption process. The largest bubbles could be derived by nucleation in the magma chamber well before eruption. Smaller sized bubbles might nucleate during magma ascent and rapid explosive ejection. Most bubbles are small (tens to hundreds of microns in diameter) and nucleation densities are very high in the range 10^{12} to 10^{14} m^{-3}. These data can be compared with experimental studies of degassing (Hurwitz and Navon, 1994) and theoretical models (Sparks, 1978; Proussevitch et al., 1993).

(3) Pumice commonly contain elongated vesicles which in extreme form are known as tube or woody pumice. Unfortunately there are no systematic studies of the proportions of such material in different kinds of deposit or any attempt to quantify the degree of elongation. The elongated vesicles are typically circular in cross-section implying strong extensional strain rather than shear.

There are some major conceptual difficulties with interpretation of pumice and volcanic ash. At the observed vesicularities (commonly in the 0.7 to 0.8 range) magma containing a few percent water will have typical internal pressures of several MPa and residual water contents in the melt phase of 1 to 2 wt % at equilibrium (Fig. 2a; Sparks, 1978). The expansion to atmospheric pressure conditions has been interrupted by fragmentation, but the mechanism is far from clear. Sparks (1978) proposed that these vesicularities represent a limit to expansion due to interference between expanding bubbles. He made an analogy with the close-packing of solid particles. However, in high viscosity magmas very high vesicularities can be achieved (Thomas et al., 1994) which are not prevented by a "packing limit". Another possible explanation is high viscosity which slows down expansion. Silicic magmas typically have viscosities in the range 10^9 to 10^{10} Pas at the observed eruption vesicularities (Fig. 2b). Bubble growth models (Proussevitch et al., 1993; Barclay et al., 1995) are discussed in more detail in a later section. Empirically it is clear that pumice is formed by disintegration of a magmatic foam followed by rapid quenching as it is ejected into the cold atmosphere. However the physical significance of the typical magmatic foam vesicularities at disintegration still requires explanation.

Another related problem is how the remaining gas escapes from pumice given the high internal gas pressures at equilibrium. Does the pumice become permeable by rupture of vesicle walls? Although laboratory analyses confirm that pumice is almost entirely permeable with few or no closed vesicles (Whitham and Sparks, 1986) it does not follow that this is the case at magmatic temperatures. It could be argued that permeability is acquired during cooling and after deposition. Another question is why pumice is preserved at all? What distinguishes magmatic foam that disintegrates to form fine volcanic glass shards from that which remains coherent as discrete pieces of pumice containing very large numbers of bubbles?

These issues inevitably focus on the mechanism of magmatic foam disintegration, which still remains one of the most poorly understood aspects of explosive volcanism. There is the question of whether magma fragments by brittle fracture or is disrupted into a vesicular spray in the liquid state. One final observation is relevant and is in fact highly misleading. Many pumice fall deposits contain predominantly angular lapilli which demonstrate brittle fracture. However many if not all these fractures are the consequence of impact breakage (e.g., Sparks et al., 1981) and are unrelated to primary disintegration mechanisms. Original shapes are more often irregular with ragged outline, although this is another area where detailed measurements and documentation are lacking.

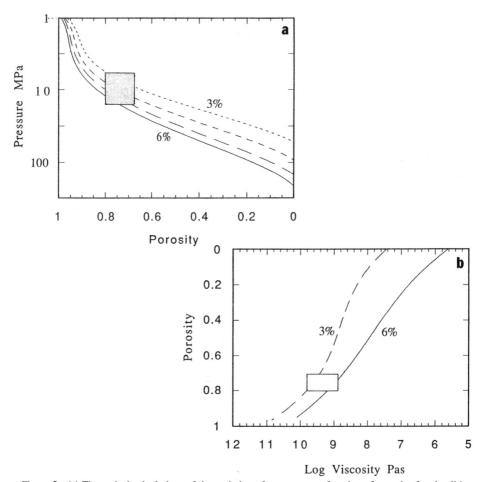

Figure 2. (a) Theoretical calculations of the variation of pressure as a function of porosity for rhyolitic magma at 700°C for different initial water contents (6, 5, 4 and 3 wt %). Calculations assume equilibrium partitioning of H_2O into the melt phase. The shaded area shows the typical range of mean pumice vesicularities in plinian deposits. (b) Theoretical calculations of the variation of viscosity with porosity for rhyoliitic magma at 700°C with an initial water contents of 3 wt % and 6 wt %. The box shows the typical range of mean pumice vesicularities in plinian deposits.

BUBBLE FORMATION

Nucleation theory

We first consider the case of homogeneous nucleation of a gas phase from a supersaturated melt. Classical theory (Zettlemoyer, 1969) envisages formation of small clusters of gas molecules due to local fluctuations in concentration. Energy is required to form the interface of the bubble embryo. Whether an embryo grows or shrinks depends on whether the decrease in free energy associated with formation of the separate gas phase is greater than the interfacial energy associated with creation of the surface of the bubble. The latter energy decreases relative to the former as the bubble embryo radius increases. A critical radius, r_c, is reached where the addition of one extra molecule results in growth (Hirth et al., 1970; Swanger and Rhines, 1972) and is given by

$$r_c = 2\sigma/\Delta P \tag{2}$$

where σ is the interfacial tension in N/m and ΔP is the supersaturation pressure defined as the difference between the gas pressure in the melt and the ambient pressure. For a water-rich silicic melt the only experimental data (Epel'baum et al., 1973) suggests a value of $\sigma = 0.06$ N/m. For example a critical radius of 0.1 micron requires a supersaturated pressure of 10 MPa for spontaneous growth. Such a bubble is still very large in comparison to molecular dimensions and at 1000°C will contain of order 10^6 molecules of H_2O! It is hard to envisage a process which would allow such a large cluster of molecules to form so it must be assumed that initial bubble embryos are much smaller than this.

Another approach to considering the kinetics of nucleation is from nucleation rate theory. Hirth et al. (1969) give the nucleation rate as

$$J = J_0 \exp [-\Delta G/kT] \tag{3}$$

where J_0 is a parameter related to the statistical distribution of gas molecules in the melt, T is temperature, and k is the Boltzmann constant. ΔG is the free energy of formation of a critical nucleus and is given as

$$\Delta G = 16 \pi \sigma^3/3\Delta P^2 \tag{4}$$

Nucleation becomes rapid when ΔG becomes comparable to or smaller than kT. Hurwitz and Navon (1994) have compared degassing of water from magma to evaporation processes and have estimated typical values of J_0 and ΔG in Equations (3) and (4). For a significant nucleation rate of 10^{10} m^3 s^{-1} they estimated that supersaturation pressures of about 60 MPa are required for significant nucleation which could account for the observed bubble densities in pumice after 100 s or so.

If crystals are present in the melt then nucleation can occur heterogeneously. If the gas phase can wet the surface of the crystal the supersaturation pressures required for spontaneous growth will be much smaller. The activation energy is now given by (Landau and Lifshitz, 1980)

$$\Delta G = 16 \pi \sigma^3 \emptyset/3\Delta P^2 \tag{5}$$

and is thus reduced by a factor

$$\emptyset = (2 - \cos\theta) (1 + \cos\theta)^2/4 \tag{6}$$

where $\quad \cos\theta = \sigma_{cg} - \sigma_{cm}/\sigma_{mg} \tag{7}$

and where σ is the surface energy and the subscripts cg, cm and mg stand for crystal-gas, crystal-melt and melt-gas interfaces respectively and θ is the contact angle. If the gas phase wets the crystal $\theta = 180°$ and $\emptyset = 0$ and no supersaturation is required. If the melt wets the crystal completely $\theta = 0°$ and $\emptyset = 1$ and there is no reduction in supersaturation pressure compared to the homogeneous case. Hurwitz and Navon (1994) have used this analysis to show that for $\theta > 150°$ there is a substantial decrease in supersaturation pressure on a flat surface (Fig. 3) and, following Sigbee (1969), that no supersaturation pressure is required for nucleation at a 90° corner if $\theta > 135°$.

Another explanation of facilitating nucleation is the presence of surfactants which reduce surface tension substantially (Sparks, 1978). Mechanisms have been identified in metallurgical systems, but as yet have not been recognised in magmatic systems.

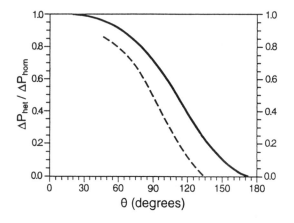

Figure 3. Ratio of the supersaturation pressure for heterogenous nucleation to that for homogeneous nucleation is plotted against wetting angle θ for a smooth surface (solid line) and a corner (dashed line) at a melt-crystal interface (after Hurwitz and Navon, 1994).

Experiments on nucleation

There have been few experimental studies of bubble nucleation (Murase and McBirney 1973; Hurwitz and Navon, 1994). The important study by Hurwitz and Navon (1994) involved dissolution of water in natural rhyolitic obsidians at temperatures of 780° to 850°C at high pressure followed by decompression. The study demonstrated the importance of heterogeneous nucleation and indicated that the large supersaturations predicted by homogeneous nucleation theory are correct. When microlites were still present (Fig. 4) nucleation occurred heterogeneously along microlite-rich bands and large nucleation densities were observed at only modest supersaturations. Decompressions of more than 5 MPa caused very large nucleation densities (10^{12} to 10^{14} m^{-3}). In those samples heated to sufficient temperature ($\geq 800°$C) to remove visible microlites no nucleation was observed for supersaturations less than 10 MPa, modest nucleation (10^9 to 10^{11} m^{-3}) was observed in the supersaturation range 10 to 70 MPa, and very high nucleation densities were seen above 70 MPa supersaturation. Even in these experiments Hurwitz and Navon (1994) considered nucleation occurred heterogeneously on unseen sites (Fig. 4a) or due to a heterogeneous melt phase. Even decompressions of 130 MPa failed to produce homogeneous nucleation. They also observed faster nucleation rates (10^{12} m^{-3} s^{-1}) with microlites present than when they were absent (10^{11} m^{-3} s^{-1}).

Another significant observation of Hurwitz and Navon (1994) concerned the propensity of different crystals to act as nucleation sites. Some mineral phases such as FeTi oxides, biotite, zircon and apatite have high wetting angles and therefore act as important nucleation sites, whereas other phases such as plagioclase have low wetting angles and do not influence nucleation.

Hurwitz and Navon (1994) propose that heterogeneous nucleation is the main mechanism of bubble formation in silicic magmas. This hypothesis seems reasonable in magma chambers where cooling and crystallization rates are slow and supersaturation pressures will increase gradually. Suspended phenocrysts are likely to act as the main nucleation sites and supersaturation pressures will remain low provided the concentration of suspended crystals is sufficient. The implications of the experiments for bubble

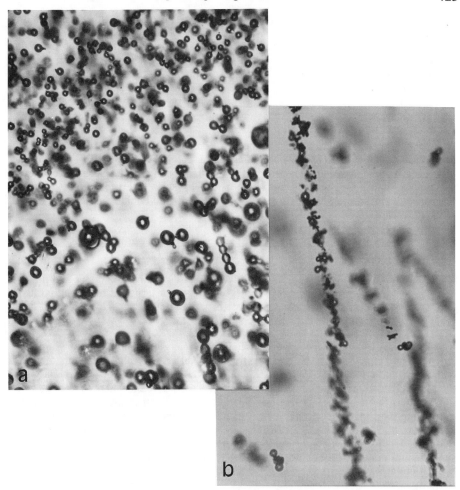

Figure 4. Photomicrographs of experiments by Hurwitz and Navon (1994). (a) The photograph shows a sample in which a saturated rhyolitic melt at 780°C decompressed by 30 MPa. The bubble sizes and number densities show marked heterogeneities. (b) The photograph shows a rhyolitic sample that was decompressed by 5 MPa at 780°C. The bubbles have nucleated on tiny microlite rich flow bands.

nucleation in phenocryst-free magmas and during eruptions are less clear. The tiny microlites in the experimental obsidians formed within supercooled lavas and have very high nucleation densities. There is no compelling evidence to suggest that comparable crystal densities or microlites exist in magma chambers or in ascending silicic magmas during explosive eruptions. Typical phenocryst concentrations in magmas range from 10^6 to zero m^{-3} which is orders of magnitude lower than the bubble nucleation densities observed in pumice (10^{10} m^{-3} and above; Whitham and Sparks, 1986; Toramaru, 1990). Nucleation on phenocrysts can only account for a few large bubbles which may form in the chamber prior to eruption (Sparks and Brazier, 1982). The experiments with microlite-free rhyolite may be closer to conditions in explosive eruptions. However these are problematic as the cause of heterogeneous nucleation is unclear. It is possible that dissolution of the microlites had occurred but the experimental time was insufficient to homogenise the melt so that remaining heterogeneities in the melt caused the heterogeneous behavior. In this case it can be argued that the behavior was determined by a crystallization and dissolution history unlikely to be realised in a natural system. Irrespective of the resolution of the problem the experiments demonstrate that large supersaturations can be developed in silicic magmas. This fact has significant implications for the occurrence of vesiculation in explosive eruptions.

Bubble growth

Bubble growth in erupting magma is a consequence of decompression. Two components of growth can usefully be distinguished: the first is due to diffusion of gas out of a supersaturated melt and the second is due to expansion of existing gas in the bubble as pressure is reduced. Bubble growth is opposed by surface tension effects, viscous resistance of the surrounding melt and inertia. The internal bubble pressure, P_b, can be written (Scriven, 1959)

$$P_b = P_a + \frac{2\sigma}{r} + 4\mu\frac{\dot{r}}{r} + \rho\left(\frac{3}{2}\dot{r}^2 + r\ddot{r}\right) \tag{8}$$

where P_a is the ambient pressure, μ is the dynamic viscosity, r is the radius and ρ is the density of the melt surrounding the bubble. The last term turns out to be negligible in almost all cases of volcanological interest (Sparks, 1978), except perhaps for lunar eruptions into vacuum, and is usually dropped.

There are two approaches to understanding bubble growth in magmas. Numerical calculations (e.g., Sparks, 1978; Proussevitch et al., 1993) allow a wide range of parameters and effects to be investigated, but have the disadvantage that interpretation of the results of a complex numerical code is not necessarily easier than interpretation of observations on the natural system (see Sparks, 1994). Another approach is to develop much simplified analytical models which have the advantage that the physical controls are very clear, but the disadvantage that the simplifications may make the calculations less realistic. In this review some simple analytical models are developed to elucidate the fundamental controls on bubble growth and some of the key results of the numerical simulations are summarised.

Diffusive growth. A simple approximate analysis of the diffusive growth of a bubble in a magma at large supersaturations can be made as follows. Consider a bubble of radius R with a thin region of gas depleted melt around the outside of thickness b. We assume that b << R and develops as a diffusional boundary layer with the saturated gas concentration at the bubble interface and the supersaturated gas concentration at R + b.

The thin boundary layer approximation gives

$$3 R^2 b = \phi R^3 \tag{9}$$

where ϕ is the ratio of gas density in the bubble to gas concentration in the melt, yielding

$$b = \frac{\phi R}{3} \tag{10}$$

The volumetric flux of gas, dV/dt, into the bubble can be approximated by assuming a linear gradient of concentration across the diffusive boundary layer and is given by

$$dV/dt = \frac{4\pi(C_s - C_m)DR^2}{b\rho} \tag{11}$$

where D is the diffusion coefficient, C_s is the equilibrium concentration of gas at the bubble interface, C_m is the supersaturated concentration of gas in the melt and ρ is the density of the gas in the bubble. Using Equations (10) and (11) and making the substitution $dV = 4\pi R^2 dR$ produces

$$dR/dt = 3D(C_s - C_m)/\rho\phi b \tag{12}$$

Integration gives

$$R = (6D(C_s - C_m)/\rho\phi)^{1/2} t^{1/2} \tag{13}$$

This result is the classic parabolic growth law (Scriven, 1959) and illustrates the strong dependence of growth on the supersaturation pressure, which controls the driving concentration difference, $(C_s - C_m)$, and diffusivity. The derivation assumes that bubble internal pressure and external pressure are constant. As is apparent from Equation (8) the bubble pressure can only be regarded as constant if surface tension, viscous and inertial effects are negligible. Furthermore, external pressure can change rapidly in a volcanic eruption. Finally the linear approximation to the boundary layer may be too crude in some situations, particularly where growth is very rapid and melt advection may be significant. Despite these caveats the parabolic law provides a useful paradigm.

Two substantial sets of numerical simulations have been carried out on the diffusive growth of bubbles in magmas (Sparks, 1978; Proussevitch et al., 1993). In the earlier study a growth rate coefficient was determined from experimental studies of CO_2 bubbles in silicone oil (Gale, 1966). The coefficient was related to the supersaturation of the gas dissolved in the liquid and, at a given supersaturation, was used as a constant in the parabolic growth equation (compare Eqn. 13). Apart from this empirical aspect, the equations that describe bubble growth were solved (e.g., Eqns. 1, 8, 13) by a simple finite difference scheme. The calculations included decompression effects due to magma ascent and to rise of individual bubbles in low viscosity melts. The calculations only concerned growth of individual bubbles in an infinite melt.

The major conclusions of Sparks (1978) are now summarised. Several results were as expected. Bubbles in magmas with high diffusivities and high water contents grow faster than bubbles in magmas with lower values of these parameters, and the bubbles achieve a larger size on reaching surface conditions during progressive decompression (Figs. 5d and 5e). Faster magma ascent rates result in smaller bubbles because there is less time for growth (Fig. 5a). A less obvious result was that onset of nucleation has little effect on final bubble size (Fig. 5b). These calculations involve some assumption about the initial supersaturation pressure. For the same ascent rate, water content and magma properties, a bubble that nucleates late at high supersaturation pressures almost catches up with a bubble that nucleates earlier at low supersaturation. Although a bubble nucleated at a high degree of supersaturation has less time for growth, the bubble growth is very rapid in its early stages (i.e., when it is small) because of the larger supersaturation driving growth (i.e., the larger the supersaturation the more rapid the growth rate). This effect almost, but not quite compensates for the late nucleation. This result takes on some significance in the light of the results of Hurwitz and Navon (1994) who established that quite large supersaturations can be developed during decompression in the absence of nuclei. If bubbles nucleate at high supersaturations then there is the potential for explosive growth rates.

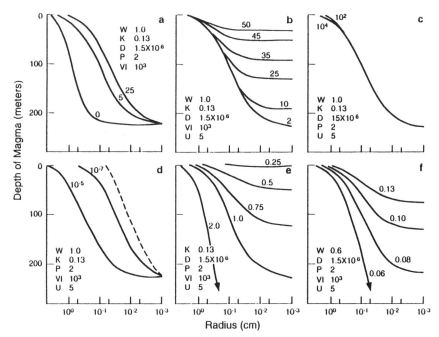

Figure 5. Bubble growth paths for bubbles in ascending magma (after Sparks, 1978). Each set of curves shows variations as a function of one of six controlling parameters, namely water content (W), viscosity (VI), gas diffusivity (D), initial supersaturation pressure (P), ascent velocity (U) and solubility constant (K). The curves are: (a) ascent velocity (cm/s); (b) supersaturation pressure (bars); (c) viscosity (poise); (d) gas diffusivity (cm^2/s); (e) water content (wt %); (f) solubility constant (bars $^{-1/2}$).

The calculations of Sparks (1978) investigated the influence of viscosity on growth. For a wide range of conditions he showed that the dynamic viscous pressure term in Equation (8) is small to negligible. However the calculations did not extend to viscosities above 10^7 Pas and did not consider decompression rates above about 0.1 MPa s^{-1}. At the time the paper was written the model of an explosive eruption consisted of fragmentation of a static overpressurised foam, which influenced the range of values investigated. It is however now evident that much larger decompression rates are possible, perhaps as high as 10 MPa s^{-1} in plinian eruptions and virtually instantaneous in vulcanian eruptions. Thus, the issue of viscous resistance still needs evaluation for higher viscosities and higher decompression rates. New calculations on this topic, after Barclay et al. (1995), are summarised below.

The calculations of Proussevitch et al. (1993) provided a more sophisticated numerical study which involved solution of the full diffusion equations for the bubble growth in a shell of melt of fixed volume. The calculations were an advance in theoretical terms on Sparks (1978) as they did not involve any empirical element. The calculations considered instantaneous decompression under a range of parameter conditions. Some important features of magma bubble growth were revealed.

The general form of the results (Fig. 6) shows a period of accelerating growth followed by a period of steady growth and then a period of decelerating growth. The initial period of growth was interpreted as a "time delay" related to the effects of surface tension. Viscous effects also retard growth and diminish with increasing bubble radius.

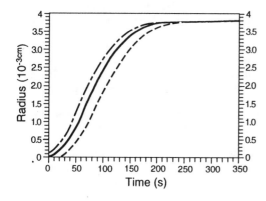

Figure 6. Typical result of numerical calculations carried out by Proussevitch et al. (1993) showing sigmoidal growth paths for bubble shells expandind following a pressure drop. The curves show runs ds-5 (dash-dot), ds-1(solid) and ds-3 (dash). The calculations involve decompression of a rhyolite with a viscosity of 10^6 Pas, a diffusivity of water of 10^{-11} m^2s^{-1} and a water content of 0.5 wt %. The three curves are for three different values of the initial radius: 4.5×10^{-7} m (ds-3), 10^{-5} m (ds-1) and 10^{-4} m (ds-5).

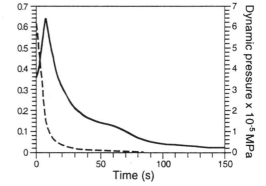

Figure 7. Variation of viscous and surface tension pressure terms in a typical numerical calculation (after Proussevitch et al., 1993). The run is ds-1 and the conditions are given in the caption to Figure 6.

Calculations by Proussevitch et al. (1993) indicate that the viscous pressure term is typically larger than the surface tension pressure (Fig. 7). The time delay and accelerating regime can therefore be interpreted as a consequence of viscosity, an interpretation supported by further calculations (Proussevitch et al., 1994). The period of steady growth occurs when the surface tension and viscosity are negligible and corresponds closely to the simple parabolic law. Finally, growth decelerates rapidly as gas is exhausted from the fixed volume shell.

The numerical calculations of Proussevitch et al. (1993) also demonstrated that for the viscosity range 10^5 to 10^8 Pas the viscous term is never large compared to the supersaturation pressure. They explored instantaneous pressure drops and higher viscosities than Sparks (1978) and found viscous pressures could reach up to about 1 MPa in the earliest stages of growth under the most extreme conditions. This effect is sufficient to explain the period of growth retardation.

Not all the results of Proussevitch et al. (1993) can be easily interpreted. The increase in the time delay with decreasing diffusivity is not obviously explicable by viscous or surface tension effects. A constructive discussion and response on the interpretation of these calculations is given in Sparks (1994) and Proussevitch et al. (1994).

Decompression of bubbles in high viscosity magma. The role of viscosity in bubble growth has been investigated by development of analytical models for instantaneous and linear decompressions (Barclay et al., 1995). The analytical models require considerable simplification to make the problem tractable, but have the merit that the results are easier to interpret than multiparameter numerical models. The analytical models neglect surface tension, inertial and diffusive effects on growth and simply consider the growth of a bubble in a viscous melt subjected to either an instantaneous pressure drop or a linear decrease in pressure. The mass of gas remains constant with no changes due to diffusion. The results allow a complete range of viscosities and pressure changes representative of any conceivable eruption condition. The influence of viscosity on bubble growth and excess internal bubble pressure have been explored in parameter ranges not examined by published numerical models of Sparks(1978) and Proussevitch et al.(1993).

Barclay et al. (1995) have developed models for a single bubble growing in an infinite melt and a thin-walled bubble modelled as a shell of melt of fixed volume. For instantaneous decompression the following solutions for bubble radius, R, as a function of time, t, were found for a bubble in infinite melt (14) and a thin shell (15)

$$R = R_0 \left(\frac{P_0}{P_e} \left(1 - \exp\left(\frac{-3P_et}{4\mu}\right) \right) + \exp\left(\frac{3P_et}{4\mu}\right) \right)^{1/3} \tag{14}$$

$$R = R_0 \left[\frac{P_e}{P_0} + \left(1 - \frac{P_e}{P_0} \right) \exp\left(\frac{3P_0R_0t}{4\mu r_0}\right) \right]^{-1/3} \tag{15}$$

where R_0 is the initial bubble radius, P_0 is the initial pressure, P_e is external pressure, μ the viscosity and r_0 is the initial radius of the shell. Analytical solutions for the case of linear decompression of a thin viscous shell can also be found

$$R = \left[\frac{R_0{}^3}{\left\{ 1 - \beta - \frac{\gamma}{\alpha} \right\} e^{\alpha t} + \beta + \frac{\gamma}{\alpha} - \gamma t} \right]^{1/3} \tag{16}$$

where $$\alpha = \frac{3P_0R_0}{2\mu r_0} \tag{17a}$$

$$\beta = \frac{\rho g h_0}{P_0} \tag{17b}$$

$$\gamma = \frac{\rho g w}{P_0} \tag{17c}$$

and ρ is the density of the magma, g is gravitational acceleration, h_0 the initial depth of the bubble and w is the velocity of magma ascent.

Figure 8 shows the variations of bubble radius and internal gas pressure with time for a viscosity of 10^9 Pas for the case of an instantaneous pressure drop of 30 MPa (equivalent to decompression from about 900 m depth). The infinite melt model (IMM) and shell model are both illustrated. The bubbles expand to the equilibrium radius and display a characteristic sigmoidal curve with some resemblance to the results of Proussevitch et al. (1993). The initial accelerating phase relates to the diminishing viscous effects as the bubble expands. The internal bubble pressure decreases as the bubble expands (Fig. 8b). Figure 9 shows the time taken to reach 99% of the final bubble size as a function of viscosity for viscous shells with different initial shell wall thickness to bubble radius ratios. The time-scale for viscous effects to be important can be readily seen.

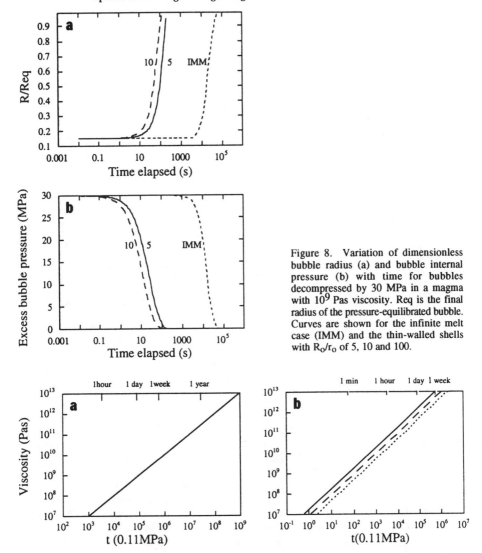

Figure 8. Variation of dimensionless bubble radius (a) and bubble internal pressure (b) with time for bubbles decompressed by 30 MPa in a magma with 10^9 Pas viscosity. Req is the final radius of the pressure-equilibrated bubble. Curves are shown for the infinite melt case (IMM) and the thin-walled shells with R_0/r_0 of 5, 10 and 100.

Figure 9. The time taken to decompress to a pressure of 0.11MPa as a function of viscosity for a decompression of 30 MPa to 0.1 MPa. Calculations are shown for (a) an infinite melt and (b) thin-walled shells with R_0/r_0 of 5 (dotted line), 10 (dashed line) and 20 (solid line).

The linear decompression model can also be applied to investigate what viscosities and decompression rates result in viscous effects becoming important. Equation (16) has been applied to look at internal bubble pressure when the pressure has dropped to 0.1 MPa. For a viscous shell with $R_0/r_0 \sim 5$ (Fig. 10) the viscous effects become important for large decompression rates and high viscosities.

The results can best be interpreted in conjunction with Figure 1 where viscosity of rhyolitic melt is shown as a function of dissolved water content. Clearly, silicic magmas containing a few percent dissolved water have viscosities which are too low to retard

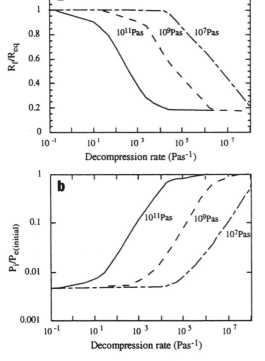

Figure 10. Calculations of final bubble radius (a) and internal pressure (b) of bubbles at an external pressure, Pe, of 0.1 MPa after different decompression rates. The calculations are for an initial pressure, Pi, of 1 MPa. The bubble radius is normalised by the equilibrium bubble radius, Req, at the external pressure. The pressure is normalised to the initial pressure so that when P_i/P_e =1 then the internal bubble pressure is equal to the initial starting pressure. Results for three different viscosities are shown for a shell bubble with an initial R_o/r_o of 10.

growth significantly and expansion time-scales are negligible. Magmas at fragmentation are partially degassed with vesicularities in the range 0.7 to 0.8 (Fig. 2b) and viscosities of 10^9 to 10^{10} Pas. The calculations of both instantaneous and linear decompression indicate the time-scales and decompression rates at which viscous effects will significantly inhibit growth. For a viscosity of 10^9 Pas the response time-scale for a bubble in an infinite melt is hundreds of seconds and of order 100 s in a viscous shell ($R_o/r_o = 10$). The time-scales are of the same order as the time-scales of magma ascent and fragmentation in explosive eruptions. This strongly suggests that there is a viscosity limit to expansion. The concept of a "viscosity quench" has been proposed by Thomas et al. (1994) based on an analysis of foam expansion; this is discussed further below.

Strongly degassed silicic magmas, such as those that occur in lava domes, can have viscosities in the range 10^{10} to 10^{13} Pas. The calculations (Figs. 9 and 10) indicate that viscous effects can be substantial with expansion time-scales of hours, days or even weeks. These calculations therefore might provide support for the concept (Sato et al., 1992) that silicic lava domes could maintain high internal pore pressures for periods of time comparable to their slow extrusion. Sato et al. (1992) proposed that in the Mount Unzen lava dome in Japan pore pressures were close to the tensile strength of the lava and this led to spontaneous disintegration when part of the lava dome collapsed to form fine-grained pyroclastic flows. There is however a logical difficulty. It is only fully degassed lava that can sustain high pore pressures, but the lava cannot be regarded as fully degassed if the pore pressures are high and therefore cannot have the viscosities necessary to sustain the high pressures. For example, if the tensile strength of lava is of order 10 MPa the equilibrium water content in the melt phase would be 1.3% and the viscosity at 800°C would be 10^9 Pas. This is too low a viscosity to allow large excess internal pressures to be maintained over the long periods of dome growth.

Decompression of high viscosity foams.   Thomas et al. (1994) have completed an analysis of the decompression of fragments of foam which has close analogies to the calculations of individual bubbles. Neglecting surface tension and diffusion they derive the following term for internal pressure variation

$$P_b = P_e + \frac{4}{3}\mu\frac{1}{\varepsilon}\frac{d\varepsilon}{dt} \tag{18}$$

where P_b is the internal pressure, P_e is the external pressure and ε is the void fraction. They introduce a dimensionless parameter, B, which is defined

$$B = \frac{4\mu}{3\tau\Delta P} \tag{19}$$

where τ is a time-scale for decompression, and ΔP is the pressure drop. For small B, viscous effects can be neglected, while for large B, viscous effects prevent foam expansion. They present results for an exponential pressure decrease to explore viscous effects. Equation (19) can be used in a simple way to explore under what conditions viscosity will start to inhibit growth. For example if $\tau \sim 100$ s (a representative value for an explosive eruption) then viscosities higher than 10^9 Pas will inhibit growth for a typical pressure drop of 30 MPa.

The calculations of Thomas et al. (1994) are complementary to those of Barclay et al. (1995). They indicate that in explosive eruptions there is a strong 'viscosity quench' in the region of 10^9 Pas for decompression rates and time-scales typical of explosive eruptions. Inspection of Figure 2b shows that this 'quench' effect occurs when vesicularities are in the range 0.7 to 0.8 and thus provides a good explanation of why pumice clasts have such narrow vesicularities and have not expanded further.

Growth of stretching bubbles. There are two observations that suggest that bubbles are deformed as they grow. First, new experiments using shock tubes (Mader et al., 1994 and described in a later section) suggest that when foams expand under explosive conditions the expanding system accelerates and bubbles could become elongated along the shock tube length. Second, tube pumice with strongly elongated cylindrical vesicles is a widespread product of volcanic eruptions.

An analysis of diffusive growth can be made for an ellipsoidal bubble with major semi-axis of length L, minor semi-axes of lengths r and eccentricity e = L/r. This shape corresponds to a bubble that is elongated along the axis of a conduit in an accelerating flow. An new analysis of diffusive growth in a strteching bubble is given in the appendix using the approximation of a thin diffusional boundary layer (compare Eqns 9 and 13). This analysis does not consider the relative motion of the melt and gaseous phase caused by bubble expansion. For an ellipsoidal bubble of constant eccentricity the minor semi-axis, r, major semi-axis, L, and equivalent spherical diameter, R, are given by

$$r = \left\{\frac{3\pi^2 D(C_s - C_m)}{8\rho\Phi}\right\}^{1/2} t^{1/2}$$

$$L = \left\{ \frac{3\pi^2 D(C_s - C_m)}{8\rho\Phi} \right\}^{1/2} et^{1/2} \tag{20}$$

$$R = \left\{ \frac{3\pi^2 D(C_s - C_m)}{8\rho\Phi} \right\}^{1/2} e^{1/3} t^{1/2}$$

where D is the diffusivity, $(C_s \ C_m)$ is the concentration difference between the bubble wall and melt interior and ϕ is a parameter defined in Equation (A1). If the eccentricity remains constant the growth still follows the parabolic law. The rate of growth of an ellipsoidal bubble will exceed that of a spherical bubble when

$$R = \left\{ \frac{3\pi^2 D(C_s - C_m)}{8\rho\Phi} \right\}^{1/2} e^{1/3} > \left\{ \frac{6D(C_s - C_m)}{\rho\Phi} \right\}^{1/2} \tag{21}$$

which is when

$$e > \left(\frac{4}{\pi^2} \right)^{3/2} = 1.273$$

In the case of a bubble growing by diffusion with the length L increasing at constant rate, L=at, where a is the proportionality constant, by stretching in an accelerating flow the eccentricity and equivalent spherical radius increases as

$$e = \left(\frac{a}{K^{1/2}} \right) t^{1/2} \quad \text{and} \quad R = (Ka)^{1/3} t^{2/3} \tag{22}$$

where K is a constant defined in the Appendix (Eqn. A11). This result shows that an ellipsoidal bubble stretching in an accelerating flow field has an enhanced rate of growth $(\alpha t^{2/3})$ in comparison to a spherical bubble of equivalent volume $(\alpha t^{1/2})$. We conclude from this analysis that there is positive feedback between degassing rates and expansion rates due to bubble deformation.

EXPERIMENTAL DEGASSING

A recent approach to understanding magma degassing has been to develop analogue experiments. Two aspects of degassing phenomena have been investigated. First, studies of the accumulation of gas bubbles to form a foam at the top of a magma chamber show that the foam goes unstable and discharges up a conduit (Jaupart and Vergniolle, 1989). These experiments indicate that for a steady flux of gas through a magma reservoir, unsteady and cyclic discharges of foam can occur up a conduit. The results show good correspondence to the behavior of the Puo-oo vent system on Hawaii (Vergniolle and Jaupart, 1990). Second, shock tube facilities have been developed at Bristol University and Caltech for the study of explosive degassing. These latter studies will be the main topic of this section.

Two sets of experiments have been carried out to study the behavior of aqueous solutions at high supersaturations of dissolved CO_2 (Mader et al., 1994). High supersaturations were achieved in two different ways. In the Caltech experiments a pyrex test cell (Fig. 11a) contains CO_2-saturated water at pressures up to 0.7 MPa. In an

a   **b**

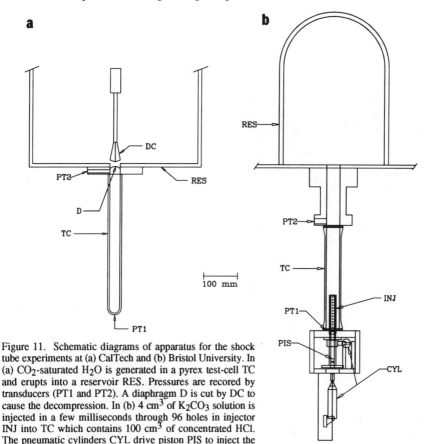

Figure 11. Schematic diagrams of apparatus for the shock tube experiments at (a) CalTech and (b) Bristol University. In (a) CO_2-saturated H_2O is generated in a pyrex test-cell TC and erupts into a reservoir RES. Pressures are recored by transducers (PT1 and PT2). A diaphragm D is cut by DC to cause the decompression. In (b) 4 cm^3 of K_2CO_3 solution is injected in a few milliseconds through 96 holes in injector INJ into TC which contains 100 cm^3 of concentrated HCl. The pneumatic cylinders CYL drive piston PIS to inject the K_2CO_3 solution.

experiment a diaphragm is cut and the solutions decompress into a reservoir at about 7 kPa pressure. Viscosity of the water is varied by adding small amounts of organic polymer. In the Bristol experiments (Fig. 11b) a concentrated solution of K_2CO_3 is injected in a few milliseconds, through 96 jets, into concentrated HCl solutions. CO_2 is formed in solution by chemical reaction with theoretical supersaturations of several MPa. The explosive flows were recorded by high speed photography and transducers recorded pressure variations within the shock tubes. Photographs of typical experiments are shown in Figure 12.

In the Caltech experiments explosive vesiculation was observed. Velocities up to 14 ms^{-1} and accelerations of 200g were achieved. The flows involve a sudden pressure drop and expand at almost constant acceleration (Fig. 13a). A strong positive correlation was observed between acceleration and the ratio of test cell pressure to reservoir pressure. A weak negative correlation was observed between viscosity and acceleration. Bubbles were observed to form in a single nucleation event. A corollary of this observation and the constant acceleration is that bubble volume is proportional to t^2. Growth of some individual bubbles was observed and they were approximately spherical implying radial growth followed a $t^{2/3}$ power law rather than the parabolic $t^{1/2}$ power law expected for diffusive growth. Van Wijngaarden (1967) has shown that bubble movement at a constant

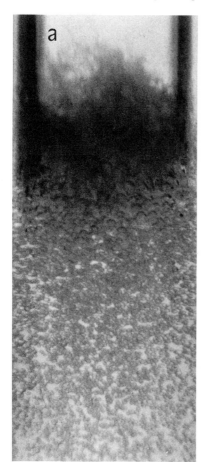
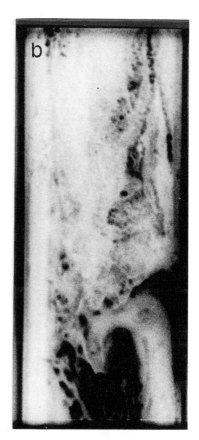

Figure 12. Photographs of the early development of simulated eruptions. The views are back lit. Light areas are bubble-free liquid or drop-free vapor. Large bubble or droplet density scatters light making these areas dark. (a) Flow generated by decompression. Image of the test-cell from 4 to 13 cm above its base. The photo shows the test solution, initially 10 cm deep, 7.1 ms after depressurization. Bubbles nucleate uniformly through the solution (about 10^9 m^{-3}) with negligible nucleation thereafter. Bubbles near the the top grow faster than those below. Above 10.5 cm the flow becomes finely fragmented while between 9.5 and 10.5 cm the bubbles have merged into a rapidly accelerating foam. (b). Flow generated by chemical reaction. Image of the test cell is 29 to 38 cm above the base and shows the test solution (initially 9.7 cm depth) 34 ms after the initiation of the reaction. Buuble growth begins after an incubation period of o(5) ms. The foam is highly heterogeneous with fragmentation occurring in the regions of most violent gas release.

velocity can also result in growth rates following a $t^{2/3}$ law. This could also be explained by the surface tension, viscous and inertial effects which retard growth but diminish in influence as bubble radius increases, resulting in growth rate which increase with time.

In the Bristol experiments velocities up to 30 ms^{-1} and accelerations of nearly 150g were achieved (Fig. 13b) when saturated K_2CO_3 (6M) was injected into concentrated HCl (12M). The maximum values of acceleration were observed to increase with larger theoretical supersaturations (Fig. 14). These flows accelerate approximately linearly with

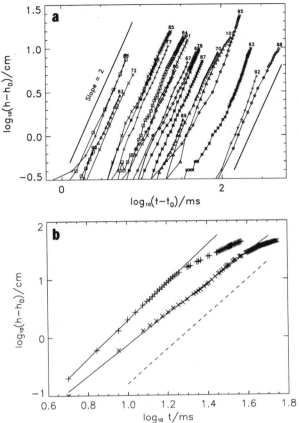

Figure 13. (a) The expansion of systems of CO_2 dissolved in aqueous solutions at high pressure after a sudden pressure drop. Data is shown as the height of the flow front versus time on a log-log plot. Origin of time was adjusted for each run to optimise the linearity of the plot. Results of 22 runs at different saturation pressures, ambient pressure, fluid viscosity and fill depth are shown. Details can be found in Mader et al. (1994). The line of slope 2 can be compared to the experimental results which all show evidence of constant acceleration.

(b) The expansion of foams created by violent reaction of concentrated HCl and K_2CO_3 solution is shown on a log-log plot of front height versus time. Top curve (6 M K_2CO_3 into 12 M HCl) and bottom curve (6 M K_2CO_3 into 12 M HCl). Each curve is a composite of data from 5 films viewing different parts of the test cell with the origin of time for each film chosen to ensure continuity of plotted data. The dashed slope on the right shows a slope of 3.

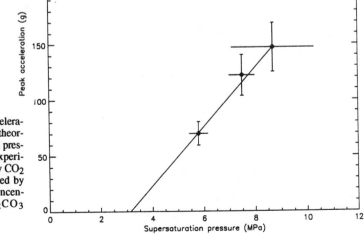

Figure 14. Peak acceleration plotted against theoretical supersaturation pressure, observed in experiments where strongly CO_2 solutions are produced by rapid mixing of concentrated HCl and K_2CO_3 solutions.

time, probably as a consequence of continuous mixing of reactants. The heterogeneities in the system may play an important role in disintegration of the foams due to the presence of weak, gas-rich zones. In both sets of experiments, foam acceleration precedes disintegration of the foam into a spray.

Application of these experiments to volcanic systems has been discussed by Mader et al. (1994). The experiments scale to volcanic systems in several important respects, notably Reynolds number, pressure drop, velocities, accelerations and the wide range of scales present in both nature and experimental simulations. Although the viscosities are many orders of magnitude less than for silicic magmas we have already established that viscosity does not inhibit bubble or foam expansion in volcanic systems, provided the visosity does not become too large. Below we will also present a conceptual model that indicates that the large pressure drops experienced in the experimental systems are comparable to those experienced in explosive eruptions.

The experiments reveal several aspects of the physics of violent degassing. First, bubble growth rates can depart from the parabolic diffusion law. This departure is most likely a consequence of advective effects, but other factors such as retardation due to viscous, surface tension and bubble distortion effects can also cause departures from the parabolic law. Bubbles can be strongly elongated in the accelerating flows and this factor might enhance degassing rates by creation of extra surface area. Second, the experiments do not support the concept of a simple downward-propagating front in an almost static foam with vesicularities of 0.7 to 0.8. Foam acceleration precedes fragmentation rather than the reverse. Third, although the precise mechanism of fragmentation is unknown, it is clear that the foam disintegrates into a spray entirely in the liquid state. Foam disintegration is controlled in this case by viscosity and surface tension as liquid films and ligaments are stretched and distorted in the flow. However magmas are viscoelastic materials and the strain rates achieved in these extreme conditions may be large enough for britttle fragmentation to occur.

CONDUIT FLOW MODELS

A larger scale approach to understanding volcanic eruptions and degassing involves modelling the overall flow of magma from a chamber through a conduit system to the surface. Examples of model numerical calculations include Wilson et al. (1980) and Dobran (1992). Such models that involve explosive degassing recognise two regimes of flow. Below the fragmentation level, the flow can be treated as a laminar viscous flow with low Reynolds number. Above the fragmentation level the flow consists of a two-phase gas-particle mixture flowing under turbulent conditions at very high Reynolds number. The latter regime can involve flows approaching, or in suitably shaped conduits, exceeding, the speed of sound where compressibility effects become substantial.

These models have provided some important constraints and insights into flow conditions in explosive eruptions. However they make various assumptions regarding the degassing processes and flow boundary conditions which mean that the models may not be wholly realistic. Below the fragmentation surface the flows are assumed to be isoviscous with gas content evolving at equilibrium. These assumptions allow simple pressure distributions to be invoked and provide estimates of how magma vesicularity will vary with fluid pressure. Wilson et al. (1980) assumed that pressure was always lithostatic and, with an isoviscous assumption, this results in a fairly uniform pressure gradient within the viscous part. Dobran (1992) relaxes the assumptions to calculate more general results in which pressure can depart from lithostatic, but retains the assumption of constant viscosity and equilibrium. The criterion for achieving fragmentation is always empirical at the observed vesicularities of 0.7 - 0.8. No physical model of why disintegration should occur at these vesicularities is involved. The validity of equilibrium, constant viscosity and fragmentation at a vesicularity threshold thus need further evaluation.

The assumption of constant viscosity is clearly incorrect. As shown in Figure 1, viscosity increases by orders of magnitude during degassing. The viscosity decrease will be most substantial in the region of fragmentation. For example, the viscosity of rhyolitic magma containing 5% H_2O at 700°C increases from 10^5 Pas to 10^9 Pas on achieving a vesicularity of 0.8 at 15 MPa (Fig.2). Large vertical viscosity gradients must therefore exist below the fragmentation level in explosive eruptions. A series of numerical calculations by Jaupart and Sparks have been carried out to investigate the effects of vertical viscosity variations. The calculations demonstrate that the pressure gradients in the magma are, as a consequence of the viscosity gradients, fundamentally different to the geostatic pressure gradients.

Before considering the detailed numerical results, the principles involved can be illustrated by a simple two layer system in which a low viscosity layer of magma containing dissolved gas is overlain by a high viscosity layer of vesicular, partially degassed magma with a much higher viscosity. The pressure at the top of the system (i.e., the level of fragmentation) is fixed. There is an excess magma chamber pressure which is greater than the hydrostatic pressure of the magma column. Mass conservation requires at steady state for the mass flux to be constant at all levels and can be described by the usual Poiseuille flow law for a cylindrical conduit of radius, r, (see Stasiuk et al., 1993)

$$M = \frac{\pi \rho r^4}{8\mu} \frac{dP}{dZ} \qquad (23)$$

where M is the mass flow rate, ρ is the density, μ is the viscosity and dP/dZ is the vertical pressure gradient. For the isoviscous case the simple result is that the pressure gradient is constant at all heights. However if there are two layers with different viscosity present and the rising magma increases its viscosity due to degassing on moving through the boundary between them, then the pressure gradient must be larger across the more viscous layer in proportion to the viscosity ratio

$$\left(\frac{dP}{dZ}\right)_2 = \frac{\mu_2}{\mu_1}\left(\frac{dP}{dZ}\right)_1 \qquad (24)$$

where subscripts 1 and 2 refer to the lower and upper layer respectively. The pressure distribution will be substantially different as a consequence of vertical viscosity variation. For large ratios of viscosity almost all the pressure drop will occur over the high viscosity, partially degassed upper layer. Also the dynamic pressure can be substantially different to the lithostatic and hydrostatic pressures.

The simple two layer model illustrates the principles but is inadequate to quantify the influence of viscosity and density changes due to degassing. Jaupart and Sparks (unpublished) have therefore carried out numerical calculations to couple relationships between viscosity and gas content with flow conditions. Simplifying assumptions still have to be made. The system is assumed to be in equilibrium so that the masses of gas in solution and in the bubbles is a function of pressure only. The assumption of equilibrium can be investigated a posteriori by comparison of time-scales for flow and decompression in the model with time-scales for bubble growth that have been discussed already. Another assumption that must be made is the pressure at fragmentation. Values in the vesiculation range of pumice (0.7 to 0.8) have been chosen. The Pouseuille flow solutions are assumed to apply in regions where there are very rapid vertical variations of viscosity. This may become invalid for extreme viscosity gradients due to the requirement for the flow to adjust to new conditions. However the model is sufficiently realistic to demonstrate that very large pressure gradients must occur immediately below the fragmentation level.

Figure 15 shows a typical result for a viscosity of 2×10^6 Pas, water content of 4 wt %, conduit length of 5 km and conduit radius of 50m. The viscosity is related to melt water content by a law which follows the curve for viscosity versus water content shown for rhyolite at 700°C (Fig 1). The fragmentation condition is specified as 0.70 void fraction. The calculations assume viscous stresses below the fragmentation level and turbulent conditions and high Reynolds number above the fragmentation level, and are in this sense similar to previous modelling assumptions (e.g., Wilson et al., 1980). The vent exit condition is the choking velocity (speed of sound). The numerical calculation finds a steady state solution for these boundary conditions with the new aspect being the strong viscosity variation below the fragmentation level which is strongly coupled to void fraction and pressure. In this case a mass flux of 2.95×10^7 kg s^{-1} is found which is a reasonably typical value for plinian eruptions. It was found that the exit pressure for this particular set of conditions was close to atmospheric. The particular calculation shows the fragmentation condition being reached at considerable depth (3.6 km).

The important features of this calculation in the present context are as follows. The flow pressure starts deviating from the lithostatic pressure dramatically upon degassing. A large break in slope occurs just above 1400 m height (at 3.6 km depth), which is where the fragmentation condition occurs. The rate of pressure variation is very large immediately below the fragmentation region and is a direct consequence of the large viscosity variation that occurs during degassing. The pressure drop over the 50 m below the fragmentation level is from 50 MPa to 12.8 MPa, which for comparison is the equivalent to the lithostatic pressure of 1.5 km of rock. Figure 16 shows the variation of gas volume fraction with height and shows that almost all the degassing is confined to a region less than 50 m in height with the vesicularity changing from 0.2 to 0.7 over this distance. The velocity at fragmentation is 5.2 m s^{-1} and thus almost all the bubble exsolution occurs in a time of 14 s. The acceleration at the level of fragmentation is 8 m s^{-2}. The time it takes for the two-phase mixture to move from the fragmentation level to the surface will be about 60 s. As the vesicularity at fragmentation increases above 0.7 the distance over which most degassing occurs decreases, the time-scale shortens and the acceleration increases.

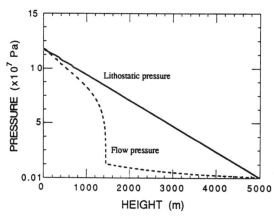

Figure 15. Numerical calculations of flow pressure variation with height in a conduit during an explosive eruption with the lithostatic gradient shown for comparison. The calculation is for eruption of rhyolitic magma containing 4 wt % water in a 50 m conduit with choked exit condition and a mass flux of 2.95×10^7 kgs^{-1}. The pressure curve shows a lower strongly curved region up to a sharp break in slope where fragmentation occurs and an upper region of slow pressure variation in the region of turbulent two-phase flow above the fragmentation level.

A CONCEPTUAL MODEL FOR EXPLOSIVE VOLCANISM

The various strands of theory, experimental data and observational constraints can now be drawn together to develop a conceptual model for explosive degassing. Several important aspects of the process remain poorly understood and so the model must be regarded as a series of working hypotheses with many questions still to be answered. We focus attention on sustained plinian eruptions as the paradigm case of explosive volcanism.

A crustal magma chamber is assumed in which pressure increases with time either due to crystallization (Tait et al., 1989) or a replenishment event (Sparks et al., 1977). Eventually the overlying rock fails and an eruption begins when the chamber develops an overpressure of order of the tensile strength of the surrounding rocks (Stasiuk et al., 1993). The chamber may be oversaturated or undersaturated in volatiles and may be homogeneous or compositionally stratified. We will explore the consequences of these different situations further below. A few bubbles probably form well before most eruptions as a consequence of slow crystallization, or saturation in CO_2 or influx of gas-rich magma during replenishment. The results of the experiments by Hurwitz and Navon (1994) suggest that the abundance of bubbles is likely to be related to the concentration of certain phenocryst phases which promote heterogeneous nucleation. The concentration of pre-existing bubble nuclei in the ascending magma is assumed therefore to be comparable or much less than the abundance of phenocrysts ($\sim 10^5$ to 10^7 m^{-3}) and orders of magnitude less than the vesicularites of erupted pumice and ash.

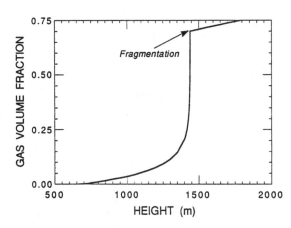

Figure 16. Numerical calculation of the variation of gas fraction with height for the same conditions as given in Figure 15. Note that almost all the gas exsolution occurs over a very narrow region with the vesicularity changing from 0.2 to 0.7 over 50 m.

Once the eruption has been initiated the system consists of three regions. In the lower region pressure is sufficiently high in the conduit that the magma can either remain undersaturated in H_2O or supersaturated. The pressure declines gradually with height and viscosity remains constant, because no gas has yet exsolved. The experimental data of Hurwitz and Navon (1994) indicates that substantial supersaturation pressures can be sustained in the ascending magma without major nucleation taking place. The middle region consists of a zone of rapid (explosive) degassing, in which the magma viscosity increases by three or four orders of magnitude as vesiculation occurs. At the base of this region the large supersaturations required for explosive nucleation rates are established. Numerical calculations (e.g., Figs. 15 and 16) suggest that this region shows large pressure gradients, is typically only tens of metres thick, and involves bubble growth over times of order 10 s. It is likely that this region is characterised by large supersaturations although this is not considered in the numerical model. The middle region is where explosive bubble growth, foam acceleration and disintegration takes place. The working hypothesis is that this region is very thin and is characterised by a large pressure drop. The upper region of conduit flow occurs above the fragmentation level where high velocity two-phase flow is achieved. In this region, shock tube studies indicate that large heterogeneties can exist (Anilkumar et al., 1993; Mader et al., 1994).

Much remains to be understood about exactly what happens in the middle region of explosive expansion and fragmentation. The shock tube experiments (Mader et al., 1994) simulate the explosive expansion of a foam in analogue systems and demonstrates that the foam disintegrates as a consequence of acceleration. The concept of a static high pressure foam that disintegrates and then accelerates is not supported by the calculations or experiments. Bubbles should grow faster in these explosive flows as they are stretched by acceleration along the conduit. Bubble elongation has been shown here to enhance degassing rates due to the increase in surface area production compared with spherical bubble growth. There may therefore be positive feedback mechanisms operating to enhance degassing rates in this region. The processes documented in the laboratory simulations might be described as a variety of wave model in which an exsolution wave propagates down into the magma column.

The mechanism of fragmentation remains poorly understood. In the shock tube experiments the accelerating foams disintegrate into a spray entirely in the liquid state. In detail these flows are heterogeneous and complex deformation of the liquid elements of the foam are sufficient to tear the foam into fragments. The shock tube experiments demonstrate that brittle failure is not essential for fragmentation. However caution is necessary in that the huge differences in viscosity may make the liquid-state disintegration processes inappropriate in magmas. At magmatic temperatures strain rates have to be sufficiently high that the timescale of deformation is shorter than the viscous relaxation timescale (Dingwell and Webb, 1989). In the numerical calculations presented in Figures 15 and 16 the strain rates approach those necessary to cross the glass transition at the top of the degassing region and thus brittle fracturing could be a consequence of the extreme flow conditions found in the thin region of explosive degassing.

Another conundrum is what significance should be placed on the observed vesicularities of pumice. Calculations on viscosity and diffusive bubble growth (Sparks, 1978; Proussevitch et al., 1993; Thomas et al., 1994; Barclay et al., 1995) indicate that even for the large pressure drops and characteristic time-scale of order 10 s the viscosity does not inhibit explosive bubble growth and foam acceleration until viscosities of around 10^9 Pas are achieved, when further expansion will be strongly inhibited. The vesicularities of pumice in the range 0.7 to 0.8 seem therefore to be a viscosity quench. One view would be that the expanding foam reaches this viscosity threshold before fragmentation and the foam expansion diminishes. A fragmentation wave then propagates into the foam. Another view is that expansion is uninhibited and the foam accelerates and disintegrates before the viscosity threshold is reached and the pumice fragments continue expansion and degassing in the upper region of two-phase flow. This later view is supported by the reasoning given in Thomas et al. (1994) where they observe that vesicularities of pumice deposits from the Taupo eruption are almost identical for different phases with discharge rates that differed by more than an order of magnitude. They take this to mean that, before fragmentation, decompression rates are not large enough for bubble expansion to be retarded by viscous effects. The numerical calculations in Figures 15 and 16 show that high strain rates are characteristic of the region of 0.7 to 0.8 vesicularity and approach the conditions for brittle fragmentation. Interesting issues then emerge about whether the preserved vesicularities represent the condition of the magma where the flow acceleration results in strain rates that cross the glass transition or represent a viscosity quench, either prior to or just after fragmentation.

Three important pressures can be usefully distinguished in an eruption. The saturation pressure for H_2O is at the deepest level and can migrate up and down in the conduit system. It can occur close to the surface in a gas-poor, undersaturated magma chamber, whereas it occurs in the chamber for a saturated magma. The depth of saturation can migrate upwards in a zoned chamber characterized by diminishing volatile content with depth, and it can migrate downwards due to the overall decrease in chamber pressure with time or due to the level of fragmentation moving downwards, which reduces the hydrostatic load on the chamber. The pressure of explosive nucleation occurs where the supersaturation is sufficient for large nucleation rates and, from the experimental data of Hurwitz and Navon (1994), will occur at supersaturations of 10 to over 70 MPa depending on the availability of sites for heterogeneous nucleation. Even gas-rich magmas may nucleate bubbles at quite shallow levels if large supersaturations are indeed developed. Fragmentation appears to occur where hydrodynamic conditions result in large accelerations and disintegration of the magmatic foam. Empirical evidence indicates that this occurs where vesicularities are in the range 0.7 to 0.8 at pressures of a few MPa.

Can this working hypothesis provide an explanation of what is observed? An important implication of this model is that pressures drop from about a few tens of MPa to a few MPa in a few short distance, probably of order of tens of metres. This pressure drop results in the explosive bubble formation and foam acceleration required for production of pumice and ash by either ductile or brittle fragmentation. The disintegrating foam can expand to vesicularities that are limited by the large viscosity increases during the later stages of degassing. The very large nucleation densities observed in pumice (10^{12} to 10^{14} m^{-3}) are consistent with the large and rapid decompressions. Tube pumice also provides evidence for the extreme accelerating flow conditions that lead to fragmentation.

ACKNOWLEDGMENTS

Research on vesiculation at Bristol is supported by grants from the Natural Environmental Research Council (GR3/8630) and the European Community (PL910499). We are grateful to Oded Navon for supplying the photographs for Figure 4. Brad Sturtevant provided many valuable comments on the manuscript and is also thanked for his contributions to developing some of the concepts in this paper. Detailed reviews by Mike Carroll, John Holloway and Megan Morrissey improved the paper.

APPENDIX

The diffusive expansion of an elongate ellipsoidal bubble is analysed using the approximation of a thin boundary layer

$$b = \frac{4}{3\pi} r\phi \qquad (A1)$$

where the symbols have the meanings previously defined in the main text (see Eqns. 9 through 13). The volume flux into the bubble is given as

$$q = \frac{dV}{dt} = \frac{D(C_s - C_m)}{\rho b} \pi^2 Lr \tag{A2}$$

The total derivative of $V(r,L)$ is formally given by

$$dV = \frac{\partial V}{\partial r} dr + \frac{\partial V}{\partial L} dL \tag{A3}$$

As $V = (4/3)\pi L r^2$ for an ellipsoid which has both of its semi-minor axes equal to r

$$dV = \frac{8}{3} \pi L r \, dr + \frac{4}{3} \pi r^2 \, dL \tag{A4}$$

Thus, the volume flow rate is given by

$$\frac{dV}{dt} = \frac{8}{3} \pi L r \frac{dr}{dt} + \frac{4}{3} \pi r^2 \frac{dL}{dt} = \frac{D(C_s - C_m)}{\rho b} \pi^2 Lr \tag{A5}$$

Substituting in for b by using Equation (A1) and rearranging gives the general differential equation

$$r \, dr + \frac{r^2}{2L} dL = \frac{(3\pi)^2 D(C_s - C_m)}{32 \rho \phi} dt \tag{A6}$$

This equation has three independent variables. If the dependence of r on L or L on t is known we can get a solution. We consider two special cases.

If the eccentricity $e = L/r$ is constant then (A6) becomes

$$\frac{3}{2} r \, dr = \frac{(3\pi)^2 D(C_s - C_m)}{32 \rho \phi} dt \tag{A7}$$

and integrating gives Equations (20) to (21) in the text.

For a bubble that grows in length linearly with time with $L = at$ where a is a constant

$$r \, dr + \frac{r^2}{2t} dt = K \, dt \tag{A8}$$

where

$$K = \frac{(3\pi)^2 D(C_s - C_m)}{32 \rho \phi}. \tag{A9}$$

Rearranging gives

$$\frac{dr}{dt} + \frac{r}{2t} = \frac{K}{r} \tag{A10}$$

A solution $r(t)$ can be found. The exact solution to general differential equations of the form

$$\frac{dy}{dx} + P(x)y = Q(x)y^n \tag{A11}$$

is due to Bernoulli and is given by

$$y^{(1-n)} e^{(1-n)\int P \, dx} = (1-n) \int Q e^{(1-n)\int P \, dx} dx + c \tag{A12}$$

If $y=r$ and $x=t$ in Equation (A13) then $P(x)=1/2x$, $Q(x)=K$ and $n=-1$, the solution is

$$r = (Kt)^{\frac{1}{2}}$$ (A13)

From Equation (A13) we get Equation (22) in the main text.

REFERENCES

Andres RJ, Rose WI, Kyle PR, de Silva S, Francis PW, Gardeweg M, Moreno, H. (1991) Excessive sulphur dioxide emissions from Chilean Volcanoes. J Volcanol Geotherm Res 46:323-329

Anilkumar AV, Sparks RSJ, Sturtevant B (1993) Geological implications and applications of high-velocity two-phase flow experiments. J Volcanol Geotherm Res 56:145-160

Barclay J, Riley D, Sparks RSJ (1995) Analytical models for bubble growth during decompression of high viscosity magmas. Bull Volcanol (in preparation)

Brey G (1976) CO_2 solubility and solubility mechanisms in silicate melts at high pressures. Contrib Mineral Petrol 57:215-221

Burnham W (1975) Water and magmas; a mixing model. Geochim Cosmochim Acta 39:1077-1084

Carey S, Sigurdsson H, Gardner JE, Criswell, W. (1990) Variations in column height and magma discharge during May 18, 1980 eruption of Mount St. Helens. J Volcanol Geotherm Res:43, 99-112

Carroll MR and Rutherford MJ (1985) Sulfide and sulfate saturation in hydrous silicate melts. J Geophys Res 90:C601-C612

Carroll MR, Rutherford MJ (1987) The stability of igneous anhydrite: experimental results and implications for sulfur behavior in the 1982 El Chichon trachyandesite and other evolved magmas. J Petrol 28:781-801

Cashman KV, Mangan MT (this volume) Physical aspects of magmatic degassing II. Constraints on vesiculation processes from textural studies of eruptive products

Criswell W (1987) Chronology and pyroclastic stratigraphy of the May 18, 1980 eruption of Mount St. Helens. J Geophys Res 92:10237-10266

Delaney JR, Karsten JL (1981) Ion microprobe studies of water in silicate melts: concentration-dependent diffusion in obsidian. Earth Planet Sci Lett 52:191-202

Dingwell DB, Webb SL (1989) Structural relaxation in silicate melts and non-Newtonian melt rheology in geologic processes. Phys Chem Minerals 16:508-516

Dixon JE, Stolper E, Delaney JR (1988) Infrared spectroscopic measurements of CO_2 and H_2O in Juan de Fuca Ridge basaltic glasses. Earth Planet Sci Lett 90:87-104

Dobran F (1992) Non equilibrium flow in volcanic conduit and applications to the eruptions of Mt. St. Helens on May 18, 1980 and Vesuvius in AD79. J Volcanol Geotherm Res 49:285-311

Dunbar NW, Kyle PR (1993) Lack of volatile gradient in the Taupo plinian-ignimbrite transition: evidence from melt inclusion analysis. Am Mineral 178:612-618

Eichelberger JC (1989) Are extrusive rhyolites produced from permeable foam eruptions? A reply. Bull Volcanol 51:72-75

Eichelberger JC, Carrigan CR, Westrich HR, Price RH (1986) Non-explosive silicic volcanism. Nature 323:598-602

Epel'baum MB, Bababashov IV, Salova TP (1973) Surface tension of felsic magmatic melts at high temperature and pressures. Geokhimiya 3:461-464

Fine G, Stolper E (1986) Dissolved carbon dioxide in basaltic glasses: concentration and speciation. Earth Planet Sci Lett 76:263-278

Gale RS (1966) The nucleation and growth of bubbles in supersaturated solutions of gases in viscous liquids. PhD dissertation, University of London

Head JW, Wilson L (1987) Lava fountain heights at Pu'u 'O'o, Kilauea, Hawaii: indicators of the amount and variation of exsolved magma volatiles. J Geophys Res 92:13715-13719

Heiken G, Wohletz K (1985) Volcanic ash. University of California Press, Berkeley, CA, 246 p

Hess PC (1980) Polymerisation model for silicate melts. In: Physics of Magmatic Processes, p 2-48, Princeton University Press, Princeton, New Jersey

Hirth JP, Pound GM, St Pierre GR (1970) Bubble nucleation. Metall Trans 1:939-945

Holloway JR (1976) Fluids in the evolution of granitic magmas: consequences of finite CO_2 solubility. Geol Soc Am Bull 87:1513-1518

Houghton BF, Wilson CJN (1989) A vesicularity index for pyroclastic deposits. Bull Volcanol 51:451-462

Hurwitz S, Navon O (1994) Bubble nucleation in rhyolitic melts: experiments at high pressure, temperature, and water content. Earth Planet Sci Lett 122:267-280

Jaupart C, Allegre CJ (1991) Gas content, eruption rate and instabilities of eruption regime in silicic volcanoes. Earth Planet Sci Planet Sci Lett 102:413-429

Jaupart C, Vergniolle S (1989) The generation and collapse of a foam layer at the roof of a basaltic magma chamber. J Fluid Mechanics 203:347-380

Karstens JL, Holloway JR, Delaney JR (1982) Ion microprobe studies of water in silicate melts: temperature-dependent water diffusion in obsidian. Earth Planet Sci Planet Sci Lett 59:420-428.

Landau LD, Lifshitz EM (1980) Statistical physics. Pergamon Press, New York

Mader HM, Zhang Y, Phillips JC, Sparks RSJ, Sturtevant B, Stolper E (1994) Experimental simulations of explosive degassing of magma. Nature (in press)

Mangan MT, Cashman KV, Newman S (1993) Vesiculation of basaltic magma during eruption. Geology 21:157-160

Murase T, McBirney AR (1973) Properties of some common igneous rocks and their melts at high temperatures. Geol Soc Am Bull 84:3563-3592

Neri A, Dobran,F. (1994) Influence of eruption parameters on the thermofluid dynamics of collapsing volcanic columns. J Geophys Res 99:11833-11857

Oppenheimar C, Francis PW, Rothery DA, Carlton WI (1993) Infrared image analysis of volcanic thermal features: Lascar Volcano, Chile, 1984-1992. J Geophys Res 98:4269-4286

Proussevitch AA, Sahagian DL, Anderson AT (1993) Dynamics of diffusive bubble growth in magmas: isothermal case. J Geophys Res 98:22283-22308

Proussevitch AA, Sahagian DL, Anderson AT (1994) Reply to Sparks RSJ (1994) Comment on 'Dynamics of diffusive bubble growth in magmas: isothermal case' by Proussevitch AA, Sahagian DL, Anderson AT (J Geophys Res 98:22283-22308, 1993) J Geophys Res (in press)

Sahagian DL (1985) Bubble migration and coalescence during the solidification of basaltic lava flows. J Geol 93:205-211

Sahagian, DL, Anderson AT, Ward B (1989) Bubble coalescence in basalt flows: comparison of numerical model with natural examples. Bull Volcanol 52:49-56

Sato H, Fujii T, Nakada S (1992) Crumbling of dacite dome lava and generation of pyroclastic flows at Unzen Volcano. Nature 360:664-666

Scriven LE (1959) On the dynamics of phase growth. Chem Eng Sci 10:1-13

Self S, Sparks RSJ, Booth B, Walker GPL (1974) The 1973 Heimaey Strombolian Scoria deposit, Iceland. Geol Mag 111:539-548

Shaw HR (1972) Viscosities of magmatic liquids: an empirical method of prediction. Am J Sci 272:870-893

Shaw HR (1974) Diffusion of water in granitic liquids, I. Experimental data. II. Mass transfer in magma chambers. In: Geochemical Transport Kinetics, Hoffmann AW, Gileth BJ, Yoder HS, Yund RA, eds, Carnegie Inst Washington Publ 634:139-170

Sigbee RA (1969) Vapor to condensed phase heterogeneous nucleation. In: Nucleation. Zettlemoyer AC (ed) 151-224, Marcel Dekker, New York

Sparks RSJ (1978) The dynamics of bubble formation and growth in magmas. J Volcanol Geotherm Res 3:1-37

Sparks RSJ (1986) The dimensions and dynamics of volcanic eruption columns. Bull Volcanol 48:3-15

Sparks RSJ (1994) Comment on 'Dynamics of diffusive bubble growth in magmas: isothermal case' by Proussevitch AA, Sahagian DL, Anderson AT (J Geophys Res 98:22283-22308, 1993). J Geophys Res (in press)

Sparks RSJ, Brazier S (1982) New evidence for degassing processes during explosive eruptions. Nature 295:281-220

Sparks RSJ, Sigurdsson H, Wilson L (1977) Magma mixing: mechanism of triggering explosive acid eruptions. Nature 267:315-318

Sparks RSJ, Sigurdsson H, Wilson L (1981) The pyroclastic deposits of the 1875 eruption of Askja, Iceland. Phil Trans Royal Soc A 299:241-273

Sparks RSJ, Wilson L (1976) A model for the formation of ignimbrite by gravitational column collapse. J Geol Soc London 132:441-451

Stasiuk MV, Jaupart C, Sparks RSJ (1993) On the variations of flow rate in non-explosive lava eruptions. Earth Planet Sci Lett 114:505-516

Stolper E (1982) Water in silicate glasses: an infrared spectroscopic study. Contrib Mineral Petrol 81:1-17

Stolper E (1989) Temperature dependence of the speciation of water in rhyolitic melts and glasses. Am Mineral 74:1247 -1257

Stolper E, Holloway JR (1988) Experimental determinations of the solubility of carbon dioxide in molten basalt at low pressure. Earth Planet Sci Lett 87:397-408

Swanger LA, Rhines WC (1972) On the necessary conditions for homogeneous nucleation of gas bubbles in liquids. J Crystal Growth 12:323-326

Swanson DA, Fabbi BP (1973) Loss of volatiles during fountaining and flowage of basaltic lava at Kilauea volcano, Hawaii. J Res U S Geol Survey 1:649-658

Tait S, Jaupart C, Vergniolle S (1989) Pressure, gas content and eruption periodicity of a shallow, crystallizing magma chamber. Earth Planet Sci Lett 92:107-123

Thomas N, Jaupart C, Vergniolle S (1994) On the vesicularity of pumice. J Geophys Res 99:15633-15644

Toramaru, A. (1990) Measurements of bubble size distributions in vesiculated rocks with estimation for quantitative estimation of eruption processes. J Volcanol Geotherm Res 43:71-90

Van Wijngaarden L (1967) On the growth of small cavitation bubbles by convective diffusion. Int'l J Heat Mass Transfer 10:127-134

Vergniolle S, Jaupart C (1986) Separated two-phase flow and basaltic eruptions. J Geophys Res 91:12842-12860

Vergniolle S, Jaupart C (1990) Dynamics of degassing at Kilauea volcano, Hawaii. J Geophys Res 95:2793-2809

Walker GPL (1981) Plinian eruptions and their products. Bull Volcanol 44:223-240

Walker GPL, Croasdale R (1972) Characteristics of some basaltic pyroclastics. Bull Volcanol 35:303-317

Watson EB (1981) Diffusion in magmas at depth in the earth: the effects of pressure and dissolved H_2O. Earth Planet Sci Lett 52:291-301

Whitham AG, Sparks RSJ (1986) Pumice. Bull Volcanol 48:209-223

Wilson CJN (1985) The Taupo eruption, New Zealand. II. The Taupo ignimbrite. Phil Trans R Soc London A 314:229-310

Wilson L. (1980) Relationships between pressure, volatile content and ejecta velocity in three types of volcanic explosion. J Volcanol Geotherm Res 8:297-314

Wilson L, Sparks RSJ, Walker GPL (1980) Explosive volcanic eruptions - IV. The control of magma properties and conduit geometry on eruption column behaviour. Geophys J R Astron Soc 63:117-148

Zettlemoyer AA (1969) Nucleation. Marcel Dekker, New York

Zhang Y, Stolper EM, Wasserburg, GJ (1991) Diffusion of water in rhyolitic glasses. Geochim Cosmochim Acta 55:441-456

Chapter 11b

PHYSICAL ASPECTS OF MAGMATIC DEGASSING II.

Constraints on vesiculation processes from textural studies of eruptive products

Katharine V. Cashman

Department of Geological Sciences
University of Oregon
Eugene, Oregon 97403-1272 USA

Margaret T. Mangan

U.S. Geological Survey
Hawaii Volcano Observatory
P.O. Box 51
Hawaii National Park, Hawaii 96718 USA

INTRODUCTION

"The little fountains scattered about looked very beautiful. They boiled, and coughed, and spluttered, and discharged sprays of stringy red fire—of about the consistency of mush, for instance—from ten to fifteen feet in the air ... The smell of sulphur is strong, but not unpleasant to a sinner." Mark Twain (1872)

It has long been understood that volatile exsolution is an important component of volcanic activity, and that the products of volcanic eruptions provide clues to their mode of origin. As early as the late eighteenth century, French geologist Dolomieu recognized that pumice was gas-bubble-bearing obsidian (Krafft, 1993). British gentleman scientist George Scrope was the first to recognize from pyroclasts that the expansive force of volatiles ('elastic aeriform fluids') could provide the driving force for volcanic eruptions (Lewis, 1982). Recognition that volcanic ash and pumice were solidified 'glass foam' came as part of the British Royal Society's investigation of the 1883 eruption of Krakatau (Simkin and Fiske, 1983). More recently, textures of volcanic pyroclasts have been measured and interpreted quantitatively. Just as the size and shape of crystals in a rock provides insight into its cooling history (e.g., Cashman, 1990; 1993), the extent and style of vesiculation in volcanic rocks can be related to eruption style and duration. In this chapter we review available data on the vesicle structure of volcanic rocks and the relationship between their preserved vesicle structure and the degassing and eruptive history of the magma from which they formed.

Quantitative data on the size, shape and connectedness of vesicles in volcanic products are limited, although "the information stored in glassy pyroclasts seems to offer a rich data base for future interpretation, and this interpretation seems basic to understanding eruptions" (Rose, 1987). Detailed examination of the structure of vesicular material can constrain parameters (vesicularity, bubble number density, bubble size, shape, and wall thickness) necessary for models of magma ascent in and eruption from volcanic conduits (e.g., Woods, 1988; Vergniolle and Jaupart, 1986; 1990; Jaupart and Allegre, 1991; Dobran, 1992; Thomas and Sparks, 1992; Papale and Dobran, 1994). Important volcanological questions that may be addressed through textural studies include (1) conditions of bubble nucleation and growth in magmas; (2) bubble size and shape parameters that affect the rheology of bubbly liquids and foams; (3) mechanisms and conditions of closed versus open system degassing, including controls on the transition from explosive to effusive volcanism; (4) the state of magma at the time of fragmentation, and (5) the importance of post-eruption expansion and exsolution in the generation of pyroclasts. Of particular interest is the role played by vesiculation in the 'fragmentation' or 'disruption' of magma; that is, in the transformation of magma from a dispersion of bubbles in a continuous liquid phase, to liquid particles entrained in a gas phase. In this paper, we first outline the structural elements particular to gas-melt mixtures, and then summarize the existing literature on the relationship between processes of vesiculation during eruption, and structural features related to vesiculation that are preserved in the eruptive products.

BUBBLE SUSPENSIONS AND FOAMS

Insight into gas-related magmatic textures can be derived from the engineering literature on laboratory and industrial foams wherein nomenclature, structural attributes, and the principles of foam generation and maturation are set forth. Useful summaries of the work done on engineering foams are provided in de Vries (1957), Bikerman (1973), Cheng and Natan (1986), Kraynik (1988), Wilson (1989), Aref and Herdtle (1990), Klemper and Frisch (1991). Applications to magmatic systems are given by Proussevitch et al. (1993), and Mangan and Cashman (1994). Below we summarize the nomenclature, topology, and processes affecting the structure of magmatic suspensions and foams.

Nomenclature and structure

Strictly speaking, a *foam* contains a dispersed gas phase of >74% by volume. This volume fraction represents the maximum packing that can be achieved with uniform spheres. At gas contents less than this value, the system is considered a *bubble suspension*. This general distinction is important from a rheologic standpoint. Foam is a cellular fluid where the volume of liquid is small compared to that of the gas phase. As such, it exhibits a complex rheology that is governed largely by interfacial phenomena. The rheology of a suspension, particularly a dilute suspension, is a good deal simpler and is controlled for the most part by the interaction of discrete gas "particles" with the external flow field.

A gas phase dispersed in a volume of liquid adopts a spherical shape as a consequence of surface tension, which acts to minimize the surface area enclosing the gas volume. A homogeneous distribution of uniform (monodisperse) spheres thus provides a basic conceptual model for the structure of bubble suspensions and foams. As stated above, rigid spheres attain maximum packing 74.05% filled space. The coordination number (number of nearest-neighbors) of each sphere in this configuration is 12. For a system of multi-sized spheres (polydisperse) the coordination number may be as high as 14, and the percentage of filled space at maximum packing approaches 85%. In fluid systems, the limits on maximum packing can be exceeded by sphere deformation. The shape adopted is governed by the minimization of surface energy, and results in the formation of 12- to 14-sided polygons with predominantly pentagonal and hexagonal faces. Figure 1 shows the most commonly assumed model forms.

To keep the mathematics tractable, the evolution of foam structure is generally modeled in two-dimensions only. A two-dimensional, ideal foam consists of closest-packed circular sections (*spherical foams*), or closest-packed hexagons (*polygonal foams*). The areas of contact between bubbles form lamella called *films*. In a *closed-cell foam* the films are intact. In an *open-cell foam* the films have burst while still fluid, and the framework of cell

Model Cells of Polygonal Foams

2 hexagons
12 pentagons

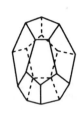

12 pentagons

6 squares
8 hexagons

Figure 1. Cell morphology of ideal polygonal foams. From **Mangan and** Cashman (1994).

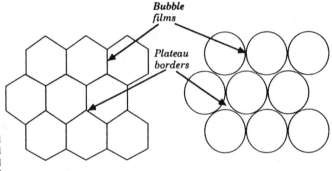

Figure 2. Two-dimensional representations of ideal spherical and polygonal foams. From Mangan and Cashman (1994).

Polyhedral Foam   *Spherical Foam*

boundaries remains. The geometry of packing requires that these boundaries of three films meet at dihedral angles of 120° forming channels called *plateau borders* (Fig. 2).

Real foams are most often polydisperse and therefore only approximate the well-ordered structure depicted in Figure 2. Several statistical measures of disorder have been developed for 2-D foams. The most widely used is based on the coordination number of bubbles (e.g., Weaire and Kermode, 1983, 1984; Glazier et al., 1990; Herdtle and Aref, 1992). Bubbles in ideal 2-D foams have coordination numbers of 6 from topological constraints (i.e., hexagonal close packing). Real foams display a distribution of coordination numbers given by $\rho(n_c)$. For a highly disordered network $\rho(n_c)$ is a broad distribution in which coordination numbers of 5 tend to be most abundant. In more orderly systems, $\rho(n_c)$ will peak at a coordination number of 6 (Fig. 3). The ratio of $\rho(5)/\rho(6)$ thus provides a simple measure of system disorder, with higher values [$\rho(5)/\rho(6) \sim 1$] implying a more disordered state, and low values [$\rho(5)/\rho(6) \sim 0.1$] a more ordered condition. Another measure of system disorder (Glazier et al., 1990) is the second moment of the $\rho(n_c)$ distribution given by

$$\mu_2 = \langle (n_c\text{-}6)^2 \rangle \tag{1}$$

which is the root-mean-square of the $\rho(n_c)$ distribution. A large μ_2 manifests disorder whereas small values of μ_2 indicate higher order.

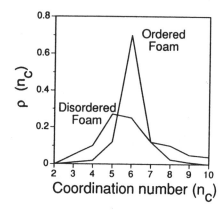

Figure 3. Coordination number distributions $\rho(n_c)$ characteristic of ordered versus disordered foams. Plots use data from numerical simulations by Glazier et al. (1990) of 2-D soap froths. From Mangan and Cashman (1994).

Processes that influence structure

Capillary drainage is initiated by the non-uniform surface curvature of bubble walls. Variations in surface curvature induce a pressure gradient in the interstitial liquid according to the well-known law of Young and Laplace

$$\Delta P_c = \sigma (1/R_1 + 1/R_2) \qquad (2)$$

where ΔP_c is the jump in capillary pressure across the gas-liquid interface, σ is the surface tension, and R_1 and R_2 are the mutually perpendicular principal radii of curvature at a particular point along the interface (e.g., Bikerman, 1973; Aref and Herdtle, 1990). The pressure differential that arises in the liquid induces capillary flow away from the center of bubble films towards the bounding plateau borders. Films are progressively thinned as a result. Gravitational flow can also lead to thinning of bubble films but has been shown to be of little consequence in magmatic systems (Proussevitch et al., 1993).

Film rupture. Film thinning, a necessary consequence of foam expansion and capillary drainage, can lead to *film rupture* . The rupture of a liquid film produces either (1) partial retraction of the film toward the bounding plateau borders to create an open pore between cells, but with each cell retaining its original structure, or (2) complete coalescence of adjoining cells, and the development of a new, larger structural unit. In extreme cases, film rupture can lead to the failure of plateau borders and ultimately to the collapse of the entire foam structure. Film retraction rates are virtually instantaneous and are independent of liquid viscosity (Cheng and Natan, 1986), so that the rate of thinning and the initiation of rupture are rate-determining in the process (Frankel and Mysels, 1969; de Vries, 1972). Brittle film rupture is also common in the solid state, and is particularly characteristic of closed-cell foams where cooling contraction is significant.

Rupture initiation is contingent on thinning to a "critical film thickness" below which local thermal, mechanical, or compositional perturbations (compounded by van der Waals forces) lead to instantaneous rupture (de Vries, 1972; Narsimhan and Ruckenstein, 1986a,b). The rate of film thinning for 'rigid' films ($V_{Re} = d\delta/dt$) is described as flow between two parallel plates (Sonin et al., 1993):

$$V_{Re} = (\delta^3/3\eta R_c^2) \Delta P \qquad (3)$$

where η is the melt viscosity, δ is film thickness, and R_c is the radius of the circle of contact between adjacent bubbles. ΔP is the pressure differential across the film. If capillary forces dominate, ΔP is that described in Equation (1). In the case of expanding foams (the more relevant case for applications to volcanic systems), the appropriate ΔP value is that describing the volume expansion related to pressure release (Mangan and Cashman, 1994). Film 'mobility' resulting from variations in film thickness will increase film thinning rates relative to this limiting value.

Coalescence of bubbles may occur as a static or dynamic phenomenon. Static coalescence occurs in evolving foams, and results from thinning and rupture of adjacent bubble walls (cf. Eqn. 3). In bubble suspensions, coalescence is controlled by bubble collisions due to variable buoyancies of bubbles of different sizes (e.g., Sahagian, 1985; Sahagian et al., 1989). In this case, the process may be viewed as dynamic, and the time scales for coalescence will depend on the initial bubble size distribution, bubble number density, and time available for bubble rise. Important in models of dynamic coalescence is the sweeping of very small bubbles around larger bubbles in the 'escape flow' such that the optimal size for bubble 'collection' is *not* the smallest bubble size but instead an intermediate size (Berry and Reinhardt, 1974). Both forms of bubble coalescence will serve to decrease the number density of bubbles and increase the mean bubble size with time.

Ostwald ripening can alter the original size distribution, number density, and packing of foams by the diffusive transfer of gas between bubbles. Ripening is driven by the pressure excess inside bubbles, which according to the Young-Laplace Law is high for small bubbles and low for large bubbles (cf. Eqns. 2,3). Foams generally contain bubbles of different sizes, thus internal pressures will be variable and gas will diffuse from regions of high to low pressure. As a consequence, large bubbles will grow and small bubbles will shrink and eventually disappear altogether. These structural changes decrease the total interfacial area, thus 'stabilizing' the material in a thermodynamic sense (lowering the overall free energy of the system).

In 1957 de Vries derived an expression for the rate of shrinkage of a single, small bubble by combining a form of Fick's law with Equation (2) to give

$$r_0^2 - r^2 = (4RTDS\sigma / P\delta) \, t \tag{4}$$

where r_0 is the original bubble radius, R is the gas constant, T is temperature, P is the ambient pressure, t is time, δ is the average distance between bubbles (i.e., film thickness), and D and S are the diffusivity and far-field solubility of the gas species, respectively. As ripening proceeds, the foam coarsens according to Equation (4) and the total number of bubbles declines. In polyhedral 2-D foams the predicted change in the bubble size distribution is encapsulated in von Neumans Law (von Neumann, 1952)

$$dA/dt = k(n_c - 6) \tag{5}$$

where the rate of change of the area, A, of an individual cell is proportional to the actual coordination number minus the ideal number (6 for hexagonal packing). In Equation (5), k is a diffusion constant. The implication of von Neumann's Law is that bubbles with fewer than six neighbors will shrink and those with greater than six will grow. Bubbles in 6-fold coordination will remain the same size. Numerical simulations and laboratory experiments have verified Equation (5), but only in terms of $<dA/dt>$, the average rate of change (see Glazier et al., 1990). Strict adherence to the law is impossible in a dynamic system because as bubbles shrink, they gain or lose neighbors accordingly. Hence, the rate of area change of an individual bubble fluctuates.

TEXTURAL CHARACTERIZATION OF PYROCLASTS

Textural interpretation of volcanic pyroclasts first requires that the structure be quantified. In this section we outline techniques for determining textural parameters aimed specifically at the interpretation of pyroclast structure in the context of the studies of industrial foams presented above.

Vesicularity

Pyroclast vesicularity is a fundamental measure of volatile expansion experienced by vesiculating magma. If vesicles are relatively undeformed, vesicularity may be measured in two dimensions using standard point counting techniques on sample thin sections. Rapid acquisition of areal data is now possible using digitized images obtained directly from a petrographic or stereoscopic microscope (for coarsely vesicular samples) or backscattered electron (BSE) images (for finely vesicular samples). If vesicle orientation is random, the area fraction of vesicles observed in two-dimensions equals the volume fraction in three-dimensions, and thus gives the vesicularity. Sample crystallinities may be similarly determined, using microscope images for phenocrysts and BSE images for microlites (Klug and Cashman, 1994).

Pyroclast vesicularity may also be measured using whole (3-dimensional) samples. The most commonly used technique is based on Archimedes' principle, with samples treated by spraying with silicone oil, weighed in air, then weighed in water (e.g., Houghton and Wilson, 1989; Hoblitt and Harmon, 1993). This technique yields an accurate measure of clast *density*, which can be converted to clast *vesicularity* if the density of the solid is known. Use of a multivolume helium pycnometer provides a direct measure of clast vesicularity (Klug and Cashman, 1994). The helium pycnometer measures the solid volume (glass + crystals + isolated vesicles) of the sample. Samples are then crushed and remeasured, such that the measured solid volume represents only the volume of glass + crystals. The latter measure of solid volume is used in conjunction with measured clast volume to determine clast vesicularity. A comparison between the He-accessible volume of whole and powdered samples yields a measure of the volume fraction of isolated pores; sample weight gives a direct measure of solid density.

Vesicle size, shape and spatial distribution

The size, shape and spatial distribution of vesicles in a sample may all be measured using image analysis techniques on thin sections (or cut slabs and outcrop surfaces for very large vesicles). After images are obtained, there are many programs available that allow both processing and analysis of images (e.g., the shareware program NIH Image for MacIntosh systems). Most important is the ability to generate a binary image of vesicles (Fig. 4a, white) and glass + crystals (Fig. 4a, black), from which individual features may be characterized. An overview of image processing techniques is provided by Russ (1988).

Vesicle shape preserves information on both the extent and style of bubble coalescence and deformation. Shape elongation relative to sphericity is most easily measured by a simple aspect ratio (length/width), which may yield information on the extent of deformation experienced by a pyroclast sample (e.g., Orsi et al., 1992). Another commonly used shape factor is $4\pi A/P^2$ (where A is vesicle area and P is vesicle perimeter, thus the shape factor of 1 for a circle decreases as the shape becomes more irregular).

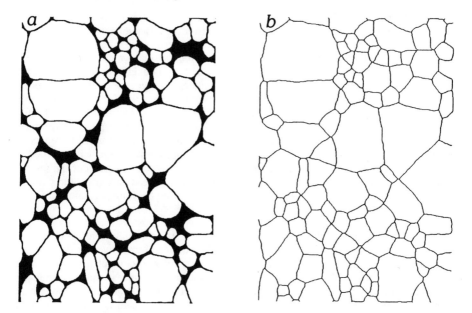

Figure 4. (a) Binary image of a typical Kilauea (Hawaii) basaltic tephra. This image was obtained from a thin section by capturing a video image, then 'threshholding' that image such that all vesicles appear white and all glass appears black. (b) 'Skeletonized' version of the image in (a). Here all areas of glass (black areas) have been 'eroded' back to a single pixel in width. This process helps to define the spatial relationship of the vesicles, and is used for determination of sample coordination number. From Mangan and Cashman (1994).

Vesicle *surface area* (S_V) will change with coalescence and ripening, and may thus reflect stages in foam evolution (Herd and Pinkerton, 1993). A measure of surface area adapted from metallurgical studies involves measurement of the number of particle boundary intersections per unit line length (P_L; $S_v = 2P_L$, Russ, 1986). Surface area may also be calculated using measured cell areas combined with model vesicle shape (Mangan and Cashman, 1994).

Spatial distribution. The spatial distribution of vesicles represents the original spatial arrangement of nucleation sites, with modifications resulting from bubble segregation, coalescence, ripening and deformation. As discussed above, the spatial distribution (coordination number) of bubbles in a foam yields a direct measure of structural order. Mangan and Cashman (1994) determine the coordination number of vesicles by 'skeletonizing' binary images of samples to generate the polyhedral framework (Fig. 4b). The number of neighbors of each cell may then be measured to yield $\rho(n_c)$ distributions. Another indication of cell ordering is increasing uniformity in average nearest-neighbor distance (Shehata, 1989).

Vesicle size and number density reflect the kinetics and timescales of vesiculation, as well as modifications by processes such as bubble loss and coalescence (e.g., Toramaru, 1989). The most useful measure of vesicle size is the *equivalent vesicle diameter L*, calculated from measured vesicle areas by assuming that vesicles are approximately spherical. Vesicle number density (number per unit area) may be obtained directly. Conversion of this data to number per unit volume is described below.

Vesicle size distributions (VSDs) measured in two dimensions (number per equivalent diameter per area, N_A) must be converted to a 3-dimensional distribution (number per equivalent diameter per volume, N_V). Two errors are introduced in making this conversion. The first source of error stems from the fact that a measurement plane is unlikely to intersect a vesicle at its maximum cross-section ('cut effect'). In addition, the 2-dimensional measurement plane is more likely to intersect large vesicles than small ones. Two different methods have been used to circumvent these errors. The most commonly used conversion (Sarda and Graham, 1990; Toramaru, 1990) is that of Saltykov (1967; described in Russ, 1986). Here a matrix of values has been calculated assuming a distribution of sphere sizes such that the largest vesicle measured is assumed to represent the maximum sphere diameter in the sample, the next largest size category is assumed to contain a fraction of sections from the largest size class, plus some vesicles of the second largest size class, etc. Problems with this approach include data partitioning into a maximum of 15 bins, the assumption of spherical forms, and the introduction of errors in the smallest size classes.

An alternative approach is given in Cheng and Lemlich (1983; similar to that of Pareschi et al., 1990). Here N_A of a given size class is divided by the average diameter of that size class, thus increasing the number of small vesicle per volume relative to the large vesicles. This method is theoretically equivalent to using more measurement planes per unit volume for the small vesicles and progressively fewer for the large vesicles, so that all vesicles are measured once and only once. The 'cut effect' is then accounted for by a multiplicative factor for each size class determined by comparing the calculated vesicle volume with measured vesicularities (Mangan et al., 1993). This permits utilization of the added constraint of measured vesicularity. Sample vesicle volume distributions based on these conversion schemes are compared in Figure 5a. Volume distributions are commonly normal or log normal in form, and the peak of the volume distribution yields a volume-based measure of average vesicle size. Plots of volume distributions also allow assessment of the degree of bubble coalescence and ripening that has affected the population (Sahagian et al., 1989; Orsi et al., 1992; Cashman et al., 1994a; Klug and Cashman, 1994).

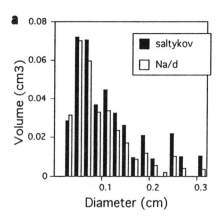 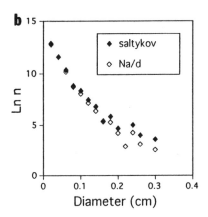

Figure 5. Comparison of methods used to convert measured vesicle number densities in two dimensions to true values in three dimensions. Input was a typical measured vesicle distribution from basaltic lava flows (from Cashman et al., 1994a); output was generated using the methods of Saltykov (1967) and Mangan et al. (1993; referred to in the plots as Na/d method). (a) Vesicle volume distributions. (b) Vesicle size distributions (VSDs); that is, plots of ln (n) vs. L (cf. Eqn. 6).

Size distributions may also be presented as population density functions (or vesicle size distributions; VSDs) in the fashion of Randolph and Larson (1988), Marsh (1988), and Cashman and Marsh (1988). For a steady-state, conservative system

$$n = n^o \exp (-L/G\tau) \tag{6}$$

where n is the volumetric number density of particles per size class (number per volume melt per bubble diameter), n^o is the volumetric number density of particle nuclei, G is the mean growth rate, and τ is the time scale of nucleation and growth. Thus a plot of ln (n) against L is linear with a slope of $-1/G\tau$ and an intercept n^o. Moments of Equation (6) yield information on the length, area, and volume distribution of the population. Thus the integrated form of the zeroth moment gives the total number density (N_T) per volume melt:

$$N_T = n^o G\tau \tag{7}$$

The maximum in the first, second, and third moments, respectively, yield the dominant diameter (L_D), area (L_A) and volume (L_V) sizes:

$$L_D = G\tau \tag{8a}$$

$$L_A = 2G\tau \tag{8b}$$

$$L_V = 3G\tau \tag{8c}$$

An example of VSD representation of the volume distributions shown in Figure 5a is illustrated in Figure 5b. Kinetic information can be derived directly from the slope and intercept of log-linear VSD plots, provided that a relevant time scale can be independently obtained (Sarda and Graham, 1990; Mangan et al., 1993; Cashman et al., 1994a; Klug and Cashman, 1994; Mangan and Cashman, 1994).

VESICULATION IN BASALTIC SYSTEMS

True foams form through extensive gas expansion in basaltic fire fountains (Mangan and Cashman, 1994), while bubble suspensions (vesicularities generally less than 70%) characterize vent samples from effusive eruptions (Mangan et al., 1993). Lava vesicularities and bubble populations are then modified substantially through transport, emplacement and solidification. Textural studies designed to elucidate processes that affect vesicle populations in subaerial basalts can aid in interpretation of both older solidified subaerial flows (Cashman et al., 1994a,b) and of the volatile content and vesicle structure of their submarine counterparts (Sarda and Graham, 1990).

Fire fountains

Basaltic fire fountains result from the fragmentation of coherent magma into an explosive spray of molten clasts that are propelled hundreds of meters above the vent. The concentration of clasts is highest in the center of the fountain and diminishes toward its exterior. The thermal profile through the fountain follows that of the clast concentration (e.g., Head and Wilson, 1987). Interior clasts undergo minimal cooling and fall to the ground while still molten. Those traversing the margins of the fountain are quenched before deposition. Their textures thus record conditions within the eruptive column, and do not bear the overprint of post-eruption transport and cooling. The most thoroughly quenched varieties, basaltic tephra and reticulite, are true magmatic foams with vesicularities between ~75 and 99%. Descriptions of these clast types are provided in Dana (1891), Wentworth and Williams (1932), Wentworth and MacDonald (1953), Heiken and Wohletz (1985), Takahashi and Griggs (1987), and quantitative structural analysis and interpretation are provided in Mangan and Cashman (1994), which we summarize below.

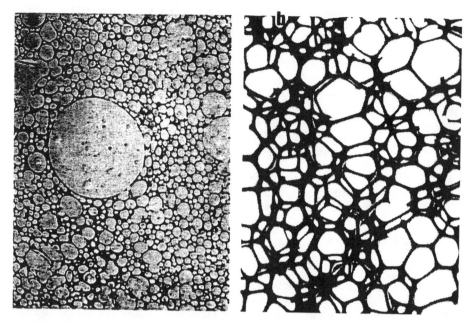

Figure 6. Typical rapidly-quenched basaltic pyroclasts from Kilauea Volcano. Field of view is 8.75 × 6.5 mm. (a) Basaltic tephra in thin section. Note that most of the vesicles are close to spherical in shape, and that there is a wide range of vesicle size. (b) Basaltic reticulite, three-dimensional sample. Note that cells are polygonal and that the structure is open. Coordination numbers of 5 and 6 are most common. From Mangan and Cashman (1994).

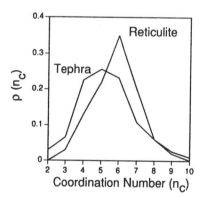

Figure 7. Coordination number distributions $\rho(n_c)$ typical of the tephra and reticulite samples shown in Figure 6. From Mangan and Cashman (1994).

Tephras are quenched before they are fully expanded and exhibit the features of a closed-cell, spherical foam. Measured vesicularities are generally less than that of maximum packing for polydisperse systems (≤85%) and vesicles are relatively undeformed, although mild flattening occurs along the planes of contact between nearest neighbors (Fig. 6a). The broad distribution of vesicles sizes leads to a structure that is relatively disordered, as illustrated by broad $\rho(n_c)$ distributions (Fig. 7).

VSDs of tephras are linear at small (<0.1 cm) vesicles sizes, and indicate that vesicles preserved in basaltic tephras represent a single population of bubbles produced by [tephra] and ash." The recent work of Mangan and Cashman (1994) supports this assertion. The structural variations among the quenched products provide "snapshots" of various stages in the vesiculation history of the samples; i.e., from a disordered, closed-cell spherical foam (tephra) to an open-celled, polygonal foam (reticulite). Textural differences are derived from the post-fragmentation thermal history of an individual clast.

Questions that remain include (1) where in the system is the foam evolving, and (2) what is the relation between fragmentation and clast structure? Magma fragmentation has always been an elusive topic. Some investigators maintain that magmatic foams evolve by the gradual accumulation of bubbles in a layer at the reservoir roof (Vergniolle and Jaupart, 1986;1990; Proussevitch et al., 1993). Fragmentation in this model is assumed to accompany physical collapse of the foam layer. Others envision foam formation as a dynamic process occurring in the conduit as the magma nears the surface (Sparks, 1978; Wilson and Head, 1981; Gerlach, 1986; Greenland et al., 1988), with fragmentation a consequence of the bubble population reaching maximum packing. Curiously, the VSDs from Mangan and Cashman (1994) are not compatible with either hypothesis. The extreme bubble nucleation and growth rates indicated by the tephra VSDs imply an intense vesiculation burst, at very high degrees of volatile supersaturation. This inference contrasts markedly with the equilibrium-controlled gas release proposed by Mangan et al. (1993) for effusive eruptions. Mangan and Cashman (1994) speculate that the "runaway" nucleation and growth rates characterizing fire fountain episodes are an indication that the fragmentation event and the vesiculation burst occur simultaneously, without extensive low-pressure exsolution beforehand. The vesiculation burst contributes to the fragmentation of magma and generates an immature foam. Evolution to a mature, polyhedral foam occurs during subsequent ascent in the fire fountain. Clasts that travel upwards along the central axis of the fountain are the most likely to pass through the full evolutionary sequence. Further aspects of these questions are discussed in Part I of this Chapter.

Effusive eruptions

During effusive eruptions magma wells up in the conduit and flows out of the vent without fragmenting. The erupting lavas commonly have vesicularities of ~60 to 70% (Einarsson, 1949; Swanson and Fabbi, 1973; Lipman and Banks, 1987; Mangan et al., 1993), and thus most closely approximate bubbly suspensions rather than actual foams. At these relatively low vesicularities the suspended bubbles are not constrained by the laws of closest-packing. Bubbles are fairly isolated and maintain a spherical shape (Fig. 9).

Mangan et al. (1993) use VSDs in vent-lava from Kilauea volcano to quantify vesiculation kinetics and assess coalescence and gas separation during the passive rise of magma in volcanic conduits. The results are summarized in a plot of vesicle number density versus diameter in Figure 10. The regular, exponential decrease in the number of vesicles with increasing vesicle size (diameters ≤ 0.2 cm) is an expected consequence of continuous nucleation and growth in a system without additions or losses due to bubble segregation. Perturbation of the tail of the VSD (diameters ≥ 0.2 cm) suggests that the original size distribution was modified by minor bubble coalescence (the tail represents only ~2 % of the population by number, or <25 % by volume). Mean bubble diameters are on the order of 0.02 to 0.03 cm and bubble-number densities typically reach $10^3/cm^3$ melt. Exsolution models for Kilauea combined with measured ascent rates indicate vesiculation time scales of ~110 s, leading to VSD-derived nucleation and growth rates of ~36 events/cm^3s and 3×10^{-4} cm/s, respectively. These rates are consistent with diffusion-controlled growth plus decompression (e.g., Sparks, 1978) and indicate that, in contrast to continuous nucleation and growth processes (Fig. 8). Bubble coalescence is evidenced by the perturbed tail of the VSD. While the number of coalesced bubbles is small (<5 % of the total number), the impact on the volume distribution of gas can be substantial, affecting between 25 and 50% of the total population. Mean vesicle diameters (Gτ) are less than 0.02 cm and vesicle number densities are high, ~10^5/cm^3 melt. The vesiculation time scale, estimated from fountain height and ascent velocity data, is ~10s during fountaining events. This implies nucleation and growth rates of ~2 × 10^4 events/s cm^3 and 9 × 10^{-4} cm/s, respectively. The derived nucleation rate is approximately three orders of magnitude greater than that calculated for effusive activity, and the rate of bubble growth exceeds effusive estimates by a factor of three (Mangan et al., 1993).

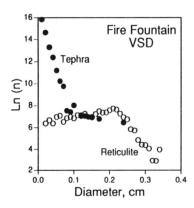

Figure 8. VSDs typical of rapidly quenched fire fountain lava such as those illustrated in Figure 6. Closed circles for tephra data and open circles for reticulite data. From Mangan and Cashman (1994).

Reticulite expands to >95% vesicularity before quenching. Its structure is a classic example an open-celled, polyhedral foam consisting of a network of predominantly 12- to 14-sided polygonal cells (Fig. 6b), including the model forms given in Figure 1. Most of the films between adjoining bubbles attained their critical thickness, ruptured, and retracted back to the plateau borders during expansion, so that virtually no closed space exists in the network. Preserved films commonly show evidence of brittle fracture, most likely the result of contraction on quenching. Reticulite structures are well-ordered relative to those preserved in tephras, as illustrated by the pronounced peak at a coordination number of 6 in the $\rho(n_c)$ distribution in Figure 7.

Reticulite VSDs (Fig. 8) show a unique "kinked" profile diagnostic of Ostwald ripening (see also Cashman and Ferry, 1988; Cashman et al., 1994a). Ripening theory predicts a gradual decline in the total number of bubbles and an overall coarsening of the population as large bubbles grow at the expense of smaller ones (cf. Eqn. 4). As a consequence, VSDs of ripened foams exhibit depletion of the smaller size classes manifested by the presence of a downturned profile with an inflection point at an intermediate size class. As a result, reticulites are substantially coarser in structure than tephras, with cells generally 0.2 to 0.3 cm in diameter and characteristic number densities of ~10^3 / cm^3 melt. The absence of a ripening signature in the tephra VSDs is consistent with their lower packing density, which serves to increase the diffusion length scale (nearest-neighbor distance).

Despite their differing characteristics, it is clear that the two clast types represent a textural continuum generated during progressive gas exsolution and expansion. Wentworth and MacDonald intuited this relation in their 1953 treatise on the features of basaltic rocks, wherein they state that "In a genetic series reticulite lies between pumice

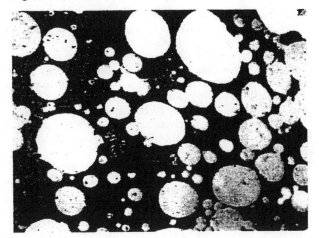

Figure 9. Typical rapidly-quenched Kilauea vent lava as described in Mangan et al. (1993). Note that the vesicularity is less than that of the products of basaltic fire fountains, and that vesicles are generally large and spherical. Field of view: 8.75 × 6.5 mm.

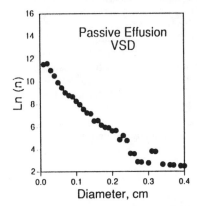

Figure 10. VSD of Kilauea rapidly-quenched effusive vent lava sample shown in Figure 9. VSD is linear at vesicle diameters <2 mm; change in slope at sizes >2 mm is the result of bubble coalescence prior to quench. From Mangan et al. (1993).

the fire fountain scenario, vesiculation maintained an equilibrium assemblage of melt + bubbles during ascent. These results contrast with predictions of Bottinga and Javoy (1991) that H_2O exsolution in basaltic magmas can occur only at very high supersaturations (i.e., far from equilibrium), and point to the need for truly predictive theoretical models of bubble nucleation in silicate melts.

Active lava tubes and surface flows

After eruption, the bubble population of lavas may be significantly modified during transport and solidification. Identification of processes responsible for these changes in active systems is integral to the interpretation of vesicle structures in solidified flows. Moreover, rates and styles of surface degassing play a crucial role in controlling basaltic lava flow rheology and morphology (Sparks and Pinkerton, 1978; Peterson and Tilling, 1980; Kilburn, 1981;1990; Rowland and Walker, 1987;1990; Dragoni and Tallarico, 1994; Crisp et al., 1994). Finally, determination of degassing rates from preserved vesicle structures is also of major interest in determining climate effects of flood basalt eruptions (e.g., Coffin and Eldhom, 1994).

Most of the work on vesicularity variations in active flows has been on pahoehoe flows, although limited data are available for a'a flows. Lipman and Banks (1987) measured density changes from 350 kg/m^3 (vesicularity of >85%) at the vent to >2400 kg/m^3 (<20% vesicularity) at the flow terminus (a distance of 10 to 15 km) in a'a flows emplaced during the 1984 eruption of Mauna Loa Volcano, Hawaii. Einarsson (1949) made similar observations a'a lava flows produced during the 1947-1948 eruption of Hekla Volcano in Iceland. In both cases, the extremely high vesicularity of the near-vent lavas significantly *decreased* the apparent viscosity of the flows, an effect that Lipman and Banks (1987) attributed to the easily deformable, very thin bubble walls. This observation is consistent with the behavior of bubble suspensions at high shear rates (e.g., Valko and Economides, 1992). Apparent viscosity *increased* downstream, largely as the result of crystallinity increases in response to cooling and extensive degassing (Crisp et al., 1994).

Tube-fed pahoehoe lava flows also decrease in vesicularity with transport distance away from the vent. Swanson and Fabbi (1973) documented an overall down-tube vesicularity decrease in samples collected from an active lava tube system established during the 1969-1974 Mauna Ulu eruption of Kilauea Volcano, Hawaii (Fig. 11). Based on concurrent decreases in measured volatiles in the glass, they attributed down-tube changes in vesicularity to continued exsolution of a volatile phase. Similar patterns of vesicularity decrease with increasing distance from the eruptive vent have been observed for the current (1983-present) eruption of Kilauea Volcano (Cashman et al., 1994a; Fig. 11). Constant H$_2$O and S contents of quenched glass in lava tube samples and in gas emitted from the tube system indicate that open system, "passive" rise and escape of larger bubbles to the lava surface probably accounts for much of the down-tube vesicularity change. Such gas loss from the tube system results in the output of 1.2×10^6 g/day SO$_2$, which represents an addition of approximately 1% to overall volatile budget calculations for the current eruption (e.g., Gerlach and Graeber, 1985). A steady increase in bubble number density in tube-transported lava with downstream distance, however, demonstrates that limited volatile exsolution is also occurring. Estimated bubble nucleation rates of 7 to 8 cm^{-3}s^{-1} are ~20% those of vent samples (see above).

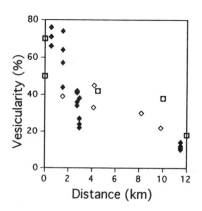

Figure 11. Lava vesicularity as a function of transport distance through the Wahaula (1991; filled diamonds) and Kamoamoa (1993; open diamonds) lava tubes, Kilauea Volcano, Hawaii. Open squares denote samples collected from the Mauna Ulu lava tube system and described in Swanson and Fabbi (1973). Replotted from Cashman et al. (1994a).

Emergence of lava from the tube system further modifies the bubble population, and may result in a preserved vesicle population that differs significantly from that erupted at the vent (Cashman et al., 1994a; Hon et al., 1994). Bubble expansion on release from pressurized tubes and resultant coalescence of adjacent bubbles combine to dramatically decrease the overall number density of bubble populations and to increase the mean bubble size and sample vesicularity (Fig. 12). Minor expansion accompanied by bubble size increases and number density decreases continues on cooling within pahoehoe toes (Walker, 1989). Modifications probably result from the combined effects of static (wall

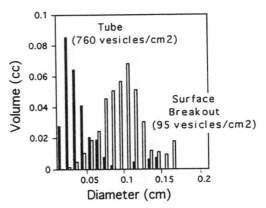

Figure 12. Comparison of vesicle volume distributions measured in rapidly-quenched samples collected directly from the Kamoamoa lava tube (Kilauea Volcano, Hawaii) and from a nearby surface breakout. Note that the vesicles in the surface breakout are larger and less numerous than those in the lava tube sample. Replotted from Cashman et al. (1994a).

thinning and rupture) coalescence and ripening of the vesicle population through interbubble diffusion of volatiles under closed-system conditions provided by the cooling surface crust (Cashman et al., 1994a). Documentation of these changes suggests caution in directly inferring vent vesicle populations (including number densities) from solidified flows (e.g., Vergniolle and Jaupart, 1990; Bottinga and Javoy, 1991).

Solidified lava flows

The vesicle structure of solidified lava flows may record dynamic processes related to flow emplacement, as well as static processes related to flow solidification. For example, vesicles preserved in a'a tend to be large and deformed (Einarsson, 1949; Fisher, 1968; MacDonald, 1972; Peterson and Tilling, 1980; Walker, 1989), reflecting the high viscosities and shear stresses developed at a'a flow boundaries. In contrast, thin fluid pahoehoe flows often have a 'spongy' texture (S-type pahoehoe of Walker, 1989; Wilmoth and Walker, 1993), the solidified equivalent of surface breakout samples from active tube systems (e.g., Fig. 12). Changes in the vesicle population inward from outer quenched surfaces can be directly related to the time available for post-emplacement ripening and coalescence (Walker, 1989; Cashman et al., 1994a). In neither case is the preserved vesicle population directly representative of the vent lavas from which the flows formed.

Inflated sheet flows. Hon et al. (1994) demonstrated that thin (<0.5m) pahoehoe sheets emplaced on near-horizontal slopes usually inflate during emplacement to thicknesses in excess of 5 m. During inflation, flow tops solidify while lava continues to intrude into the interior. Thus solidification of inflated sheet flows is a dynamic process, with new lava continually fed to the central lava tube. Recognition of inflation during flow emplacement has important consequences for the interpretation of vesicle structure, and may require some reinterpretation of previous work.

A direct correlation between preserved vesicle structure and the inflation process has been made by Cashman et al. (1994b) in their study of tumuli (former lava tubes) in an inflated sheet flow emplaced at Kalapana, Hawaii, in 1990. Their results are summarized here. The sheet flow can be divided into three vesicularity units: an upper vesicular zone, a dense interior, and a thin vesicular base. This classification is similar to those of Aubele et al. (1988) and Sahagian et al. (1989) for basalt flows of unspecified origin.

The thickness of the upper vesicular zone is a direct function of both duration of inflation and the time during which the central lava tube continues to supply new lava to the

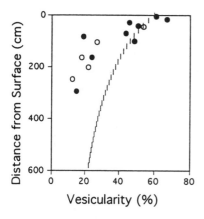

Figure 13. Vesicularity measured as a function of distance below the flow surface (upper vesicular zone only) from inflated sheet flows, Kalapana, Hawaii. Filled circles are vesicularities measured in the axial crack of the central tumulus from the Woodchip Tube, active July 1990 to February 1991. Open circles are measurements from a tumulus exposed in a coastal sea cliff, Kalapana, an ocean entry active in September and October, 1990. Short vertical lines represent the decrease in vesicularity resulting solely from the overburden of the growing crust, calculated for a lava that was initially 60% vesicular. The offset in the calculated and measured values most likely results from ~0.2 MPa overpressures within the growing tumulus. Replotted from Cashman et al. (1994b).

sheet. Vesicularity within the upper vesicular zone decreases with increasing depth at a rate significantly greater than that predicted by the increasing pressure imposed by the growing crust (Fig. 13). While gases certainly rise and escape through axial cracks over the lava tube (Walker, 1991), it is likely that the extreme and repeatable decrease in vesicularity records overpressuring of the flow during active inflation (overpressures at ~0.2 MPa are indicated by Fig. 13). Imposed on this general structure may be local perturbations reflecting transient changes in flow occurring within the lava tube. For example, horizontal bands of small vesicles record pressure fluctuations within the lava tube that accompanied changing supply rates from the vent.

Vesicular bases of inflated sheet flows are thin (commonly <0.5 m) regardless of the final flow thickness or duration of emplacement (Cashman et al., 1994b). Uniform basal vesicular zone thicknesses are most likely a consequence of decreasing cooling rates inward, and mark the point at which bubble rise rate exceeds the rate of lower crust formation. Flow bases may be pipe-vesicle bearing.

The dense interior zone forms through (1) solidification from the flow base during emplacement, and (2) solidification of stagnant lava remaining in the tube after flow ceases. Evidence that bubble-poor melt exists in active tubes lies in pahoehoe flows that emerge under pressure from the base of active tube systems. These flows have been broadly described as 'dense' (Swanson, 1973; Hon et al., 1994) or 'blue glassy' (Cashman et al., 1994b). They are characterized by a dense glassy rind, indicating that the lava was initially depleted in bubbles. These flows commonly have pipe vesicles generated at an inner selvage (P-type pahoehoe of Walker, 1989; Wilmoth and Walker, 1993), vesicles that form through elongation and inward forcing of bubbles by the solidification front (see Walker, 1987; Philpotts and Lewis, 1987 for different variants of this basic model). Bubble redistribution is also likely within the molten core remaining after flow ceases. Redistribution may occur both by single bubble rise and coalescence (Sahagian, 1985; Sahagian et al., 1989) and continued volatile exsolution at the lower crystallization front (e.g., Aubele et al., 1988; McMillan et al., 1987; 1989). Extensive late-stage differentiation and vesiculation at the lower boundary layer may also generate Rayleigh-Taylor instabilities, similar to the vesicle-rich diapirs observed in Kilauea Iki lava lake ('VORBs' of Helz et al., 1989) and the vesicle cylinders commonly observed in inflated sheet flows (Goff, 1977; Cashman et al., 1994b).

Submarine lava flows

Experimental volatile solubility constraints (e.g., Stolper and Holloway, 1988) combined with gas composition and volume and size distributions of vesicles in submarine basalts yield information on original volatile contents (MacPherson, 1984), composition (Kurz and Jenkins, 1981; Marty and Ozima, 1986; Javoy and Pineau, 1991), vesiculation kinetics (Dixon et al., 1988; Sarda and Graham, 1990; Blank et al., 1993; Pineau and Javoy, 1994), and complex degassing histories of rift-transported magmas (Dixon et al., 1991). The vesicularity and mean vesicle size in glassy rims of pillow basalts generally decrease with increasing eruption depth (Moore, 1965; 1970; Moore and Schilling, 1973; MacPherson, 1984; Fig. 14). Eruption depths of MORBs are commonly in excess of 2 km. At this depth CO_2 is the primary exsolving phase for magmas with low H_2O contents, and vesicularities are <5 vol % (e.g., Kurz and Jenkins, 1981; Marty and Ozima, 1986; Dixon et al., 1988). Significantly higher vesicularities reported in MORB 'popping rocks' (Sarda and Graham, 1990; Pineau and Javoy, 1994) reflect high CO_2 contents (4000 ppm) that may indicate primary MORB values. The unusually high vesicularities measured in back-arc basin basalts (e.g., Dick, 1980; Gill et al., 1990) are most likely the result of anomalously high water contents.

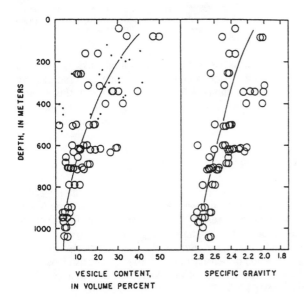

Figure 14. Vesicularity vs. depth relations for basalt samples from the Reykjanes Ridge. Vesicularity data from subglacial pillow basalts is shown in small dots. [Used by permission of *Contributions to Mineralogy and Petrology*, from Moore and Schilling (1973), Fig. 2, p. 110.]

Recent measurements of combined vesicularity and volatile contents of submarine basalt glasses suggest caution in the interpretation of vesicularity as directly reflecting eruption depth, particularly when CO_2 is the primary exsolving phase. CO_2 contents of basaltic glasses erupted on the Juan de Fuca ridge show an inverse correlation with depth, and no clear relation with sample vesicularity (Dixon et al., 1988; Blank et al., 1993). These vesicle characteristics result from incomplete degassing of CO_2 in regions where magma storage time was minimal. Conversely, the rarity of vesicular MORBs suggests that pre-eruptive degassing from magma chambers is common (Pineau and Javoy, 1994). Tholeiitic glasses from the submarine portion of the east rift zone of Kilauea Volcano show CO_2 contents consistent with the eruption depth, but display a wide range of SO_2 and H_2O contents (Dixon et al., 1991) produced by low-pressure degassing during transport through

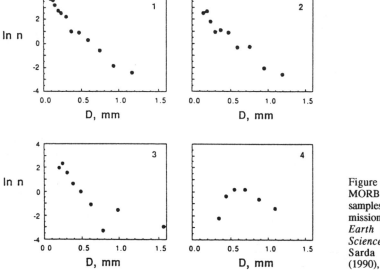

Figure 15. VSDs of MORB 'popping rock' samples. [Used by permission of the editor of *Earth and Planetary Science Letters*, from Sarda and Graham (1990), Fig. 2, p. 273.]

the subaerial rift zone. Variable shapes of VSDs in MORB 'popping rocks' (Sarda and Graham, 1990; Fig. 15) also suggest that the vesicle populations of submarine samples are frequently modified from original vent populations before solidification. In this regard, estimates of bubble number densities in magma chambers from measured vesicle number densities in MORBs must be used with caution (e.g., Bottinga and Javoy, 1990; Vergniolle and Jaupart, 1990).

VESICULATION IN SILICIC MAGMAS

Significantly more data exist on the structure of silicic pyroclasts than for their basaltic counterparts. Juvenile products of silicic explosive eruptions may also form true foams with vesicularities in excess of 75%. Preserved vesicle shapes are commonly spherical or cylindrical (in fibrous pumice), as recorded by the shapes of glass shards (representing remnant films and plateau borders) and by structure preserved in pumice clasts. Physical properties of pumice are described in terms of density (vesicularity) and the shapes, sizes and connectedness of vesicles (Sparks and Brazier, 1982; Whitham and Sparks, 1986; Heiken, 1987; Klug and Cashman, 1991; 1994; Orsi et al., 1992). Textures preserved in pumice clasts from explosive eruptions will reflect vesiculation during eruption plus expansion and coalescence in eruption columns prior to quenching. The mechanisms and role of volatile exsolution in effusive eruptions of silicic lava is more controversial (e.g., Eichelberger et al., 1986; Eichelberger, 1989; Fink et al., 1992). Characteristically low volatile contents preserved in obsidian require volatile loss and equilibration at atmospheric pressures, but where and how that loss occurs is a matter of debate. Additional discussion of degassing and fragmentation of silicic magmas may be found in Part I of this Chapter.

Explosive eruptions

Explosive silicic eruptions are characterized by extremely high eruption velocities, column heights that reach tens of kilometers into the atmosphere, and the production of large quantities of both ash and pumice. While theoretical studies have yielded important insights into processes occurring during explosive eruptions (e.g., Sparks, 1978; 1986; Woods, 1988; Thomas and Sparks, 1992), textural analysis of the products of these eruptions provides an important complement to the models. Heiken and Wohletz (1985) generated a remarkable compendium of volcanic ash textures for all types of eruptions, including documentation of eruptive history, chemical composition and ash morphology. In this section we do not attempt to review their work, but instead focus on information provided by recent studies of pumice clast morphology.

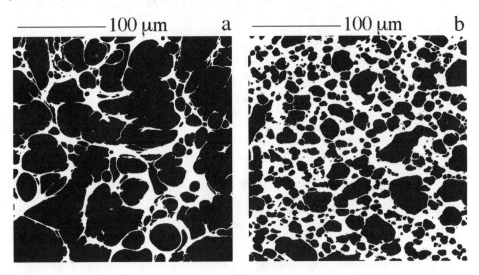

Figure 16. Binary BSE images of Mount St. Helens 1980 pumice. Here vesicles are black, interstitial glass + crystals is white. (a) 'White' pumice; average vesicularity 81%, crystallinity 4%, no microlites. Vesicles are highly coalesced, although in many cases wall rupture was only partial, and wall remnants remain. (b) 'Gray' pumice; average vesicularity 58%; crystallinity 22% (including 15% microlites). Vesicles are less spherical than in the microlite-free white pumice, and coalescence, while evident, is less ubiquitous. Replotted from Klug and Cashman (1994).

Vesicularity of silicic pumice from plinian columns is commonly 70 to 80%, and clast density is correspondingly low (<1000 kg/m^3). According to the nomenclature presented above, pumice may thus be classified as transitional between a bubbly suspension and a spherical foam, although significant deviations from spherical shapes are common (Fig. 16). Sparks (1978) postulated that pressure increases occur when bubble expansion is impeded after bubbles attained spherical packing limits, ultimately resulting in fragmentation of the magma (see also Part I). Houghton and Wilson (1989) found support for this theory in the uniformity of clast density in magmatic ('dry') eruptions, which they interpreted to preserve the state of the magma at the time of fragmentation. There are some problems, however, with this interpretation. As in the basaltic fire fountains described above, pumice clasts in plinian deposits may have both lower and higher vesicularities than the proposed fragmentation limits of 75 to 83 %. For example, pumice from the Plinian phase of the Bishop Tuff eruption ranges in bulk density from 510 to 820 kg/m^3, representing vesicularities of 60 to 77% (Gardner et al., 1991), while densities of 180 to 330 kg/m^3 reported by Whitham and Sparks (1986) for the Askja and Minoan eruptions imply vesicularities of 86 to 92% (assuming glass densities of 2300 kg/m^3). Thus, any model of vesiculation-induced fragmentation of magma must account for this variability.

Additionally, silicic 'reticulites' are extremely rare, although typically high water contents (>2 to 3 wt %) in silicic magmas (Chapter 8, this volume) dictate that eruptive products should show vesicularities in excess of 99% (ash) if equilibration to atmospheric pressures occurred under closed system conditions.

Vesicularities <70% result from incomplete clast expansion. Studies of both vesicularity and shard morphology in hydrovolcanic deposits indicate that 'premature' quenching of expanding magma within the conduit can occur through interaction with external water (Sheridan and Wohletz, 1983; Wohletz, 1983; 1987; Heiken, 1987; Houghton and Wilson, 1989; Heiken and Wohletz, 1991). Water introduced deep in the magmatic system would thus be expected to produce higher density clasts than water introduced near the surface. Vesicularities <70% are also found in products of purely magmatic eruptions, particularly when magma crystallinities are high. For example, microlite-rich pumice from the 1980 eruption of Mount St. Helens has significantly lower vesicularity (60%; Fig. 16b) than concurrently-erupted microlite-free pumice (81%, Fig. 16a). Sequentially-erupted microlite-free and microlite-bearing pumice from the 79 AD eruption of Mt. Vesuvius show similar variations in vesicularity (Carey and Sigurdsson, 1987). In both cases, the presence of microlites apparently increased the bulk viscosity of the magma sufficiently to retard bubble expansion.

High vesicularities (>80%) can also occur in silicic pumice. In contrast to the polyhedral cell formation seen in basaltic pyroclasts, high vesicularities in silicic pumice are achieved primarily through bubble coalescence and stretching. *Coalescence* occurs by expansion of smaller bubbles into larger neighbors, as a consequence of higher internal pressures in smaller cells (cf. Eqn. 2). Limited coalescence increases the bubble size range by generating a population of large bubbles without completely eliminating small bubbles, and thus allows more efficient packing of spheres. *Stretching* of vesicles in 'fibrous' or 'long-tube' pumice into cylindrical shapes (probably through shearing) allows a further increase in overall vesicularity without transformation into a polyhedral foam by removing packing constraints for spheres (Klug and Cashman, 1991). Measured vesicularities up to 93% in fibrous, or 'woody' pumice (Kato, 1987; Klug and Cashman, 1991) approach packing limits for a size range of cylinders (Burk and Apte, 1987). Finally, 'breadcrusting' of larger clasts is common in silicic eruptions, and indicates that internal clast expansion has continued after quenching of the outer rind.

Vesicle volume distributions in silicic pumice are commonly polymodal. Sparks and Brazier (1982) first made this observation based on Hg-porosimetry data (Fig. 17). Their preliminary interpretation that the three vesicle populations resulted from three different episodes of vesiculation was later reinterpreted by Whitham and Sparks (1986), who noted that the smallest size population of Figure 17 (<8 μm) represented connections between larger vesicles, and not a separate vesicle population. The smaller true vesicle population (12 to 20 μm) most likely represents syn-eruptive vesiculation during magma ascent and fragmentation. The larger vesicle population (>100 μm) has been variously interpreted as pre-eruptive bubble growth in magma chambers (Sparks and Brazier, 1982; Whitham and Sparks, 1986) or post-fragmentation expansion and coalescence in eruption columns (Orsi et al., 1992; Klug and Cashman, 1994). It is important to note that pumice samples from many different eruptions exhibit this bimodal volume distribution (Whitham and Sparks, 1986; Toromaru, 1990; Orsi et al., 1992; Klug and Cashman, 1994). This suggests that (1) the preserved vesicle population in silicic pumice rarely (if ever) results from a single, steady state bubble nucleation and growth event without later modification, and (2) determination of vesiculation kinetics from VSDs depends on the origin assumed for the different vesicle populations.

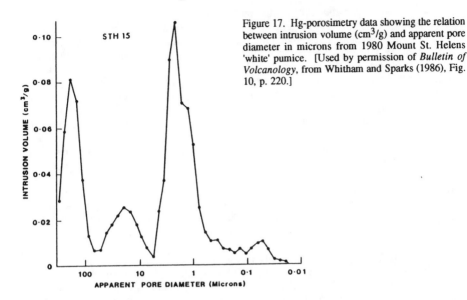

Figure 17. Hg-porosimetry data showing the relation between intrusion volume (cm^3/g) and apparent pore diameter in microns from 1980 Mount St. Helens 'white' pumice. [Used by permission of *Bulletin of Volcanology*, from Whitham and Sparks (1986), Fig. 10, p. 220.]

As an example of the use of vesicle size information for determination of vesiculation kinetics we briefly review work by Klug and Cashman (1994) on pumice from the May 1980 eruption of Mount St. Helens. Vesicle volume distributions from 'gray' pumice (microlite-bearing; Fig. 16b) are shown in Figure 18. Here both raw and 'decoalesced' volume distributions are shown, where the latter were obtained by manually separating partially coalesced vesicles (those with partially intact walls). The peaks in the decoalesced volume distributions provide the closest obtainable approximation of mean bubble size at the time of magma fragmentation. Using calculated ascent times of ~1000 to 4000 s (following Wilson et al., 1980), a mean size of 15 μm yields bubble growth rates of 1 to 5×10^{-7} cm/s. Growth rates are an order of magnitude slower than those calculated assuming a simple diffusive + decompressive growth model (Sparks, 1978; Mangan et al., 1993). Possible explanations for this discrepancy include viscosity-limited growth, hindered bubble expansion because of interference with surrounding bubbles, or overestimation of vesiculation times.

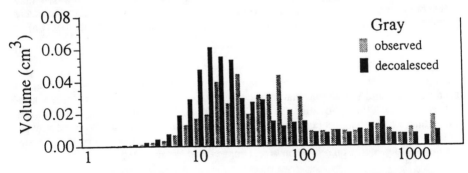

Figure 18. Vesicle volume distributions measured in 'gray' 1980 Mount St. Helens pumice (as shown in Fig. 16b). 'Decoalesced' distributions were made by drawing lines of minimum thickness between vesicles where partial walls remained. The decoalesced distributions are the closest available representation of the bubble population at the time of magma fragmentation. Replotted from Klug and Cashman (1994).

Vesicle number densities for the gray pumice are extremely high (>10^9), resulting in nucleation rates of 3 to 10×10^5 cm^{-3}s^{-1}. Again, these rates are predicated on times calculated assuming that the system maintained equilibrium during ascent. Recent experimental work by Hurvitz and Navon (1994) provides a limited basis for comparison. Nucleation rates estimated by Klug and Cashman span the range observed by Hurvitz and Navon for bubble nucleation in water-saturated rhyolite at large pressure drops (>80 MPa). However, vesicle number densities in the pumice exceed those in the experimental charges by an order of magnitude (even for experimental overpressures of >100 MPa). High measured vesicle number densities in the Mount St. Helens pumice may indicate that vesiculation occurred at very high supersaturations (that is, large ΔP). Alternatively, it is possible that the high vesicle number densities are related to the prevalent microlite population, although the evidence is ambiguous. Experimental studies on obsidian have shown that the presence of microlites may significantly increase bubble nucleation rates (Bagdassarov and Dingwell, 1992; Hurvitz and Navon, 1994). Although vesicle number densities in microlite-free pumice from Mount St. Helens are slightly lower than those in microlite-bearing samples, this difference can be explained by the high degree of bubble coalescence in microlite-free samples (Klug and Cashman, 1994).

Vesicle shapes in pumice commonly record stages of bubble coalescence (Heiken, 1987; Orsi et al., 1992; Klug and Cashman, 1994). Evidence for inter-bubble expansion preserved in silicic pumice (Fig. 19a,b) indicates clast expansion times less than times required for complete rupture of melt films and retraction to spherical shapes. Instead, bubble walls thin unevenly, and breakage occurs where walls reach thicknesses of <1 μm (Fig. 19c). Film thinning times resulting from capillary pressure may be estimated from the integrated form of Equation (3). An initial film thickness of 10 μm will thin to 1 μm over an area of radius 8 μm (from Fig. 19a) in 400 to 4000s (given μ = 10^5 to 10^6 Pa s, assuming that ΔP can be estimated from the size difference between coalescing bubble; cf. Eqn. 2). These times are excessive given times available for expansion in eruption columns (Thomas and Sparks, 1992), although the time constraints are weak. However, magmatic pressures at the vent are well in excess of atmospheric (Wilson et al., 1980) and, as in basaltic fire fountains, expansion resulting from decompression in eruption columns must contribute significantly to the thinning process.

Vesicle shapes also record stretching of bubbles in response to shear during transport. Formation of cylindrical vesicles may involve coalescence in both longitudinal and radial directions (Fig. 20) in addition to simple stretching, thus complicating shape interpretation. However, when stretching alone is recorded, vesicle aspect ratio as a function of size may yield constraints on the shear strain experienced by the magma during transport. The extent of bubble deformation that occurs during shear is measured by the capillary number (Ca)

$$Ca = (d\varepsilon/dt)r\mu_{liq} / \sigma \qquad (9)$$

where $(d\varepsilon/dt)$ is the shear strain rate, r is bubble radius, μ_{liq} is the liquid viscosity and σ is the surface tension. A Ca number of 1 represents the maximum shear strain rate due to shear flow around the bubble that may occur without deforming the bubble to a non-spherical shape. From Equation (9) it is apparent that large bubbles will deform more easily than smaller bubbles (a common observation in pumice) and that if applied stress is not maintained through the glass transition, strained bubbles may return to spherical geometries (depending on relaxation time; Dingwell and Webb, 1989).

Combined bubble coalescence and deformation will increase magmatic gas permeability, which will in turn increase rates of volatile escape, heat transfer, and clast cooling (Thomas and Sparks, 1992). Permeable gas loss will also permit continued

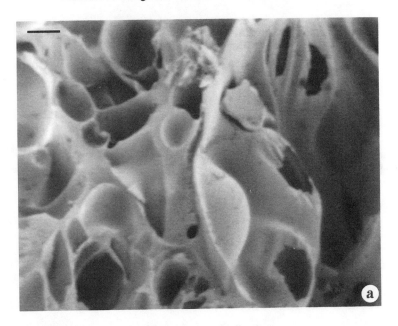

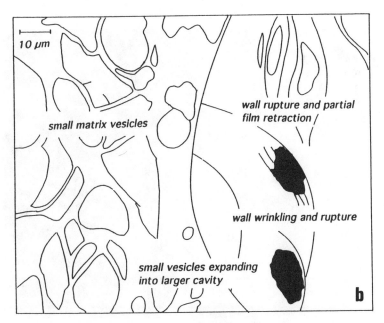

Figure 19. Characteristic film rupture structures in climactic fall pumice from the 6800 ybp eruption of Mt. Mazama, Oregon. Scale bars are 10 μm. (a) SEM image showing expansion of small bubbles into larger cavities, and plastic 'wrinkling' around rupture that demonstrates that film breakage occurred prior to quenching. (b) Line drawing of the image shown in (a) with important features labeled.

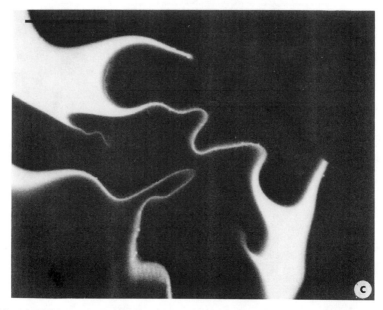

Figure 19. (c) BSE image from thin section of a transect across a vesicle similar to those in (a) with wrinkle structures. Note that bubble wall wrinkling occurs at thicknesses <1 μm; 1 μm is therefore considered an upper thickness limit for film failure.

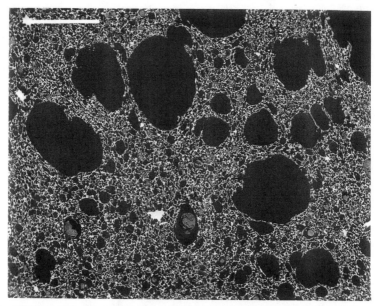

Figure 20. BSE image of a thin section cut perpendicular to the direction of elongation of cylindrical vesicles in a sample of the Wineglass Welded Tuff, 6800 ybp eruption of Mt. Mazama, Oregon. Note the wide range in vesicle size, and the evidence for radial bubble coalescence. Scale bar is 500 μm.

volatile exsolution without further affecting foam structure. Very few measurements of pumice permeability exist, and none have been made at eruptive temperatures. Eichelberger et al. (1986) measured an increase in permeablity of pumice in the Inyo rhyolite from 10^{-16} m^2 at <60% porosity up to 10^{-12} m^2 at 75% porosity. Thomas and Sparks (1992) estimate permeabilities of 10^{-9} to 10^{-12} m^2 from the Hg-porosimetry measurements of Sparks and Brazier (1982) on highly vesicular (>75%) pumice. The prevalence of brittle fractures in pumice samples indicates that these measurements yield permeabilities that are high relative to permeability in the liquid state. However, evidence of non-brittle wall rupture in pumice is common (Fig. 19), suggesting that at least part of the measured permeability was attained prior to quench. Cylindrical vesicles developed in 'long-tube' pumice will also drastically increase magmatic gas permeability. Well-developed cylindrical vesicles (Fig. 20) commonly have radii on the order of 0.1 to 1 mm at spacings averaging of 0.1 to 5 mm. Permeabilities of the large cylindrical vesicles alone will be 10^{-9} to 10^{-8} m^2 (using the permeability model of Berryman and Blair, 1987).

The above observations on the vesicle structure of silicic pumice have important consequences for investigations of magmatic fragmentation. Pumice from magmatic ('dry') eruptions may have vesicularities significantly *less* than the theoretical 75% packing limit, supporting explanations for magmatic disruption that are not controlled exclusively by bubble packing (e.g., Heiken, 1987; Dingwell and Webb, 1989; Mangan and Cashman, 1994). Moreover, significant *post*-eruption expansion requires re-examination of interpretations of quenched pumice vesicularity as a direct measure of fragmentation vesicularity (Klug and Cashman, 1994). Based on a comparison with the behavior of materials in basaltic fire fountains, we suggest that 70 to 80% vesicularity may represent the easily attainable expansion limit (not necessarily achieved at the time of fragmentation). The high viscosities of silicic melts (see Chapter 9) severely limit overall rates of foam expansion (cf. Eqn. 3). Continued volatile exsolution and expansion beyond spherical packing limits must proceed by increasing permeability through local wall rupture (coalescence) and bubble deformation. Much of this expansion probably occurs in the eruption column after magmatic disruption. Thus final pumice vesicularity will be controlled by the relative rates of vesiculation, magma rise, and location of quenching. As seen in basaltic fire fountains, this model allows a range of vesicularities to be produced from a single event, depending on the time an individual clast has to equilibrate with atmospheric pressure before quenching (see also Part I).

Effusive silicic eruptions

Rates, styles and mechanisms of magmatic degassing are central to the controversy surrounding the origin of obsidian flows. Obsidian flows are generally produced following explosive eruptions of volatile-rich rhyolite. The effusive eruption style coupled with typically low measured volatile contents have led workers to assume that obsidian represents the eruption of a volatile-poor magma under closed-system conditions; the implication is that these eruptions are the result of tapping of magma chambers strongly zoned in volatiles (e.g., Eichelberger and Westrich, 1981). An alternative explanation for obsidian is offered by Eichelberger et al. (1986). They suggest that the explosive-to-effusive transition represents a change in system behavior from 'closed' to 'open', with volatile escape in the latter case permitted through generation of a 'permeable foam' (open-celled foam in the terminology presented earlier). Closed-system degassing requires that the melt stay in contact with volatiles during exsolution and transport to the level of magmatic fragmentation, as appears to be the case in high-velocity plinian eruptions (Jaupart and Allegre, 1991). Open-system degassing requires separation of volatiles from the melt before or during magma ascent (Newman et al., 1988; Taylor, 1991).

Macroscopic textural data on glassy and vesicular rhyolite flows are reviewed by Eichelberger et al. (1986) and Fink et al. (1992). The complexity of textural features of silicic lava flows often belies the simple permeable versus non-permeable foam distinction (Fink, 1983; Fink and Manley, 1987; 1989; Manley and Fink, 1987; Anderson and Fink, 1989). The development of distinct vesicular zones is common to rhyolitic lava flows (Fig. 21). Flow top vesicularities (25 to 40%) result from primary vesiculation ('effervescence') with equilibration of the flow surface to atmospheric pressures. Bands of coarsely vesicular (~20%) pumice exist *below* the upper vitrophyre as the result of upward migration and concentration of volatiles after flow ceased. Coarsely vesicular pumice layers ofter rise buoyantly through the upper part of the flow to the flow surface. Extensive bubble coalescence and expansion during rise can result in high vesicularities (70 to 80%; Manley and Fink, 1987), or even 'giant' bubbles (~3 to 4 m diameter) such as those preserved in the Big Obsidian Flow in Newberry caldera, Oregon (Jensen, 1993). Post-intrusion volatile migration may also lead to high overpressures and the potential for endogenic pyroclastic flow generation from the fronts of advancing lavas. Finally, devitrified rhyolite commonly contains vesicle bands generated by local concentration of volatiles during crystallization of anhydrous minerals ('second boiling').

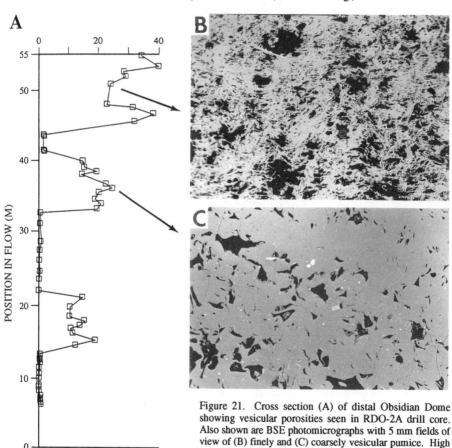

Figure 21. Cross section (A) of distal Obsidian Dome showing vesicular porosities seen in RDO-2A drill core. Also shown are BSE photomicrographs with 5 mm fields of view of (B) finely and (C) coarsely vesicular pumice. High porosity zone in the lower third of the flow was formed by late-stage crystallization and resultant 'second boiling' after the lava stopped advancing. From Fink et al. (1992).

Open-system volatile loss. Questions that arise from the textual data revolve around where and how quickly volatiles are lost from magmas. Open-system volatile loss, either before or during magma extrusion, is required for interpretation of measured $H_2O-\delta D$ correlations in many obsidian-forming eruptions (e.g., Eichelberger et al., 1986; Newman et al., 1988; Taylor, 1991; Fink et al., 1992; see also Chapter 8). Decreasing volatile contents recorded in a stratigraphic sequence of plinian deposits may represent the *pre-eruptive*, depth-dependent, volatile content of the quenched margin of the conduit/feeder system (Newman et al., 1988). Extensive gas loss preceding eruptive activity is observed in many active systems (Westrich and Gerlach, 1992; Stix et al., 1993; Hoblitt and Harmon, 1993) and suggests open system behavior within the magma reservoir itself. Degassing *between* eruptive episodes is also common (reviewed in Fink et al., 1992), although actual mechanisms of gas migration out of these magmas are unknown. Volatile loss during magma ascent requires that volatiles are able to escape through the magma, which in turn requires that the magma is relatively permeable to gases. Jaupart and Allegre (1991) assume that the flow rate of volatiles out of magma is equal to the flow rate through the country rock, although the nature of the boundary layer between rising magma and country rock is poorly understood. They find that for low initial ascent velocities (small magma chamber overpressures), permeabilities of 10^{-13} to 10^{-15} m^2 will result in magma ascent dominated by gas loss to the country rock. From the limited permeability measurements of Eichelberger et al. (1986), these values require porosities approaching 60%. Clearly more detailed analysis of permeability development is required for a thorough analysis of volatile loss in volcanic conduits.

CONCLUSIONS

Vesicle shapes and sizes in basaltic lavas reflect the location and style of volatile exsolution. Explosive basaltic eruptions are characterized by extremely rapid vesiculation and fragmentation. Continued expansion occurs in the thermal core of fire fountains, with evolution to highly expanded open cell foams (Mangan and Cashman, 1994). Effusive eruptions show near-equilibrium rates of volatile exsolution, as controlled by diffusion and decompressional expansion in volcanic conduits (Mangan et al., 1993). Bubble populations in lavas at volcanic vents are significantly modified during transport and solidification by passive bubble loss, coalescence, and continued nucleation (Swanson and Fabbi, 1973; Sahagian, 1985, McMillan et al., 1987; 1989; Aubele et al., 1988; Sahagian et al., 1989; Walker, 1989; Wilmoth and Walker, 1993; Cashman et al., 1994a,b).

Silicic pumice formation is less well understood, with questions persisting concerning conditions of pumice preservation, and when and how volatiles are lost (see also Part I, and Chapter 8). The former question is important in the context of understanding the nature of magmatic fragmentation, the latter in modeling rates of gas loss during ascent in the conduit, during upward expansion of the eruption column, and during transport within pyroclastic flows. Crystal-poor magmatic eruptions commonly produce pumice with vesicularities of 70 to 80% (Houghton and Wilson, 1989). As in basaltic eruptions, the final texture preserved in pumice is related to the time available for clast expansion. Timescales of eruption are related to eruption intensities, which in turn are ultimately related to the size of the magma chamber being tapped (Carey and Sigurdsson, 1989). Development of magma permeability is also central to the controversy surrounding the origin of effusively erupted obsidian flows (Eichelberger et al., 1986; Fink et al., 1992). Here chemical evidence for 'open-system' behavior of gases is compelling (Newman et al., 1988; Taylor, 1991), but the timing of gas loss is uncertain. Textural evidence for post-emplacement vesiculation (Fink et al., 1992) suggests that rates of volatile equilibration to atmospheric pressure are low.

ACKNOWLEDGMENTS

This work was supported by funding through the U.S. Geological Survey Volcano Hazards Program (to MTM) and by National Science Foundation grants EAR9017364 and EAR9218908 (to KVC). We appreciate thoughful reviews by Steve Carey, Michael Carroll, Roger Denlinger, Jon Fink, Grant Heiken and Jim Kauahikaua. We also thank Caroline Klug for many helpful discussions, as well as for the use of her unpublished SEM images of pumice from Mt. Mazama.

REFERENCES

Anderson SW, Fink JH (1989) The development and distribution of surface textures at the Mount St. Helens dome. IAVCEI Proc Volcanol 2:25-46

Aref H, Herdtle T (1990) Fluid networks. In: Moffat HK (ed) Topological Fluid Mechanics. p 745-764

Aubele JC, Crumpler LS, Elston WE (1988) Vesicle zonation and vertical structure of basalt flows. J Volcanol Geotherm Res 35:349-374

Bagdassarov NS, Dingwell DB (1992) A rheological investigation of vesicular rhyolite. J Volcanol Geotherm Res 50:307-322

Berry EL, Reinhardt RL (1974) An analysis of cloud droplet growth by collection: Part II. Single initial distributions. J Atmos Sci 31:1825-1831

Berryman JG, Blair SC (1987) Kozeny Carmen relations and image processing methods for estimating Darcy's constant. J Appl Phys 62:2221-2228

Bikerman J (1973) Foams. Springer-Verlag, New York, 337 p

Blank JG, Delaney JR, DesMarais DJ (1993) The concentration and isotopic composition of carbon in basaltic glasses from the Juan de Fuca Ridge, Pacific Ocean. Geochim Cosmochim Acta 57:875-888

Bottinga Y, Javoy M (1990) Mid-Ocean Ridge basalt degassing: Bubble nucleation. J Geophys Res 95:5125-5131

Bottinga Y, Javoy M (1991) The degassing of Hawaiian tholeiite. Bull Volcanol 53:73-85

Burk RC, Apte PS (1987) A packing scheme for real size distributions. Am Ceram Soc Bull 66:1389-1392

Carey S, Sigurdsson H (1987) Temporal variations in column height and magma discharge rate during the 79 A.D. eruption of Vesuvius. Geol Soc Am Bull 99:303-314

Carey S, Sigurdsson H (1989) The intensity of plinian eruptions. Bull Volcanol 51:28 - 40

Cashman KV (1990) Textural constraints on the kinetics of crystallization of igneous rocks. In: Nicholls J, Russell JK (ed) Modern Methods of Igneous Petrology: Understanding Magmatic Processes. Rev in Mineral 24:259-314

Cashman KV, (1993) Relationship between crystallization and cooling rate—insight from textural studies of dikes. Contrib Mineral Petrol 113:126-142

Cashman KV, Ferry JM (1988) Crystal size distribution (CSD) in rocks and the kinetics and dynamics of crystallization III. Metamorphic crystallization. Contrib Mineral Petrol 99:401-415

Cashman KV, Marsh BD (1988) Crystal size distribution (CSD) in rocks and the kinetics and dynamics of crystallization II. Makaopuhi lava lake. Contrib Mineral Petrol 99:292-305

Cashman KV, Mangan MT, Newman S (1994a) Surface degassing and modifications to vesicle size distributions in Kilauea basalt. J Volcanol Geotherm Res 61:45-68

Cashman KV, Kauahikaua JP, Pallon JE (1994b) Emplacement of inflated sheet flows II. Vesicle structure. Bull Volcanol (in review)

Cheng H, Lemlich R (1983) Errors in measurement of bubble size distribution in foam. Industr Eng Chem Fund 22:105-109

Cheng HC, Natan TE (1986) Measurement and physical properties of foam. In: Cheremisinoff NP (ed) Encyclopedia of Fluid Mechanics. Gulf Publ Co, Houston, Texas, p 3-25

Coffin MF, Eldholm O (1994) Large igneous provinces: crustal structure, dimensions, and external consequences. Rev of Geophys 32:1-36

Crisp J, Cashman KV, Bonini JA, Hougen SB, Pieri DC (1994) Crystallization history of the 1984 Mauna Loa lava flow. J Geophys Res 99:7177-7198

Dana JD (1891) Characteristics of Volcanoes. Dodd, Mead & Co, New York, 399 p

deVries AJ (1957) Foam Stability. Rubber-Stitching, Delft, Netherlands, 1-77

deVries AJ (1972) Morphology, coalescence, and size distribution of foam bubbles. In: Lemlich R (ed) Adsorptive Bubble Separation Techniques. Academic Press, New York, 7-31

Dick HJB (1980) Vesicularity of Shikoku Basin basalt: A possible correlation with the anomalous depth of back-arc basins. Init Rep Deep Sea Drilling Proj Leg 58:895-901

Dingwell DB, Webb SL (1989) Structural relaxation in silicate melts and non-Newtonian melt rheology in geologic processes. Phys Chem Minerals 16:508-516

Dixon JE, Stolper E, Delaney JR (1988) Infrared spectroscopic measurements of CO_2 and H_2O in Juan de Fuca Ridge basaltic glasses. Earth Planet Sci Lett 90:87-104

Dixon JE, Clague DA, Stolper EM (1991) Degassing history of water, sulfur and carbon in submarine lavas from Kilauea Volcano, Hawaii. J Geol 99:371-394

Dobran F (1992) Nonequilibrium flow in volcanic conduits and application to the eruptions of Mt. St. Helens on May 18, 1980, and Vesuvius in AD 79. J Volcanol Geotherm Res 49:285-311

Dragoni M, Tallarico T (1994) The effect of crystallization on the rheology and dynamics of lava flows. J Volcanol Geotherm Res 59:241-252

Eichelberger JC (1989) Are extrusive rhyolites produced from permeable foam eruptions? A reply. Bull Volcanol 51:72-75

Eichelberger JC, Westrich HR (1981) Magmatic volatiles in explosive rhyolitic eruptions. Geophys Res Lett 8:757-760

Eichelberger JC, Hays DB (1982) Magmatic model for the Mount St. Helens blast of May 18, 1980. J Geophys Res 87:7727-7738

Eichelberger JC, Carrigan CR, Westrich HR, Price RH (1986) Non-explosive silicic volcanism. Nature 323:598-602

Einarsson T (1949) The flowing lava. Studies of its main physical and chemical properties. In: Einarsson T, Kjartansson G, Thorarisson S (eds) The Eruption of Hekla 1947-1948. Visindafelg Islendinga, Reykjavik, 71 p

Fink JH (1983) Structure and emplacement of a rhyolitic obsidian flow: Little Glass Mountain, Medicine Lake Highland, northern California. Geol Soc Am Bull 94:362-380

Fink JH, Manley CR (1989) Explosive volcanic activity generated within advancing silicic lava flows. IAVCEI Proc Volcanol 1:169-179

Fink JH, Anderson SW, Manley CR (1992) Textural constraints on effusive silicic volcanism: Beyond the permeable foam model. J Geophys Res 97:9073-9083

Fisher RV (1968) Puu Hou littoral cones, Hawaii. Geol Rundschau 57:837-864

Frankel S, Mysels KJ (1969) The bursting of soap films. II. Theoretical considerations. J Phys Chem 73:3028-3038

Gardner JE, Sigurdsson H, Carey S (1991) Eruption dynamics and magma withdrawal during the plinian phase of the Bishop Tuff eruption, Long Valley Caldera. J Geophys Res 96:8097-8111

Gerlach TM (1986) Exsolution of H2O, CO2, and S during eruptive episodes at Kilauea volcano, Hawaii. J Geophys Res 91:12177-12185

Gerlach TM, Graeber EJ (1985) Volatile budget of Kilauea Volcano. Nature 313:273-277

Gill and others (1990) Explosive deep water basalt in the Sumisu backarc rift. Science 248:1214-1217

Glazier JA, Anderson MP, Grest GS (1990) Coarsening in the two-dimensional soap froth and the large-Q Potts model: a detailed comparison. Phil Mag B 62:615-645

Goff FE (1977) Vesicle cylinders in vapor-differentiated basalt flows. Univ Calif Santa Cruz, unpubl PhD dissertation, 181 p

Greenland LP, Okamura AT, Stokes JB (1988) Constraints on the mechanics of the eruption. US Geol Surv Prof Paper 1463:155-164

Head JW, Wilson L (1987) Lava fountain heights at Pu'u O'o, Kilauea, Hawaii: Indicators of amount and variations of exsolved magma volatiles. J Geophys Res 92:13715-13719

Heiken G (1987) Textural analysis of tephra from a rhyodacitic eruption sequence, Thira (Santorini), Greece. In: Marshall JR (ed) Clastic Particles. Van Nostrand Reinhold Co, New York, 67-78

Heiken G, Wohletz K (1985) Volcanic Ash. Univ of California Press, Berkeley, 246 p

Heiken G, Wohletz K (1991) Fragmentation processes in explosive volcanic eruptions. SEPM Spec Publ 45:19-26

Helz RT, Kirschenbaum H, Marinenko JW (1989) Diapiric transfer of melt in Kilauea Iki lava lake, Hawaii: A quick efficient process of igneous differentiation. Geol Soc Am Bull 101:578-594

Herd RA, Pinkerton H (1993) Bubble coalescence in magmas. 24th Lunar Planet Sci Conf Abstr XXIV:641-642

Herdtle T, Aref H (1992) Numerical experiments on two-dimensional foam. J Fluid Mech 241:233-260

Hoblitt RP, Harmon R.S. (1993) Bimodal density distribution of cryptodome dacite from the 1980 eruption of Mount St. Helens, Washington. Bull Volcanol 55:421-437

Hon K, Kauahikaua JP, Denlinger R, McKay K (1994) Emplacement and inflation of pahoehoe sheet flows - observations and measurements of active lava flows on Kilauea Volcano, Hawaii. Geol Soc Am Bull 106:351-370

Houghton BF, Wilson CJN (1989) A vesicularity index for pyroclastic deposits. Bull Volcanol 51:451-462

Hurwitz S, Navon O (1994) Bubble nucleation in rhyolitic melts: Experiments at high pressure, temperature, and water content. Earth Planet Sci Lett 122:267

Jaupart C, Allegre CJ (1991) Gas content, eruption rate and instabilities of eruption regime in silicic volcanoes. Earth Planet Sci Lett 102:413-429

Javoy M, Pineau F (1991) The volatiles record of a "popping" rock from the Mid-Atlantic Ridge at 14°N: chemical and isotopic composition of gas trapped in the vesicles. Earth Planet Sci Lett 107:598-611

Jensen RA (1993) Explosion craters and giant gas bubbles on Holocene rhyolite flows at Newberry Crater, Oregon. Oregon Geol 55:13-19

Kato Y (1987) Woody pumice generated with submarine eruption. J Geol Soc Japan 93:11-20

Kilburn CRJ (1981) Pahoehoe and aa lavas: a discussion and continuation of the model by Peterson and Tilling. J Volcanol Geotherm Res 11:373-389

Kilburn CRJ (1990) Surfaces of aa flow-fields on Mount Etna, Sicily: Morphology, rheology, crystallization and scaling phenomena. In: Fink JH (ed) Lava Flows and Domes: Emplacement Mechanisms and Hazard Implications. Springer-Verlag, Berlin, 129-156

Klempner D, Frisch KC (1991) Handbook of Polymeric Foams and Foam Technology. Hanser Publ, Munich, 442 p

Klug C, Cashman KV (1991) Effects of vesicle size and shape on the vesicularity of silicic magma fragmentation. EOS Trans Am Geophys Union 72:312

Klug C, Cashman KV (1994) Vesiculation of May 18, 1980 Mount St. Helens magma. Geology 22:468-472

Krafft M (1993) Volcanoes: Fire From the Earth. Harry N. Abrams, Inc, New York, 206 p

Kraynik AM (1988) Foam flows. Ann Rev Fluid Mech 20:325-357

Kurz MD, Jenkins WJ (1981) The distribution of helium in oceanic basalt glasses. Earth Planet Sci Lett 53:41-54

Lewis TA (1982) Volcano. Time-Life Books, Virginia, 176 p

Lipman PW, Banks NG (1987) Aa flow dynamics, Mauna Loa, 1984. US Geol Surv Prof Paper 1350:1527-1567

Macdonald GA (1972) Volcanoes. Prentice-Hall, New Jersey, 510 p

Macpherson DW (1984) A model for predicting the volumes of vesicles in submarine basalts. J Geol 92:73-82

Mangan MT, Cashman KV (1994) Character and evolution of magmatic foams formed during basaltic fire-fountaining. J Geophys Res (in review):

Mangan MT, Cashman KV, Newman S (1993) Vesiculation of basaltic magma during eruption. Geology 21:157-160

Manley CR, Fink JH (1987) Internal textures of rhyolite flows as revealed by research drilling. Geology 15:549-552

Marsh BD (1988) Crystal size distributions (CSD) in rocks and the kinetics and dynamics of crystallization I. Theory. Contrib Mineral Petrol 99:277-291

Marty B, Ozima M (1986) Noble gas distribution in oceanic basalt glasses. Geochim Cosmochim Acta 50:1093-1097

McMillan K, Cross RW, Long PE (1987) Two-stage vesiculation in the Cohassett flow of the Grande Ronde Basalt, south-central Washington. Geology 15:809-812

McMillan K, Long PE, Cross RW (1989) Vesiculation in Columbia River basalts. Geol Soc Am Spec Pap 239:157-167

Moore JG (1965) Petrology of deep-sea basalt near Hawaii. Am J Sci 263:40-52

Moore JG (1970) Water content of basalt erupted on the ocean floor. Contrib Mineral Petrol 28:272-279

Moore JG, Schilling J-G (1973) Vesicles, water, and sulfur in Reykjanes Ridge basalts. Contrib Mineral Petrol 41:105-118

Murase T, McBirney AR (1973) Properties of some common igneous rocks and their melts at high temperatures. Geol Soc Am Bull 84:3563-3592

Narsimhan G, Ruckenstein E (1986a) Hydrodynamics, enrichment and collapse in foams. Langmuir 2:230-238

Narsimhan G, Ruckenstein E (1986b) Effect of bubble size distribution on the enrichment and collapse in foams. Langmuir 2:494-508

Newman S, Epstein S, Stolper E (1988) Water, carbon dioxide and hydrogen isotopes in glasses from the ca. 1340 A.D. eruption of the Mono Craters, California: Constraints on degassing phenomena and initial volatile content. J Volcanol Geotherm Res 35:75-96

Orsi G, Gallo G, Heiken G, Wohletz K, Yu E, Bonani G (1992) A comprehensive study of pumice formation and dispersal: the Cretaio tephra of Ischia (Italy). J Volcanol Geotherm Res 53:329-354

Papale P, Dobran F (1994) Magma flow along the volcanic conduit during the plinian and pyroclastic flow phases of the May 18, 1980, Mount St. Helens eruption. J Geophys Res 99:4355-4373

Pareschi MT, Pompilio M, Innocenti F (1990) Automated evaluation of volumetric grain-size distribution from thin-section images. Comp Geosci 16:1067-1084

Peterson DW, Tilling RI (1980) Transition of basaltic lava from pahoehoe to aa, Kilauea Volcano, Hawaii: field observations and key factors. J Volcanol Geotherm Res 7:271-293

Philpotts AR, Lewis CL (1987) Pipe vesicles—an alternate model for their origin. Geology 15:971-974

Pineau F, Javoy M (1994) Strong degassing at ridge crests: The behavior of dissolved carbon and water in basalt glasses at 14°N, Mid-Atlantic Ridge. Earth Planet Sci Lett 123:179-198

Proussevitch AA, Sahagian DL, Kutolin VA (1993) Stability of foams in silicate melts. J Volcanol Geotherm Res

Randolph AD and Larson MA (1988) Theory of Particulate Processes. Academic Press, New York, 251 p

Rose WI (1987) Active processes studied with scanning electron microscopy. In: Marshall JR (ed) Clastic Particles. Van Nostrand Reinhold Co, New York, 136-158

Rowland SK, Walker GPL (1987) Toothpaste lava: Characteristics and origin of a lava structural type transitional between pahoehoe and aa. Bull Volcanol 52:631-641

Rowland SK, Walker GPL (1990) Pahoehoe and aa in Hawaii: volumetric flow rate controls the lava structure. Bull Volcanol 52:615-628

Russ JC (1986) Practical Stereology. Plenum Press, New York

Russ JC (1988) Computer Assisted Microscopy. The measurement and analysis of images. North Carolina State University, Raleigh, NC

Sahagian D (1985) Bubble migration and coalescence during the solidification of basaltic lava flows. J Geol 93:205-211

Sahagian DL, Anderson AT, Ward B (1989) Bubble coalescence in basalt flows: comparison of a numerical model with natural examples. Bull Volcanol 52:49-56

Saltykov (1967) The determination of the size distribution of particles in an opaque material from the measurement of the size distribution of their sections. In: Elias H (ed) Proc 2nd Int'l Congress for Stereology. Springer-Verlag, New York, 163 p

Sarda P, Graham D (1990) Mid-ocean ridge popping rocks: implications for degassing at ridge crests. Earth Planet Sci Lett 97:268-289

Shehata MT (1989) Applications of image analysis in characterizing dispersion of particles. Mineral Assoc Can Short Course 16:119-132

Sheridan MF, Wohletz KH (1983) Hydrovolcanism: basic considerations and review. J Volcanol Geotherm Res 17:1-29

Simkin T, Fiske RS (1983) Krakatau 1883: The Volcanic Eruption and Its Effects. Smithsonian Inst Press, Washington, DC, 464 p

Sonin AA, Bonfillon A, Langevin D (1994) Thinning of soap films: The role of surface viscoelasticity. J Coll Inter Sci 162:323-330

Sparks RSJ, Pinkerton H (1978) The effect of degassing on the rheology of basaltic lava. Nature 276:385-386

Sparks RSJ (1978) The dynamics of bubble formation and growth in magmas. J Volcanol Geotherm Res 3:1-37

Sparks RSJ, Brazier S (1982) New evidence for degassing processes during explosive eruptions. Nature 295:218-220

Stix J, Zapata JA, Calvache M, Cortes GP, Fischer TP, Gomez D, Narvaez L, Ordonez M, Ortega A, Torres R, Williams SN (1993) A model of degassing at Galeras Volcano, Columbia, 1988-1993. Geology 21:963-967

Stolper E, Holloway JR (1988) Experimental determination of the solubility of carbon dioxide in molten basalt at low pressure. Earth Planet Sci Lett 87:397-408

Swanson DA (1973) Pahoehoe flows from the 1969-1971 Mauna Ulu eruption, Kilauea Volcano, Hawaii. Geol Soc Am Bull 84:615-626

Swanson DA, Fabbi BP (1973) Loss of volatiles during fountaining and flowage of basaltic lava at Kilauea volcano, Hawaii. J Res, U.S. Geol Surv 1:649-658

Takahashi TJ, Griggs JD (1987) Hawaiian volcanic features: A photoglossary. US Geol Surv Prof Paper 1350:845-902

Taylor BE (1991) Degassing of Obsidian Dome rhyolite, Inyo volcanic chain, California. Geochem Soc Spec Publ 3:339-353

Thomas RME, Sparks RSJ (1992) Cooling of tephra during fallout from eruption columns. Bull Volcanol 54:542-553

Toramaru A (1989) Vesiculation process and bubble size distribution in ascending magmas with constant velocities. J Geophys Res 94:17523-17542

Toramaru A (1990) Measurement of bubble size distributions in vesiculated rocks with implications for quantitative estimates of eruption processes. J Volcanol Geotherm Res 43:71-90

Valko P, Economides MJ (1992) Volume equalized constitutive equations for foamed polymer solutions. J Rheol 36:1033-1055

Vergniolle S, Jaupart C (1986) Separated two-phase flow and basaltic eruptions. J Geophys Res 91:12842-12860

Vergniolle S, Jaupart C (1990) Dynamics of degassing at Kilauea volcano, Hawaii. J Geophys Res 95:2793-2809

von Neumann J (1952) Metal Interfaces. Am Soc Metals, Cleveland, Ohio, 108-110

Walker GPL (1989) Spongy pahoehoe in Hawaii: a study of vesicle-distribution patterns in basalt and their significance. Bull Volcanol 51:199-209

Walker GPL (1991) Structure, and origin by injection of lava under surface crust, of tumuli, "lava rises," "lava rise pits", and "lava-inflation clefts" in Hawaii. Bull Volcanol 53:546-558

Weaire D, Kermode JP (1983) Computer simulation of a two-dimensional soap froth I. Method and motivation. Phil Mag 48:245

Weaire D, Kermode JP (1984) Computer simulation of a two-dimensional soap froth II. Analysis of results. Phil Mag B 50:379-395

Wentworth CK, Williams H (1932) The classification and terminology of the pyroclastic rocks. National Res Council Bull 89:19-53

Wentworth CK, MacDonald GA (1953) Structures and forms of basaltic rocks in Hawaii. US Geol Surv Bull 994, 98 p

Westrich HR, Gerlach TM (1992) Magmatic gas source for the stratospheric SO_2 cloud from the June 15, 1991, eruption of Mount Pinatubo. Geology 20:867-870

Whitham AG, Sparks RSJ (1986) Pumice. Bull Volcanol 48:209-223

Wilmoth RA, Walker GPL (1993) P-type and S-type pahoehoe: a study of vesicle distribution patterns in Hawaiian lava flows. J Volcanol Geotherm Res 55:129-142

Wilson A (1989) Foams: Physics, Chemistry and Structure. Springer-Verlag, London

Wilson L, Head JW (1981) Ascent and eruption of basaltic magma on the earth and moon. J Geophys Res 86:2971-3001

Wilson L, Sparks RSJ, Walker GPL (1980) Explosive volcanic eruptions - IV. The control of magma properties and conduit geometry on eruption column behavior. Geophys J R Astron Soc 63:117-148

Wohletz KH (1983) Mechanisms of hydrovolcanic pyroclast formation: Grain-size, scanning electron microscopy, and experimental studies. J Volcanol Geotherm Res 17:31-63

Wohletz KH (1987) Chemical and textural surface features of pyroclasts from hydrovolcanic eruption sequences. In: Marshall JR (ed) Clastic Particles. Van Nostrand Reinhold, New York, 79-97

Woods AW (1988) The fluid dynamics and thermodynamics of Plinian eruption columns. Bull Volcanol 59:169-191

Chapter 12

EARTH DEGASSING AND LARGE-SCALE GEOCHEMICAL CYCLING OF VOLATILE ELEMENTS

Albert Jambon

Laboratoire MAGIE, CNRS URA1762
Université P. et M. Curie
75252 Paris Cedex 05, France

INTRODUCTION

Volatile elements are of specific geochemical importance because of their potentially large influence on a wide range of geological processes: e.g., the evolution and differentiation of mantle and crust, the geophysical and rheological properties of earth materials, the formation of the atmosphere and oceans, the appearance and evolution of life with changing environment of the early Earth, and the nature of sedimentary processes and external geochemical cycles. The dynamics of the Earth, as a system of chemical reservoirs interacting with one another, has been termed chemical geodynamics (Allègre, 1982). It was initially founded on data for both trace elements and radiogenic isotopes (e.g., Sr, Nd, Pb). These are mostly lithophile elements whose external reservoir is the continental crust and they are therefore not appropriate to provide satisfactory constraints for use in describing properly all geochemical cycles. Studying volatile elements whose major exospheric reservoirs are different provides valuable and complementary information. For instance, noble gases are stored chiefly in the atmosphere, water in the ocean, and carbon in sediments. These reservoirs are loosely connected to crustal evolution and have had their own history. Because of their major impact on terrestrial environments, understanding the exchange of volatile elements between the mantle and the external reservoirs is of utmost importance. In addition, water may have had a dramatic impact on terrestrial evolution through its role in continental crust formation (Campbell and Taylor, 1983) and its influence on mantle rheology. Less abundant volatiles can be used as tracers, e.g., the noble gases and their radiogenic isotopes. The least abundant helium isotope, ^{3}He, is of particular interest: as this element is primordial (no production after accretion) and as helium is continually lost to space from the atmosphere, its source must be pristine deep material. Therefore, element fluxes from the mantle are scaled to that of ^{3}He, and because of its very short residence time in the atmosphere its abundance is a good index of instantaneous mantle dynamics. Other examples are iodine, which is suggested to be a tracer of organic matter, fluorine, a companion tracer for water, and chlorine, a potential indicator of magma-seawater interaction. Sulfur appears to have a short residence time in the exosphere and therefore provides information on the recent dynamics of mantle and crust. Our present knowledge of the nitrogen cycle is fragmentary and its abundance and isotope composition in the interior of the Earth are controversial. Because of its very low concentration in mantle materials, analytical difficulties are great. The absence of reliable data concerning fluxes makes further discussion difficult and therefore they will not be included in this chapter.

In order to quantitatively model Earth-degassing and geochemical cycling of volatile elements, three sets of information are required. First, one must decide what is the topology of the cycle—what are the reservoirs and how do they communicate. Second, the mass of the reservoirs and the concentrations of volatile elements in them are required in order to calculate the stored volatile masses; these data are summarized in Table 1. Finally, exchange fluxes between reservoirs must be estimated. This information allows assessment of an instantaneous (present day) or a time independent (steady state model) cycle. The secular evolution requires some additional time information.

Table 1.

Summary of volatile abundance in the Earth and major reservoirs (gm).

	Mass	^{40}Ar	CO_2	H_2O	S	F	Cl	I
	10^{27}	10^{18}	10^{22}	10^{22}	10^{22}	10^{22}	10^{22}	10^{18}
Bulk silicate Earth *	4.0	144	92	440	80	10.6	8	60
Degassed mantle **	2.0	3-4	24	60	40	4.0	1	1.6
Exosphere ***	0.02	66	22	160	1.1	1.3	3.0	30
Major reservoir		atmosphere	sediments	ocean	mantle	crust	ocean	organic matter

* *Bulk silicate Earth is equivalent to the primitive mantle, i.e., after core formation; it is the sum of mantle and exosphere.*

** *The degassed mantle is assumed to be 50% of the whole mantle, in agreement with the ^{40}Ar budget. The enriched mantle which results from recycling is not considered here since its mass and mean concentrations are unknown.*

*** *Exospheric values are fairly well known, others are model dependent and discussed in relevant sections.*

The choice of the reservoirs is not the same for all the elements. For instance, if we consider argon, relevant reservoirs are the mantle and the atmosphere. The continental crust is only a subsidiary, minor reservoir. There is no need to distinguish sediments from the continental crust and ocean from the atmosphere because of the small size and similar characteristics of these reservoirs. One could subdivide the mantle into degassed (depleted?) and undegassed (undepleted?) parts, but this requires additional assumptions which need to be substantiated by the data. The masses of the main reservoirs are usually known with a reasonable accuracy, but the concentration in the reservoirs is often difficult to obtain because of heterogeneity (e.g., continental crust) or the difficulty of sampling without bias (e.g., mantle). The bulk Earth abundance is a useful reference as it permits cross-checking the data using mass balance constraints. Unfortunately, the chondritic reference cannot be used for volatile elements, and the bulk Earth abundance of volatiles is subject to significant uncertainty.

The transfer paths between reservoirs are also important to specify properly. For instance, the hydrothermal exchange between seawater and oceanic crust was discovered only recently. This permitted a better balancing of the flux of magnesium. The same would be true for water and carbon dioxide among others. The flux among reservoirs is also difficult to estimate. The short term cycle may be much more important than the long term one which we try to estimate here. For example, the transfer of water from the continents to the oceans as a result of alteration and erosion cannot be measured directly. The anthropogenic contribution to the input of sulfur into the atmosphere is on the order of 90% of the total input and has increased strongly over the last decades (Bluth et al., 1993).

Fluxes may be at the same time localized (at the Earth scale) and diffuse (at human scale) making direct evaluation of a mean value difficult. Episodic input, for instance by explosive volcanism, may also be important and variations on longer time scales may occur as well.

The evolution of the terrestrial system with time is an important aspect of its dyna)ficult to approach quantitatively. There is abundant geological and geochemical evidences that the Earth evolves at an irregular pace. This can be traced for recent periods (e.g., dating the ocean floor and recording Sr isotopes in oceanic carbonates) but is out of reach for most of the Earth history (beyond 500 Ma). Useful approaches to this problem include: (1) comparison of the present day input flux with the amount in the reservoir; in a closed system, the integrated input flux should equal the reservoir content. In the case of mismatch and if the system is shown to be closed, then the flux must have changed with time; (2) the noble gases possess radiogenic isotopes stemming from parents with different half-lives and thus studies of their isotopic ratios and their differences between reservoirs provide valuable time information. Unfortunately, the more abundant volatiles do not possess radiogenic isotopes, thus limiting possible inferences.

It appears therefore that three classes of volatiles elements can be distinguished: (1) helium, a unique tracer of the exhalation of primordial volatiles from the mantle; (2) noble gases in general, and their radiogenic isotopes, which provide secular information (e.g., mean age of reservoirs); (3) major volatiles, which permit description of present-day cycles (rate of exchange between and mass of the reservoirs). In this review we will examine successively the different elements and their isotopes of interest. We will not give for each volatile a complete view of the degassing history and cycle, but rather present in each case the more specific information that can be obtained and specify the major unresolved issues.

NOBLE GASES AND THE DEGASSING HISTORY OF THE EARTH

The study of noble gases in meteorites started long before the study of terrestrial samples (see Ozima and Podosek, 1983, for references). As the noble gases possess radiogenic isotopes, their isotopic composition has been the subject of many works and a complex picture has emerged over the years. We will summarize the major conclusions as they can be of interest to the understanding of Earth dynamics.

Helium: a non-conservative element

Helium in meteorites. For all refractory elements on earth, two reference reservoirs are commonly invoked, the solar and chondritic references. This is true for the volatile elements as well; their terrestrial abundance are significantly lower but elemental and isotope ratios can be usefully compared. In the case of the rare gases, the cosmogenic proportions can be obtained from observations of solar flares (Cameron, 1973). From the study of meteorites several components can be distinguished, in particular spallation, radiogenic and two trapped components, solar and planetary (Signer and Suess, 1963; Pepin and Signer, 1965) each of which has a specific abundance and isotope pattern. The solar pattern is interpreted to represent implanted atoms from the solar wind. Carbonaceous chondrites, the most volatile-rich meteorites, provide evidence for the planetary component (Mazor et al., 1970). Chondrites quite commonly contain a mixture of the two components and this could also be the case for the Earth. Further details of the isotopic subdivisions of these components can be found in Ozima and Podosek (1983), and references therein.

The atmospheric budget and He residence time in the atmosphere. The two isotopes of helium, ^{3}He and ^{4}He, have an atmospheric isotopic ^{3}He/^{4}He ratio (R_A) of 1.4×10^{-6} (Mamyrin et al., 1969). The continuous escape of helium to space yields a short residence time in the atmosphere of ~10^6 years (Kockarts and Nicolet, 1962; Axford, 1968; Banks and Holzer, 1968). The amount of He in the atmosphere is therefore nearly zero and no rock will ever be contaminated by atmospheric helium. The atmospheric isotopic composition results from a balance between degassing by volcanism (mostly ocean ridges), continental degassing (liberation of radioactive production by erosion and crustal gas escape), and cosmogenic and solar input into the atmosphere. Early attempts to estimate a terrestrial budget suggested a significant cosmogenic input (Johnson and Axford, 1968; Kockarts, 1973), but following the discovery of mantle outgassing at ridges (Clarke et al., 1969; Mamyrin et al., 1969), this source need not be significant, although it is still poorly estimated. High ^{3}He/^{4}He ratios are found in meteorites in both the solar component (300 R_A) and the planetary component (100 R_A). The bulk Earth isotope ratio is not known, but this value is probably irrelevant as most of the primordial and radiogenic helium have been lost to space. The major source of He on Earth is from the radiogenic decay of ^{238}U, ^{235}U and ^{232}Th, to yield ^{4}He, with present-day contributions of 50, 2 and 48% respectively. ^{3}He can be produced in very small amounts by neutron bombardment on Li. As the major source of neutrons is from ^{238}U spontaneous fission, the production ratio of ^{3}He to ^{4}He in the continental crust is on the order of 10^{-8} and depends on Li and U abundances. Another source of ^{3}He is from the decay of tritium (solar or anthropogenic). This may be a significant contributor of ^{3}He in near surface groundwater as a decay product of bomb tritium, but is otherwise negligible.

Cosmogenic helium. Cosmogenic He can be implanted in minerals, especially at high altitudes. The accumulation of He with a high ^{3}He/^{4}He ratio depends on the exposure age of minerals and the ability of He to diffuse out after implantation. At low altitude and below the surface, this contribution is negligible (Kurz, 1986a,b; Craig and Poreda, 1986; Lal et al., 1987, 1989; Porcelli et al., 1987; Kurz et al., 1990).

The flux of cosmogenic and solar noble gases carried by cosmic dust into the atmosphere can be estimated from analysis of cosmic dust and magnetic particles in arctic ice and oceanic sediments (Fukumoto et al., 1986; Amari and Ozima, 1985, 1988; Takayanagi and Ozima, 1987; Matsuda et al., 1990; Nier and Schlutter, 1990; Nier et al., 1990; Maurette et al., 1991). Diverse approaches have estimated the cosmic dust input at ~10^{10} g/yr (Esser and Turekian, 1988; Robin, 1988) corresponding to 1300 mol/yr of Ne and 10 to 100 mol/yr ^{3}He. This may be compared to the outgassing flux at Mid-Ocean Ridge (see below) of about 400 mol/yr for both ^{3}He and Ne. As He and Ne are the most abundant noble gases in the cosmic dust, it has been suggested that subduction of extraterrestrial components might account for the mantle noble gas abundances. This point is discussed below

He in the ocean: the MORB source and He flux from the mantle. He in Mid-Oceanic Ridge Basalts displays a ratio of about 8 R_A (Krylov et al., 1974; Craig and Lupton, 1976; Kurz and Jenkins, 1981; Kurz et al., 1982a,b; Allègre et al., 1983; Marty and Ozima, 1986; Poreda et al., 1986; Graham et al., 1987). This high ratio must be inherited from primordial helium (≥ 100 R_A) with the addition of some radiogenic helium. As such a high ratio cannot be produced in surface reservoirs, it must originate from the mantle itself, indicating that the present day mantle still contains primordial gases (not

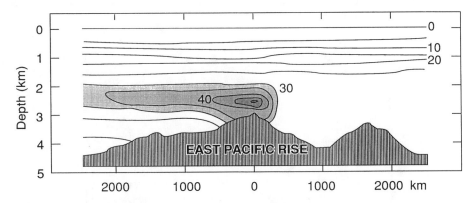

Figure 1. Helium flux at ridges. Numerous measurements of ^{3}He/^{4}He in the oceans (over 100 stations) enabled to map the helium influx to the ocean. This is illustrated here for the East Pacific Rise at 15°S (after Lupton and Craig, 1981) as δ^3He (= (R/Ra – 1) × 1000). Helium is clearly exhalated from the ridge where magmatism and hydrothermal activity reach their maximum. When using a circulation model of the ocean (Stuiver et al., 1983) these data permit calculation of a ^{3}He ridge flux of 400 mol/yr (Craig et al., 1975; Jean-Baptiste, 1992).

recycled atmosphere). This is particularly obvious when looking at the He composition in deep Pacific waters (Fig. 1; Lupton and Craig, 1981) which shows a plume originating from the EPR (East Pacific Rise) crest and moving westward. A simple model of oceanic circulation coupled with the measured gradient of helium isotope ratio in the water column, enabled Craig et al. (1975), to estimate the ^{3}He outgassing flux at ~10^3 mol/yr. Further analyses of He in the ocean (1700 measurements at 103 stations) and more precise modelling of oceanic circulation based on ^{14}C abundance (Stuiver et al., 1983) suggest a downward revision of this flux to 400±100 mol/yr (Jean Baptiste, 1992) . This agrees with a simple estimate of 20 km^3/yr degassing of basalt containing on average 0.7×10^{-14} mol/g of ^{3}He (1.4×10^{-5} cm3/g of He with 8 R_A). The ^{3}He flow from hot-spots is estimated to be less than 20% of the ridge flow (Torgersen, 1989).

Estimating the He concentration in the parent magma of oceanic basalts before eruption is difficult. Carbon dioxide in most oceanic basalts exceeds its solubility at eruption depth and consequently CO_2-rich vesicles are formed. As all noble gases partition favorably into the gas phase, a small vesicle fraction (< 1 vol.%) may trap most (> 90%) of the initial helium, and more of the heavier noble gases (Jambon and Shelby, 1980; Kurz and Jenkins, 1981; Jambon et al., 1986). The vesicles, once formed, may migrate and eventually escape from the melt. Indeed, He concentration in oceanic basalts varies widely even if care is taken to avoid breaking the vesicles prior to analysis. It is usually assumed that samples with the largest abundance of vesicles are the least degassed but this ignores the possibility of vesicle fractionation and accumulation (see Sarda and Graham, 1990). A mean abundance of ^{4}He in the range of 5 to 20×10^{-10} mol/g is a reasonable estimate for MOR basalts.

The helium:heat-flow relationship. The ^{3}He to heat flow relationship at ridges provides an independent means of estimating the oceanic helium flux; the measured heat flow at ridges is significantly lower than that predicted by a model of conductive cooling of the ocean floor. Hydrothermal circulation through the ocean crust is assumed to be responsible for this difference (Wolery and Sleep, 1976). As both heat and helium have

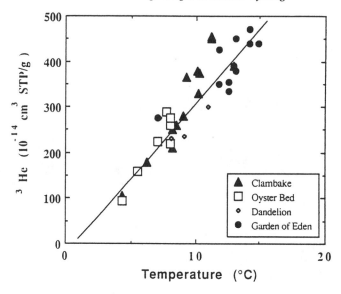

Figure 2. [3]He:heat-flow relationship. [3]He and temperature are correlated in hydrothermal vents as shown on these measurements from the Galapagos submarine waters (Jenkins et al., 1978). This confirms that both heat-flow and [3]He find their origin in ridge magmatism. Measurements at several stations showed that hydrothermal vents carry a variable [3]He/heat-flow ratio of $17\pm13 \times 10^7$ at/Joule. Regional estimates exhibit less variability at 0.9 ± 0.4 10^7 at/Joule, a significantly lower value, as much of the heat is transferred by diffusion and not by the vents.

their source in the generation of new sea-floor, a correlation between the two parameters is expected. This can be investigated on a regional or a local scale (Jenkins et al., 1978; Lupton and Craig, 1981; Welhan and Craig, 1983; Merlivat et al., 1984, 1987; Craig et al., 1987; Rosenberg et al., 1988; Lupton et al., 1989; Jean Baptiste et al., 1991; Jean Baptiste, 1992). For each basin a global [3]He/heat ratio of $\sim 0.9\pm0.4 \times 10^7$ atom/Joule can be estimated. Sampling of hydrothermal vents reveals a correlation between [3]He concentrations and temperature (Fig. 2); a ratio of $17\pm13 \times 10^7$ atom/Joule shows that despite the larger variability this ratio is significantly higher than the regional estimate, indicating that He outgassing is focused by hydrothermal circulation while a significant part of the heat is transferred by diffusion. This relationship yields an average [3]He flow derived from the heat flow at ridges that is in fair agreement with the direct measurements.

Helium from arc magmas. The helium isotopic composition of arc magmas shows more variability than observed in MORB, with reported values of [3]He/[4]He ranging from 8 R_A to about 1 R_A (Poreda, 1985; Sano et al., 1987; Hilton and Craig, 1989; Marty et al., 1989; Poreda and Craig, 1989; Trull et al., 1989; Hilton et al., 1992). The very low ratios cannot be generated by mixing mantle helium with radiogenic helium produced within the subducted oceanic crust because the time between crust generation and subduction is too short and thus the incorporation of radiogenic helium from the continental crust or subducted sediments is required.

Helium flow from the continental crust. The high concentration of U and Th in the continental crust yields a significant flux of radiogenic He, especially in those regions where continental heat flow is high (Sano et al., 1986; Oxburgh and O'Nions, 1987;

O'Nions and Oxburgh, 1988; Torgersen, 1989). The flux of ^{3}He from the continental crust is negligible compared to ridge flux, but a significant contribution of ^{4}He to the atmosphere is from radiogenic crustal helium (with ^{3}He/^{4}He ~10^{-2} R_A). The estimates given by the above authors confirm that ^{3}He flux is negligible while that of ^{4}He is significant. From the He/heat flow ratio a flux on the order of 2 to 7×10^6 atoms/cm^2-s has been estimated (Sano et al., 1986). A comparable estimate of radiogenic argon flux is not possible at present because of the overwhelming atmospheric contribution in crustal waters.

Helium model age of the mantle and outgassing. For a MORB ^{4}He abundance of 2×10^{-5} cm^3/g (1.8×10^{-9}mol/g) and U and Th concentrations of 71 and 187 ppb respectively (Hofmann, 1988), and assuming that He, U and Th are highly incompatible, one obtains from the closed system assumption, a mantle age of about 1.2 Ga. This age would be larger if the amount of He is underestimated (e.g., vesicle loss or degassing before eruption or lesser incompatibility of helium). If the ^{4}He abundance is overestimated by a factor of 2, the age is overestimated by a similar factor but this does not alter the argument. An age of 1.2 Ga can seem paradoxical as the MORB source is nearly completely degassed and contains only a few percent of total terrestrial ^{4}He (less than 4% of ^{4}He produced over the earth lifetime). This is easy to understand if degassing of ^{4}He and U and Th extraction are coherent. The formation of the continental crust by extraction from the mantle occurred with a mean age of about 2.5 Ga, implying that extraction of He after crustal formation was not very important or was coherent with U and Th extraction. The ratio of strongly incompatible elements (like U and Th) over compatible elements (like Si) in the depleted mantle has changed with time by a factor of about four as a result of crustal formation. The model age reflects the differential extraction of He and U+Th from the mantle; if they were extracted similarly, the model age would be the earth age. This is in agreement with (1) complete recycling of the crust formed before 4 Ga (but not He), (2) absence of significant recycling of U and Th in the depleted mantle and (3) gentle continuous degassing of He at ridges from the depleted mantle at present.

The ^{3}He/^{4}He vs ^{87}Sr/^{86}Sr relationship. Helium isotopic variations in oceanic basalts correlate with other radiogenic isotopes (Kurz et al., 1982a,b). N-MORB (i.e., normal or depleted MORB) display a mean value of 8 ± 0.5 R_A while E-MORB (i.e., enriched MORB) and some OIB (Oceanic Island Basalts like those of Gough, Tristan da Cunha...) exhibit a negative correlation between R/R_A and ^{87}Sr/^{86}Sr (Fig. 3). A limited number of hot spots (e.g., Hawaii, Réunion, Iceland, Samoa) display much higher values (up to 30 R_A in Loihi), without clear correlation with other radiogenic isotopes (Kaneoka and Takaoka, 1980; Kurz et al., 1983, 1991; Rison and Craig, 1983; Staudigel et al., 1984; Kurz, 1987; Vance et al., 1989; Farley et al., 1990; Graham et al., 1990; Staudacher et al., 1990; Kurz and Kammer, 1991; Poreda and Craig, 1992). Isotope ratios can be interpreted as resulting from subducted U and Th with either altered oceanic crust and/or sediments to form an enriched source (high ^{4}He). A simple calculation can be done to check whether this interpretation is realistic or not. If one assumes that the U is increased by a factor of 50 relative to the primitive mantle (taken here as a reference) in a component devoid of ^{4}He and ^{3}He initially, and if this component is mixed with 2/3 of a MORB type end member, it follows that about one Ga is necessary to decrease the He ratio from 8 R_A (MORB value) to the value of 6 R_A observed in Tristan da Cunha. For the observed ^{87}Sr/^{86}Sr ratio, a duration of about 1 to 2 Ga is necessary depending on the Rb/Sr of the enriched source and the initial Sr isotopic value. These calculations confirm that this process is of appropriate scale for generating the observed ^{4}He excess; this is also consistent with the Pb isotope interpretation as giving an age of U/Pb fractionation (Chase, 1981). The high He ratios (Loihi type) are interpreted as revealing a contribution from

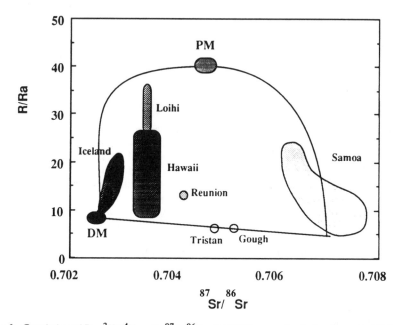

Figure 3. Correlation of R = ^{3}He/^{4}He with ^{87}Sr/^{86}Sr. N-MORB carry the depleted mantle (DM) signature with a well defined (normalized to atmospheric ratio $R_A = 1.4 \times 10^{-6}$) R/R_A of 8 and little variability. Kurz et al., (1982) have shown that E-MORB and some oceanic islands, define a correlation to high ^{87}Sr/^{86}Sr and low ^{3}He/^{4}He as expected from crustal recycling (straight line from DM through Gough and Tristan). Islands with high ^{3}He/^{4}He (Hawaii, Réunion, Iceland, Samoa) suggest a primitive end member with a ^{87}Sr/^{86}Sr of about 0.7048 and R/R_A of about 40. Mixing lines between this primitive end member and degassed mantle (the low R/R_A line) do not reproduce the fields observed at the Reykjanes ridge or Samoa islands among others. Here are presented two mixing lines for the purpose of illustration : one with He_{PM}/He_{DM} = 100/1 and one with He_{PM}/He_{DM} = 10/1 (DM = Depleted Mantle; PM = Primitive Undegassed Mantle) (towards Samoa). A similar picture would emerge with other radiogenic isotopes (Nd or Pb). The mixing process must be more complex than a simple binary mixing. (Data from Kurz et al., 1982; Poreda and Craig, 1992.)

some undegassed part of the primitive mantle as no other feasible mechanism to produce significant amounts of ^{3}He can be envisioned. Some islands like Réunion island have very homogenous ^{3}He/^{4}He values (Staudacher et al., 1990; Graham et al., 1990), while Hawaii, on the other hand, displays variability in both space and time (Kurz et al., 1983; Kurz, 1987; Kurz and Kammer, 1991). Evolution in space is well illustrated by data from Iceland and the Reykjanes Ridge which are understood to represent mixing of MORB and plume components (Kurz et al., 1982a,b, 1985; Condomines et al., 1983; Poreda et al., 1986; Hilton et al., 1989). In a ^{3}He/^{4}He vs ^{87}Sr/^{86}Sr diagram, two degassed end members can be identified: the MORB source and the source of OIB with low ^{3}He/^{4}He (like Tristan da Cunha). One can assume that samples with high He ratios are generated by mixing of a primitive (undegassed) component with a degassed component. This latter component can be a mixture between the two identified degassed end members. For a mixing process between primitive and depleted mantle, the primitive component is expected to have a high ^{3}He/^{4}He (~40 R_A) and a ^{87}Sr/^{86}Sr near 0.7048 (bulk Earth value). The depleted mantle component is expected to have a ^{3}He/^{4}He of 8 R_A and a ^{87}Sr/^{86}Sr ~0.7025. Curvature of the mixing line depends on k = (^{4}He$_{PM}$/^{4}He$_{DM}$)/(Sr$_{PM}$/Sr$_{DM}$), where PM stands for primitive (undegassed) mantle and DM for depleted (degassed) mantle. As the depleted mantle is more than 90% degassed (in ^{4}He), and Sr is mildly incompatible, we estimate $k \geq 20$. The calculated mixing lines are not compatible with the trends observed for specific islands, e.g., Iceland and the Reykjanes Ridge, Hawaii and Loihi, Samoa, or Lau Basin; in particular, the observed curvature (Samoa or Reykjanes Ridge: Kurz et al., 1982a,b; Poreda and Craig, 1992) is in the wrong direction.

The above results can be summarized as follows:

1. The terrestrial mantle is still degassing some of its primordial gases.

2. The degassing flux of primordial ^{3}He at ridges is about 0.4×10^3 mol/yr (with an uncertainty of probably 25%), slightly less than the first estimates of Craig et al. (1975).

3. Data from OIB provide evidence for the depleted mantle (8 R_A), the enriched and degassed mantle (R < 8 R_A), and the undegassed mantle (R >30 R_A).

Argon and catastrophic versus continuous degassing

Argon, the third most abundant gas in the atmosphere consists of three isotopes, ^{36}Ar, ^{38}Ar and ^{40}Ar, of which only ^{40}Ar is of radiogenic origin (^{40}K decay). The ratio of ^{36}Ar to ^{38}Ar is essentially the same in the atmosphere and in the solar and planetary components identified in meteorites. The primordial ^{40}Ar/^{36}Ar ratio is observed to be nearly zero (in meteorites) and terrestrial ^{40}Ar is essentially of radiogenic origin. The abundance of argon in meteorites is variable but correlates with carbon, suggesting that carbonaceous matter is the principal carrier of primordial argon during accretion (Otting and Zähringer, 1967). When we divide the amount of ^{36}Ar in the atmosphere by the mass of the earth, we obtain about 0.9×10^{-12} mol/g, a value within the range of that measured in chondrites. Unfortunately, this is a weak constraint as to the origin of gases; carbonaceous chondrites contain about hundred times more ^{36}Ar than ordinary chondrites whose ^{36}Ar content is quite variable. The addition of 1% of carbonaceous chondrites to an Earth severely depleted in argon would incorporate the same amount as accreting homogeneously ordinary chondrites only.

Argon isotopes in terrestrial reservoirs. Argon is characterized by the ratio of radiogenic to primordial argon, ^{40}Ar/^{36}Ar. Argon composition in the continental crust varies from nearly atmospheric (e.g., groundwaters) to "purely" radiogenic for old rocks devoid of fluid inclusions. Rock samples always contain radiogenic argon. In the K/Ar dating technique, radiogenic ^{40}Ar is used to measure the age of the rock; ^{36}Ar is assumed to be the result of contamination by atmospheric argon (e.g., adsorption or alteration by meteoric water) and therefore the corresponding amount of ^{40}Ar, according to the atmospheric ^{40}Ar/^{36}Ar ratio, is subtracted before calculating an age. After subtraction of this component, zero age samples usually contain no significant amount of radiogenic ^{40}Ar. When, they do, this inherited argon is called excess argon and it may contribute to erroneous age estimation. Excess argon has been discovered in oceanic basalt glasses (Dalrymple and Moore, 1968; Fisher, 1971; Dymond and Hogan, 1978) and is measured now routinely.

Glassy rims of MORB have revealed very high ^{40}Ar/^{36}Ar in excess of 20,000 (Allègre et al, 1983; Jambon et al., 1985; Sarda et al., 1985; Sarda and Graham, 1990), i.e., seventy times the atmospheric ratio (^{40}Ar/^{36}Ar = 295). At this level, argon can be considered as purely radiogenic because the amount of ^{36}Ar is usually near blank variability. Lower ratios are usually interpreted to reflect atmospheric contamination (e.g., adsorption, incorporation of altered crust or seawater: Jambon et al., 1985; Fisher, 1971, 1985; Patterson et al., 1990). It is now widely accepted that the MORB source is strongly degassed (very little or no ^{36}Ar left), with a mostly radiogenic composition. There have been several measurements of argon in glasses, phenocrysts, xenoliths from Loihi seamount, south of Hawaii (Kaneoka and Takaoka, 1980; Kaneoka et al., 1983; Allègre et al., 1983; Kurz et al., 1983; Rison and Craig, 1983; Vance et al, 1989). These samples have the highest ^{3}He/^{4}He ratios so far observed and are therefore considered to hold a primitive mantle component. The measured argon isotopic composition is nearly atmospheric and this result has been interpreted in two radically opposite ways :

   1. A nearly atmospheric composition is what is expected for an un-degassed-undepleted (primitive) mantle. The small difference observed corresponds to the small amount of radiogenic argon concealed in the depleted mantle and in the continental crust. One can estimate that in this model, the primitive mantle value should be about 400 (Sarda et al., 1985).

   2. The measured, nearly atmospheric value results from atmospheric contamination (Fisher 1985, 1987; Patterson et al., 1990). This is supported by the observation that all Loihi samples have suffered significant degassing before sampling and are therefore sensitive to minor contamination of atmospheric origin. Finding xenoliths with high ^{3}He/^{4}He and high ^{40}Ar/^{36}Ar ratio is not a final proof either, as samples may have exchanged with some magma derived from the undegassed mantle during the upwelling of a plume of deep origin. Recycling of atmospheric ^{36}Ar during subduction seems however unlikely (Staudacher and Allègre, 1988).

This difficult problem awaits further clues and is still open to discussion.

^{40}**Ar** *and long term terrestrial outgassing***:** As for all volatile elements, the amount of primordial argon is unknown. Fortunately however, because of the very low primordial ^{40}Ar/^{36}Ar (about 10^{-4}) , ^{40}Ar can be considered to be zero in the beginning (unlike ^{4}He) and this provides some fundamental information about Earth outgassing.

From the bulk Earth K abundance (about 245ppm for a K/U ratio of 1.2×10^4; Jochum et al. 1983), the total amount of radiogenic argon in the Earth-atmosphere system is calculated to be 3.6×10^8 mol ^{40}Ar. This indicates that ~46% of ^{40}Ar is presently in the atmosphere, ~2% is in the continental crust and ~2% in the depleted mantle source of MORB. This leaves ~50% of total radiogenic argon still concealed somewhere in the mantle, presumably the primitive (undegassed) mantle, or a combination of degassed and undegassed primitive mantle. A mass balance calculation on lithophile elements between crust and mantle indicates that 30% of the present mantle could be of depleted character; accordingly one simple model would consist of 50% primitive undegassed mantle, 20% degassed undepleted mantle and 30% depleted and degassed mantle. The relative fraction of depleted and undepleted degassed mantle is uncertain and model dependent. This evaluation is quite robust as the amount of ^{40}Ar in the atmosphere is accurately known and ^{40}Ar in the depleted mantle and continental crust is inconsequential in this evaluation. The amount of degassed mantle depends chiefly on bulk Earth K estimate. As about half of ^{40}Ar is in the atmosphere, a 10% error on K abundance corresponds to an error of ±5% on the undegassed mantle fraction.

It is important to note that the ^{40}Ar abundance in the atmosphere provides little information about an early outgassing event, because it had near zero abundance in the beginning, and the ^{40}K half life is 1.3 Ga. For the same reason, ^{40}Ar abundances indicate that a significant portion of the mantle did not participate in the long term degassing process. It is also interesting to compare the present day flux to the steady state flux, that is obtained when the radiogenic production exactly balances the outgassing flux. The present day flux amounts to about 75% of the production in the degassed (depleted+undepleted) mantle. If this value is correct, it implies that the mantle activity has decreased significantly with time. As atmospheric ^{40}Ar represents more than 10 times the amount of ^{40}Ar presently in the degassed mantle, one should expect a steady or near steady state to prevail.

In order to derive information about a possible catastrophic degassing one must use primordial isotopes of argon, i.e., ^{36}Ar and the variations of $^{40}Ar/^{36}Ar$ between the reservoirs that can be sampled. Degassing models must be developed and compared with available data on the present day flux of ^{36}Ar and ^{40}Ar, the amounts of ^{40}Ar and ^{36}Ar in the atmosphere and the $^{40}Ar/^{36}Ar$ in the degassed mantle (Damon and Kulp, 1958; Turekian, 1959; Ozima, 1975; Hamano and Ozima, 1978; Hart et al., 1979; Sarda et al., 1985; Turner, 1989; Zhang and Zindler, 1989; Azbel and Tolstikhin, 1989). All models differ in some details and we can summarize them as follows:

1. The atmosphere is derived from mantle outgassing.

2. There is no recycling of atmospheric components to the mantle.

3. The very high $^{40}Ar/^{36}Ar$ in the depleted mantle (>20,000) relative to the atmospheric value (295.5) requires that early outgassing was far more important than it is presently. All models favor an early catastrophic outgassing which cannot be dated accurately using argon data alone.

Important questions that remain unresolved are:

1. Is the whole mantle degassed of its ^{36}Ar or was only half of the mantle affected as for ^{40}Ar? This clearly translates into the case of a present-day primitive mantle having either a high or a nearly atmospheric $^{40}Ar/^{36}Ar$.

2. Is the early event of atmospheric formation a process of mantle out-gassing or is it the result of degassing as a consequence of impacts during the accretion process? In the case of argon, this distinction is rather academic because the time scale for accretion is very short compared to the half-life of ^{40}K (about 1.3 Ga), but he same is not true for Xe as will be discussed next.

Further inferences from neon and xenon

In addition to helium and argon, neon and xenon display interesting isotopic anomalies which can be used to constrain volatile geodynamics.

Constraints from terrestrial xenology. Xenon has nine stable isotopes. All are primordial and some are also produced by a variety of nuclear reactions (in space, in meteorites, or on Earth). ^{130}Xe is usually taken as the reference isotope (like ^{36}Ar) as no secondary production is expected after nucleosynthesis. The major isotopic anomalies are for ^{129}Xe, produced by ^{129}I decay (half life 17 Ma), and ^{134}Xe and ^{136}Xe, produced by spontaneous fission of ^{244}Pu (half life 82 Ma) and ^{238}U. Excesses of these Xe isotopes observed in chondritic meteorites have been used to constrain their accretion timescales (Podosek, 1970). ^{129}Xe excesses of up to 20% (relative to the atmospheric $^{129}Xe/^{130}Xe$ of 6.48) have been found in some terrestrial samples (Phinney et al., 1978 Kaneoka and Takaoka, 1980; Staudacher and Allègre, 1982; Allègre et al, 1983; Marty, 1989). The excess ^{129}Xe correlates with ^{134}Xe and ^{136}Xe excesses and this is interpreted to indicate that MORB Xe records some mixing between an atmospheric component and an upper mantle component with significant excess of ^{129}Xe, ^{134}Xe, ^{136}Xe. If ^{134}Xe and ^{136}Xe excesses (denoted by *) can be fission products of either ^{244}Pu or ^{238}U, the ratio of ^{4}He*/^{136}Xe* as measured in MORB is of the correct order of magnitude to be generated by ^{238}U fission; the short lived (now extinct) ^{244}Pu, another parent to ^{134}Xe* and ^{136}Xe*, is not necessary. The magnitude of Xe isotopic excesses is therefore consistent with the abundance of ^{4}He in the depleted mantle.

The observed excess of ^{129}Xe may only be produced by the decay of the short lived ^{129}I in a system having a sufficiently high I/Xe ratio (i.e., a xenon-poor system). In a model of atmospheric formation by mantle outgassing, this implies that most of the atmospheric xenon was already degassed from the mantle before the complete decay of ^{129}I. Subsequent radioactive decay of the remaining ^{129}I, generated a high ^{129}Xe/^{130}Xe in a xenon impoverished mantle. Because of the short half life of ^{129}I, this constrains atmospheric formation to within a few half lives of ^{129}I after accretion. At variance with this simple interpretation is an explanation by heterogenous accretion, in which mantle and atmospheric xenon were accreted independently (heterogeneous accretion model) and never mixed together (Ozima et al., 1985; Marty; 1989). In this scenario the atmosphere is not extracted by catastrophic mantle outgassing, and the ^{129}Xe anomaly puts some constraint on accretion history (a few half lives after nucleosynthesis) but not on the degassing time.

Neon. Neon has three isotopes with atmospheric ratios of ^{20}Ne/^{22}Ne = 9.80 and ^{21}Ne/^{22}Ne = 0.029. Excess ^{20}Ne and ^{21}Ne (relative to atmospheric isotopic composition) of up to 30% and 230% respectively have been found in MORB (Poreda and Radicati, 1984; Staudacher, 1987; Sarda et al., 1988), in Hawaiian basalts (Honda et al., 1991) and in diamonds (Honda et al., 1987; Ozima and Zashu, 1988). A mixing line between atmospheric composition and these radiogenic samples (which display high ^{40}Ar/^{36}Ar ratio) most certainly reflects late atmospheric contamination as observed for argon (see above). Data from Loihi Seamount plot very close to the atmospheric point. Ne isotopes are thus similar to argon in many respects. The excess ^{21}Ne can be explained by nuclear reactions within the mantle and according to Kyser and Rison (1982), the observed excess is of the right order of magnitude. More difficult however is explaining the high ^{20}Ne/^{22}Ne ratio observed. Nuclear reactions can decrease this ratio (Wetherill, 1953) and mass fractionation also could decrease the ^{20}Ne/^{22}Ne ratio in the residual degassed mantle, but there is no simple way to increase ^{20}Ne/^{22}Ne. Allègre et al. (1993) propose that solar-like Neon in the MORB source (i.e., neon with high ^{20}Ne/^{22}Ne) is derived from cosmic dust deposited in sediments and subducted. According to the estimated input rate of cosmic dust (Takanayagi and Ozima, 1987; Robin, 1988; Esser and Turekian, 1988; Robin, 1988; Maurette et al., 1991), this model can provide the right amount of ^{20}Ne excess if a small fraction of particles is subducted. An unexpected but important consequence of this model is that the maximum input of ^{3}He into the mantle corresponding to the cosmic ^{20}Ne is negligible when compared to the observed ^{3}He (Allègre et al., 1993). Honda et al. (1991), from their analyses of Hawaiian basalts, argue that mantle neon, unlike atmospheric neon, has a solar ^{20}Ne/^{22}Ne; in this interpretation the atmosphere could not be derived from the mantle.

CO_2, A MAJOR VOLATILE WITH MAJOR PROBLEMS

Major volatiles like CO_2 and H_2O are far more abundant than noble gases but, because of the absence of suitable radiogenic isotopes of either C or H, their history is not well constrained. In the present discussion, CO_2, is taken to represent all forms of carbon including carbonates and reduced (organic) carbon. Information that can be obtained for these species pertains essentially to the present-day cycle, and therefore we will focus on

All Chondrites

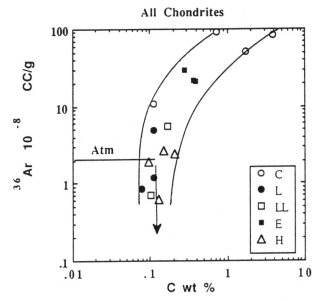

Figure 4. ^{36}Ar versus C in chondrites. An important carrier of noble gases is the carbonaceous matter in chondrites. More specifically, the relationship between ^{36}Ar and C in diverse types of chondrites (average values calculated from the data of Otting and Zähringer, 1967) permit to approach an average C content for the Earth. Assuming that the atmospheric ^{36}Ar represents more than 99% of terrestrial ^{36}Ar enables to derive a carbon concentration of 0.11±0.05 wt %. Another possibility consists in assuming that most of ^{36}Ar is from about 1 to 3% of carbonaceous chondrites: this would place the minimal terrestrial carbon value at the low end of the range defined previously.; as the remaining 97% (ordinary chondrites) could still be a significant contributor for C, the lower bound can hardly be smaller. Should atmospheric ^{36}Ar represent a fraction (at least 50%) of the overall terrestrial amount, then the terrestrial carbon should be increased accordingly (up to a factor 2).

determining the carbon content of the various terrestrial reservoirs and the dynamics of exchange among them. As will be seen below, the answers are fragmentary as neither the amounts nor the fluxes can be established very accurately. The contribution of magmatic CO_2 to the external, short-term carbon cycle is marginal.

The bulk Earth carbon abundance

C/^{36}Ar and bulk Earth carbon. One strategy to estimate the terrestrial carbon abundance considers the composition of meteorites. Even though volatiles are depleted to various degrees in planetary objects it is observed that the ratios of volatile elements of similar volatility are often similar in chondrites and planets. The C/^{36}Ar ratio among chondrites varies from 2 to 20×10^7 mol/mol (Fig. 4) (Otting and Zähringer, 1967) from which a terrestrial mean C abundance of 0.11±0.05 wt % C ($5.5\pm2.5 \times 10^{23}$ mol) can be derived, assuming that almost all ^{36}Ar is now in the atmosphere. If only 50% of ^{36}Ar is now in the atmosphere—see above—then the total C amount is increased to 6 to 12×10^{23} mol. The correlation with Ne leads to the same conclusion, and the reconstructed C/^{3}He used by Marty and Jambon (1987) relies on a similar approach.

The mean terrestrial abundance of ^{36}Ar is in the same range as that of ordinary chondrites and a factor of 30 to 50 below that of carbonaceous chondrites. It has been therefore suggested that terrestrial volatiles, including C, were accreted at a late stage, from about 3% of a component similar to carbonaceous chondrites. If correct, the amount of terrestrial carbon would be 2×10^{23} mol, which is near the minimal value calculated previously or about twice this value if the previously accreted material contained some carbon. If the approach taking benefit of the $C/^{36}Ar$ correlation is valid then the terrestrial carbon abundance is well constrained. The incorporation of carbon upon accretion is a subject of controversy. Carbon may have been lost in early stages of accretion, when the earth gravity field was insufficient to retain CO_2 volatilized on impact (Lange and Ahrens, 1986; Matsui and Abe, 1986). It has also been argued (O'Neil, 1991) that heterogenous accretion predicts incorporation of reduced carbon in early stages: under this form carbon is refractory and not prone to volatilization. Oxidized carbon was introduced in the late stages of accretion with more oxidized material and it is not clear whether it was effectively retained or not.

Terrestrial inventory. A second strategy consists of evaluating the amount of carbon in the core, the mantle, the crust, the ocean and the atmosphere. The atmospheric CO_2 is important because of the biological and climatic implications but the corresponding mass of carbon is negligible (2.4×10^{18} mol). The amount of carbon dissolved in seawater is greater (3.3×10^{18} mol) but still a minor amount in the external reservoirs. Most of the carbon is stored in the crust and more precisely in sedimentary carbonates which have been estimated to represent about 15% of the sedimentary mass, i.e., 3.6×10^{21} mol of C; organic carbon represents a subsidiary component of lesser importance and carbon of metamorphic and magmatic rocks add up to about 5×10^{21} mol of C (Hoefs, 1965; Javoy et al., 1982).

The carbon concentration in the mantle is difficult to evaluate. Ultramafic nodules have variable amounts, from negligible abundances to carbonated nodules (Nadeau et al. 1990; Pineau and Mathez, 1990; Trull et al., 1993; Hauri et al., 1993), reflecting a depleted source on one hand or a metasomatized source on the other (Berg, 1986). Ultramafic massifs are very heterogeneous and may contain reduced carbon (graphite or diamond; Nixon et al., 1991), as well as CO_2 fluid inclusions, making derivation of mean abundances difficult.

The amount of CO_2 in the MORB source can be derived from analyzing oceanic basalts which, when quenched at sea floor pressures may retain most of their carbon as either dissolved or gaseous CO_2 (Killingley and Muenow, 1975; Pineau et al., 1976; Moore et al., 1977; Delaney et al., 1978; Muenow et al., 1979, 1980, 1991; Garcia et al., 1979, 1989; Byers et al., 1983, 1984, 1985, 1986; Harris and Anderson, 1983, 1984; Pineau and Javoy, 1983; 1994; DesMarais and Moore, 1984; Mattey et al., 1984, 1989; Sakai et al., 1984; Exley et al., 1986; Jambon and Zimmermann, 1987; Aggrey et al., 1988a,b; Dixon et al., 1988, 1991; Gerlach, 1989; Javoy and Pineau, 1991; Blank et al., 1993; Macpherson and Mattey, 1994; see also Chapters 5 and 8). The results are summarized in the histogram of Figure 5, which indicates a mean CO_2 concentration of 1300 ppm for MORB. The variability ($\sigma = 600$ ppm) could be interpreted as for other trace elements as resulting from variable degrees of partial melting and possibly fractional crystallization, or as a reflection of differences in magma composition. From such analyses, the carbon content of the source mantle can be reconstructed if the partition coefficient of carbon is known. The simplest approach is to assume that C is incompatible, which requires that neither carbonate nor diamond (or graphite) remains in the residue following melt extraction. This assumption has the advantage of simplicity but relies on the closed system assumption (see below). The significant variability in MORB carbon abundance (from 300 to 3000 ppm CO_2) could be interpreted as the signature of

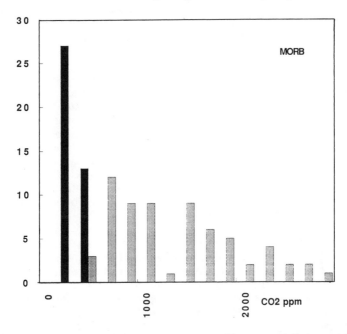

Figure 5. CO_2 abundance in oceanic basalt glasses: bulk CO_2 content (glass + vesicles). A mean concentration of 1300 ppm is obtained (mean of 8.04% MgO) with a variability of $\sigma = 600$. The black bars correspond to CO_2 dissolved in glass (Dixon et al., 1988) mean of 150 ppm CO_2, and illustrates the importance of vesicle CO_2. N- and E-MORB mean values differ by 60 ppm only. The major difficulty consist in deciding whether the observed scatter reflects source composition or late vesicle loss and degassing. (Data from Aggrey et al., 1988a; Byers et al., 1983, 1984, 1985, 1986; Delaney et al., 1978; Dixon et al., 1988, 1991; Exley et al., 1986; Garcia et al., 1979, 1989; Jambon and Zimmermann, 1987; Javoy and Pineau, 1991; Killingley and Muenow, 1975; Mattey et al., 1984, 1989; Muenow et al., 1979, 1980, 1991; Pineau and Javoy, 1983; 1994; Pineau et al., 1976.)

incompatibility, as the most incompatible elements in MORB have the most variable concentrations (e.g., Hofmann, 1988).

One serious difficulty with the previous approach is related to the vesiculation of oceanic basalts. Are vesicular glassy margins representative samples of the magma before degassing? Some vesicles are broken and lose their gas (before sampling or during lab preparation). Do the observed variations reflect deep processes (source composition and degree of melting) or a late stage degassing before or during eruption? This point is controversial and some authors consider that only a few percent carbon remain, because of a distillation process (e.g., Javoy et al., 1982; Pineau and Javoy, 1983, 1994) while others consider that gas rich samples keep nearly all their magmatic CO_2. Basaltic magmas may be gas saturated at depth and loose a significant amount of CO_2 in a magma chamber. If so the total amount of CO_2 in sea-floor samples would be influenced by magma ascent and degassing processes and not necessarily directly related to the abundance of CO_2 in the source mantle. CO_2 loss cannot be discounted but it is believed not to completely erase the primary melt signature. For the purpose of illustration, consider a mean CO_2 value in MORB of ~1300 ppm; 10% partial melting of a mantle source, with carbon being incompatible, corresponds to ~130 ppm CO_2 in the source. This is a minimal abundance assuming that no significant CO_2 loss occurred before eruption (e.g., in a magma chamber); if so, the mean CO_2 concentration in bulk MORB glasses can be interpreted as the amount of CO_2 which is soluble at the time the melt resided in a magma chamber. 1300 ppm CO_2 corresponds to the solubility at about 2400 bars, about 7 km under the sea floor (Pan et al., 1991). Variable CO_2 concentrations along MORs could then represent different depth of magma chambers. The problem of CO_2 loss could be explored using the carbon isotopic composition as carbon isotopes are believed to be fractionated upon degassing (Pineau and Javoy 1983; 1994; see Chapter 5). Isotopic results are however difficult to interpret because of several complications. Reduced (organic) carbon has been shown to occur in some samples with a low $^{13}C/^{12}C$ and its origin is controversial: primary or secondary or contamination (e.g., Des Marais, 1986). The melt-gas fractionation factor for carbon isotopes is also not accurately known, and both kinetic effects and equilibrium thermodynamic fractionation may be important. Isotopic variability between glass and vesicles, and even among vesicles can be observed only when specific procedures are used (leaching, stepwise crushing, stepwise heating). Because of these serious analytical problems, it is still difficult to obtain a clear picture of CO_2 abundance and isotopy and their variability in relation with other geochemical parameters.

Another approach to estimating MORB carbon abundances has been suggested by Jambon et al. (1985), based on the observation that the $^4He/^{40}Ar$ in the mantle is controlled by the (U+Th)/K ratio, which is nearly constant. It follows that a radiogenic ratio of 2(±1) is expected. As the solubility constants for He and Ar differ by nearly one order of magnitude (Jambon et al., 1986; Lux, 1987), argon is lost more rapidly than He when vesiculation proceeds. As He and CO_2 do not fractionate significantly upon degassing of basaltic melts (Jambon and Zimmermann, 1987; Marty and Jambon, 1987; Javoy et al., 1986) the fraction of He loss is similar to that of CO_2. For the purpose of illustration a MORB glass with a $^4He/^{40}Ar$ of about 6 corresponds to about 50% loss of the initial He, and hence CO_2. Usually, bulk MORB glasses exhibit $^4He/^{40}Ar$ ratios in the range 1 to 10 when vesicle gas is not lost. This result illustrates how degassing can be a serious problem and, in the case of MORB glasses, a fractional loss of 50% is quite common; a factor of ten loss is however unusual. Some samples such as those from Loihi however appear to be severely degassed and the He/Ar ratio differs too much from the radiogenic value to enable an appropriate correction. This suggests that the source magma was very rich in volatiles (both CO_2 and H_2O) and significant gas loss has occurred before eruption on the sea floor.

Concerning CO_2 in MORB basalts, the helium age (as calculated above) suggests a similar conclusion. An age of 2.3 Ga (corresponding to 1300 ppm CO_2 and $CO_2/^4He = 2.4 \times 10^4$)) will be increased by a factor of 2 if ^4He is increased by a similar value. Increasing the age further would make serious difficulty with our present understanding of mantle dynamics. As the CO_2/He ratio does not fractionate significantly, the *mean* amount of CO_2 can hardly exceed 2000 ppm in N-MORB before degassing. Enriched basalts with higher (U+Th) concentrations might have slightly (but not drastically) higher CO_2 concentrations because the CO_2/He ratio in MORB is observed to be rather constant (Marty and Jambon, 1987). Hawaiian primary magmas have been proposed to contain as much as 5000 ppm CO_2 (Gerlach and Taylor, 1990) and this could possibly be the rule for OIB.

The oceanic helium flux enables to calculate a primary He content of 1.5×10^{-5} cm^3/g corresponding to a primary CO_2 concentration of 700 ppm. This rather low value suggests that either the He flux is underestimated or that the oceanic crust is not completely degassed at depth.

To summarize this discussion: the CO_2 concentrations measured in MORB samples (from 300 to 3000 ppm CO_2) sometimes underestimates the amount of CO_2 in the primary magma. Based on He flux and a mean age of 2 Ga (corrected from the He/Ar ratio) a mean primary concentration of 1200 ppm CO_2 is estimated. Age considerations give an upper bound of 2600 ppm CO_2. Variability due to partial melting and fractional crystallization is probably less than a factor of 2 about the mean value, further variations relate to source composition (enrichment, metasomatism).

Estimating CO_2 in the MORB source provides only part of the information needed to characterize the terrestrial C budget because some parts of the mantle may not be sampled during MORB production (as discussed for argon). The total amount of CO_2 in the mantle depends on whether some primitive mantle remains and whether some carbon recycling operates. If 50% of the mantle is primitive, 50% degassed (and complementary to the outer reservoirs) mass balance calculations indicate 4.6×10^{23} g CO_2 in the primitive mantle (230 ppm CO_2), compared with 2.4×10^{23} g CO_2 for the depleted mantle (120 ppm CO_2); this corresponds to 9.2×10^{23} g CO_2 for the silicate Earth. Notice that a degassed mantle value as low as 60 ppm CO_2 would reduce the silicate Earth abundance by 25% only, which gives an idea of the uncertainty introduced in our assessment. The amount of C obtained in this way (0.2×10^{23} mol) is more than ten times smaller than that derived above from the $C/^{36}Ar$ ratio (5.5×10^{23} mol). This discrepancy has been explained in several ways:

(1) the reference to chondrites is not appropriate as the terrestrial accretion history is different from that of meteorites. Still, in this case, one expects that carbon on Earth should be less depleted than argon which is more volatile, the opposite of what is actually observed.

(2) C during partial melting is not incompatible, e.g., if the C carrier is a reduced phase, the amount dissolved depends on the mantle oxygen fugacity. If carbonates are the residual carbon phase, then CO_2 solubility is in equilibrium with carbonates. The stability of carbonates under mantle conditions (Brey et al., 1983; Canil and Scarfe, 1990; Katsura and Ito, 1990; Biellmann et al., 1993) does not imply that they can survive the fraction of partial melting that prevail during the generation of oceanic basalts.

(3) Most of terrestrial carbon might reside in the core. This has been suggested several times and discussed in detail recently by Wood (1993). As some light elements amount to 10% of the core mass, there is the possibility that part of it is carbon, which would then explain the apparent deficiency in the outer layers. The excess carbon calculated from the C–^{36}Ar correlation (6×10^{24} g) corresponds to about 3200 ppm C were it dissolved in the whole core.

The $C/^3He$ ratio and the flux of CO_2.

One first consideration about the carbon cycle takes advantage of its isotopic composition. The estimated isotope ratio ($^{13}C/^{12}C$) of the exosphere is similar to the mantle and MORB value, suggesting that the mantle/exosphere system has reached a steady state. This inference is correct within the accuracy of the estimates of mean isotopic compositions. However, as ridge output and subduction recycling are probably very close to one another in both amounts of carbon and isotopic composition, the isotopic similarity is not a strong constraint.

Jambon and Zimmermann (1987) and Marty and Jambon (1987) have shown that glassy MORB samples have a nearly constant $C/^3He$(~2×10^9) in both glass and vesicles. This implies that the two species do not fractionate significantly during magmatic evolution and degassing. This observation is important as it permits scaling of the C flux to the ^3He

flux (400 mol/yr) which yields a CO_2 flux of 3.5×10^{13} g/yr. The independent estimate of 150×10^{13} g/yr (Javoy et al., 1982) illustrates the difficulty of the problem. They assumed, based on their carbon isotope measurements, that MORB glasses are strongly degassed before eruption. Further, the estimate based on the ^{3}He flux makes no hypothesis as to the efficiency of degassing of the newly formed crust. An independent argument against such a high CO_2 flux has been presented by Walker (1982) who pointed out that such a massive injection of CO_2 into the ocean would result in a severe imbalance of the cations (Ca, Mg) necessary to neutralise seawater.

The $C/^3$He ratio of MORB is much larger than the reconstructed exospheric value (from the exospheric C and a planetary He/Ne ratio) but falls close to the chondritic value. The $C/^3$He ratio at hot spots appears to be slightly higher than that of MORB (Marty and Jambon, 1987; Trull et al., 1993). This should be interpreted with caution since hot spot glasses with high ^{3}He/^{4}He are usually strongly degassed and the $C/^3$He ratio could therefore be fractionated. The analysis of nodules confirms these higher ratios, however, even though the range is quite large. The larger $C/^3$He ratio could simply result from carbon recycling to the hot spot source (the lower mantle). According to Trull et al. (1993), the higher $C/^4$He of hot spots compared to MORB can be explained by recycling of carbon to the lower mantle. Actually, the contribution of a small amount of primitive mantle is necessary in the source of high ^{3}He/^{4}He hot spots, otherwise trace elements and radiogenic isotopes (especially Pb) would be significantly affected. If both primitive and depleted mantle have a similar $C/^3$He ratio, then recycling of C to either of these sources will increase their $C/^3$He.

The $C/^3$He ratio in arc volcanics and arc gases is definitely higher than that of MORB but a wide range of values is observed (Marty and Jambon, 1987; Varekamp et al., 1992, and references therein). The negative correlation between $C/^3$He and ^{3}He/^{4}He ratio corresponds to a mixing trend between a MORB-like mantle and a component extracted from the subducted slab. Despite numerous measurements, it is still difficult to estimate quantitatively the CO_2 flux at arcs and therefore to estimate how much carbon is effectively recycled and how much is expelled with arc magmas. Analysis of CO_2 and H_2O in oceanic arc glasses (e.g., Stolper and Newman, 1994) indicates significant degassing of submarine arc magmas and that CO_2 is lost first; the primary CO_2 content of such magmas is thus difficult to assess. For a primary magma with 3700 ppm CO_2 (about 3 times the estimated MORB value), one predicts a flux on the order of 9.2×10^{13} g CO_2/yr, compared to a ridge flux of about 3.5×10^{13} g/yr. According to Staudigel et al. (1990), trapping of hydrothermal carbonates in oceanic crust involves $\sim 1.1 \pm 0.2 \times 10^{14}$ g/yr, a value significantly larger than the ridge flux. It is anticipated that most of this carbon as well as some sedimentary carbon will be outgassed at arcs, but this is of little help to balance the carbon flux quantitatively and we must conclude that the significance of C recycling in subduction zones is presently unknown.

The importance of C recycling in subduction zones has been the subject of numerous works and we summarize briefly the most important points. Carbonates may be stable in the mantle in the absence of melt and they could survive the subduction process for some slab P-T-t paths (Wyllie and Huang, 1976; Ellis and Wyllie, 1980; Canil and Scarfe, 1990; Becker and Altherr, 1992; Edwards, 1992). The temperature of the slab is controlled by its age (older plates are colder), the rate of subduction (slower yields greater cooling), and the slope of the descending slab (geothermal gradient) (Abbott and Lyle, 1984; Staudigel and King, 1992; McCulloch, 1993). It is important to note that these conditions have undoubtedly changed with time, in particular the thermal gradient in the mantle and the mean age of the plates. Carbonates were possibly less recycled in the past than they are presently.

An additional complication with the carbon cycle is the variable oxidation state of carbon. Both oxidized and reduced carbon phases (graphite or diamond; Nixon et al. 1991) are found in rocks of mantle origin but this provides little constraint on the relative proportions of reduced and oxidized forms and their abundance (Ballhaus, 1993; Blundy, 1991).

All of these observations concerning global cycles of carbon can be interpreted in several ways:

1. He is easily degassed while C is not (compatible ?). The increase of the $C/^3He$ in the outgassing flux, indicated by the difference in MORB and exospheric ratios, results from extensive outgassing of the mantle He. In this model, a bulk earth $C/^3He$ cannot be derived as the ratio and the amount of C in the mantle are not known.

2. The low $C/^3He$ in the exosphere results from massive C recycling to the mantle. Both C and He are incompatible and degassed at a similar rate, hence a $C/^3He$ similar to the primitive chondritic-like value. Recycled C is not mixed back with the MORB source. The $C/^3He$ of the depleted mantle is identical, but the whole mantle C abundance depends on the size of the enriched mantle.

3. C was strongly recycled when both the continental crust and ocean were absent and is not anymore today because of efficient degassing at arcs. The similarity of MORB and chondrite $C/^3He$ is coincidental. A similar explanation relates to the catastrophic degassing event which affected He but not C (may be C was rapidly incorporated back to the mantle). Most of the missing C, relative to chondrites, is now concealed in the core.

H_2O: A MARGINALLY VOLATILE SPECIES

Water in the exosphere

The mass of water in the ocean amounts to 1.4×10^{24} g. The water in the continents amounts to a smaller but non-negligible amount of about 0.2×10^{24} g. The oceanic crust is significantly altered during hydrothermal interaction at ridges. Ito et al. (1983) considered this problem in great detail, taking into account the various degrees of alteration of the different rock types. They arrive at an average oceanic crust (6.5 km thick) containing 1.5 ± 0.5 wt % H_2O, which amounts to 8.8×10^{14} g H_2O/yr for 3.0 km^2/yr of subducted crust.

Water in H_2O:MORB and the H_2O:ridge flux

Analyses of water in MORB glasses are now reliable and a number of workers using different techniques have obtained consistent results showing that the amount of water varies from about 1000 to 6000 ppm (Moore, 1970; Moore and Schilling, 1973; Delaney, et al., 1978; Byers et al., 1983, 1984, 1985, 1986; Kyser and O'Neil, 1984; Poreda, 1985; Poreda et al., 1986; Jambon and Zimmermann , 1987, 1990; Aggrey et al., 1988a; Dixon et al., 1988; Michael, 1988). OIB and E-MORB contain generally more water than N-MORB but there is a significant overlap between the different types of basalts (Fig. 6). A mean value of 3000 ppm is reasonable with a variability (one σ) of about 1000 ppm. Unlike CO_2, these H_2O contents generally represent the magmatic water and are not influenced by vesiculation. Equilibrium calculations, as well as direct measurements (Javoy and Pineau, 1983), confirm that water in the vesicles of deep erupted samples is on

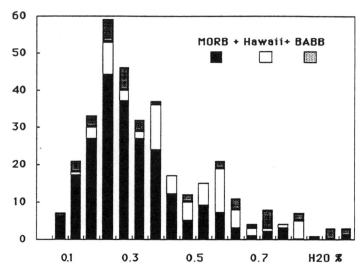

Figure 6. H_2O abundance in oceanic glasses. The observed variability can be interpreted in the following ways: (1) N-MORBs display low values (usually less than 0.4%) which vary according to the degree of partial melting and the depleted character of the source. (2) E-MORB in addition are often enriched and have been subject to fractional crystallization. Mean H_2O for all MORB is 0.33%; MgO = 7.5%. (3) Hawaii data include differentiated samples from a source rich in H_2O (mean H_2O = 0.45%; MgO = 6.8%). (4) BABB samples vary from very low to high water concentrations. Fractional crystallization may also be significant as well as source composition (mean H_2O = 0.71%; MgO = 6.7%) resulting from recycled water. (Data from Aggrey et al., 1988a; Byers et al., 1983, 1984, 1985, 1986; Delaney, et al., 1978; Dixon et al., 1988; Jambon and Zimmermann , 1987, 1990; Kyser and O'Neil, 1984; Michael, 1988; Moore, 1970; Moore and Schilling, 1973; Poreda, 1985; Poreda et al., 1986.)

the order of 1% of the total water. This is not the case for samples erupted at very shallow depth (less than 500 m) where water starts to exsolve significantly and for arc magmas (may be some OIB) which contain water at the per cent level. The mean water content of oceanic basalts corresponds to a flux of less than 2×10^{14}g/yr, which is small compared with e.g., the annual discharge of fresh water to the oceans (7×10^{19} g/yr) or the amount of water carried by the subducting plates (8.8×10^{14} g/yr; Ito et al., 1983).

Water and subduction

All of the water removed by subduction cannot remain in the mantle, or the world ocean would disappear completely after 2 Ga. As no evidence of such large variations of the ocean mass appear in the geological record, one conclusion must be that the exospheric water cycle is nearly steady state and water is most probably returned to the exosphere by arc magmas. It appears therefore that the key to the water cycle is the behavior during subduction, a question which is not properly answered at present. Water is released from sediments and oceanic crust as P and T increase during subduction, and it is generally accepted that slab water induces melting of the mantle wedge to produce arc magmatics (Gill, 1981). Still unknown is how much of this water is released from slab, when this occurs and how much of this water is carried further down to the mantle.

Estimating the amount of water released by arc volcanism is difficult. Firstly, the mass of magmatic rocks formed every year is uncertain with estimates ranging from 0.5 to

10×10^{15} g/yr (see discussion in Ito et al., 1983). Secondly, subaerial magmas are subject to degassing and assessing water contents in the primary magma is difficult (see also Chapter 8). Few samples quenched at ocean depth have been recovered and analyzed. Still it appears that some back arcs basalts cannot be distinguished from N-MORB (Garcia et al., 1979; Muenow et al., 1979; Harris and Anderson, 1983; Poreda, 1985; Aggrey et al., 1988b; Hochstaedter et al., 1990; Jambon and Zimmermann, 1990) while others may have up to several percent dissolved water (e.g., Danyushevsky et al., 1993; Stolper and Newman, 1994).

Additional evidence supporting this picture includes: (1) the observed progressive dehydration of eclogitic rocks affected by subduction; (2) the water abundance in enriched basalts (e.g., OIB) showing independent evidence of recycling. Jambon and Zimmermann (1990) argued that the K_2O/H_2O ratio of N-MORB is nearly constant as expected for incompatible species subject to partial melting and fractionation. Enriched samples on the other hand display a marked K_2O enrichment with K_2O/H_2O up to five times higher than in N-MORB. Because of the high water content of the slab initially, very low K_2O/H_2O ratio would be observed if significant amounts of water were not lost before returning to the mantle. A recent study of boron and its isotopes in oceanic basalts (Chaussidon and Jambon, 1994) has shown that less than 2% at most of slab boron is recycled to the mantle. In order to keep the B/H_2O ratio nearly constant, it is expected that about 20% of the slab water is recycled, which is in agreement with a model of quasi steady state for water cycle: the remaining 80% is added to the mantle wedge and induces melting and production of arc magmas.

SULFUR: AN EXAMPLE OF MASSIVE RECYCLING

Long term sulfur cycle

As for the other volatile elements, the budget of sulfur can be evaluated by considering the concentration in and transfer between the major reservoirs of continental crust, oceanic crust and mantle. The continental inventory amounts to 1.5×10^{22}g (Holser and Kaplan, 1966). The long term run-off by rivers amounts to 0.6×10^{14}g/yr (Holser and Kaplan, 1966), which yields a calculated residence time of about 250 Ma for sulfur in the continental crust and suggest a rapid turnover of continental sulfur. The accuracy of this figure is however limited because of the strong anthropogenic overprint to the natural cycle.

The major control of sulfur mobility is its oxidation state. Reduced sulfur forms insoluble sulfides (mostly Fe-sulfides), but during subsurface weathering, these may be transformed to soluble sulfates which are leached away to the ocean. During hydrothermal alteration at ridges, marine sulfate is converted to sulfide and trapped in altered oceanic crust. It is carried down with the subducting slab and a significant part of this sulfide returns to the continental crust via arc magmatism. This picture results from numerous works which conjugate the study of sulfur concentration and isotopic composition (Albarède and Michard, 1986; Alt et al., 1989, 1993; Alt and Anderson, 1991; Alt and Burdett, 1992; Anderson, 1974; Hochstaedter et al., 1990; Kanehira et al., 1973; Kellog et al., 1972; Mathez, 1976; Moore and Fabbi, 1971; Peach et al., 1990; Puchelt and Hubberten, 1980; Rye et al., 1984; Sakai et al., 1982, 1984; Ueda and Sakai, 1984; Von Damm et al., 1985; Woodruff and Shanks, 1988).

According to continental run off to the ocean, seawater sulfur (1.3×10^{21}g) is renewed with a time constant of about 20 Ma. The study of hydrothermal vents at mid-ocean ridges has shown that a major sink of oceanic sulfates correspond to their reduction and entrapment in the oceanic crust (Edmond et al., 1979). From the ^{3}He/heat relationship, this corresponds to a sulfur flux of ~1.3×10^{14}g/yr into the oceanic crust (Edmond et al., 1979; Von Damm et al., 1985), about twice the input flux by rivers. If the ^{3}He flux reappraisal by Jean-Baptiste (1992) is to be preferred over Craig et al. (1975) value, then the hydrothermal uptake by the oceanic crust becomes 0.52×10^{14} g/yr. If 30% of the sulfur supplied by continental run off goes to sediments, the remainder (0.42×10^{14} g/yr) should be taken up by the oceanic crust, a value identical within error to the hydrothermal estimate.

A characteristics of the external sulfur cycle is its short residence time in the exosphere relative to the Earth age. This is a strong argument in favor of a steady state for the exchange of sulfur between the mantle and the exosphere. In the steady state hypothesis one needs a major input to the continental crust in order to balance the sulfur budget This can be provided only by arcs volcanics which are known to deliver large amounts of sulfur, with δ^{34}S in the range 0 to 10‰, as expected from the composition of recycled altered oceanic crust (Alt et al., 1989, 1991). The amount of sulfur provided by this process is difficult to evaluate accurately even though many workers have attempted to reach reasonable figures for this return flux. There have been several attempts to infer the volcanic flux of SO_2 to the atmosphere (Stoiber and Jepsen, 1973; Cadle, 1975; Le Guern, 1982; Rose et al., 1982; Berresheim and Jaeschke, 1983; Stoiber et al., 1987; Lambert et al., 1988; Bluth et al., 1993). This represents only a small fraction of the sulfur carried by arc magmatics (about 10% or less; see Chapter 1).

Sulfur in oceanic basalts

Sulfur in MORB glasses occurs at an average concentration of 1060 ppm S ($\sigma = 230$ ppm) (Kanehira et al., 1973; Moore and Schilling, 1973; Mathez, 1976; Delaney et al., 1978; Byers and Muenow, 1983; Sakai et al., 1984; Byers et al., 1984, 1986; Aggrey et al., 1988a; Peach et al., 1990; Chaussidon et al., 1991) (Fig. 7); there is no significant difference between N- and E-MORB. The limited variability results from the compatible character of sulfur and variations in degree of melting do not induce significant variations in the S concentrations in mantle-derived magmas. Fractional crystallization has a small effect on sulfur concentration; in the case of the Reykjanes Ridge, for example (Moore and Schilling, 1973), sulfur increases by ~80 ppm for each 1 wt % decrease in melt MgO content, while for samples from the EPR it increases by ~240 ppm for each 1 % decrease in MgO (Mathez, 1976). This behavior can be explained by variations in sulfur solubility in sulfide saturated melts undergoing differentiation (see Chapter 7). Sulfur therefore appears to be a compatible element during partial melting, with abundance buffered by coexisting iron sulfide. The abundance of sulfur in MORB cannot be used to determine accurately the abundance of sulfur in the mantle but a lower bound of 100 ppm is obtained by assuming ~10% partial melting for an incompatible element; the actual mantle abundance must be significantly higher.

Hawaii is the only oceanic island for which numerous analyses of sulfur in submarine basaltic glasses exist (Killingley and Muenow, 1975; Muenow et al., 1979; Sakai et al., 1982; Byers et al., 1985; Garcia et al., 1989; Dixon et al., 1991). The mean value of 1020 ppm S (±300 ppm; mean MgO of 6.5%) is not significantly different from the MORB concentration, and is in agreement with the compatible character of sulfur and a small increase of sulfur concentration with fractional crystallization (Reykjanes Ridge type).

In recent years several studies of sulfur in submarine arc and back arc samples have documented differences between subduction and MOR environments (Garcia et al., 1979;

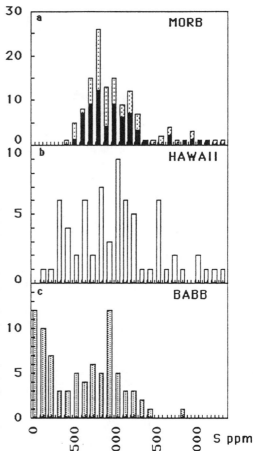

Figure 7. Sulfur abundance in oceanic basalts.

(a) N- and E-MORB exhibit similar concentrations (mean 1060 ppm, MgO = 7.7%). This is the result of S compatibility upon both partial melting and fractional crystallization (N = black; E = white).

(b) The hawaiian data are more scattered partly because of some fractional crystallization. The mean S content is not significantly different from the MORB value as low sulfur samples may be resulting from some sulfur degassing (not observed in MORB).

(c) BABB and arc samples display a bimodal distribution. The high sulfur peak is similar to MORB abundance. The low sulfur samples have been subject to significant loss of sulfur, a characteristics of arc magmas. (Data from Kanehira et al., 1973; Moore and Schilling, 1973; Mathez, 1976; Delaney et al., 1978; Byers and Muenow, 1983; Sakai et al., 1984; Byers et al., 1984, 1986; Aggrey et al., 1988a; Peach et al., 1990; Chaussidon et.al., 1991.)

Muenow et al., 1980, 1991; Aggrey et al., 1988b; Hochstaedter et al., 1990; Alt et al., 1993; Nilsson and Peach, 1993). Most notable in arc and back arc basalts, is the bimodal distribution of sulfur contents with one maximum near 1000 ppm (like MORB) and a second maximum near zero; the latter is interpreted to be the result of degassing during differentiation of arc magmas. This aspect has been discussed thoroughly by Alt et al., (1993), who conclude that S isotope variations (from 0 ‰ to 10 ‰) result chiefly from degassing. In the arc context, because of the abundance of water and of the oxidation state, S appears to be quite mobile (volatile). This difference is likely a result of the generally more oxidized nature of arc basalts which yields greater amounts of sulfur in the magmatic vapor phase (Carroll and Rutherford, 1985, 1987). Because of this degassing effect, the primary S content of the melt is difficult to assess. The analysis of melt inclusions in phenocrysts provides a better idea of the primary melt composition, especially where volatiles are concerned. Several studies of lavas from Etna and Vulcano (Italy) (Clocchiatti et al., 1994; Metrich and Clocchiatti, 1989) have shown that bulk lavas are strongly degassed (about 120 ppm S) while melt inclusions display variable concentrations from about 3000 ppm for the least differentiated liquids down to values similar to the matrix for evolved melts. This confirms that unlike the differentiation of MORB, arc magmas loose significant sulfur as crystallization proceeds and also indicates that the primary magmas may contain large amounts of S (~3000 ppm for 8 to 10% MgO liquids).

One could conclude that the low sulfur content in arc magmas is not a primary feature, but results from differentiation. If one assumes that sulfur cycle is at steady state, arc magmas bring upwards about 0.5×10^{14} g S/yr. For a concentration in the primary magma of 1000 ppm, it corresponds a mass of arc magmas of 5×10^{16} g/yr. As mentioned above, arc magma production (see discussion in Ito et al., 1983) is in the range of 0.05 to 1.0×10^{16} g/yr with a preferred range of 0.5 to 1.0×10^{16} g/yr. If these have a mean 5% MgO, they correspond to primitive magmas of 1.5 to 3×10^{16} g/yr: this is about a factor of two less than the amount obtained when assuming a primary sulfur concentration of 1000 ppm. In the arc context, analyses of melt inclusions in crystals reveal sulfur contents up to ~3000 ppm which would indicate a mass of primitive magma of 1.7×10^{16} g/yr in agreement with the direct estimate. This value permits balancing of the sulfur budget and constrains the mass flux at arcs. The oxidized state of sulfur from arc magmas may be an explanation of the very rapid turnover of sulfur in the continental crust (~250 Ma) compared to other volatile species (e.g., water or carbon).

Mantle sulfur

The mantle should be the major reservoir of terrestrial sulfur. Because sulfur appears to be compatible in mantle-derived basaltic magmas, implying melting under sulfide-saturated conditions, little variation in the sulfur content of the mantle is expected. The study of ultramafic rocks derived from the mantle provide better information if no secondary alteration has affected their sulfur abundances (Lorand, 1990). From sulfur in orogenic lherzolite and the correlation with Cu concentrations, a mean mantle concentrations of about 200 ppm can be derived (Lorand, 1991; O'Neil, 1991). Assuming a constant sulfur concentration for the whole mantle, the calculated global abundances indicate that ~97% of terrestrial sulfur is in the mantle ~1.3% is in the continental crust, ~1.0% is in the oceanic crust and 0.1% is in the ocean.

HALOGENS

The halogens: F, Cl, Br and I can be divided into three groups according to their different geochemical behavior. Fluorine is not very soluble in water and because of this, its external cycle and interactions with the oceanic crust are quite simple. Chlorine and bromine are mainly stored in seawater and their external cycles are dominated by their high solubility in water. Iodine has a very peculiar geochemistry because it resides specifically in the organic matter of sediments being too large an anion to substitute into any common minerals. The halogens thus provide different information concerning geochemical cycling and mantle-exosphere interactions.

Fluorine

The fluorine content of oceanic basalts has been determined on pillow interiors (Rowe and Schilling, 1979; Schilling et al., 1980), lava flows (Sigvaldasson and Oskarsson, 1986), crystalline rocks (Aoki et al., 1981), and glassy pillow rims (Garcia et al., 1979, 1989; Muenow et al., 1979, 1980, 1991; Byers et al., 1983, 1984, 1985, 1986; Aggrey et al., 1988a,b; Michael and Schilling, 1989). According to the correlations presented by Sigvaldasson and Oskarsson (1986) and Rowe and Schilling (1979) fluorine is not significantly degassed from magmas, even at low pressure. A more serious problem is the variable and unpredictable loss of fluorine upon crystallization (Schilling et al., 1980). Seawater contamination of oceanic samples is unlikely because of the low fluorine concentration in seawater (1.3 ppm) and because of the good correlations observed between fluorine and other geochemical tracers not prone to seawater contamination. During the crystallization of oceanic basalts, F behaves incompatibly until late stages of fractionation. Accordingly, correction of shallow level fractionation permits evaluation of a mean F concentration of melts at the conventional 8% MgO composition (Fig. 8). One obtains 140 (±50) ppm for N-MORBs and ~200 (±80) ppm for E-MORB. This difference originates in mantle heterogeneity as the degree of partial melting appears similar for the two populations.

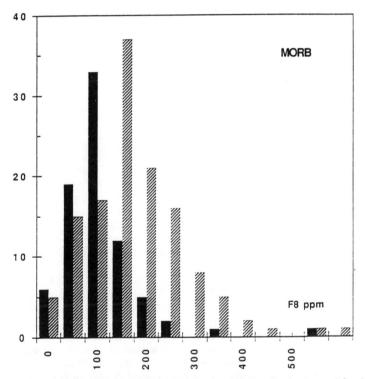

Figure 8. Fluorine abundance in oceanic basalts. The fluorine concentration is corrected for shallow level fractionation at a conventional 8% MgO. N-MORB (black bars) display lower concentrations (mean 140 ppm) than E-MORB (mean 200 ppm) which suggests that some fluorine is recycled with the subducting slab. (Data from Aggrey et al., 1988a,b; Byers et al., 1983, 1984, 1985, 1986; Garcia et al., 1979, 1989; Michael and Schilling, 1989; Muenow et al., 1979, 1980, 1991.)

The H_2O/F ratio of MORB varies in the range of 10 to 20 for both E- and N-MORB with a strong overlap between the two populations (Fig.9). It is significantly lower than the exospheric ratio (ocean+ sediments +continental crust) of about 125. OIB data are mostly from Hawaii (Muenow et al., 1979; Byers et al., 1985; Garcia et al., 1989). Fluorine concentrations vary widely and the H_2O/F ratios exhibit much scatter without any apparent logic. It is not possible to discuss these data further until we have better understanding of the origin of the wide range of observed values. Arc and Back Arc Basin Basalts (BABB) (Garcia et al., 1979; Muenow et al, 1980, 1991; Aggrey et al.,1988b)

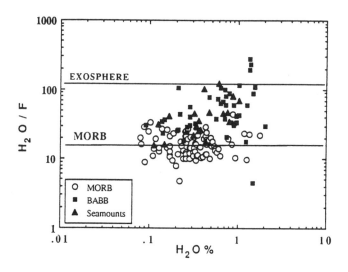

Figure 9. H_2O/F ratio in oceanic basalts. The H_2O/F ratio of both N- and E-MORBs is nearly constant at about 16, independently of water concentration. BABB exhibits a larger ratio which reflects some interaction with the exosphere (recycling of altered crust). The higher exospheric ratio is the expression of the lesser degassing of F. Seamounts may have been contaminated by some altered crust incorporation. Same data source as Figure 8.

display similar F concentrations to those observed in ridge basalts. Their H_2O/F ratios therefore vary from MORB-like to significantly higher values, according to their water content: up to 40 at Vulcano and Etna (Clocchiatti et al., 1994; Metrich et al., 1993).

The above results can be interpreted in the following way: (1) As both F and H_2O are incompatible upon mantle partial melting, the H_2O/F ratio of MORB is also that of the mantle. (2) F has been transferred less efficiently than H_2O to the exosphere, hence the different H_2O/F ratios in the MORB source and the exosphere. (3) Recycling of water and fluorine is such that the mantle H_2O/F ratio is not significantly affected. If this were not the case, E- and N- MORB would exhibit different compositions as is observed for many other trace elements. The mean F and H_2O concentrations in N-MORB is slightly less than in E-MORB but the overlap is such that the two populations cannot be distinguished on these parameters. (4) As water is significantly expelled from the slab during subduction, arc and, to a lesser extent, back arc basalts are enriched in H_2O. Fluorine-transfer at this stage is minimal so that the left-over after devolatilization of the slab has a mean H_2O/F ratio nearly similar to that of the depleted mantle.

New oceanic crust brings to the surface about 1.0×10^{13} g/yr of F. If all of this fluorine were transferred to the exosphere then it would take 1.5 Ga to supply all the exospheric F. This implies that a significant fraction of this fluorine is not delivered to the exosphere but rather recycled to the mantle with the oceanic crust. We have seen above that a significant amount of water is trapped by the oceanic crust upon hydrothermal alteration. Water is outgassed at arcs and not at ridges and little fluorine is released either at ridges or arcs. This behavior of fluorine relative to water is in agreement with the greater stability of fluorinated phases compared to hydroxilated equivalents and explains the quite high H_2O/F ratio in the exosphere.

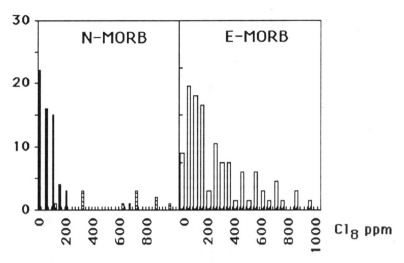

Figure 10. Chlorine abundance in MORBs. N-MORB (black) display very low concentrations once corrected for fractional crystallization (Cl_8 = Cl corrected at 8% MgO). N-MORB from the Galapagos Spreading center display exceedingly large chlorine abundances (and Cl/H_2O ratio). E-MORB exhibit both low and large abundances, an evidence of seawater derived chlorine (possibly brine). (Data from Schilling et al., 1980; Michael and Schilling, 1989; Byers et al., 1983, 1984, 1986; Aggrey et al., 1988a.)

Chlorine and bromine

Chlorine contents have been determined for crystalline basalts, pillow interiors or lava flows (Sigvaldasson and Oskarsson, 1976; Rowe and Schilling, 1979; Schilling et al., 1980; Aoki et al., 1981). Unni and Schilling (1978) and Schilling et al., (1978) neither list their data nor provide information about the crystallinity of the samples analysed, the only substantial results being presented graphically. Other works (Schilling et al., 1980; Michael and Schilling, 1989; Byers et al., 1983, 1984, 1986; Aggrey et al., 1988a) deal with glassy samples (Fig. 10).The most substantial information concerning the chlorine cycle are found in Michael and Schilling (1989) and Ito et al. (1983).

The first important observation is that the Cl/Br ratio in both N- and E-MORB is constant at about 400±100 which is nearly the seawater ratio of 290. There are few Br measurements for OIB (Hawaii) and BABB (Déruelle et al, 1991) but they confirm the good coherence between chlorine and bromine over a range of concentrations from 20 to 1000 ppm Cl (Fig. 11). According to chlorine data for the Reykjanes ridge and the good correlation between Cl abundance and latitude it has been inferred that the plume component present under Iceland is very rich in chlorine (about 200 ppm in the magma) (Unni and Schilling; 1978). They further interpreted the decreasing chlorine values North of the Reykjanes Ridge and over Iceland to reflect degassing at shallow depths as is the case for CO_2 and H_2O. This interpretation is however untenable as chlorine concentrations correlate with K_2O (Sigvaldasson and Oskarsson, 1976) and anticorrelate with MgO, which can in no way be the result of degassing. In addition, the degassing of magmas at Etna and Vulcano (Clocchiatti et al., 1994; Metrich and Clocchiatti, 1989) as observed from the melt inclusions in crystals, is significant for sulfur as fractionation proceeds but not for chlorine which increases to levels of up to 4000 ppm. More recent studies have fueled a significant data base for all oceans but do not simplify the picture.

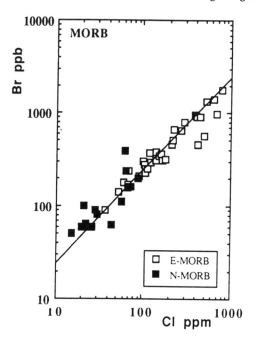

Figure 11. Cl/Br ratio in oceanic basalts. The Cl/Br ratio (~400) in MORBs is constant. (Data from Schilling et al., 1980; Déruelle et al., 1991.)

One puzzling characteristics of chlorine in oceanic basalts is its somewhat erratic variation in abundance. It behaves as an incompatible element upon fractional crystallization but displays no reasonable correlation with any of the conventional incompatible trace elements, including H_2O and K_2O (Fig. 12). One possible explanation for this particular feature is magma interaction with seawater. It has been observed that a majority of fresh oceanic MORB (E- and mostly N-MORB) exhibit a H_2O/Cl ratio of 50±20 which is similar to the seawater ratio. This is expected in a model of ocean formation by mantle outgassing and does not call for interaction with seawater unless unusually high concentrations are encountered. High concentrations of water in fresh MORBs (> 4000 ppm) are believed to reflect an H_2O-enriched source rather than late interaction with seawater; a number of such samples exhibits normal H_2O/Cl ratios. Byers et al. (1983), Aggrey et al. (1988a) and Michael and Schilling (1989) have analysed samples with exceedingly low H_2O/Cl ratios, as low as about 3. This is a regional characteristics of the Galapagos Spreading Center at 85°W, and some parts of the EPR. EPR seamounts (Aggrey et al., 1988a) appear dramatically enriched in chlorine over H_2O (Fig. 12). This can be obtained by incorporation of brine by the magmas, as brines are the only component having a sufficiently low H_2O/Cl ratio (Michael and Schilling, 1989). The increasing chlorine enrichment with fractional crystallization substantiates this interpretation as it exceeds what crystallization could induce. Fractionated magmas are more contaminated than primitive ones (or contaminated magmas fractionate more). The Reykjanes Ridge chlorine data and the Cl–MgO correlation could be explained by the same kind of evolution. If this interpretation is correct, then the H_2O/Cl ratio of the mantle is within the range of 50±20. In the previous interpretation of Unni and Schilling (1978) and Schilling et al., (1978), the high chlorine values were considered as the plume signature, with major consequences concerning the internal chlorine cycle. In the interpretation of a late contamination, chlorine like fluorine and water is only marginally reinjected into the deep mantle; exceedingly high chlorine concentrations are late phenomena.

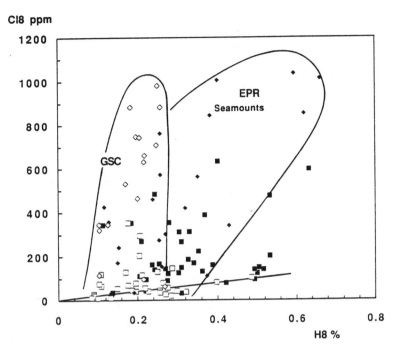

Figure 12. Cl versus H_2O in oceanic basalts. As both water and chlorine are incompatible elements, their concentrations can be corrected for shallow level fractionation. Samples from the Galapagos Spreading Center (GSC) display huge enrichment in chlorine. The same is observed for the EPR seamounts, with Cl/H_2O of up to 0.5 compared to seawater ratio of 0.02. The only possible contaminant with such high ratio is brine generated by hydrothermal activity. A number of N-MORB samples have ratios nearly identical to the seawater ratio. Most E-MORBs exhibit larger ratios which are interpreted as the result of interaction with altered crust; they are, on the average more fractionated than E-MORB. If this interpretation is correct then the mantle Cl/H_2O ratio should be about the same as that of seawater. (Data from Byers et al., 1983; Aggrey et al., 1988a; Michael and Schilling, 1989; Déruelle et al., 1991.)

Iodine

Although the external chemistry of iodine is well documented (Ullman and Aller, 1985; Kennedy and Elderfield, 1987a,b), little was known concerning the internal iodine abundance and cycle until the recent work by Déruelle et al. (1992). Iodine is the largest of all monatomic ions. As such, it cannot fit in the structure of any common mineral on Earth and it is found intimately associated with organic matter. This specificity is well demonstrated for the behavior of iodine during early diagenesis (Kennedy and Elderfield, 1987a,b) where it is found to follow organic carbon quite closely. Estimation of mantle iodine concentration from ultramafic nodule analysis (Wänke et al., 1984) does not appear to be a reliable approach, because concentrations of mobile trace elements are dramatically affected by metasomatism and alteration.

The iodine concentrations in all types of oceanic basalts cluster around 8 ppb. Some enriched basalt (but not all) display large iodine enrichment (up to 260 ppb). This has been interpreted as the signature of organic matter incorporation which is not well correlated with the recycling of altered crust or sediments.

One specific feature of iodine is its very strong concentration in the exosphere (1500 ppb) (Wänke et al, 1984; Déruelle et al., 1992). After the rare gases, it is the most enriched element, significantly more than other halogens and the major volatiles like H_2O and CO_2. Depending on whether the whole mantle or only 50 to 30% of the mantle has transferred its iodine, the exospheric iodine represents from 97 to 29% of Bulk Earth Iodine. Another possibility is that Iodine was never incorporated in the mantle and represents the remainder of a late accretional veneer (Wänke et al., 1984) which, because of the specific iodine chemistry, remained in the sediments.

CONCLUSIONS

The present-day volatile cycle can be described as the exchange between two major reservoirs, the mantle and the exosphere (continental crust, sediments, atmosphere and ocean). Because of its short life, the oceanic crust can be viewed as an interface between these two major reservoirs with major interactions at mid-ocean ridges and at subduction zones. The size of the reservoirs and the mass of volatiles therein are fairly well known and are summarized in Table 1 according to the above discussion.

One early stage of major outgassing is necessary to explain the specific isotope ratios of noble gases in the mantle relative to atmospheric composition. This catastrophic outgassing event could also correspond to some addition during the final accretion stage, without calling for an homogenous primitive Earth to start with (a more simple model used previously); this would allow explanation of the solar composition of neon isotopes in the mantle.

Present day outgassing fluxes at ridges are minor: at present rate, the H_2O, CO_2 and noble gas inventory in the exosphere could not have been supplied within the Earth lifetime. This point is even stronger when considering that some volatile elements are recycled to the mantle. This argument supports the contention that major volatiles were affected by the early catastrophic outgassing event. The outgassing flux may have decreased significantly with time (as indicated by radiogenic argon) but this effect does not suffice to explain the noble gas data. It seems that about half of the mantle was affected by long term degassing as only 50% of radiogenic argon is contained in the exosphere + depleted mantle.

^{3}He flux at ridges is calibrated using an oceanic circulation model in agreement with the ^{3}He/Heat flow correlation and the sea-floor spreading rate. Helium flux is used to calibrate other element fluxes, in particular CO_2 and noble gases. Other elements which are not exsolved at depth in the ocean (H_2O, S, halogens) are constrained by correlations with other elements (e.g., trace elements, stable and radiogenic isotopes). The present day ridge flux appears to be a rather small flux compared with hydrothermal exchange. The altered oceanic crust traps large amounts of water, CO_2, halogens and sulfur which are now reasonably estimated. The process of recycling is controlled by arc magmatism. The subducted slab delivers some of its volatiles to the overlying mantle and arc magmas, the remainder being recycled to produce enriched mantle. Assessing how much is recycled and how much is released is a difficult problem. One major constraint stems from the minor ridge input. A minor subduction output is expected as a severe imbalance of the exospheric budget is not substantiated by observations; near steady-state is likely for the long term exospheric cycle.

The object of future work will be to better estimate subduction fluxes. In addition to the volatile elements discussed in this chapter, additional tracers (stable and radiogenic isotopes, trace elements like boron) will aid more quantitative estimates. A number of outgassing models for both noble gases and major elements have been proposed but they are either too simple when they do not take recycling into account or too poorly constrained when they specify the physical mechanism which controls outgassing. A combination of (1) transfer from the mantle to the surface (mid-ocean ridges and hot spots), (2) interaction with the exosphere by hydrothermal alteration and sedimentation, and (3) complex processes during subduction (accretion, recycling, devolatilization, partial melting, fractional crystallization), regulates the Earth outgassing and cycling of volatile elements.

ACKNOWLEDGMENTS

The author is indebted to M.R. Carroll, B. Déruelle, P. Jean-Baptiste, J.P. Lorand, F. Pineau, and J.C. Varekamp for discussion and helpful comments.

REFERENCES

Abbott D, Lyle M (1984) Age of oceanic plates at subduction and volatile recycling. Geophys Res Lett 11:951-954

Aggrey KE, Muenow DW, Batiza R (1988a) Volatile abundances in basaltic glasses from seamounts flanking the East Pacific Rise at 21°N and 12-14°N. Geochim Cosmochim Acta 52:2115-2119

Aggrey KE, Muenow DW, Sinton JM (1988b) Volatile abundances in submarine glasses from the North Fiji and Lau Back Arc Basins. Geochim Cosmochim Acta 52:2501-2506

Albarede F, Michard A (1986) Transfer of continental Mg, S, O and U to the mantle through hydrothermal alteration of the oceanic crust. Chem Geol 57:1-15

Allègre CJ (1982) Chemical geodynamics. Tectonophysics 81:109-132

Allègre CJ, Staudacher T, Sarda P, Kurz M (1983) Constraints on the evolution of Earth's mantle from rare gas systematics. Nature 303:762-766

Allègre CJ, Sarda P, Staudacher T (1993) Speculations about the cosmic origin of He and Ne in the interior of the Earth Earth Planet Sci Lett 117:229-233

Alt JC, Anderson TF, Bonnell L (1989) The geochemistry of sulfur in a 1.3 km section of hydrothermally altered oceanic crust, DSDP Hole 504b. Geochim Cosmochim Acta 53:1011-1023

Alt JC, Anderson TF (1991) Mineralogy and isotopic composition of sulfur in layer 3 gabbros from the Indian Ocean, Hole 735b, Proc ODP Sci Results 118:113-125

Alt JC, Burdett J (1992) Sulfur in Pacific deep-sea sediments, ODP Leg 129, and implications for cycling of sediments in subduction zones, Proc ODP Sci Results 129:283-294

Alt JC, Shanks Iii WC, Jackson MC (1993) Cycling of sulfur in subduction zones: The Geochemistry of Sulfur in the Mariana Island Arc and Back-Arc Trough. Earth Planet Sci Lett 119:477-494

Amari S, Ozima M (1985) Search for the origin of exotic helium in deep-sea sediments. Nature 317:520-522

Amari S, Ozima M (1988) Extra-terrestrial noble gases in deep-sea sediments. Geochim Cosmochim Acta 52:1087-1095

Anderson Jr. AT (1974) Chlorine, sulfur and water in magmas and oceans. Bull Geol Soc Am 85:1485-1492

Aoki K, Ishiwaka K, Kasisawa S (1981) Fluorine in rocks from continental and oceanic regions and petrogenetic application. Contrib Mineral Petrol 76:53-59

Axford WE (1968) The polar wind and the terrestrial helium budget. J Geophys Res 73:6855-6859

Azbel IY, Tolstikhin IN (1989) Geodynamics, magmatism and degassing of the earth. Geochim Cosmochim Acta 54:139-154.

Ballhaus C (1993) Redox State of Lithospheric and atmospheric upper mantle. Contrib Mineral Petrol 114:331-348.

Banks PM, Holzer TE (1968) The polar wind. J Geophys Res 73:6846-6854

Becker H, Altherr (1992) Evidence from ultra-high-pressure marbles for recycling of sediments into the mantle. Nature 358:745-748

Berg GW (1986) Evidence for carbonate in the mantle. Nature 324:50-51

Berresheim H, Jaeschke W (1983) The contribution of volcanoes to the global atmospheric sulfur budget. J Geoph Res 88C6:3732-3740

Biellmann C, Gillet P, Guyot F, Peyronneau J, Reynard B (1993) Experimental evidence for carbonate stability in the Earth's lower mantle. Earth Planet Sci Lett 118:31-41

Blank JG, Delaney JR, Des Marais DJ (1993) The concentration and isotopic composition of carbon in basaltic glasses from the Juan de Fuca Ridge, Pacific Ocean. Geochim Cosmochim Acta 57:875-887

Blundy JD (1991) Carbon-fluid equilibria and the oxidation state of the upper mantle. Nature, 349:321-324

Bluth GJS, Schnetzler CC, Krueger AJ (1993) The contribution of explosive volcanism to global atmospheric sulphur dioxide concentrations. Nature 366:327-329

Brey G, Brice WR, Ellis DJ, Green DH, Harris KL, Ryabchikov ID(1983) Pyroxene-carbonate reactions in the upper mantle. Earth Planet Sci Lett 62:64-74

Byers CD, Muenow DW, Garcia MO (1983) Volatiles in basalts from the Galapagos Spreading Center 85° W To 86°W. Geochim Cosmochim Acta 47:1551-1558

Byers CD, Christie DM, Muenow DM , Sinton JM (1984) Volatile contents and ferric-ferrous ratios of basalt ferrobasalt andesite and rhyodacite glasses from the Galapagos 95°W propagating rift. Geochim Cosmochim Acta 48:2239-2245

Byers CD, Garcia MO, Muenow DW (1985) Volatiles in pillow rims glasses from Loihi and Kilauea volcano, Hawaii. Geochim Cosmochim Acta 49:1887-1896

Byers CD, Garcia MO, Muenow DW (1986) Volatiles in basaltic glasses from East Pacific Rise 21°N: Implications for MORB sources and submarine lava flow morphology. Earth Planet Sci Lett 79:9-20

Cadle RD (1975) Volcanism emissions of halides and sulfur compounds to the troposphere and stratosphere. J Geoph Res 80:1650-1652

Cameron AGW (1973) Abundances of the elements in the solar system. Space Sci Reviews15:121-146

Campbell IH, Taylor SR (1983) No water, no granites-No oceans, no continents. Geophys Res Lett 10:1061-1064

Canil D, Scarfe CM (1990) Phase relations in peridotite + CO2 systems to 12 GPa: Implications for the origin of kimberlite and carbonate stability in the Earth's upper mantle. J Geophys Res 95:15805-15816

Carroll MR, Rutherford MJ (1985) Sulfide and sulfate saturation in hydrous silicate melts. J Geophys Res 90C:601-612

Carroll MR, Rutherford MJ (1987) The stability of igneous anhydrite: experimental results and implications for sulfur behavior in the 1982 El Chichon trachyandesite and other evolved magmas. J Petrol 28:781-801

Chase GC (1981) Oceanic Islands Pb: two stage histories and mantle evolution. Earth Planet Sci Lett 52:277-284

Chaussidon M, Sheppard SMF, Michard A (1991) Hydrogen, sulphur and neodymium isotope variations in the mantle beneath the EPR at 12°50'N. Geochem Soc Spec Publ 3:325-337

Chaussidon M, Jambon A (1994) Boron content and isotopic composition of oceanic basalts: Geochemical and cosmochemical implications. Earth Planet Sci Lett 121:277-291

Clarke WB, Beg MA, Craig H (1969) Excess [3]He in the sea: evidence for terrestrial primordial helium. Earth Planet Sci Lett 6:213

Clocchiatti R, Gioncada A, Mosbah M, Sbrana A (1994) Possible deep origin of sulfur output at Vulcano (Southern Italy) in the light of melt inclusion studies. Acta Volcanologica (in press)

Condomines M, Gronvold K, Hooker PJ, Muehlenbachs K, O'Nions RK, Oskarsson J, Oxburgh ER (1983) Helium, oxygen, strontium and neodymium relationships in icelandic volcanics. Earth Planet Sci Lett 66:125-136

Craig H , Clarke WB, Beg MA (1975) Excess [3]He in the deep water on the East Pacific Rise. Earth Planet Sci Lett 26:125-132

Craig H, Lupton JE (1976) Primordial neon, helium and hydrogen in oceanic basalts. Earth Planet Sci Lett 31:369-385

Craig H, Poreda RJ (1986) Cosmogenic [3]He in terrestrial rocks: the summit lavas of Maui. Proc Nat'l Acad Sci 88:1970-1974

Craig H, Craig VH, Kim KR (1987) Papatua expedition I, Hydrothermal vent survey in back arc basins: the Lau, N. Fiji, Woodlark and Manus Basins, and the Havre trough. EOS 68:100

Dalrymple GB, Moore JG (1968) Argon-40 excess in submarine pillow basalts from Kilauea Volcano, Hawaii. Science 161:1132-1135

Damon PE, Kulp JL (1958) Inert gases and the evolution of the atmosphere. Geochim Cosmochim Acta 13:280

Danyushevsky LV, Falloon TJ, Sobolev AV, Crawford AJ, Carroll M, Price RC (1993) The H2O content of basalt glasses from Southwest Pacific back-arc basins. Earth Planet Sci Lett 117:347-362

Delaney JR, Muenow DW, Graham DG (1978) Abundance and distribution of water carbon and sulfur in the glassy rims of pillow basalts. Geochim Cosmochim Acta 48:581-594

Déruelle B, Dreibus G, Jambon A (1991) Halogene abundances in worldwide oceanic basalts and their petrogenetic implications. Terra abstracts 3:456

Déruelle B, Dreibus G, Jambon A (1992) Iodine abundances in oceanic basalts: implications for earth dynamics. Earth Planet Sci Lett 108:217-227

Des Marais DJ (1986) Carbon abundance measurements in oceanic basalts: the need for a consensus. Earth Planet Sci Lett 79:21-26

Des Marais DJ, Moore JG (1984) Carbon and its isotopes in mid-oceanic basaltic glasses. Earth Planet Sci Lett 69:2433-2441

Dixon JE, Clague DA, Stolper EM (1991) Degassing history of water, sulfur and carbon in submarine lavas from Kilauea volcano, Hawaii. J Geol 99:371-394

Dixon JE, Stolper EM, Delaney JR (1988) Infrared spectroscopic measurements of CO_2 and H_2O in Juan de Fuca Ridge basaltic glasses. Earth Planet Sci Lett 90:87-104

Dymond J, Hogan L (1978) Factors controlling the noble gas abundance patterns in deep sea basalts. Earth Planet Sci Lett 38:117

Edmond JM, Measures C, McDriff RE, Chan LH, Collier R, Grant B, Gordon LI, Corliss JB (1979) Ridge Crest hydrothermal activity and the balances of the major and minor elements in the ocean: the Galapagos data. Earth Planet Sci Lett 46:1-18

Edwards G (1992) Mantle decarbonation and Archean high-Mg magmas. Geology 20:899-902

Ellis DE, Wyllie PJ (1980) Phase relations and their petrological implications in the system $MgO-SiO_2-H_2O-CO_2$ at pressures up to 100 kbar. Am Mineral 65:540-556

Esser BK, Turekian KK (1988) Accretion rate of extraterrestrial particles determined from osmium isotope systematics of pelagic clay and manganese nodules. Geochim Cosmochim Acta 52:1383-1388

Exley RA, Mattey DP, Clague DA, Pillinger CT (1986) Carbon isotope systematics of a mantle "hot Spot": a comparison of Loihi Seamount and MORB glasses. Earth Planet Sci Lett 79:21-26

Farley K, Natland J, McDougall JD, Craig H (1990) He, Sr, and Nd isotopes in Samoan basalts: evidence for enriched and undepleted mantle components. EOS (Trans Am Geophys Union), 71:1669

Fisher DE (1971) Incorporation of Ar in east Pacific basalts. Earth Planet Sci Lett 2:321-324

Fisher DE (1985) Noble gases from oceanic island basalts do not require an undepleted mantle source. Nature 316:716-718

Fisher DE (1987) Reply to comment by Staudacher, Sarda. J Geophys Res 92:2813-2817

Fukumoto H, Nagao K, Matsuda JI (1986) Noble gas studies in the host phase of high $^3He/^4He$ ratios in deep-sea sediments. Geochim Cosmochim Acta 50:2244-2253

Garcia MO, Liu NWK, Muenow DW (1979) Volatiles in submarine volcanic rocks from the Mariana island arc and trough. Geochim Cosmochim Acta 43:305-312

Garcia MO, Muenow DW, Aggrey KE (1989) Major element, volatile, and stable isotope geochemistry of hawaiian submarine tholeiitic glasses. J Geoph Res 94 B8:10525-10538.

Gerlach TM (1989) Degassing of carbon dioxide from basaltic magma at spreading centers: II Mid-oceanic ridge basalts. J Volc Geoth Res 39:221-232.

Gerlach TM, Taylor BE (1990) Carbon isotopes constraints on degassing of carbon dioxide from Kilauea volcano. Geochim Cosmochim Acta 51:2051-2058.

Gill JB (1981) Orogenic Andesites and Plate Tectonics, Springer Verlag, New York, 390 p

Graham DW, Jenkins WJ, Kurz MD, Batiza R (1987) Helium isotope disequilibrium and geochronology of glassy submarine basalts. Nature 326:384-386.

Graham D, Lupton J, Albarède F, Condomines M (1990) Extreme temporal homogeneity of helium isotopes at Piton de la Fournaise, Réunion Island. Nature 347:545-548.

Hamano Y, Ozima M (1978) Earth-atmosphere evolution models based on Ar isotopic data. In: Terrestrial Rare Gases, Alexander EC Jr , Ozima M (eds), Japan Scientific Soc Press, Tokyo

Harris DM, Anderson Jr. AT (1983) Concentrations sources and losses of H_2O, CO_2 and S in Kilauean basalts. Geochim Cosmochim Acta 47:1139-1150

Harris DM, Anderson Jr. AT (1984) Volatiles H_2O, CO_2, S and Cl in a subduction zone basalt. Mineral Petrol 87:120-128

Hart RA, Dymond J, Hogan L (1979) Preferential formation of the atmosphere-sialic crust system from the upper mantle. Nature 278:156-159

Hauri EH, Shimizu N, Dieu JJ, Hart SR (1993) Evidence for hotspot-related carbonatite metasomatism in the oceanic upper mantle. Nature 365:221-227

Hilton DR, Craig H (1989) A helium isotope transect along the Indonesian archipelago, Nature 342: 906-908

Hilton DR, Gronvold K, O'Nions RK, Oxburgh ER (1990) Regional distribution of ^3He anomalies in the Icelandic crust. Chem Geol 88:53-67

Hilton DR, Hoogewerff JA, VanBergen MJ, Hammerschmid K (1992) Mapping magma sources in the east Sunda and Banda arcs: constraints from He isotopes. Geochim Cosmochim Acta 56:851-859

Hochstaedter AG, Gill JB, Kisakabe M, Newman S, Pringle M, Taylor B, Fryer P (1990) Volcanism in the Simisu rift, I. major element, volatile, and stable isotope geochemistry. Earth Planet Sci Lett 100:179-194

Hoefs J (1965) Ein Beitrag zur Geochemie des Kohlenstoffs in magmatischen und metamorphen Gesteinen. Geochim Cosmochim Acta 29:399-428

Hofmann AW (1988) Chemical differenciation of the Earth: the relationship between mantle continental crust and oceanic crust. Earth Planet Sci Lett 90:297-314

Holser WT, Kaplan IR (1966) Isotope geochemistry of sedimentary sulfates. Chem Geol 1:93-115

Honda M, Reynolds JH, Roedder E, Epstein S (1987) Noble gases in diamonds: occurrences of solar-like helium and neon. J Geophys Res 92:12507-12521

Honda M, McDougall I, Patterson DB, Doulgeris A, Clague DA (1991) Possible solar noble gas component in Hawaiian basalts. Nature 357:680-684

Ito E, Harris DM, Anderson AT (1983) Alteration of oceanic crust and geologic cycling of chlorine and water. Geochim Cosmochim Acta 47:1613-1624

Jambon A, Shelby JE (1980) Helium diffusion and solubility in obsidians and basaltic glass in the range 200-300°C. Earth Planet Sci Lett 51:206-214

Jambon A, Weber HW, Begemann F (1985) Helium and argon from an Atlantic MORB glass: concentration, distribution and isotopic composition. Earth Planet Sci Lett 73:255-267

Jambon A, Weber HW, Braun O (1986) Solubility of He, Ne, Ar, Kr and Xe in a basalt melt in the range 1250-1600°C. Geochemical implications. Geochim Cosmochim Acta 50:255-267

Jambon A, Zimmermann JL (1987) Major volatiles from a North Atlantic MORB glass and calibration to He ; a size fraction analysis. Chem Geol 62:177-189

Jambon A, Zimmermann JL (1990) Water in oceanic basalts: evidence for dehydration of recycled crust. Earth Planet Sci Lett 101:323-331

Javoy M, Pineau F, Allegre CJ (1982) Carbon Geodynamic Cycle. Nature 300:171-173

Javoy M, Pineau F, Delorme H (1986) Carbon and nitrogen isotopes in the mantle. Chem Geol 57:41-61

Javoy M, Pineau F (1991) The volatile record of a "popping" rock from the Mid-Atlantic Ridge at 14°N: Chemical and isotopic composition of the gas trapped in the vesicles. Earth Planet Sci Lett 107:598-611

Jean Baptiste P, Charlou JL, Stievenard M, Donval JP, Bougault H, Mevel C (1991) Helium and methane measurements in hydrothermal fluids from the mid-Atlantic Ridge: the Snake Pit site at 23°N. Earth Planet Sci Lett 106:17-28

Jean Baptiste P (1992) Helium-3 distribution in the deep world ocean, in Isotopes of noble gases as tracers in environmental studies. Proc IAEA Vienna, 219-240

Jenkins WJ, Edmond JM, Corliss V (1978) Excess [3]He and [4]He in Galapagos submarine hydrothermal waters. Nature 272:156-158

Jochum KP, Hofmann AW, Ito E, Seufert HM, White WM (1983) K, U and Th in Mid-Ocean Ridge basalt glasses and heat production, K/U and K/Rb in the mantle. Nature 306:431-436

Johnson HE, Axford WE (1968) Production and loss of [3]He in the earth's atmosphere. J Geophys Res, 306:431-436

Kanehira K, Yui S, Sakai H, Sasaki A (1973) Sulphide globules and sulphur isotope ratios in the abyssal tholeiite of the Mid-Atlantic Ridge near 30°N latitude. Geochem 7: 89-96

Kaneoka I, Takaoka N (1980) Rare gas isotopes in Hawaiian ultramafic nodules and ultramafic rocks. Science 208:1366

Kaneoka I, Takaoka N, Clague D (1983) Noble gas systematics for coexisting glass and olivine crystals in basalts and dunite xenoliths from Loihi seamount. Earth Planet Sci Lett 66:427-437

Katsura T, Ito E (1990) Melting and subsolidus phase relations in the $MgSiO_3$ system at high pressures: implications for evolution of the Earth's atmosphere. Earth Planet Sci Lett 99:110-117

Kellog WW, Cadle RD, Allen ER, Lazrus AL, Martel EA (1972) The Sulfur cycle. Science 175:587-596

Kennedy HA, Elderfield H (1987a) Iodine Diagenesis in pelagic deep-sea sediments. Geochim Cosmochim Acta 51:2489-2504

Kennedy HA, Elderfield H (1987b) Iodine Diagenesis in pelagic deep-sea sediments. Geochim Cosmochim Acta 51:2505-2514

Killingley JS, Muenow DW (1975) Volatiles from Hawaiian submarine basalts determined by dynamic high temperature mass spectrometry. Geochim Cosmochim Acta 39:1467-1473

Kockarts G (1973) Helium in the terrestrial atmosphere. Space Sci Reviews 14:723-757

Kockarts G, Nicolet M (1962) Le problème aéronomique de l'hélium et de l'hydrogène neutres. Annales Geophys 18:262-290

Krylov A Ya, Mamyrin BA, Khabarin LA, Mazina TI, Silin YI (1974) Helium isotopes in ocean floor bedrock. Geokhimiya 8:1220

Kurz MD (1986a) Cosmogenic helium in a terrestrial igneous rock. Nature 320:435-439

Kurz MD (1986b) In situ production of cosmogenic terrestrial helium and some applications to geochronology. Geochim Cosmochim Acta 50:2855-2862

Kurz MD (1987) Temporal helium isotopic variations within Hawaiian volcanoes: basalts from Mauna Loa and Haleakala. Geochim Cosmochim Acta 51:2905-2914

Kurz MD, Jenkins WJ (1981) The distribution of helium in oceanic basalt glasses. Earth Planet Sci Lett 53:41-54

Kurz MD, Jenkins WJ, Hart SR (1982a) Helium isotopic systematics of oceanic islands: implications for mantle heterogeneity. Nature 297:43-47

Kurz MD, Jenkins WJ, Schilling JG, Hart SR (1982b) Helium isotopic variations in the mantle beneath the North Atlantic Ocean. Earth Planet Sci Lett 58:1-14

Kurz MD, Jenkins WJ, Hart S, Clague D (1983) Helium isotopic variations in Loihi Seamount and the island of Hawaii. Earth Planet Sci Lett 66:388-406

Kurz MD, Meyer P, Sigurdsson H (1985) Helium isotopic variability within the neovolcanic zones of Iceland. Earth Planet Sci Lett 74:291-305

Kurz MD, Colodner D, Trull TW, Moore RB, O'Brien K (1990) Cosmic ray exposure dating with in situ produced cosmogenic ^{3}He: results from young Hawaiian lava flows. Earth Planet Sci Lett 97:177-189

Kurz MD, Kammer DP (1991) Isotopic evolution of Mauna Loa Volcano. Earth Planet Sci Lett 103:257-269

Kurz MD, Le Roex AP, Dick H (1991) Isotope geochemistry of the Bouvet hot spot. Geochim Cosmochim Acta 55:2269

Kyser TK, O'Neil JR (1984) Hydrogen isotope systematics of submarine basalts. Geochim Cosmochim Acta 48:2123-2133

Kyser TK, Rison W (1982) Systematics of rare gas isotopes in basic lavas and ultramafic xenoliths. J Geophys Res 87:5611-5630

Lal D, Nishiizumi K, Klein J, Middleton R, Craig H (1987) Cosmogenic ^{10}Be in Zaire alluvial diamonds: implication for ^{3}He contents of diamond. Nature 328:139-141

Lal MD, Craig H, Wacker JF, Poreda R (1989) ^{3}He diamonds: the cosmogenic component. Geochim Cosmochim Acta 53:569-574

Lambert G, Le Cloarec MF, Pennisi M (1988) Volcanic output of SO_2 and trace metals: A new approach. Geochim Cosmochim Acta 52:39-42

Lange MA, Ahrens E (1986) Shock induced CO_2 loss from $CaCO_3$: implications for early planetary atmospheres. Earth Planet Sci Lett 77: 409-418

Le Guern (1982) Les débits de CO_2 et de SO_2 volcaniques dans l'atmosphère. Bull Volcan 45-3:39-42

Lorand JP (1990) Are spinel Lherzolite xenoliths representative of the abundance of sulfur in the upper mantle ? Geochim Cosmochim Acta 54:1487-1492

Lorand JP (1991) Sulphide petrology and sulphur geochemistry of orogenic Lherzolites: a comparative study of the Pyrenean Bodies (France) and the Lanzo Massif (Italy). J Petrol Spec Lherzolite issue, 77-95

Lupton JE, Craig H (1981) A major helium-3 source at 15°S on the East Pacific Rise. Science 214:13-18

Lupton JE, Baker ET, Massoth GJ (1989) Variable ^{3}He/heat ratios in submarine hydrothermal systems: evidences from two plumes over Juan de Fuca Ridge. Nature 337:161-164

Lux G (1987) The behavior of noble gases in silicate liquids: solution, diffusion, bubbles and surface effects, with applications to natural samples. Geochim Cosmochim Acta 51:1549-1565

Macpherson C, Mattey D (1994) Carbon isotope variations of CO_2 in central Lau Basin basalts and ferrobasalts. Earth Planet Sci Lett 121: 263-276

Mamyrin BA, Tolstikhin IN, Anufriev GS, Kamenskii IL (1969) Isotopic analysis of terrestrial helium on a magnetic resonance mass spectrometer. Geochem Int 6:517-524

Marty B (1989) Neon and xenon isotopes in MORB: implications for the earth-atmosphere evolution. Earth Planet Sci Lett 94:45-56

Marty B , Ozima M (1986) Noble gas distribution in oceanic basalt glasses. Geochim Cosmochim Acta 50:1093-1097

Marty B, Jambon A (1987) C/^{3}He in volatile fluxes from the solid earth: implications for carbon geodynamics. Earth PlanetSci Lett 83:16-26

Marty B, Jambon A, Sano Y (1989) Helium isotopes and CO_2 in volcanic gases of Japan. Chem Geol 7:25-40

Mathez EA (1976) Sulfur solubility and magmatic sulfides in submarine basalt glass. J Geoph Res 81:4269-4276

Matsuda JI, Murota M, Nagao K (1990) He and Ne isotopic studies in the extraterrestrial material in deep-sea sediments. J Geophys Res 95:7111-7117

Matsui T, Abe Y (1981) Formation of a magma-ocean on the terrestrial planets due to the blanketing effect of an impact induced atmosphere. Earth Moon Planets 34: 223-230

Mattey DP, Carr RH, Wright IP, Pillinger CT (1984) Carbon isotopes in submarine basalts. Earth Planet Sci Lett 70:196-206

Mattey DP, Exley RA, Pillinger CT (1989) Isotopic composition of CO_2 and dissolved carbon in glass. Geochim Cosmochim Acta 53:2377-2386

Maurette M, Olinger C, Christophe Michel-Levy M, Kurat G, Pourchet M, Brandstätter F, Bourot-Denise M (1991) A collection of diverse micrometeorites recovered from 100 tonnes of Antarctic blue ice. Nature 351:44-47

Mazor E, Heymann D, Anders E (1970) Noble gases in carbonaceous chondrites. Geochim Cosmochim Acta 34:781-802

McCulloch MT (1993) The role of subducted slabs in an evolving earth. Earth Planet Sci Lett 115:89-100

Merlivat L, Andrie C, Jean-Baptiste P (1984) Distribution des isotopes, de l'hydrogène de l'oxygène et de l'hélium dans les sources hydrothermales sous-marines de la ride Est-Pacifique à 13°N. C R Acad Sci Paris 299, II:17

Merlivat L, Pineau F, Javoy M (1987) Hydrothermal waters at 13 N on the East Pacific Rise: isotopic composition and gas concentration. Earth Planet Sci Lett 84:100-108

Metrich N, Clocchiatti R (1989) Melt inclusion investigation of the volatile behaviour in historic alkali basaltic magmas of Etna. Bull Volcanol 51:185-198

Metrich N, Clocchiatti R, Mosbah M, Chaussidon M (1993) The 1989-1990 activity of Etna magma mingling and ascent of H_2O-Cl-S rich basaltic magma. Evidence from melt inclusions. J Volcanol Geotherm Res 59:131-144

Michael PJ (1988) The Concentration, Behavior and storage of H_2O in the suboceanic upper mantle: implication for mantle metasomatism. Geochim Cosmochim Acta 52:555-556

Michael PJ, Schilling JG (1989) Chlorine in Mid-Ocean Ridge magmas: evidence for assimilation of seawater-influenced components. Geochim Cosmochim Acta 53:3131-3143

Moore JG (1970) Water content of basalt erupted on the ocean floor. Contrib Mineral Petrol 28:272-279

Moore JG, Fabbi BP (1971) An estimate of the juvenile sulfur content of basalt. Contrib Mineral Petrol 33:118-127

Moore JG, Schilling JG (1973) Vesicles water and sulfur in Reykjanes ridge basalts. Contrib Mineral Petrol 41:105-108

Moore JG, Batchelder J, Cunningham C (1977) CO_2-filled vesicles in mid-ocean ridge basalt. J Volc Geoth Res 2:308-329

Muenow DW, Garcia MO, Saunders AD (1980) Volatiles in submarine volcanic rocks from the spreading axis of the East Scotia Sea back-arc basin. Earth Planet Sci Lett 47:272-278

Muenow DW, Graham DG, Liu NWK (1979) The abundance of volatiles in Hawaiian tholeitic basalts. Earth Planet Sci Lett 42:71-76

Muenow DW, Perfit MR, Aggrey KE (1991) Abundances of volatiles and genetic relationships among submarine basalts from the Woodlark Basin, Southwest Pacific. Geochim Cosmochim Acta 55:2231-2239

Nadeau S, Pineau F, Javoy M, Francis D (1990) Carbon concentrations and isotopic ratios in fluid-inclusions-bearing upper mantle xenoliths along the northwestern margin of North America. Chem Geol 81:271-297

Nier AO, Schlutter DJ (1990) Helium and Neon isotopes in stratospheric particles. Meteoritics 25:263-267

Nier AO, Schlutter DJ, Brownlee DE (1990) Helium and Neon isotopes in deep Pacific ocean sediments. Geochim Cosmochim Acta 54:173-182

Nilsson K, Peach CL (1993) Sulfur speciation, oxidation state, and sulfur concentration in back-arc magmas. Geochim Cosmochim Acta 57: 3807-3813

Nixon PH, Pearson DG, Davies GF (1991) Diamonds: the oceanic lithosphere connection with special reference to Beni Bousera, North Marocco. In: Peters TJ et al. (eds) Ophiolite Genesis and Evolution of the oceanic lithosphere: 275-289

O'Neil HSC (1991) The origin of the Moon and the early history of the Earth. A chemical model. Part I: the Earth. Geochim Cosmochim Acta 55:1159-1172

O'Nions RK, Oxburgh ER (1988) Helium volatile fluxes and the development of continental crust. Earth Planet Sci Lett 90:331-347

Otting W, Zähringer J (1967) Total carbon and primordial rare gases in chondrites. Geochim Cosmochim Acta 31:1949-1965

Oxburgh ER, O'Nions RK (1987) Helium loss, tectonics and the terrestrial heat budget. Science 237: 1583-1588

Ozima M (1975) Ar isotopes and Earth atmosphere evolution models. Geochim Cosmochim Acta 39:1127-1134

Ozima M, Podosek FA (1983) Noble Gas Geochemistry. Cambridge University Press, London

Ozima M, Podosek FA, Igarashi G (1985) Terrestrial xenon isotope constraints on the early history of the Earth. Nature 315:471

Ozima M, Zashu S (1988) Solar-type neon in Zaire cubic diamonds. Geochim Cosmochim Acta 52:19-25

Pan V, Holloway JR, Hervig RL (1991) The pressure and temperature dependence of carbon dioxide solubility in tholeiitic basalt melts. Geochim Cosmochim Acta 55:1587-1595

Patterson B, Honda M, McDougall I (1990) Atmospheric contamination: a possible source for heavy noble gases in basalts from Loihi seamount, Hawaii. Geophys Res Lett 17:705-708

Peach CL, Mathez EA, Keays RR (1990) Sulfide melt-silicate melt distribution coefficients for noble metals and other chalcophile elements as deduced from MORB: Implications for partial melting. Geochim Cosmochim Acta 54:3379-3390

Pepin RO, Signer P (1965) Primordial rare gases in meteorites. Science 149:253-265

Phynney DJ, Tennyson J, Frick U (1978) Xenon in CO_2 well gas revisited. J Geophys Res 83:2313-2319

Pineau F, Javoy M (1983) Carbon isotopes and concentrations in mid-ocean ridge basalts. Geochim Cosmochim Acta 46:371-379

Pineau F, Javoy M (1994) Strong degassing at ridge crests: the behaviour of dissolved carbon and water in basalt glasses at 14°N, Mid-Atlantic Ridge. Earth Planet Sci Lett 123:179-198

Pineau F, Javoy M, Bottinga Y (1976) $^{13}C/^{12}C$ ratios of rocks and inclusions in popping rocks of the mid-Atlantic ridge. Their bearing on the problem of isotopic composition of deep seated carbon. Earth Planet Sci Lett 29:413-421

Pineau F, Mathez E (1990) Carbon isotopes in xenoliths from Hualalai Volcano Hawaii and the generation of isotopic variability. Geochim Cosmochim Acta 54:217-227

Podosek FA (1970) Dating of meteorites by the high temperature release of iodine-correlated ^{129}Xe. Geochim Cosmochim Acta 34:341-365

Porcelli DR, Stone JOH, O'Nions RK (1987) Enhanced $^{3}He/^{4}He$ ratios and cosmogenic helium in ultramafic xenoliths. Chem Geol 66:89-98

Poreda R (1985) Helium-3 and Deuterium in back-arc basalts: Lau Basin and the Mariana trough. Earth Planet Sci Lett 73:244-254

Poreda R, Radicati F (1984) Neon isotope variations in Mid-Atlantic Ridge Basalts. Earth Planet Sci Lett 69:277-289

Poreda R, Schilling JG, Craig H (1986) Helium and hydrogen isotopes in ocean-ridge basalts north and south of Iceland. Earth Planet Sci Lett 78:1-17

Poreda R, Craig H (1989) Helium isotope ratios in circum-Pacific volcanic arcs. Nature 338:473-478

Poreda R, Craig H (1992) He and Sr isotopes in the Lau Basin mantle: depleted and primitive mantle components. Earth Planet Sci Lett 113:487-493

Puchelt H, Hubberten HW (1980) Preliminary result of sulfur isotope investigations on Deep Sea Drilling Project cores from lets 52 and 53. Initial Rept DSDP 51-52-53/2:1145-1148

Rison W, Craig H (1983) Helium isotopes and mantle volatiles in Loihi seamount and Hawaiian island basalts and xenoliths. Earth Planet Sci Lett 66:407-426

Robin E (1988) Des poussières cosmiques dans les cryocomites du Groënland: nature, origine et applications. Dissertation, Univ Paris-Sud, France

Rose WI, Stoiber RE, Malinconico LL (1982) Eruptive gas composition and fluxes of explosive volcanoes: budget of S and Cl estimated from Fuego volcano, Guatemala. In: Andesites, Thorpe RS (ed), Wiley, New York, 669-676

Rosenberg ND, Lupton JE, Kadka D, Collier R, Lilley MD, Pak H (1988) Estimation of heat and chemical fluxes from a sea floor hydrothermal vent field using radon measurements. Nature 334:604-607

Rowe EC, Schilling JG (1979) Fluorine in Iceland and Reykjanes Ridge Basalts. Nature 279:33-37

Rye RO, Luhr JF, Wassermann MD (1984) Sulfur and Oxygen isotopic systematics of the 1982 eruptions of El Chichòn volcano, Chiapas, Mexico. J Volcanol Geotherm Res 23:109-123

Sakai H, Casadevall TH, Moore JG (1982) Chemistry and isotope ratios in basalts and volcanic gases at Kilauea volcano, Hawaii. Geochim Cosmochim Acta 46:229-738

Sakai H, Des Marais DJ, Ueda A, Moore JG (1984) Concentrations and isotope ratios of carbon, nitrogen and sulfur in ocean-floor basalts. Geochim Cosmochim Acta 48:371-379

Sano Y, Wakita H, Ching-Wang H (1986) Helium flux in a continental Land area estimated from $^{3}He/^{4}He$ in northern Taiwan. Nature 323:55-57

Sano Y, Wakita H , Giggenbach WF (1987) Island arc tectonics of New Zealand manifested in helium isotope ratios. Geochim Cosmochim Acta 50:2429-2432

Sarda P, Graham D (1990) Mid-Ocean ridge popping rocks: implications for degassing at ridge crests. Earth Planet Sci Lett 97:268-289

Sarda P, Staudacher T, Allegre CJ (1985) $^{40}Ar/^{36}Ar$ in MORB glasses: constraints on atmosphere and mantle evolution. Earth Planet Sci Lett 72:357

Sarda P, Staudacher T, Allegre CJ (1988) Neon Isotopes in submarine basalts. Earth Planet Sci Lett 91:73-88

Schilling JG, Bergeron MB, Evans R (1980) Halogens in the mantle beneath the North Atlantic. Phil Trans R Soc London A297:147-178.

Schilling JG, Unni CK, Bender ML (1978) Origin of chlorine and bromine in the oceans. Nature 273:631-636.

Signer P, Suess HE (1963) Earth Sciences and Meteoritics. North-Holland, Amsterdam, 241 p

Sigvaldasson E, Oskarsson N (1976) Chlorine in basalts from Iceland. Geochim Cosmochim Acta 40:777-789.

Sigvaldasson E, Oskarsson N (1986) Fluorine in Basalts from Iceland. Contrib Mineral Petrol 94:263-271

Staudacher T (1987) Upper mantle origin for Harding County well gases. Nature 325:605

Staudacher T, Allegre CJ (1982) Terrestrial Xenology. Earth Planet Sci Lett 60:389-406

Staudacher T, Allegre CJ (1988) Recycling of oceanic crust and sediments: the noble gas subduction barrier. Earth Planet Sci Lett 89:173-183

Staudacher T, Sarda P, Allegre CJ (1990) Noble gas systematics of Reunion island, Indian Ocean. Chem Geol 89:1-17

Staudigel H, King SD (1992) Ultrafast subduction: the key to slab recycling efficiency and mantle differenciation? Earth Planet Sci Lett 109:517-530

Staudigel H, Hart SR, Schmincke HU, Smith BM (1990) Cretaceous ocean crust at DSDP sites 417 and 418: carbon uptake from weathering versus loss by magmatic outgassing. Geochim Cosmochim Acta 63:3091-3094

Staudigel H, King SD (1992) Ultrafast subduction: the key to slab recycling efficiency and mantle differenciation? Earth Planet Sci Lett 109:517-530

Stoiber RE, Jepsen A (1973) Sulfur dioxide contributions to the atmosphere by volcanoes. Science 182:577-578

Stoiber RE, Williams SN, Huebert B (1987) Annual contribution of sulfur dioxide to the atmosphere by volcanoes. J Volcan Geotherm Res 33:1-8

Stolper E, Newman S (1994) The role of water in the petrogenesis of Mariana trough magmas. Earth Planet Sci Lett 121:293-325

Stuiver M, Quay P, Ostlund G (1983) Abyssal water carbon-14 distribution and the age of the world oceans. Science 219:849-851

Takanayaji M, Ozima M (1987) Temporal variation of 3He/4He ratio recorded in deep-sea sediments cores. J Geophys Res 92:12531

Torgersen T (1989) Terrestrial helium budget: implications with respect to the degassing processes of continental crust. Chem Geol (Isotope Geoscience Section) 78:1-14

Trull TW, Perfit MR , Kurz MD (1989) He and Sr isotopic constraints on subduction contributions to Woodlark Basin volcanism. Geochim Cosmochim Acta 54:441-453

Trull T, Nadeau S, Pineau F, Polve M, Javoy M (1993) C-He systematics in hotspot xenoliths: implications for mantle carbon contents and carbon recycling. Earth Planet Sci Lett 118:43-64

Turekian KK (1959) The terrestrial economy of Helium and Argon. Geochim Cosmochim Acta 17:37

Turner G (1989) The outgassing history of the Earth's atmosphere. J Geol Soc Lond 146:147-154

Ueda A, Sakai H (1984) Sulfur isotope study of quaternary volcanic rocks form the japanese islands arcs. Geochim Cosmochim Acta 48:1837-1848

Ullman WJ, Aller RC (1985) The geochemistry of iodine in near shore carbonate sediments. Geochim Cosmochim Acta 49:967-978

Unni CK, Schilling JG (1978) Cl and Br degassing by volcanism along the Reykjanes Ridge and Iceland. Nature 272:19-23

Vance D, Stone JOH , O'Nions RK (1989) He, Sr, and Nd isotopes in xenoliths from Hawaii and other oceanic islands. Earth Planet Sci Lett 96:147-160

Varekamp JC, Kreulen R, Poorter RPE, Van Bergen MJ (1992) Carbon sources in arc volcanism, with implications for the carbon cycle. Terra Nova 4:363-373

Von Damm KL, Edmond JM, Measures CI, Grant B (1985) Chemistry of submarine hydrothermal solution at Guyamas Basin, Gulf of California. Geochim Cosmochim Acta 49:2221-2237

Walker JCG (1982) Carbon geodynamic cycle. Nature 303:730-731

Wänke H, Dreibus G, Jagoutz E (1984) Mantle Chemistry and accretion history of the Earth. In Archean Geochemistry, Kröner A et al. (eds), Springer-Verlag, Berlin, 1-23

Welhan JA, Craig H (1983) Methane, hydrogen and helium in hydrothermal fluids at 21°N on the East Pacific rise. In: Hydrothermal Processes at Sea floor Spreading Centers, NATO Advanced Research Institute, Plenum Press, New York

Wetherill GW (1953) Spontaneous fission yields from uranium and thorium. Physical Rev 92:907-912

Wolery TJ, Sleep NH (1976) Hydrothermal circulation and geochemical flux at mid-ocean ridges. J Geol 84:249-275

Wood BJ (1993) Carbon in the core. Earth Planet Sci Lett 117:593-607

Woodruff LG, Shanks Iii WC (1988) Sulfur isotope study of chimney minerals and vent fluids from 21°N, East-Pacific Rise: Hydrothermal sulfur sources and disequilibrium sulfate reduction. J Geophys Res 93:4562-4572

Wyllie PJ, Huang WL (1976) Carbonation and melting reactions in the system $CaO-MgO-SiO_2-CO_2$ at mantle pressures with geophysical and petrological applications. Contrib Mineral Petrol 54:79-107

Zhang Y, Zindler A (1989) Noble gas constraints on the evolution of the Earth's atmosphere. J Geophy Res 94 B10:13719-13737